# Electric Process Heating

## Technologies/Equipment/Applications

By

**Maurice Orfeuil**

Chief, Electricity Applications Department
Electricité de France

Preface by

**Albert Robin**

Directeur Général Adjoint
Electricité de France

BATTELLE PRESS
Columbus • Richland

Cover Photo
· Infrared furnace for aerospace
simulations. Centre des Mureaux

Translation by: TRADUCTOR, Montreal, Quebec

Printed in the United States of America.

**Library of Congress Cataloging-in-Publication Data**

Orfeuil, Maurice.
Electric process heating.
Translation of: Electrothermie industrielle.
Includes bibliographies and index.
1. Electric heating. 2. Electric furnaces.
I. Title.
TK4601.07313 1981 621.402'8 86-3531
ISBN 0-935470-26-3

Originally published in French under the title, "Electrothermie Industrielle: Fours et équipements thermiques électriques industriels". All Rights Reserved. Authorized translation from French language edition published by Bordas Dunod, Paris 1981.

# Preface

The work of Maurice Orfeuil fills a gap and responds to a need.

First, it fills a gap by synthesizing the many ways that one can heat a material electrically — solid, liquid, or gaseous — a synthesis heretofore lacking.

Many of these processes have been known since the advent of industrial electricity and have led to the development of traditional electrothermal technologies. Their preferred field of application has been the working, forming, and thermal treatment of metals. Resistance furnaces, arc furnaces and induction furnaces, in their diverse variations, have long constituted the core of the panoply at the disposal of the practitioner of these technologies.

But gradually, technological evolution has made industrially usable phenomena, long known in principle but relatively little practiced (dielectric hysteresis, plasma, electron beam), or newly discovered ones, such as the laser, which was unknown in laboratories less than 20 years ago.

It was essential that a work be written that would give students, engineers and managers sufficient exposure to the state of the art of industrial process heating technologies to enable them to evaluate their possibilities. As we will see, even where the whole scale of industrial needs is filled, electricity would still open avenues not easily accessed by other forms of energy.

Second, M. Orfeuil's work responds to a need. A deep understanding of the possibilities that electricity offers for thermal and ore treatment applications is already essential. It will become even more so in the near future for those who make or prepare investment decisions in the industrial sector. Two principal factors will motivate an increased recourse to electricity:

a) The evolution of relative costs of energy, making electricity more attractive than it was 10 or 20 years ago;

b) The necessity for rational, i.e., economical utilization of the available energy forms.

To be competitive, one must use the least possible energy and have recourse to sources of energy which are most available, i.e., lowest expenditure of foreign exchange; stable costs; assured supply. Contrary to opinion, still too widely held, satisfying thermal needs by the "electricity" route brings not only important economies in the use of primary energy, but also can produce foreign exchange savings. This is true, even if all the electricity used derives from heavy fuel fired thermal plants, which no longer is the case.

Presently, France's basic production is assured by coal, nuclear and hydroelectric energy. Thus, the foreign exchange cost of electricity for fuel oil is rapidly declining and is practically nonexistent during several summer months.

The gap which the book fills and the need to which M. Orfeuil's work responds are sufficient reasons for it to be of interest, but there are additional ones, notably:

— the preciseness of the documentation of the technology;
— the clarity of the presentation;
— the methodical indication of implemented applications, or those which one can envisage on short notice both in the technical and economic area.

Although this is essentially a technical work, the author, a mining engineer and an MBA, does not forget economic and financial considerations which make a technology industrially acceptable or not. In the course of the book, the author maintains the reader's interest by his didactic style and the mastery of the subjects covered. It is a mastery which he acquired during his studies and his years as an engineer at Division Industries Électricité du Service commercial of the E.D.F., where his task was to promote the development of the technology which he presents in this volume.

Thanks to this technology, industry enjoys a basic opportunity to be seized without hesitation. In the delicate situation to remain competitive and to reduce the risk of partial paralysis due to energy shortage, it is necessary to convert industry to increased utilization of available electric processes as well as to production of new materials.

Albert Robin
Director of General Management of Électricité de France

# Foreword

**Electrothermie Industrielle** was first published in French in 1981 and attempted to provide an overall view of the state of the art in electric process heating techniques throughout the world. Since then, no major revolution has occurred in this field, but rather a continuous flow of innovations presenting various degrees of importance, all having a positive effect on the efficiency and the competitiveness of these techniques. The most prominent ones mainly involve the use of existing techniques for new applications, with the emergence of a few really new technologies and the development of computerized design methods.

As for this last point, two main types of **Computer-Aided Design** systems are now available. The first family is based on complex mathematical models which provide a very accurate representation of the physical phenomena involved. These models use sophisticated numerical techniques to solve the corresponding equations and generally require powerful computers. They are mainly used in research, but manufacturers of equipment and industrial companies now sometimes rely on them to study problems which cannot be solved by conventional methods. The second family includes simplified design tools which can often be implemented on microcomputers. Their main goal is to solve the bulk of repetitive and often classical problems encountered in the design and engineering of equipment in a more accurate, efficient and less expensive way. However the frontier between these two families tends to vanish with the increased speed of microcomputers, thus offering more and more powerful tools to equipment designers and users.

The most interesting programs which have been recently released mainly concern induction, resistance, microwaves and high frequency dielectric heating. Some of them can solve problems in a space at 2 or 3 dimensions and cope with both electrical and thermal aspects (distribution of electromagnetic fields, heat transfers, etc. as a function of time and space).

To list all the innovations regarding the different technologies, the corresponding equipment, their components and their applications, would be too long and tedious. Therefore, this foreword will emphasize the most important innovations and outline a few major trends for the future.

Three major evolutions in **Induction Heating** should be mentioned. First, solid state generators have displayed a consistent growth which should continue. For medium frequencies, up to 10 kHz, motor generators have almost disappeared in new equipment; thyristor generators have become the standard frequency converters, and single modules of up to 5 MW should come to the market shortly. In the range 10 kHz — 100 kHz, thyristor or transistor generators have started to replace tube generators, and solid state generators operating up to 100 kW at 100 kHz probably will be available soon. The second evolution concerns coils: transverse flux coils, multilayer coils, coils with opened or closed magnetic core, wide rectangular coils and others. These have started to leave the laboratory for the shop floor, even if their use is still limited. Third, a substantial extension of the applications of induction heating can be foreseen, both in its traditional field, the metal industries (new heat treatments and reheating processes, strip edge heating, elimination of skidmarks), and in other sectors such as chemical, food, paper, and textiles industries (heating of reactors, vessels, screws, cylinders, concentrators). Induction heating may also become a great success for cooking in the commercial and domestic sectors.

**Resistance Furnaces** are highly praised, but face strong competition from equipment using other fuels, since they must prove their value on the basis of energy cost comparisons. However, theoretical studies and measures on furnaces in industrial operation have shown that the efficiency of resistance furnaces is generally far higher and decreases less with a reduction of the furnace load. The most interesting innovation is probably the development of a range of high power density furnaces (up to 80 $kW/m^2$ of furnace wall at 900°C) which can compete in terms of productivity with gas or oil furnaces and sometimes exceed their performances. This is mainly achieved through the use of solid state control, improved heating elements and a better design increasing heat transfer between the elements and the load.

Two other developments could be of a great significance. First, the implementation of discrete time multivariable self adaptative controllers on resistance furnaces will give them a considerable competitive edge, permitting an increase in quality of manufactured goods and the automation of production lines. These systems may be extended later to the optimal control of other electric heating processes. Second, the development of high power and high temperature (up to 30 MW and 900°C, and perhaps to 1,300°C) air and process gas heaters has started to open new markets to resistance heating in many industries, often in **Dual Energy Systems** (electricity is only used when the cost of the heat it provides is lower than the cost of the heat produced from fossil fuels).

Incidentally, it should be noted that **Dual Energy Systems** are displaying fast growth in countries with abundant resources of cheap nuclear or hydroelectricity. They are based on diversified equipment, such as gas-resistance furnaces; hot gases generators using, alternatively or simultaneously, resistance or plasma heating and fuel firing; heat pump-fossil fuel boilers systems; and, of course, electric and fossil fuel boilers. Two innovative electric boilers are, for example, now coming to the market. The first has incorporated in the same pressure shell the immersion electric heaters and the gas heated tubes. The second, based on induction heating, is a 3-phase transformer. This avoids the use of another transformer to connect it to the line supply. The primary is the induction coil, the secondary, wound around the same magnetic core, is an electrically shorted steel tube through which the fluid or gas to be heated is passed.

**Infrared** emitters and ovens have not changed basically, but infrared heating has gained wider acceptance and its applications keep diversifying, mainly in the food, paper and textiles industries. The same trend can be observed with **Radiofrequency and Microwave** heating; moreover better control of the energy transfer from the source to the product to be heated has been achieved.

**Lasers** are experiencing a fast growth for materials processing. The electronics industry depends almost entirely on lasers for such vital operations as scribing and trimming of components, but more conventional mass producing industries have started to use lasers for welding, machining and heat treating. Large sums of money are presently spent in research to develop higher output power sources and promote the use of lasers in highly automated and flexible production lines. Lasers will, however, stay a limited market in terms of electricity consumption.

**Electron Beam** furnaces have recently regained interest for the melting of high purity metals and electron beam welding and heat treating are most often used in industry. This technique will, nevertheless, also have a limited contribution to the electrification of industry.

A few **Direct Current Arc Furnaces** have been put into operation for steel making, but it is too large to know if this technique will replace AC arc furnaces on a large scale.

The use of **Arc Plasma** heating have been researched for many years, but this technology is just beginning to mature into a more fully understood industrial process. Recent studies undertaken in the U.S.A., Canada, Japan, Sweden, France and other countries have shown that increased implementation of plasma could be one of the most attractive possibilities to extend the electrification of manufacturing industries. Moreover, the flexibility of combining electricity, in the form of plasma, and other fuels, notably coal, still adds to the interest of this technique.

In the short term, the most promising areas lie within the iron, steel and ferroalloys industries for upgrading conventional systems, such as blast furnaces, cupolas and even arc furnaces, or replacing some of them by new processes. Thus, a few plasma systems have recently reached the stage of the pilot plant or of commercial demonstration. For example, an existing blast furnace producing ferrochrome has been equipped with nine torches each rated at 2 MW to increase its productivity and reduce its coke consumption. The feasibility of plasma fired cupolas and their ability to melt machine chips at high productivity levels has been demonstrated. The use of plasma torches to supply hot reducing gas by injecting coal into the hot air stream has been tested succesfully on a blast furnace in industrial operation, thus showing that electricity and low-quality coal can replace a high percentage of the coke. Pilot plants have also been built for direct reduction of iron ore, ironmaking, ferrochrome production, and recovering zinc from steelmaking dust. A few new routes for iron and steel making can be imagined including such other miscellaneous utilisations as induration of pellets, heating of soaking pits and rolling-mill furnaces, temperature adjustments during ladle refining and laddle preheating.

In the long term, plasma processing, which has been used since the late thirties to produce acetylene in Germany, could provide new opportunities in the chemical industries for the processing of coal and hydrocarbons, the synthesis of special materials such as titanium dioxide, silica and refractory powders or the destruction of the wastes generated by many industries. This will however require an important research and development effort to overcome the technical problems. Further penetration of this technology will also remain largely dependent on the economic situation and on the price of electricity as compared to that of other energies.

On the whole, the future of electric process heating is bright and it should be remembered that imagination is what tends to become real.

Maurice Orfeuil
Paris, June 1986

# Acknowledgements

We acknowledge with sincere appreciation the contributions made by the Electric Power Research Institute (EPRI) and the Canadian Electrical Association (CEA) in funding the translation of this book. In particular, we are indebted to Mr. Thomas G. Byrer, Director of the EPRI Center for Metals Fabrication, for bringing the original version to our attention and to Dr. Edward M. Ezer, CEA Director of Research and Development, for overseeing the French to English translation. Special thanks are extended to the author, Mr. Maurice Orfeuil of Electricité De France and to Mme. Maryvonne Vitry of Bordas Dunod for arranging a publication grant from the French Ministry of Culture.

The author would like to express his heartfelt gratitude to all those who helped and advised him in the task of writing this book. Thanks are due especially to:

— Mr. J. Gosse, professor of industrial Thermal Science at the Conservatoire des Arts et Métiers, who offered him a teaching post in electrotechnology and encouraged him to start writing this book;

— Mr. A. Busson, professor of Electricity at the Conservatoire des Arts et Métiers, and technical and scientific advisor to the President of the Union technique de l'Électricité who, years ago, stimulated his interest in writing and encouraged him throughout this undertaking;

— Mr. A. Robin, managing director at Électricité de France, who agreed to preface this book and to place his objectives in the context of today's short- and long-term energy trends;

— Mr. R. Robin, assistant director in the Distribution Department, Électricité de France-Gaz de France, who encouraged the project and provided all the necessary facilities;

— Mr. L. Hallot, assistant director, Méthodes Fonderie de la Régie nationale des Usines Renault, who shared with him his vast experience and supplied valuable background information.

Also to be thanked are the Institut français de l'Énergie and, in particular, the director of his school of Thermal Science for having allowed him to teach industrial electrotechnologies for so many years, and thus enrich his experience through the multiple contacts he has made with continuing-education students.

The following pages owe much to the comprehensive documentation made available to the author by:

— the Comité français d'Électrothermie, particularly Mr. P. Michel, the Delegate General;

— manufacturers of electroheating equipment, thanks to whom the author was able to amply illustrate his text;

— users of electroheating equipment, who agreed to share the results of their experience;

— research centres involved in innovation in the area of electrotechnologies.

Finally, as they went their way through these pages, his colleagues, engineers and researchers at Électricité de France, especially the personnel in the Industrie-Électricité Division under the leadership of Mr. C. Médan, who, in their joint endeavor to further the various electrotechnologies, will recognize the author's indebtedness to them. Their unstinting support is gratefully acknowledged.

# Table of contents

# Electric Process Heating

## Technologies/Equipment/Applications

# Chapter 1

# Electrothermal Technology and Industrial Processes

## 1. THERMAL EQUIPMENT AND ELECTROTHERMAL TECHNOLOGY

To invest is to have confidence in the future. This, for an economic entity (an individual, a business, a nation, ...) is to be able to supply future products and services. An analysis of an investment program has the following independent aspects:
— technical: capacity of production, technology, financial product features;
— economic: market analysis, production costs (labor, capital, energy, raw materials, etc.), competition;
— finances: financial available resources (owners' finances and self financing, external short and long term resources) and their cost, overall cost effectiveness;
— social: work place conditions, effect on the environment, pollution.

Most investment decisions require thorough initial analysis because they involve events that take place over a long period of time, and play an important role in the life of an enterprise or even a nation. This is particularly true for furnaces and other industrial heating equipment, which, in many cases, require a long time to install.

The main objective of this book is to help all those who must make decisions in selecting industrial heating equipment. Electrical energy can be used for heating in many different ways. The chapters that follow will answer questions concerning:
— types of heating equipment presently available;
— construction and operating principles of this equipment;
— relative advantages of different electrical heating techniques, and their limitations;

— technical and economic data. In particular, data on energy consumption of industrial electric heating equipment;
— current and future technological developments;
— energy saving capabilities.

It would be presumptuous to attempt a presentation of all industrial applications of electrothermal technology in one book. Examples found in different chapters are thus presented primarily as analogies. A technique proven successful in one area can often be transferred to other, totally different areas with some modifications.

The following observation is at the root of our reasons for writing this book: the teaching of electrothermal technology is, at present, in its infancy. This is so at all educational institutions, including universities.

Thus, the development of our discipline should start by collecting the basic concepts in one place with numerous references for in-depth study.

Another, and more ambitious goal is to promote understanding between the electrical and heat engineering specialists. The electrical and heat engineering disciplines were born at different times and under different circumstances. These have developed independently and in complete isolation from each other. Electrothermal technology is at the confluence of these two disciplines. A remarkable synergistic effect could result from their coming together.

## 2. ELECTROTHERMAL TECHNOLOGY AND INDUSTRIAL PROCESSES

Since the time of Vulcan's forge, considerable progress has been made in industrial heating, and thermal processes have become very diversified (tempering, annealing, polymerization, drying, fusion, lyophilization, etc.). However, a fundamental fact remains: the progress of industry is closely linked with the availability of energy, choices of energy sources, scarcity of alternative energy and technological innovations which modify the relationship between different energy sources.

Beyond mere considerations of cost and availability which justify its use in industrial heating, electrothermal technology owes not a small part of its development to the following characteristics:
— ease of measurement, control and regulation;
— ease of confinement to a specific area;
— high energy conversion efficiency;
— possibility of obtaining high temperature levels;
— high energy density;
— ease of decentralization and modular design;
— ease of automation and duplication of heating conditions;
— control of work atmosphere;
— high product quality;
— decreased pollution and improved working conditions.

Above all, the most remarkable property of electricity as a source of thermal energy is its adaptability. It can be used for heating in many different ways. With

respect to energy transmission, electric heating processes are often divided into two major categories:
— indirect heating — the transfer of energy from heat source to heated object obeys the conventional heat transfer principles;
— direct heating — the electric current flows through the heated object where energy is released. Heat transfer then follows the common thermodynamic laws.

Electrothermal processes are separated into direct and indirect heating

| Direct heating | Indirect heating |
|---|---|
| — Direct resistance heating (conduction)<br>— Induction heating (electromagnetic)<br>— Dielectric heating: (microwave & radiowaves)<br>• high frequency<br>• ultra high frequency (microwave)<br>— Electron beam heating<br>— Laser heating<br>— Electric arc direct heating | — Indirect resistance heating<br>— Indirect heating by electric arc<br>— Infrared heating<br>— Plasma heating |

**Fig. 1**

The table in Figure 1 lists the different electrothermal processes by separating them into two categories.

For any industrial process, this fundamental difference in basic techniques can be used to determine the best possible electrothermal procedure or combination of procedures. In particular, when a problem arises involving a choice between electricity and fossil fuel as a source of energy, care must be taken not to consider electrothermal equipment as a furnace where burners are simply replaced by electrical resistance heaters. It is imperative to seek the most efficient electrothermal process. When resistance heating is selected, equipment should not be a simple replica of a fuel-fired furnace, but must be designed for the maximum possible efficiency. Several choices are available: decentralized heating, assessment of energy density, heat recovery system efficiency, etc.

To facilitate this type of analysis, practically the whole classification of electrothermal processes, given above, should be considered. In fact, each chapter of this book is dedicated to one method of electric heating. However, to better account for their industrial applications, direct and indirect arc heating are presented in the same chapter. For the same reasons, indirect resistance heating is given, albeit somewhat arbitrarily, in two separate chapters (actually in three chapters, since infrared radiation heating, traditionally considered as separate from resistance heating, is, in fact, a special case of the latter):
— resistance heating — application to electric furnaces;
— heating with enclosed electrical resistance.

Each chapter has the same structure, except for some variations required by differences in each technique:
— principles and theory of heating methods under consideration;
— characteristics of this method of heating;
— general features of industrial installations;
— applications and limitations of the technique;
— economic considerations, in most cases;
— references.

Each chapter is handled as a unit, as independently as possible from the remainder of the book, but comparisons between different methods are made frequently. This modular concept involves some overlap, but enhances the consistency of each chapter.

Besides the above-mentioned heating methods, other techniques which sometimes make use of electrothermal processes exist; in particular the heat pump and ultraviolet radiation, even though electricity is not the heat source. In the heat pump, electricity provides mechanical energy only, not heat energy, since it drives an engine enabling a fluid following a thermodynamic cycle to transfer heat from a cold source to a warm source.

Appendix 2 of this chapter briefly presents the principle and industrial applications of the heat pump.

Ultraviolet radiation's effect on material is radiochemical and not thermal. Its industrial applications are few, and consist mainly of accelerated hardening of inks and varnishes, chemical synthesis and sterilization of food and pharmaceutical products (5), (7), (18).

The table in Figure 2 gives an estimate of nominal power rating and electricity consumption of electrothermal installations in France since 1978, with the exception of infrared radiation and laser heating, for which no data are available.

## 3. ELECTROTHERMAL TECHNOLOGY AND ELECTROMAGNETIC RADIATION

Among electrothermal techniques, electromagnetic radiation occupies a particularly important place. Its various applications include:
— high — or ultra-high frequency dielectric (hysteresis) heating;
— infrared radiation heating;
— induction heating;
— electron beam heating;
— laser heating;
— indirect resistance heating (heat transfer by radiation).

In reality, all these techniques form a cohesive physical whole. In all applications of these processes, energy transfer involves bombardment by "energy particles", which are either "parts of matter" (electrons) or "matter-free particles" (photons) associated with the electromagnetic field waves. These rays (for example, light) are characterized in a vacuum by wavelength $\lambda$ and frequency $f = c_o/\lambda$, where $c_o$ is the speed of electromagnetic wave propagation in a vacuum, or about $3 \times 10^8$ m/sec.

Rated power and electricity consumption
of electrothermal installations in France (34)

| Method of heating | Power (MW) | | Annual consumption (GWh) | |
|---|---|---|---|---|
| Furnaces and resistance heating equipment: | 2,785 | | 5,945 | |
| • melting of metals | | 26 | | 51 |
| • heat treatment of metals | | 1,402 | | 3,294 |
| • glasses and ceramics | | 111 | | 198 |
| • food processing | | 867 | | 1 832 |
| • other applications | | 379 | | 570 |
| Furnaces and induction heating equipment: | 766 | | 1 537 | |
| — Low frequency (under 60 Hz): | | | | |
| • melting of metals | | 338 | | 700 |
| • other applications | | 62 | | 147 |
| — Intermediate frequency (60 — 10 kHz): | | | | |
| • melting of metals | | 155 | | 313 |
| • other applications | | 104 | | 213 |
| — High frequency (over 10 kHz) | | 107 | | 164 |
| Furnaces and arc heating equipment: | 1,911 | | 6,755 | |
| • production of steel and castings | | 1,200 | | 2,800 |
| • production of ferroalloys | | 550 | | 3,000 |
| • production of calcium carbide | | 40 | | 300 |
| • other applications | | 121 | | 655 |
| Furnaces and conduction heating equipment: | 21 | | 90 | |
| • including fusion under electroconductive slag | | 11 | | 28 |
| High frequency dielectric heating | 37 | | 70 | |
| Microwave heating | 2 | | 3 | |
| Electron beam | 5 | | 3 | |
| Plasma heating | 33 | | 51 | |
| Heat pumps: | 64 | | 204 | |
| • manufacturing processes | | 40 | | 177 |
| • space heating | | 24 | | 27 |
| Igneous electrolytic furnaces | 408 | | 3,520 | |
| TOTAL | 6,032 | | 18,178 | |

**Fig. 2**

The frequency f is associated with energy quanta (photons). A single quantum of energy is W = hf where h is Planck's constant, $6.6 \times 10^{-34}$ J/sec. The emission of these rays and their absorption by different bodies is of fundamental importance in electrothermal processes. These different points are treated in the chapter for each type of heating, since each radiation type has specific heating characteristics.

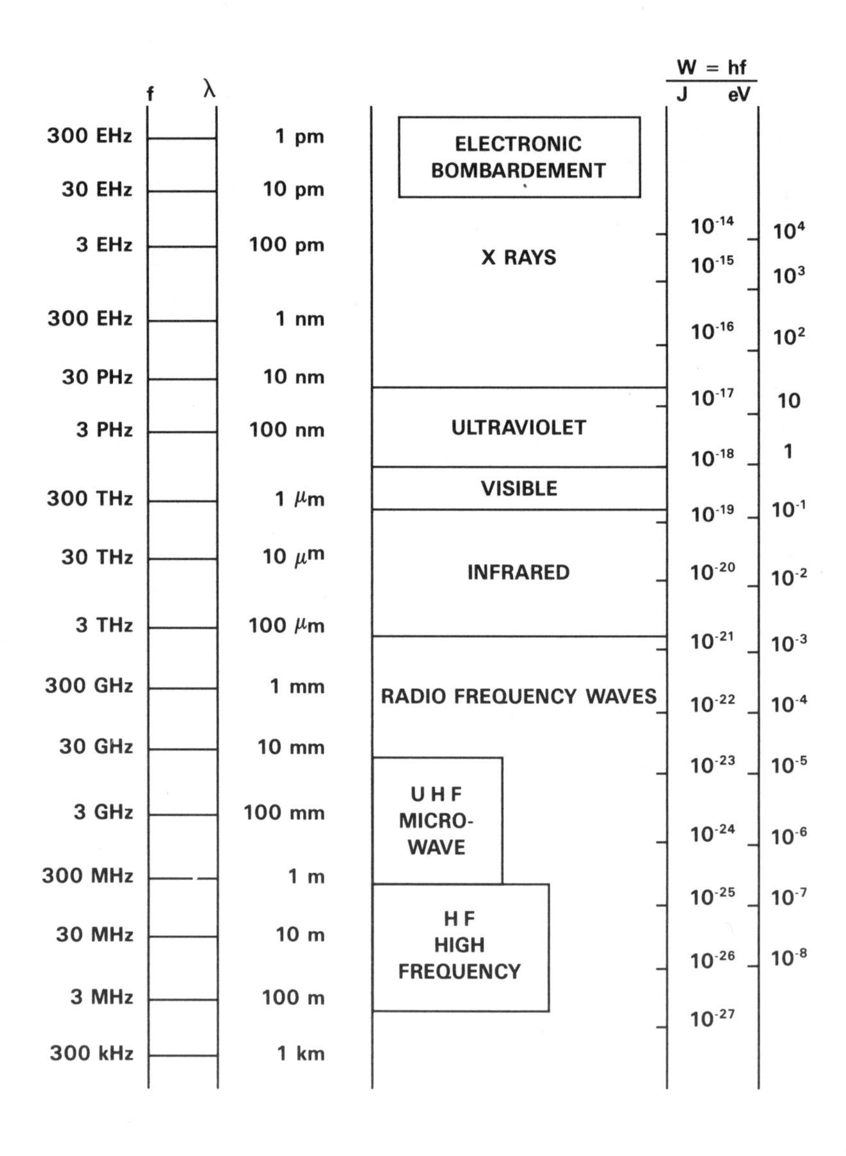


**Fig. 3.** Electromagnetic spectrum — Logarithmic scale (24) 1 eV = $1.6 \times 10^{-19}$ J.

## 4. DEVELOPMENT OF INDUSTRIAL ELECTROTHERMAL TECHNOLOGY

Since the turn of the century, the use of electricity for industrial heating has continued to grow even with the occasional lowering of fossil fuel prices. This growth should accelerate in the future, but will involve new developments and major innovations. This evolution tends to follow some definite directions (25), (30), (31), (32), (33a, b, c), (37), (38), (39).

### 4.1. Increasing power densities

In industrial heating, productivity increases are often synonymous with an increase in power density. Several factors contribute to the increase of power density in electrothermal processes, but two trends are apparent:
— improving conventional processes;
— recourse to special processes.

#### *4.1.1. Improving conventional processes*

In this category are the increase of power density in heating by infrared, induction, arc or resistance. An increase of the power density in resistance heating is brought about essentially by:
— improvement of metal alloys used in the manufacture of electric resistance elements, with respect to their maximum operating temperature;
— progress in the development of instruments, in particular temperature gauges with low thermal inertia and also in thyristor applications;
— development of special, non-metallic resistance heaters operating at high temperature;
— better understanding of heat transfer between electrical resistance heaters and heated objects.

Combining these methods can, in many instances, double or even triple the achievable specific power. For example, the heat density at the wall of a furnace containing a load at 900°C can be increased from 15 — 20 kW/m$^2$ to 30-40 kW/m$^2$.

The same considerations apply to infrared heating because emitters are electrical resistors. However, other types of advances are involved in this method of heating: improvements in emitter reliability, increase in emission temperature (short wavelength infrared), introduction of radiation reflectors, protection of materials in the event of conveyor stoppage or cooling of the emitters (with energy recovery for increased efficiency). Thus, while conventional infrared heaters give power densities of 15-80 kW/m$^2$ at the furnace wall, the most advanced units produce densities of 200-300 kW/m$^2$, and have the same operating life span.

In induction heating, increases in power density are obtained principally by using thyristor solid state static generators, and from a better knowledge of electromagnetic and thermal phenomena.

Thus, for example, the alternating frequency generators with inverters used in melting and heating metals before casting and in a host of other applications permit, because of frequency adjustment to optimum value, increases of 5-15% in the heating rate while, thanks to higher efficiency and less heat loss, reduce specific consumption by 10-20%. In channel-type induction furnaces, improvement of the knowledge of electromagnetic and thermal phenomena results in the design of new types inductors and an increase in power density exceeding 25%. Power densities encountered in induction heating present a wide range of values (50 — 50,000 kW/m$^2$) because of constraints imposed by certain phenomena. For example, in heating metals, an excessively high power density can lead to surface oxidation before casting.

The development of design and operation know-how for electric arc furnaces lead to the UHP (Ultra High Power, or Ultra High Productivity) concept, which can provide specific power 1.5 to 2 times greater than that possible with conventional furnaces. Increase in specific power is obtained by the better understanding of electrical and thermal phenomena characterizing electric arcs and the introduction of new furnace control techniques (electronic regulation of electrode positions, computer-based operation control, etc.). In UHP furnaces, power density can reach 2,500 kW/m$^2$ at the bath surface. (This value is given only as an illustration; it is rarely used in relation to arc furnaces. Usually the power density is expressed in kilovolt-Amperes per ton of capacity).

These data demonstrate that the recently attained or expected gains in power density are very substantial, and that vast reserves of productivity exist in conventional electrothermal processes.

### *4.1.2. Recourse to special processes*

Heating methods which twenty years ago were practically never used are now beginning to appear in large-scale applications. The majority of these electrothermal methods provide very high power densities. This is especially evident in the case of plasma, electron beam and laser heating which give 5 x $10^6$, $10^{10}$ and $10^{16}$ kW/m$^2$ power densities respectively. These densities are, in most cases, available only for small surface areas. The concentration of very high power in a small area is an advantage in some industries where great operating precision is required. The minimum surface areas heated by these areas are of the order of $10^{-3}$, $10^{-7}$ and 5 X $10^{-8}$ cm$^2$, which makes these methods very useful in electronics and micromechanics.

Dielectric, high frequency or ultra high frequency (microwave) heating enables higher power densities to be obtained. While not as spectacular as the preceding methods, these generally provide densities of between 30 and 100 kW/m$^2$ but compare favorably with densities obtainable by conventional procedures used in many large-scale processes which give densities in the 5-20 kW/m$^2$ range (e.g. convection heating).

In addition, the latter two methods can provide very large power volume densities giving an important advantage.

#### *4.1.3. Range of power densities*

The application areas and conditions of different electrothermal processes are very different, and techniques are not always interchangeable, but their wide power density ranges allow their use in many situations.

Power density of electrothermal processes

| Heating technique | Power density ($kW/m^2$) |
|---|---|
| Indirect resistance | 5 -60 |
| High frequency dielectric | 30 -100 |
| Infrared | 10 -300 |
| Ultra high frequency (microwave) | 50 -500 |
| Induction | 50 $-5.10^4$ |
| Conduction | 100 $-10^5$ |
| Arc | $10^3-5.10^5$ |
| Plasma | $10^3-5.10^6$ |
| Electronic beam | $10^4-10^{10}$ |
| Laser | $10^5-10^{16}$ |

**Fig. 4**

### 4.2. The "energy saving" requirement

The evolution of electrothermal technology will have to reflect the increasing concern about energy conservation. Because of its importance, an entire section (5) is dedicated to this subject.

### 4.3. Development of electronics

In the paragraph describing the power density increases, it was stated several times that the progress was linked to the introduction of electronic systems. The importance of electronics is not limited to only this aspect.

Electrothermal technology, like many other areas, is actually undergoing a revolution due to the arrival of electronic systems in their many forms. The areas of modern electronics development are often interdependent, but several directions can be recognized:
— electronic regulation;
— solid state static converters;
— automation
— microprocessors, microcomputers and optimization.

#### *4.3.1. Electronic regulation*

Traditionally, the control of electrothermal processes is based on electromechanical systems. These systems, in general, provide more accurate process regulation than can be achieved in heating with fossil fuels. Their performance, however, is always limited by characteristics common to all electromechanical devices, especially the inertia of their mechanical components which impose constraints on their response times. Electronic modules have been progressively substituted for electromechanical devices to provide comparison and signal amplification, but most switching relays continue to be electromechanical with much greater inertia than that of electronic elements. Thus, for example, in resistance furnaces, the current supply to heating elements is provided via electromechanical switches, and power modulation is provided by the "On-Off" system (i.e. the resistor current is turned off) or "On-partly on-off" systems (current flows through an electromechanical switch to a delta, and then to a star connected resistance circuit. For a more detailed description, see the chapter on resistance furnaces).

Therefore, the development of solid state switches represents a major innovation: it enables entirely electronic control systems to be designed. These switches consist of thyristors (basic relay consists of two head-to-tail, parallel-connected thyristors) which provide power modulation at speeds comparable to one alternating current cycle (phase angle regulation) or with several cycles (wavetrain regulation). Inertia of these control systems is very low because the power modulation is carried out at the level of a fraction of a sinusoidal current wave at 50 Hz, or, at most, at a level of a few cycles. Combined with temperature sensors with low thermal inertia, these systems provide very fine temperature regulation with precision determined only by that of the sensors. Thus, for example, it becomes possible to design resistance furnaces that reach temperature accuracis of $\pm$ 0.5°, thanks to this new control system.

Thus, novel applications and new products, such as semiconductors and other technical devices become possible.

This type of current supply regulation is not limited to temperature adjustment and to resistance furnaces. For example, these systems can be used to:
— smooth out the load curve (electric power demand) of the current-using apparatus;
— use entirely static devices, without any moving parts, that have long service life and require infrequent and simple maintenance procedures;
— increase the maximum temperature obtainable and the durability of heating elements in resistance furnaces, and increase power density and enhance performance of certain resistor types (silicon carbide, molybdenum disilicide);
— improve the performance of glass melting furnaces heated directly by electric current;
— simplify the construction of hot water or steam boilers with electrodes having rather complex electromechanical control systems;
— design static systems for suppressing voltage surges and the electric arc flicker phenomenon (statocompensators).

### *4.3.2. Frequency converters*

Heating by electromagnetic induction requires higher frequencies than are available from the power grid; therefore, frequency converters are needed. For intermediate frequency generation, electromechanical systems (motors and alternators) are commonly used. They are durable equipment but have limitations, and in particular, lack output flexibility. Thyristor systems help overcome these limitations by increasing power density and furnace output rate, and at the same time provide high energy conversion (see chapter on heating by electromagnetic induction).

### *4.3.3. Automation*

The automation of electrothermal processes advances constantly, responding very well to control. Automation is used to regulate temperature cycles, to manipulate products and to adjust various process variables such as chemical composition of solutions, duration of immersion of a work piece in a bath, adjustment of a coating thickness, checking the mechanical characteristics of products, controlling deviation of product quality from specifications, etc.

Automation often increases productivity, eliminates repetitive operations which are often unhealthy and boring, saves raw materials and energy, and makes it possible to develop products which previously could not be made.

Automation makes use of an ever-increasing number of electronic components. Microcomputer applications and performance should expand considerably over the next few years.

### *4.3.4. Microprocessors, microcomputers and optimization*

Mainframe computer-based data processing has been successful for many years in monitoring and controlling electrothermal processes, as for example in the operation of a large metal remelting furnace with either induction or arc heating. However, such equipment is still relatively rare and is used only by very large factories.

As in other areas, micro-based data processing should soon expand very rapidly. A microprocessor can be defined as an integrated microelectronic circuit contained on a silicon chip only a few square millimeters in size, and which has programmable logic and memory. Silicon chips continue to grow in number as integration techniques progress. Most microprocessors function as the central processing unit (CPU) of a microcomputer. Other functions (memory, input-output interfaces, etc.) are performed by other components. After receiving input signals, a microprocessor carries out calculations by combining several programs and delivering the output commands.

There are two completely different types of microprocessors:

— a digital controller, consisting essentially of a microprocessor, memory and input-output modules. Instructions can be provided externally or produced internally by a program stored in memory. The digital controller is based on algorithms (P.I.D. type) or uses numerical calculation capability;

— automatic programmable controller, an on-off device similar to a conventional relay. Its use is spreading rapidly because of the considerable savings in its installation and operational flexibility. Modifications are implemented by changing the program and not the wiring. Some advanced versions of programmable controllers have input-output functions enabling their use as monitors and regulators.

Soon, these systems will be installed in an ever growing number of electrothermal units, not only in new industries such as electronics, but also in established industries such as the metal processing or food industries. These will place the data processing science in the service of electrothermal units of moderate size, for which mainframe computing would not be economical, and will lead to optimization of electrothermal manufacturing processes.

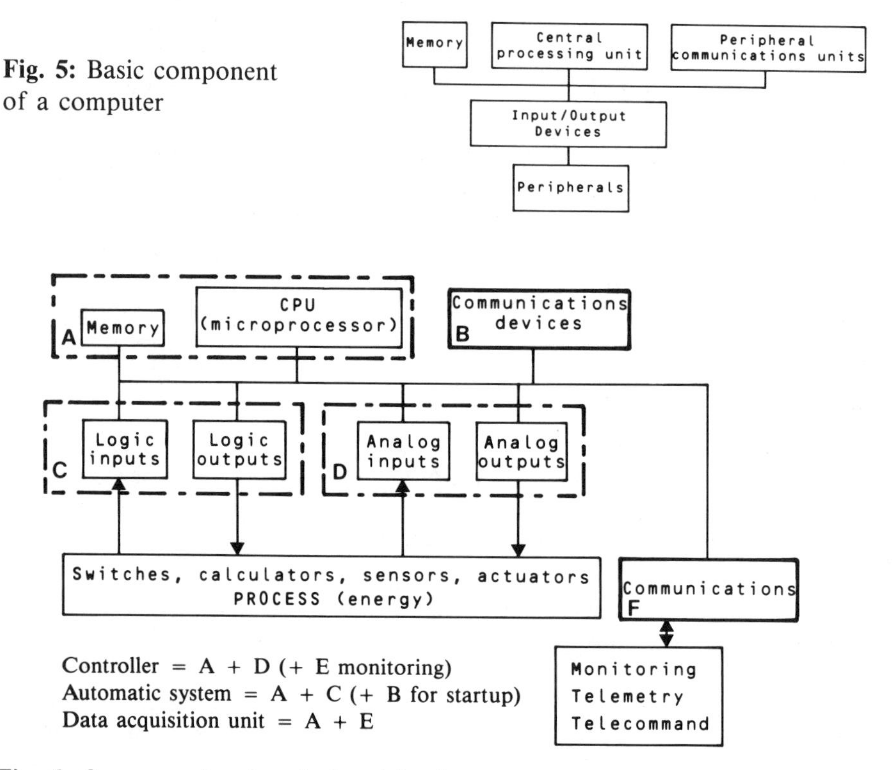


**Fig. 5:** Basic component of a computer

**Fig. 6.** Components of an industrial microcomputer (31)

### 4.4. Development of electric power costs and electrothermal technology

The cost of electric power has always had a profound effect on electrothermal manufacturing processes. It is evident that the cost of electricity, to some extent, determines how competitive these processes are in relation to processes using fossil fuels, but this is not the only factor to consider. Electrothermal techniques can offer advantages which compensate for any energy price differences (raw material economy, product quality improvement, fewer rejects, new product development, increase of production rate due to greater power density or more efficient energy transmission, improvement of working conditions, pollution reduction, etc.).

The price structure has considerable influence, not only on the ability of electrothermal processes to compete, but also on the selection of a particular process. For a long time, for example, price discounting during off-peak hours (which without going into the detail of electricity rate schedules, corresponds to the night hours when generators and electricity distributing equipment idle) prompted energy storage (for example, large capacity water tanks of 100,000 liters or more, melting metals at night and pouring in daytime, heating work space by tiles preheated at night with immersed electric elements, etc.). In France, this trend will no doubt be enhanced by the development of nuclear energy and hot water storage methods by large industrial consumers.

These developments shall be accompanied by innovations such as the storage of hot water combined with energy recovery, or harnessing "new" energy sources such as solar radiation, decentralization of energy-producing facilities to bring them closer to the consumer (in contrast to fuel-fired boilers, the combined energy production-distribution efficiency of which is rather low). The monitoring of water heating during off-peak hours should further lower the average per-kilowatt cost to industries.

Profound changes can be expected in France from nuclear plant electricity production. The power cost differences, now based on the time of the day, will gradually be replaced by seasonal differences. These exist even now but to a limited extent only. Electric power delivered at high-to-moderate voltage should be cheaper during the "summer" (actually during six to seven months of the year, including the conventional summertime). This seasonal variation of electric power costs could bring about new solutions to satisfy the heating requirements of industries. Solutions involving multi-energy sources could thus appear. Electrothermal equipment, providing its installation costs were acceptable, could be used only during certain months of the year as a substitute for fossil fuel; for example, during part of a winter day or during low rate periods. These solutions could be attractive to industries consuming large quantities of hot water and steam. This idea of multi-energy sources is a fertile research area and a beginning for various optimization schemes.

### 4.5. Introduction of electrothermal technology into new areas

For a long time, electrothermal processes have been flourishing in industries that require high temperatures such as metal production/ processing and ceramics. Electric furnaces and other heating systems giving a rapid temperature increase are easy to control at high temperatures and provide high temperature yields and advantages specific to each process.

On the other hand, electrothermal technology is used little in industries such as food processing, textiles or chemicals, where temperatures required are moderate, generally below 250°C. In these industries, heating by heat transfer media, such as hot water or steam, is required and the economic appeal of electrothermal technology is not so apparent. Despite high energy conversion efficiency, the savings provided by electrothermal heating are not substantial enough to overcome the cost difference. However, this situation is now undergoing far-reaching change because of the following factors:

— price shifts favoring electricity;

— innovations such as the heat pump, high frequency dielectric heating or microwave heating, and also improvements in conventional methods such as infrared and resistance heating;
— energy recovery designs based on electrothermal techniques;
— introduction of decentralized heating (hot water storage, small electric boilers, etc.);
— development of two-energy and multi-energy systems.

Electrothermal techniques appear to be well adapted to the needs of industries which will develop rapidly in the future: Electronics, bio-engineering, fine chemical production and the food industries should provide new applications.

## 5. ELECTROTHERMAL TECHNOLOGY AND ENERGY CONSERVATION

Energy conservation is an absolute necessity even in the areas of electric power and electrothermal processing. Conservation involves two areas:
— lowering the consumption of a given electrothermal process, most often in a plant;
— selection of the most power-efficient electrothermal process.

### 5.1. Electricity conservation in an electrothermal process and in plants

In most industries, electricity accounts for only a few percentage points of the overall production costs. Optimum use of electricity, which is easily attainable, only rarely preoccupies plant management. There are many occasions when electricity can be conserved profitably. Each plant is a special case, but systematic analysis of the following items should help define a conservation plan, not only for electrothermal types of equipment, but also for most electrical systems (8), (26), (28), (30), (32 a, b, c, d), (33 c), (34), (36) and (38):
— electricity distribution in plants;
— electric motor efficiency;
— lighting;
— electrothermal equipment;
— rates.

#### *5.1.1. Distribution*

From its point of entry, electricity is led to the various user stations. Well planned distribution can provide appreciable savings:

1. Choice of the most economical cables. Electrical cables are often sources of excessive heat loss. This is caused by two factors: progressive, and often thoughtless increase in current flow as the plant expands, and inefficient use of high performance cables (better insulation provides an increase in electric current carried). It is important to optimize cable dimensions.

2. High voltage distribution using substations, instead of low voltage distribution, where the same electric power is transported by a 20,000 V cable and a 380 V cable; the former carries only one fiftieth as much current as the latter. Using the 380 V cable enables cable size to be reduced by a factor of ten,

and energy losses to be cut by a factor of a hundred. The current transport costs are reduced by several percent.

3. Correct placing of capacitors. A high overall power factor, cos $\varphi$, reduces electricity costs because it lowers distribution losses. However, compensation of overall cos $\varphi$ masks the real situation in the plant, since a low power factor is due to equipment characteristics and use. Therefore, insertion of capacitors just before the equipment lowers cos $\varphi$ at this point instead of at the head of the distribution system and provides savings of up to 10%.

### *5.1.2. Electric motor efficiency*

Using an electric motor with a higher power rating than necessary and, of course, prolonged operation with the clutch disengaged, cause the following:

— lower efficiency: a motor running at 30% of its rated power provides an efficiency of about 70%, while at full load, this reaches 90%;

— lower power factor; the cos $\varphi$ of a motor operating under the above-mentioned conditions is only 0.46, while at full load, it is 0.87.

Excessive supply line voltage can produce a significant decrease in cos $\varphi$ in a small motor of a few kilowatts, i.e. about 0.2 for a motor operating at a voltage 10% too high.

Correct sizing of motors and their intelligent operation are therefore significant in energy conservation. Innovations, such as variable speed motors (electronic speed variation) and high efficiency motors facilitate improvements in energy conservation.

### *5.1.3. Lighting*

Good illumination in work places ensures security, improves productivity and enhances comfort, if:

— a correct illumination level is selected;

— lighting is provided only where it is needed;

— the system is well maintained;

— natural lighting is taken into consideration;

— efficient lamps are installed.

Thus, to provide illumination of 1,000 lm, it is possible to use:

— an incandescent lamp of 77 W;

— a white fluorescent tube of 16 W;

— a high pressure sodium lamp of 15 W.

Savings are also achieved by regular maintenance of the lighting system: light sources should be cleaned often and replaced near the end of their nominal life span. Loss due to deposited dirt leads to installation of lamps with powers 1.5 to 2.0 times that required for adequate illumination. The problem of lighting should be examined from the point of view of industrial building optimization. Natural lighting should be taken into consideration during the thermal design of the building.

### *5.1.4. Electrothermal equipment*

Electricity is widely used in melting furnaces, heat treatment, reheating, calcining, drying, polymerization, etc.

Numerous factors affect the energy performance of these operations, but a few points should always be taken into consideration:

1. *Insulation*: It is necessary to keep insulation in good condition, to keep doors sealed, and to use new low thermal momentum materials, such as ceramic-fibre- refractories (refer to the "Resistance furnaces" chapter for evaluation of possible savings). For example, placing polypropylene balls on the surface of baths decreases heat loss due to convection, radiation and evaporation by about 20 to 50%.

2. *Regulation*: precise regulation is synonymous with energy conservation. New thyristor regulators often provide the best solution in these cases. For example, thyristor controllers installed in a 1,500 kW high power resistance furnace make it unnecessary to use mixing turbines; this is an excellent example of the substitution of "know-how" for energy.

3. *Increase in furnace operating time*: better organization of production brings about substantial reductions in specific energy consumption (programming of production, decrease of idling and unnecessary downtime, use of large batch and campaign-type manufacturing, coordination of outside contract work). Thus, for example, when an induction melting furnace operates at 40% instead of 70% normal capacity, the specific energy consumption in kilowatthours per ton of poured metal increases by 15%.

4. *Energy recovery*: energy recovery is not, as some tend to think, associated only with fossil fuel-fired furnaces. Recovery is possible in all cooling systems and liquid or gas evacuations, such as cooling water from presses, organic materials eliminated during firing of ceramics, collection of sludges from industrial baths, as well as the cooling of solids (ceramic products, forged or heat-treated parts, etc.).

Numerous examples are given in the different chapters.

### *5.1.5. Rates*

A better use of electricity rates (reduced usage during peak periods, compensation of cos $\varphi$, management of demand surges, etc.) results in financial savings, not only for the enterprise, but also for the nation.

At the electricity generation level, the most efficient power plants should be producing all the time, and the less efficient only during the peak demand periods.

To conclude, despite small savings achievable by correcting any one of the above-mentioned items, when combined these can amount to up to 15% of a plant's energy costs. In most instances, these corrections can be achieved without large investment; the capital recovery period should rarely exceed two or three years.

### 5.2. Electrothermal technology and energy conservation

Evaluation of an investment project should include, as was indicated at the beginning of this chapter, an analysis of technical aspects, economics (in particular energy availability and costs), financing and social impact. Therefore, at each plant, a complete evaluation should be carried out before any investment. On the other hand, in most countries, changes in the energy situation prompt authorities to adopt legislation favoring energy conservation, and thus promote competition between the various energy production and utilization techniques. Electricity, that "secondary" energy which is often generated from other energy sources, finds itself in a peculiar situation. It is necessary to define an equivalence coefficient between electricity and "primary" energy sources, such as coal or petroleum. A disadvantage which is sometimes mentioned is that, when electricity is used for heating, the primary energy is wasted because one of the fundamental laws of thermodynamics, usually called the Carnot cycles, limits the efficiency of thermal power plants to about 35%. For the comparison of energy efficiency, it is common to assume that for each 1 kWh delivered to an industrial enterprise, the power plant consumes 10.46 MJ (2.5 th) of fuel ([1]).

This equivalence is acceptable, despite being much less than ideal, since electric power is also supplied by nuclear and hydraulic power plants. However, it completely ignores energy conversion within a plant where electric heating is favored over conventional heating.

An objective evaluation should begin with the final product and proceed backward, adding up all the energy consumed by its source along the energy chain, without neglecting the real yields in each step.

The preceding equivalence is often misleading; only a thorough analysis of all the pertinent industrial operations gives a true picture of the real energy efficiency. This analysis would show that when 1 kWh is required for an electrical process, frequently 3 or 4 th are needed when competing processes are used. How can this surprising result be explained? From reading thermodynamic texts too rapidly! Substitution of electricity for fossil fuels does not mean that we waste electricity.

It would not be wise to produce steam on a very large scale in electric boilers without taking advantage of the decentralization possible with electricity. Practice shows that, in the chemical industry, electric boilers placed close to the point of use can successfully compete with some of the central systems.

The changeover from fossil fuels to electricity in many instances requires adaptation of other work methods, other fabrication procedures and, above all, another approach to energy selection. Electricity requires no heat transfer fluid; it is changed into heat at a point where it is needed by the method best suited to a particular situation. In fact, electricity can be used in any of the following ways:

---

([1]) The therm (th) has recently disappeared as a conventional energy unit. The equivalences to be used are th = 4.185 MJ = 1.16 kWh. For primary energy, 1 ton-equivalent petroleum (t.e.p.) = 1.5 ton-equivalent coal (t.e.c.) = 10,000 th.

— joule effect;
— electromagnetic induction;
— infrared, ultraviolet, high and ultra high frequency radiation;
— electric arc;
— plasma;
— force modes (heat pump, separation by semi-permeable diaphragms, compression);
— electron beam.

Each of these techniques offers advantages that can result in energy savings:
— direct heating by the joule effect (i.e., by current flowing through the heated object) dissipates energy directly inside a piece of material, avoiding all problems associated with energy transmission from a heat source to the heated object;
— induction heats a metal (or a conductor of electricity) directly with induced currents;
— infrared transfers energy from the emitting source to the heated body without heating the intervening space, provides high power density concentration and has low thermal momentum;
— ultraviolet radiation substitutes radiochemistry for conventional thermal effects;
— heat pumps transfer energy from a cold source to a warm source, expending relatively little energy;
— microwaves, strategically located in a production chain, speed up heat transfer considerably and heat the core of objects that are poor heat conductors.

Many studies of furnaces and industrial heating systems demonstrate that the installation of electrothermal devices in many cases provides energy savings: examples are given in several chapters (8), (10), (22), (23), (26), (28), (3O), (321 c, d, g and h), (33 c, f and g), (34), (35).

## 6. APPENDIX 1. HEAT TRANSFER AND HEATING OBJECTS

Many technical papers deal with the physical laws governing thermal effects. This appendix is only a brief summary of fundamental ideas; specifics for each electrothermal technique are described in the corresponding chapters (1), (2), (3), (4), (6), (7), (9), (21).

### 6.1. Relation between heat and temperature

Temperature and heat are linked by two phenomena:
— when an object receives a certain amount of heat, its temperature increases or a transition occurs (fusion, vaporization, sublimation);
— when two objects are at different temperatures, the hotter one loses its heat to the colder and their temperatures tend to equalize.

Heat transfer involves three mechanisms:
— conduction;
— convection;
— radiation.

However, when heat is supplied by electrothermal means, the above transfer modes can be bypassed because other phenomena are involved, such as the passage of current or electromagnetic waves through the heated object.

## 6.2. Heat transfer

The three heat transfer modes correspond to different physical mechanisms. In industrial practice, these all occur simultaneously, but their relative importance varies as the application and technology change.

### *6.2.1. Conduction*

Thermal conduction (not to be confused with heating by electrical conduction or direct heating of an object by electric current passing through it; see chapter on "Heating by conduction") is propagation of heat from molecule to molecule in a body or in several solid, contiguous bodies, without any movement of the medium or any intervention of movement in the heat transfer. This transfer mode is characteristic of heat transmission in solids, or between solids in contact. In liquids and gases, thermal conduction also occurs, but, except in very viscous media, its effect is marginal when compared to that of convection.

#### *6.2.1.1. Fourier's law*

The analysis of thermal conduction is based on Fourier's law, which deals with the relationship between heat flux and temperature gradient in each part of a body:

$$\vec{D} = -\lambda \vec{G},$$

Where:
$\vec{D}$ is the density gradient of thermal current (heat flux per unit of surface area);
$\vec{G}$ is the temperature gradient;
$\lambda$ is specific thermal conductivity of a given material.
This relationship can be expressed for a one-dimensional medium or in cases where the temperature, for reasons of symmetry, depends on only one coordinate (plane, cylinder, sphere), as follows:

$$\boxed{\frac{d\Phi}{ds} = -\lambda \frac{d\theta}{dx},}$$

$\Phi$ heat flux in watts
$\theta$ temperature in degrees Celsius
$s$, where heat transfer occurs
$\chi$, coordinate of the point in meters
$\lambda$, thermal conductivity in watts per meter per degrees Celsius.
The above vector relationships represent two laws:
— direction of heat flow coincides with that of the temperature gradient;
— heat flux per unit of surface area is proportional to the temperature gradient.

The — sign indicates that the heat flow is in the direction of temperature decrease, i.e. from the warmer body to the cooler body.

Application of Fourier's law to a body with dimensions dx, dy, dz, and the use of integral calculus enable us to calculate the quantity of heat transferred by conduction through a body of any shape (this relation is usually called the "heat equation", in which t represents time):

$$\frac{\partial \theta}{\partial t} = \frac{\lambda}{c\gamma}\left(\frac{\partial^2 \theta}{\partial x^2} + \frac{\partial^2 \theta}{\partial y^2} + \frac{\partial^2 \theta}{\partial z^2}\right),$$

$c$ and $\lambda$ are the specific heat and density; $\lambda/c\gamma$ is known as thermal diffusivity. The expression presupposes $\lambda$ to be temperature independent. This is true only in the first approximation; $\lambda$ varies with temperature and moisture content of materials; these should be considered when high precision is required. When a system is in equilibrium, $\partial\theta/\partial t = 0$, the preceding relation becomes $\Delta\theta = 0$, $\Delta$ where is the Laplace operator.

Fourier's law enables thermal conduction in systems at equilibrium or in the dynamic state to be studied. We can determine heat losses through furnace walls, rates of temperature increase in heated bodies, temperature gradients in heated bodies and optimum heat insulation of equipment.

#### *6.2.1.2. Permanent modes*

When a solid is subjected to constant limit conditions, after a certain length of time (determined by study of variable mode) it reaches a permanent state characterized by temperature equilibrium at each point. The paragraphs that follow give some results obtained while studying permanent modes of objects with simple geometric shapes. Thermal conductivity is considered constant.

##### *6.2.1.2.1. Plane composite wall*

Walls of resistance furnaces, or tanks containing loads heated by immersion heaters, are rarely constructed of a single material. Usually, these are made of several materials, each for a specific purpose (refractory, thermal insulator, corrosion-resistant coating, etc.). When these walls have flat parallel planes, it is easy to determine heat transfer through them, providing the surface is assumed to be an infinite plane (in practice, end effects have to be considered).

Each wall component is traversed by the same flux, which is represented, where there are n plates with faces kept at temperatures $\theta_1$ and $\theta_{n+1}$, by:

$$\Phi = \frac{\theta_1 - \theta_2}{e_1/\lambda_1 S} = \frac{\theta_2 - \theta_3}{e_2/\lambda_2 S} = \ldots = \frac{\theta_n - \theta_{n+1}}{e_n/\lambda_n S},$$

or by adding:

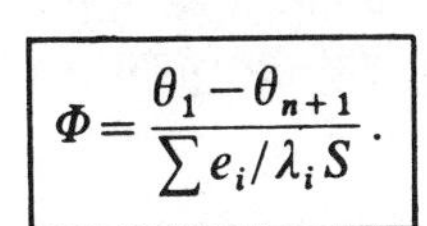

$$\Phi = \frac{\theta_1 - \theta_{n+1}}{\sum e_i/\lambda_i S}.$$

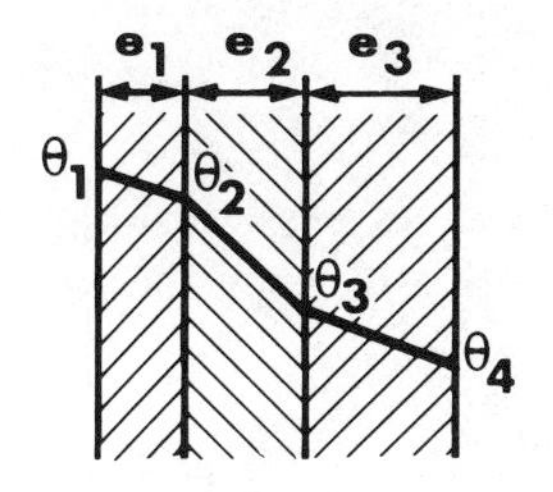


**Fig. 7:** Heat flux in a composite wall.

The expression R = $e_i/\lambda_i S$ is, by analogy with electricity, known as the thermal resistance. Some authors use this term to designate the thermal resistance per surface unit, $e_i/\lambda_i$). The thermal resistance of a plane composite wall is equal to the sum of the thermal resistances of its components.

The temperature difference across a plate is proportional to its thermal resistance:

$$\theta_i - \theta_{i+1} = \Phi \frac{e_i}{\lambda_i S}.$$

In the plate of row $i$, the temperature gradient is given by:

$$\theta_x = \theta_i - \frac{\theta_1 - \theta_{n+1}}{RS} \frac{x}{\lambda_i}.$$

where $x$ is the distance of the row $i-1$ plate from the wall surface.

The temperature distribution graph is therefore a series of straight lines with slopes inversely proportional to the thermal conductivity coefficients of the materials from which the different wall components are made.

This analysis demonstrates the importance of the conductivity coefficient λ in heat transfer by conduction. Low values indicate thermal insulators, and high values good heat conductors. The λ values are provided by material producers. The reference temperature and λ = f($\theta$) curve are very important in selecting these materials. The table in Figure 8 gives the thermal conductivity values of some representative materials.

*6.2.1.2.2. Cylindrical composite wall*

Many thermal units have cylindrical shapes (ducts, pipes, circular furnaces). For two concentric surfaces of radii $r_1$ and $r_2$, kept at temperatures $\theta_1$ and $\theta_2$ respectively, the heat flux per unit length is given by:

$$\Phi = 2\pi\lambda \frac{\theta_1 - \theta_2}{l_n(r_2/r_1)}.$$

The thermal resistance per unit length of the tube is then:

$$R = \frac{l_n(r_2/r_1)}{2\pi\lambda}.$$

This can be represented by an expression similar to that found for the plane wall, where R = $(r_2 - r_1)/\lambda S$, and when the following condition is satisfied:

$$S = \frac{2\pi(r_2 - r_1)}{l_n(r_2/r_1)} = \frac{S_2 - S_1}{l_n S_2 - l_n S_1},$$

i.e., when S is the logarithmic average of the interior and exterior surfaces.

Thermal conductivity of selected materials in Watts per meter per degree Celsius at 20°C (the international standard λ is given at 23.9°C).

| Material | λ | Material | λ |
|---|---|---|---|
| Copper | 394 | Heating conductor supports | 1.0 to 2.5 |
| Aluminum | 208 | Refractory clay (1,000°) | 1 to 1.3 |
| Slag | 109 | Magnesia | 0.8 |
| Iron-steel | 46 to 56 | Refractory casing | 0.5 |
| Chrome-nickel | 12 | Sillimanite | 1.4 |
| Granite | 2.5 to 3.5 | Slag wool | 0.03 to 0.06 |
| Glass | 1 to 1.5 | Glass wool | 0.04 |
| Sandstone | 0.9 | Asbestos | 0.1 to 0.2 |
| Porcelain | 0.7 to 2 | Diatomaceous earth | 0.08 |
| Soil | 0.5 | Cork crumbs | 0.04 |
| Water | 0.6 | Fine sand | 0.06 |
| Oil | 0.17 | Air | 0.025 |
| Bakelite | 0.25 | Hydrogen | 0.160 |
| Cardboard | 0.1 to 0.4 | Water vapor (100°) | 0.020 |
| Soft rubber | 0.1 to 0.2 | | |
| Rubber foam | 0.05 | | |
| Leather | 0.12 to 0.15 | | |
| Loose cotton | 0.05 | | |
| Wool | 0.04 | | |
| Oakwood | 0.15 to 0.20 | | |

**Fig. 8**

For a cylindrical tube composed of n layers, the heat flux expression becomes:

$$\Phi = \frac{\theta_1 - \theta_{n+1}}{R},$$

Where

$$R = \sum \frac{l_n(r_{i+1}/r_i)}{2\pi\lambda_i}.$$

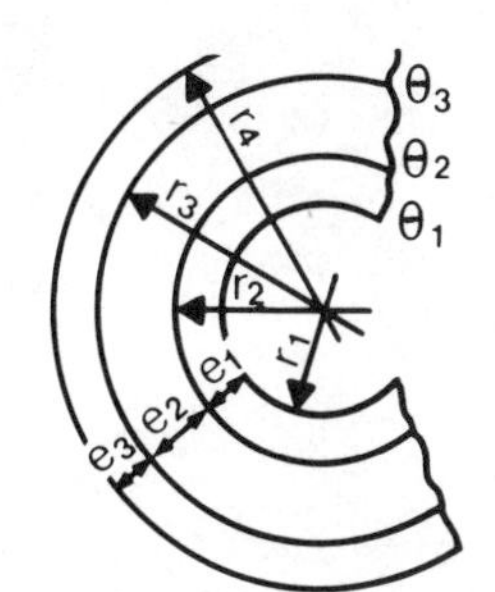


**Fig. 9** Heat Flux in a Cylindrical Composite Wall

The temperature gradient inside the row *i* layer is given by:

$$\theta_r = \theta_i - \frac{\theta_1 - \theta_{n+1}}{R} \frac{l_n(r/r_i)}{2\pi\lambda_i},$$

or

$$\theta_r = \theta_i - \frac{\theta_1 - \theta_{n+1}}{RS} \frac{e}{\lambda_i} \quad \text{with} \quad S = \frac{2\pi(r - r_i)}{l_n(r/r_i)} \quad \text{and} \quad e = r - r_i.$$

*6.2.1.2.3. Spherical composite wall*

For a solid enclosed by two concentric spheres with radii $r_1$ and $r_2$ and kept at temperatures $\theta_1$ and $\theta_2$ respectively, the heat flux is given by:

$$\Phi = \frac{4\pi\lambda(\theta_1 - \theta_2)}{(1/r_1) - (1/r_2)}.$$

The thermal resistance of a spherical vessel is then:

$$R = \frac{(1/r_1) - (1/r_2)}{4\pi\lambda}.$$

This can be expressed by a formula similar to that found for the plane wall, i.e. $R = (r_2 - r)/\lambda S$ and $S = 4\pi r_1 r_2 = \sqrt{S_1 S_2}$, i.e. S is now the geometric average of the internal surface $S_1$ and the external surface $S_2$.

For a composite wall of *n* layers, the thermal flux expression becomes:

$$\Phi = \frac{\theta_1 - \theta_{n+1}}{R} \quad \text{with} \quad R = \Sigma \frac{(1/r_i) - (1/r_{i+1})}{4\pi\lambda_i}.$$

The temperature gradient inside a layer of row $i$ is given by:

$$\theta_r = \theta_i - \frac{\theta_1 - \theta_{n+1}}{R 4\pi \lambda_i}\left(\frac{1}{r_i} - \frac{1}{r}\right),$$

or

$$\theta_r = \theta_i - \frac{\theta_1 - \theta_{n+1}}{RS}\,\frac{e}{\lambda_i} \quad \text{with} \quad S = 4\pi r_i r \quad \text{and} \quad e = r - r_i.$$

*6.2.1.2.4. General case*

Industrial thermal systems are often of relatively complex shape. Furnaces in general are parallepiped in shape. The heat losses of furnaces and other equipment are calculated per unit surface area; thus, the total heat loss of a uniform and homogeneous surface is obtained by multiplication. The heat flux penetrates the internal surface and flows through expanding sections to the external surface which has the maximum surface area. Different authors (Langmuir, Trinks, Paschkis, Heiligenstadt, etc.) have tried to find an equivalent average surface taking wall thicknesses and furnace dimensions into consideration. These procedures were published and are mentioned here for reference only. In the most general case of a surface with any shape, it is necessary to turn to the fundamental equations of heat conduction and use analog or numerical methods to solve the problem. In most cases, the equations given above are sufficiently accurate.

***6.2.1.3. Variable modes***

Heat conduction in a variable mode occurs during heating of the charge in a furnace, heating of furnace walls and heat recovery. It is necessary to return to the general equation:

$$\frac{\partial\theta}{\partial t} = \frac{\lambda}{c\gamma}\Delta\theta.$$

Now, the significance of $a = \lambda/c\gamma$, the thermal diffusivity, becomes clear. The specific heat c and mass $\gamma$ impede heating of an object; this decreases as c increases, while the rate of temperature increase is proportional to the coefficient of thermal conductivity.

Mathematical study of these variable modes is, in general, extremely complex. An example of this is the semi-finite body, bounded on one side by a plane surface, the temperature of which is initially uniform ($\theta$ — $\theta_0$). At time t = 0, the surface of this body is rapidly brought to temperature $\theta_1$ = 0 and kept at this temperature. Integration of the preceding equation and application of the initial and boundary conditions gives the temperature gradient as a function of time t and distance x from the surface:

$$\boxed{\frac{\theta}{\theta_0} = 2G\left(\frac{x}{\sqrt{2at}}\right),}$$

where G is the Galton function (the integral of a Gaussian curve) defined by:

$$G(X) = \frac{1}{\sqrt{2\pi}} \int_0^X e^{-u^2/2}\, du.$$

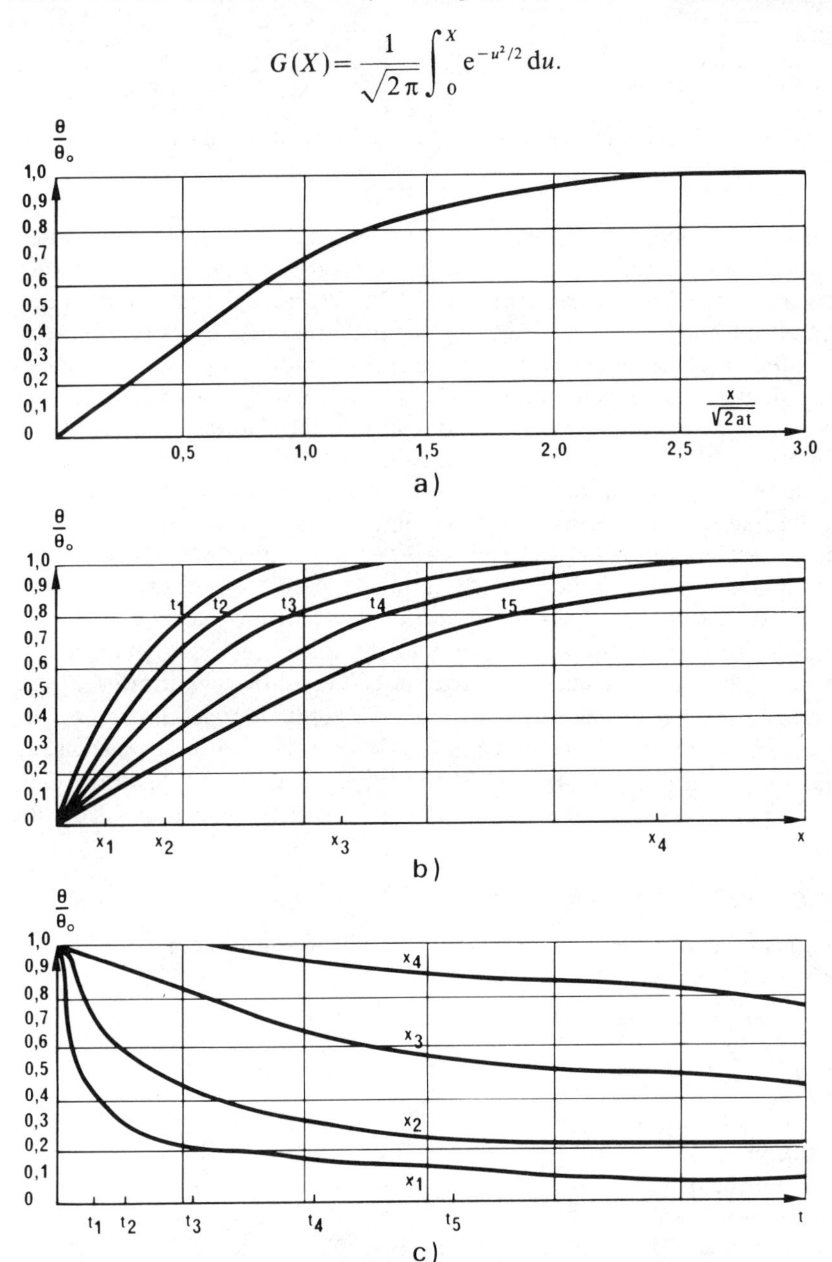


**Fig. 10.** Temperature gradient in a semi-finite mass ($\theta_1 = 0$);
(a) variation of $\theta/\theta_0$ as a function of $\chi/\sqrt{2at}$
(b) temperature distribution at any given instant
(c) temperature variation with time at a given distance x from the surface

If the body surface temperature becomes $\theta_1$, the temperature gradient equation becomes:

$$\frac{\theta-\theta_1}{\theta_0-\theta_1}=2G\left(\frac{x}{\sqrt{2at}}\right).$$

Many methods have been proposed for study of more general cases.

### *6.2.2. Convection*

Convection is characteristic of heat propagation in a fluid, gas or liquid, where the molecules are in motion. This phenomenon is basic to the study of thermal effects in fluids and in fluids touching solids. When a solid is immersed in a moving fluid, the heat is exchanged by thermal conduction between the solid and the fluid particles touching it. Because of fluid motion, a point on the solid's surface does not remain in contact with a fluid particle for long. The fluid particles constantly collide with the solid body which exchanges heat successively with different fluid particles. Similarly, inside the fluid, heat is transferred by conduction between neighboring but constantly changing particles. The elementary heat transfer mechanism is conduction, but the overall phenomenon of heat transfer, known as convection, is the result of convection and fluid movement obeying specific laws. The heat transfer becomes more intense as fluid agitation increases. Two types of convection are generally recognized:
— natural convection, in which the fluid motion is caused by temperature differences between the solid and the fluid, or between different parts of the fluid;
— forced convection, in which the fluid motion is produced by mechanical means, independent of thermal phenomena and used to speed up the heat transfer.

The study of heat transfer by convection is therefore related to fluid flow. Only the most important principles of this extremely complex branch of science are given here.

#### *6.2.2.1. Coefficient of convection*

The heat flux exchanged between a solid and its surroundings through an interface is given by:

$$\boxed{\mathrm{d}\Phi=\alpha\,\mathrm{d}S(\theta_a-\theta_s),}$$

$\Phi$, heat flux transferred in watts;
$S$, exchange surface in $m^2$;
$\theta_a$, ambient temperature, °C;
$\theta_s$, temperature of solid, °C;
$\chi$, convection coefficient in watts per $m^2$ per °C.

The thermal convection coefficient is not an absolute constant, but depends on many factors, such as:
— fluid characteristics: conductivity coefficient, specific heat, mass, viscosity, velocity, temperature;

— flow type: turbulent or laminar;
— wall characteristics: shape, dimensions, roughness, temperature, cleanliness;
— respective position of the wall and the fluid; the angle of attack of the fluid.

This does not imply that the heat flux transmitted should be proportional to the temperature. The study of convection is largely an experimental science and uses many dimensionless numbers which characterize fluids from the physical and thermal viewpoints (Nusselt, Biot, Prandtl, Reynolds, Grashoff, Margoulis, Stanton, Peclet numbers etc.). The purpose is to express heat transfer by accessible values and explain by analogy (in geometrically similar heat transferrs, heat transfer conditions are the same when these dimensionless numbers have the same values). Numerous empirical expressions exist such as the following:

$$N_u = \frac{\alpha d}{\lambda} = C P_r^m R_e^n G_r^p \left(\frac{L}{d}\right)^q,$$

$N_u$ = $\alpha d/\lambda$ Nusselt number;
$P_r$ = $v/a$ = $\eta c/\lambda$ Prandlt number;
$R_e$ = $wd/v$ Reynolds number;
$G_r$ = $(d^3 g\beta(\theta_a — \theta_s))/v^2$ Grashoff number;
$\alpha$, convection coefficient in Watts per square meter per degrees C;
$\lambda$, conduction coefficient in Watts per meter per degrees C;
$d$, characteristic dimension, for example diameter of a tube or of a sphere in meters;
$L$, length, for example of a tube, in meters;
$a$, thermal diffusivity equal to $\lambda/c\gamma$ in square meters/second;
$c$, specific heat in Joules per kg per °C;
$\gamma$, mass of fluid in kg/m$^3$;
$\nu$, kinematic viscosity in m$^2$/sec;
$\eta$, dynamic viscosity, equal to $\nu\gamma$, in kg per meter per second;
$w$, fluid velocity in meters per second;
$g$; acceleration due to gravity in meters per second squared;
$\beta$, volume expansion coefficient, equal to $(1/V)(\mathrm{d}V/\mathrm{d}T)$ where
$\beta = 1/T$ for gas, in $(°\mathrm{C})^{-1}$;
$T$, absolute temperature, degrees K;
$\theta_a — \theta_s$, temperature difference between a fluid and a wall, in degrees C;
$C$, constant
$m$, $n$, $p$, $q$, experimental indexes.

The Reynolds number characterizes the conditions of fluid flow. Expressed as a function of dynamic viscosity, this becomes:

$$R_e = w^2 \gamma / \frac{\eta w}{d}.$$

It therefore represents a ratio between the forces of momentum and friction. A turbulent flow favors heat transfer (Reynolds number larger than 2,500-3,000).

The Prandtl number characterizes the physical and thermal aspects of the fluid itself.

The Grashoff number represents perturbations in the fluid produced by stresses due to temperature differences in a current. These stresses are negligible when forces causing a fluid to flow predominate, but become high, for example, when an enclosed furnace wall is cooled.

The L/d ratio describes effects which occur in fluid flow (for example at the point of turbulence, near the entrance to a tube). Causes of repetitive or permanent turbulence, such as sharp projections from the walls, should be analysed and evaluated experimentally.

Extensive research has been conducted in determining, the convection $\alpha$ coefficient, which is a factor in equations describing many industrial processes.

### *6.2.2.2. Heat transfer by convection*

The convection coefficient varies considerably:
— for gases, from 2 to 100 W/m$^2$°C;
— for liquids not undergoing a phase transition, from 100 to 200 W/m$^2$°C;
— for liquids which evaporate or condense: 1,000 — 50,000 W/m$^2$°C.

The convection coefficient in gases is very small. Therefore, to accelerate thermal exchange, forced and not natural convection is used in most industrial applications; the effective convection coefficient then increases and becomes:
— for calm air, 3 to 15 W/m$^2$°C;
— for moving air, 10 — 150 W/m$^2$°C.

A good approximation for the convection coefficient in calm air is given by:

$$\alpha = A\,(\theta_a - \theta_s)^{0.25},$$

where:
A = 1.8 for vertical walls;
A = 1.3 for downward inclining horizontal walls;
A = 1.5 for upward inclining horizontal walls.

For tubes, the convection coefficient in calm air is given by:

$$\alpha = 1.32\,\frac{(\theta_a - \theta_s)^{0.25}}{d}.$$

For objects located in moving fluids, methods presented in the preceding paragraph must be used.

### *6.2.3. Radiation*

Heat transfer by radiation involves two objects at different temperatures, separated by a space transparent to the radiation. Thermal radiation is an electromagnetic phenomenon.

Heat transfer by radiation between surfaces of two solids obeys the Stefan-Boltzmann law. Its physical importance and general acceptance are due to its agreement with natural phenomena and its accuracy which is greater than that of other heat transfer laws. The Stefan-Boltzmann law has the following form:

$$\Phi = \varepsilon F \sigma S\,(T_1^4 - T_2^4),$$

$\Phi$, heat flux transmitted in Watts;
$T_1$, temperature of the emitting surface in degrees K;
$T_2$, temperature of the receiving surface in degrees K;
$\sigma$, Stefan constant, equal to 5.73 x $10^{-8}$W/m$^2 \cdot$ K$^4$;
$S$, surface of the emitting body
$\epsilon$, reciprocal radiation coefficient;
$F$, angular coefficient of the receiving surface with respect to the emitting surface.

The reciprocal radiation coefficient depends mostly on the surface properties of bodies with respect to the emission and absorption of radiation, but also on their shape and relative position. The angular coefficient is simply a function of the surface shapes and their respective positions. Heat transmission by radiation is of the utmost importance in industrial electrothermal technology. Infrared heating is, in fact, based on this energy transfer mode, and plays an important role in resistance furnaces. Therefore, it is interesting to present the principles of thermal radiation in some detail and to show relationships between theory and practical applications. The most important developments are included in the chapters "Heating by infrared radiation" and "Resistance heating". These show how a thorough understanding of radiation laws, combined with electronic monitoring and control technology, provide considerable improvements in the economy and performance of these furnaces.

### *6.2.4. Heat transfer through a wall dividing two fluids*

Many problems of industrial heat transfer concern transmission of heat from one fluid to another through a wall or series of walls and various media (heat losses from wall to furnace or other heat apparatus, heat transfers, etc.).

In the case of planar surfaces, or surfaces that can be mathematically expressed as planes, the total heat flux through a composite wall is often expressed as:

$$\boxed{\Phi = KS\,(\theta_2 - \theta_1),}$$

$\Phi$, heat flux in Watts;
$S$, exchange surface in m$^2$;
$\theta_2$ and $\theta_1$, temperature of two fluids in degrees C;
$K$, coefficient of total heat transmission W/m$^2$°C.

The two wall faces are not at the same temperature as the two fluids in contact with them. The heat transmission is due to:

— contact of the fluid with the wall by convection, or radiation and convection. The relative importance of these two modes depends on the medium and emission type (convection and conduction with liquids, radiation and convection in furnaces at high temperature);

— through the walls, essentially by conduction.

Coefficient K is expressed as:

$$\frac{1}{K} = \frac{1}{K_1} + \sum \frac{e_i}{\lambda_i} + \frac{1}{K_2},$$

or:

$$R = R_{c_1} + \sum R_i + R_{c_2}.$$

A composite wall surrounded by fluids is thus equivalent, from the thermal viewpoint, to several electrical resistances connected in series. The total thermal resistance ($R = 1/KS$) is equal to the sum of the wall resistance and the two interface resistances. This method is often used in designing heating systems for buildings. In a high temperature furnace, the effect of $R_{c_2}$ on total thermal resistance is small because the exchange coefficient is very large, and in practical calculations, $\theta_2' = \theta_2$. On the other hand, $\theta_1'$ can be very different from the ambient temperature $\theta_1$, and the exchange coefficient becomes very high. In reality surface temperature is not discontinuous, as shown in Figure 11, but changes very rapidly in fluids touching the wall.

In the case of non-planar surfaces, it is also possible to define an equivalent total resistance.

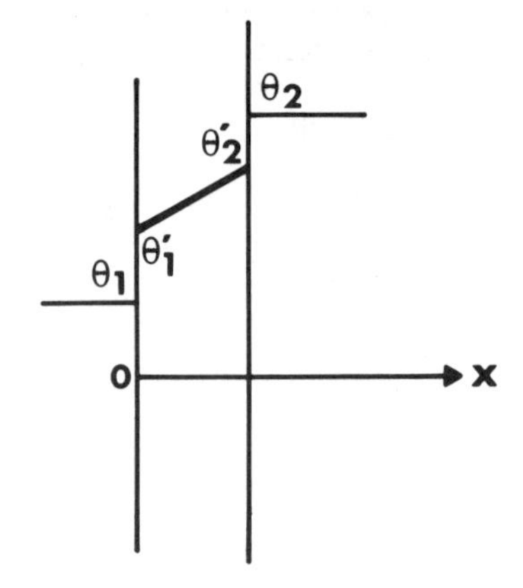


**Fig. 11.** Heat transfer through a wall dividing two fluids

Many methods have been proposed to calculate heat losses from furnace walls; the phenomena are complex, and calculations are largely based on experimental data.

### 6.3. Heating of objects

The combined heat transfer by radiation, convection and conduction constitutes a thermal process which raises the temperature of an object. It is not always possible to express this complex phenomenon by any one equation, or even by a set of equations. However, in many instances, an analysis based on the laws given above, given certain approximations and simplifications, enables problems to be solved with reasonable precision. Examples of these methods are given in a chapter dealing with encased electrical resistances when used for heating baths and moulds. The energy transmission, however, often occurs independent of heat transfer; it is important to take this into account when analysing the heating of objects.

The energy required to heat a body is equal to the sum of:

— sensible heat, which is expressed as $mc\ (\theta_2 - \theta_1)$, c, specific heat; $m$, mass to be heated; $\theta_2 - \theta_1$, temperature increase;

— latent heat of a phase change or of a chemical reaction, usually expressed as $Lm$ where $L$ is the latent heat or heat of reaction and m is the mass to be heated (in case of an exothermic reaction, $Lm$ is subtracted from the heat required).

Pressure and volume changes have hardly any effect on the specific heat of solids and liquids, but not so with gases. For gases, it is necessary to distinguish between the specific heat at constant volume $c_v$ and the specific heat at constant pressure $c_p$.

When a body to be heated contains several states of matter, the heat required for its heating is the sum of the sensible and latent heats.

These notions enable us to determine heat requirements, to calculate the power to be installed, to evaluate efficiency and to predict energy consumption. Examples are given in several chapters, particularly in the chapter on resistance furnaces.

## 7. APPENDIX 2: HEAT PUMPS AND THEIR INDUSTRIAL APPLICATIONS

### 7.1. Principle

Modern heat pumps are compressors operating like refrigerators but with an inverted thermodynamic cycle. These withdraw energy from a low temperature source (outside air, gas, liquid, etc.) and transfer it at a higher temperature.

Basically, a heat pump consists of:

— a compressor;

— two heat exchangers (evaporator and condenser);

— heat transfer fluid, which passes through a thermodynamic cycle by circulation between the heat exchangers, the compressor and the expansion chamber.

The economic and theoretical interest of this system lies in the fact that it can take out a quantity of heat $Q_f$ from the environment at a low temperature $T_f$, where it is available, and transfer a quantity of heat $Q_c$ at a higher temperature $T_c$, where it can be used for a relatively small amount of electricity corresponding to work $W$ of the compressor.

More precisely, thermodynamics states that:

$$\frac{Q_c}{Q_f} = \frac{T_c}{T_f} \qquad \text{(T in degrees Kelvin with TK} = \theta\ ^\circ\text{C} + 273\text{).}$$

$$Q_c = Q_f + W \quad \text{(principle of energy conservation).}$$

The theoretical performance coefficient of the system is therefore:

$$\text{Cop}_t = \frac{Q_c}{W} = \frac{T_c}{T_c - T_f} = \frac{\theta_c + 273}{\theta_c - \theta_f}.$$

The practical performance coefficient is:

$$\mathrm{Cop} = m\ \mathrm{Cop}_t = \mathrm{m}\frac{T_c}{T_c - T_f} = m\frac{\theta_c + 273}{\theta_c - \theta_f}.$$

The factor m is generally larger than 0.5. This is because the real performance efficiency is always lower than that expected from a perfect Carnot cycle, without losses or irreversibility.

The performance coefficient increases as the temperature difference between the cold and hot sources ($T_c$ — $T_f$) decreases. In practice, Cop is no larger than 3, providing $T_c$ — $T_f$ is lower than about 50°C.

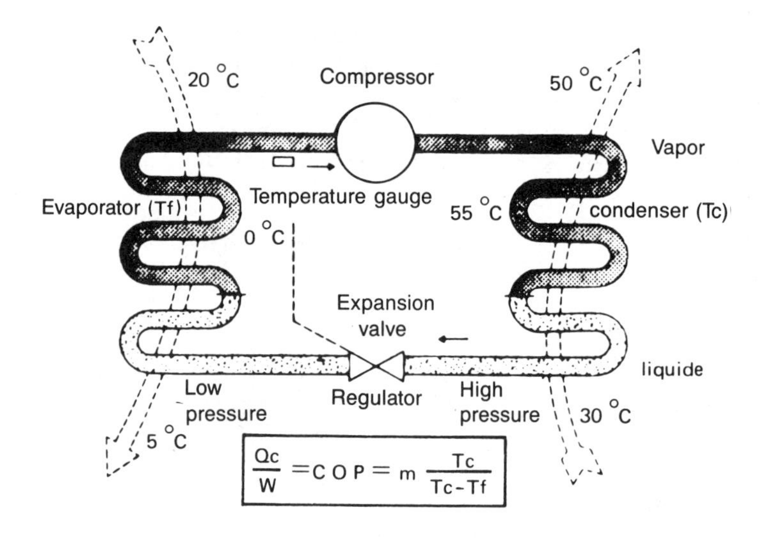


**Fig. 10.** Principle of a heat pump

Frequently, when this coefficient is more than 3, the use of a heat pump driven by an electric motor becomes uneconomical, even if the electricity could be provided by a power plant producing it from fossil fuel with an efficiency coefficient equal to 1, and transferred to the user with the same efficiency coefficient of 1, which is impossible.

Saving primary energy with the heat pump appears to be achievable only when the performance coefficient is 2, or preferably 1.5.

The heat pump could thus prove effective for:

— unit operations involving mass transfer (drying, concentration, evaporation, distillation, etc.) where latent heat lost can be recovered;

— heat transfers, expecially when there is need for heating and cooling.

Thermodynamics and industrial practice differentiate two types of heat pumps:

— *The indirect heat pump*

A heat transfer fluid undergoing a thermodynamic cycle travels in a closed loop; it serves only as an energy transporting agent. Such fluid is generally a halogenated hydrocarbon. The performance coefficients obtainable with this family of heat pumps are usually 2 — 5, which do not encourage more extensive use. However, it can be used to extract energy from any available heat source providing the temperature difference between the cold and hot sources are sufficiently small, and if the high initial investment can be cost-justified. The maximum possible temperature of the heat source is about 100°C, due to compressor technological limitations (13 a, b), (14 a), (15 b, d), (34).

**Fig. 13.** Diagram of an air dehumidification wood dryer (indirect heat pump)

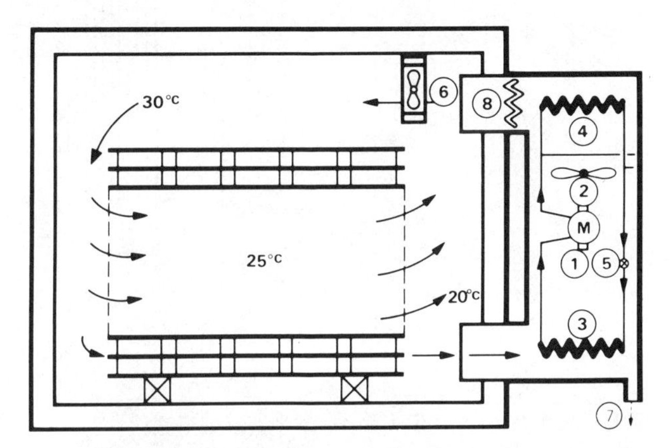


1. Compressor for heat transfer fluid
2. Blower circulating air inside the dehumidification chamber
3. Evaporator (cold element of the apparatus) where warm, humid air releases its water.
4. Condenser (hot element of the apparatus) where the dry, cold air is heated
5. Expansion valve for the heat transfer fluid
6. Blower circulating air in the dryer
7. Discharge of water extracted from wood
8. Electrical heating resistances for use if required

— *The direct heat pump*

In this type of heat pump, the heat transfer fluid is absent. Fluid to be evaporated, often (but not exclusively) water, takes its place.

In a conventional evaporator, hot vapor is condensed producing heat which, on the other side of the heat transfer surface, evaporates the fluid. In direct heat pumps (often called heat pumps with mechanical vapor recompression), vapors from the evaporator are recompressed by an electric compressor for use at higher pressure and temperature as the heating medium for evaporation.

For this system to be economically viable, the temperature difference between the evaporator and the condenser (there is only one heat exchanger, while the indirect heat pump requires two exchangers) should be small: lower than 20°C. In industrial installations, this difference is often less than 10°C, and the practical performance coefficient is very high, generally between 10 and 20 (13, a, c, d), (17), (19), (27), (29), (33e), (34).

Concentrated fluid

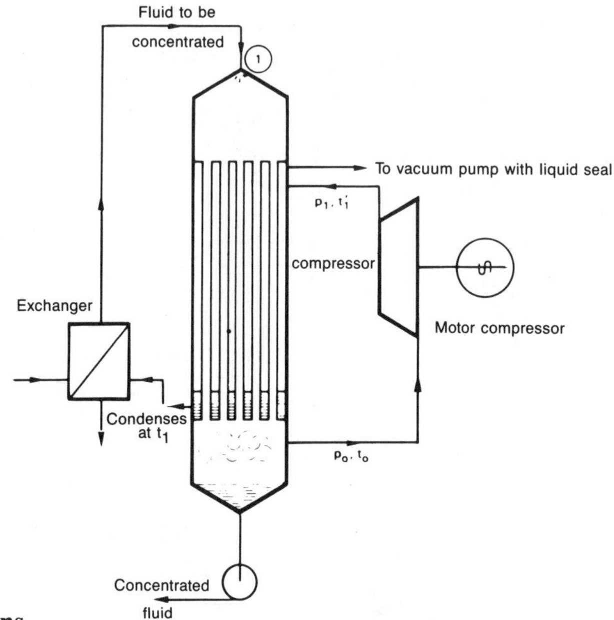


**Fig. 14.** Principle of an evaporator with mechanical recompression (27).

### 7.2. Industrial applications

The potential applications of heat pumps are numerous because they involve most of the heat transfer processes (drying, evaporation, concentration, distillation, etc.). By transferring heat from a colder to a warmer place, the heat pump can also be used to recover heat energy from hot liquids which would normally be wasted.

High investments involved in heat pump systems have, in the past, limited their use, especially with no other reason than conservation of energy. The application of heat pumps in drying wood or hides for leather manufacture is justified more by gains in quality and a decrease in the number of rejects than by the energy savings. However, high energy costs are beginning to make heat pumps more attractive. The return on investment is now somewhere between one and four years for most suitable industrial processes.

The following examples can be cited from existing heat pump applications:
— *the indirect heat pump* (11), (12 a, b, c, d, e), (13 b, d), (14 a, b, c), (15 a, b, c, d), (20), (33 d), (45), (35):

- drying wood, hides and furs,
- drying food products (hams, sausages, etc.)
- drying meat packing by-products (intestines, bladders, etc.)
- drying heat sensitive products,
- drying special ceramics,

• heat recovery from cooling, especially in rubber and plastic manufacturing, or food and agricultural product processing (most of these industries require heating as well as cooling) and metal surface treatment industries,
• drying photographic emulsions,
• production of hot water, making use of auxiliary heat pump condensers used in drying,
• drying plasterboard,
• heating buildings, especially by withdrawing heat from plastic strata (mixed electrothermal-geothermal application).

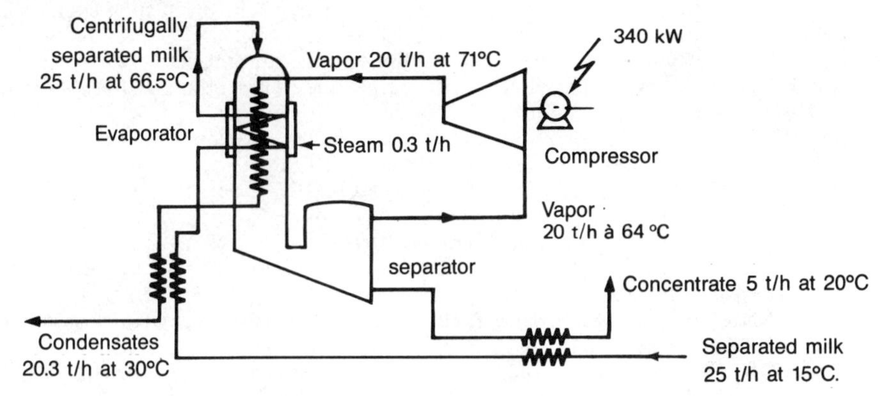


**Fig. 15.** Diagram of an installation for concentration of milk from which cream has been removed by evaporation with mechanical recompression (34).

— *for the direct heat pump* (13 c, d, e), (16), (17), (19), (27), (29), (32 e, f), (33 d, e), (34), (35):
• concentration of whey and other dairy products,
• concentration of distillery slops for upgrading,
• distillation of concentration of chemical products,
• concentration of nuclear effluents,
• concentration of brines,
• concentration of sugar syrups and apple juice,
• concentration of waste effluents for yeast-making,
• concentration of black liquors from paper mills,
• concentration of sodium aluminate and zinc sulphate,
• concentration of glycerol in soap-making,
• concentration of various meat industry products (cooking juices, effluents, etc.)
• drying sheet products (textiles, paper, etc.).

## 8. BIBLIOGRAPHY

[1] V. PASCHKIS, *Les fours électriques industriels*, Dunod, 1952.
[2] R. LOISON, *Chauffage industriel*, École nationale supérieure des Mines de Paris, 1956.
[3] W. TRINKS, *Les fours industriels*, Dunod, 1957.
[4] W. H. MAC ADAMS, *Transmission de la chaleur*, Dunod, 1961.
[5] F. LAUSTER, *Manuel d'électrothermie industrielle*, Dunod, 1967.
[6] W. HEILIGENSTAEDT, *Thermique appliquée aux fours industriels*, Dunod, 1971.
[7] KEGEL *et al.*, *Elektrowärme, Theorie und Praxis*, Verlag W. Girardet, Essen, 1974.
[8] W. L. HARRISON, Electroheat. Conserving energy in the production engineering industry, *Elektrowärme International*, n° 32, décembre 1974.
[9] J. GOSSE, *Rayonnement thermique*, Éditions Scientifiques Riber, 1975.
[10] J. BOUCHET et P. CLÉMENT, Crise énergétique et électrothermie, *Revue générale de Thermique*, n° 167, novembre 1975.
[11] P. PAUMIER, *Les séchoirs dans l'industrie de la viande*, Comité français d'Électrothermie, Journées du Touquet, 12-13 mai 1976.
[12] COMITÉ FRANÇAIS D'ÉLECTROTHERMIE, Journées de Strasbourg, 3-4 novembre 1976 :
*a*) R. CAPPA, *Le séchage en charcuterie industrielle;*
*b*) R. ROUSSEL, *Le chauffage de l'Usine Gardy par récupération sur les condenseurs de l'entrepôt frigorifique voisin, la STEF;*
*c*) C. FENIGER, *Quelques exemples de climatisation et de vinification sous régimes dirigés dans les caves de Champagne;*
*d*) L. JOST, *Récupération sur condenseurs de groupes frigorifiques industriels;*
*e*) J. CONAN, *Récupération de chaleur sur les condenseurs de groupes frigorifiques pour le séchage du malt.*
[13] COMITÉ FRANÇAIS D'ÉLECTROTHERMIE, Colloque de Versailles, 21-22 avril 1977 :
*a*) A. ROUX, *Principe de fonctionnement des pompes à chaleur industrielles;*
*b*) P. MANNONI, M. ORFEUIL et N. VION, *Applications industrielles de la pompe à chaleur dans quelques processus industriels;*
*c*) E. LEGENDRE et F. BONDUELLE, *Application de la pompe à chaleur à la concentration par évaporation et au séchage dans les industries agro-alimentaires;*
*d*) P. LAGUILHARRE, *L'évaporation à compression mécanique dans les industries alimentaires;*
*e*) G. DEBERON, *Possibilité d'amélioration du bilan thermique des unités d'extraction au solvant par l'utilisation de pompes à chaleur.*
[14] BULLETIN PAC-INDUSTRIE N° 5, EDF, Direction Générale, avril 1977 :
*a*) J. H. MORGAT, *Un séchoir industriel de carreaux de plâtre par pompe à chaleur;*
*b*) M. BATAILLE, *Exploitation du prototype de séchoir de carreaux de plâtre équipé d'une pompe à chaleur;*
*c*) N. TECULESCU, *Le séchage des carreaux de plâtre par pompe à chaleur : réalisme économique et faisabilité.*
[15] COMITÉ FRANÇAIS D'ÉLECTROTHERMIE, Journées de Grenoble, 17-18 mai 1977 :
*a*) M. CURT, *Chauffage d'ateliers et de bureaux par récupération sur process;*
*b*) M. BEUZIT, *Récupération de chaleur dans une fromagerie;*
*c*) P. PAUZE *et al.*, *Utilisation de la pompe à chaleur sur nappe phréatique dans une usine de presse;*
*d*) M. COMBETTE *et al.*, *Chauffage et rafraîchissement d'ateliers par pompe à chaleur sur nappe phréatique.*
[16] A. BERTAY et A. BONNELIE, La concentration du lactosérum par évaporation avec recompression mécanique des vapeurs, *Industries agricoles et alimentaires*, n° 9-10, septembre-octobre 1977.

[17] R. LELEU, L'évaporation par pompe à chaleur, *Bulletin PAC-Industrie*, n° 7, EDF, Direction générale, janvier 1978.
[18] F. GOUBET, Les applications de l'électricité dans l'industrie, *Travaux communaux*, février-mars 1978.
[19] J. C. DURAND, Installations d'évaporation par recompression mécanique des vapeurs, *Bulletin PAC-Industrie*, n° 8, EDF, Direction générale, avril 1978.
[20] A. YOT, Deux exemples d'utilisation de pompes à chaleur dans l'industrie des matières plastiques, *Bulletin PAC-Industrie*, n° 9, EDF, Direction générale, juillet 1978.
[21] J. SCADURA *et al.*, *Initiation aux transferts thermiques*, Technique et Documentation, 1978.
[22] J. VASSAUX, Le choix d'un matériel thermique dans la transformation du caoutchouc et des matières plastiques et les économies d'énergie, *Revue technique du Bâtiment*, n° 71, mars-avril 1979.
[23] P. MANNONI, *Rendements énergétiques et problèmes de choix d'équipement dans l'entreprise*, Comité français d'Électrothermie, Journées de Dinard, 9-10 mai 1979.
[24] A. BUSSON, *Les rayonnements électromagnétiques et leurs applications industrielles*, Exposé à la Société des Ingénieurs et Scientifiques de France, Journées d'Études « L'Électricité dans l'Industrie », 30 mai 1979.
[25] T. GAUDIN *et al.*, *Premiers éléments pour un programme national d'innovation*, Ministère de l'Industrie, 1979.
[26] M. ORFEUIL et J. M. PERIANI, Utilisation de l'électricité et économies d'énergie dans l'industrie, *La Technique moderne*, juillet-août 1979.
[27] G. ABECASSIS, L'évaporation avec recompression mécanique des buées en industrie laitière, *Bulletin PAC-Industrie*, n° 12, EDF, Direction générale, juillet 1979.
[28] C. MEDAN et M. ORFEUIL, Électricité et économies d'énergie dans l'industrie, *Revue française de l'Électricité*, n° 267, 1979.
[29] J. HUCHON, *La concentration par évaporation à recompression mécanique des buées dans le secteur laitier*, Journées du Comité français d'Électrothermie, Deauville, 23-24 octobre 1979.
[30] G. PIETRE-CAMBACEDES, Les applications de l'électricité dans l'industrie : progrès récents et économies d'énergie, *Travaux communaux*, janvier-février 1980.
[31] B. SARRETTE, *Électrothermie et microprocesseurs, applications actuelles et perspectives*, Journées d'Études du Comité français d'Électrothermie, Versailles, 6-7 mars 1980.
[32] REVUE GÉNÉRALE D'ÉLECTRICITÉ, n° 3, mars 1980, consacré au rôle de l'électricité dans les économies d'énergie :
*a*) J. BOUCHET, *L'électricité, facteur d'économies d'énergie et de matières premières dans l'industrie;*
*b*) A. BOURMAULT, *Moteurs électriques asynchrones triphasés à faibles pertes;*
*c*) J. MALHERBE, *Économies d'énergie dans la transmission de la force motrice et dans le réglage des processus;*
*d*) P. PAQUETEAU, *Importance de l'analyse des consommations énergétiques : gestion d'une centrale des fluides;*
*e*) J. HUCHON, *La concentration par évaporation à recompression mécanique des buées dans le secteur laitier;*
*f*) L. MULLER, *Distillation et recompression mécanique des vapeurs : optimisation technico-économique;*
*g*) J. L. MINGAUD et J. SOLA, *Analyse et modélisation du bilan énergétique des fours de traitement thermique;*
*h*) J. HEURTIN et J. GAULON, *Les fours à résistances et les économies d'énergie.*
[33] CONGRÈS DE L'UNION INTERNATIONALE D'ÉLECTROTHERMIE, Cannes, 20-24 octobre 1980 :
*a*) G. VAN DICK et H. SAVONET, *Évolution des applications électrothermiques industrielles au cours des dernières années;*
*b*) A. P. ALTGAUZEN *et al.*, *Einfluss des technischen Fortschritts auf die Entwicklung der Elektrowärme;*

*c*) P. MICHEL, R. THOMASSIN et M. ORFEUIL, *Facteurs économiques et techniques susceptibles de favoriser la pénétration de l'électricité dans l'industrie au cours de la prochaine décennie;*
*d*) P. MANNONI, M. PAYEN et J. VASSAUX, *Le séchage par pompe à chaleur dans l'industrie;*
*e*) R. FREYTAG et P. VIALLIER, *Un circuit original de récupération de chaleur par recompression de vapeur. Application au séchage en continu de matériaux en nappes;*
*f*) A. BERTAY *et al.*, *Consommations énergétiques comparées entre systèmes électriques et systèmes utilisant un combustible. Intérêt des mesures énergétiques;*
*g*) W. AYLOT *et al.*, *Energy saving and process heat recovery in electroheat plants.*
**[34]** DOCUMENTATION EDF, Paris, 1980.
**[35]** DOCUMENTATION ELECTRICITY COUNCIL, Londres, 1980.
**[36]** DOCUMENTATION APAVE, Paris, 1980.
**[37]** M. KIMURA *et al.*, *Flexible manufacturing system complex provided with laser*, août 1980, Tokyo et Cité des Sciences de Tsukuba.
**[38]** G. SEGUIER, *L'électronique de puissance*, Dunod, Paris.
**[39]** P. MICHEL, J. LHERMITTE et M. ORFEUIL, *New trends in electroheat technology*, Conférence au Congrès « Energy Use Management », Tucson, USA.

# Chapter 2

# Resistance Heating

## Electrical furnace applications

The resistance furnace is probably the most widely known and most frequently encountered electrothermal apparatus. Its use in fact dates from the early 1920's, and its technology has been improving ever since.

## 1. PRINCIPLE OF RESISTANCE FURNACE HEATING

### 1.1. Power dissipation in a resistance

The heat produced by resistance furnaces is based on Joule's law: any electrically conductive substance through which an electric current flows gives off heat. The electric power converted into heat is expressed by the equation:

$$P = UI = RI^2 = \frac{U^2}{R},$$

$P$, watts (W),
$U$, volts (V),
$I$, amperes (A),
$R$, ohms (Ω)

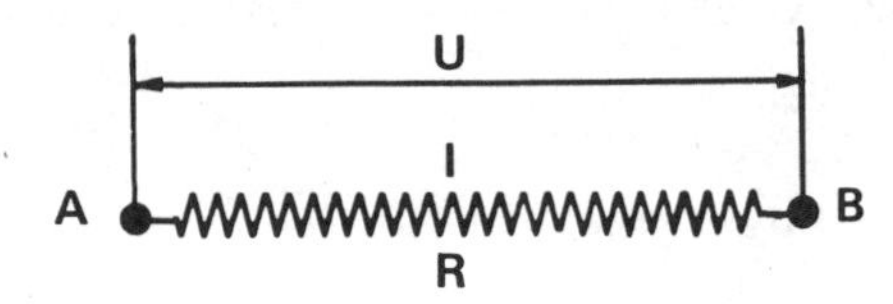


**Fig. 1.** Resistance heating principle

Ohm's law states the relationship between voltage, resistance and current.

$$U = RI.$$

If the same electric power is used for time t, the electrical energy converted into heat will be:

$$Q = Pt = RI^2 t,$$

where Q is expressed in joules and t in seconds.

In industrial applications, power is generally expressed in kilowatts, and energy in kilowatthours (kWh) representing the energy dissipated in one hour by a circuit absorbing a constant power of 1 kW (1 kWh therefore represents 3,600 kJ).

This assumes the various physical values are constant; if not, it is necessary to calculate the energy emitted as:

$$Q = \int_0^t U_t I_t \, dt.$$

In practice, for alternating current, constant physical values are used for each cycle. A load supplied with an alternating voltage $U_t$, through which current $I_t$ flows, with a phase angle $\varphi$ with respect to the voltage, absorbs active power P equal to:

$$P = UI \cos \varphi,$$

*U* and *I* are root mean square values (rms) of the voltage and the current over one cycle.

The current and voltage are practically in-phase in a resistance, and the previous expression can therefore be reduced to the simplified form of Joule's law, as stated above.

Some electronic regulation systems (phase angle operating thyristors), however, use fractions of a cycle to modulate the emitted power.

Present-day supply lines almost exclusively use three-phase alternating current; the power value depends on the resistor circuit (see paragraph 3.2.).

## 1.2. Heating a resistance furnace

The principle involved in resistance furnaces is extremely simple; schematically, it consists of a chamber heated by electrical resistances. This chamber, which is often known as a heating chamber or furnace, must be insulated to reduce thermal losses as much as possible. The charge to be heated is placed in this chamber. The generic term "furnace" is used to designate this type of heating equipment, but other names are often used in special cases. For example, in most cases, ovens are resistance furnaces operating at low or medium temperatures, less than 600 — 700°C. The name can also apply to ovens used as dryers or warming hoppers. Also, some equipment is similar to furnaces but differs in shape or application, as is the case with resistance hot air generators.

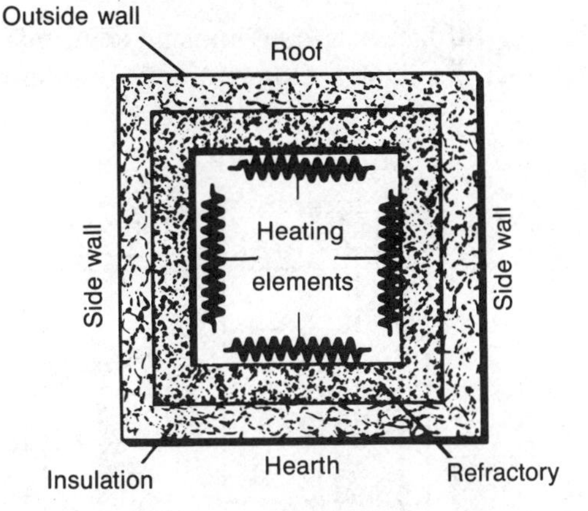


**Fig. 2.** Diagram of a resistance furnace

Resistance furnaces are indirect heating devices (as opposed to conduction heating or resistance heating by direct current flow: see "Heating by conduction"). The heating elements produce calorific energy P. This energy is then transferred to the substance to be heated and to the walls of the furnace. The charge absorbs a part, $P_1$, of the energy dissipated in the resistances (effective energy), but another part, $P_2$, remains unused and constitutes thermal loss. Therefore, the study of a furnace involves analysis of the following points:

— heat production,
— heat transfer,
— heat utilization.
  • absorption of heat by the substance to be treated
  • thermal losses and furnace efficiency.

Resistance furnaces are quite similar to conventional fossil fuel furnaces, although for maximum efficiency, they must often differ in certain ways. Resistance furnaces use the three standard thermal laws which determine heat transmission by radiation, convection and conduction, and which are dealt with in numerous publications. Therefore, in this chapter, we are more concerned with the specific characteristics of resistance furnaces: type and performance of various heating element resistances, implementation of these resistances, thermal transfer between heating elements and charge, resistance furnace applications, etc.

To simplify the presentation of this subject, encased element resistance heating, as well as conduction heating, are covered in a separate chapter.

Resistance furnace applications are extremely numerous in industry, and there is hardly an industrial sector which does not use them. This is due primarily to the ruggedness and reliability of such equipment, the ease with which they can be implemented, their maintainability and high efficiency.

## 2. CONSTITUTION AND CHARACTERISTICS OF RESISTANCE FURNACES

Essentially, resistance furnaces consist of:

— a heating chamber, which receives the charge usually made of refractory or semi-refractory materials. Insulating materials placed against the refractory materials provide furnace heat-containment; some refractory materials are also excellent thermal insulators (for example, refractory fibre-based ceramic products).

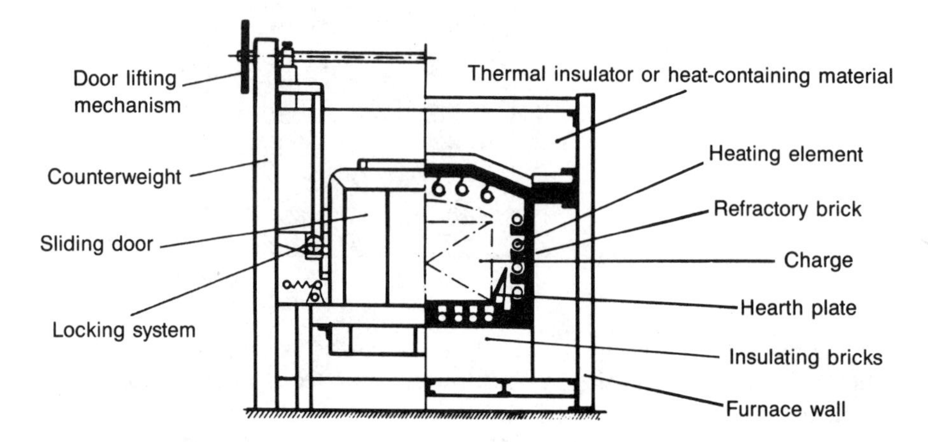


**Fig. 3.** General structure of a resistance furnace.

— a structural casing, generally metal;
— electric resistances forming the heating elements and their power supply system;
— charge temperature regulation and handling systems;
— auxiliary devices, air recycling turbines for forced convection furnaces or air mixing fans, cooling systems for some parts of the furnace (for example, handling rollers in high temperature furnaces), special blowers, etc.

### 2.1. Heat transfer in a resistance furnace

In a resistance furnace, heat transfer is made according to the three laws mentioned in the introduction:
— convection,
— radiation, and
— conduction.

Throughout a complete treatment cycle in an electric furnace, the three categories of heat transfer are encountered:
— transfer of heat from the heating element to the charge and the furnace (refractory, insulation and metal structure): heating of the furnace, both for temperature rise and for normal sustained operation;
— transfer of heat from the heated charge to the body of the furnace, and from the body to the exterior; i.e. natural cooling;
— special heat transfers: load cooling by forced convection, cooling of handling systems and some resistance terminations.

In general, heating is caused by heat transfer between the resistances and the charge, but a major role is also played by the walls, since they reflect most of the energy received from the heating elements (1), (2), (3), (12), (14), (15), (18), (32), (59), (79), (82).

### 2.1.1. *Conductive heat transfer*

Except in special cases, conductive heat transfer plays only a small role in heat transfer between the resistances and the charge. Conversely, thermal conduction controls:
— transfer of energy inside the charge and equalization of its temperature, and
— thermal losses through the furnace walls.

The furnace walls must be composed of materials which reduce thermal losses to a minimum. Thermal conduction inside the load can, especially for products with poor thermal conductivity or complex shapes, limit thermal flux and thus reduce the thermal gradient in the charge, decreasing the risks of distortion or local overheating.

### 2.1.2. *Radiant heat transfer*

Radiant heat transfer is fundamental in high temperature electrical furnaces operating at more than 500°C. Beyond this temperature, heat transfer between the resistances and the charge consists basically of radiation, and beyond 750°C, of radiation only.

Heat transfer takes place either directly between the resistances and the charge, or indirectly by reflection or re-emission from the furnace walls at high temperatures.

Thermal exchanges between the resistances, the load and the walls obey all the radiation laws, and in particular the Stefan-Boltzmann, Planck and Wien laws. With the usual thermal calculations, these laws can be used to calculate the characteristics of resistance furnaces. Paragraphs 3.6 and 5.1 are devoted to calculation of resistances and high power density resistance furnaces and highlight special applications of these laws, in particular the Stefan-Boltzmann law:

$$\boxed{\Phi = \varepsilon_{RC}\, SF\, \sigma (T_R^4 - T_c^4),}$$

$\Phi$, thermal flux exchanged between the resistances and the charge,
$T_R$ and $T_c$, resistance and charge surface absolute temperatures,
$\epsilon_{RC}$ coefficient allowing for emissivities $\epsilon_R$ and $\epsilon_c$ of the resistances and the charge,
$F$, furnace global shape factor,
$\sigma$, Stefan's constant,
$S$, area of emitting surface.

This expression shows the importance of using a heating system which emits at high temperature to obtain high power densities.

### 2.1.3. *Convective heat transfer*

When heat transfers between the resistance and the load use this heat transfer method, two techniques are implemented:

— *natural convection*: the temperature differences in the heating chamber move the internal fluid and heat transfers from hot areas (resistances) to the cold areas (charge);

— *forced convection*: fluid movements created by a mechanical process independent of thermal phenomena (mixing, recirculating turbines, etc.).

Convection plays an important role especially in furnaces and ovens operating at low temperature. Within this temperature range, convection, and in particular forced convection, can provide much higher power densities than radiation (although the use of infrared radiation offers even higher power densities) and excellent temperature homogeneity.

There are two types of forced convection:

— simple mechanical mixing of the air surrounding the charge, intended essentially to distribute temperature;

— forced air aimed directly at the resistances and then the charge so as to accelerate heat transfer, while providing excellent temperature distribution.

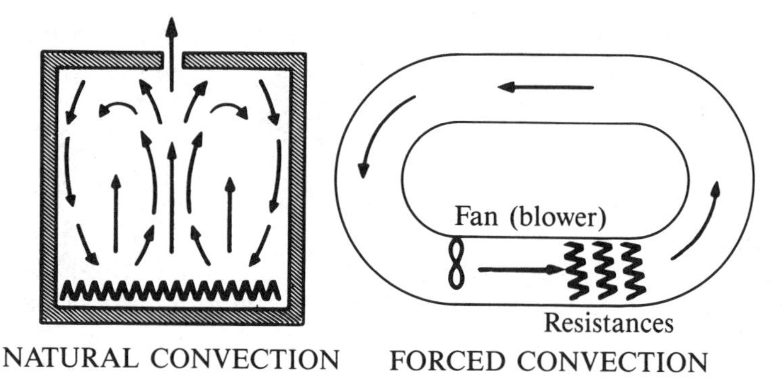


**Fig. 4.** Diagram showing natural and forced convection

Forced convection is used in all low temperature furnaces where high productivity and precise temperatures are required.

This transfer method is, of course, not applicable to vacuum furnaces which must use radiative heat transfer.

### *2.1.4. Overall heat transfer in a resistance furnace*

In practice, radiative and convective heat transfer take place simultaneously with ambient air or internal furnace air serving as the convection fluid.

However, Figure 5 shows that, in a resistance furnace, radiation is still important. Thus, for low or medium temperature furnaces, where high temperature accuracy or rise are required, the use of forced convection suffices (this represents up to 90 — 95% of the total energy transferred).

Moreover, in low temperature furnaces, it is sometimes necessary (drying, calcination, polymerization, etc.) to vent vapors and sludge. Partial recourse to convection then becomes necessary to facilitate elimination of products emitted by the charge.

Also, in the low and medium temperature range, metal and fluidized beds are used (e.g. fluidized sand beds) to heat the products to be treated more quickly by immersion (thermally, a fluidized bed is similar to a liquid), since the convection coefficient and therefore the heat transfer are much higher for liquids than for gases, with better temperature distribution.

Conversely, in high temperature furnaces, radiation dominates providing high power densities only if the resistances are correctly used. However, simple atmospheric mixing is used in some furnaces to improve temperature distribution, but above all to improve the controlled atmosphere used to treat the charge.

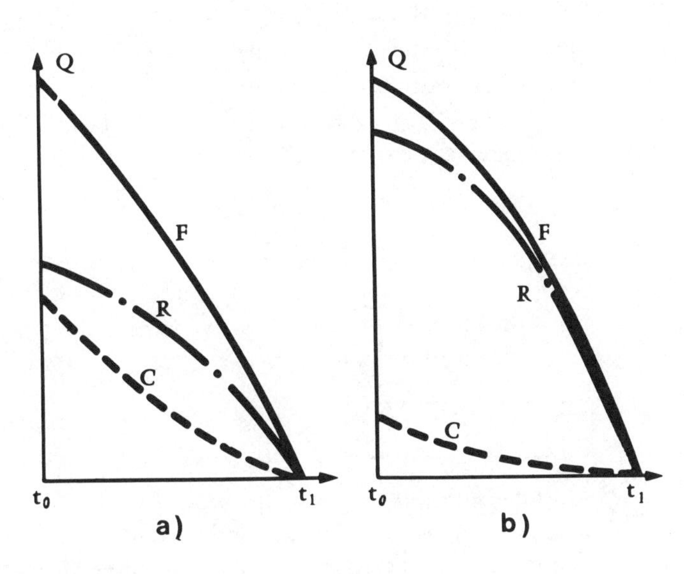


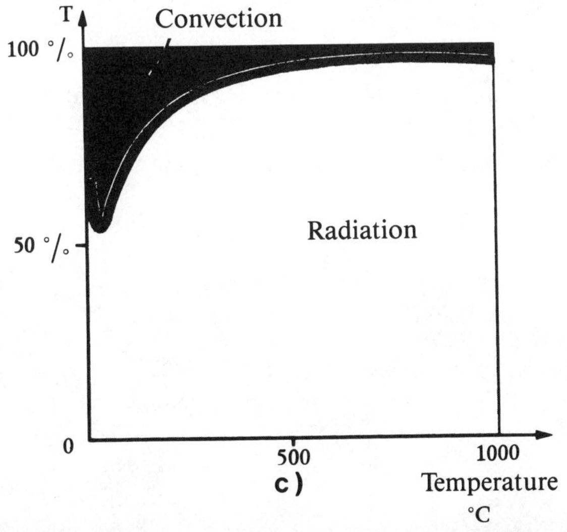


**Fig. 5.** Exchange in a resistance furnace (12). F, total heat flux; R, flux due to radiation, C, flux due to natural convection;
a) chamber held at constant temperature (170°C), charge inserted at ambient temperature
b) furnace held at constant temperature (750°C), charge inserted at ambient temperature;
c) distribution of thermal flux between radiation and natural convection as a function of temperature

Therefore, from the heat transfer point of view, there are two major categories of resistance furnaces using different construction technologies and, despite some overlap, used in different temperature ranges:

— *radiation furnaces* for high temperatures (more than about 600°C);

— *forced convection furnaces*, for low and medium temperatures of about 700 — 750°C. Natural convection is used only rarely for very low temperature chambers of less than 100 to 150°C when the required thermal performances are low (or, in special cases, where liquids are used as an intermediary).

Mixed convection-radiation furnaces also exist but due to their construction technology, generally belong to the forced convection furnace family.

## 2.2. Calculation of resistance furnace power

The energy emitted by the heating system is used to:

— raise the charge temperature as required;

— heat the materials forming the chamber;

— compensate for furnace thermal losses.

The simplified thermal balance and power to be installed in a resistance furnace therefore are very easy to approximate.

### *2.2.1. Energy absorbed by the load*

The quantity of heat to be applied to a substance is generally equal to the sum of:

— the sensible heat $Q_1$ necessary to increase the temperature from $\theta_1$ to $\theta_2$, without a change of state. This energy can be expressed as:

$$Q_1 = mc(\theta_2 - \theta_1),$$

$Q_1$, in kilowatthours;
$c$, mean specific heat between $\theta_1$ and $\theta_2$ (rigorous calculation should make use of the law of the variation of $c_0$ as a function of temperature, using integral calculus) in kWh per kg per °C;
$m$, charge weight in kilograms;

$\theta_1$ and $\theta_2$, initial and final temperatures of the bodies to be heated in degrees Celsius;

— the latent heat $Q_2$ required for eventual changes of state (fusion, vaporization):

$$Q_2 = m_2 L,$$

$Q_2$, in kilowatthours;
L, latent heat of change of state in kilowatthours per kilogram;
$m_2$, weight of part of charge undergoing change of state in kilograms.

If several changes of state occur, these must be taken into account, and the sum of the sensible heats on the various intervals without changes of state and latent heats must be calculated.

When exothermal and endothermal chemical reactions take place in the furnace, the corresponding energy is subtracted from or added to the above values.

Numerous tables are available providing the thermal characteristics of a wide range of substances; for special products (special plastics or ceramics), thermal characteristics can generally be obtained from the suppliers.

### 2.2.2. *Energy absorbed by the walls*

Energy absorbed by the walls can be calculated from the thermal capacity $m_i c_i$ of each of the materials forming the walls and from the temperature profile of these when thermal balance is attained (also refer to paragraph 2.5.5).

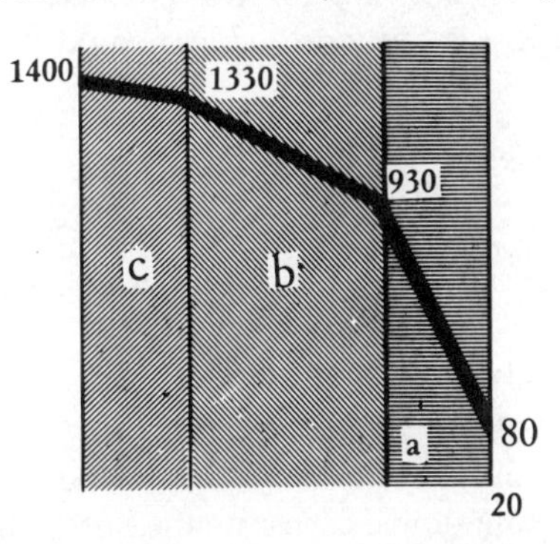


**Fig. 6.** Temperature distribution in the composite wall of a resistance furnace.

The heat stored in the walls provides a heat inertia which enhances temperature uniformity throughout the enclosure and prevents sudden temperature variations.

During furnace cooling, the energy accumulated in the walls is either partially or totally lost (in the event of prolonged shutdowns). For intermittently or discontinuously operating furnaces, this energy has an important effect on the thermal balance and practical efficiency of the furnace. Conversely, in furnaces where shutdowns are infrequent, the effect of this energy on thermal balance and efficiency is generally negligible, being distributed over a long production run. In this case, thermal losses play an important role in the thermal efficiency of the furnace.

### 2.2.3. *Thermal losses in furnaces*

— *Losses due to thermal conduction through the walls*

Because of the thermal gradient between the furnace enclosure and the environment, a heat flux passes through the furnace walls. Heat transfer is caused by conduction in the walls, then by radiation, conduction and convection from the outside walls of the furnace.

Calculation of these losses is made using the standard heat transfer laws.

Thermal insulation can reduce these losses substantially.

— *Losses through openings*

These losses consist of:

• losses due to door opening; these can be high for treatment processes requiring frequent openings (e.g. enamelling furnaces),

• losses through slots (cracks, misfitting doors, bad sealing) and the passages reserved for power conduits.

Careful design and regular maintenance reduce these losses in resistance furnaces. Such losses cannot be accurately evaluated (except when due to door openings), since these often occur due to furnace ageing, lack of maintenance or incorrect application.

— *Losses through volatile product venting conduits*

Some thermal operations call for the evacuation of sludges, vapors and solvents, etc. These losses are more or less equal to the energy stored in the products vented.

— *Miscellaneous losses*

These consist of:

• losses through the exposed surface of baths (metals, salts, aqueous solutions, etc.),

• losses due to inadvertent cooling (seals, handling equipment, etc.),

• losses through exposed furnace parts(e.g. certain forging furnaces).

The above losses can generally be greatly reduced by a rigorous analysis of their causes and adapting simple corrective measures (thermal insulation, furnace maintenance, leakage checks, reduction in the dimension of certain openings and door hatches, etc.). Efficiency, which is already high in resistance furnaces, can be noticeably improved by these methods.

### *2.2.4. Power requirements*

Approximate calculation of the power to be installed is very simple; it differs slightly for continuous and discontinuous furnaces.

#### *2.2.4.1. Intermittent or discontinuous furnaces*

The power produced by these furnaces is expressed as:

$$\boxed{W_1 = \frac{C_1 + C_2}{t} + a\,D,}$$

$W_1$, power to be installed in kilowatts;
$C_1$, heat required to raise the charge temperature and, where necessary, to obtain its change of state, in kilowatthours;
$C_2$, heat accumulated in furnace walls, in kilowatthours;
D, thermal losses at furnace final temperature, in kilowatts
a, coefficient allowing for the mean loss value during temperature rise; a is generally between 0.6 and 0.8;

t, temperature build-up time in hours.

In calculating $C_1$, the various components of the charge, the products to be heated supports, and products eliminated during heating (water, solvents, etc.) must be taken into account.

Coefficient a can be calculated more precisely by constructing a model of the furnace and following its heating cycle. The calculations, which make use of the general laws of heat, are relatively complex. For normal furnaces, the mean losses during temperature rise are generally 60% to 80% of their value at final temperature.

For example, for furnaces operating between 800 and 1,000°C, the accumulated energy in the walls when using modern insulating and refractory materials (e.g. ceramic fibres) is between 10 and 50 kWh/m$^2$ of wall area (for the furnace hearth, these values must be increased 20 to 30%).

The power value obtained is simply an order of magnitude. In some cases, so as to make use of a power reserve, the installed power is 10 to 30% higher (production peaks, difficulty in estimating some thermal losses, etc.). Conversely, in other cases this is less; for example, when cooling between two heating cycles is relatively limited. Only a thorough analysis of the operating conditions can enable a final selection of the amount of power to be installed.

### *2.2.4.2. Continuous furnaces*

For furnaces in which products are treated continuously, it is not always necessary to allow for the furnace heat-up energy when calculating power requirements. Since charge treatment often begins only after the furnace has reached thermal stability, products are not in the furnace during temperature rise, and the installed power is generally sufficient to ensure that the walls are heated rapidly. In this case, the power requirement becomes:

$$W_2 = P_1 + D,$$

$W_2$, power requirement in kilowatts;
$P_1$, energy absorbed in hourly production, in kilowatts,
D, thermal losses in kilowatts.

In practice, installed power is 15 to 30% higher, so as to provide a power reserve.

If the power is not sufficient for rapid heating (if the charge is contained in the furnace during this phase), the expression giving $W_1$ may be used. However, if the furnace is used infrequently, this expression should not be used since the furnace is then "overpowered" (electricity contract at fixed prime cost, higher furnace investment costs and higher maintenance expenses, etc.). Efficient production planning (e.g. furnace automatic starting at off-peak rates) helps avoid overpowering.

## 2.3. Efficiency of resistance furnaces

It is imperative that efficiency be clearly defined in order to take advantage of the thermal and economic performance of resistance furnaces (31).

Numerous fruitless discussions, and even misunderstandings, have occured when comparing resistance furnaces with other types of furnaces.

### 2.3.1. *Electrical efficiency*

Electrical efficiency is expressed as $\eta_1 = Q_1/Q$ where $Q_1$ is the electrical energy converted into thermal energy in the heating chamber, and $Q$ is the electrical energy supplied to the furnace. The difference $Q - Q_1$ is equal to the sum of the electrical losses in the power lines, voltage transformers and voltage regulators.

In industrial furnaces, this electrical-to-thermal conversion ratio is generally very high, about 95%.

### 2.3.2. *Thermal efficiency*

Thermal efficiency is expressed as $\eta_2 = Q_2/Q_1$ where $Q_2$ is the thermal energy stored in the treated charge at the end of the thermal operation, and $Q_1$ is the electrical energy converted into heat in the chamber. The difference $Q_1 - Q_2$ gives furnace losses.

$\eta_2$ is an operating efficiency, being basically a function of:

— furnace construction (furnace type, refractory and insulating materials, regulation, shape, dimensions, etc.).

— furnace use (weight, physical, thermal and chemical characteristics of the materials to be heated, treatment time, furnace charge factor, type of treatment, operating mode, labor quality, etc.).

Some of these criteria can also be used to define special thermal efficiencies, for example efficiency at full load or percentage load, efficiency for certain types of product, etc.

### 2.3.3. *Practical or industrial efficiency relative to specific consumption*

This efficiency depends not only on the product of $\eta_1 \times \eta_2$, but also on its variation in time; in particular, this efficiency includes furnace charge variations over time, and can be estimated from specific quantities of energy consumption over long periods (e.g. month, quarter or year). This therefore represents the mean elecrothermal efficiency value $\eta_1 \times \eta_2$ over long periods. During industrial acceptance of electrothermal equipment (as compared to equipment using another form of energy), it is generally the thermal efficiency which is used to check design specifications.

Conversely, this constitutes the economic efficiency (i.e. the mean specific consumption over a long period) and is significant when assessing the performance of equipment.

Now, although related, these values sometimes differ widely. For example, Figure 7 compares specific consumption of steel or cast iron product enamelling furnaces (treatment temperature 850°C) measured in a variety of furnaces performing comparable treatments (medium power fossil fuel furnaces are not equipped with regenerators). The difference between efficiencies is not at all the same at maximum load and under the usual production conditions (specific consumptions must then be weighted using the respective costs of energy, and the overall economy obtained from an evaluation of the cost of all production factors).

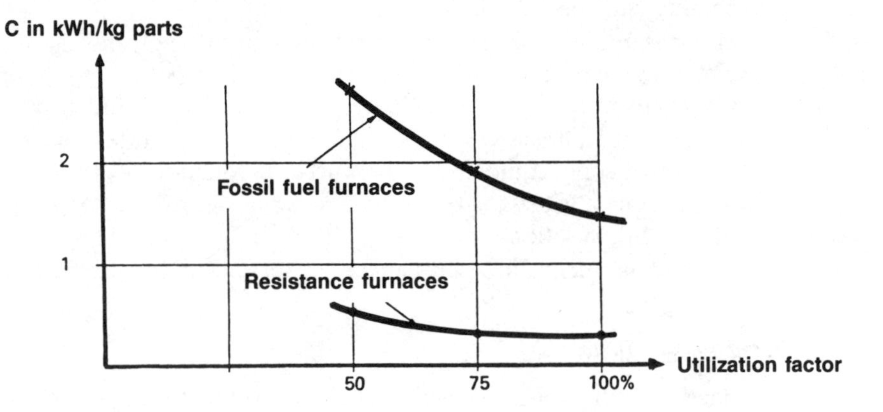


**Fig. 7.** Comparison of energy consumption in resistance and fossil fuel furnaces used for enamel baking (79).

The graph in Figure 7 above obviously has no normative value. However, it demonstrates that the energy cost criteria alone, before practical efficiencies are taken into account, cannot be compared (89), (90). Only a detailed analysis (technical, economic and social factors) can lead to the proper decision.

It is difficult to give practical efficiency values for resistance furnaces in absolute terms. However, experience shows that, for the usual industrial operations, this efficiency is generally between 50 and 75% — very high efficiency values providing significant energy savings.

## 2.4. Different types of resistance furnaces

Resistance furnaces can be classified according to numerous criteria:

— *Type of resistance*: metal, cermet (ceramic-metal) resistances, non metallic resistances, solid, liquid, granular or sectional, wire or ribbon, moulded, radiating tube resistances, etc. The choice depends basically on the temperature level and the power density sought, the atmosphere and the type of application.

— *Furnace utilization method*: continuous (conveyor heating) or discontinuous (fixed or static charge). The operating method depends essentially on the production required, the shape and nature of loads, duration of thermal cycle, etc.

— *Handling method*: continuous conveyor, vibration, helical screw, moving spars, etc.

— *Environment*: normal atmosphere (air), controlled atmosphere, vacuum, liquid, fluidized bath, etc. These atmospheres protect the material to be treated or play a chemical role.

— *Heat transfer method*: radiation, convection, etc. The heating method is basically a function of the desired temperature, but also depends on the accuracy

of the temperature required and the charge characteristics (in particular, thermal conductivity and thermal fragility).
— *Position of resistances with respect to charge*: direct radiation, indirect (muffled furnace, crucible or pot furnaces, forced convection with battery of resistances, heating chamber ventilation, etc.)
— *Furnace application*: heating, heat treatment, calcination, polymerization, stabilization, melting, drying, oven furnaces, etc.
— *Furnace shape*: chamber, moving hearth, well, crucible, elevator, drum, bell.

The combinations are numerous and resistance furnaces are often custom constructed, although some equipment is produced in volume.
Figures 8 to 11 illustrate a wide range of resistance furnaces, without, however, mentioning all current variations.

The furnaces described in the following paragraphs are some of the most widely used.

### *2.4.1. Fixed hearth furnaces*

These furnaces, which are also known as chamber or box furnaces, are the most simple and most common. Loading is manual or mechanical, being a function of the volume and weight of the products to be heated. These are discontinuous furnaces.

Fixed hearth furnaces are widely used and are encountered most often in the mechanical, metallurgical, electrical, ceramic and food industries, and also in laboratories.

However, the volume of the heating chamber is generally limited to a few cubic meters, since beyond this, it is preferable to use furnaces enabling easier charge handling.

### *2.4.2. Movable hearth furnaces*

Movable hearth furnaces are fixed discontinuous operation furnaces, in which the hearth, whether heated or not, is mobile and can be removed from the furnace. The charging and discharging operations are facilitated being performed outside the furnace with normal handling facilities: block and tackle, overhead cranes, etc.

To increase furnace productivity, it is possible to use two movable hearths, saving charge and discharge time. For long duration treatments, including slow cooling down periods, cooling of the charged hearths can be done using "muffles" outside of the furnace which are then used only for heating operations.

The applications field is more or less the same as that of chamber furnaces (with the exception of laboratories), but for bulkier charges, ranging from a few tons to tens of tons.

### *2.4.3. Bell furnaces*

Generally, the furnace consists of a fixed hearth and a circular or rectangular cross-section bell. The resistances are attached to the bell walls and, where necessary, to the roof and hearth. In some furnaces where only limited power and accuracy are required, the resistances are installed in the hearth only.

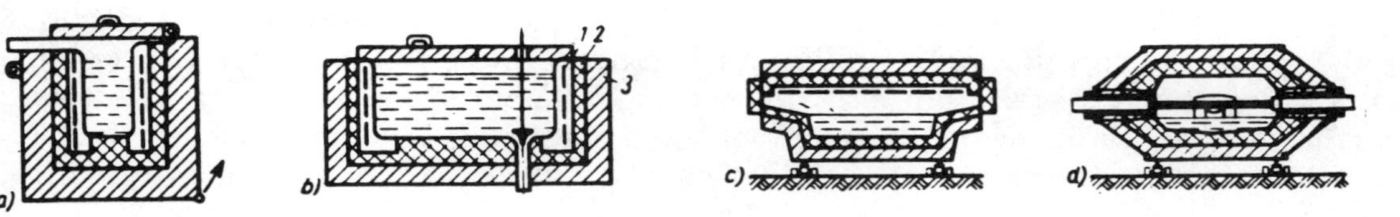


**Fig. 8.** Melting resistance furnaces: a) crucible furnace; b) tank furnace; c) hearth furnace (radiating basin or roof); d) radiating bar-type rotary furnace (or radiating resistor) (7).

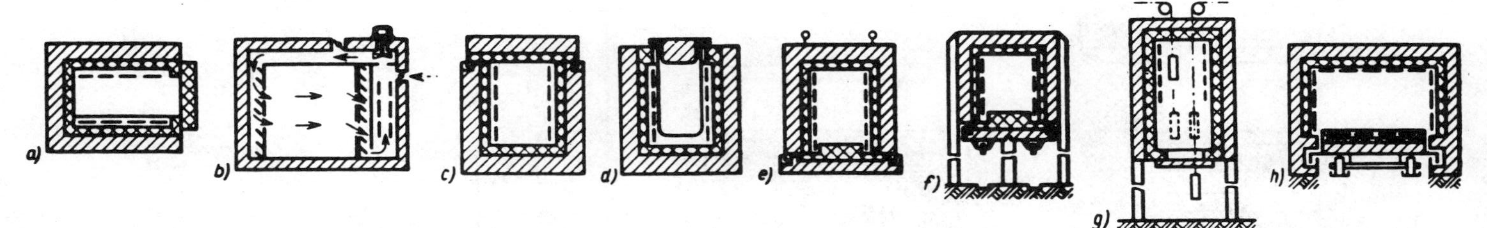


**Fig. 9.** Resistance furnaces for heating fixed loads (7): a) chamber furnace (fixed hearth or box); b) chamber furnace with air circulation; c) tank (or well) furnace; d) pot furnace (muffled); e) bell furnace; f) elevating hearth furnace; g) vertical furnace; h) moveable hearth furnace.

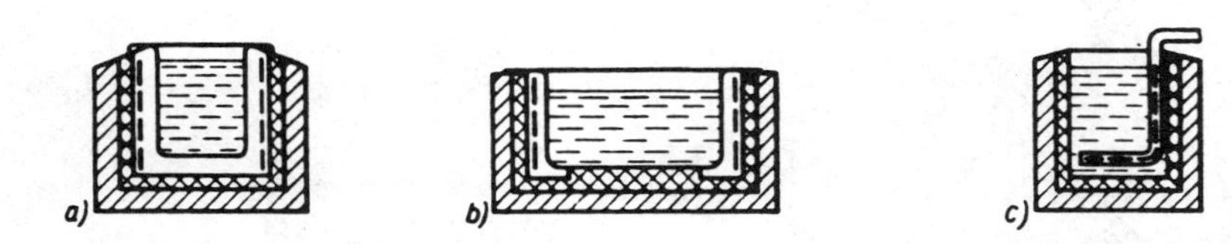


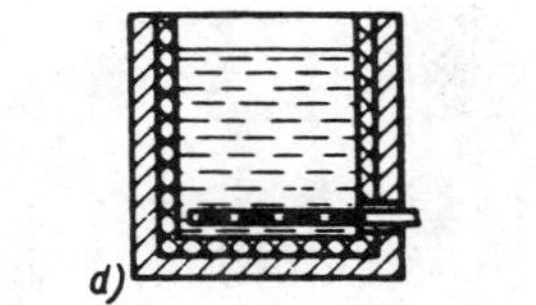


**Fig. 10.** Resistance furnaces for heating liquids: a) bath furnace with crucible and external heating; b) bath furnace with tank and external heating; c) bath furnace with crucible and internal heating (immersion heater); d) bath furnace with crucible and bottom heating (immersion heater).

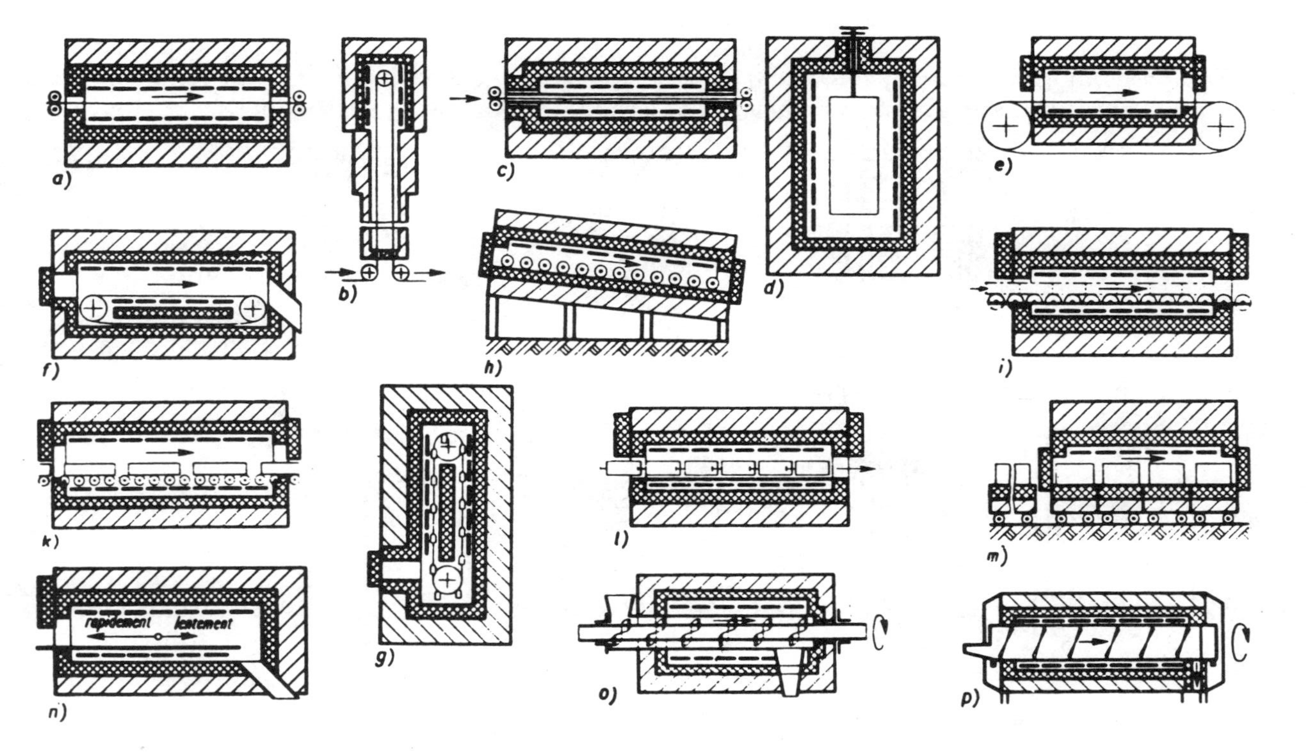


**Fig. 11.** Resistance furnaces for conveyor type heating (7): a) continuous furnace; b) vertical continuous furnace; c) muffle-equipped continuous furnace; d) monorail continuous furnace; e) chain or belt conveyor furnace; f) chain or belt conveyor furnace with return to furnace interior; g) swing-tray furnace; h) roller furnace; i) roller pusher furnace; k) roller hearth furnace l) pulling furnace; m) moving hearth tunnel furnace (wagon furnace); n) vibrating hearth furnace; o) drum furnace, fixed, wormscrew; p) rotating drum furnace.

These discontinuous furnaces obviously require an overhead crane or some handling system powerful enough to lift the bell. This emphasizes the attractiveness of fibrous ceramic refractory materials, which reduce total bell weight by a factor of 2 to 4.

A furnace bell can be used with several bases, enabling productivity to be increased when slow cool-down times are required. The furnace is then often muffled, providing slow cooling periods in a vacuum or in special treatment atmospheres (see paragraph 7.1.2.1.1.).

The capacities of such furnaces are generally high, ranging from a few tons to a hundred tons.

Circular cross-section furnaces equipped with muffles are widely used for heat treatment of metal coils, strips or wires. Rectangular cross-section and sometimes circular cross-section furnaces are most often used to heat treat rough parts (forgings, castings, weldings, bars, tubes, etc.) in the manufacturing and metallurgical industries, treatment of large parts in electrical construction, firing normal or special ceramic products, etc.

### *2.4.4. Elevating hearth furnaces*

This type of furnace chamber rests on pillars, forming a fixed bell, allowing the hearth to be moved vertically. The hearth is lowered for charging and mechanically raised for treatment. The furnace does not require a high-power overhead crane since the hearth-lifting mechanism is built into the furnace, thus minimizing the floor area required.

This type of furnace is used mostly in the manufacturing, metallurgical and electrical construction industries for heat treatment of rough finished parts.

### *2.4.5. Rotating hearth furnaces*

These furnaces are circular and tunnel-shaped. The ring-shaped hearth is mobile and rotates around a race located outside the heating chamber. The hearth is rack-and-pinion driven.

These furnaces offer adequate efficiency since the movable hearth remains in the heating chamber. This offers economies in labor, since only one worker is required to charge and discharge the furnace (however, this is hard physical work). Also, production organization is simplified.

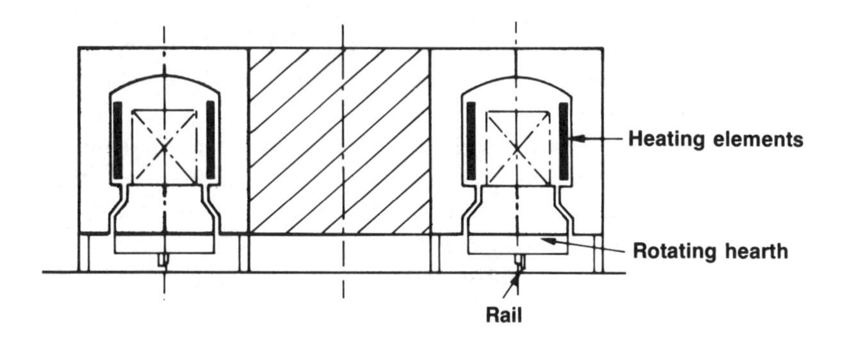


**Fig. 12.** Diagram of rotating hearth furnace

This type of furnace is used most often for heating steel prior to shaping (forging, stamping, etc.), and metal heat treatment. It is rarely used in other industries except occasionally in the ceramics industry.

### *2.4.6. Rotating plate furnaces*

In these furnaces, the hearth receiving the charge consists of a metal plate mounted on a rotating shaft traversing the bottom wall of the furnace. Depending on the temperatures, the hearth is made from ordinary steel or refractory steel. These furnaces offer excellent efficiency.

These furnaces are used mostly in the metal industries for heating light alloy batches prior to stamping (the furnace features internal ventilation) and for heating prior to hardening of small parts. Some ovens (plastic transformation industries) use a similar design.

### *2.4.7. Inclined hearth furnaces*

In inclined hearth furnaces, products contained in the furnace are moved along by gravity. These furnaces are mostly used for heating metal parts which can roll such as billets, tubes, and cylindrical products prior to forming (e.g. extrusion of light alloy rods).

### *2.4.8. Vibrating and reverberatory furnaces*

Product conveyance is achieved:
— in reverberatory furnaces, by moving the hearth backwards and forwards;
— in vibrating furnaces, by vibrating the hearth through an electromagnetic vibrator.

Mostly, these furnaces are used in metal conversion industries for bulk treatment of small parts, being quite often continuous and sometimes completely automated.

### *2.4.9. Chain and wire conveyor furnaces*

Chain conveyor furnaces are continuous furnaces in which product conveyance is provided by one or more chains. The chains may be load carrying and thus fully heated, or simply used for traction where they are located in relatively cold areas and drive the parts either through pawls (chains located beneath the furnace hearth), or by beams (chains located in the furnace roof).

The drive sprockets are located either in or out of the furnace. Having the sprockets outside where the complete mechanical system operates cold, makes it easier to supervise and maintain. Conversely, chains cool as they exit the furnace and adversely affect efficiency.

Wire conveyor design is similar to that of the chain type. The parts are conveyed on cables controlled by external drums.

Chain and wire conveyor furnaces are used mostly in the metal conversion industries for heating billets, reheating or treating half-finished products, drying relatively light products prior to enamelling, etc. Their main advantage is low conveyor weight compared to that of the parts.

### *2.4.10. Continuous hearth furnaces using a belt*

The parts or products are carried through these furnaces on a conveyor belt which may or may not be metal, depending on operating conditions. The conveyor can consist of:

— an endless refractory steel belt;

— a belt, consisting of articulated sheet common plane or refractory steel elements;

— a cast refractory steel belt, consisting of small elements transversely assembled with small bars;

— a common or refractory steel wire belt, consisting of woven strands assembled using transverse bars;

— a belt of non-metallic material, when oven operating temperature is low (e.g., when processing certain food products).

The resistances are located only above the belt, if temperature accuracy is not critical, or both above and beneath the belt for faster and more even heating. In the latter case, the belt must be metallic in order to enhance heat transfer and prevent damage. The atmosphere can be mixed to accelerate thermal exchange and improve temperature uniformity. Belt furnaces also exist in which the heat transfer uses forced convection only. Belt furnaces can operate in a controlled atmosphere or be muffled.

These furnaces are very good for integral automation of industrial heating operations. The conveyor drive system can be located inside or outside the furnace; the internal type is highly efficient.

These furnaces are widely used throughout all sectors of industry. They are constructed in a wide range of dimensions and power requirements varying from a few to several hundred kilowatts.

For high production runs, the furnace hearth is fitted with controlled or non-controlled rollers, which rotate at the conveyor speed to decrease undesirable pulling.

These furnaces are common in the manufacturing and metallurgical industries for many continuous heat treatment operations and reheating prior to forming some products, in electronic and electrical component treatment, in the construction materials industry for drying and baking, in the ceramics industry for drying or firing, in the foodstuff industries (bread, biscuit, cakes, prepared products such as pasta, pizzas, cooked dishes, etc.), in the wood industry (drying, polymerization, etc.) and in the plastic industries (gelling, polymerization, etc.)

### *2.4.11. Tunnel furnaces*

In tunnel furnaces, the products to be heated are placed on wagons which form a movable hearth. To improve efficiency (the high amount of energy accumulated in the hearth is practically lost), these furnaces are often fitted with heat recovery devices, either forced air systems (e.g., blowers between furnace exit and entry) or by the use of a double tunnel in which the wagons travel in opposite directions. When the temperature is not too high (less than 1,000°C), it is also possible to construct furnaces using light trolleys where the load is supported only by a beam traversing the hearth. These furnaces are most widely used in the ceramic industry, but may also be suitable for other applications.

**Fig. 13.** Forced convection furnace or oven, with heating resistance battery (A).

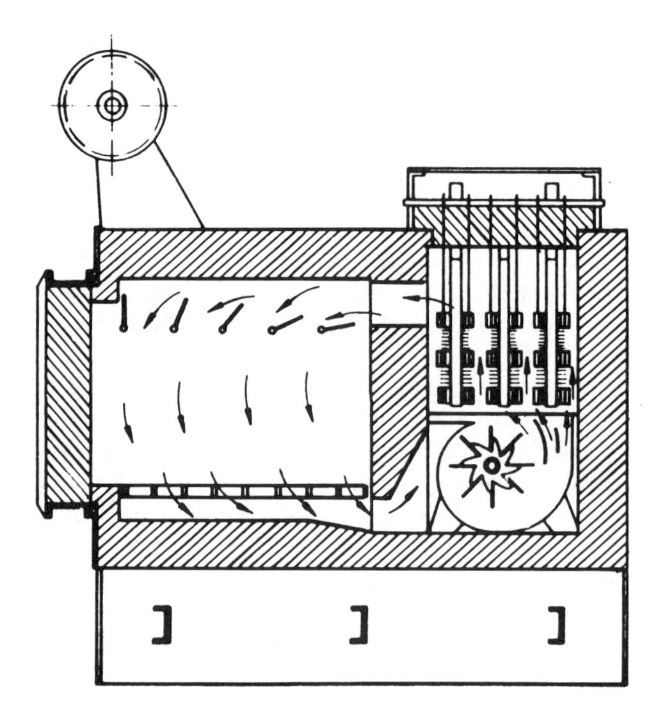

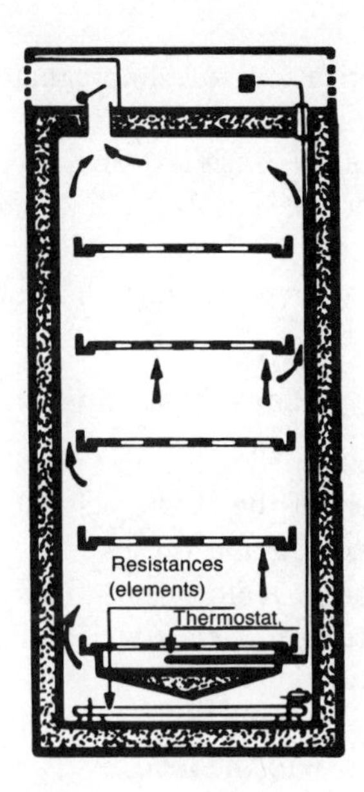


**Fig. 14.** Resistance-heated natural convection oven.

### *2.4.12. Electric resistance ovens*

This term applies to electric ovens operating at low or medium temperatures (less than 600°C), in which the heat transfer is effected by forced convection or, more rarely, natural convection.

The resistances are generally combined to form batteries (see paragraph 3.5.1.) over which the air (or treatment atmosphere) is driven by a blower. Heat transfer can also be obtained by natural convection.

The power of these ovens ranges from a few kilowatts to several hundred kilowatts. Generally they are discontinuous, but when continuous, their design is similar to that of a rotating plate or hearth furnace as described above (monorail ovens are also used for baking surface coatings such as paint, together with swing-tray furnaces).

In discontinuous ovens, the charges are located on trolley-mounted shelves.

Particular attention must be paid to the design of hot air distribution ovens, which must provide an even temperature throughout the chamber. Performance analysis on some ovens in fact often shows temperature stratification, whether vertical or in the air flow direction, calling for inverted air flow; however, the best solution is generally to allow for charge losses with high air speeds and improved distribution.

Electric resistance ovens are widely used throughout industry:

— drying and cooking paints, varnishes, plasters on varied surfaces (metal, wood, construction materials, plastics, fabrics, paper, etc.);

— heating plastic materials;

— drying and baking mould cores;

— drying wood;

— drying ceramic products;
— drying and polymerization of impregnated materials in electrical manufacturing;
— processing pharmaceutical, chemical or bacteriological products;
— bonding various materials;
— drying powder products;
— drying and baking meat preserves, miscellaneous meat products, bread and pastries, cooked dishes, etc.

## 2.5. Refractories and insulation: the heating chamber

An oven or furnace must provide the best possible conditions for the charge heating cycle. The purpose of the heating chamber is to:
— *PREVENT* heat dissipation and thus:
• enable the desired charge temperature to be obtained;
• provide optimum thermal efficiency, which calls for the lowest possible wall heat capacity and thermal conductivity:
— *ESTABLISH* a thermal reserve to complement the heating elements, to:
• maintain high heat flow to the load,
• provide better heat distribution using heat from the walls.
The latter two conditions require that the chamber absorb a certain quantity of heat and contradicts the previous point; therefore, an optimum solution must be found (see paragraph 2.5.5.);
— if necessary, *MAINTAIN* the appropriate atmosphere in the furnace, and
— *ENSURE* the structural integrity of the furnace.
Refractory and insulating materials are an essential part of furnace walls playing two distinct roles.

### *2.5.1. Refractories*

Conventionally, refractory products are non-metallic mineral materials, the melting point (or, to be more precise, the softening point, defined from the heat resistance of the material using standard tests) being greater than 1,500°C. These materials line the hot surfaces of furnace walls (1), (2), (12), (14), (15), (74), (75), (BQ), (BR), (BS), (BT), (BU), (BV), (BW).
The optimum qualities required for refractories are as follows:
— *thermal*: the softening and melting temperatures must be high, much higher than the furnace operating temperature; in addition, heat capacity and, above all, thermal conductivity, must be as low as possible;
— *mechanical*: materials must offer excellent mechanical strength and resiliency at operating temperature; they must withstand compression and traction forces, thermal shocks resulting from sudden temperature variations whether cyclic or not, and enable resistance coupling. Material shrinkage must be low;
— *chemical*: materials must offer high resistance to corrosive atmospheres, loads or materials forming the resistances and conversely, must not corrode the resistances;

— *electrical*: the electrical insulation resistance must be high, even at high temperatures;
— *morphological*: these materials must be capable of being produced in highly varied shapes and dimensions;
— *economical*: material costs must, of course, be kept as low as possible and maintain furnace profitability.

The composition, and the physical, chemical and electrical qualities of a given refractory material must be as constant as possible within its operating temperature range.

These materials are available as prepared (bricks, shaped pieces) and non-prepared refractories (concretes, mixes, slurries, etc.). The physical and thermal characteristics (conductivity, diffusivity, density, mechanical strength, etc.) vary with form, consistency and chemical composition.

Characteristics of some current refractory products

| Refractory | Base | Melting point (°C) |
|---|---|---|
| Silica | $SiO_2$ | 1,700 |
| Clay | $2\ SiO_2.Al_2O_3.2H_2O$ | 1,700 |
| Sillimanite | $SiO_2.Al_2O_3$ | 1,800 |
| Alumina, bauxite | $Al_2O_3$ | 2,050 |
| Chrome-plated iron, chromite | $FeO.Cr_2O_3$ | 2,000 |
| Carbon | C | 3,000(1) |
| Silicon carbide | SiC | 2,500 |
| Quicklime | CaO | 2,200-2,570 |
| Magnesia | MgO | 2,800 (2) |
| Dolomite | $CO_3Ca.CO_3Mg$ | 2,300 |
| Zircon | $SiO_4Zr$ | 2,000 |
| Zirconia | $ZrO_2$ | 2,400-2,500 |

(1) Flash point
(2) Softening at 1,300°C if impure

**Fig. 15**

There are three major families of refractory materials:
— *Alumino-silicate products* (sometimes known as "acid products") ranging, in increasing order of performance, from pure silica to pure alumina and including alumina silicates (refractory clays, fire clay, sillimanite, cyanite, mullite, gibbsite, corundum, etc.);
— *basic products*, consisting essentially of single or combined basic oxide; mainly, these oxides are magnesia (MgO), chromites ($FeO\text{-}Cr_2\ O_3$), lime (CaO) and calcinated dolomite (CaO-MgO);
— *special products*, covering materials of diverse chemical nature (zirconia, zircon, silicon carbide, graphite, etc.).

A product type is determined by its element content, its mineral characteristics, impurity content and physical characteristics (porosity).

A detailed study of refractory materials exceeds the range of this work. When employed in electric resistance ovens, it is obviously necessary to ensure that the products used meet the above requirements.

### *2.5.2. Thermal insulation*

These materials, used on the furnace surfaces, essentially provide thermal insulation, i.e. reduce thermal losses. Thermal conductivity must therefore be very low. To meet this requirement, these materials have a porous or fibrous structure, with small air or vacuum "pockets" to prevent convection (conventionally, an insulating material should have a porosity greater than 45%). Thermal insulation sometimes stores a certain amount of heat which is not transferred to the material to be heated. The resulting thermal losses, while negligible in a continuously operating furnace, considerably reduce the efficiency of a discontinuous furnace (1), (2), (12), (14), (15), (29), (74), (BS), (BT), (BU), (BV).

For low temperatures, it is possible to use organic substances (cork, wool, sawdust, etc.). However, modern industrial installations generally use mineral materials (glass, rock or slag wools, expanded clay, lightened refractory materials, vermiculite, alum-silicate products, manganese, silicon, asbestos, etc.), for both low and medium temperatures.

Figure 16 provides the characteristics of some insulators used in industrial furnaces and ovens.

### *2.5.3. Refractory insulation*

Refractory and insulating materials as described above must first offer excellent temperature adaptability and heat resistance, and second have high insulating capability, to reduce thermal losses.

With the development of refractory materials having adequate insulating properties (a refractory material is considered to be an insulator when its porosity is at least 45%), i.e. low thermal conductivity and reduced heat capacity, this distinction tends to disappear. Insulating refractories can be subdivided into four major groups:

— formed insulating refractories (bricks, shaped parts);

— rigid insulating refractory concretes;

— fibrous insulating refractory materials;

— fibrous insulating concretes and composite fibrous products.

The main interest of such products is the substantial energy savings obtained, in particular with discontinuous furnaces (35), (46), (74), (BQ), (BR), (BS), (BT), (BU), (BV), (BW).

Of these products, fibrous insulating refractories merit special attention. In fact, these have a thermal conductivity 20 to 40% lower than that of other insulating refractories such as bricks or concrete, but, above all, are six to ten times less dense.

Average characteristics of some thermal insulators

| Composition | Conductivity at 0°C (W/m°C) | Specific weight (kg/m$^3$) |
|---|---|---|
| Asbestos | 0.14 | 576 (1) |
| Magnesium carbonate | 0.056 | 250 (2) |
| Slag wool | 0.03-0.06 | 120-250 |
| Kieselguhr (fired bricks) | 0.08 | 180-250 (4) |
| Cellular concrete | 0.06-0.2 | 260-900 |
| Glass wool | 0.056 | 220 |
| Cork (grains) | 0.04 | 80-100(3) |
| Agglomerated cork | 0.06 | 250-300 (3) |
| Wool | 0.037 | 136 |
| Kapok | 0.034 | 18 |
| Felt | 0.031 | 120 |
| Water | 0.50 | 1 000 |
| Moving dry air | 0.120 | 1.3 |
| Still dry air | 0.02-0.035 | 1.3 |

(1) Withstands 800°C. Asbestos: 4 $SiO_2$ . 3 MgO . CaO.
(2) Up to 300°C.
(3) Below 120°C.
(4) Kieselguhr is a natural hydrated silicate.

**Fig. 16**

These materials can provide spectacular energy savings with intermittent operation furnaces, ranging from 30 to 60% (see paragraph 5.2.).

Most ceramic fiber refractory materials are made from mixtures of arc furnace-melted alumina and silica. The maximum operating temperature is about 1,450°C on the heated surface for 60% alumina refractory materials. For temperatures higher than 1,700°C, fibres consisting of more than 80% alumina or zirconia must be used.

Fibrous materials can include bulk fibres, layers, felts, cords, beads, papers, pre-shaped parts and composite materials (concretes, shaped parts).

### *2.5.4. Refractories and resistances*

Resistance mountings are often made from refractory materials which must be highly resistant to:
— corrosion
— temperature variations
— current flow and be a good dielectric

The materials used are high alumina-content alumino-silicate products (more than 45%).

These supports have a wide range of shapes (see paragraph 3.5.) and are a controlling factor in the service life of resistances (see paragraph 3.5.) and therefore require careful attention.

### *2.5.5. Optimum wall thickness*

Thicker furnace walls lower thermal losses for a given type of material. Conversely, energy accumulated in the walls increases, and is partially or totally lost in discontinuous or intermittent operation. Therefore wall thickness can minimize total furnace losses but is a function of furnace operating mode (1), (2), (12), (14), (30), (59), (79).

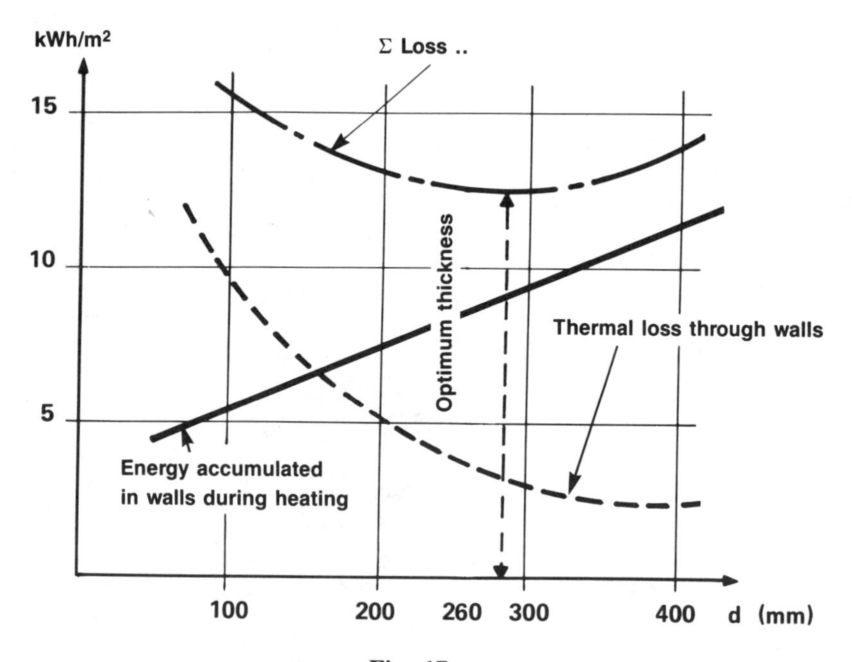


**Fig. 17**
Furnace wall thickness offering minimum total thermal loss (79).

However, this analysis is not complete because it neither allows for the respective costs of energy and construction materials, nor for the possibility of using different materials varying in price and performance. By integrating all these variables, it is possible to define an optimum wall thickness minimizing overall cost — not just thermal losses (this calculation must also allow for the heat rise rate and therefore productivity).

Figure 18 shows the effect different materials have on both losses and furnace productivity. Complemented by a cost examination, this type of analysis can be used to determine the economic optimum mentioned above (also refer to paragraph 5.2.).

Comparison of thermal performances of various composite walls for a furnace operating at 950°C (79).

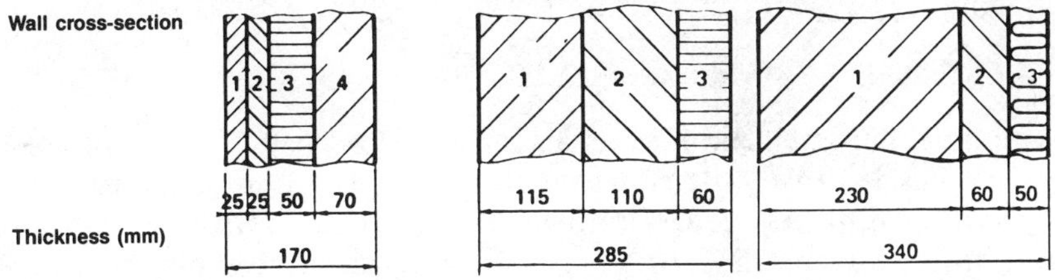

| Wall cross-section | 1, 2, 3, 4 | 1, 2, 3 | 1, 2, 3 |
|---|---|---|---|
| Thickness (mm) | 25, 25, 50, 70 (170) | 115, 110, 60 (285) | 230, 60, 50 (340) |
| Type | I Ceramic fibres (2) | II Insulating brick | III Compact bricks (1) |
| Wall composition, per layer | Density (kg/m$^3$) | Density (kg/m$^3$) | Density (kg/m$^3$) |
| No. 1 | Fibrous 1 refractory 128 | Insulating brick 500 | Compact brick 2160 |
| No. 2 | Fibrous 2 refractory 96 | Insulator I 350 | Insulator I 400 |
| No. 3 | Insulator 1 96 | Insulator 2 144 | Insulator 2 350 |
| No. 4 | Insulator 2 144 | | |
| Weight per square meter (kg/m$^2$) | 20.5 | 104.6 | 538.3 |
| Heating time (thermal balance at 950°C)(h) | 3 | 15 | 70 |
| Accumulated energy at a given instant (kWh/m$^2$) | 2.7 | 17.5 | 125 |
| Accumulated energy after 10 hours of heating (short cycle)(kWh/m$^2$) | 2.7 | 15.8 | 85.3 |
| Outside wall temperature (°C) | 90 | 62 | 106 |
| Losses due to transfer (W/m$^2$) | 670 | 340 | 900 |
| Natural cooling time to 300°C inside the furnace (h) | 17 | 120 | 180 |

(1) Compact bricks are used very rarely today.

(2) This furnace could have been better insulated (e.g. 220 mm instead of 170) to limit operating mode losses.

**Fig. 18**

Figure 18 shows the effect different materials have on both losses and furnace productivity. Complemented by a cost examination, this type of analysis can be used to determine the economic optimum mentioned above (also refer to paragraph 5.2.).

## 3. ELECTRIC RESISTANCE ELEMENTS

Continual improvement in materials intended for the production of heating elements has led to the development of electrical resistance heating. It has enhanced ruggedness, longevity and resistance to corrosion. Operating temperatures have increased with two essential consequences: high density, and therefore high productivity furnaces; very high operating temperatures which have provided new industrial processes and products [13], [39], [53].

### 3.1. Resistance and resistivity

The resistance $R$ of a heating body of length $l$, uniform cross-section $S$ and resistivity $\rho$ is given by the expression:

$$\boxed{R=\rho\, l/S,}$$

$R$, in ohms;
$\rho$, in ohm-meters;
$l$, in meters;
$S$, in square meters;

The inverse of the resistance $C = 1/R$ is conductance. Resistivity varies as a function of temperature:

$$\boxed{\rho_\theta=\rho_0(1+\alpha\theta),}$$

$\rho_\theta$, resistivity at temperature $\theta$;
$\rho_0$, resistivity at 0°C;
$\theta$, temperature in degrees Celsius;
$\alpha$, resistivity temperature coefficient in °C$^{-1}$.

Coefficient $\alpha$ is generally positive for metals, especially those used in the manufacture of electric resistances. This may be practically zero (e.g. constantan) or negative (graphite below 500°C).

### 3.2. Connecting resistances

Resistances can be assembled and connected in different ways, the most common being:

— series, where total resistance is equal to the sum of the individual resistances;
— parallel, where total conductance is equal to the sum of the individual conductances;
— mixed, combining both the above methods;
— series-parallel which, by operation of a selector switch, can vary the power dissipated by a ratio of 1 to 4;

— and star-delta connection for three-phase AC supply, enabling a power ratio of 1 to 3 through a selector switch.

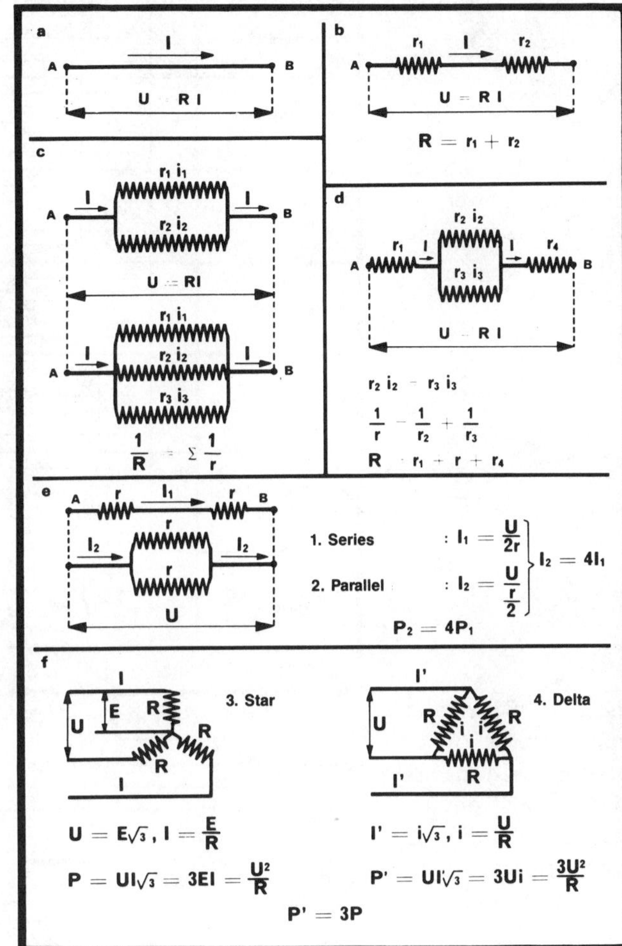


**Fig. 19.** Equivalent resistance of various connections of resistors [13].

Electromechanical selector switches are used to switch from one setup to another, and therefore to vary furnace power instantaneously. For three-phase AC, a delta connection is used during temperature build-up, followed by a star configuration during temperature stabilization. Changing from one type of configuration to another also enables resistance to be adjusted for some heating elements where temperature is a factor (e.g. molybdenum disilicide resistances).

Electronic continuous power modulation systems (thyristor power supplies) are rapidly replacing the power variation systems discussed above (*see* paragraph 4.3.).

| Description | No of filaments | Circuit diagram | Voltage | Power |
|---|---|---|---|---|
| Single-phase AC or DC | 2 | U | U | $P = UI$ |
| Single-phase AC or DC | 3 | U E | $U = 2E$ | $P = UI = 2EI$ |
| Three-phase AC and neutral (star) | 4 | U E | $U = E\sqrt{3}$ | $P = UI\sqrt{3} = 3EI$ |
| Three-phase AC without neutral (star) | 3 | U | $U = E\sqrt{3}$ | $P = UI\sqrt{3}$ |
| Three-phase AC (delta) | 3 | U | U | $P = UI\sqrt{3}$ |
| Two-phase AC (3-wire) | 3 | U' U U | $U' = U\sqrt{2}$ | $P = 2U'I$ |
| Two-phase AC (4-wire), no neutral | 4 | U U | U | $P = 2UI$ |
| Two-phase AC (5-wire) with neutral | 5 | U E U' U' | $U = 2E$<br>$U' = E\sqrt{2}$ | $P = 2UI = 4EI$ |

U, E, U' = Voltage I = Current P = Power

**Fig. 20.** Methods of supplying power to heating resistance [13].

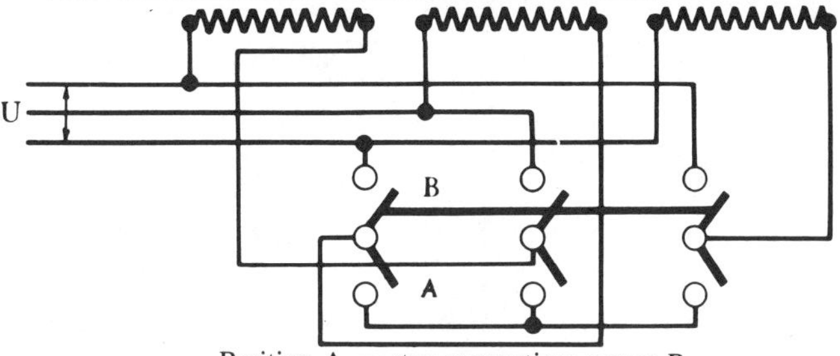


Position A = star connection; power $P_e$
Position B = delta connection; power $P_t$
$P_t = 3\ P_e$

**Fig. 21.** Schematic representation of an electromechanical switching device enabling changeover from star to delta connection and vice versa [12].

### 3.3. Criteria for selecting resistances

Resistance selection is fundamental in optimizing investment cost and service life. In particular, it seems that some of the failures encountered in practice are due to incorrect evaluation of the latter. The first characteristic to be considered in choosing a resistance is its maximum operating temperature; this, in fact, controls the maximum temperature of the charge placed in the furnace and the resistance surface charge, and therefore furnace productivity. Expansion constraints, risks of softening or localized melting also depend on operating temperature.

Other characteristics such as susceptibility to vibrations or resistance to chemical action also play a role but are easier to compensate for.

— *Heating system temperature*

The softening temperature and, most certainly, the melting point of the materials forming the resistances must be much higher than the maximum temperature reached during normal operation to prevent damage to the elements.

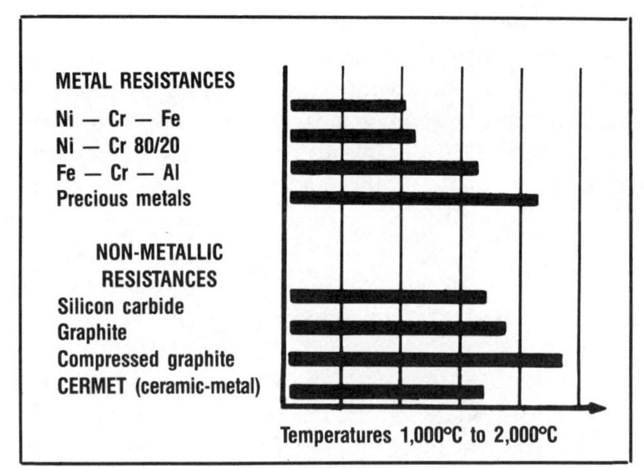


**Fig. 22.** Operating temperatures of some materials used in resistances

To obtain high heating element service life, it is necessary to predict the maximum operating temperature, i.e. taking into consideration operation as a whole (furnace, regulation, heating). This temperature is reached at the end of the heating cycle when the regulator turns the system off. If the temperature sensor measures furnace temperature only, there is a risk of overheating the elements, requiring very high safety margins, and limiting furnace performance. Due to the use of regulation systems, modern furnaces are spared this constraint, and productivity is improved (*see* paragraph 5.1.).

— *Resistivity*

For constant power, a relatively high resistivity at a given voltage reduces the heating system dimensions and, for fixed dimensions, limits current.

This current limitation enhances the service life of the cold outlets and contacts, where heating is then only due to thermal conduction. Low thermal conductivity and heat capacity prevent heat loss at the extremities, and facilitate cold outputs.

The temperature coefficient of resistivity $\alpha$ must therefore be low, positive and more or less constant over the furnace operating range. A low constant coefficient dispenses with the requirement for expensive starting and regulation devices. In addition, a positive temperature coefficient lets the dissipated power regulate itself since, in the event of an accidental increase in temperature, the resistance increases and the dissipated power decreases.

— *Mechanical strength*

A heating system must be mechanically strong, not only when installed, but also at maximum operating temperature. Strength must not decrease with time, especially after rapid heating and cooling which result in successive expansion and contraction.

A low expansion coefficient generally favors mechanical strength, but the heating element must also withstand shocks and vibrations.

The method used to attach and support heating elements also plays a major role in element mechanical strength and service life.

The resistance material must be uniform, flexible and maintain structural integrity up to the operating temperature. Adequate ductility enables diverse shapes, promotes the best possible fit between surfaces and provides optimum heat transfer. When materials are easy to weld, repairs and connection to cold parts can be made rapidly and simply.

— *Resistance to chemicals*

The materials used in the resistances must withstand corrosive vapors or gases in the natural or artificial atmosphere of the furnace, emanations from materials being treated or the heating system refractory supports. Resistances used in vacuum furnaces must not evaporate at working temperature.

In addition, materials used in the manufacture of industrial heating resistances must be relatively inexpensive and offer the numerous qualities required to ensure long service life:
— high softening and melting temperatures;
— relatively high resistivity and low, positive constant temperature coefficients;
— high mechanical strength;
— resistance to chemical agents, in particular oxidation;
— stability under operating conditions;
— ease of use.

It is, of course, difficult to find all these qualities in a single material, but several families come close to meeting all industrial requirements.

All the characteristics stated in the description of the various types of resistances are given for reference only, since only manufacturers can provide the exact characteristics of their products.

## 3.4. Different types of resistances

Resistive heating elements are generally divided into three major families:
— metallic resistances;
— cermets;
— non-metallic resistances.

### *3.4.1. Metallic resistances*

Numerous alloys are used to make metallic resistances. In practice, these can be broken down into three categories:

— iron-nickel-chromium or nickel-chromium alloys;
— iron-chromium-aluminum alloys;
— special molybdenum, tungsten, tantalum, platinum, niobium, iron-nickel-chromium-aluminum resistances.

### *3.4.1.1. Iron-nickel-chromium or nickel-chromium alloy resistances*

#### *3.4.1.1.1. Characteristics*

These alloys are produced in electrical induction or arc furnaces and undergo special heat treatment, generally in resistance furnaces. These are by far the most widely used for the production of resistances intended for industrial furnaces, since they offer the qualities required at a relatively low cost [13], [31], [39], [69], [73], [*BG*], [*BH*], [*BI*], [*BK*], [*BN*].

*— Principal characteristics*

Figure 23 provides examples of the characteristics of a family of nickel-chromium and iron-nickel-chromium alloys as a function of nickel and chromium content. The remainder consists of iron and additives which account for less than 2% of the total composition.

Resistivity variation with temperature is very low; moreover, these alloys are completely stable, since they are not subject to temperature-induced expansion or contraction.

Therefore, nickel-chromium alloy resistances can be used at constant voltage without requiring special starting devices. From the mechanical viewpoint, these offer adequate rigidity and strength and excellent ductility. All nickel-chromium based alloys used in the manufacture of resistances are non-magnetic.

It is often necessary to perform welding operations on these resistances, either to obtain heating elements of adequate length or for repair, or again to attach current entries (cold outputs). Preferably arc-argon welding is used, but oxyacetylene welding can also be used.

*— Power densities*

For ordinary furnaces, the surface power usually varies from 0.5 to 2.0 W/cm$^2$. Such a load is, for comparison purposes, from 0.4 to 0.6 W/cm$^2$ for a furnace temperature of 1,100°C, from 0.6 to 0.8 W/cm$^2$ for 1,000°C, 1.0 W/cm$^2$ for 900°C, 2.0 W/cm$^2$ for 700-800°C, but could be much higher in certain conditions (refer to section 5.13).

In general, a distinction is made between:
— iron-nickel-chromium resistances, and
— nickel-chromium resistances.

#### *3.4.1.1.2. Iron-nickel-chromium resistances*

These ternary (i.e. three-part) alloy resistances contain iron in proportions ranging from 15 to 70%; all are used as heating elements in industrial furnaces, except for that richest in iron which is used only for low temperature applications (e.g. electrical convectors). Additional substances can be used to improve

## Characteristics of a family of iron, nickel and chromium alloy resistances

| Composition | 1 | 2 | 3 | 4 | 5 | 6 | 7 | 8 |
|---|---|---|---|---|---|---|---|---|
| Characteristics | Ni80-Cr20 | Ni70-Cr30 | FeNi60-Cr15 | FeNi45-Cr23 | FeNi36-Cr18 | FeNi32-Cr20 | FeNi19-Cr25 | FeNi10-Cr18 |
| Density | 8.35 | 8.15 | 8.20 | 8.00 | 8.00 | 7.90 | 7.90 | 7.90 |
| Melting point (°C) | 1,400 | 1,400 | 1,400 | 1,400 | 1,390 | 1,390 | 1,380 | 1,380 |
| Mean coefficient of expansion in $10^{-6}$/°C between 0°C and: | | | | | | | | |
| — 500°C | 15.5 | 15.0 | 15.5 | 16.0 | 16.0 | 16.0 | 16.3 | 18.0 |
| — 1,000°C | 17.5 | 17.5 | 17.5 | 17.8 | 18.0 | 18.0 | 18.5 | |
| Specific heat at 20°C (Wh/kg x °C) | 0.12 | 0.13 | 0.13 | 0.14 | 0.14 | 0.14 | 0.14 | 0.14 |
| Resistivity at 20°C ($\mu\Omega$ x cm) | 108 | 118 | 112.2 | 112 | 105 | 104 | 95 | 74 |
| Mean temperature coefficient of resistivity ($10^{-6}$/°C) | 52 | 58 | 120 | 158 | 250 | 250 | 400 | 330 |
| Maximum operating temperature (°C) | 1,200 | 1,250 | 1,100 | 1,150 | 1,050 | 1,050 | 950 | 500 |
| Usual maximum load in Watts per square centimeter at: | | | | | | | | |
| — 800°C | 2.2 | 2.40 | 2.0 | 2.0 | 1.8 | 1.8 | 1.0 | |
| — 900°C | 1.7 | 1.90 | 1.5 | 1.5 | 1.3 | 1.3 | 0.5 | |
| — 1,000°C | 1.3 | 1.50 | 1.1 | 1.1 | 0.9 | 0.9 | | |
| — 1,100°C | 0.7 | 0.85 | 0.5 | 0.5 | | | | |
| — 1,150°C | 0.5 | 0.65 | | | | | | |

Fig. 23

resistance characteristics. A chromium oxide coating prevents most resistance oxidation which forms at the surface and prevents, because of oxygen tightness, a further oxidation. Because iron and iron oxide are present, this coating however more or less adheres, oxygen tightness is not perfect, and the formation of some iron oxide is still allowed at high temperatures.

These resistances are thus progressively subjected to intense oxidation accelerated by vibration and thermal cycles and, in some cases, aggravated by oxidizing atmospheres. The effective cross-section of the resistance decreases, creating hot points, and finally breaks down.

The operating temperature is therefore relatively limited, around 1,100°C at the resistance itself, and the service life of ternary alloy elements operating in oxidizing atmospheres is shorter than that of binary alloy elements. Conversely, costs are lower, and only a thorough analysis of a given problem enables the proper selection of either element. However, binary elements are most widely used in oxidizing atmospheres.

When used in reducing atmospheres of hydrogen or cracked ammonia, these alloys offer excellent resistance to corrosion, if the atmosphere used is purified (no residual sulphur or ammonia).

If temperature remains below 1,000°C, these alloys, especially 45% nickel and 23% chrome alloys, withstand carbonizing, carbon monoxide, carbon dioxide and hydrocarbon atmospheres. However, if generated from hydrocarbons, such atmospheres often contain traces of sulphur which is especially corrosive to nickel-rich alloys. Under such conditions, it is better to use an alloy containing at least 30% iron but it lowers the maximum operating temperature.

Basic refractory materials react with chromium to produce chromates and thus erode the protective oxide coating. The refractory materials used are therefore silicon and alumina based.

#### *3.4.1.1.3. Nickel-chromium resistances*

These heating elements consist essentially of two families:

— alloys containing more or less 80% nickel, 20% chromium (sometimes known as 80/20's);

— very high chromium content alloys consisting of 70% nickel, 30% chromium (often known as 70/30's).

These alloys contain less than 2% of other elements; manganese and magnesium (rolling and drawing facilitating metals), silicon (deoxidant) and various additives (rare earths) intended to increase maximum operating temperature and service life.

These heating elements use the same protection against corrosion as ternary alloy elements, but in the absence of iron oxide, the chromium oxide layer adheres well and is oxygen-proof. The coating has more or less the same expansion coefficient as the alloy, and therefore provides good protection. Since oxidation is more active as temperature increases, the alloy must contain extra chromium (however, the chrome content is limited to 30% approximately; more than 35% makes the alloy unstable).

The maximum operating temperature is about 1,200°C for 80/20 and 1,250°C for 70/30 nickel-chromiums. These alloys do not become brittle, even when held at high temperatures for long periods, and at 1,100°C in a normal oxidizing atmosphere their service life can reach ten years if not subjected to mechanical shock.

While nickel-chromium alloys offer excellent resistance to oxidation, they are sensitive to corrosive gases such as sulphur dioxide ($SO_2$), hydrogen sulphide ($H_2S$), sulphuric oxide ($SO_3$), water vapor, oil vapors, aluminum vapor, cyanhydric vapor and carbonizing atmospheres. Corrosion due to sulphur or carbon can cause rapid destruction of elements (a phenomenon known as green rot). In carbonizing atmospheres containing sulphur, alloys seem to disintegrate due to carburization of the chrome and sulphuration of the nickel, carburization being followed by partial or complete oxidation [69], [BG].

The most expensive alloys, such as 80/20 or 70/30, are not necessarily best suited for caustic atmospheres, and the use of ternary alloys may be advantageous, especially in carbonizing atmospheres if the temperature is not prohibitive. Otherwise, radiating tubes or other resistances must be used.

As for ternary alloys, the basic refractory materials must be avoided, and silicon and alumina based materials used.

#### *3.4.1.1.4. Heating element forms*

Nickel-chromium or iron-nickel-chromium alloys are available in:

— *circular cross-section products:* wires or rods;

— *semi-elliptical cross-section products:* strips obtained by cold rolling circular cross-section wires;

— *rectangular cross-section products:* strips obtained by shearing from flat wider products.

Some alloys are also used in the form of cast or moulded resistances (e.g. grades 70/30 and 60/15) or tubes (grade 80/20).

Resistances are formed by the alloy manufacturer, the user or specialists. Since these alloys are easy to work, heating elements are available in a wide range of forms.

The dimensions of numerous products such as wires, rods, strips and, more recently, moulded or cast resistances have become standard (U.T.E. (Union Technique de l'Électricité) publications). Quality standards also exist for these alloys.

Therefore, standard diameters provide at least forty values ranging from 7.34 to 0.079 mm. In some cases, manufacturers can provide non-standard sizes.

#### *3.4.1.1.5. Applications*

Due to their many qualities (adequate mechanical strength, high resistance to oxidation, long service life, malleability and repairability, relatively low cost, etc.), these alloys are very widely used, not only in the production of furnace heating elements but also for the production of encased heating resistances (immersion heaters, heating cartridges, finned elements, etc.).

These applications meet a wide range of requirements in low temperature ovens, and also in the medium-to-high temperature range. However, in the case of temperatures above 1,100-1,150°C requiring very high power densities, other resistances must be used.

### 3.4.1.2. *Iron-chromium-aluminum alloy resistances*

These alloys contain 20 to 30% chromium, 2 to 6% aluminum, the remainder being iron with small amounts of additives (for example, traces of cobalt increase resistance to corrosion).

#### *3.4.1.2.1. Characteristics*

Figure 24 provides the essential characteristics of the main iron-chromium-aluminum alloys used as resistances [39], [BH], [BG].

Characteristics of a range of iron-chromium-aluminum resistances [*BH*].

| Characteristics | Composition | | |
|---|---|---|---|
| | 1<br>Cr22-A15.5-Fe | 2<br>Cr22-A15.0-Fe | 3<br>Cr22-A14.5-Fe |
| Density | 7.1 | 7.15 | 7.25 |
| Permanent mode maximum operating temperature (°C) | 1,375 | 1,330 | 1,280 |
| Resistivity at 20°C ($\mu\Omega$ x cm) | 145 | 139 | 135 |
| Mean temperature coefficient of resistivity ($10^{-6}$/°C) | 40 | 60 | 70 |
| Melting point (°C) | 1,500°C approximately | | |
| Specific heat at 20°C (Wh/kg x °C) | 0.13 | | |

**Fig. 24**

Below the Curie point (about 730°C), these alloys are magnetic.

— *Element surface protection and resistance to aggressive atmospheres.*

The protection of iron-chromium-aluminum resistances is provided by a coating of oxides which form due to oxidation of the metals forming the alloys (chromium and aluminum oxides). This layer of oxides adheres strongly to the surface of heating elements and prevents further oxidation. At temperatures higher than 1,100°C, protection is due essentially to an alumina film. Therefore, these elements offer excellent protection against corrosive oxidation, enabling them to be used in oxidizing atmospheres up to 1,375°C for the highest quality grade. The surface power may therefore be higher than with nickel-chromium alloys, and higher furnace operating temperatures may be used. However, these alloys are more costly and, in some furnaces, are only used where the charged temperature must exceed 1,000°C.

In atmospheres containing sulphur (hydrogen sulphide $H_2S$ and sulphur dioxide $SO_2$), these alloys offer adequate protection against corrosion if they are used in an oxidizing environment.

In endothermally or exothermally controlled atmospheres, iron-chromium-aluminum alloys offer adequate protection against corrosion if the element surfaces are oxidized beforehand at more than 1,050°C, and then repeated periodically. However, the oxide film does not provide as good protection in such atmospheres as in air, and some corrosive impurities contained in gas mixtures (e.g. chlorine) can cause damage. All the halogens are corrosive to these resistances.

These alloys are not attacked by pure hydrogen and also withstand pure cracked ammonia quite well, but the resistance to partially burned ammonia is not so good.

Very dry, oxygen-poor nitrogen causes formation of aluminum nitride where the maximum temperature is 950°C. Conversely, if the nitrogen atmosphere contains a little oxygen, this prevents nitride forming, and the resistance to corrosion is better.

High quantities of water vapor can reduce element service life; this phenomenon is liable to occur in high temperature firing operations, but is easy to remedy by first eliminating most of the water.

For reference purposes, Figure 25 provides the operating conditions for iron-chromium-aluminum resistances in different atmospheres.

Alkaline and halogen metal salts, together with nitrates, silicates and boron composites, prevent oxidation and therefore promote corrosion in these alloys. This also applies to oxides of heavy metals such as copper, lead and iron. In fusion, some metals such as zinc, aluminum and copper, rapidly destroy iron-chromium-aluminum alloys.

The refractory materials supporting the heating elements must be carefully chosen; in particular, they must not contain free silicon and their ferric oxide content must be as low as possible. It is better to use products having a high alumina content. The supports for these resistances are generally produced from sillimanite or steatite.

— *Power density*

Figure 26 provides the most common specific powers for iron-chromium-aluminum resistances. These of course vary with furnace design, the arrangement and shape of the resistances and the type of charge.

— *Mechanical strength*

Unlike austenitic and nickel-chrome alloys, iron-chrome-aluminum alloys have a ferritic structure. These alloys crystallize when hot. After prolonged heating

Maximum temperature of heating elements as a function of the type of atmosphere [*BH*].

| Atmosphere | Cr22 Al5.5-Fe | Cr22 Al4.5-Fe |
|---|---|---|
| Dry air | 1,375 | 1,280 |
| Wet air | 1,300 | 1,100 |
| Pure hydrogen | 1,375 | 1,280 |
| Oxygen-poor dry nitrogen | 950 | 900 |
| Cracked ammonia | 1,200 | 1,100 |
| Carbon monoxide | 1,100 | 1,000 |
| Protective gas A* | 1,100 | 1,100 |
| Protective gas B* | 1,150 | 1,000 |
| Protective gas C* | 1,050 | 1,000 |

Approximate percentage composition of protective gasses

| | A* | B* | C* |
|---|---|---|---|
| CO | 10 | 10 | 20 |
| $CO_2$ | — | 10 | — |
| $H_2$ | 15 | 10 | 40 |
| $CH_4$ | 0.5 | 1 | 1 |
| $N_2$ | Remainder | | |

**Fig. 25.**

Maximum power in Watts per square centimeter for a furnace temperature in degrees Celsius, as a function of the type of alloy

| Alloy | Furnace (oven) temperature in degrees Celsius | | | | | | | |
|---|---|---|---|---|---|---|---|---|
| | 600 | 700 | 800 | 900 | 1,000 | 1,100 | 1,200 | 1,300 |
| Cr22-Al5.5-Fe | | | | | 4.0 | 3.0 | 2.0 | 1.5 |
| Cr22-Al5.0-Fe | | | | 4.0 | 3.0 | 2.2 | 1.6 | |
| Cr22-Al4.5-Fe | 3.9 | 3.5 | 3.0 | 2.4 | 1.5 | 1.0 | | |

**Fig. 26**

at high temperature, crystallization is maintained. Therefore, these offer lower mechanical strength when hot, and are rather fragile when cold after prolonged use. Moreover, plasticity increases so that the material can elongate in time due to its own weight when the operating temperature exceeds 1,100°C. Therefore, elements of this type must be carefully supported, and all precautions taken to allow for expansion.

When welding these elements, arc welding with a graphite electrode or, better yet with a tungsten electrode and protective gas (argon), is recommended. The welded area must be annealed to relieve internal tension.

— *Service life*

Service life may exceed five years of continuous operation at 1,250°C under normal operating conditions. Mechanical defects or corrosive atmospheres can, however, noticeably reduce the service life of such resistances depending upon other operating conditions.

*3.4.1.2.2. Heating element forms*

Iron-chromium-aluminum alloy resistances consist of "corrugated" strips and spring-wound wires on ceramic supports.

As for nickel-chromium alloy elements, the available product dimensions are many and subject to standards (U.T.E. (Union Technique de l'Électricité) publications).

*3.4.1.2.3. Applications*

These alloys are used primarily to produce resistances for high temperature furnaces, higher than can be reached with nickel-chromium alloys; in comparable temperature ranges, they can also increase power density.

Encased resistances also use these elements where high temperatures are required (e.g. sheathed ceramic resistances for heat treatment of weld beads).

### *3.4.1.3. Iron-nickel-chromium-aluminum alloy resistances*

These have appeared only very recently. Their approximate composition is 5% aluminum, 35% nickel and 20% chromium, the remainder consisting of iron (the percentages indicated represent the maximum content for each element). Additives, in particular rare earths, improve resistance characteristics.

The high aluminum and chrome contents offer excellent resistance to oxidation, and in oxidizing atmospheres, enable maximum operating temperatures of approximately 1,350°C.

Compared to nickel-chromium resistances, these alloys offer the advantage of higher temperatures, and therefore higher productivity. They also have better mechanical strength than iron-chromium-aluminum alloys, which makes them easier to install and increases their service life.

Up to now, these alloys consisted of moulded resistances but strips and wires are rapidly becoming available [90], [BG].

### *3.4.1.4. Other types of metal resistances*

To obtain high temperatures, it is necessary to use special metals and alloys with high melting points: nickel, chromium, iridium, platinum, platinum/rhodium, rhodium, molybdenum, tungsten, tantalum, niobium, etc. For the most part, these metals are very costly and thus are only used for laboratory furnaces or special industrial furnaces [13], [BJ], [BO].

*3.4.1.4.1. Platinum, platinum/rhodium and rhodium*

The main characteristics of these metals are given in Table 27.

Resistivity increases rapidly with temperature, and a starting system limiting power at low temperature is required. Generally, the resistance consists of a wire wound around a refractory tube free of impurities such as silicon. The vapor tension is high at high temperature and must be taken into account in the furnace atmosphere.

Platinum and platinum/rhodium resistance characteristics [13]

| Properties | Pt | Pt-Rh at 10% Rh | Pt-Rh at 20% Rh | Rh |
|---|---|---|---|---|
| Density | 21.5 | 20 | 18.75 | 12.48 |
| Melting point (°C) | 1,769 | 1,850 | 1,884 | 1,985 |
| Resistivity at 0°C ($\mu\Omega$ x cm) | 9.81 | 18.4 | 20.4 | 4.3 |
| Resistance maximum operating temperature (°C) | 1,400 | 1,500 | 1,600-1,700 | 1,850-1,900 |
| temperature coefficient of resistivity between 0°C and operating temperature ($10^{-3}$/°C) | 31.62 | 14.22 | 12.10 | 5.96 |
| Maximum power density at operating temperature (W/cm$^2$) | 2 | 2 | 2 | 2 |

**Fig. 27.**

Due to the price of platinum, this type of resistance is gradually disappearing since much cheaper resistances (e.g. molybdenum disilicide) can be used to reach these temperatures.

Resistances using pure sintered rhodium also exist and enable temperatures at the heating element of almost 1,900°C.

The power obtained with furnaces equipped with these types of resistance remains less than 10 kW.

### *3.4.1.4.2. Molybdenum, tungsten, tantalum and niobium*

The main characteristics of molybdenum, tungsten and tantalum are provided in figure 28 below.

The temperature coefficients of resistivity are rather high, and it is generally necessary to provide starting systems which limit low-temperature current surges.

Essentially, these resistances are used as heating elements in vacuum furnaces or controlled atmosphere furnaces in the heavy industries (aviation, micromechanical, aerospace, nuclear, electronic components, electrical and mechanical construction, tool-making, etc.), and for brazing, sintering, bonding, heat treatment, outgassing, monocrystal growth and test operations.

Characteristics of molybdenum, tungsten and tantalum resistances [13].

| Properties | Mo | W | Ta |
|---|---|---|---|
| Density | 9.6-10.28 | 19.32 | 16.65 |
| Melting point (°C) | 2,610 | 3,410 | 3,000 |
| Vaporization temperature (°C) | 4,800 | 5,930 | 4,100 |
| Resistivity at 0°C ($\mu\Omega$ x cm) | 5.17 | 5.5 | 12.4 |
| Mean temperature coefficient of resistivity between 0°C and 2,000°C ($10^{-3}$/°C) | 5.5 | 5.5 | 3 |
| Resistance maximum operating temperature (°C) | 1,500-1,700 | 2,200-2,800 | 2,400 |
| Normal surface power (W/cm$^2$) | 10-20 | 10-20 | 10-20 |

**Fig. 28.**

— *Molybdenum resistances*

Molybdenum is used in the form of spring-wound wires, assembled bundles of wires (cage consisting of links), rods, bands or strips.

This metal oxidizes starting at 400°C. The oxidation rate increases very quickly at 600°C since molybdenum oxide $MoO_3$ becomes volatile and no longer protects the resistance.

Molybdenum also oxidizes at 800°C in the presence of water vapor, and at 1,000°C in carbon dioxide $CO_2$. It carbonizes at 1,000°C in the presence of carbon monoxide CO and hydrocarbons. In a high vacuum ($10^{-4}$ torr), evaporation becomes high at about 1,700°C.

At high temperatures, molybdenum has to be protected by an inert or gas-reducing atmosphere ($N_2$, dissociated $NH_3$, He, Ar, etc.). A limit vacuum (of between $10^{-4}$ and $10^{-2}$ torr) is also highly appropriate.

The specific surface power often reaches 15 to 25 W/cm$^2$ for operation at 1,600°C. In-service furnace powers range up to 100 kW approximately.

— *Tungsten resistances*

Since its melting point is much higher, tungsten is more widely used than molybdenum. The evaporation rate is also lower.

Tungsten begins to oxidize in air at 500°C, and therefore an inert atmosphere or a vacuum ($10^{-4}$ to $10^{-6}$ torr) must be used at high temperatures.

In most cases, this metal consists of round bars, strips, wires, tubes and grids (cage consisting of element bundles).

The power range extends up to 200 kW approximately.

— *Tantalum resistances*

Tantalum resistances are used in vacuums or inert gases. In air, oxidation occurs at 500°C and nitriding at 700°C. This metal also withstands high vacuums, up to $10^{-6}$ torr at 2,000°C.

Tantalum is used in the form of helical wires, tubes, sheets and grids. It is extremely costly. In general, furnace power does not exceed 60 kW.

— *Niobium resistances*

This metal has melting and boiling points of 2,500°C and 3,700°C respectively. It is more expensive than tantalum, and is almost never used in heating elements.

### 3.4.2. Cermets

The term "cermet" is a contraction of the expression "ceramic-metal". The materials forming these resistances are composites of metals and metalloids in the form of high temperature sintered ceramics. In the type usually encountered, the basic material is approximately 95% molybdenum disilicide $MoSi_2$. Resistances consisting of other materials have appeared recently, but their use is confined to special applications [9O], [BH], [BL], [BN].

#### 3.4.2.1. *Molybdenum disilicide resistances*

##### 3.4.2.1.1. *Characteristics*

— *Density and expansion*

The density of molybdenum disilicide resistances, which is approximately 5.6, is between that of non-metallic resistances (about 2 for carbon and 2.6 for silicon carbide) and the usual metal resistances (approximately 8.2 for nickel-chromium and 7.2 for iron-chromium-aluminum). The coefficient of linear expansion is about 7.5 x $10^{-6}$/°C.

— *Resistivity*

Resistivity is high, but varies with temperature. It ranges from 35 $\mu\Omega$ x cm when cold to 350 $\mu\Omega$ x cm at 1,500°C, calling for self-regulation of the supply current and special starting devices. Slight variations in resistivity exist between the various grades, but are generally within 10%.

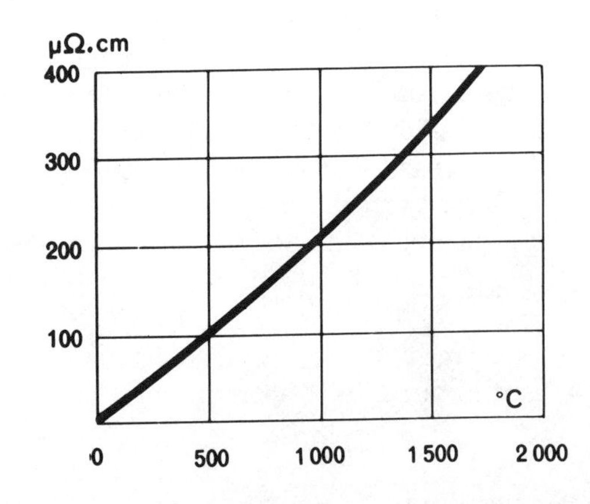


**Fig. 29.** Change in resistivity of molybdenum disilicide heating elements as a function of temperature [13].

Resistivity varies little with time; this decreases slightly over the first few hours of use for some varieties and increases very slightly for others. Unlike silicon carbide (*see* paragraph 3.4.3.2.), there is therefore no aging of the resistances, enabling new and old elements to be connected together in series or in parallel.

— *Element surface protection*

Molybdenum disilicide MoSi2 oxidizes forming a layer of silica $SiO_2$ on the surface of the element starting at 980°C,and very rapidly at temperatures above 1,400°C. This highly adherent film protects the element from corrosion and oxidation produced during element curing. Breaks in the silica film are sometimes unavoidable in industrial use, for example during sudden cooling. To conserve this protective film, it is important to expose the elements to a temperature of 1,400°C periodically in an oxidizing atmosphere.

From 1,600°C, the silica film tends to adhere to the furnace refractory covering causing a rupture of the element on cooling. To prevent contact between the resistance and the refractory material, elements are suspended vertically.

— *Operating temperature*

Depending on the grade used, the maximum operating temperature at the resistance is 1,700 or 1,800°C. The maximum temperature obtainable in a direct radiating furnace is therefore between 1,650° and 1,750°C. However, these are upper limits; lower operating temperatures, and consequently lower power densities, improve element service life.

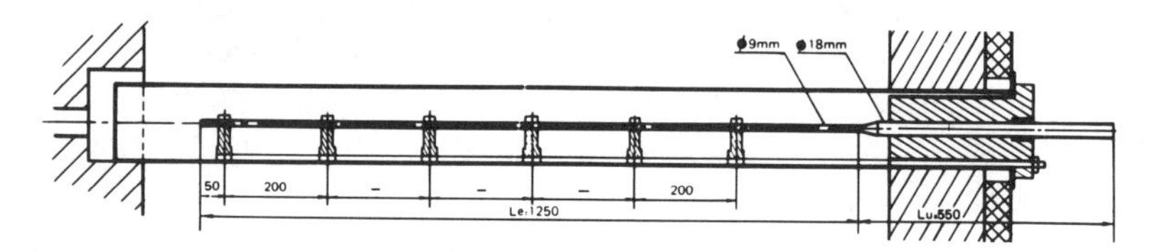


**Fig. 30.** Radiating tube with molybdenum disilicide heating element [BH].

For horizontally mounted elements, the maximum operating temperature can be much lower. In fact, the element begins to lose its rigidity near 1,300°C. However, it is possible to operate at much higher temperatures if the heating element is supported, while allowing it to move horizontally during elongation or shrinkage after heating and cooling. Some furnaces are fitted with horizontal radiating tubes, each containing a molybdenum disilicide heating element. The ceramic radiation tube is sillimanite, and the element supports are zirconium silicate (basic materials are not to be used as supports). To reduce overheating, the surface area touching the supports must be very small. Under such conditions, the maximum operating temperature can be about 1,500°C.

— *Resistance to corrosion*

The maximum operating temperatures mentioned above apply to resistances used in air. This temperature may vary as a function of furnace atmosphere.

However, due to the protective silica surface film, resistance to chemical agents is generally excellent. The reducing effect of some atmospheres on silica $SiO_2$, however, calls for limitation of the element operating temperature, as shown in Figure 31. Also, with reducing atmospheres, superficially reoxidizing the element by admitting air into the furnace and operating it at a temperature higher than 1,400°C to regenerate the protective film is recommended.

Maximum operating temperature of a molybdenum disilicide heating element as a function of the type of atmosphere [BH].

| Atmosphere | Maximum temperature in element (°C) |
|---|---|
| Air | 1,800 |
| Oxygen | 1,700 |
| Nitrogen, nitrogen oxide (NO2) | 1,600 |
| Nitrogen protoxyde (NO) | 1,600 |
| Sulphur dioxide (SO2) | 1,600 |
| Carbon dioxide (CO2) | 1,600 |
| Argon, helium, neon | 1,550 |
| Carbon monoxide (CO) | 1,500 |
| Hydrogen, dew point 15°C | 1,460 |
| Cracked, partially burned ammonia (8% $H_2$) | 1,400 |
| Dry hydrogen | 1,350 |
| Methane | 1,350 |
| Usually encountered controlled atmospheres | 1,350-1,500 |

**Fig. 31.**

Molybdenum disilicide elements are not suitable for vacuum furnaces. The maximum temperature at the resistance varies with residual pressure; therefore,

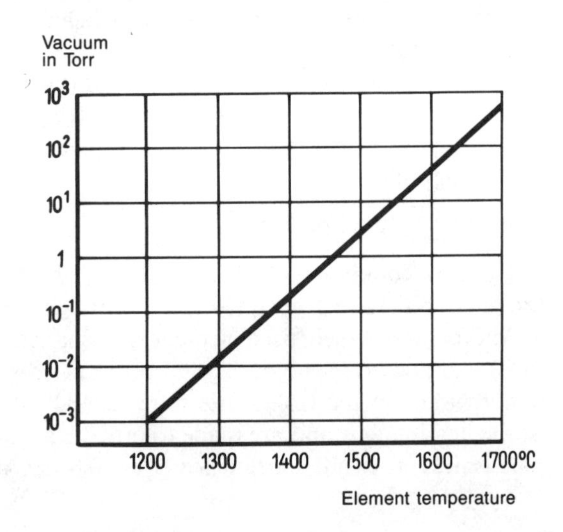


**Fig. 32.** Molybdenum disilicide heating element maximum operating temperatures in vacuum furnaces [BL].

graphite, molybdenum or tungsten resistances are used in industrial vacuum furnaces.

— *Surface power*

The surface power to be adopted depends on the number of variables, furnace charge temperature, maximum element operating temperature, type of atmosphere, heating element arrangements, etc. The surface power generally used is 10-12 W/cm2 for a furnace temperature of 1,500°C, and 6-10 W/cm2 for a temperature of 1,600°C.

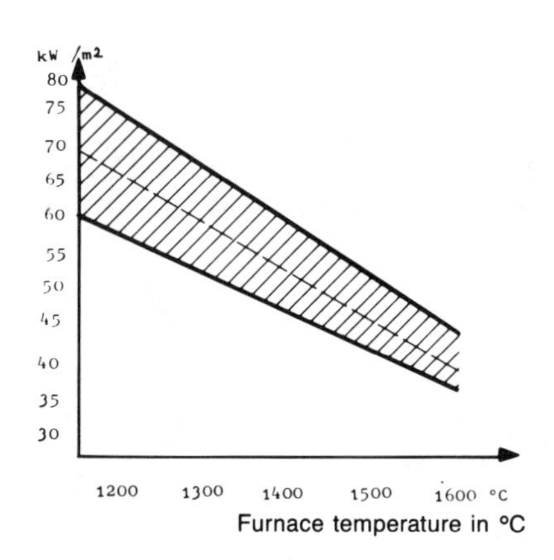


**Fig. 33.** Surface power per square meter of wall [88].

When the furnace temperature is low, the surface power used is much greater up to 20 W/cm$^2$. These specific powers are much higher than those obtained with metal resistances, enabling low or medium temperature furnaces to be constructed (temperatures less than 1,100°C) with very high power densities. Figure 33 gives the power that can be installed for each square meter of furnace wall as a function of temperature.

With commonly used metal resistances (nickel-chromium 80/20), the power density in furnaces operating around 950°C can barely exceed 20 kW/m$^2$ of wall. With this type of resistance, values three to four times higher can be reached (*see* paragraph 5.1.3.).

### *3.4.2.1.2. Types of heating elements*

The most common molybdenum disilicide heating elements are U-shaped, with the leads generally suspended freely through the furnace roof.

Another less widely used element is W-shaped (or, to be more precise, double "U") and is used in the horizontal position only. The heating part has a small 6 to 9 mm diameter. Cold outlets are generally equal to or double the diameter of the heating part and are made to order. The maximum length of the heating part is 1.25 m, while current leads are 0.8 m. At the other end of the range, very small dimension elements having a diameter of 0.4 mm, heating length of 1.5 cm and an overall length of 3.6 cm exist.

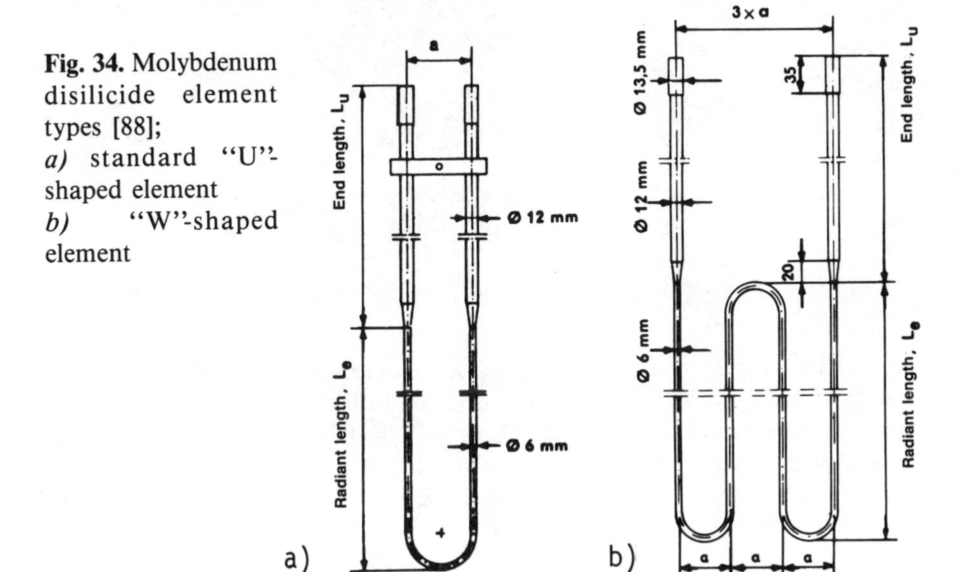


**Fig. 34.** Molybdenum disilicide element types [88]; *a)* standard "U"-shaped element *b)* "W"-shaped element

To facilitate installation, the cold outlets of some elements are not a continuation of the heating part, but form an angle of up to 90° with it.

The unit power of the most common elements varies between 4 and 10 kW.

*3.4.2.1.3. Source of electricity*

Heating elements are supplied with three-phase or single-phase AC via several resistances (generally 4 to 12) connected in series. The supply voltage is usually between 100 and 400 V.

Generally, power is controlled by regularly separated resistance zones. Power feeds of the "full, medium, and off" type are often provided in furnaces. The power control system must prevent heavy overcurrents that often occur with cold starting due to low resistivity at low temperature; generally, it is recommended to use a starting voltage approximately one-third that of the normal operating voltage. Power supplies similar to those mentioned above (changeover power supply to the resistance groups using star, delta, series, parallel connection systems) help solve this problem. Other cold starting techniques exist. In addition, temperature and power control is provided by thyristor systems. The increase in resistivity at high temperature offers a certain degree of self-regulation.

Due to heating element shape and high currents flowing through them, strong magnetic fields often occur between connections; the forces exerted may exceed the breaking limit. When designing a furnace, it is very important then to analyze this problem and use heating connection arrangements, power supplies and electrical circuits which reduce it. A simple way to reduce magnetic fields is to connect consecutive elements so that current flows through adjacent heating parts in opposite directions.

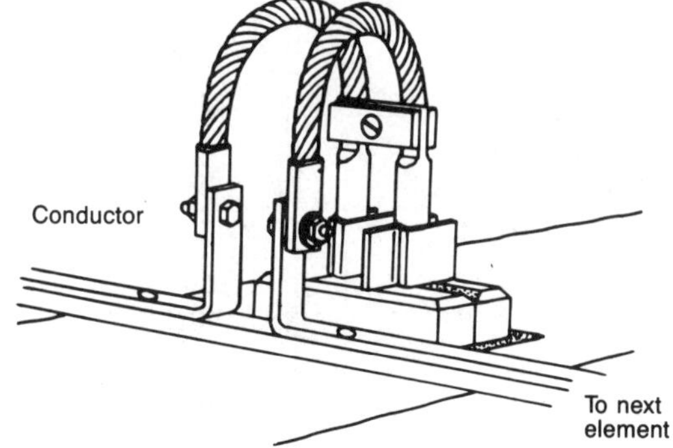


**Fig. 35.** Electrical connection of molybdenum disilicide elements [BL].

Installation of elements and their electrical connection is generally very simple. The cold outlets of each heating element are, in fact, inserted in a refractory plug consisting of two special ceramic half-shells; this plug is inserted with the element it supports in a space prepared in the roof of the furnace. The elements are connected by means of aluminum cables or braids. Some manufacturers provide elements with terminals welded to the cold outlets.

*3.4.2.1.4. Useful life of elements*

Molybdenum disilicide is a sintered ceramic, with some brittleness when cold. Therefore, precautions must be taken when handling and installing this material. The heating element is most often coupled to a refractory head which decreases the risk of mechanical damage during installation. From 1,200°C, this, however, offers excellent plasticity, enabling it to withstand vibrations or mechanical shock.

The service life of elements is affected by the number of heating and cooling cycles which eventually make it brittle. Therefore, in continuous furnaces, service life is very long and, in the absence of other destructive factors, may reach ten years. However, this is lowered (by some years) in discontinuous furnaces with frequent cool-downs to ambient temperature. For discontinuous furnaces, and also continuous furnaces, it is recommended the standby temperature to kept at 600°C. Under such conditions, service life can be extended further.

The longest service life is obtained in oxidizing atmospheres (air, oxygen, water vapor, carbon dioxide). However, elements resist reducing gases well only if the temperature limits mentioned above are respected, and superficial reoxidization at a temperature exceeding 1,400°C is carried out periodically. Elements also offer adequate heat resistance to iron oxides which do not adhere to the elements. Conversely, elements adhere to refractory materials at temperatures above 1,600°C, but this drawback can be remedied by simply suspending the elements vertically without allowing them to touch the refractory materials.

Finally, aging of such elements, if some precautions are respected in furnace operation, is very low and, in most cases, element breakdown is due to mechanical damage only. The service life of molybdenum disilicide resistances is therefore very long if mechanical stress and shocks are avoided. Service life in industries having rugged operating conditions, such as steel-making, heavy mechanical engineering or glass manufacturing, vary between three and ten years.

*3.4.2.1.5. Advantages and limitations of molybdenum disilicide resistances*

The main interest in these resistances is due to their ability to operate at high temperatures which provide high power densities and widens the field of application of resistance furnaces; they also offer excellent resistance to corrosion and have a long service life.

Conversely, their cost is rather high and their brittleness calls for precautions when handling and in furnace operation.

*3.4.2.1.6. Applications*

Conventionally, these resistances were reserved for furnaces operating at high temperatures:
— ceramic industries for production of refractory materials, porcelain, special ceramics (e.g. sillamnite), sanitary ceramics and ferrites;
— metal industries for construction of Pits furnaces, forge furnaces, heat treatment (with or without a special atmosphere), brazing and heating of billets, etc.;
— glass industries for melting (pot furnaces, crystalwork furnaces), holding, refining and heating of the drawing area, reheating of blanks, etc.;
— electronic industries, for production of semiconductors;
— laboratory furnaces.

At present, molybdenum disilicide resistances are also used in medium temperature furnaces (up to 1,000°C approximately) with very high power densities and therefore less high volume production (metal heat treatment furnaces and ceramic furnaces), [13], [27], [57], [68], [83], [88], [BH], [BL].

*3.4.2.2. Lanthanum chromite resistances*

These resistances are composed mainly of chromium oxide and lanthanum oxide (lanthanum chromite). This compound offers the uniqueness, which is rare for oxides, of being highly refractory since it melts near 2,500°C, and is also a conductor of electricity above room temperature. Resistivity depends on composition and temperature: between 1 and 40 Ohms x cm cold; it also varies between 0.08 and 1 Ohm-cm at 1,800°C. This property is used in designing heating elements. They consist of rods or tubes with a central heating area of average resistivity and low resistivity non-heating terminations which enable very high temperatures in the central area and a cold outlet temperature of less than 300°C.

These resistances are composed of oxides that are inert in air or pure oxygen. Conversely, the lower the partial oxygen pressure in a gas mixture, the more the operating temperature is limited and service life shortened.

The maximum operating temperature is about 1,850°C, but service life is then very short, a matter of several hundred hours maximum. When used at temperatures of less than 700°C, these elements have much longer service life:

several thousand hours. Basically, these resistances are used for heating small laboratory furnaces [BM].

#### *3.4.2.3. Zirconia resistances*

These sintered ceramic resistances consist of tubes or rods having a heating part in stabilized (e.g. lime) zirconia $ZrO_2$ and lanthanum chromite-based cold outlets. This design enables very high temperatures to be obtained in the heating part, while keeping electrical connections at a temperature of less than 300°C. Other elements consist of thin blades of hollowed out zirconia, cut from a mass. A part of this blade is used only to couple the resistance.

These heating elements require preheating using molybdenum disilicide or lanthanum chromite resistances, since below 1,100°C, zirconia is a very bad conductor of electricity [BM].

These resistances can be used to construct laboratory furnaces operating up to 2,200°C in an oxidizing atmosphere. These have only appeared recently, and it is difficult to estimate their service life.

### *3.4.3. Non-metallic resistances*

This class of resistances consists of heating elements made from carbon and silicon carbide.

#### *3.4.3.1. Carbon resistances*

##### *3.4.3.1.1. Characteristics*

Two allotropic varieties of carbon are used:
— amorphous carbon;
— synthetic graphite.

Both materials can be produced with a very high degree of purity, of the order of 99% carbon for artificial graphite.

Amorphous carbon or graphite heating elements can be divided into two main families: aggregates (consisting of pieces or grains of coal, coke or graphite, placed on supports such as ducts or channels) and molded elements. The former is very primitive and only provides coarse temperature control and, in spite of some advantages (low cost, protective atmosphere favorable for certain applications), is becoming obsolete. Conversely, molded elements are being used more and more in high temperature furnaces.

Amorphous carbon parts are generally produced by extruding a petroleum coke or powdered anthracite paste under pressure, mixed with tar which acts as an agglomerant; they are then calcinated at high temperature. Many forms are possible such as bars, tubes, coils, etc.

Graphite is produced from amorphous carbon in high temperature electric furnaces (*see* paragraph entitled "Manufacture of graphite electrodes" in the "Heating by conduction" chapter). During graphitization, impurities such as silicon, aluminum and iron oxides are eliminated.

Graphite and carbon can be machined or molded to obtain many shapes and dimensions.

Characterisics of carbon resistances

| Characteristics | Amorphous carbon | Synthetic graphite |
|---|---|---|
| Real density | 1.8 | 2.25 |
| Apparent density | 1.1 | 1.56 |
| Specific heat (Wh/kg x °C) | 0.20 to 0.23 at 20°C; 0.33 to 0.55 between 1,000 and 2,000°C | |
| Resistivity ($\mu\Omega$ x cm) | 7,000 to 8,000 | 800 to 1,050 |
| temperature coefficient of resistivity $\alpha$ | Slightly negative | Negative from 0 to 500°C ($-4 \times 10^{-4}$/°C) Positive from 500 to 1,200 ($3.6 \times 10^{-4}$/°C) Slightly positive, then ($1.3 \times 10^{-4}$/°C) |
| Boiling point (°C) | 3,600 | |
| Distilling point (°C) | 2,200-2,800 (highly intense in vacuum) | beyond 2,700 |
| Maximum operating temperature (°C) | 2,200-2,300 | 2,600 in vacuum 2,800 in argon 3,000 in helium |
| Initial oxidation temperature (°C) | 400 | 500 |

**Fig. 36.**

Basically, synthetic graphite is used in the production of heating elements since it offers constant practical resistivity and improved resistance to oxidation. Figure 36 gives the main characteristics of carbon resistances.

Carbon heating elements have high resistivity. Graphite offers the advantage of little temperature variation with temperature increase. However, graphite heating elements are sometimes supplied with low voltage, since cross-sections are large (generally to ensure adequate mechanical strength) resulting in low resistance.

Carbon is one of the most refractory elements in existence. It can be used to design furnaces operating at high temperature. Unfortunately, it oxidizes in air at a relatively low temperature. At 500°C, graphite loss in continuous operation is about 1% per day. To reduce such wear, some furnaces have inert or confined atmospheres. For example, high temperature crucible furnaces use heating elements placed around the crucible, and permanently immersed in nitrogen. Recently designed radiating tubes contain heating elements consisting of graphite pins placed in ceramic tubes; the leakproof tubes, contain a special confined atmosphere and slow down oxidation [H], [AO].

*3.4.3.1.2. Element forms and applications*

Carbon elements are available in many forms: coal or coke grains and pieces, blocks or bricks, bars, rods, tubes, crucibles (*see* "Direct current flow crucible melting furnaces" in the "Conduction heating" chapter), pins, coils, graphite deposits on glass fabric (*see* "Other types of encased resistance" in the chapter entitled "Heating with encased electric resistances") and powdered carbon.

**Fig. 37.** Carbon resistance-heated Pits furnace ("Elpit" furnace) [90].

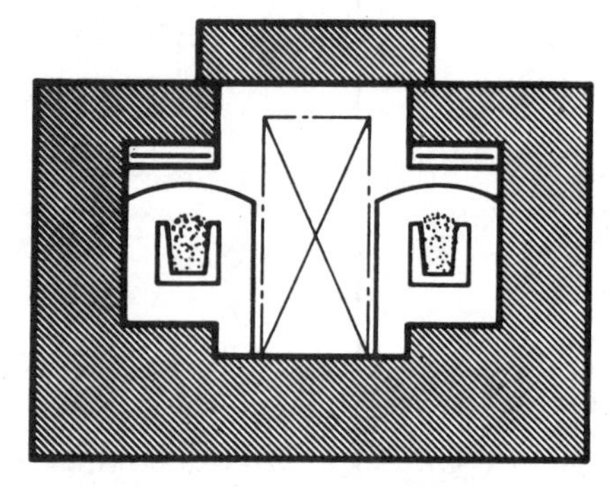

The structure of furnaces using carbon resistances is sometimes closely related to the shape of the resistances. Examples of these are:

— *the Pits furnace*, with a capacity of 20 to 160 t and a power of 600 to 2,000 kW, using petroleum coke pieces placed in silicon carbide troughs ("Elpit" furnaces). Coke consumption is approximately 0.5 kg per ton of steel. One of the advantages of these furnaces is that a protective atmosphere is produced automatically (carbon monoxide base) which effectively protects the metal against oxidation and decarbonization;

— *"kryptol" furnaces*, using a mixture of coke or graphite pieces, silicon carbide and silicates. This mixture, located around the heating chamber (muffle, tube, crucible), acts like a heating element radiating through the roof, or by conduction due to a crucible. The current is applied through embedded graphite electrodes;

**Fig. 38.** Carbon grain-based resistance (kryptol furnaces) [13].

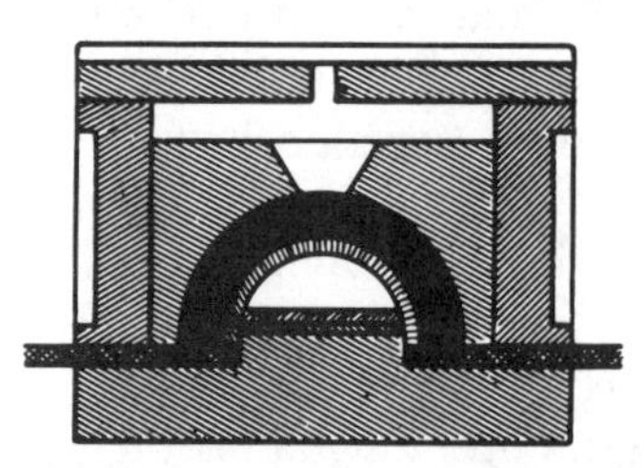

— *horizontal radiating resistor melting furnaces*. These are oscillating horizontal furnaces using a graphite rod radiating directly onto the melt. These furnaces are employed for melting pig iron, steel, bronze and silica.

While some cases offer appreciable advantages, most of these furnaces have been abandoned for other techniques such as arc, induction or resistance furnaces using other heating elements.

Conversely, the use of molded graphite resistances is rapidly expanding. New elements are enabling temperatures of approximately 3,000°C. In particular, this development is related to that of vacuum furnaces in which graphite resistances are commonly used as heating elements. In most cases, the elements consist of bars of various sizes, and also pins or coiled resistances, with cold outlets located at the same end or at opposite ends [20], [33], [34], [45], [47].

### *3.4.3.2. Silicon carbide resistances*

Carbon oxidizes rapidly in air. Two solutions have been studied to overcome this drawback. The first involves the use of carbon resistances, the surfaces of which have been superficially converted into silicon carbide; this solution seems to have been abandoned. The second consists of using very pure silicon carbide resistances which offer adequate resistance to oxidation. This material is produced in electric arc furnaces, and first appeared at the beginning of the 1930's, improving ever since. Various crystalline varieties are used to produce resistances. In some cases, the heating part is in $\alpha$ silicon carbide, with cold ends in ß silicon carbide. In other cases, the heating part is in ß silicon carbide. Crystallization affects some properties of the resistance such as the maximum operating temperature or resistivity [13], [BL], [BP], [BQ], [BR], [CA].

#### *3.4.3.2.1. Characteristics*

— *density*

The mean density of the silicon carbide used for the production of heating elements is approximately 2.6, but for some high temperature recrystallized varieties, it can reach 2.9 (for 80/20 nickel chromium it is 8.3-8.4).

— *resistivity*

Resistivity is very high, varying between 0.15 and 0.5 ohm-cm at ambient temperature versus 110$\mu$ohm-cm for 80/20 nickel-chromium. Figure 39 gives change in resistivity versus temperature for several heating elements.

Resistivity initially decreases as the temperature approaches 650°C. Above this temperature, resistivity increases to normal furnace operating conditions approaching the initial value. The lower portion of the curve represents an average trend, since traces of impurities have a marked effect on low temperature resistivity. Conversely, at high temperature, this phenomena disappears, and the rated resistance values can be used in calculating dissipated power. The ascending curve, under normal conditions, prevents runaway increases in absorbed power.

The resistivity variation law differs for elements of various composition, and can only be provided by the manufacturer. Tolerances for silicon carbide industrial resistances are generally between ±10 and 15%, but for elements of special shapes, can reach ±20%. Narrower ranges can, however, be obtained for simply shaped elements.

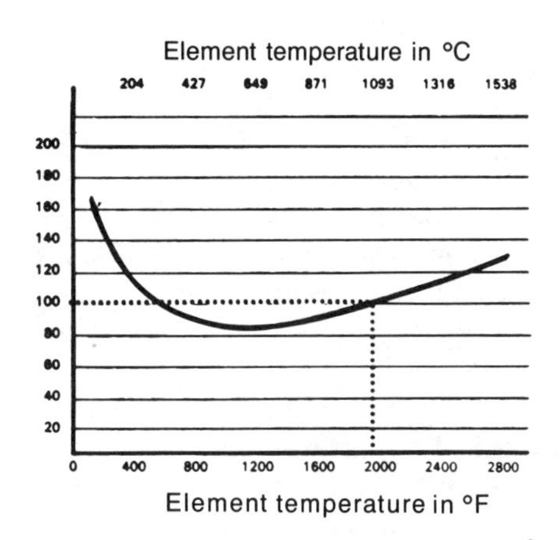


**Fig. 39.** Variation in resistivity for a family of elements, expressed as a percentage of the rated value at 1,960°F (1,071°C) versus temperature [BK].

— *expansion*

The coefficient of linear expansion between 20 and 1,500°C is generally between 5 x 10-6/°C and 6 x 10-6/°C and must be taken into account.

— *operating temperature*

The maximum operating temperature depends on many factors: furnace atmosphere, resistance charge, shape, crystalline struture, etc. In general, the maximum operating temperature for elements varies between 1,500 and 1,650°C, enabling maximum temperatures of about 1,600°C in a furnace radiating directly onto the charge in a normal atmosphere.

— *surface power*

Maximum surface power depends on the type of heating elements, operating temperature, atmosphere, furnace composition and its optimum temperature. As an example, Figure 40 gives the maximum permissible value for furnace operating temperatures between 1,100 and 1,400°C, i.e. 1,500°C on the rod in a normal atmosphere.

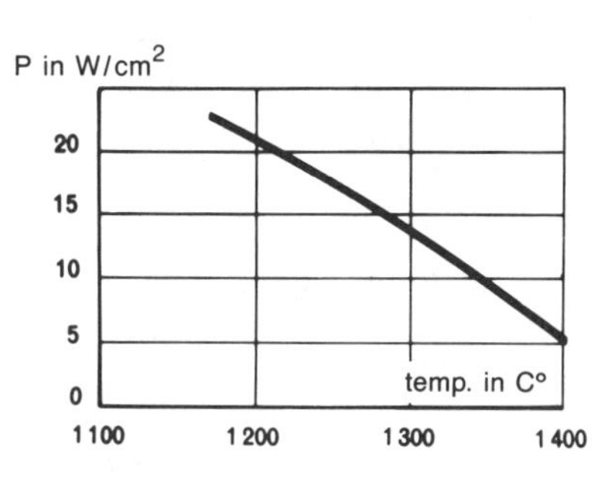


**Fig. 40.** Surface power as a function of oven temperature

In a reducing atmosphere, power density must be much lower so that heating element temperature is never within the silicon carbide chemical transformation zone. With most elements, a hydrogen atmosphere causes destruction of silicon carbide, CSi, at 1,300°C and the formation of silicon Si; at 1,300°C, the specific level must not exceed 3 Watts/cm$^2$ (compared to 14 W/cm$^2$ in air).

In general, and to increase service life, it is recommended that silicon carbide rods be used at a surface power slightly lower than maximum. The benefit of a level lower than the limit is, however, related to the quality of the temperature measurement and regulation system.

— *resistance to corrosion*

At high temperature, silicon carbide reacts whenever its two components, carbon and silicon, can act as reducers. Therefore, silicon carbide heating elements oxidize in air. Since the elements are porous, silica ($SiO_2$) is formed due to the effect of oxygen, increasing resistivity (aging phenomenon). This aging causes a large increase in resistance, which for the most common elements, is between 50 and 100% of the initial value during the first hundred hours. Subsequently, the phenomenon continues more slowly and, after a year and a half to two years, resistivity can be four times higher than at the beginning (*see* paragraph 3.4.3.2.4.). Carbon dioxide $CO_2$ also causes aging of silicon carbide rods.

In hydrogen, cracked ammonia or other atmospheres containing hydrogen, the maximum operating temperature is 1,300°C, which involves limiting power density to prevent this temperature being exceeded at the heating elements. Beyond this temperature, the reducing effect of hydrogen increases rapidly and causes decomposition of the elements. In a high-nitrogen-content atmosphere or one containing free nitrogen, temperatures of 1,400°C must not be exceeded, since at 1,450°C formation of silicon nitrite begins.

When the atmosphere contains carbon monoxide or hydrocarbons, silicon carbide elements become enriched in carbon; resistivity then decreases gradually and element current increases. Air must be periodically admitted into the furnace to burn off excess carbon.

Water vapor also adversely affects elements. With the reducing gas mixtures used in the heat treatment of some metals, gradual aging of the elements also occurs. To prevent an excessively rapid increase in resistivity, it is necessary to reduce the specific load of the elements and to use a lower dew point. Silicon carbide elements also react with basic materials such as alkalines, earth-alkalines, vaporous heavy metal oxides or molten oxides and with silicates and borates.

Therefore, when drawing up a technical specification, it is necessary to define accurately the type of atmosphere in which silicon carbide elements are to be used.

Recently, there has been much progress in the production of these elements, enabling improvements in performance, in particular in resistance to corrosion. For example, high density silicon carbide rods offer much lower porosity than those usually encountered, and the chemical effect of corrosive atmospheres is therefore much slower and limited. Other elements are subjected to special surface impregnation treatment and deposit of a silicon carbide coating. This fills the pores between the silicon carbide crystals, rendering the surface impervious, and delaying the aging phenomenon. The service life of elements can be increased from 20 to 300%, depending on operating conditions [BL], [BP], [BQ], [BR], [CA].

Currently available silicon carbide heating elements can therefore be used in most of the atmospheres encountered in industry, if carefully chosen and the operating constraints are respected.

*3.4.3.2.2. Heating element types*

— *design*

Silicon carbide heating elements are available in a wide range of shapes, as shown in Figure 41. The rods may be solid (*a,b,c*), hollow (*d,e,f*) or U or W shaped (*g*).

**Fig. 41.** Silicon carbide heating elements

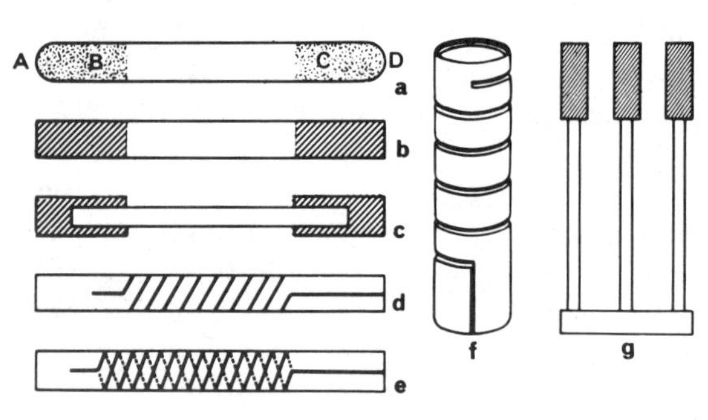


**Fig. 42.** Silicon carbide coaxial element

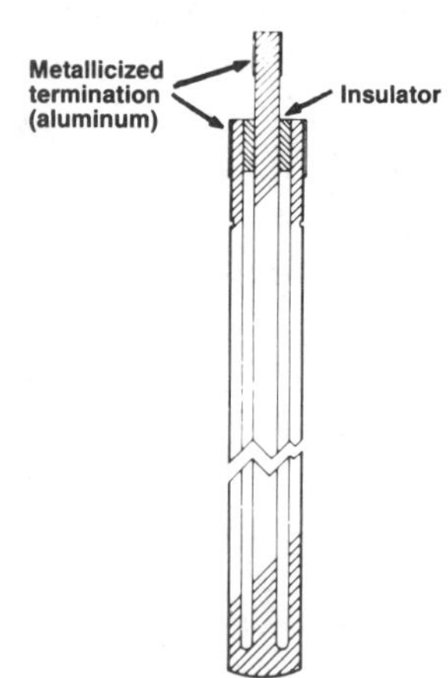


Other silicon carbide elements have been developed, such as coaxial elements joined at one end with the current inlets located at the other end and only the external tube heated (Figure 42).

Heating element cold outlets are created by various techniques:

— the cold ends consist of material other than that forming the heating zone (*a,b*); thus, some resistances have an $\alpha$ silicon carbide heating zone with cold ends in $\alpha$ silicon carbide;
— the ends are reinforced by matching sleeves, or of a material having lower resistivity (*c*);
— the rod consists of a single material, but the diameter of the heating part is much smaller than the ends (solution similar to *c* above, but in this case, the element consists of a single piece);
— the element is made from a thin-walled tube composed of one material (in general, ß silicon carbide), with the heating part produced by cutting a spiral in the tube. The two un-cut ends of the tube are used as a current inlet to the heating part (Figure 43);

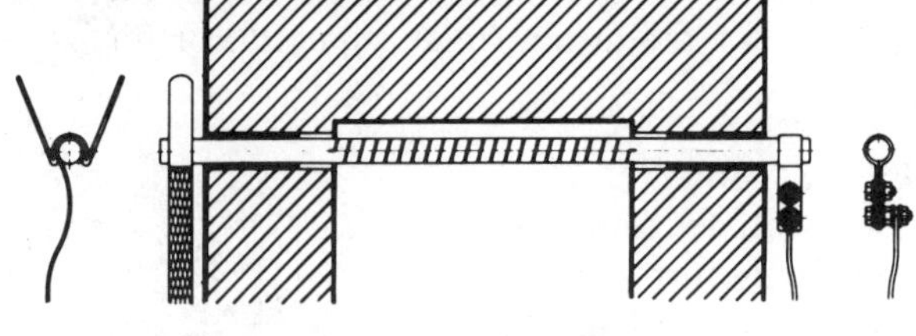

**Fig. 43.** Single spiral element with cold termination at each end of the element [13].

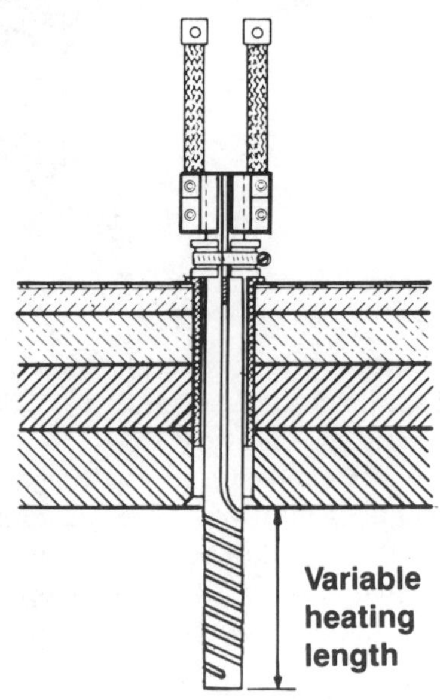


**Fig. 44.** Double spiral element roof suspended, with cold terminations located at the same end (horizontal or vertical wiring are equally also possible).

— the element is made from a single piece and the heating part consists of a double spiral providing current flow in both directions (bayonet element). The cold outlet is divided into two parts, coupled to the two spirals. This type of element offers the advantage of having the connections located at one end of the resistance, which reduces wiring costs and offers a wide choice of applications (Figures 41 and 44). Combinations of the above methods or other techniques can be used to obtain cold outlets. In order to decrease the contact resistance with current inlets, the ends of the cold outlets are usually plated with aluminum.

The connections consist of clamps with aluminum braid ribbons, sheet steel clamps or clips.

— *installation*

Rods may be positioned horizontally or vertically. Arrangements, however, vary with the type of element. Single spiral elements, when vertically mounted, must not be suspended but rested on their bottom end. Conversely, double spiral single cold end elements offer installation flexibility. They can be suspended vertically or placed horizontally, since the end corresponding to the junction of the two spirals is not supported (short lengths), or they can rest on a blind hole (long lengths) or even rise vertically from the furnace hearth.

Some hollow rods have a large inside diameter and can surround a tube or crucible for producing small quantities or for laboratory tests.

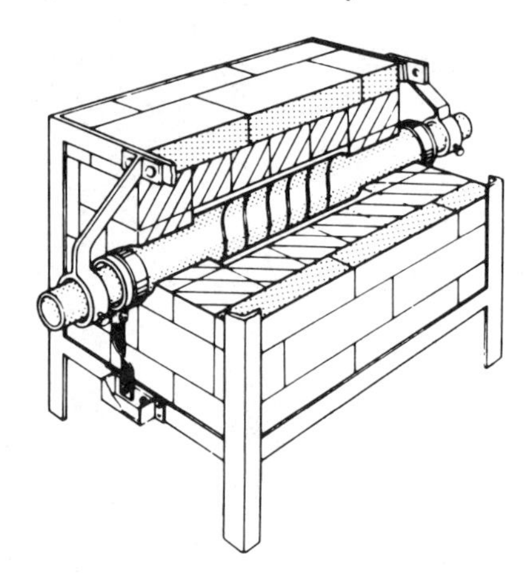

**Fig. 45.** Small furnace with resistance completely surrounding the tubular chamber [BR].

Silicon carbide heating elements can also be placed in silicon carbide envelopes and used as immersion heaters (*see* paragraph 5.11.1. of the chapter on "Enclosed electrical resistance heating").

Silicon carbide rods are generally single phase. For use with three-phase, the heating elements must be connected correctly. However, there are three-phase elements consisting of three vertical rods, one end of which bears a small sleeve; the other three ends are interconnected, either through a part of the same type, or a highly conductive ceramic part.

— *element range*

An extremely wide range of elements exists, both in terms of dimensions and unit power. For example, a heating length of a range of bar elements (Figure 41*b*) varies between 102 and 1,829 mm (special rods can be up to 2,800 mm), with powers of between 0.5 and 45 kW approximately. The supply voltage must be chosen so that the maximum surface powers are respected; lower powers of course can also be achieved. Overall lengths and diameters of the corresponding elements vary from 279 to 2,667 mm and from 7.9 to 54 mm respectively. For double spiral elements, the heating lengths and maximum powers range from 102 to 610 mm and 0.8 to 20 kW respectively.

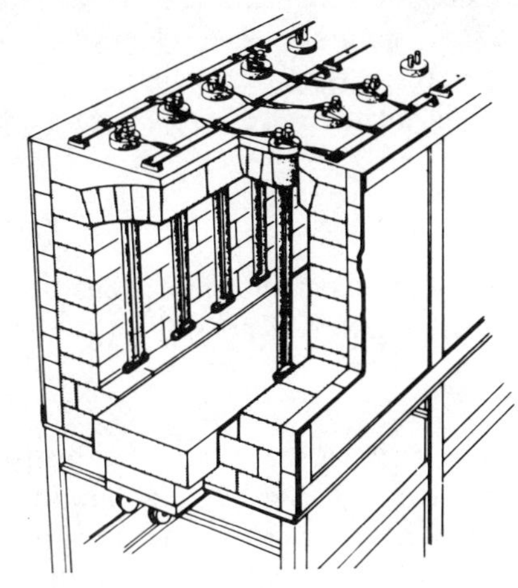

**Fig. 46.** Movable hearth or wagon furnace with U-shaped resistances [BR].

*3.4.3.2.3. Useful life of elements*

This is a cost factor, since high performance silicon carbide heating elements are much more expensive, for equal power, than the normal metal resistances. If silicon carbide element-equipped furnaces are constructed and controlled respecting manufacturers' and constructors' instructions, service life can be very long for continuous operation furnaces. For intermittent operation, it is generally lower. Moreover, manufacturers sometimes recommend that a temperature of about 800°C be maintained in intermittent operation with brief production shutdowns (for example shutdown between two successive working days).

The main items affecting element service life are:

- installation of elements in the furnace and furnace design,
- furnace temperature and element regulation and specific load,
- the type of atmosphere and load,
- element heat-up and furnace operating conditions,
- rod aging and type of power supply.

— *furnace design and element installation*

In general, sufficient play for expansion and contraction of rods (these phenomena occur most often in intermittent operating furnaces) is generally provided by a system of flexible connections (contacts consisting of clamps or clips supplied through flexible braids). In furnace wall feed-throughs, the elements must be free to expand and be protected by sleeves consisting of an excellent dielectric, which does not react chemically with silicon carbide. In general, these consist of high alumina content materials, 60 to 90%, depending on temperature and atmosphere, and in which fibrous materials are being more and more widely used. When two cold outlet elements are mounted vertically, they rest on insulating ceramic supports. Finally, these elements are sensitive to mechanical shock, and must therefore be handled carefully.

— *furnace temperature*

The most common elements enable maximum temperatures of about 1,450°C in the furnace. The most recent elements, using recrystallized silicon carbide, enable temperatures of approximately 1,600°C. However, below these maximum temperatures, element service life is increased. Regulation should prevent element overheating.

— *furnace atmosphere*

The effect of special atmospheres has been described above. In the case of corrosive atmospheres, it is possible to use a muffle furnace or radiating tubes.

— *heating up*

The furnace is sometimes heated gradually up to 800°C, when full power is then applied. It would seem that, with modern elements, these precautions are no longer necessary. Conversely, thermal shock must be avoided, and continuous operation offers a much longer service life for silicon carbide rods.

— *aging*

As previously mentioned, rod resistance increases with time. For some rods, this increase is initially rapid (60 to 100% minimum over 100 hours), then very slow. For other rods, and in particular the most recent, aging may be very gradual and the increase in resistivity less pronounced.

This aging phenomenon is due to two basic causes:

— oxidation of silicon oxide (caused by either air or other atmospheres) producing silica and increasing element resistance;

— disintegration due to physical (successive contraction and expansion) and chemical causes.

For a given voltage, the dissipated power becomes very low, the furnace heatup time is increased, and it is sometimes impossible to reach normal operating temperature. With older elements, the value of the resistance can be multiplied by approximately four over its operating period. Conversely, for more recent elements, this value increases by only 50 to 100% during the operating period (high density recrystallized elements and special impregnated elements). To maintain that power, it is necessary to double the voltage in the first case, and to multiply it by a factor of between 1.25 and 1.5 in the second case.

Various solutions have been proposed to compensate for the aging of silicon carbide resistances and to regulate furnace power — use of high power rods (as aging progresses, the power comes close to the actual power required), compensation resistances, compensation or variable inductances, variable resistances (rheostats), autotransformers, induction regulator (for high power furnaces or to obtain fine temperature settings).

The most widely used and best adapted technique seems to be the multitapped transformers which have up to twelve secondary tappings providing wide flexibility, and voltage ratios of 1 to 4. The supply voltages must, however, remain less than 250 V. The use of thyristor power supplies in new furnaces should soon appear [54], [84].

For temperatures of the order of 1,300 to 1,400°C in the heating chamber, the service life elements, for intermittent operating furnaces is, on average, 2,000 hours approximately. In continuous operation (or intermittent, but with a permanent holding temperature of at least 800°C), the service life can, under certain operating conditions, be 2 to 5 times higher.

#### *3.4.3.2.4. Electrical installation and replacement of elements*

Heating elements can be connected in series, in parallel or series-parallel. However, parallel connection is generally recommended. When two heating elements of identical resistance are connected in series and the resistance varies with time, the element offering the highest resistance absorbs more power, further accelerating aging. This unbalance becomes accentuated (for constant overall power), and the element having the largest resistance is rapidly destroyed. Conversely, with parallel coupling, each heating element is effectively supplied at the same voltage. The rod offering the lowest resistance therefore absorbs the highest power and ages more rapidly (its resistance increases) which tends to restore the balance between originally identical elements, and ensures resistance stability.

In practice, it is therefore preferable to connect heating elements in parallel. However, when this leads to an excessively high feed current, two elements of equal resistance are connected in series (connection of more than two must be avoided), and these pairs of elements are then connected in parallel.

In numerous cases, replacement of the rods can be performed without interrupting furnace operation. Precautions must be taken when replacing elements; for example, new elements must be installed in parallel with old elements if these have a resistance exceeding by 15 to 20% that of the new elements. If a break occurs in a group of elements having operated over a long period, the defective element must be replaced by an old element, or all elements of the group must be replaced. If the failure occurs in series-connected elements, it is better to replace the elements connected in series, which can be reused. To keep silicon carbide heating element costs to a minimum, spare parts inventory must be carefully managed.

#### *3.4.3.2.5. Advantages and limitations of silicon carbide elements*

The main attraction of these elements is their capacity to operate at high temperature, provide high power densities, and their ability to withstand corrosion.

Conversely, they are quite expensive and require a special power supply to combat early aging. Their brittleness also calls for handling precautions.

#### *3.4.3.2.6. Applications*

Silicon carbide elements are most often used in high temperature furnaces in:

— ceramic industries for treatment of metals, such as ferrites, titinates, steatites, refractories, grinding stones, electrotechnical porcelain, kitchen articles, potteries, tiles, fluorescent powders, miscellaneous powders, etc.;
— metal industries, for heat treatment, heating prior to forming, sintering, brazing, melting and holding, mould heating, production of metal-glass compounds, enamelling, etc.;
— glass industries for melting (pot furnaces), holding, refining and feed crucible heating;
— semiconductor industries for resistance furnaces requiring high temperatures.

Air heaters permitting temperatures of 1,200°C have also been designed.

Heating elements can also be used in low or medium temperature furnaces (down to 1,000°C) with very high power densities, and therefore high productivities (for example, aluminum melting and metal heat treatment furnaces) [50], [66], [71], [77].

### 3.5. Form and arrangement of elements in furnaces

Elements are produced in a wide range of forms, wires, rods, strips, ribbons, sheets, plates, bars and clips. Also, tubular and moulded resistances are available. Some materials (molybdenum disilicide, silicon carbide, etc.) require special forms while others, such as the standard metal alloys, permit elements of highly varied form.

The builder's choice depends on many technical or economic factors which are sometimes contradictory. Therefore, a compromise between the following must be made:
— required electrical power;
— heating chamber volume;
— volume or area available to house heating elements (ease of installation and replacement; position; hearth, roof, walls, doors);
— power supply type (AC or DC: single-phase, 2-phase or 3-phase) with or without phase balancing requirement;
— supply voltage;
— type of treatment to be performed: protection of heating elements against shocks and gases, vapors or special atmospheres, heat distribution in the heating chamber;
— temperature: value, required accuracy, variation during treatment time (choice of heat transfer mode: radiation, natural or forced convection, conduction);
— mechanical strength of heating elements;
— replacement of elements without furnace shutdown;
— element cost;
— type of regulation.

Thermal and electrical calculations, tests conducted in laboratories and industrial experience permit more rational choices to be made. The following paragraphs concern the use of metal resistances, since other resistance types are often subject to more stringent installation requirements as mentioned above.

Analysis of the use of resistances has generally been divided into two parts according to the heat transfer method used in the furnace: forced convection or radiation.

### *3.5.1. Forced convection furnace elements*

In forced convection furnaces, the heating elements are often combined in batteries of rigid strips or wires, mounted on refractory supports (up to 800°C) or finned shielded resistances (up to 500°C approximately). However, as for resistances used in radiation furnaces, these can be distributed along the walls of the furnace, the charge being protected from direct radiation by a muffle, or more rarely, submitted to radiation from the elements. The arrangements used are then similar to those encountered in radiation furnaces; however, the resistances are sometimes placed in protective tubing due to high air speeds and at low temperature, enclosed elements can be used.

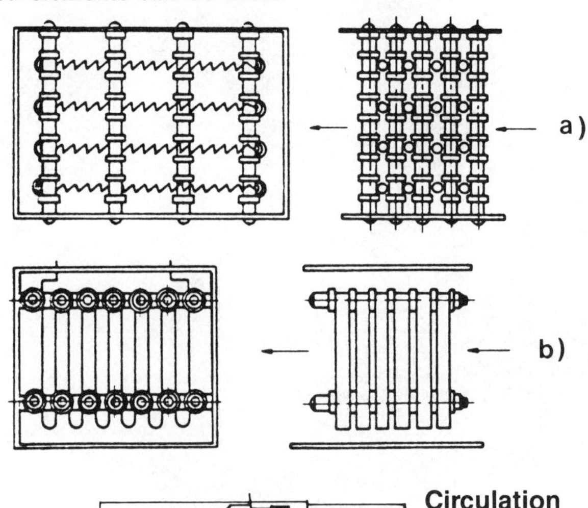


**Fig. 47.** Batteries of heating elements for forced convection furnaces [90]:
*a*) wire resistances
*b*) ribbon resistances

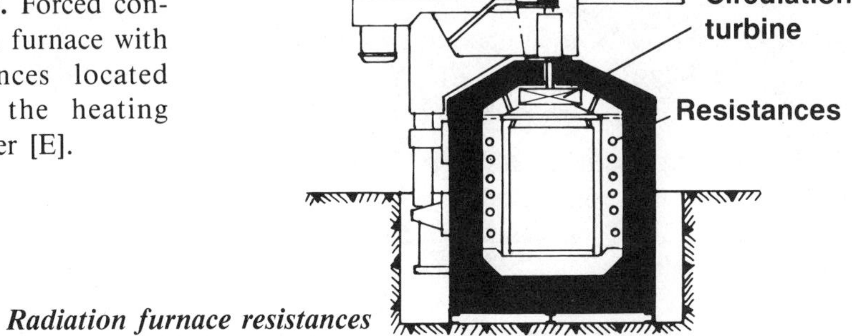


**Fig. 48.** Forced convection furnace with resistances located along the heating chamber [E].

### *3.5.2. Radiation furnace resistances*

The most widely used heating elements in radiation furnaces are:
— coil resistances;
— ribbon resistances;
— thick wire resistances;
— tubular resistances;
— cast resistances.

### *3.5.2.1. Coil elements*

A coil element consists of a helically wound round wire. These elements are distributed along the walls of the furnace and supported on refractory brackets or placed in grooved bricks. When located on the hearth, resistances are generally protected from falling debris, refractory materials or other elements by sheets of refractory steel.

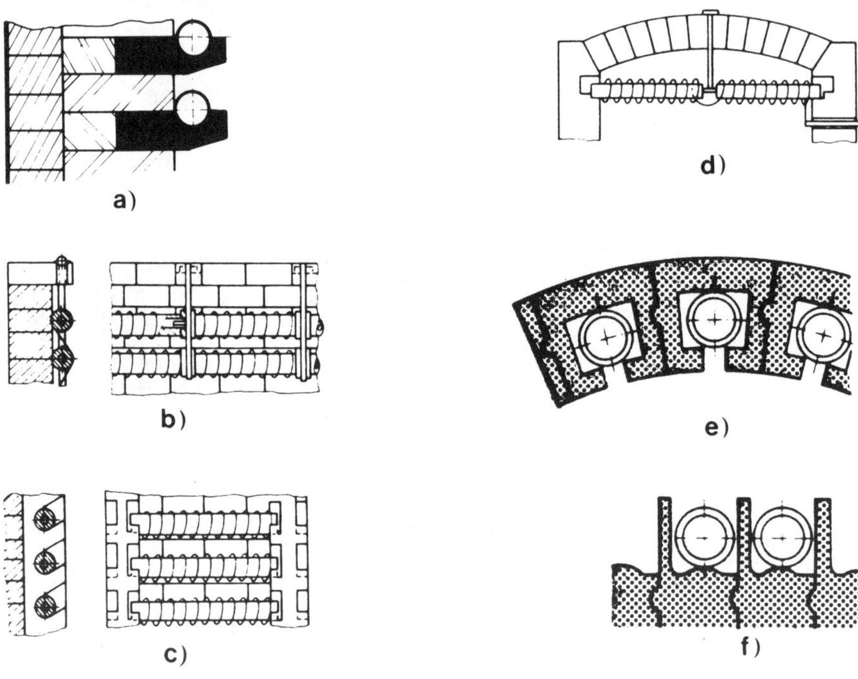


**Fig. 49.** Coil elements: *a*) supported by shaped bricks (in wall); *b*) and *c*) on ceramic mountings (sillimanite); *d*) suspended from the roof; *e*) located in the roof; *f*) located in the hearth [24], [90].

The composition of materials used to produce the supports has an important effect on resistance service life; these materials must be free of iron oxide and silica, since both have a chemical effect on some metals used in the production of resistance alloys. Such chemical action is also a function of furnace temperature and atmosphere.

When working coil elements, the following rules must be respected:

— to ensure correct mechanical behavior, the coil diameter must not be more than 10 times the wire diameter;

— to obtain free radiation, the center-to-center distance between two coils must not be less than three times the wire diameter.

Basically, this type of element is used with iron-nickel-chromium, nickel-chromium and iron-chromium-aluminum alloys. This design, however, limits direct radiation from the resistances onto the charge to be heated. For nickel-chromium alloys, the maximum operating temperature is about 1,000°C, but very well designed shaping bricks can enable 1,100°C to be reached. For iron-chromium-aluminum alloys, this setup does not permit temperatures higher than 1,300°C to be radiated directly onto the charge.

### *3.5.2.2. Ribbon elements*

A ribbon resistance arrangement is often preferred to a coil resistance arrangement. The resistance itself is flattened, and offers a highly favorable radiating surface — cross-section ratio. Two devices are used to attach these resistances to the walls of the furnace.

**Fig. 50.** Ribbon resistances:
*a*) refractory mounting
*b*) metal plug mounting
*c*) ceramic parts offering better element dimension stability [24], [90].

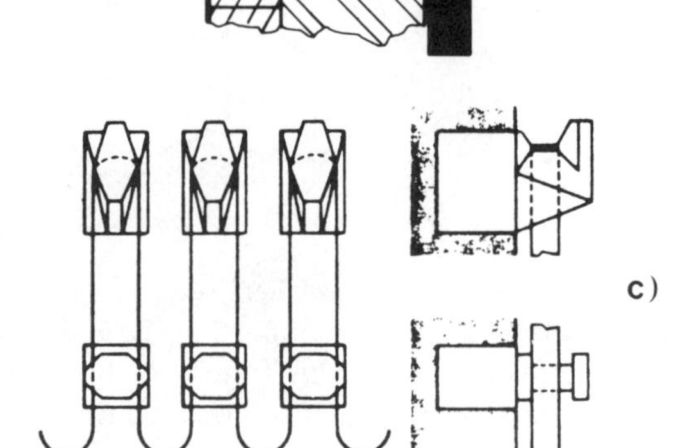


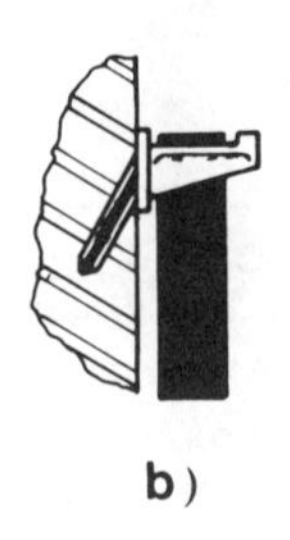


The first consists of a special refractory part, consisting of 65% alumina; the second consists of a 35% nickel, 15% chromium refractory steel hook attached in the same manner as a plug in the laboratory wall (this arrangement is commonly used with light refractories).

Both suspension methods offer free resistance radiation, and therefore prevent overheating.

To obtain adequate mechanical characteristics, a maximum element height of 7 to 8 times the ribbon width is recommended. In general, the thinner the ribbon, the shorter the element height should be.

In numerous furnaces, and especially in the movable hearth type, heating elements are located directly on the hearth, protected by refractory steel. Sections of the resistance are separated by 65% alumina refractory "combs".

Ribbon resistances are used up to a maximum of 1,150°C for nickel-chromium alloys (installation on hooks or brackets) and 1,320°C for iron-chromium-aluminum alloys (bracket mounted).

### *3.5.2.3. Thick wire elements*

This heating element arrangement is similar to that used with ribbon resistances, but the lengths consist of thick wires. Figure 51 shows a support and anchoring system for this type of resistance made from fibrous ceramic refractory materials.

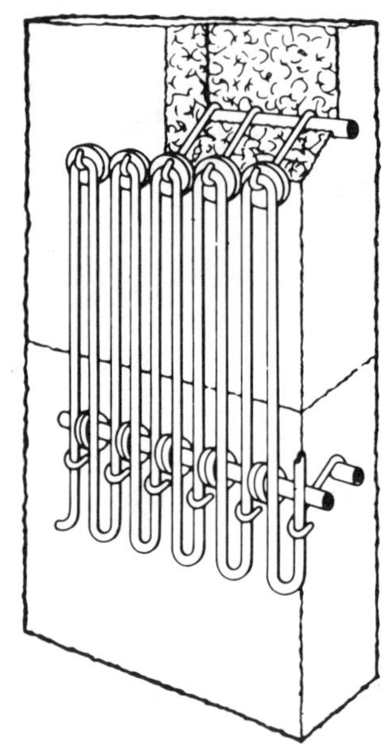

**Fig. 51.** Thick wire element mounting system in a fibrous-ceramic refractory material furnace.

The anchoring bars and washers separating the resistances, are made from sintered ceramic. The hook supporting the resistances and those keeping it in place are refractory steel. This system enables free vertical expansion of the heating elements, while reducing horizontal movement and risk of deformation. This system can be used during construction of new furnaces or during renovation, especially when a conventional refractory lining is converted into a fibrous ceramic refractory lining.

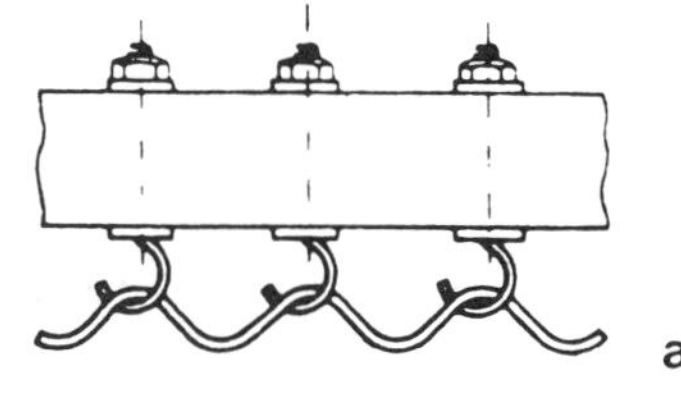


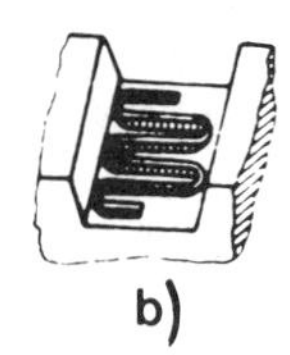


**Fig. 52.** Wire elements [90]:
*a)* in roof;
*b)* in hearth.

### *3.5.2.4. Tubular elements*

Tubular elements of the type shown in Figure 53 *a)* can be placed horizontally or vertically in new or renovated furnaces, either in the walls, the roof or the hearth.

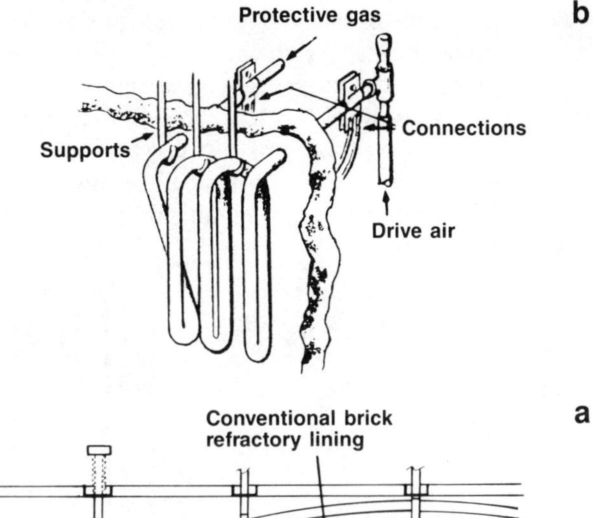


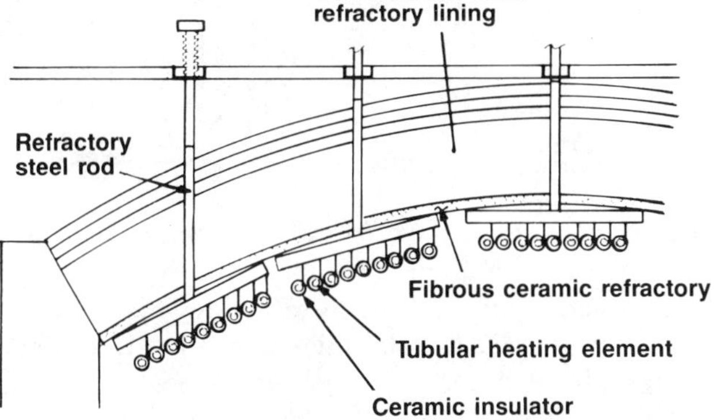


**Fig. 53.** Tubular elements:
*a)* Renovation of a furnace using fibrous ceramic refractories and tubular heating elements [R];
*b)* Radiating tube consisting of a rectangular coiled tube, through which a decarbonizing fluid flows [S].

Sintered ceramic insulators suspended from the furnace walls by refractory steel rods support the tubes and keep them away from the inside surface of the furnace, while permitting free expansion. This type of installation is particularly useful in fibrous ceramic refractory-lined furnaces. Heating elements can easily be placed in the roof.

Other types of tubular resistance consist of conventional large diameter radiation tubes forming a rectangular coil. These modules are connected in equal numbers to each phase and supplied through a voltage step-down transformer.

### *3.5.2.5. Cast resistances*

Cast resistances are rough cast, solid heating elements, consisting of alternating U-sections.

These have high cross-sections, of the order of 200 mm$^2$, and thus must be supplied with low voltage in low capacity, limited power furnaces, but offer the advantage of being free standing. This characteristic dispenses with miscellaneous supports (brackets, ceramic tubes, pillars) which cause local overheating in the resistances. The large cross-section and absence of support prevent free radiation from the resistance, enabling operation at approximately 50°C above the usual safety limits for drawn alloys. For example, 70/30 nickel-chromium wire is used up to 1,150-1,200°C maximum, while a cast resistance, poured from the same alloy, can operate up to 1,200°C permanently and reach peaks of 1,250°C. Due to radiation laws (emission increasing to the fourth power of the absolute temperature), a temperature gain of several tens of degrees noticeably increases furnace performance [13], [62*b*], [78], [79], [82], [86].

In addition, the porosity of cast resistances improves inherent emissivity $\epsilon_r$, and therefore heat transfer. Lastly, cast resistances provide an excellent furnace overall form factor, since the wall lining coefficients can be very high (*see* paragraph 5.1.1.).

**Fig. 54.** Cast resistance diagram [90].

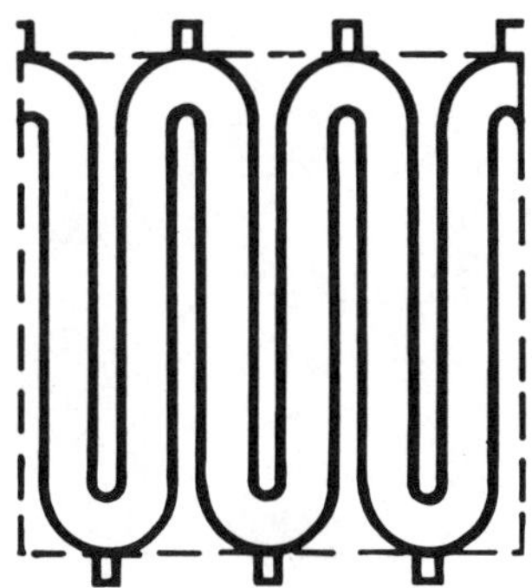

These elements therefore offer excellent thermal qualities, enabling maximum performance from a given metal alloy.

Cast resistances are solid and therefore very rugged, with very good resistance to oxidation (minimum metal thickness: 4 mm). Conversely, the supports or couplings provided for this type of element must support their weight: approximately 750 g per section. This problem is easy to resolve with refractory brick walls (metal hook or ceramic support). Conversely, when the wall consists of alumino-silicate ceramic fibre layers, with very low mechanical strength, the problem becomes much more delicate. Several solutions are at present undergoing long-term tests; in all, the mechanical forces are born by the furnace casing. Other solutions using compressed refractory parts are also being developed. All these systems enable suspension of resistances from the roof and from the walls.

Cast resistances are generally made from 70/30 nickel-chromium, 60/15 iron-nickel-chromium and iron-nickel-chromium-aluminum alloys, but other alloys can also be used.

### *3.5.2.6. Other forms of elements*

Metal elements also exist in other forms: sheets, corrugated plates, grids, meshes, which are, however, not currently used in furnaces.

### *3.5.2.7. Electrical radiation tubes*

No resistance material can withstand all atmospheres. Therefore, it is often necessary to protect heating elements with a sealed refractory or ceramic envelope. This phenomenon is often seen in high temperature radiation furnaces. The protective tube together with the heating element-protective tube assembly, is then known as a radiation tube. In particular, this problem is encountered in metal transformation industries, where very corrosive atmospheres are used (endothermically or exothermically controlled atmospheres, etc.) but may also be encountered in other sectors (fabrication of special ceramics, etc.).

There are many types of radiation tubes. Most use metal resistances, but recently, non-metallic resistance radiation tubes enabling very high temperatures, have appeared.

*3.5.2.7.1. Metallic resistance radiation tubes*

*3.5.2.7.1.1. Coil resistance radiation tubes*

Figure 55 shows a coiled resistance radiation tube assembly. The heating element consists of a wire, generally 80/20 or 70/30 nickel-chromium, single-phase connected on insulating supports made of steatite. The ceramic inserts are kept in place by a central rod or two ceramic lateral rods. The heating element forms a whole which can be easily replaced. Most importantly, it is possible to replace an element without shutting down the furnace.

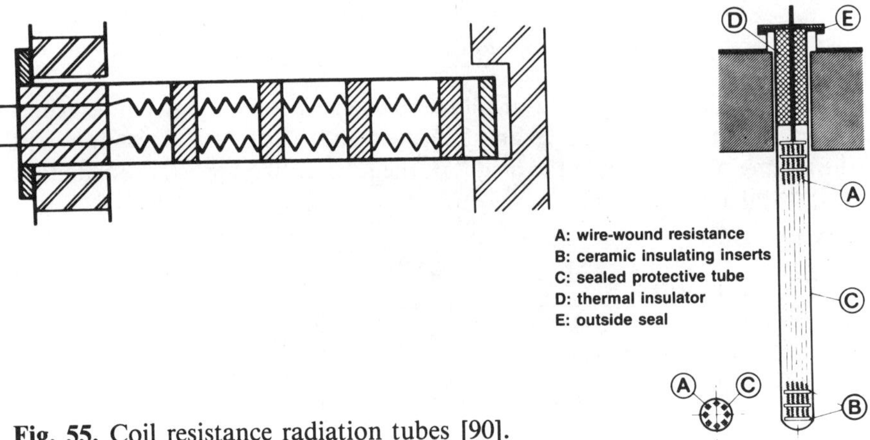


**Fig. 55.** Coil resistance radiation tubes [90].

With this arrangement, the heating elements tend to radiate onto each other. The operating temperature is therefore limited to about 970°C with 80/20 nickel-chromium wire (900°C in furnace) and 1,020°C with 70/30 nickel-chromium (950°C in furnace). Under these conditions, the temperatures of the tube itself are 940 and 990°C respectively.

Due to inherent wire strength limits, radiation tubes must be horizontally mounted; however, where operating temperature is low and brief they can be mounted vertically.

Because of the limited thermal and mechanical performances of coiled resistance radiation tubes, they are generally preferred over other types of radiation tubes, edge-mounted spiral ribbon tubes and coil resistance tubes.

*3.5.2.7.1.2. Spiral ribbon radiation tubes*

In spiral ribbon radiation tubes, the heating element consists of an 80/20 or 70/30 nickel-chromium alloy steel ribbon, helically edge-wound at a given pitch, and supported by a central ceramic rod.

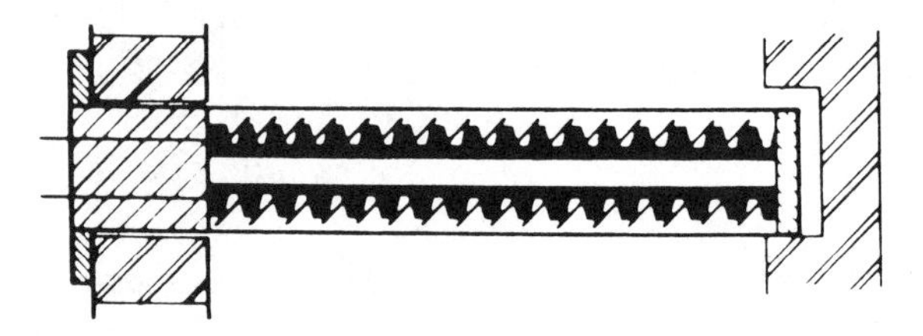

**Fig. 56.** Edge-rolled ribbon radiation tube [24].

The resistance can be mechanically supported by a screw-shaped insulator (however, this limits resistance radiation) or simply by ceramic insulators located at regular intervals (this offers better resistance radiation).

Edge-wound ribbon radiation tubes can be mounted vertically or horizontally. With the standard alloys, furnace temperature can reach 1,000°C. After systematic laboratory tests, this type of radiation tube seems to offer the best thermal and reliability performance [90].

*3.5.2.7.1.3. Thick wire radiation tubes*

The heating elements consist of folded thick cross-section wire connected in series; each section is suspended and free to expand. The sections are supported by insulating plates running the height of the tube. Ribbon resistance radiation tubes of this type also exist.

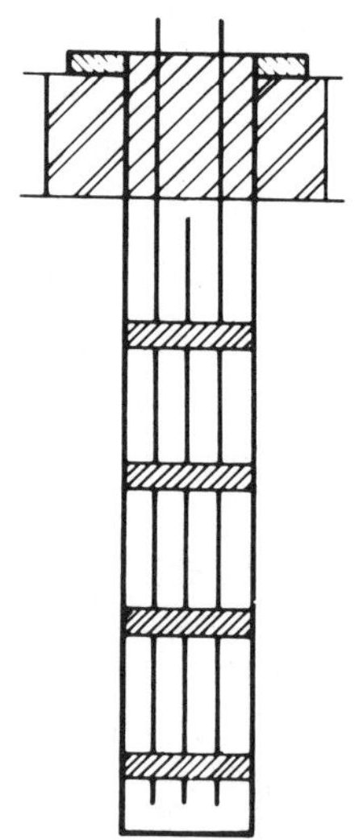

**Fig. 57.** Thick wire radiation tube [24].

Generally, this type of tube is installed vertically. Performance is between that of coil resistance tubes and edge-wound ribbon tubes [90].

*3.5.2.7.1.4. Other radiation tubes with metal resistances*

Barrel-mounted resistance radiation tubes are sometimes used (*see* presentation of barrel resistances in paragraph 3.4 "Heating using enclosed resistances"). Free radiation of the resistances onto the tube is strongly affected with this setup being suitable for temperatures less than 800°C only. Therefore, it is preferable to use radiation tubes.

New types of resistances adapted to controlled atmospheres, especially carbonizing atmospheres, have also recently appeared. Neither conventional resistances nor standard radiation tubes, these elements consist of rectangular coiled refractory alloy tubes (*see* paragraph 3.5.2.4.). The decarbonizing fluid flows permanently through the tubes, ensuring long service life in carbonizing atmospheres. These heating elements are suitable for all radiation heating purposes up to 1,100°C, but are best for tempering and gas state carbonitrizing furnaces.

*3.5.2.7.2. Non-metallic resistance radiation tubes*

Non-metallic resistance radiation tubes have appeared recently. Basically, there are three types: molybdenum disilicide, graphite and silicon carbide.

Molybdenum disilicide resistance radiation tubes can be used in the horizontal or vertical position (*see* paragraph 3.4.2.1.). Due to the temperature level reached (more than 1,500°C at the tube), the tube must be constructed from ceramic material.

Graphite resistance radiation tubes consist of a graphite section, with a ceramic envelope. Generally, these are used in the vertical position, and the resistance operating temperature can be very high (*see* paragraph 3.4.3.1.) enabling extremely high power densities. Silicon carbide resistance radiation tubes are provided with a silicon carbide or ceramic envelope.

The use of radiation tubes is still limited, but may be developed for high temperature furnaces, or where very high power densities are desired.

*3.5.2.7.3. Protective tubes*

The most widely used protective tubes are metal, being thinner than 4 mm to facilitate heat treatment from the resistances to the charge, since these tubes actually form heat shields.

In general, metallic radiation tubes consist of welded, rolled sheet steel (inconel-type stainless steel, 36/8, conventional nickel-chromium alloys, etc.). These tubes offer the advantage of relatively low cost and good resistance to corrosion. However, creep becomes high beyond 1,000°C, rendering them unsuitable for high temperatures. Under these conditions, the performance limiting factor is no longer the heating element, but the enveloping tube. Centrifugally cast refractory steel tubes (with rather high carbon content to provide adequate rigidity) should enable higher temperatures and therefore permit the construction of high power density radiation tube furnaces.

Ceramic protective tubes have also provided good results, although the power density, using the standard metal resistances, is lower than that obtained with metallic tubes.

Silica protective tubes are also possible for production, but have limitations due to high cost and a certain amount of brittleness (*see* "Infrared radiation heating" chapter).

These tubes offer excellent sealing and therefore efficient protection of the heating elements and long service life.

The unit power for radiation tubes generally varies from a few to about 20 kilowatts.

#### *3.5.2.8 Radiating panels*

A new type of radiation heating element has recently appeared, consisting of resistive wires embedded in fibrous ceramic refractory material close to the surface (*see* "Enclosed resistance heating", paragraph 3).

These panels enable easy construction of furnaces and heating chambers that withstand temperatures up to 900°C approximately.

## 3.6. Calculation of resistances

### *3.6.1. Principle of calculation*

The two characteristic equations used to calculate resistances are those which give the heat transfer between the heating elements, the load and the walls, on the one hand, and on the other, the electrical power dissipated in the resistance.

Where resistances are essentially radiating (the calculation principle remains the same even if heat transfer is due to convection, but in this case the laws concerning this heat transfer method must be used) the thermal flux exchanged between the heating elements and the load is expressed as (Stefan-Boltzmann law):

$$\Phi = \varepsilon_{RC} F \sigma (T_R^4 - T_C^4),$$

$\Phi$, thermal flux exchanged between the resistances and the charge in Watts per square meter;

$\epsilon_{RC}$, coefficient (less than 1) allowing for emissivities $\epsilon_R$ and of $\epsilon_C$ the resistances and the load;

$F$, global form factor (less than 1), in particular allowing for filling the wall with resistances and their position with respect to the load.

$\sigma$, Stefan's constant, equal to $5.73 \times 10^{-8}$ W/m$^2$ X K$^4$;

$T_R$ and $T_C$, absolute temperatures of the resistances and the charge surface in degrees Kelvin.

Paragraph 5.1, concerning the evolution of resistance furnaces towards higher power densities, provides a more detailed analysis of the significance of quantities $\epsilon_{RC}$ and $F$, and the consequences of the value of the product $\epsilon_{RC} \times F$ on furnace performance.

Electrically, the equation that must be compared with that defining thermal flux is Joule's law, expressed as:

$$P_W = \frac{U^2}{R},$$

$P_W$ in watts,
$U$ in volts
$R$ in ohm-meters.

The characteristic element used for calculation of the resistance is the limit temperature $T_{RM}$. Knowing $\epsilon_{RC}$ x $F$, it is possible to calculate permissible superficial thermal flux and the dimensions of the heating element; in balanced conditions, all the electrical power dissipated must be contained in the thermal flux emitted by the resistance:

$$\Phi . S = \frac{U^2}{R},$$

$S$, heating element external radiating surface.

### 3.6.2. *Example*

The power to be installed in an independently regulated area of a furnace operating at 900°C is 10 kW per phase (voltage 220 V). A nickel-chromium resistance of resistivity 100 $\mu\Omega$-cm at 1,000°C is used. This is a wire resistance, and product $\epsilon_{RC}$ x $F$ is equal to 0.6.

— *Case of a 4 mm diameter wire*

• For each phase, resistance is $R = U^2/P = 4.84$ ohms.
• Allowing for resistivity, the length $l$ of wire to be used is: $l = (R \times S_2)/\varrho = 60$ m, $S_2$ wire cross-section.
• The outside surface of the wire $S$ is: $S = \pi \times d \times l = 7{,}539$ cm$^2$.
• The resistance load factor is then $\Phi = P/s_1 = 1.33$ W/cm$^2$.
• The resistance temperature can be determined from the thermal flux expression given above; the calculation gives $\theta_R = 956$°C.
• The weight of metal required to produce the resistance is then (106.18 g/m): 6.37 kg.

— *Case of a 3 mm diameter wire*

• The resistance value remains the same; the length of wire to be used becomes $l = 34.2$ m.
• The outside surface of the wire is only: $S = 3{,}223$ cm$^2$.
• The resistance load factor is then $\Phi = 3.10$ W/cm$^2$.
• The resistance temperature now reaches $\theta_R = 1{,}020$°C.
• The weight of metal required to produce the resistance (59.73 g/m) is only: 2.04 kg.

Both temperatures ([1]) determined above are fully compatible with the use of 80/20 nickel-chromium resistances if the ambient atmosphere is suitable, and especially in the second case, that a temperature limiter is provided at the resistances, to ensure thermal safety.

This example shows the effect of permissible surface power density on heating element investment costs. When the surface power density increases, cost decreases, but service life may be decreased since the resistance is used more. Therefore, an optimum between the service life of the elements and their investment costs must be found.

Moreover, this optimum is not always easy to determine, since its evaluation must allow for regulation quality and the temperature measurement system, together with the resistance arrangement (as power density increases, hot points related to incorrect arrangement of resistances, especially in corners where free radiation is limited, become important together with the reliability of the temperature sensors located as safety devices at the resistances).

— *Effect of factor* $\epsilon_{RC}$ x $F$

The above calculations were conducted using factor $\epsilon_{RC}$ x $F$ equal to 0.6. This value presupposes that the wire can radiate freely (wire simply supported from hooks). With such wire diameters mechanical strength is relatively limited, and it is often necessary to form the heating elements into coils supported by sillimanite bars. In this case, and due to mutual radiation from the elements, factor $\epsilon_{RC}$ x $F$ is much lower, about 0.3 (in addition, certain thermal, mechanical or electrical aspects of the turns and their diameter must be respected) [61], [73].

With the above wire lengths and specific charges, the temperature of the heating elements becomes:

— for 4 mm diameter wire: 1,005°C;

— for 3 mm diameter wire: 1,114°C.

These temperatures are still fully acceptable for 80/20 nickel-chromium alloys, but in the second case (3 mm diameter wire), it is absolutely necessary to control correctly the temperature at the resistance using a low thermal inertia sensor, carefully placed to prevent incorrect arrangement of resistances in corners.

### *3.6.3. Practical calculation of resistances*

In practice, the most delicate part of resistance calculation is the evaluation of factor $\epsilon_{RC}$ .F, since this is subject to important variations. Thus, for ribbon resistances, it is 10 to 40% more than that for coiled wires made from the same alloy, as a function of dimensions.

These values are simply averages, and correct arrangement of the resistances in angles and corners, together with their support, is of primary importance. Moreover, the temperature measurement and regulation quality greatly control resistance performances, and therefore furnace productivity.

Calculation of resistances are derived from both thermal and electrical calculations, with the diverse experience of constructors and laboratory experiments enabling a better understanding of the phenomena involved. Graphs

(1) *N.B.* Several degrees of liberty exist for calculation of resistances which may be performed according to different methods, for example by limiting length, specific power density and arrangement.

and tables facilitating resistance design are available, and, more recently, computer programs for electrical heating elements have been developed by resistance or furnace manufacturers, and by certain research laboratories, which now offer the user more security [61], [90].

Recent research has also noticeably improved the performance of resistance ovens (*see* paragraph 5.1.), due to a better understanding of the phenomena involved and technological advances with certain components.

However, whenever high heat transfer performance is not absolutely required (which is most common due to the need for precise thermal cycles, including gradual temperature increases), it is best to use large heating elements, since the corresponding investment costs can be rapidly recouped.

## 4. RESISTANCE FURNACES AND CONTROL

Resistance furnaces offer accurate temperature control. Temperature uniformity depends upon distribution and type of heating element, atmosphere mixing (especially at low and medium temperatures) and regulation. The first three categories reflect the thermal design of the furnace and provide precise, even load temperatures (or heterogeneous, where necessary). However, control is most important for fine temperature adjustment.

Electric furnaces generally feature two types of control: electrical power and temperature (load and resistance temperatures), which are complementary [10], [11], [12], [17], [18], [32], [38], [40], [42], [43], [65], [78], [82], [91].

### 4.1. Control systems for resistance furnaces

A control system consists basically of:

— *One or more sensors measuring furnace (or resistance) temperature* which produce an equivalent electrical value.

Generally, this device consists of a pyrometer consisting of a temperature sensor — also known as a probe or pyrometric rod — a thermocouple (the use of optical or infrared pyrometers is rare, while mechanical systems have almost disappeared) and a galvanometer complete with amplifier for the thermocouple signal if required. This galvanometer enables temperature display, since the emf produced and therefore the current, is a function of the difference in temperature between the hot and cold parts of the thermocouple. Digital displays are gaining wide acceptance.

The thermoelectric couples most widely used are iron-constantin, up to a temperature of 900°C, chromel-alumel up to 1,200°C, and platinum-platinum-rhodium up to 1,600°C. These temperature sensors must be chosen very carefully, especially where they are used as temperature limiters at the resistances. An analysis of the sensors available on the market and used in resistance furnaces has, in fact, demonstrated that their response time at 95% (that is, taking into account thermal inertia and reaction time under given experimental conditions, a temperature equal to 95% of the real temperature) can vary between 3 seconds and 1 hour.

Response time in seconds of a sensor sample used in resistance furnaces [82].

| Sensor number | 1 | 2 | 3 | 4 | 5 | 6 | 7 |
|---|---|---|---|---|---|---|---|
| Response time at 63% (s) | 1 | 2 | 7 | 40 | 50 | 60 | 1,200 |
| Response time at 95% (s) | 3 | 6 | 20 | 120 | 250 | 180 | 3,600 |

**Fig. 58.**

It is obvious that some of these sensors are not necessarily adapted to resistance furnace control and, when employed as temperature limiters at resistances, give poor results. The design of high power density electric furnaces necessitates low response-time sensors (*see* paragraph 5.1.).

The position of the probes in a furnace is fundamental for both load temperature accuracy and for resistance protection. The probe controlling the load temperature must, in radiation furnaces, receive radiations directly from the load without producing deposits that hinder heat transfer, and the probe protecting the resistances must be near the hottest point in the resistances. With heat transfer by convection or conduction, the probe must be closest to the heating element (a second probe can be placed against or inside the load, to detect complete heating).

— *A device for comparison of measured temperature to reference temperature*

The electrical value measured by the pyrometer is compared to a reference value or set point which gives a difference or error signal $\epsilon$. When the furnace or its resistances are at the desired temperature, the difference is zero, and positive or negative if the reference value deviates.

The comparison device may be mechanical (clamp, pressure or gravity detectors), photoelectric or inductive. All these systems are based on the same principle: determination of the temperature indicator pointer position, and therefore are essentially position detectors. These are gradually being replaced by more accurate electrical and electronic comparators, which are more reliable and simple, and enable direct difference $\epsilon$ or error signal calculation.
The set point temperature which has been assumed to be constant, can vary with time. Therefore, the set point must be changed. In electromechanical systems, this change is easily obtained with a profiled cam driven by a small induction motor [17].

— *A power control and variation system*

The error signal is applied to the correction device which then controls the electrical energy absorbed by the furnace.

With mechanical comparison systems, control is generally obtained via an electromechanical system (with judicious combinations of cams, motors, levers, springs, toothed-drive disks, etc.) controlled by the pointer position detector, which then open or close contacts controlling power to the heating elements. With photoelectric or induction systems, contact control is generally electromechanical or electronic.

With electrical and electronic comparators, control of the resistance power supply uses either conventional electromechanical relays, opening and closing of which are controlled by electromagnetic, electromechanical or electronic devices (transistors), or static switching devices (thyristors).

Furnace power can be controlled with other techniques.

## 4.2. Power variation devices

From the expression of power as a function of voltage and resistance $P = U^2/R$, methods of adjusting the power produced in the furnace can result by varying voltage or resistance. New fully electronic control systems (static thyristor control) also enable easy power control at the level of the AC sine wave itself, by modulating the number of cycles permitted in the charge, and by applying only portions of each cycle.

— *Action on voltage value*

The power value can, for example, be modified by varying the voltage using a rheostat, an inductance, a transformer or an induction regulator, the latter being very rarely used.

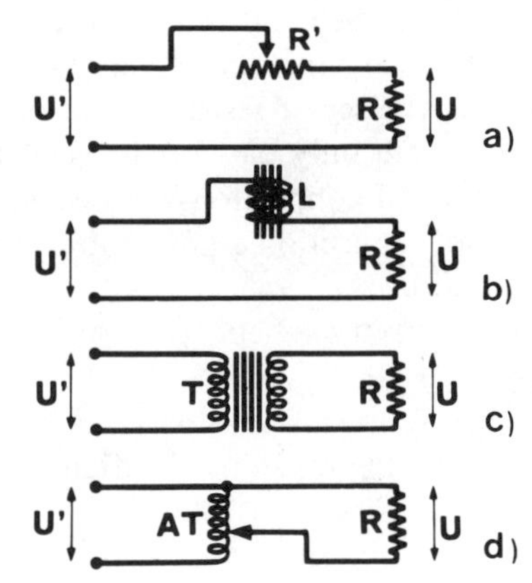


**Fig. 59.** Voltage variation [13]:
*a)* rheostat;
*b)* variable inductance;
*c)* multi-tapping transformer;
*d)* auto-transformer.

• *rheostat*

By inserting a rheostat (variable resistance) in the heating element supply circuit, and supplying the assembly with constant voltage *U'*, it is possible to vary voltage *U*. Power loss in the rheostat is, however, high, and precludes industrial applications.

• *inductance*

A variable inductance is located in the heating element circuit. The control can then, for example, be obtained as follows: the thermocouple is connected to a pyrometer which detects the difference between the furnace temperature and the set point. An electronic amplifier amplifies the error or difference signal and varies the saturated coil inductance, which for a constant supply voltage *U'*, modifies voltage *U* across the heating elements, and therefore the absorbed power (inductance variation is obtained by more or less saturating the core with a DC current, or by varying core position with a motor).

In this manner, the heating elements have a continuous voltage applied which makes it possible to realize continuous regulation (e.g. P.I.D. type, *see* paragraph below concerning thyristor control), which offer excellent temperature accuracy.

The drawback with this system is that the power factor decreases as power demand decreases, together with efficiency. This system has therefore not been very widely used.

• *multi-tap transformer and auto-transformer*

These solutions are often used in furnaces using heating elements with a high resistivity variation between ambient temperature and operating temperature (molybdenum disilicide resistances and, to a lesser degree, carbon silicide), to cancel out all power surges and any risk of damage to the elements during cold starting, or aging due to resistivity increase with time (silicon carbide resistances).

Conversely, these solutions are rarely used for continuous control in normal operation.

— *Action on resistance values*

The use of the different resistance connections described in paragraph 3.2. can be used to vary the equivalent resistance, and therefore power:

• switching the set of resistances in or out: "on-off" adjustment (it is also possible to switch only part of the elements in or out)

• series-parallel connections: power variation of 1 to 4

• delta-star connections: power variation of 1 to 3

• combinations.

Changeover from one setup to another is provided by electromechanical relay control systems.

Due to their simplicity, reliability and low cost, control systems featuring power control by changing resistance configurations, and in particular "on or off" or "full-medium-off" systems (star-delta coupling), are the most widely used. For all the reasons mentioned, and despite the availability of full electronic control (thyristor control), these systems are still very widely used.

## 4.3. Thyristor regulation

### *4.3.1. Principle*

Two- or three-position, discontinuous control systems generally provide sufficient temperature accuracy for normal applications. However, the design of such systems is such that a certain control temperature latitude must be accepted (at high temperature, generally between ±5 and ±10°C) to prevent excessive chattering and rapid deterioration (dead zone, obtained by closing down the furnace at a temperature slightly higher than the set point temperature, and starting at slightly lower temperatures. In conventional furnaces, this dead zone varies between one and a few minutes). Moreover, it does not enable optimum resistance operation (*see* high power density furnaces, paragraph 5.1.).

Conversely, continuous control offers very high accuracy and enables an increase in the maximum operating temperature of the resistance. Continuous power modulation control systems using thyristors now make this possible but without the drawbacks inherent in variable inductance or electromechanical devices [28], [49], [51], [54], [63], [65], [78], [91].

Thyristors are semi-conductors consisting of four p-n-p-n layers, and can be thought of as three series-connected diodes, the polarity of one inverted with respect to the other two. Therefore, current is normally blocked in both directions. The center layers are generally less doped than the outer layers, and behave as a grid which, when biased, enables current to flow (this part of the thyristor is known as the gate).

The theory of thyristor operation is beyond the scope of this work, and only those elements of interest to electric furnace control, and in particular resistance furnaces, are described.

A thyristor control system is, in principle, extremely simple. Thyristors are located in the heating resistance power supply circuit and, as a function of the error or difference signal emitted by the regulator, control current. Therefore, the thyristors behave as static relays whose response times, which are much shorter than the AC mains cycle, enable continuous modulation of the power applied to the resistances.

### *4.3.2. Thyristor control modes*

There are two main types of thyristor control:

— *Pulsed multiple cycle control*

Control is effected over pulses of complete cycles. The thyristors are triggered only when the AC voltage supplying the load passes through zero.

Under these conditions, a time base $T$ corresponding to a whole number of cycles must be used. As a function of the error signal providing the difference between the load temperature and the set point, the thyristor control system allows only a whole number of cycles equal to or less than that characterizing the time base to pass through the load during this reference period. Conduction time and power utilization are proportional to the value of the signal emitted by the regulator.

In this system, the time base is generally a few seconds, which is much longer than that of one cycle but largely sufficient considering the inertia of the temperature sensors used in resistance furnaces.

Power variation is linear with the control voltage, and switching of the thyristors during passage of the voltage through zero prevents risks of interference. The control circuit is more complex than for phase angle control.

— *phase angle control*

In this system, for each cycle the error signal causes the thyristor to switch with a certain delay proportional to passage through zero of the voltage. As a function of this phase angle, the rms power for one cycle is proportionally higher or lower.

This control method, which enables cycle chopping, can provide high accuracy since its response time is extremely short. Conversely, this system generates interference (creation of harmonics) due to a sudden variation in thyristor current when switched, and the power variation is not linear with the control voltage.

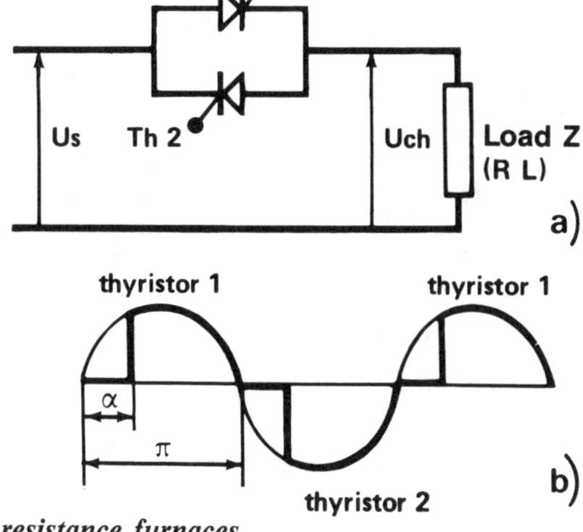


**Fig. 60.** Diagram of a phase angle controlled thyristor static switch [90]:
*a)* load supply via thyristor static switch.
*b)* thyristor switching delay with resistive load.

### *4.3.3. Thyristor control and resistance furnaces*

There are only a few resistance furnaces using the thyristor control method. However, these appear to be quite satisfactory and offer excellent temperature accuracy.

In particular, thyristor control facilitates production of P.I.D. control systems (Proportional, Integral, Differential), i.e. continuous control systems:

— *proportional*, since around set point temperature $\theta_2$ (within the proportionality band), the absorbed power increases proportionally to the difference $\epsilon$ between temperature $\theta_2$ and the charge temperature $\theta_3$ (*see* Figure 62).

The load therefore never reaches the set point temperature, although it approaches it closely, and the temperature at which the furnace stabilizes is the mode temperature $\theta_s$, determined by the point at which the heating power curve crosses the furnace loss curve $P_v$ ($\theta$).

— *integral*, since with purely proportional regulation systems, the permanent difference in temperature between the load and the set point can become too high if, for stability reasons, the proportionality band cannot be too narrow; integral action compensates for this temperature difference. This action is known as integral because the action of the corrective device is also a function of integral

$\int \epsilon \, dt$ of the temperature difference $\epsilon$, and $t$ as time.

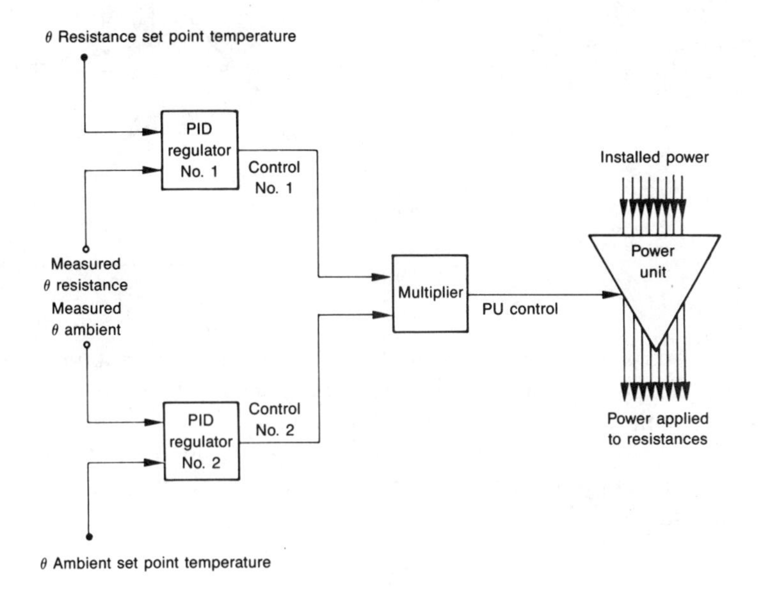


**Fig. 61.** Circuit diagram of a thyristor power supply controlled by two P.I.D. regulators (one for the load, the other for the resistances), connected in parallel [82].

As long as a temperature difference persists, the corrective device tends to equalize the temperatures. The down time effect is, however, reinforced, and stability decreases.

— *differential*, since with purely proportional regulation, oscillations can occur if the proportionality band, unlike the above, is too narrow; the differential action limits these oscillations. This action is known as differential, since the action on the corrective device is also a function of the derivative $d\,\epsilon/dt$ of the temperature difference in time (this action allows for the temperature change rate, enabling the final temperature to be reached more rapidly and reducing oscillation amplitude).

Both integral and differential compensation can be combined with proportional regulation to obtain very accurate P.I.D. regulation. Thyristor control is particularly suited to P.I.D. systems.

It would seem that, in the future, thyristor control and, in particular, the pulsed multiple cycle version, will be developed widely for the reasons already stated:

— continuous power modulation

— temperature accuracy

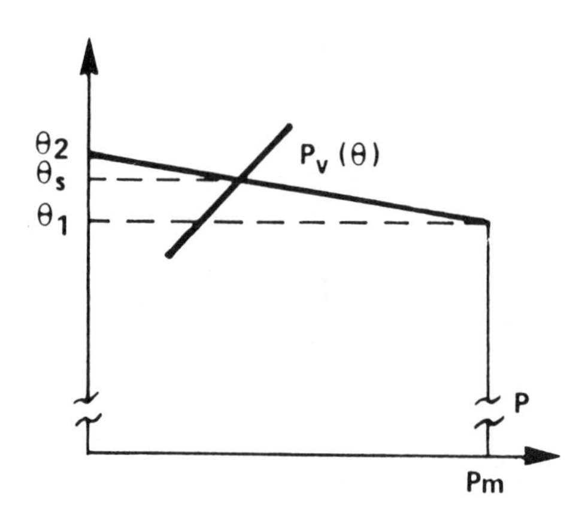


**Fig. 62.** Proportional regulation: $\theta_2$-$\theta_1$: proportionality band; *Pm* maximum power [90].

— reduction of resistance operating temperature, or increase in power density (these possibilities are only two facets of the same advantage, better control of resistance temperature and variation amplitude)
— unlike electromechanical relays, absence of wear due to very frequent power switching operations (which justifies the widely used term of static switches or static regulation).

## 5. DEVELOPMENT OF RESISTANCE FURNACES

Resistance furnaces, which have been used for a long time, are being continuously improved with the introduction of new alloys for resistances, improvement in existing alloys, intensive mechanization, finer control, etc. However, three types of development seem particularly important:
— development of high power density, high temperature resistance furnaces, both characteristics being partially related;
— development of low thermal inertia furnaces;
— development of vacuum furnaces.

This triple evolution which, of course, leads to the production of very high productivity furnaces, also provides substantial energy and raw material savings, an attractive advantage in the economic context of the 1980's.

### 5.1. High power density electric furnaces

Resistance furnaces are sometimes reproached for their "sluggishness" when compared to fossil fuel furnaces. Most of the time, this is unjustified, since numerous thermal operations require a stringent temperature cycle to be followed: excessive temperature rise rates can, under these conditions, adversely affect the quality of heated products, leading to critical deformation of parts and incorrect treatment. In some cases, the heating rate can become a major criterion, and resistance furnaces are, in such cases, inferior to fossil fuel furnaces. Recent developments are such that resistance furnaces now offer equivalent or even better performance than the fossil fuel type.

### *5.1.1. Remarks concerning heat transfer between resistance and load*

In furnaces, with atmosphere mixing or not, heat transfers take place through convection and radiation. Beyond 500-700°C, depending on the type of furnace, radiation becomes preponderant. For higher temperatures, radiation is responsible for 90% of the heat flux received by the load, whether this flux comes directly from the resistances or is reflected onto it from the furnace walls.

Essentially, increasing resistance furnace performance means increasing the energy transferred by radiation by improving transfer conditions. The thermal flux exchanged between the resistances and the load can be stated in the form (Stefan-Boltzmann law):

$$\Phi = \varepsilon_{RC} FS\, \sigma (T_R^4 - T_C^4),$$ (*see* paragraph 3.6).

To improve radiation exchange, the following factors must be controlled:
— maximum resistance temperature;
— furnace overall form factor;
— resistance emissivity

Of these three factors, maximum resistance temperature is by far the most important. However, careful choice of the furnace form, the type of resistances and their arrangement on the furnace walls also improves resistance furnace performance. Before discussing these points, it is necessary to comment ont the relationship between the heat flux applied to the load and the installed electrical power, and also on the consequences of this relationship on resistance furnace operation, sometimes neglected by users.

In normal operation, the mean power dissipated by the resistances is used to increase the load temperature and to compensate for thermal losses. However, the energy applied to the load is determined by the thermal flux expression mentioned above. Also, an increase in installed electrical power does not necessarily result in an increase inthe mean power absorbed by the load, and therefore productivity. In fact, if the flux transmitted is already at its maximum before power is increased (resistances operating almost permanently at their maximum temperature), the sole result of this power increase is to cause resistance overheating and destruction (where there is the exceptional lack of a temperature limiter at the heating elements), or, due to control, a shorter mean energizing time for heating elements.

Therefore, in numerous cases, it is an improvement in heat transfer, by one of the methods mentioned above, that effects an increase in resistance furnace performance [78], [79], [82].

#### *5.1.1.1. Influence of total resistance emissivity*

Resistance emissivity depends on the type of material used, its surface condition and temperature, etc. It is thus not generally possible to control this parameter, and the difference between the various materials or the forms under which a given material is presented is generally rather low. Due to surface roughness, cast resistances, for example, have a slight advantage, since their emission factor is approximately 5% greater than that of wire or ribbon resistances obtained from an identical alloy, having hardened surfaces which are smoother [73].

*5.1.1.2. Influence of overall furnace form factor*

The overall form factor of a furnace is a complex idea, since it reduces the multiple interactions of the resistances, load and walls of a furnace to a single factor [78], [79], [82].

— *Overall efficiency factor e of the resistance*

This factor is equal to a ratio of thermal flux emitted by the resistance, occupying a surface $S_e$, to flux emitted by the same surface when taken to a temperature equal to that of the resistances and having the same emissivity. In fact, this action covers two efficiency factors: the resistance's inherent efficiency factor $e_0$, which allows for their shape but not for the reflection of a part of the flux emitted onto the supporting wall; and the overall efficiency factor $e$, which allows for this reflection.

When these resistances are used in a furnace, i.e. located against a supporting wall, overall efficiency factor $e$ must be considered.

This value can be obtained by calculation or tests. For example, for cast resistances, the inherent efficiency factor is about 0.80, and the overall efficiency factor 0.95. Generally, for other types of resistances, this is 5 to 20% less.

— *Coverage factor g*

In general, between end elements, the resistances do not occupy all the available surface area. Effective surface area $S_u$, the sum of surface areas $S_e$ as defined above, is less than the total surface $S_R$. The coverage factor is equal to the ratio of $S_u$ to $S_R$. Heating elements, such as cast or corrugated sheet resistances provide an excellent coverage factor.

— *Angle factor $f_{CR}$*

The angle factor $f_{CR}$ of the surface of resistances $S_R$ with respect to the load surface $S_C$ is equal to that fraction of the total radiation flux emitted by $S_R$ and directly intercepted by $S_C$. These factors do not allow for the furnace walls which may be re-emissive. Determination of angle factors is generally complex, but can be easily obtained by calculation in certain cases (regular-shaped furnaces of long length compared to their cross-section, etc.). This factor results from the fact that the charge "sees" or does not "see" the resistances at all angles.

To obtain a high angle factor, it is necessary to line as much of the furnace walls as possible, while avoiding overheating in corners. As an example, for a round charge, heat transfer is maximum in a round furnace with resistances arranged around its perimeter.

— *Overall form factor*

Finally, flux emitted by reflection from non-resistance lined walls must be taken into account. The favorable effect of this reflection must be added to the angle factor.

The overall form factor allows both for resistance efficiency factor $e$, coverage factor $g$, angle factor $f_{CR}$ and re-emission from the walls.

Improvement of the overall form factor increases the thermal flux transferred from the resistances to the load, and therefore furnace thermal performance. Cast or corrugated resistances, followed by ribbon and wire resistances, give a decreasing form factor in that order.

### 5.1.1.3. *Influence of maximum resistance temperature*

Resistance temperature, its evolution during the heating cycle and its control are determining factors in the development of high power density resistance furnaces. There are two major trends:

— the use of non-metallic resistances emitting at high temperature;

— the use of improved alloy metal resistances, combined with fine control systems, enabling permanent operation at optimum temperature.

### 5.1.2. *Very high temperature, non-metallic resistance furnaces*

If heating elements withstanding very high temperatures (molybdenum disilicide or silicon carbide resistances) are not used in very high temperature furnaces, but in medium (900°C) or relatively high (up to 1,250°C) temperature furnaces, it is possible to dissipate very high power offering very short heat-up times.

For example, molybdenum disilicide resistances have maximum emission temperatures of 1,700 to 1,800°C. When used at slightly lower temperatures, of the order of 1,400 to 1,500°C, these resistances enable high power density (between 50 and 70 kW/m$^2$ on furnace walls) for charge temperatures of between 800 and 1,100°C, while guaranteeing long element service life. As a comparison, in metallic resistance furnaces, the power density does not exceed 15 to 25 kW/m$^2$ on the furnace walls. Under these conditions, an optimum must be found between operating gains and eventual prohibitive costs.

The advantage of such resistances is obviously increased temperature level. For a charge temperature of 1,300°C and above, they are the only resistances which can be used.

### 5.1.3. *High power density metallic resistance furnaces*

As a function of load temperature, Figure 63 gives the transferred heat flux per square meter of wall, between the heating elements and the load when the resistance limit temperature is 1,300°C (coefficient $\epsilon_{RC}$ x $F$ equal to 0.5). This figure is a simple graphic representation of the Sefan-Boltzmann formula given at the beginning of this paragraph, and it is possible to plot a group of curves corresponding to each resistance temperature limit value (curves $T_R$ = 630°C, $T_R$ = 1,100°C and $T_R$ = 1,300°C, on Figure 63).

In this case the maximum heat flux transferred, when the load is cold and the resistances are at their maximum temperature, is 184 kW/m² (point A''). However, if a power of this type were installed in the furnace, and as soon as the load temperature began to rise, it would be necessary to limit dissipated electrical power to prevent resistance overheating (a thermal flux of this type cannot, in fact, be exchanged between the resistances and the load since their difference in temperature would be too low) and follow curve 3 by means of the control system. Therefore, it is totally useless to seek such high powers. However, the installed powers generally encountered in metallic resistance furnaces, between 10 and 25 kW/m², seem relatively low and, in some cases, limit productivity.

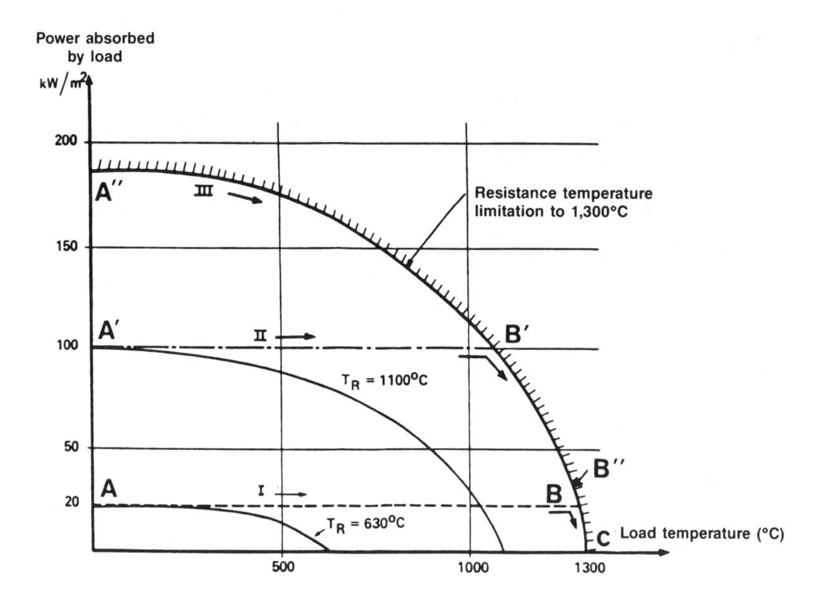


**Fig. 63.** Instantaneous power exchanged between the resistances and the load when the limit temperature is 1,300°C (for a coefficient $\epsilon_{RC} \times F = 0.5$); [79].

Recent technologies have resolved the technical problems causing such limitations, and enabled a new generation of resistance furnaces to be developed [78], [79], [82], [91].

### *5.1.3.1. Remarks on resistance furnaces of conventional design*

Conventionally designed resistance furnaces are "on or off" regulated by means of an ambient pyrometric probe located between the resistances and the load; as a function of the difference between its set point and the measured temperature from the ambient probe, the regulator or control system controls a relay which cuts off or applies electrical power to the heating elements.

The heating elements can therefore be supplied at maximum voltage only, with dissipated power equal to installed power, or cut off.

In many cases, the control system also features a resistance temperature safety device, operating a relay connected in series with the power circuit.

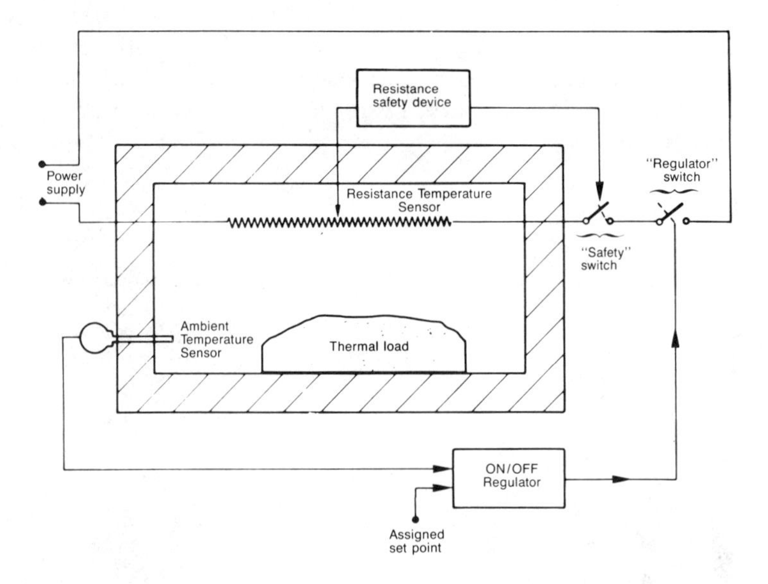


**Fig. 64.** Regulation by Classian ON/OFF Technique using electromechanical contactor [82].

During the temperature rise phase, installed power is constantly on. In the normal operating mode, the control system, operated by the ambient temperature device, periodically switches the resistance power supply. The resistance temperature changes according to a sawtooth pattern with a period $T$, around average temperature $\theta_m$; the resistances would have the same heating efficiency if they were taken to constant temperature $\theta_m$.

With this type of control, there are at least two reasons for limiting resistance specific power, and therefore power density:

— *relay service life*

To avoid excessive frequent power relay operation and to ensure adequate service life, period $T$ must not be made too low (with furnaces offering adequate temperature accuracy, of between $\pm 5$ and $\pm 10°C$, the time base being about 1 minute). The resistance temperature variations must therefore be relatively high, which proportionally reduces mean maximum temperature $\theta_m$ and therefore maximum power density.

— *Resistance temperature sensor performance*

The resistance temperature-limiter sensor has a certain thermal inertia. While the furnace temperature is rising, it indicates the real temperature of the resistance with a certain delay or error which increases as power density increases. From a critical power density value, the measurement error becomes such that, while the temperature is rising, and eventually in the normal mode, the resistance can periodically exceed the maximum temperature. Then, it is necessary either to tolerate a certain period where the resistance set point temperature is exceeded (which may shorten service life), or limit power density. It is therefore critical to use resistance temperature sensors offering minimum thermal inertia (far from being the general case in conventional furnaces) and placed as near as possible to the resistances.

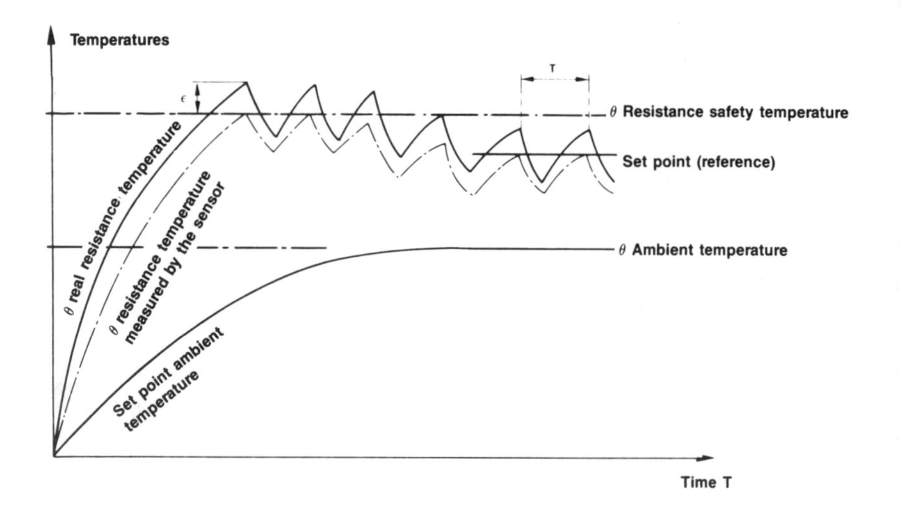


**Fig. 65.** Temperature rise in a resistance furnace equipped with conventional ambient temperature-controlled "on or off" control [82].

Although the heating elements periodically reach (or exceed) a temperature close to their maximum operating temperature under these conditions, the load only receives an energy corresponding to their average temperature $\theta_m$ by radiation, which, according to the Stefan-Boltzmann law, limits power density.

The previous remarks apply essentially to the end of temperature rise and the normal operating mode. An examination of the beginning of the temperature rise highlights one of the consequences of using low specific power density in conventional resistance furnaces to ensure long heating element service life. Figure 63 shows that, with a low power density, for example 20 kW per square meter on the walls, the resistances are often underused, since their temperature does not exceed 630°C when the charge is cold (curve 1). A much higher power density can therefore be used at the beginning of heating, if an efficient control system, which obviates the previously mentioned drawbacks at the end of temperature rise and holding temperature, is used. *It is during this temperature rise phase that the adverse effect of limited power density on productivity occurs.*

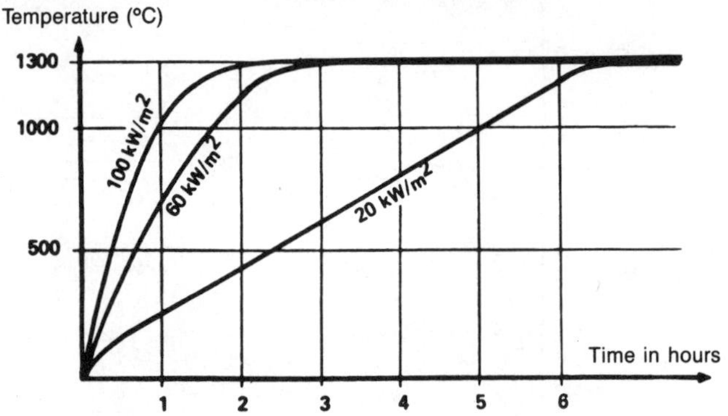


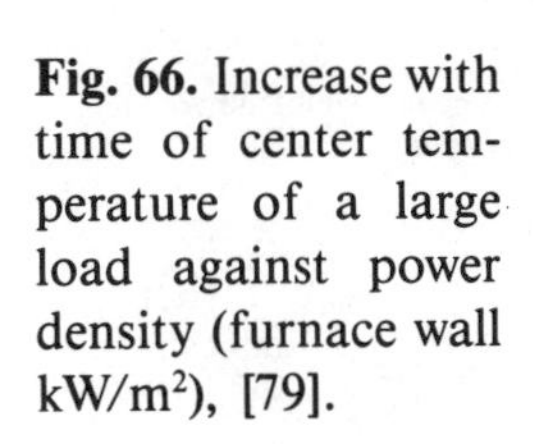

**Fig. 66.** Increase with time of center temperature of a large load against power density (furnace wall kW/m$^2$), [79].

Temperature variations in resistances are inversely proportional to wall thermal inertia (in particular, this applies to refractory ceramic fibre walls). This is another argument in favor of fine control, since it is probable that repeated thermal shock to the resistances limits service life.

To conclude, "on or off" regulation, which is fully adequate for numerous applications, together with incorrect use of certain resistance temperature sensors, prohibits installation of very high specific powers in resistance furnaces, and in particular in those using metallic resistances for which maximum temperature is limited.

### *5.1.3.2. Principle of resistance furnace with high specific power density*

To use given resistances with high power densities (the reasoning which follows, although developed for conventional metallic resistances, also applies to non-metallic resistances), it is necessary to control resistance temperature, at least until the load reaches its target temperature.

When starting the furnace, three phases are to be considered as shown in Figure 67 and graph 2 of Figure 63.:

— the maximum installed power (100 kW/m$^2$ in this case) is dissipated; the resistance temperature increases to the maximum allowed as the load temperature increases (curve A'B' of Figure 63 and zone 1 of Figure 67);

— the temperature difference between the resistances and the load is not sufficient to enable a flux equal to the installed electrical power to be transferred by radiation from the heating elements to the load. The resistance, which is continually kept at its maximum temperature, drives the control system, and the power applied to the resistances must be instantaneously equal to the limit power given by the Stefan-Boltzmann law (curve 2 between B' and B" of Figure 63 and zone 2 of Figure 67);

— the load temperature reaches the set point value and assumes temperature control. The power required to maintain the temperature becomes lower and lower as the furnace walls heat up, while the resistance temperature drops to the absolute minimum value (zone 3 of Figure 67).

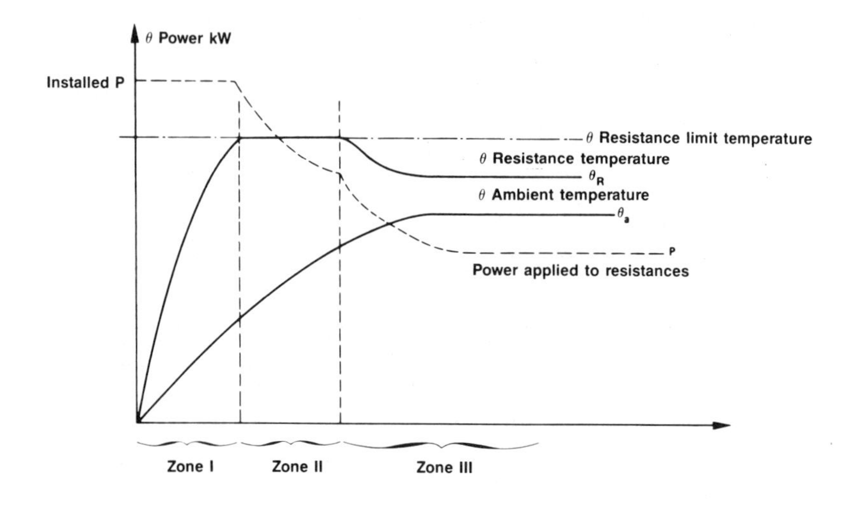


**Fig. 67.** Ideal temperature rise diagram for a resistance furnace [82]:
zone I, power demand = installed power (maximum power)
zone II, power demand permanently equal to $\sigma \times \epsilon_{RC} F(T_E^4 - T_E^4)$;
zone III, power required to maintain temperature

This furnace design implies:

— installation of two temperature sensors, one for the resistances at the hottest point, and the other for the load offering low thermal inertia; the first sensor measures a temperature very close to that of the heating elements;

— use of a control system enabling continuous modulation of dissipated electrical power; the control system must use static instead of electromagnetic relays (thyristor power supply);

— use of a double input control system.

Sensors meeting the above requirements are available, but any choice requires careful analysis since the sensor response times offered for resistance temperature control varies, according to a comparative study performed on a sample, from 1 to 1,000 (*see* paragraph 4.1.).

The control system can use various configurations but in all cases, calls for the possibility of varying power to the heating elements at 0 to 100% of its maximum value.

In general, temperature measurements are applied to P.I.D. (Proportional, Integral, Differential) regulators which produce a heating control signal. The power input is controlled by an integer pulsed multiple cycle thyristor system with a short time base $T$ (in general between 2 and 3 seconds, i.e. 100 to 150 times line frequency; to be compared with time bases of 1 minute or more for electromechanical systems). Conduction time t and area heating power utilization factor ($t/T$) are proportional to the signal emitted by the temperature regulator (phase angle control can also be used).

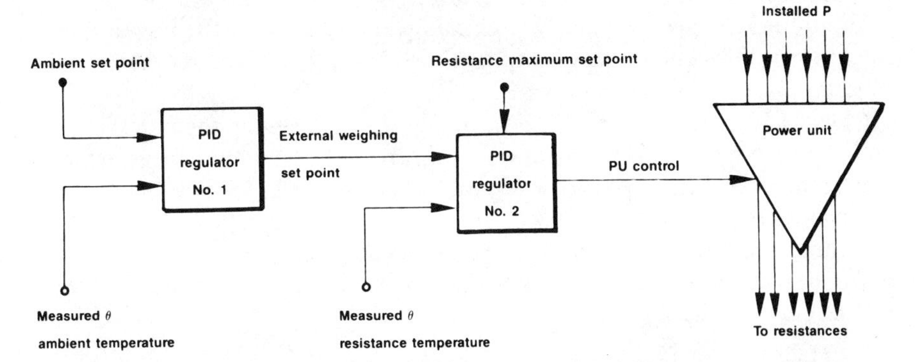


**Fig. 68.** Block diagram of a thyristor power supply controlled by two cascade-mounted PID regulators (one for the load and the other for the resistances) [82].

*5.1.3.3. Advantages and limitations*

The metallic resistance furnace design described above offers numerous advantages.

An examination of Figure 63 shows that this design enables installation, in furnaces using conventional metal resistances (but preferably offering a good form factor, i.e. cast, ribbon, thick wire and, for radiation tubes, edge wound or ribbon elements( of high specific powers between 30 and 60 kW/$^2$ at walls. This depends on the type of alloy used in the resistances, their arrangement and the need for radiation tubes, and load temperature. However, attempting to install excessively high power densities is useless, since Figures 63 and 66 show that they would be used only during very short periods, and that gain would be low.

This gain in power density is due to several separate but interdependent factors:
— operation during the temperature rise phase, with very high specific powers at the resistances without overheating of the elements;
— at end of temperature rise, the possibility of gradually decreasing the specific power by continuous modulation of the electrical power supplied;
— stable operation with a small temperature difference between the load and the resistances;
— the use of very low thermal inertia temperature sensors;
— the use of tyristor control systems.

This concept offers four major advantages, based on the possibility of:
— constructing very high productivity normal metallic resistance furnaces;
— reaching relatively high load temperatures with low cost resistances;
— operation during holding phases with lower resistance temperatures than in conventional furnaces, and therefore an increase in resistance service life;
— excellent temperature accuracy.

Moreover, this type of furnace can be used to benefit from the electricity rate structure. For example, for discontinuous operation furnaces, the temperature rise, especially if the electrical power supply system is combined with the use of low thermal inertia refractory materials such as ceramic fibres, can generally be obtained during off-peak hours, which is not always the case with conventional furnaces. Similarly, continuous modulation of the power supply can prevent unwanted excessive power demands. All these measures offer substantial energy savings.

Conversely, furnaces of this type require great care in their design and production and also during operation, to ensure that temperature measurements remain very reliable. Moreover, overheating may occur in loads which are bad conductors (e.g. certain ceramics).

Here again, a thorough analysis should dictate investment choices. The use of furnaces of this type will grow in the steel, metallurgical, mechanical, electrical and ceramic industries.

## 5.2. Low thermal inertia resistance furnaces

Electric furnace design has evolved steadily over recent years, and much progress has been made concerning resistances, construction materials and handling.

But, significant progress towards a more rational utilization of energy and an increase in productivity has been accomplished recently with the development of low thermal inertia resistance electric furnaces [21], [35], [36], [46], [58], [62*d*], [77].

For a long time, the internal lining of most furnaces, at least for high temperatures, consisted of dense refractory bricks.

In particular for discontinuous furnaces, this construction method led to high energy consumption. Over recent years, a major innovation has been the

Comparison of different refractory materials with equal insulating capability (hot surface temperature 1,000°C)

| Characteristics | Materials | | | |
|---|---|---|---|---|
| | Supercompressed brick | Insulating brick | Insulating concrete | Fibrous materials |
| Thermal conductivity (W/m x °C) | 1.4 | 0.28 | 0.27 | 0.19 |
| Density (t/m$^3$) | 2.3 | 0.8 | 1.25 | 0.1 |
| Lining thickness (m) | 1.6 | 0.32 | 0.30 | 0.21 |
| Lining weight per square meter (t/m$^2$) | 3.68 | 0.128 | 0.48 | 0.02 |
| Refractory cost index | 1 | 0.36 | 0.70 | 0.50 |

**Fig. 69.**

development of light fibrous refractory materials, combining low apparent density and a low thermal conductivity coefficient.

Although insulating brick is the cheapest material, its thermal momentum is six times higher than ceramic fibre refractories.

Due to the substantial energy savings, high investment costs can be recovered very rapidly in the case of intermittent furnaces (generally over one or two years).

Moreover, fibrous ceramic refractories offer faster heating and cooling cycles, an increase in the number of treatment operations for a given type of furnace, and better thermal control.

Furnaces of this type require the use of special resistance suspension devices, due to their low tensile strength. Numerous systems are available: tubular resistance modules (Figure 53); metal or ceramic hooks from which ribbon or coiled wire resistances (Figure 51) are suspended; cast ceramic fibre-formed pieces providing the same function (Figure 70); resistances mounted on supporting frames fitted to the furnace walls; coiled wires on ceramic rods supported by metal hooks; cast resistances attached to the roof or walls using different techniques, etc.

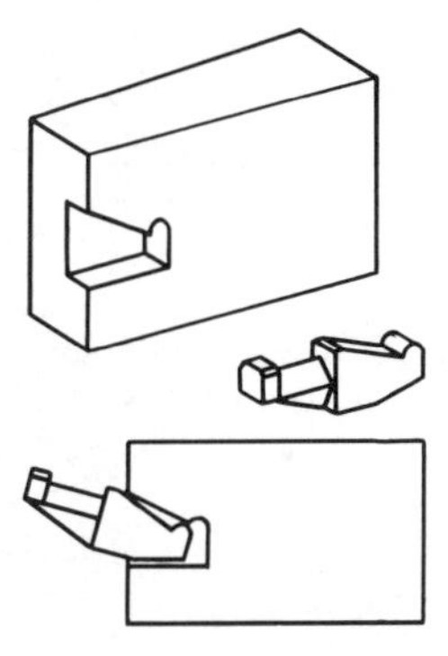

**Fig. 70.** Example showing low thermal inertia furnace ribbon resistance mountings (cast ceramic fibre refractory supports and bricks [M], [BS].

The low thermal inertia of electric furnaces of this type enables improvements in control systems due to continuous power modulation using thyristor power supplies instead of the conventional "on or off" or "full or medium" systems.

Basically, the most widely encountered ceramic fibres consist of silica and alumina, and enable temperatures of about 1,400°C to be reached. Special high alumina or zirconia content fibres enable temperatures of approximately 1,600°C.

These furnaces can be used for a wide range of applications: boilerworks, forges, foundries, mechanical and ceramic industries, etc. Easily adapted to resistance heating, these fibrous ceramic refractories also enable electricity rates to be used advantageously (temperature increased during off-peak hours), and therefore represent an important factor in increasing productivity.

## 5.3. Vacuum resistance furnaces

Industrial use of vacuum furnaces is relatively recent, dating from some twenty years ago. These furnaces are responsible for numerous advances in the special metals industry, and have also enabled production of new materials such as semiconductors and special sintered ceramics, or new techniques such as vacuum impregnation, vacuum bonding, vacuum brazing [33], [34], [60], [67], [85].

There are two main types of vacuum furnace:

— hot wall furnaces;
  • single vacuum
  • double vacuum
— cold wall furnaces.

### *5.3.1. Hot wall vacuum furnaces*

#### *5.3.1.1. Single vacuum hot wall furnaces*

These furnaces are of relatively simple design and make use of conventional materials (the standard insulating refractories and refractory steels) enabling equipment comparable in price to controlled atmosphere furnaces.

The charge is placed in a refractory steel chamber in which the vacuum is created. Generally, the heating elements are located outside the vacuum chamber, operate in air and are made from the usual metallic alloys. For relatively low temperatures (less than 500°C) and limited vacuum pressure, resistances are sometimes placed directly inside the vacuum chamber reducing thermal loss and increasing the heating rate. To decrease thermal inertia during cooling and to increase speed, the vacuum chamber and furnace itself can be separated.

**Fig. 71.** Hot wall bell-type vacuum furnace [E].

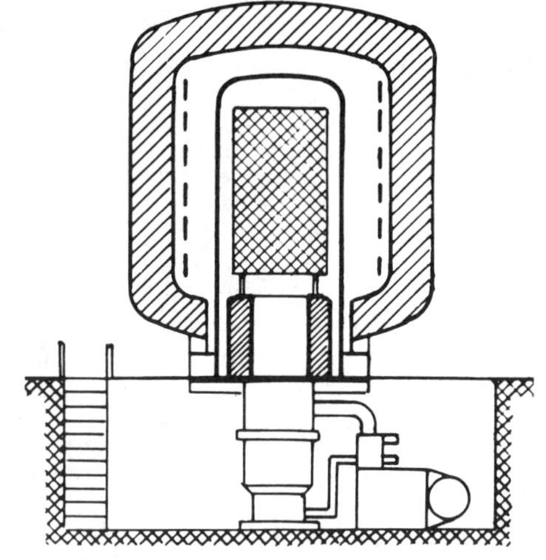

The vacuum chamber may be vertical (pot or bell furnace) or horizontal. The bell furnace configuration offers adequate vacuum chamber mechanical strength and is the most widely used since the pump unit, coupled to the base, remains fixed.

In this type of furnace, the vacuum chamber must withstand atmospheric pressure, while being taken to a temperature slightly higher than the furnace operating temperature. For reasons related to the mechanical behavior of refractory steels forming the vacuum chamber, the maximum operating temperature for such furnaces is between 800-900°C. Moreover, the vacuum is rarely less than $10^{-2}$ torr.

At present, these furnaces are not widely used but can be used as vacuum ovens, or vacuum annealing and tempering ovens.

*5.3.1.2. Double vacuum hot wall furnaces*

This type of furnace is identical to the single-wall type, but the heating elements are contained in a vacuum-sealed chamber in which the pressure is dropped to approximately 1 torr. This chamber may be the brick-lined furnace itself (bricking must be carefully performed to provide sufficient sealing to maintain a low vacuum) or a special chamber inside the refractory lining.

The load is placed in a second chamber, or converter, concentric with the first, in which the vacuum required for load protection is maintained.

The hot chamber is therefore not submitted to atmospheric pressure, enabling this type of furnace to be used up to temperatures of 1,100°C. Residual pressure can drop down to $10^{-2}$ torr approximately.

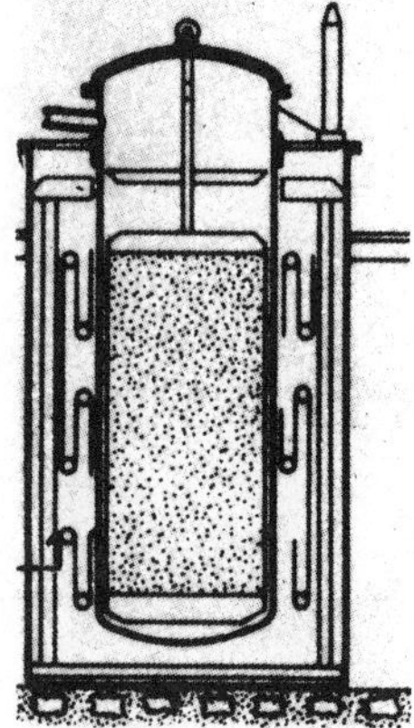

**Fig. 72.** Double hot wall vacuum furnace [T].

Productivity and thermal efficiency can be increased by using several converters so that the furnace is permanently charged (one converter heating with the second on controlled cool-down).

These furnaces can have very high payloads, ranging from a few to several tens of tons. These are used mainly in primary transformation of metals, for annealing, or heat treatment of products such as coils of ribbon or wires [E], [K], [L], [AF].

*5.3.2. Cold wall vacuum furnaces*

Temperature limits and residual pressure are the main drawbacks with hot wall vacuum furnaces. This has resulted in an important development in cold wall furnaces over recent years.

Cold wall furnaces in which the envelope is generally cooled by water circulation enable very high temperatures to be reached, together with very low residual pressures, and large furnace production [A], [C], [D], [E], [G], [H], [I], [J], [K], [L], [AG], [CC].

The heating elements are placed in the furnace heating chamber and radiate directly onto the charge. Generally, these are graphite, molybdenum, tantalum or tungsten elements (for temperatures of less than 1,000°C, nickel-chromium resistances are sometimes used). Element choice depends on the temperature level desired, the residual pressure value, cost and, where necessary, on the type of load (*see* paragraph 3.4.). These elements are supplied at voltages lower than vacuum breakdown voltage.

To limit radiation onto the furnace walls and thermal losses, the heating chamber is surrounded by successive shields reflecting the radiation. The shields are either metallic (molybdenum, tungsten, stainless steel, etc.), graphite or carbon felt shields. Generally between three and eight shields are installed. In most cases, the furnace outside wall is in thick stainless steel and water cooled.

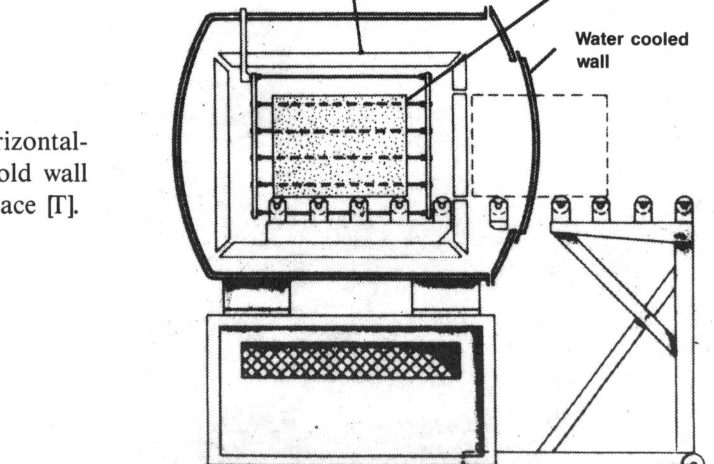


**Fig. 73.** Horizontally charged cold wall vacuum furnace [T].

The maximum operating temperature in cold wall vacuum furnaces is based on the physical behavior of the structural components and the resistors. In some extreme cases, this temperature can reach 3,000°C, but in most applications, temperatures are between 1,200 and 2,000°C.

These furnaces can withstand extremely low residual pressures, down to $10^{-6}$ torr approximately, calling for very powerful pump units. To obtain a low vacuum, down to 1 torr approximately (primary vacuum), mechanical vane or rotary piston pump mechanical systems are used. For a vacuum between 1 torr and $10^{-3}$ torr, Roots pump systems must be used. Finally, for a very high vacuum between $10^{-5}$ and $10^{-6}$ torr, diffusion pumps must be used, the various systems being generally employed sequentially.

Numerous types of vacuum furnace are now available: horizontal, bell and elevating hearth furnaces, and also top charging furnaces, hardening chamber furnaces using oil and furnaces equipped with forced convection systems for charge cooling, etc. Due to the low thermal inertia of the cold walls (no

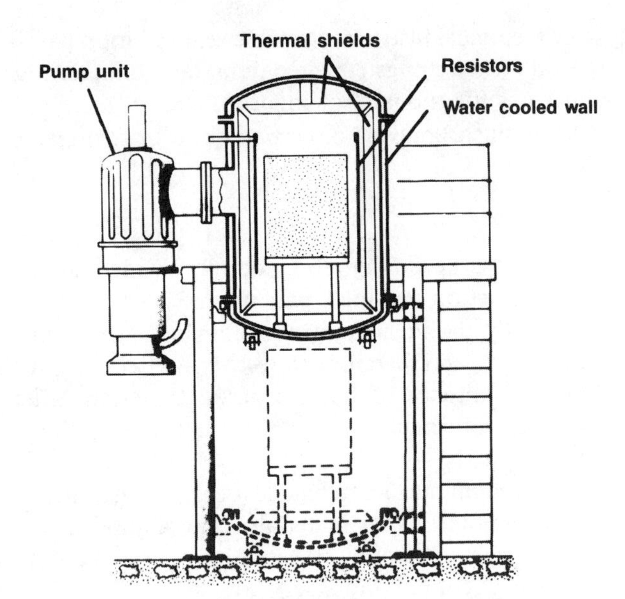


**Fig. 74.** Cold-wall-vacuum elevating furnace

brickwork), these furnaces can cool the load very rapidly (gas hardening); thermal cycles which were unavailable with conventional furnaces are now easy to obtain, for example:

— rapid temperature rise up to 1,250-2,000°C with a vacuum of about $10^{-5}$ torr;

— accelerated cooling by means of a pressure relieved protective gas in the furnace, recycled through a turbine and around a water-cooled exchanger.

These are continuous furnaces which may be a limiting factor for high production rates. Moreover, continuous or semi-continuous furnaces have been recently developed where the load is introduced and extracted through a hatch. Compared to discontinuous furnaces, the energy gain obtained is substantial.

The range of furnaces is very wide, power and capacity varying from a few kilowatts to several hundred kilowatts and a few tens of cubic decimeters to approximately 100 cubic meters respectively.

### *5.3.3. Vacuum furnace applications*

Vacuum furnaces are most widely used in advanced industries such as aviation, aerospace, nuclear, electronic, special steels and special ceramics, but their use in more conventional sectors is growing.

*— Heat treatment of metals*

Vacuum furnaces offer high quality and mechanical performances, minimum part deformation, excellent metal surface states and intensive outgassing, and also can be used to perform treatments which are difficult to obtain using other methods.

These furnaces are being more widely used for treating parts such as cutting tools, moulded parts, extrusion dies, mandrels, rams, stamping dies, extruding screws, surgical instruments, tablewear, aviation parts, ferromagnetic materials, special metals (tungsten, zirconium, tantalum, titanium, etc.) and special steels and parts for the nuclear industry, etc.

Vacuum carbonizing or carbon nitriding treatments are also being developed [20].

— *Brazing*

Brazing is becoming more and more widely used as an assembly technique (*see* "Induction heating", paragraph 5.3.).

Basically, the advantages in vacuum brazing are the elimination of an oxidizing atmosphere (and therefore protective flux), no risk of flux inclusion in the joint, and total absence of trapped gases. Generally, after brazing, no cleaning is required, deformation is very low and the mechanical characteristics of the assembly high.

The vacuum brazing technique is of particular interest for stainless steels, nickel alloys, titanium-based materials, ceramic and metal assemblies, assembly of complex parts, assembly of very high precision parts and assembly of different metals (e.g. copper and steel), etc., [67].

— *Metallization (plating)*

Vacuum evaporation plating is used in numerous industries to obtain thin films offering special product surface features. Generally, this operation is performed intermittently for treating objects such as castings and continuously for materials in bands.

The metals to be deposited — aluminum, copper, gold, silver, chromium, antimony, etc. — often form a Joule effect-heated spiral filament (this application is then more similar to conduction heating); these can also be contained in an indirectly heated crucible.

This technique is used to deposit conductive films on plastics (prior to electrolytic coating) and anti-corrosion protection layers on metals, and for coating accessories for automobile parts, water fittings, souvenirs, jewelery and plastic licence plates, and depositing thin films on optical and electronic components. Metallization of paper strip, textiles and plastic can be used to produce films for capacitors, insulating sheets, packing and decorative films, special textiles, etc. [AG], [AN], [CC].

In optics and electronics, and also in more conventional sectors, resistance heating is now subject to competition from more recent processes such as electronic beam heating (*see* corresponding chapter).

— *Sintering*

Vacuum furnaces are suitable for agglomeration of particle powders by heat (and previously applied mechanical pressure). Vacuums prevent oxidation and eliminate volatile impurities and trapped gases. The finished product density is often increased significantly since it is possible to use finer powders. Metal or non-metallic powders (for example, tungsten carbide and tantalum capacitors) can be sintered in these furnaces, enabling the sintered components to be assembled with other materials [47].

— *Vacuum bonding (or welding)*

The metal bonding-welding technique is recent. If two flat surfaces are heated together and submitted to high pressure, a metal joint, resulting from micro-deformation of the surfaces and diffusion at the interface, is created. If the surfaces are first of all submitted to a vacuum, the superficial oxide films and the trapped gases are eliminated, facilitating bonding.

This process is not widely employed at present but can be used to produce stainless steel surgical instruments and cutlery, high speed steel and precision wire drawing assemblies, and reload valve seats, etc.

— *Manufacture of semiconductors*

Drawing monocrystals sometimes calls for a high vaccum ($10^{-5}$ and $10^{-6}$ torr), and sometimes the use of normal or very high pressures (up to 100 bars).

Vacuum resistance furnaces are often used to perform this operation in conjunction with induction furnaces (*see* corresponding chapter, paragraph 5.7.). These furnaces must offer very high temperature accuracy, of about $\pm 0.5°C$, and are equipped with P.I.D. control systems.

Vacuum furnaces can also be used for applications such as outgassing, resin impregnation, mechanical tests, etc.

### *5.3.4. Advantages and limitations of vacuum furnaces*

Full vacuums do not exist. Vacuum furnaces are in fact controlled atmosphere furnaces, held at a very low pressure. Therefore, these furnaces provide a very pure, easily adjustable atmosphere, for relatively low cost since such pressures are obtained by mechanical or similar methods. Basically, the advantage of vacuum furnaces rests in qualities which each application tries to maximize. Some of these are:

— *outgassing*

Trapped gases or gases dissolved in metals or other materials are eliminated. Mechanical properties are improved giving longer service life, better resistance to fatigue and increased resilience for higher hardnesses.

— *high temperatures*

The production and treatment of numerous new materials require very high temperatures. Vacuum furnaces which enable temperatures of almost 3,000°C are excellent for such purposes.

— *reduced deformation*

Heating in a vacuum furnace is regular and even. This minimizes deformations and eliminates the requirement for straightening after treatment.

— *no oxidation, decarbonizing or carbonizing*

A vacuum is an excellent, easily obtained and controllable atmosphere. After treatment, parts are perfectly clean and free of oxidation. There are no cleaning operations required; elimination of cleaning operations is of special interest for complex-shaped parts containing blind holes, deep grooves and locations to which access is difficult.

The absence of decarbonizing or carbonizing dispenses with surface hardness differences, providing constant mechanical quality and avoiding reworking costs to remove the affected surface layer.

— *high productivity*

Due to their construction characteristics, vacuum furnaces have low thermal inertia, enabling the desired thermal cycles to be obtained rapidly; temperature homogeneity in the treatment chamber is excellent. Low thermal inertia also enables rapid cooling, improving the furnace utilization and electrical power factors and therefore better absorption of fixed costs.

— *special atmospheres*

The residual atmosphere can consist of a gaseous environment other than air. These special atmospheres can be used, for example, to obtain thermo-chemical treatments such as vacuum hardening or carbonitriding at high temperatures. Parts can be easily cooled in special atmospheres (argon, etc.).

The principal limitation in vacuum treatment is the high investment costs involved, but with total production costs not necessarily being high. The savings obtained by eliminating most of the cleaning and reworking operations, together with an increase in product quality can, in many cases, decrease unit cost.

Also, some products must be vacuum produced and treated (special metals such as titanium, tungsten, uranium or some sintered ceramics).

Vacuum furnaces are essentially used in the transformation of metals, but the possible applications are constantly increasing (impregnation of products with special resins, vacuum casting of various products, production of special ceramics, and special operations in the food industry, etc.).

## 6. ADVANTAGES AND LIMITATIONS OF RESISTANCE FURNACES

Resistances furnaces offer numerous advantages from the technical, economic and social points of view.

### 6.1. Advantages of resistance furnaces

— *Technical*

• the resistances presently available cover almost all of the temperature ranges used in industry, and progress in the science of materials constantly improves their characteristics.

• the heat transfer mode is easily adapted to product heating requirements;

• if necessary, the available control systems provide high temperature accuracy, and wise distribution of resistances in the furnace offers excellent temperature homogeneity;

• resistance furnaces use many types of atmosphere: air, vacuum, neutral, reducing or oxidizing controlled atmospheres and, in general, resistances have no effect on the furnace atmosphere;

• these furnaces can also provide heating cycles over rather wide temperature ranges, increasing their value;

• the results obtained can be reproduced from one cycle to another, and such furnaces are easily automated;

• resistances can be designed for a wide range of furnace forms;

• their cleanliness and thermal qualities enhance the likelihood of obtaining high quality products;

• furnace operation is generally very simple and safe.

— *Economic*

• investments required with resistance furnaces are generally limited, the same or often less than those of alternate solutions;

• resistance furnace efficiency is very high and drops less quickly than fossil fuel furnaces under conditions other than nominal; energy costs compare favorably in numerous cases to those of furnaces using other energies (comparisons should be made between energy costs after efficiencies have been taken into account, i.e. on effective, not gross energy costs);

• continuous measurement of the quantities of electricity consumed enables easy "energy" management;

• these furnaces require limited maintenance, therefore offering low downtime, further improving productivity;

• the heating quality reduces the number of production rejects;

• automatic operation reduces labor costs.

— *Social*

The absence of smoke and noise improves working conditions and guarantees clean premises. In addition, controlability and automation dispenses with the most difficult tasks.

Resistance furnaces are therefore simple, reliable, high performance production tools, which explains their wide use throughout industry.

## 6.2. Resistance furnace limitations

For a long time, resistance furnace development was hindered by two obstacles: the relatively high cost of electricity, even after allowing for the much higher efficiency of electrical equipment; and the sometimes lower power densities obtained compared with fossil fuel furnaces. Due to changes in the relative prices of energies, and the development of high power density electrical furnaces, these limitations no longer apply.

Resistance service life has not always lived up to user expectations and to the figures announced by manufacturers. Analyses show that the destruction of resistances is often due to operational neglect or errors, design errors being rarer and in general easier to correct. Although resistance furnaces are very rugged tools, it is very important that the minimum precautions be taken with their use to obtain a working service life comparable to their nominal service life, which is normally very high.

However, the advantages and limits of resistance furnaces are closely related to their applications.

## 7. APPLICATIONS OF RESISTANCE FURNACES

Resistance furnaces are used throughout industry. Therefore, it is very difficult to provide an exhaustive list of their industrial applications.

The examples discussed below cover a wide range of applications, and are primarily intended to give potential users a general overview.

### 7.1. Metallurgical, mechanical and electrical industries

#### *7.1.1. Melting metals*

Resistance furnaces are widely used for melting metals and, in particular, non-ferrous metals. Their application continues to increase due to the development of high power density, and therefore high productivity furnaces, and also due to the development of very high temperature furnaces.

The accent on energy conservation also has led to the development of super-insulated furnaces having very low power and specific energy consumption.

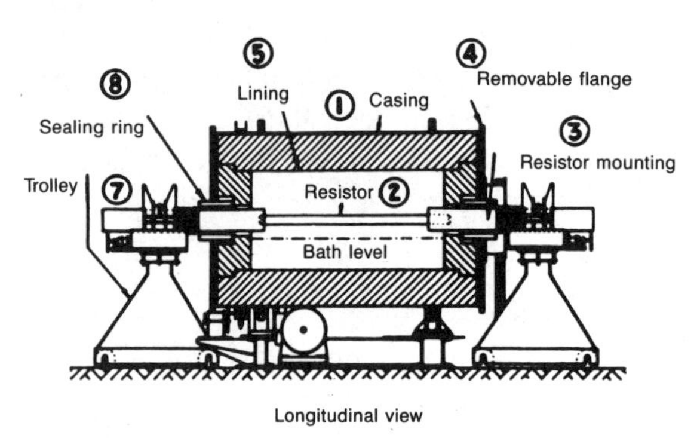


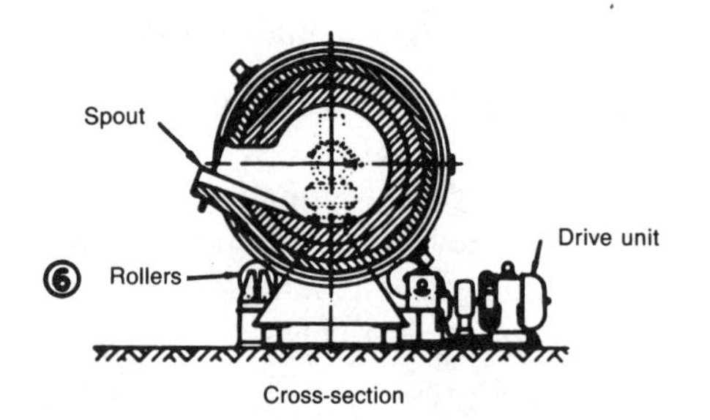


**Fig. 75.** Radiant rod furnaces [76], [AC].

### *7.1.1.1.* Radiant rod furnaces

Radiating rod furnaces are reverberating furnaces in which the heating element consists of a graphite or silicon carbide rod. Such furnaces consist of a cylinder mounted on two rails enabling alternate rotary movement, or of a tank similar to that used in a three-phase arc furnace. The resistance radiates onto the load and onto the furnace refractory lining. These furnaces are subjected to an oscillatory movement enhancing heat transfer from the refractories to the load through thermal conduction. In general, these furnaces are single-phase supplied at a voltage of 20 to 50 V, with very high currents of about 4,000 to 5,000 A flowing through the radiating rod [76], [AC] [BX].

These furnaces are used in casting foundries to produce special castings and steels, and rarely for certain copper alloys.

Radiant rod furnace range (melting time, with furnace hot)

| Power (kW) | Cast Iron | | Steel | | Copper and bronze | |
|---|---|---|---|---|---|---|
| | Capacity (kg) | Melting time (h) | Capacity (kg) | Melting time (h) | Capacity (kg) | Melting time (h) |
| 60 | 50 | 1 | 30 | 1 | 50 | 0.5 |
| 150 | 250 | 1.5 | 150 | 1.5 | 250 | 0.75 |
| 220 | 500 | 2.25 | 300 | 1.75 | 500 | 1.25 |
| 350 | 1,200 | 3 | 750 | 2.5 | 1,200 | 1.50 |
| 500 | 2,000 | 3.5 | 1,500 | 3.5 | 2,000 | 1.75 |

Cast iron, 1,550°C; steel, 1,650°C; copper alloys, 1,300°C.

**Fig. 76**

Specific energy consumption varies widely and is affected by melting frequency and the lining weight. For cast iron and steel, the usual consumptions vary respectively from 750 to 950 kWh/t and 850 to 1,250 kWh/t; for a cold start, these consumptions are much higher, between 2,000 and 2,800 kWh/t. Conversely, for high capacity furnace and in the normal temperature mode, consumption is low, and a power of 250 kW is sufficient to keep a 20 ton furnace filled with cast iron at 1,500°C.

Resistance consumption is rather high, about 2 to 5 kg of graphite per ton cast.

In spite of certain advantages such as relatively low investment costs, limited fire loss and high metal quality, these furnaces are gradually being replaced by crucible induction furnaces, offering better operating flexibility.

### *7.1.1.2.* Basin furnaces

Reverberatory furnace design is similar to that of radiant rod furnaces; however, they differ in several ways, especially in how resistances are used and arranged. The heat required for melting the metal and keeping it molten is provided by heating elements located in the roof, radiating directly onto the load.

The heating elements used differ widely. Metal elements consist mostly of cast resistances wire wound on sillimanite tubes, and ribbon or thick wire resistance radiation tubes; the best results seem to be obtained with cast resistances and edge-wound ribbon resistance tubes. For low power holding furnaces, encased resistances are sometimes used. Silicon carbide rods are the most currently used non-metallic elements. To protect the resistances, a stiff steel sheet, which also behaves as a heat distributor, is inserted between the resistances and the charge.

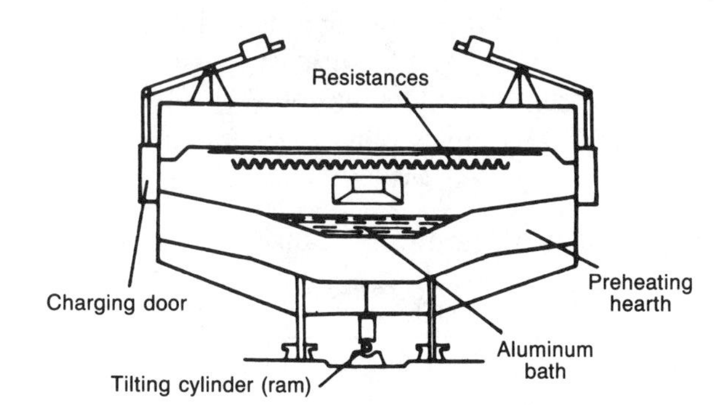


**Fig. 77.** Diagram of a preheating hearth resistance basin furnace [90].

To increase flow, tank furnace design involves a high basin area compared to the height of the molten metal. Sometimes an inclined hearth to preheat the metal to be melted is used with large basin furnaces. This device also enables higher temperatures on the hearth than in the bath, thus accelerating metal melting. Generally, these are fixed furnaces but tilting furnaces can easily be designed.

Most of the furnaces in service in foundries are of medium or low capacity (between 600 and 2,000 kg aluminum), but it is possible to construct furnaces of much higher capacity, ranging from 10 to several tens of tons of aluminum.

The progress obtained with these furnaces involves the use of resistances and their regulation (high density, therefore high production furnaces), and the refractories to increase service life (vacuum moulded fibrous refractories, special concretes, etc.).

Variations of such furnaces exist but up to now their development has been limited. For example, resistances can be located in the hearth preventing risky projections and facilitating replacement, but this limits installed power. Special immersion heaters can also be installed in the bath which increases thermal efficiency, but calls for the prevention of corrosion and solidification of the molten metal.

These furnaces are used in casting foundries for melting and temperature holding of aluminum alloys, low melting point alloys (zamak and other zinc, tin and lead alloys, etc.), but their use can be extended to copper alloys (aluminum bronze, etc.). Furnaces of a capacity of several tens of tons are also used for maintaining aluminum after production by electrolysis.

As an example, the table below gives the nominal current at 720°C of basin furnaces designed for melting aluminum alloys.

Basin resistance furnace range for melting aluminum alloys
Temperature: 750°C

| Maximum power (kW) | Holding power ([1]) (kW) | Rate (kg/hr) | Capacity (kg) |
|---|---|---|---|
| 30 | 20 | 14 | 950 |
| 60 | 22 | 85 | 1,850 |
| 90 | 24 | 160 | 2,050 |
| 120 | 26 | 235 | 2,550 |
| 150 | 28 | 280 | 2,950 |
| 200 | 36 | 420 | 4,400 |
| 250 | 40 | 545 | 5,100 |
| 300 | 44 | 655 | 5,700 |
| 350 | 48 | 780 | 6,350 |
| 400 | 52 | 910 | 7,000 |

([1]) With tank open

**Fig. 78.**

However, basin furnaces having a higher specific power exist (installed kilowatts per ton of furnace capacity); these are about 150 kW for a capacity of 2,000 kg.

The specific melting consumption is about 0.5 kWh/kg to 0.6 kWh/kg of melted metal. Consumption and normal operation depends heavily on the thermal design of the furnace. Super-insulated furnaces have recently been developed: the power necessary to keep 700 kg of aluminum at 750°C is now only 4 to 5 kW, compared to 15 to 20 kW with conventional furnaces, providing a substantial energy saving. Overall consumption (melting and holding) depends essentially on operating conditions; in aluminum casting foundries, consumption generally varies between 800 and 1,300 kWh/t.

The primary advantage in resistance basin furnaces is the absence of a crucible, very high service life of the refractory lining, very low metal gassing, limited oxidation losses, very high thermal efficiency and the option of installing several drawing stations on a single furnace. Moreover, the high capacity of these furnaces often enables melting of a high proportion of the metal required for production, during low electricity rate periods and limiting of power during peak hours [70], [C], [D], [G], [AM], [AP], [BA], [BZ], [CB].

Conversely, these furnaces are not the most flexible since alloy charging demands much time.

*7.1.1.3.* Crucible furnaces

The metal is melted in a crucible placed in a resistance lined chamber. Metal crucibles (steel or cast iron) are rugged, cheap and provide adequate thermal conductivity, but often react with the molten metals contained in them.

Graphite or silicon carbide crucibles prevent continuation of the molten metal in contact with cast iron and steel but have a rather low service life. The most common crucibles are:

Fig. 79. Diagram of a resistance crucible furnace [90].

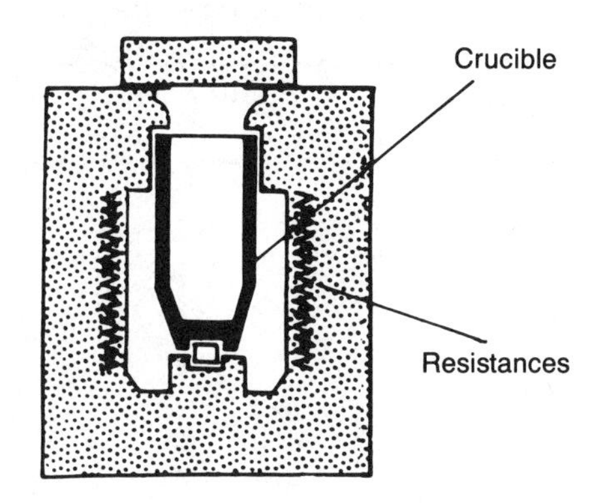


— graphite, for aluminum alloys;
— graphite or silicon carbide for copper alloys;
— steel, for magnesium alloys;
— cast iron for zinc alloys (zamak).

As required, resistance crucible furnaces can be either fixed or tilting. Some enable melting in a vacuum or special atmosphere (for example, the surface of magnesium alloy baths, which is a very oxidizable metal, is often protected with an atmosphere of sulphur hexafluoride $SF_6$) [48], [70], [C], [D], [G], [I], [W], [Y], [AO], [AQ], [BA], [BR], [BZ].

Up to a temperature of 1,100-1,150°C, in molten metal baths, resistances are generally metallic (nickel chromium or iron-chromium-aluminum, depending on temperature and power density) and consist of corrugated ribbons or wires wound around refractory supports. For higher temperatures, non-metallic resistances must be used; these are usually silicon carbide elements but molybdenum-silicide elements may also be used with fixed furnaces.

These furnaces are intended for melting of non-ferrous metal alloys in casting foundries, where average or low rates are involved:
— melting and holding aluminum alloys
— melting and holding low melting point alloys (tin, lead, zinc, etc.);
— melting and holding copper base alloys.

— *Aluminum alloys*

For aluminum alloys, specific consumption in the molten state is about 0.5 kWh/kg of melted metal at 700°C, starting with a hot furnace. For melting at 800°C, the rate is reduced by 10 to 15% and specific consumption increases in comparable proportions. The overall specific consumptions combining melting and holding depend on the operating conditions, and are generally between 0.8 and 1.4 kWh/kg of poured metal.

Crucible resistance furnace range for melting and holding of aluminum alloys, temperature: 700°C

| Capacity (kg) | Maximum power (kW) | Maximum flow rate (kg/hr) | Power required for holding | |
|---|---|---|---|---|
| | | | Cover open (kW) | Cover closed (kW) |
| 30 | 15 | 30 | 6 | 4 |
| 70 | 18 | 38 | 7 | 4.5 |
| 120 | 24 | 53 | 9 | 5.5 |
| 210 | 46 | 110 | 11 | 6.5 |
| 390 | 54 | 130 | 16 | 8 |
| 625 | 72 | 180 | 25 | 12 |

**Fig. 80**

— *Zinc alloys*

For zamak, the specific melting consumption is about 0.10-0.12 kWh/kg of molten metal, and overall consumption is generally between 0.15 and 0.22 kWh/kg of molten metal.

— *Copper base alloys*

For copper-base alloys with low pouring temperatures (for example 60/40 for brass), it is possible to use iron-chromium-aluminum metallic resistance furnaces. conversely, for pouring temperatures higher than 1,150°C (aluminum, bronze or manganese, etc.), non-metallic resistances, generally silicon carbide, must be used; these furnaces enable bath temperatures of between 1,300 and 1,400°C.

For brass which is poured at around 900°C, specific consumption is about 0.3 kWh/kg of molten metal.

Range of resistance crucible furnaces for melting and holding zamak temperature 450°C

| Furnace capacity (kg) | 200 | 300 | 385 | 500 | 650 | 1,350 |
|---|---|---|---|---|---|---|
| Installed power (kW) | 30 | 30 | 30 | 38 | 38 | 60 |
| Melting production with cover (kg/h) | 300 | 300 | 300 | 370 | 370 | 580 |
| Time for initial melting, from cold start (min) | 170 | 190 | 210 | 290 | 310 | 350 |
| Holding power (kW): | | | | | | |
| • cover fitted | 4.3 | 4.3 | 4.3 | 5 | 6 | 9.5 |
| • no cover | 5 | 5 | 5 | 6 | 7 | 11 |

**Fig. 81.**

Range of iron-chromium-aluminum alloy resistance crucible furnaces for melting and
holding brass, temperature: 900°C

| Power (kg) | Maximum power (kW) | Maximum flow rate (kg/hr) | Power required for holding (cover open) (kW) |
|---|---|---|---|
| 100 | 15 | 40 | 8 |
| 210 | 18 | 54 | 10 |
| 360 | 24 | 72 | 13 |
| 640 | 46 | 160 | 16 |
| 1,200 | 54 | 190 | 28 |

**Fig. 82**

Range of silicon carbide resistance crucible furnaces, melting temperature: copper-based alloys 1,200°C, aluminum alloys 750°C.

| Power (kW) | Copper-based alloys | | Aluminum alloys | |
|---|---|---|---|---|
| | Capacity (kg) | Maximum rate (kg/h) | Capacity (kg) | Maximum rate (kg/h) |
| 28 | 60 | 75 | 20 | 55 |
| 45 | 200 | 135 | 60 | 95 |
| 100 | 500 | 300 | 150 | 225 |
| 170 | 1,000 | 570 | 300 | 425 |
| 250 | 2,000 | 865 | 600 | 625 |

**Fig. 83**

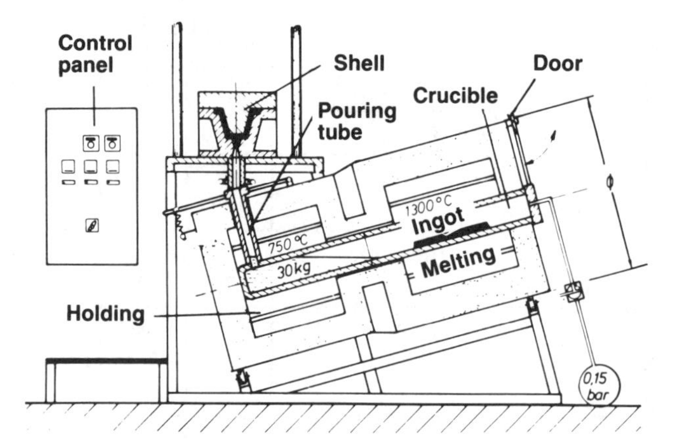


**Fig. 84.** Special low pressure casting furnace [AL].

Much higher specific powers are obtained with furnaces using silicon carbide resistances.

New furnaces which are not real crucible furnaces generally use silicon carbide resistances for melting.

These furnaces are intended for low pressure techniques. The holding zone resistances are nickel chromium wires wound on a ceramic support. A 31 kW furnace offers a maximum rate of 55 kg/hr approximately at 750°C.

Other special, very high density furnaces have recently been developed. These use graphite elements, and, to ensure long service life, a nitrogen atmosphere is maintained in the heating chamber. The installed power for a furnace of capacity 300 kg bronze, is then 150 kW, which is similar to the power densities obtained with induction furnaces with efficiency much higher than obtained with the latter. These furnaces are generally similar in structure to the standard resistance crucible furnaces; however, they are rather costly and have appeared on the market too recently to enable an accurate estimate of their performance [AO].

The main interest in resistance furnaces is due to their simple design and operation, low investment costs, limited oxidation losses and the quality of the metal provided. Crucible service life is often higher, by some 20 to 50%, than that obtained with similarly designed fossil fuel furnaces. Conversely, power density and maximum operating temperature of metallic resistance crucible furnaces are relatively limited, but it is always possible to use non-metallic resistances. Also, the crucible must be replaced in basin furnaces. Resistance crucible furnaces compare favorably in energy consumption to equivalent fossil fuel furnaces, and offer substantial energy savings for high temperatures.

### *7.1.2. Heat treatment of metals*

Metallurgists use this term to describe thermal operations which, due to controlled temperature variation and if necessary combined with an active or simply protective environment, are used to improve the mechanical and sometimes chemical characteristics of metals.

Frequently, a somewhat unclear distinction is often made between two types of heat treatment:

— Treatment of rough parts, half products and castings obtained by hot or cold mechanical formation (forging, rolling, drawing, etc.) or mechanically welded parts. Generally, heat treatments of this type are intended to restore the internal structure of metal destroyed during working (parts obtained by plastic deformation) or melting (cast or welded parts), and consist essentially of annealed products.

— Treatment of machined or ready-for-use parts, to provide special mechanical properties for mechanical components or half finished products; such treatments are numerous and, with progress in the science of metallurgy, are becoming more and more diverse. The most common are tempering, annealing, hardening, carbonitriding and nitriding.

Thermally, a heat treatment is characterized by a thermal cycle (temperature change in time) and, especially, by a maximum working temperature. These variables and the type of treatment used dictate furnace construction methods. Listed by temperature range, the most common treatments are:
— 150 to 300°C, stabilization of steels, tempering of high carbon steels, tempering and aging of light alloys, blueing of steels;
— 300 to 550°C, nitriding, treatment of light and ultra light alloys, and annealing of copper;
— from 500 to 750°C, tempering of ordinary steels, annealing of steels, welds, copper alloys and stabilization of cast iron;
— 700 to 850-900°C, carbon nitriting, carbon hardening and annealing of steels, tempering of steels, annealing of copper-nickel-zinc alloys and copper-nickel alloys;
— 900 to 1,000-1,100°C, carbon hardening, annealing of special steels, tempering of special steels, treatment of stainless and refractory steels, treatment of malleable cast iron (black and white center);
— 1,050 to 1,300-1,350°C, treatment of high-speed steels.

Resistance furnaces are widely used in heat treatment shops, and a wide variety of designs are available. In addition to the usual resistance furnaces, these shops use salt bath, fluidized bed and vacuum furnaces [20], [21], [22], [23], [24], [36], [37], [41], [44], [60], [62 *a,d*], [86], [A] to [AG], [AN], [BA], [BO].

### *7.1.2.1.* Conventional resistance furnaces

#### *7.1.2.1.1. Types of furnaces*

The most widely used furnaces are (also *see* Figures 9 to 11):
— *for half finished products* (wires, strips, tubes, bars, formed sections, etc.):
- bell furnaces
- elevating furnaces
- continuous strand furnaces (stepped)
- muffled tube furnaces (treatment of bars, tubes and wires)
- controlled roller furnaces
- pot furnaces
- fixed or movable hearth furnaces
- continuous traction furnaces (sheet, wires, etc.)
- pusher type furnaces.

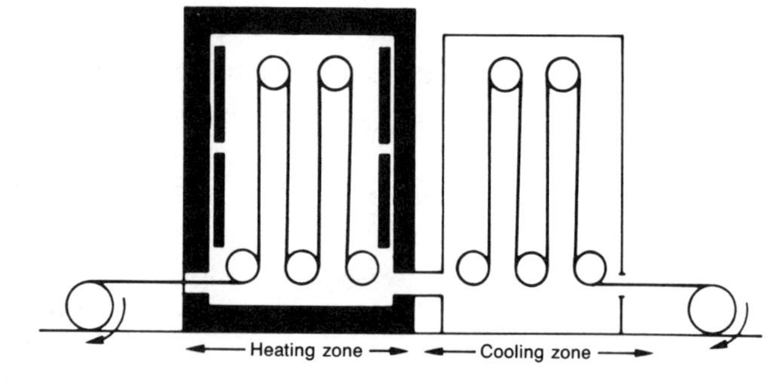


**Fig. 85.** Continuous vertical traction furnaces for treatment of strips and sheets [90].

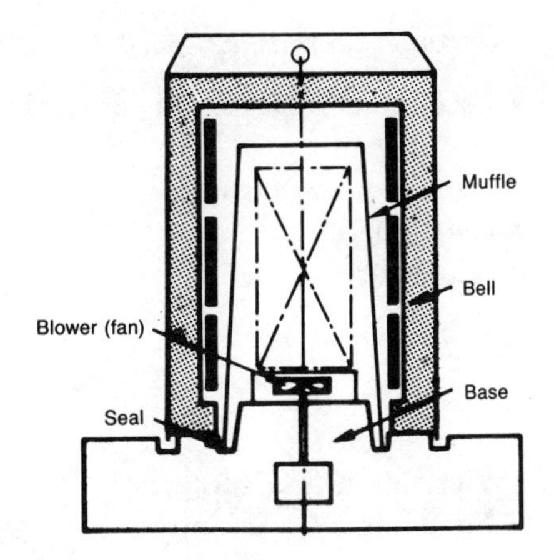


**Fig. 86.** Multibase muffled bell furnace for heat treatment of wire coils [90].

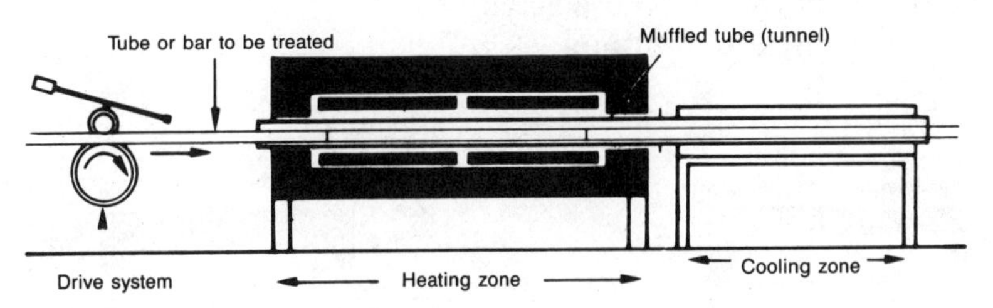


**Fig. 87.** Muffled tunnel furnace for treatment of tubes, bars, wires, etc. [90].

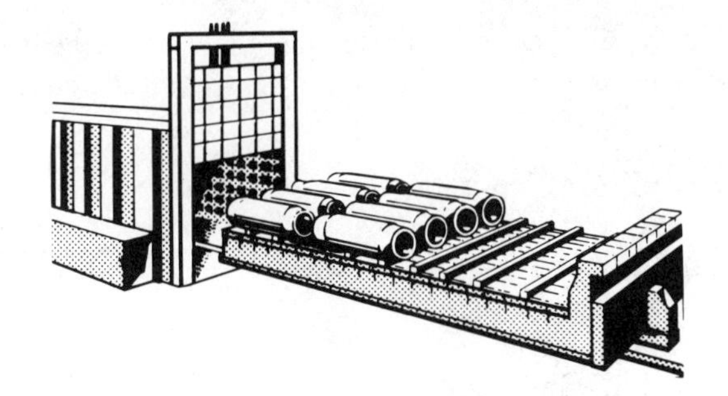

**Fig. 88.** Movable hearth furnace for treatment of rough parts [90].

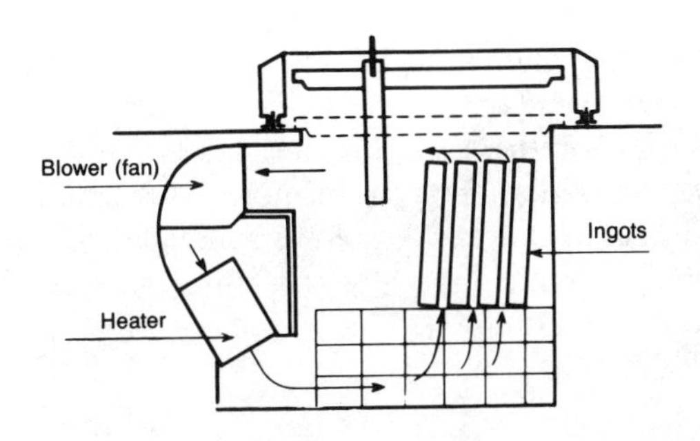


**Fig. 89.** Forced convection furnace for homogenizing aluminum ingots [90].

— *for rough (unfinished) parts* (castings, forgings, weldments):

- fixed or movable hearth furnaces
- elevating furnaces
- metallic apron furnaces
- bell furnaces
- pit furnaces

— *for mechanical parts:*

- apron furnaces
- vertical pot furnaces
- Vibrating hearth furnaces
- Continuous or discontinuous furnaces with built-in hardening tank
- forced or unforced convection chamber furnaces
- muffled or unmuffled vertical furnaces
- pit furnaces
- rotating converter furnaces.

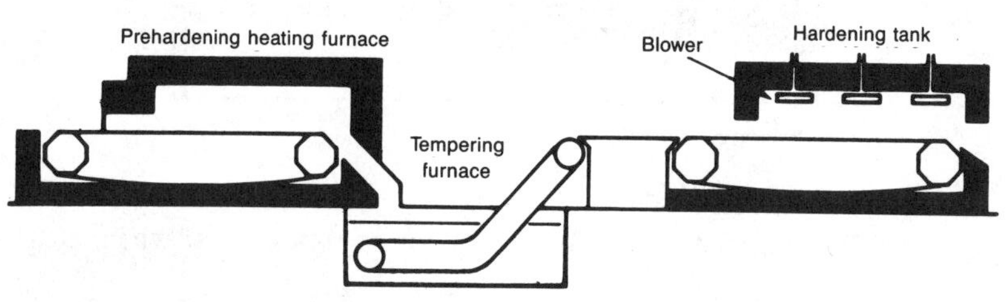


**Fig. 90.** Continuous heat treatment line for steel parts, with prehardening heating furnace and forced convection tempering furnace [90].

**Fig. 91.** Muffled pot furnace for gas state hardening, carbonitriding, gas nitriding, annealing and tempering [E].

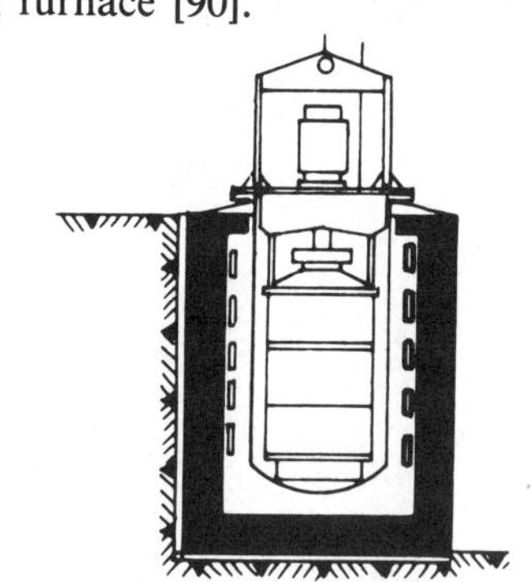

Generally, aluminum alloys are treated according to temperature level (less than 550°C) and the accuracy sought (often better than 5°C) in forced convection furnaces. For the same reasons, tempering furnaces are often of the mixed convection-radiation type. In most cases, controlled atmosphere furnaces are fitted with mixing fans or blowers intended to homogenize the atmosphere but generally operate at high temperature, with heat transfer to the charge taking place by radiation.

These furnaces are constantly being improved:

— use of fibrous ceramic refractory materials, especially for discontinuous furnaces in which rough or half-finished products are treated (*see* paragraph 5.2.);

— increased use of controlled atmospheres; the development of rugged, reliable electric radiation tubes has facilitated this change (*see* paragraph 3.5.2.7.);

— search for higher power densities, either by using non-metallic resistances or a control system (*see* paragraph 5.1.).

*7.1.2.1.2. Power and specific consumption*

The power to be installed can be calculated easily by using the method described in paragraph 2.2. When determining this power, the eventual supports for the parts (racks, etc.) must be included in the mass to be heated. A very rough but quick estimate can be obtained from the theoretical energy required to heat the load (energy stored by the products) and the temperature rise time, using an efficiency of 60 to 70% for discontinuous furnaces and 70 to 80% for

Heat stored by some metals at treatment temperature (kWh/rt)

| Temperature (°C) | Steels | | Copper | 60/40 brass | Aluminum | Nickel | Titanium |
|---|---|---|---|---|---|---|---|
| | Carbon and low alloy | High speed | | | | | |
| 200 | 25 | 22 | 23 | 23 | 58 | 20 | 35 |
| 300 | 41 | 35 | 35 | 34 | 85 | 33 | 51 |
| 400 | 56 | 48 | 46 | 45 | 113 | 47 | 68 |
| 500 | 75 | 64 | 58 | 57 | 140 | 62 | 85 |
| 600 | 96 | 80 | 71 | 70 | 167 | 77 | 102 |
| 700 | 123 | 97 | 84 | 82 | — | 93 | 120 |
| 800 | 148 | 121 | 96 | 94 | — | 111 | 138 |
| 900 | 176 | 140 | 109 | — | — | 129 | 158 |
| 1,000 | 189 | 153 | 122 | — | — | 149 | 179 |
| 1,100 | 206 | 170 | — | — | — | 169 | 200 |
| 1,200 | 224 | 187 | — | — | — | 189 | 223 |

**Fig. 92.**

continuous furnaces (generally, efficiency increases with furnace size). This rather rough method provides an initial estimate of the power to be installed; but then a much more accurate calculation is required (method described at beginning of this chapter).

As an example, Figure 93 gives the characteristics of a family of atmosphere chamber furnaces with built-in hardening tank used for hardening, carbonitriding, tempering and annealing steel products.

Range of chamber furnaces with built-in hardening tank

| Power (kW) | Heating capacity (1) (kg/h) | Maximum charge (kg) |
|---|---|---|
| 48 | 150 | 200 |
| 76 | 250 | 350 |
| 126 | 600 | 700 |

(1) Rough charge (installation included) at 850°C

**Fig. 93.**

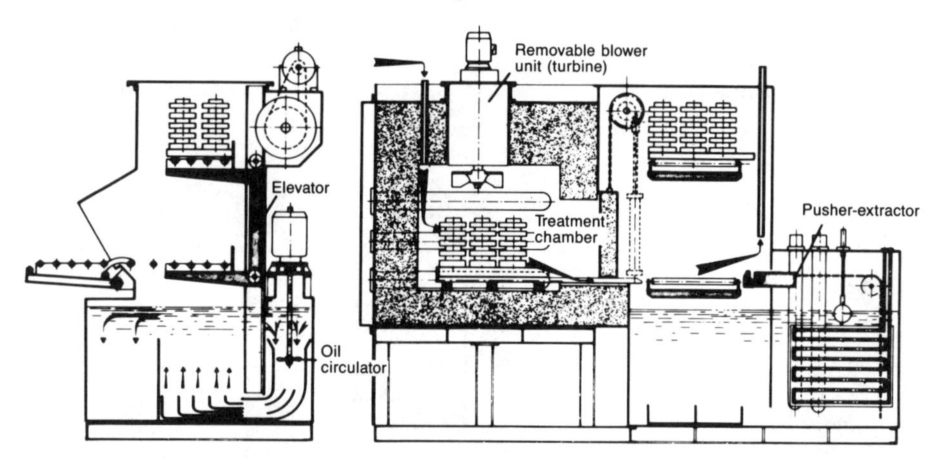


**Fig. 94.** Diagram of a radiation tube-heated furnace with built-in hardening tank intended for pre-hardening heating, hardening and carbonitriding [E].

The specific electrical consumption, compared to the mass of products treated, varies widely and depends on variables such as furnace charge, type of parts and their treatment, temperature level and cycle duration, etc. An estimate of the specific consumption is easily obtained using the thermal laws mentioned above, and cross-references can easily be made to the most commonly used values:

— gas hardening of steel parts: 400 to 1,200 kWh/t;

— carbonitriding of steel parts: 300 to 900 kWh/t;

— pre-hardening heating of steel parts: 200 to 600 kWh/t;

— tempering of steel parts: 70 to 200 kWh/t;

— annealing of steel half-finished products (tubes, bars, etc.): 180 to 300 kWh/t;

— heat treatment of aluminum alloys: 150 to 400 kWh/t;

— production of endothermal gases: 0.4 to 0.8 $kWh/m^3$.

The variation in the above figures is an accurate image of the utilization conditions of heat treatment furnaces, and only careful analysis will enable this consumption to be accurately estimated [64], [89].

Heat treatment furnaces offer the usual advantages of resistance furnaces, and in particular:
— high treatment quality due to temperature homogeneity and accuracy, to accurate control and reproducibility of results;
— high operational flexibility due to ease of control and automation, reduced monitoring and the possibility of operating over a wide range of temperatures using the same furnace;
— high energy efficiency is maintained even if the furnace charge is lowered; consumption measurements performed on numerous industrial installations show that the overall production efficiency is often 2.2 to 3 times higher than that of fossil fuel furnaces performing identical work, which in many cases, results in savings of primary energy or enables resistance furnaces to achieve equivalent performances;
— high maintainability, related to the simple design of these furnaces and widespread use of this technique.

Conversely, it is critical that requirements be clearly defined before designing the furnace, so that the heating elements are correctly selected and arranged, and to take measures to prevent mechanical and thermal shock on the elements in operation.

#### *7.1.2.2.* Vacuum furnaces

Vacuum resistance furnaces (also *see* paragraph 5.3.) are more and more widely used for heat treatment of metals. Originally, these were reserved for special metals, but are now frequently used for treatment of quality steels. They are also being substituted for salt bath furnaces for tool treatment since they prevent pollution, improve quality and generally provide energy savings. Figure 95 gives an example of the characteristics of a family of cold-wall vacuum furnaces.

Cold wall vacuum furnace range (maximum temperature 1,500°C)

| Power (kW) | Load (kg) | Cooling water (1/hr) |
|---|---|---|
| 20 | 10 | 20 to 30 |
| 55 | 80 | 50 to 90 |
| 90 | 150 | 70 to 120 |
| 150 | 300 | 100 to 250 |

**Fig. 95.**

**Fig. 96.** Vacuum furnace with outside heat exchanger [K]:
1. Outside heat exchanger
2. Mixing turbine
3. Furnace proper

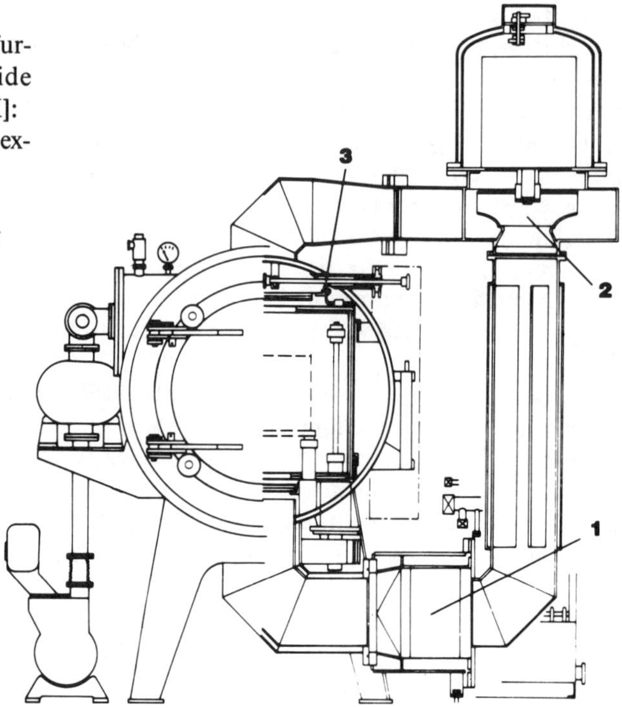

In the most recent vacuum furnaces, the heat exchanger used during cooling is now often located outside and not inside the furnace, allowing much higher surface exchanges, and therefore higher cooling speeds and productivity.

Single or double vacuum hot wall furnaces are generally used for treatment of half-finished products, while cold wall furnaces are used for mechanical parts. There is some overlap, however, and the use of continuous vacuum furnaces is gradually growing [20], [33], [60], [62*a*], [85], [D], [E], [H], [I], [K], [T], [AF], [AN], [BO], [CC].

### *7.1.2.3.* Fluidized bed furnaces

These furnaces are as yet little used. Basically, these were developed to replace salt bath furnaces which are often considered as excessively pollutive. Fluidization consists of rendering a fine sand bed loose, for example alumina, and maintaining it in suspension through a rising gas flow. Thermally, this fluidized environment can be compared to a liquid and gives a high heat transfer coefficient, enabling fast and even heating.

According to the parts treated, the fluidizing gas can be chemically active or inert. Heating is obtained by encased resistances immersed in the fluidized bath, or standard resistances located outside the muffle.

Fig. 97. Diagram of a fluidized sand bath furnace [L], [AN].

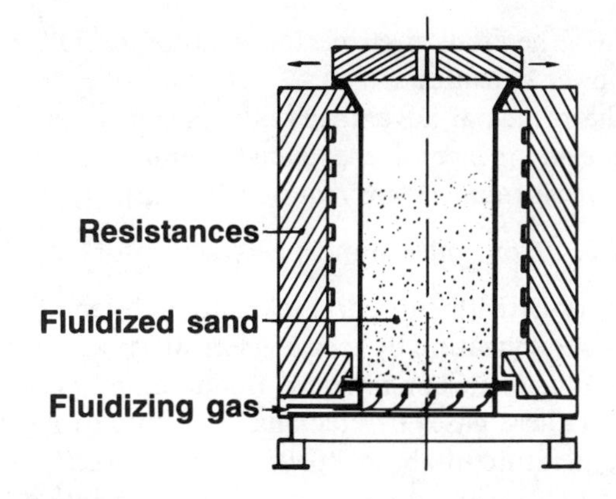


These furnaces are used for treatment of tools, hardening and tempering of mechanical parts, patenting of wires, treatment of bars, and as isothermal enclosures. The maximum operating temperatures are around 1,100°C.

Capacities vary from a few kilograms to several tons, and rates can reach 10 tons per hour.

### *7.1.2.4.* Salt bath furnaces

Load heating is provided by immersion in a mixture of melted salts. These salts can also have a chemical effect (hardening, nitriding, coloring, etc.). Two

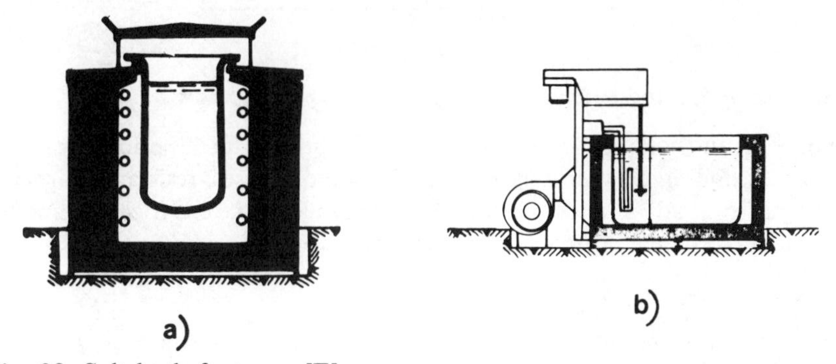


**Fig. 98.** Salt bath furnaces [E]:
*a)* radiation heated crucible
*b)* immersion heater with temperature equalization system.

types of resistance furnaces are used for this operation: externally heated crucible furnaces, similar in design to those described above for melting metals; and immersion heater furnaces, including a crucible or simple refractory lining. Other types of conduction-heated salt bath furnaces using electrodes immersed in the bath exist (*see* conduction heating).

The maximum temperature reached with immersion heater furnaces is about 500°C, and externally heated crucible furnaces do not permit temperatures over

900°C. Therefore, electrode furnaces are generally preferred for high temperatures or power densities.

The principal advantage with resistance salt bath furnaces is that they are simple and inexpensive, and are therefore widely used in installations where very high performances are required [D], [E], [I], [W].

*7.1.2.5.* Controlled atmosphere generators

In heat treatment, parts taken to high temperature are sometimes protected against oxidization by special gases; these are also used as carrier gases for case hardening or carbon nitriting. Endothermic gas generators, for example, provide one of these gases by cracking a mixture of air and gaseous hydrocarbons at a temperature of about 1,050°C. The converter containing the catalyser can be heated by electrical resistances. The main interest in this method of heating is the energy economy obtained compared to fossil fuel heating; the primary energy gain is often 60%.

Range of endothermic atmosphere generators

| Power (kW) | Rate ($m^3$/hr) |
|---|---|
| 10 | 15 |
| 15 | 30 |
| 20 | 60 |
| 25 | 80 |

**Fig. 99.**

### *7.1.3. Metal heating prior to forming*

Induction and conduction heating are used for heating metals prior to forming (*see* corresponding chapters). However, in numerous cases, resistance furnaces offer significant advantages and complete the range of the electrical techniques available in this field. In particular, these furnaces are used for:

— *Heating non-ferrous metals*

The temperature required for forming such metals is relatively low, about 500°C for light alloys, 800°C for copper-base alloys and 900°C for titanium alloys.

Metallic resistance furnaces then offer relatively low investment costs and adequate energy efficiency while providing an accurate even temperature.

The most widely used furnaces are rotating hearth, rotating plate, metal apron, inclined hearth, conveyor and chamber furnaces (*see* paragraph 2.4.). Forced convection furnaces are generally used for heating light alloys.

— *Heating of complex shaped products*

Heating through direct current passage is impossible while induction heating is difficult. Conversely, resistance heating is quite suitable for complex shape blanks.

For heating steel parts to 1,200-1,250°C, non-metallic resistances (molybdenum disilicide or silicon carbide) must be used.

Iron-chromium-aluminum resistances are, however, used in some metallic resistance high power density electrical furnaces equipped with very accurate control systems (*see* paragraph 5.1.) and should promote their use. For example, a rotating hearth atmospheric furnace, of 600 kW, equipped with molybdenum silicide resistances, is capable of producing 1.1 tons per hour of special steel parts at 1,250°C. Chamber or rotating hearth furnaces are the most widely encountered.

— *Heating in atmospheres*

With fossil fuels, atmosphere heating often requires muffled or radiation tube furnaces. Investment and operating costs are the deciding factors, and non-metallic resistance furnaces, and especially molybdenum disilicide, offer an interesting alternative in such cases. As before, rotating hearth and chamber or pusher type furnaces are the most common.

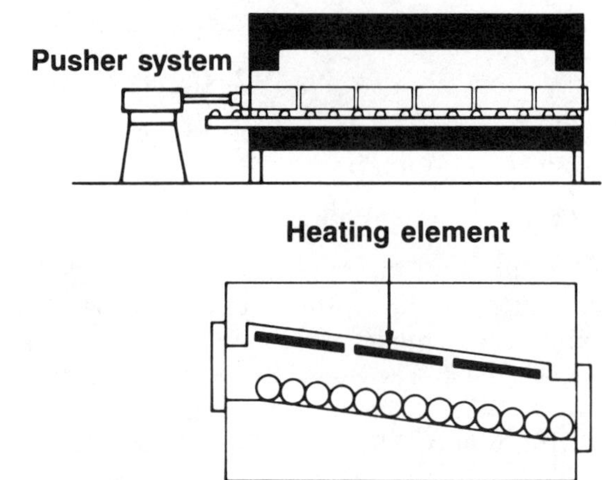


**Fig. 100.** Pusher type furnace [90].

**Fig. 101.** Inclined hearth furnace [90].

Also, high capacity Pits furnaces heated by molybdenum disilicide resistances, are used to heat special steel ingots prior to rolling in normal, inert or special atmospheres (installations of several thousand kilowatts are currently in use).

In practice, combinations of two or three of these situations — non-ferrous metals, complex-shaped products, use of atmosphere, etc. — are encountered; in such cases, resistance heating is highly profitable.

For example, for the forging of light alloy parts from blanks which are unsuitable for induction heating (e.g. disks), forced convection ovens are frequently used. Resistance furnaces are also fully adapted to heating titanium products prior to forming. Non-metallic resistance furnaces, molybdenum disilicide or silicon carbide, can also be used to heat quality steels up to temperatures of 1,300°C at a specific consumption similar to that of induction furnaces (of the order of 500 kWh/t); in addition, with resistance furnaces, oxidation losses are very low, representing an important gain with steels intended for forging or rolling (generally, this gain is between 1 and 4%) [19], [26], [66], [72].

### 7.1.4. *Galvanization furnaces*

Galvanization is a treatment applied to the surface of steels and other ferrous metals for protection against corrosion. Hot galvanization uses the reciprocal affinities of zinc and iron. After surface preparation (cleaning, stripping, fluxing in baths which can be heated by immersion heaters), the parts are immersed in a bath of molten zinc at a temperature of approximately 450°C (higher in some cases).

An intimate alloy of iron and zinc then forms in several layers on the surface providing excellent protection against corrosion.

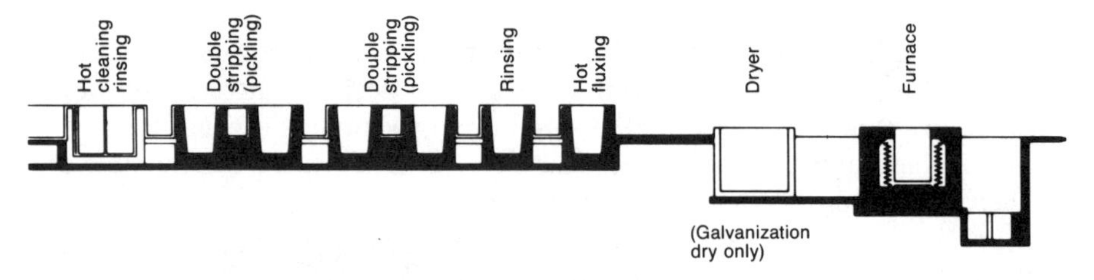


**Fig. 102.** Galvanization line [90].

Two types of resistance furnaces are used for galvanization: metal tank and ceramic tank furnaces [6], [81].

#### *7.1.4.1.* Metal tank furnaces

The furnace consists of a refractory material chamber with very good thermal insulation which keeps heat losses very low. Nickel-chromium, or more rarely, chromium-aluminum (to obtain higher power densities) resistances are used to heat the tank. The emissivity factor of nickel-chromium resistances is about 1 to 1.2 W/cm². These ribbon type elements are uniformly distributed along the vertical walls of the furnace providing several heating areas.

Each area is controlled by automatic temperature regulation, offering homogeneous and uniform heating of the bath. Application of the regulation concepts described in paragraph 5.1. should increase power density in new furnaces.

**Fig. 103.** Metal galvanization tank

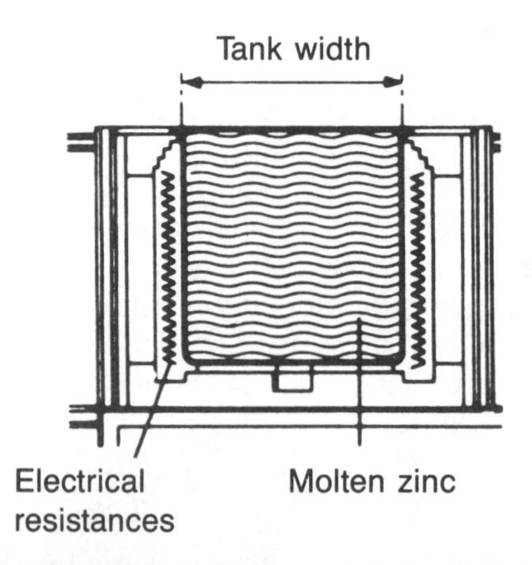

As with all ferrous products, contact of molten zinc with the tank results in the formation of iron-zinc compounds known as mattes. Formation of mattes increases with temperature and is accelerated beyond 480°C, while around 450°C the corresponding curve slope is low. Due to the excellent control obtained with resistance heating, it is easy to keep within this area and therefore increase tank service life (the service life of resistance heated tanks often exceeds five years).

### *7.1.4.2.* Ceramic tank furnaces

The tank consists of a special low porosity, high density refractory brick coating (silico-aluminous refractory), the service life of which can exceed twenty years. Bath heating is provided by nickel-chromium ribbon resistances suspended from the roof partially covering the bath; these resistances, which are independently controlled and located in several areas, radiate directly onto the bath.

The basic interest in the ceramic tank is due to high service life of the refractory lining, and thermal inertia favorable to treatment of large batches and the option of working at temperatures much higher than those currently used. Conversely, usage of extra zinc, due to its design, represents a rather heavy investment cost.

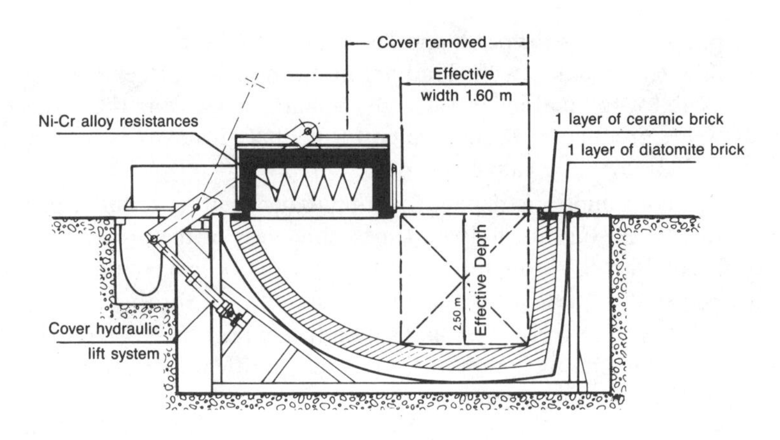


**Fig. 104.** Ceramic tank galvanizing furnace [AD].

### *7.1.4.3.* Installed Power and Specific Energy Consumptions

The data given below are intended to enable an initial estimate of installed power and, if necessary, an approximate evaluation of energy costs.

Simplified thermal balance for galvanization furnace

| Thermal balance | Hourly energy consumption |
|---|---|
| Bath surface losses (uncovered bath)(kWh/m²) | 16 |
| Bath surface losses (covered bath) (kWh/m²) | 0.8 to 1 |
| Side wall losses ([1]) (kWh/m²) | 0.7 to 1, depending on insulation quality |
| Losses through tank bottom (kWh/m²) | 1.4 |
| Load usage, quantity of energy required to heat one ton of steel to 450°C (kWh/t) | 65 |
| Added zinc ([2]), quantity of energy required to melt one ton of zinc at 450°C (kWh/t) | 82 |

([1]) Make sure to use the tank dimensions and not the overall furnace dimensions in calculations.

([2]) The quantity of added zinc represents 8 to 10% of steel weight treated.

**Fig. 105**

The power required for production is equal to the sum of the energy required to raise the temperature of the load per hour, to melt replacement zinc, and to compensate for thermal losses through the bath surface and the furnace walls. To allow for production peaks, the power actually installed is often increased by 15 to 25% over the calculated power.

Specific consumptions depend on numerous factors: ensured production, number of work stations, holding times, thickness of the protective layer, etc. Since thermal losses through the surface are high, it is recommended that the bath always be covered when not in production.

Measurements performed on tanks in industrial use have demonstrated that, under normal operating conditions, the total specific power varies between 180 and 250 kWh per ton of parts treated.

The main advantages of these resistance furnaces are high constant efficiency, temperature homogeneity, control quality and safety of use, together with reduced formation of mattes and ash.

### *7.1.5. Enamel baking furnaces*

Enamelling consists of applying one or several layers of a special mineral product to a metal surface and vitrifying them to obtain a protective, decorative covering. Enamel is a product of the glass family, consisting essentially of silica to which melting and adherence agents and dyes are added. The metals commonly treated using this process are steel, cast iron and aluminum.

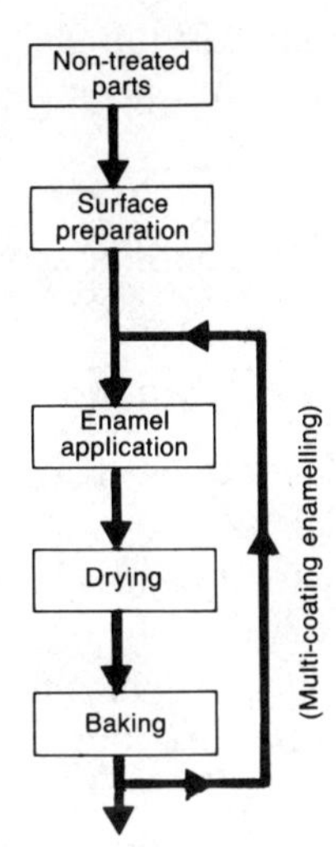


**Fig. 106.** Block diagram of an enamelling installation

After preparation of the surface (cleaning, stripping, mechanical treatments, etc., some operations can be performed in baths heated by immersion heaters), the enamel is deposited on the parts using one of several processes (dipping, spraying, powdering, electrophoresis), then generally dried in a forced convection or infrared radiation furnace; it is then baked in a resistance furnace.

Depending on the processes used, the coating consists of one or more layers applied and baked successively. The operations performed vary with the type of metals enamelled. The enamel baking temperature varies as a function of the type of metal, and is generally around 830-850°C for steel and cast iron, and 560-580°C for aluminum.

The most widely used resistance furnaces are chamber furnaces (box furnaces) and continuous furnaces. Due to the temperature level, the resistances are generally nickel-chromium alloy consisting of coiled wires on a ceramic support, or ribbons radiating directly onto the charge. Evolution of enamel baking resistance furnaces has been particularly marked by increased use of fibrous ceramic refractories and energy recovery [A], [AH], [AI].

*7.1.5.1.* Chamber furnaces (box furnaces)

These are mostly used for small and medium production runs, and have highly variable dimensions, geometry and powers. With relatively low capacity for treatment of special household articles, these furnaces can reach volumes of several cubic meters, or even tens of cubic meters for baking coatings on industrial parts. Enamelling furnaces of several thousand kilowatts also exist.

The baking time for enamelled parts is often rather low, a matter of several tens of minutes. Box furnace doors are therefore frequently opened, and to conserve high efficiency, fast charging and uncharging systems must be used. The specific consumption of enamelling chamber furnaces is generally between 0.2 and 0.5 kWh/kg of gross load for steel or cast iron products (weight of charge-supporting parts included). Power calculations can be made using the method described at the beginning of this chapter, but loading and unloading times and thermal losses due to door opening must be taken into account.

*7.1.5.2.* Continuous furnaces

The most widely used continuous furnaces are U-shaped or metal conveyor furnaces.

— *Continuous U-shaped furnaces*

This type of furnace is widely used for large industrial production runs. This furnace offers very high thermal efficiency together with constant product quality. The parts to be baked follow a U-shaped path inside the furnace which is divided into two separate parts:

• the heat transfer area, in which the baked parts give off part of their energy to parts entering the furnace; there are no heating elements in this area;

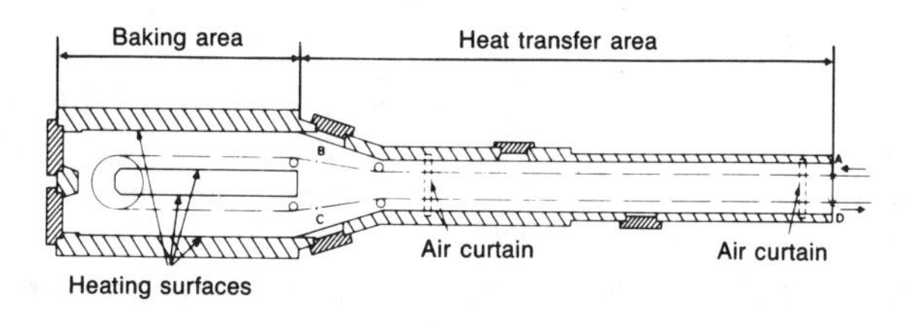


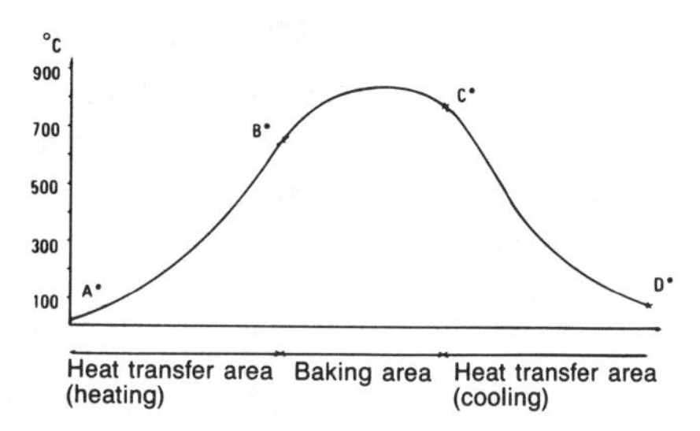


**Fig. 107.** "U"-shaped enamelling furnace with part temperature profile curve [90].

*A,B,C and D correspond to the points on the top diagram.

• the baking area is located in the U-shaped part itself, and sometimes features a median wall used to improve heating uniformity. The ribbon or coiled wire heating elements are suspended from the walls of this area.

Parts circulating through the furnace are supported by hooks attached to a chain located above the roof, which contains an opening enabling the hooks to pass. To limit heat losses without hindering operation, this opening is covered with articulated plate systems. Some installations use the heat losses through this opening to dry enamel; in such cases, the dryer is located over the furnace, and additional heating is provided by resistances. This arrangement, which saves energy, must, however, be carefully studied since it can adversely affect product quality; some noxious gases are given off during baking and can pollute products during drying causing defects. To limit thermal losses, air curtains are often located at the entry-exit of continuous furnaces, preventing convection currents which can draw hot air from the furnace. U-shaped furnaces are available in several forms: symmetrical, asymmetrical, corner and baffle.

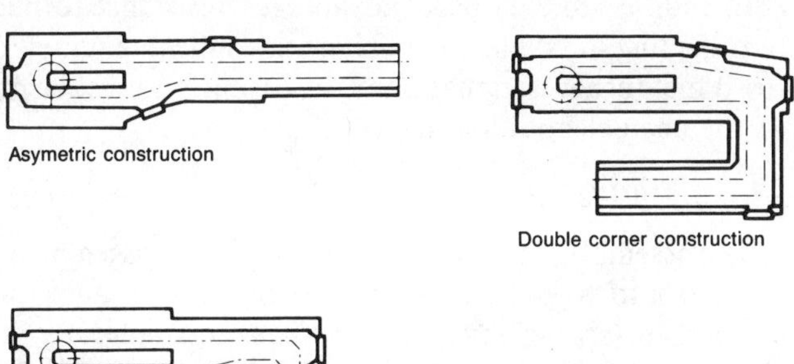


**Fig. 108.** Special construction U furnaces [90].

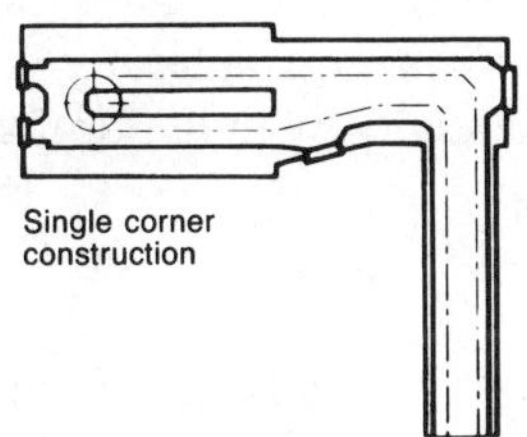


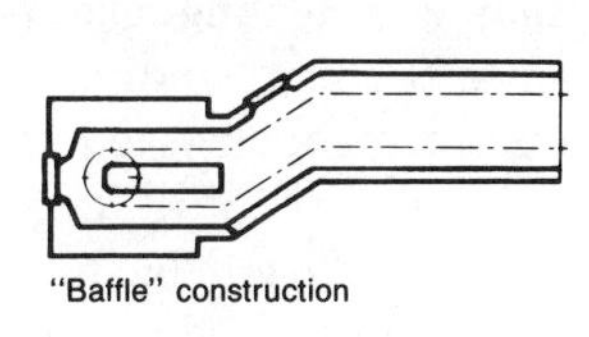


**Fig. 109.** U-shaped enamelling furnace with superimposed dryer (sectionthrough heating area) [90].

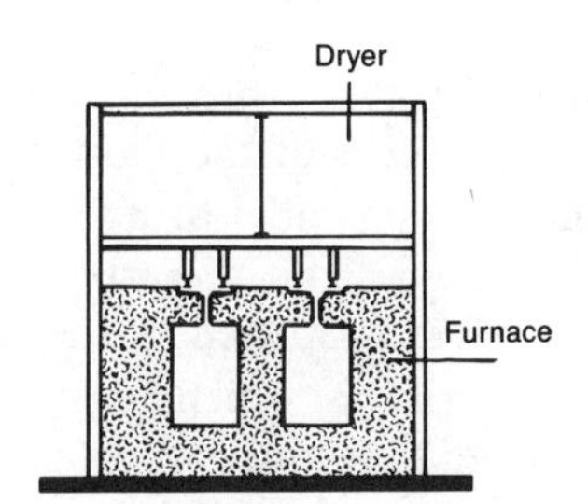


— *Continuous conveyor furnaces*

This type of furnace is widely used for baking small enamelled parts such as pans, pressure cookers and miscellaneous household articles. These furnaces often contain a preheating area where additional energy is uniquely obtained by thermal losses from the central heating area and a pulsed-air cooling area; the energy recovered in this area can be used to dry parts, bake decorations, preheat parts, or heat the building. As with U-shaped furnaces, the furnace itself can be equipped with a thermal loss-heated oven, used for drying parts or baking decorations.

*7.1.5.3. Specific consumption and advantages of resistance furnaces*

For continuous furnaces with energy recovery and fibrous ceramic refractories, the total specific consumption is generally between 200 and 300 kWh/t; without recovery, this consumption increases by 20 to 30%. For discontinuous furnaces, this consumption is generally between 400 and 600 kWh/t. Specific consumptions, however, vary widely with furnace design and operating conditions.

In addition to their usual advantages, resistance furnaces offer the advantage of providing the slightly oxidizing atmosphere required for baking enamel; with fossil fuel furnaces, radiation tubes or muffles must be used to prevent pollution of the enamel by sulphur.

### *7.1.6. Brazing*

The assembly, consisting of the parts to be assembled and the brazing metal, is heated to a given temperature in the resistance furnace in a controlled atmosphere (*see* brazing as described in paragraph 5.4. of the "Induction heating" chapter). Heating takes place essentially by radiation from the heating bodies, since the temperature is often high, up to 1,200°C. In numerous cases, the treatment quality requires complex thermal cycles which must be carefully controlled.

Generally, a reducing atmosphere such as cracked ammonia, pure hydrogen, cracked or burned hydrocarbons (exothermal gases) or fluorated mixtures is used. Vacuum brazing is also being developed. The choice of resistances depends basically on the type of atmosphere and the temperature required. The most widely used resistances are 80/20 nickel-chrome, molybdenum disilicide and silicon carbide and, in vacuum furnaces, graphite and special metal resistances [62*c*], [67], [68].

In most cases, brazing furnaces are similar to heat treatment resistance furnaces; some of the most widely used furnaces are:

— muffled chamber furnaces, mostly used for small installations in which the shape and weight of parts and production rate are irregular; the protecting atmosphere is applied to the inside of the muffle only;

— bell furnaces, which are suitable for heavy or bulky parts; the load is placed on a fixed base and covered either with a muffle and a heating bell, or directly with a heating bell; the former enables numerous bases higher than those of the heating bells to be used, and optimizes operation;

— tunnel furnaces, usually intended for production lines; these furnaces can be muffled or not, depending on the type of atmosphere and resistances (in particular, molybdenum disilicide or silicon carbide resistances enable operation directly and at high temperature in the furnace atmosphere, thus dispensing with the requirement for a muffle). A flame curtain at the furnace entry and exit, or hatch systems, prevent controlled atmosphere losses. Part handling is generally provided by a metal apron or rollers. Operation of such furnaces can be completely automated;

— vacuum furnaces, used for very high quality products; in spite of higher investment costs, these are competing more and more with the usual resistance furnaces;

— salt bath furnaces, mostly used for brazing aluminum parts; this process is not fully developed, and the furnaces used are generally electrode furnaces and rarely resistance furnaces.

Resistance brazing furnace applications are extremely diverse and are found in most of the metal transformation industries. Because of this diversification, it is difficult to provide specific energy consumption: the consumptions measured vary highly (from 400 to 2,000 kWh/t).

### *7.1.7. Sintering*

In the mechanical, electrical and electronic industries, two clearly different types of sintered parts are used: the first produced from metal powders and the second from ceramic powders. However, the furnaces used are of very similar design and are only discussed below [47], [51], [71], [87].

Generally, sintered parts are produced more or less identically. Powders of the various materials are combined to form a clearly defined mixture which also contains lubricants and binders. First of all, this mixture is compressed in presses to form relatively fragile parts which, however, can be handled; these parts are then baked in controlled atmosphere furnaces, providing them with mechanical strength. Sintering consists of agglomerating solid adjacent particles, or, in some cases, agglomerating particles through a short liquid phase. After sintering, parts normally have the shape, dimensions and structure desired, but may undergo complementary mechanical operations to improve dimensional accuracy or to alter shape (machining, hot or cold deformation, etc.) and, especially for metal parts, special heat treatments intended to modify their mechanical or physical characteristics (hardening, tempering, annealing, carbonitriding, controlled oxidation, etc.).

The atmospheres used vary widely. With ceramic powders, it is often possible to operate in a normal air atmosphere, although some products require special atmospheres (for example, highly oxidizing for sintering of ferrites). Generally, with metal powders, a reducing or neutral atmosphere must be used (endothermic gas, hydrogen, etc.). The choice of resistances is directly related to the type of atmosphere and the maximum temperatures. Temperature levels are generally high, between 1,100 and 1,700°C, and the most currently used resistances are iron-chromium-aluminum, silicon carbide, molybdenum disilicide or special metal (molybdenum, etc.) resistances. The furnaces can be muffled or not, muffles being made from metal or ceramic (alumina) parts, depending on temperature and atmosphere. Flame curtains and hatch systems prevent controlled atmosphere leakage. Some of the most widely used furnaces are:

— chamber furnaces, for low production runs;

— metal conveyor or pusher furnaces; Figure 110 shows an example of a furnace used for sintering special steels. A furnace of this type can produce 50 to 200 kg of sintered parts per hour.

The resistances are made from iron-chromium-aluminum wires wound on sillimanite tubes. Furnace consumption is about 0.8-0.9 kWh/kg of sintered parts.

For very high temperature furnaces, the resistances used in the sintering area are often non-metallic, and metallic in the pre-sintering area;

— vacuum furnaces, which are gaining wide acceptance for very high quality products.

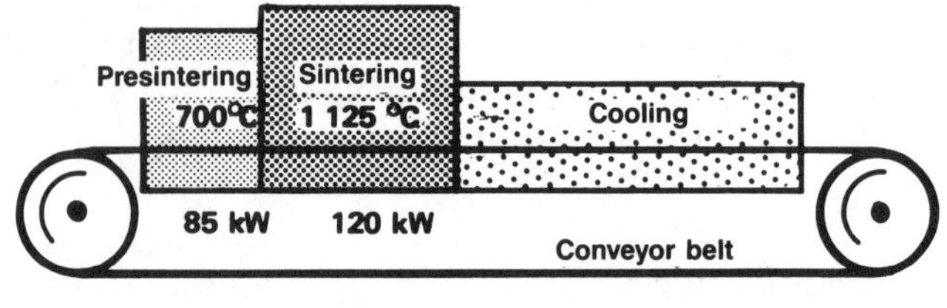


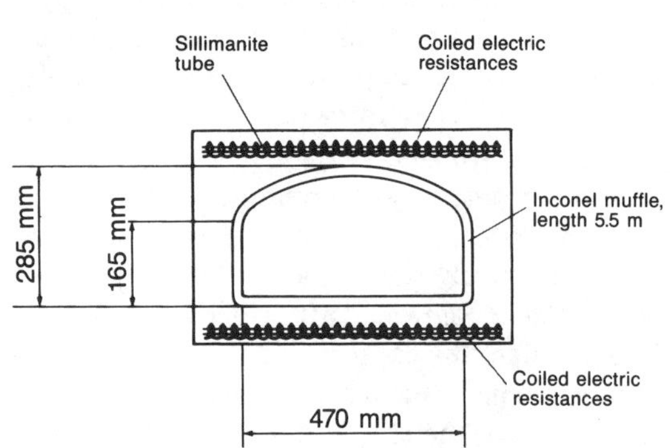


**Fig. 110.** Diagram of a conveyor type sintering furnace [90], [D].

Generally, specific consumptions vary widely according to the type of parts and the temperature level, usually between 500 and 1,500 kWh/t of sintered parts.

For example, these furnaces are used in the electrical and electronic industries for production of capacitors, resistors, ferrites, wire guides, insulators and numerous other sintered ceramic products. Many mechanical parts can also be produced by sintering.

In this field, the main advantage of resistance furnaces are temperature accuracy, high energy efficiency at high temperatures and very limited maintenance requirements.

### *7.1.8. Drying and firing organic coatings*

Resistance furnaces and ovens are widely used for drying and firing paint and varnish coatings. While discontinuous resistance ovens are widely used for limited production installations, the low cost of fossil fuels and relatively high efficiency within this temperature range (150 to 220°C approximately) have limited the use of electric ovens for very high capacity installations. Numerous medium size production lines have, however, been constructed. When product form permits, infrared radiation ovens sometimes enable higher energy efficiency and productivity (*see* infrared heating chapter).

As an example, figure 111 shows a painting line in which all thermal operations are performed by resistance heating; this installation produces approximately 650 kg of small parts over a period of eight hours (1,750 kg of untreated parts). Heating of fluids (aqueous solutions) used in the part spraying tunnel is provided by immersion heaters. The paint firing oven-tunnel is a forced convection oven, heated by batteries of resistances. The parts are placed in swing trays coupled to a conveyor. The mean power demand during production is only two-thirds of the installed power, which is common with this type of installation. It is very difficult to provide specific consumptions since they are extremely variable for this type of application, and each case must be analysed separately; however, specific consumption is high in most cases, and all possible energy recovery methods must be used.

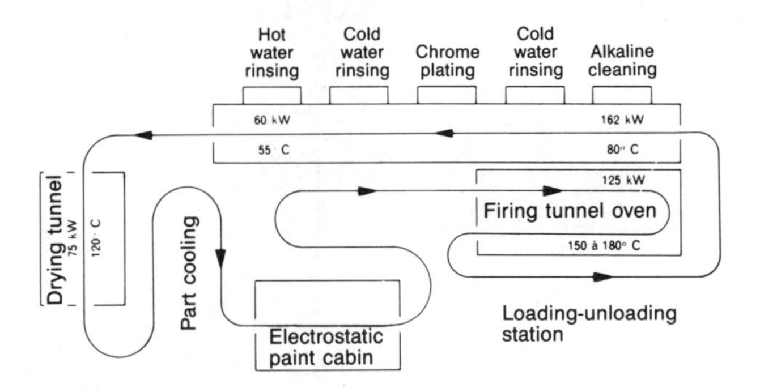


**Fig. 111.** Diagram of painting line for small mechanical parts [90].

**Fig. 112.** Monorail oven (current handling system in paint firing furnaces and ovens) [90].

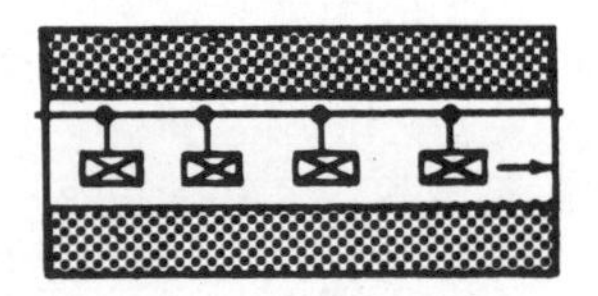

In electrical manufacturing, electric ovens are also used for drying windings, baking and preheating impregnations, preheating prior to plasticizing electrical insulation, polymerization of araldite, etc. (*see* figures accompanying paragraphs 2.4.12 and 7.5.1).

### *7.1.9. Manufacturing semiconductors and electronic components*

Because of the temperature accuracy obtained, resistance heating is widely used in the manufacture of semiconductors and products for the electronics industry (single crystal drawing ovens, sintering ovens, etc.) [34], [45], [51], [91].

For example, there are numerous methods of drawing single crystal (Czochralski, Bridgman, floating zone, etc.) in which resistance heating competes with induction heating (*see* "Induction heating" chapter, paragraph 5.7.).

Metallic or non-metallic single crystal growth ovens are generally built into more complete systems comprising:

— a supporting frame, including the devices required for opening the door and the water and gas circuits;

— crystal and crucible drive mechanisms (drawing heads);

— electronic control and spindle movement variation and regulation systems for each drawing head;

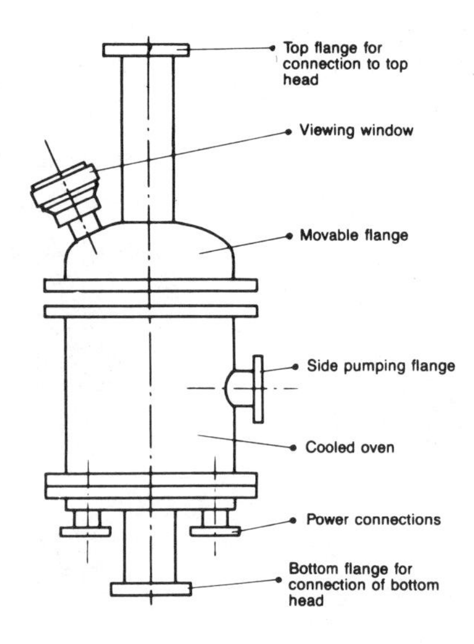


**Fig. 113.** Diagram of an oven for the growth of single-crystals [AN].

— the oven itself;
— the oven power units and control system;
— with numerous high pressure machines, a TV camera.

The furnaces used are vacuum (down to $10^{-6}$ torr), normal pressure or high pressure (up to 100 bars). Special atmospheres are often maintained in the furnace, and at the drawing head connection, sealing must be carefully maintained. Operating temperatures range from 500 to 2,800°C. Structurally, these furnaces are similar to the cold wall vacuum furnaces described above. In general, the graphite, tungsten or molybdenum heating elements are located in the heating chamber and surround the crucible which is supported by refractory steel, molybdenum or tungsten parts, depending on temperature. In many cases, the metallic heating elements consist of a grid providing excellent heat distribution. A series of tungsten, molybdenum or graphite heat shields are inserted between the heating elements and the water-cooled walls of the furnace. Due to the temperature accuracy required, which is generally better than $\pm 0.5$°C, these furnaces are fitted with P.I.D. electronic regulation systems, with thyristor power supplies and automated control systems in which microprocessors are becoming more and more widely used [34], [AN], [CC].

## 7.2. The ceramic industry

### *7.2.1. Manufacturing ceramic products*

In the ceramics industry, resistance furnaces have been used for some fifty years, and are constantly being improved. Ceramic (from the Greek "Keramos": pottery) is the art and industry of fired-clay. The field covers a wide variety of products (bricks, roofing tiles, floor tiles, vases, sanitary items, household articles and industrial products, etc.) produced using similar techniques:
— preparation of a more or less plastic paste;

— shaping this paste as required;
— firing worked objects at high temperature to provide special properties;
— where necessary, protection and decoration of objects using special coatings.

Generally, a ceramic product is produced from a mixture of mineral materials each fulfilling a different function; numerous classifications have been proposed. A detailed analysis of the various types of ceramics is outside the scope of this work. For reference, the following classification system can be used:

— *Porous paste ceramics*

These are water permeable pastes, not very compact, with an earthy consistency.

• *ordinary earthenware (terra cotta)*

These are simple colored paste ceramics (ferriferous clay, resulting in the name "red products" often given to this family) without glazing (a glaze is a transparent vitreous covering applied to ceramic products) sometimes enamelled or varnished; the main products are bricks, roofing tiles, flooring tiles, pottery and other items such as statuettes, etc.; after firing, products are often waterproof.

• *refractory terra cotta*

These are ceramic products withstanding high temperatures and consist of alumino-silicate aluminous and magnesian composites.

• *eartenwares*

Earthenwares are terra cottas which are glazed to combat porosity; earthenwares can be enamelled using an opaque glazing. A wide range of earthenware products, ranging from the ordinary to very high value, exists. Earthenware is used for roofing tiles, kitchenware, sanitary items and decorative objects.

— *Compact paste ceramics*

The paste is compact, waterproof and has a fine fracture.

• *stoneware*

These are opaque, very hard vitrified fracture pastes obtained from ordinary clays. Stoneware can be covered with other products (vitreous salt clay varnishes or lead-alkaline "coverings"). Stoneware is used to produce kitchen pottery, pipes, paving tiles, artwork ceramics, etc.

• *porcelains*

Porcelains are obtained from a white paste (using very pure clays such as kaolin) and vitrified to translucency. Basically, porcelains are used for tableware and artwork.

The chemical composition of some pastes is given for reference in figure 114:

Approximate chemical composition of some ceramic pastes

| Fine earthenware (%) | | Stone ware (%) | | Porcelain (%) | |
|---|---|---|---|---|---|
| $SiO_2$ | 69.5 | $SiO_2$ | 76 | $SiO_2$ | 60 |
| $Al_2O_3$, $Fe_2O_3$ | 29 | $Al_2O_3$, $Fe_2O_3$ | 20 | $Al_2O_3$ | 32 |
| $K_2O$, $Na_2O$ | 1.5 | $CaO$, $MgO$, $Na_2O$, $K_2O$ | 4 | $CaO$, $MgO$, $Na_2O$, $K_2O$ | 8 |

**Fig. 114.**

In addition to these products, special ceramics of highly variable chemical composition exist (ferrites and zirconium-based products) which are basically used in industry. These materials, which often offer high quality characteristics, are at present being widely developed. Some special products are produced by sintering ceramic powders (*see* paragraph 7.1.7.).

Table 115 provides the main applications of ceramics, according to product type.

Ceramic product applications.

| Product application | | Ceramic family |
|---|---|---|
| Construction materials | Building or facing bricks | Terra cotta |
| | Roofing tiles | Terra cotta |
| | Vitrified bricks | Stoneware |
| | Paving tiles | Terra cotta, stoneware |
| | Architectural products | Terra cotta, earthenware, stoneware |
| Domestic items | Tableware | Feldspathic earthenware, porcelain |
| | Cooking pottery | Varnished terra cotta, stoneware, porcelain |
| | Toilette items | Feldspathic earthenware, stoneware, porcelain |
| | Sanitary articles | Feldspathic earthenware, stoneware, porcelain |
| Industrial ceramics | Refractory products (bricks, etc.) | Refractory terra cotta (1) |
| | Thermal insulation products (bricks) | Refractory terra cotta (2) |
| | Electrical insulating products | Stoneware, porcelain, steatite (3) |
| | Abrasives (grounding stones, etc.) | Refractory terra cotta (4) |
| | Miscellaneous parts (recipients) | Stoneware |
| | Special products | |
| Artwork ceramics | All objects | Earthenware, porcelain, stoneware, terra cotta |

(1) Clays, bauxite, silica, sillimanite, mullite, silicon carbide, chromite, magnesia, dolomite, zirconia.
(2) Diatomite and products listed in (1)
(3) Magnesian silicate, prepared from steatite or one of its derivatives.
(4) Corundum or silicon carbide with vitreous binders.

**Fig. 115.**

The main thermal operations performed during production of standard ceramic products are drying, firing the paste and glazings, covering and decoration. Figure 116 provides the firing temperature for the main ceramic products.

Treatment temperatures for some ceramic product families

| Type of ceramic | Temperature °C |
|---|---|
| Tiles and ordinary bricks .... | 800-1,100 |
| Refractory bricks: | |
| silica .................... | 1,500 |
| alumina silicate........... | 1,250 minimum |
| magnesia ................ | 1,700 |
| Porous potteries: | |
| first firing ............... | 950-1,000 |
| glazing .................. | 950-1,000 |
| Common earthenware: | |
| first firing ............... | 900-1,000 |
| glazing .................. | 900-1,000 |
| Fine earthenware: | |
| bisque.................... | 1,200-1,280 |
| glazing .................. | 1,000-1,150 |
| Stoneware.................. | 1,250-1,320 |
| Hard porcelain ............. | 1,400 |
| Dental porcelain ............ | 1,300-1,400 |
| Soft porcelain .............. | 1,200-1,250 |
| Artwork ceramic............ | 900-1,300 |
| Special products ............ | up to 2,000 |

**Fig. 116.**

Firing pastes (bisques), coating (glazing, enamel, covering, varnishes) and decoration can be performed either simultaneously or successively, depending on the product. During heat treatment, products must be subjected to stringent thermal cycles during temperature rise, holding and cooling. The temperature rise is often divided into several phases where rate of increase varies (slow firing at start of cycle, then fast or full), and involves several levels enabling structural modification of the products. The type of atmosphere is also very important. For earthenwares, terra cotta and vitreous porcelain, firing must be performed completely in an oxidizing atmosphere. Conversely, for porcelain the atmosphere must vary according to the firing stage: initially oxidizing and then reducing. The total duration of firing cycles for ceramic products is generally very long, between 10 and 25 hours in most cases, but the heating time itself is often lower, 6 to 12 hours.

### *7.2.2. Resistance furnaces in the ceramic industry*

The main types of resistance furnaces used in the ceramic industry are chamber, movable hearth, bell, tunnel and wagon furnaces. Infrared radiation furnaces are sometimes used for continuous drying, but drying of wet products using energy recovered from fired products is becoming more and more widely used [5], [9], [12], [45], [51], [54], [D], [G], [Y], [AE], [AJ], [AK].

The evolution of resistance furnaces in this field has been marked by the development of fibrous ceramic refractories, an increase in accessible temperature levels in electrical furnaces and overall energy recovery.

#### *7.2.2.1. Discontinuous furnaces*

Discontinuous furnaces used in ceramic product firing consist of chamber, movable hearth and bell furnaces. These are intended for limited or highly diverse production runs, used in the production of tableware and pottery, artwork, sanitary products and technical parts (in corundum, steatite, silicon carbide and zirconia, etc.).

##### *7.2.2.1.1. Chamber furnaces*

These furnaces are used in artwork and industry. Their main advantage is operational flexibility. Generally, the resistances consist of wires wound on ceramic supports and placed on the walls, doors and sometimes in the furnace hearth, but other arrangements may be used. Depending on the temperatures to be reached, resistances are nickel chromium (up to 1,050-1,100°C) or iron-chromium-aluminum (up to 1,300°C approximately). For temperatures over 1,100-1,200°C, molybdenum disilicide or silicon carbide resistances are also widely used. It is possible to fire all ceramic products in resistance furnaces.

In many cases, furnaces are heated during off-peak hours which results in very low energy costs, especially with low thermal inertia furnaces using ceramic fiber refractories, for which the whole temperature cycle takes place within the off-peak rate period. During product cooling, hot air extracted from the furnace can be recovered for drying and preheating new charges or for local heating, considerably improving thermal efficiency. Figure 117 gives an example of the characteristics of a range of insulating refractory brick chamber furnaces with a maximum operating temperature of 1,300°C (iron-chrome-aluminum resistances). The weights loaded and specific consumption per kilogram for the whole load (ceramic products and supports) determine a well-filled furnace; consumption compared to net load is 10 to 25% greater depending on the type of supports, and increases in inverse proportion to the furnace filling factor. The total specific consumption is generally between 0.6 and 1.2 kWh/kg of parts, and depends on numerous factors (furnace filling, temperature level, thermal cycle duration, type of refractories used, etc.).

Range of chamber furnaces for firing ceramics
(specific consumptions assume a charge temperature of 1,200°C approximately)

| Power (kW) | Chamber dimensions | | | | Weight of charge (kg) | Electrical consumption | |
|---|---|---|---|---|---|---|---|
| | Volume ($m^3$) | Width (mm) | Depth (mm) | Height (mm) | | For one firing (kWh) | Per kilogram loaded (kWh/kg) |
| 81 | 1 | 800 | 1,000 | 1,260 | 500- 600 | 350- 480 | 0.65-0.75 |
| 99 | 1.5 | 900 | 1,200 | 1,390 | 650- 850 | 400- 650 | 0.6 -0.7 |
| 126 | 2.0 | 1,000 | 1,400 | 1,430 | 1,000-1,200 | 650- 850 | 0.6 -0.7 |
| 168 | 3.0 | 1,100 | 1,820 | 1,500 | 1,500-1,800 | 850-1,200 | 0.55-0.65 |
| 220 | 4.4 | 1,250 | 2,300 | 1,530 | 2,200-2,650 | 1,100-1,600 | 0.5 -0.6 |

**Fig. 117.**

The volume of chamber furnaces is limited to approximately 10 cubic meters. Products are heated primarily by radiation from the resistance onto the products and from the hottest products onto cooler products. As volume increases, the products in the center of the load become more and more difficult to heat. Low thermal conductivity of ceramic materials accentuates this phenomenon, therefore imposing an upper limit on furnace capacity.

*7.2.2.1.2. Movable hearth furnaces*

In design, these furnaces are similar to chamber furnaces but loading and unloading are facilitated by the movable hearth. The use of two hearths and a cooling unit also considerably increases productivity when charges are not disturbed by thermal shock. In such cases, it is possible to unload the charge undergoing controlled cooling from the first hearth, and to transfer this to the cooling unit, then to load the second hearth with a cold charge with the furnace still hot. This system actually reduces specific electricity consumption. However, the interest in movable hearth furnaces has decreased due to the development of low thermal ineertia furnaces and energy recovery during cooling.

*7.2.2.1.3. Bell furnaces*

Although this type of furnace facilitates charge handling, and by using several bases can increase the rate of usage and therefore productivity while reducing energy consumption, these furnaces are not used very often. System design is similar to that used in chamber furnaces. The use of ceramic fibres considerably lightens the bell rendering it easy to handle, and therefore reducing the lifting system power requirement.

*7.2.2.2. Continuous furnaces*

Wagon tunnel furnaces (*see* figure 11) and tunnel furnaces are the most widely used continuous furnaces in the ceramics industry. Rotary hearth furnaces are sometimes used, but their development has been very limited.

*7.2.2.2.1. Parallel passage furnaces*

These furnaces have been specially designed for series firing of small parts. They are used for firing of earthenware of tiles (bisque and enamel) or stoneware, but may also be used to produce other products (steatite technical parts, small household items, etc.).

**Fig. 118.** Parallel passage furnace [90].

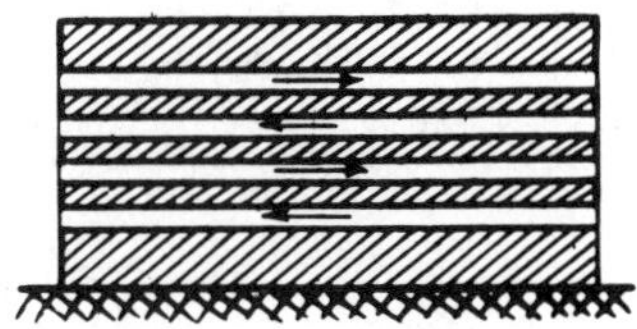

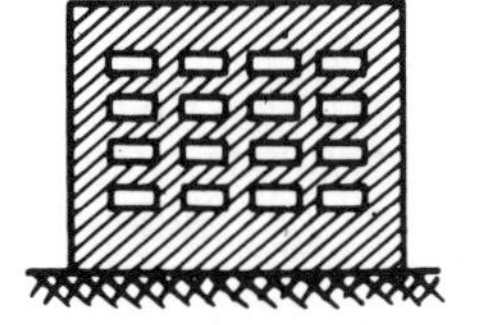

A parallel passage furnace consists of refractory masonry containing small cross-section tunnels or passages in which the products to be treated circulate. Only the central area of the furnace is heated by resistances mounted perpendicular to the passages and accessible via the sides of the furnace. The parts circulate in opposite directions in two adjacent tunnels, enabling the energy given off in the tunnel cooling air to be used to preheat products advancing in the opposite direction in the other passage. Product movement is provided by automatic loading pushers. Generally, there are between 8 and 48 passages enabling a production rate up to 70,000 tiles per day. Due to their design, these furnaces offer excellent thermal efficiency but their range of application is rather limited. For example, a 24-passage furnace with an installed power of 255 kW enables production of 50,000 10 x 10 mm stoneware tiles per day; higher capacity furnaces can be constructed.

*7.2.2.2.2. Tunnel furnaces with trucks*

*— Principle*

Truck-type tunnel furnaces are intended for mass production of a wide range of ceramic products. The trucks or wagons on which the objects to be treated are placed, pass through a uniform cross-section tunnel at constant speed. During their passage through the furnace, the objects are submitted to a predetermined temperature and, where necessary, atmosphere cycle. Only the center part is equipped with heating elements since most of these furnaces are equipped with sophisticated energy recovery systems. The resistances used in the preheating area are sometimes different from those in the heating area; in general, resistances are nickel-chromium in the preheating area, and iron-chromium-aluminum or molybdenum disilicide or silicon carbide in the final, high temperature heating area.

— *Energy recovery*

Numerous types of tunnel furnace enabling exhaustive energy recovery exist.

• *double tunnel furnaces*

The trucks circulate in opposite directions in two adjacent tunnels; in between, partitions form the high temperature area. In the remaining areas of the furnace,

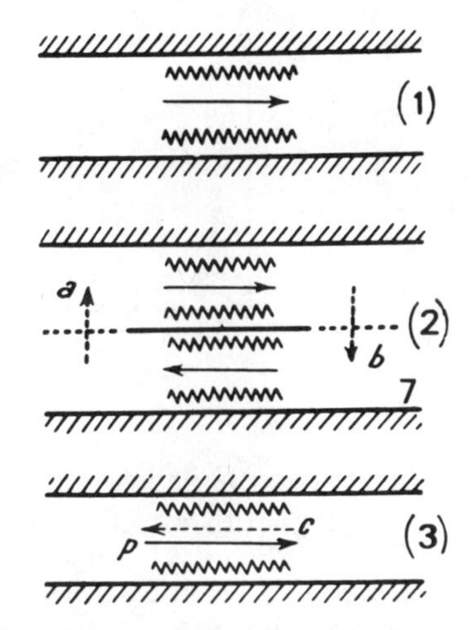


**Fig. 119.** Energy recovery principle [90]:
1. tunnel without energy recovery;
2. double tunnel with transverse heat recovery;
3. single tunnel with air counterflow longitudinal recovery.

the trucks face each other directly and heat is transferred from one tunnel to the other transversely. Therefore, both tunnels have identical but opposite temperature curves which are symmetrical with respect to an axis passing through the center of the furnace and perpendicular to the direction of truck movement. This type of furnace can be used only for firing similar products, and the production rates for both tunnels must be more or less balanced. However, it is possible to work at different temperatures with different products, but the thermal cycles are different.

• *air counterflow furnaces*

An air flow, in the opposite direction to that of the trucks, is provided by fans; heat recovery is then longitudinal. The air flow can be limited to prevent movement of dust. This arrangement dispenses with the requirement for a double tunnel, and therefore reduces investment costs (also, it is possible to combine both systems).

• *furnaces equipped with various recovery systems*

Heat recovery systems are continually being developed, and various systems can be used concomittantly. For example, numerous tunnel furnaces are equipped with forced convection systems, the hot air being recovered for various thermal applications.

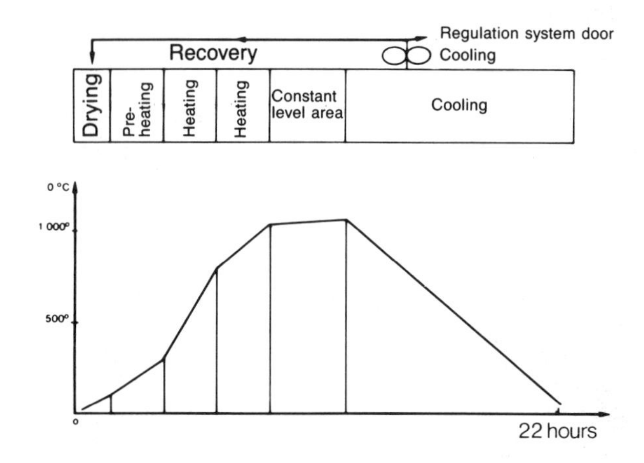


**Fig. 120.** Diagram of an earthenware bisque-firing furnace (equally suitable for enamel firing) with forced convection recovery [90], [AK].

This hot air is then used for drying, the dryer being built inside or outside the furnace, and also for preheating ceramic products. Moreover, and if the preheating area is well insulated and sufficiently long, the thermal losses in the central area of the furnace are used to heat incoming cold products. During the first firing phase, any organic materials contained in the products are burnt off in the form of gases and vapors; these are extracted for the furnace, and it is possible to place an exchanger (for example, a plate-type exchanger) to recover energy from these products; the recovered heat can be used for heating premises or product drying to limit thermal losses during charging and withdrawal operations.

Forced-convection energy-recovery single-tunnel furnaces are at present the most widely used since they offer high efficiency, while relatively simple in design, and therefore low investment costs. Also, in modern furnaces, efforts have been made to reduce the thermal capacity of the trucks by decreasing their weight and using low specific heat insulating refractory materials.

For example, a 600 kW tunnel furnace is effectively 600 mm wide, 820 mm high and 50 meters long. This furnace is equipped with heat recovery systems: a forced convection system takes up the hot air in the cooling area, enables preheating of incoming products and feeds a dryer separate from the furnace, while a plate-type exchanger in the organic material combustion gas extraction system partially heats the shops in winter. Used for firing cooking pottery (simultaneous firing of bisque, enamel and decorations at 980°C), this furnace produces eight tons of ceramic products per day, with a mean power demand of 300 kW and therefore a mean specific consumption of 0.9 kWh/kg (at the nominal production rate of 15 tons per day, this consumption is much lower). In the preheating area, the resistances used are nickel-chromium alloy, and iron-chromium-aluminum in the heating area. As designed, this furnace can reach a maximum temperature of approximately 1,250°C with a daily production rate of more than 15 tons providing high adaptability to market conditions (possibility of producing terra cottas, stoneware or earthenware, daily production rate variation, etc.).

### *7.2.2.3. Specific consumption and applications*

Specific consumption is closely related to the recovery system, the temperature level, the thermal cycle and the furnace charge. As a function of these parameters, the specific consumption is generally between 0.8 and 1.3 kWh/kg of product (net load) for bisque firing (earthenware), 0.7 and 1.3 kWh/kg for enamel and pottery or tableware products, 0.3 and 0.7 kWh/kg for earthenware tile enamel (correct use of furnace space), 0.2 to 0.3 kWh/kg for firing decorations on porcelain, 0.9 to 1.5 kWh/kg for porcelain and 0.2 to 0.3 kWh/kg for firing ceramic tiles in passage furnaces. These specific consumptions are of course cumulative when production of a product involves several runs through the furnace.

Truck furnaces are suitable for high production rates, and are employed in the manufacture of a wide range of products such as tableware and earthenware pottery, earthenware and stoneware sanitary items, tiles and floor tiles (very high production rates, much higher than those of passage furnaces), steatite products, electrotechnical porcelain, etc. In furnaces using a partial reducing atmosphere, gas has often replaced electricity for heating the furnaces due to the ease with which an atmosphere of this type is obtained (for example, high quality hard porcelain fired at 1,400°C which between 1,000 and 1,300°C requires a reducing atmosphere). Electrical furnaces can, however, operate in controlled reducing atmospheres (ready-to-use atmosphere or gas generators). Primarily economic considerations govern energy choices in this field.

Electric furnaces offer numerous advantages in the ceramics industry: excellent temperature control and accuracy, high energy efficiency, automatic operation, frequent recourse to off-peak rates, very wide temperature ranges (including special ceramics fired at very high temperatures), easy control and operation, production quality, etc.

## 7.3. The glass industry

Electrical fusion of glass is ensured in furnaces in which the current flows directly through the molten glass bath (*see* conduction furnace chapter); however, resistance indirect heating is sometimes used in special cases such as lead glass fusion in crucible furnaces or holding in forehearths and feeder channels in mechanical glass-making machines, in forebays in glassworks in which the glass is blown, and in the production of float glass. Resistance furnaces are also used for heating blown glass blanks between two blowing phases (Glory Holes), heat treatment of glass, firing decorations on glass and many other operations (bonding, composite material production, metallization operations, etc.) [25], [27], [45], [52], [CC].

### *7.3.1. Crucible fusion furnaces*

Crucible fusion furnaces (also known as pot furnaces) are intended for fusion of lead crystal where low production rates are involved. Furnaces of this type are similar to the crucible furnaces used for melting metals; however, they are generally placed in a completely closed refractory chamber which can contain one or two crucibles. Due to the temperature level, the resistances used are molybdenum disilicide (in this case suspended from the refractory roof, and located around the crucible and along the vertical walls of the chamber) or silicon carbide.

Furnaces of this type are often used to fuse and refine all the daily production during low rate hours, thus ensuring very low energy costs. The use of an entirely closed electrical furnace also avoids the release of smoke and dust containing lead which can be very noxious, thereby considerably reducing pollution and improving working conditions. Also, the quality of the glass and service life of crucibles are greatly increased. For example, crucible furnaces having a capacity of 550 kg of lead crystal using molybdenum disilicide resistances with a power of 75 kW distributed over 15 identical power elements (the maximum installable power with a crucible of this capacity is about 150 kW) enable melting of the complete charge during off-peak hours. Specific consumption for fusion and refining is between 0.7 and 1 kWh/kg of crystal; total consumption depends on the holding time. The mean service life of heating elements is between three and five years [BH].

### *7.3.2. Heating of drawing heads, feed lines and pre-furnaces*

As above, molybdenum disilicide or silicon carbide resistances are used. However, safety precautions must be taken with high fluorine content glass (more than 0.8-1%). The main interest in resistance heating is due to its high efficiency, three or four times higher than with fossil fuel heating, but also due to better working conditions especially where glass products are manually produced. In addition, conduction heating can be used profitably for this application.

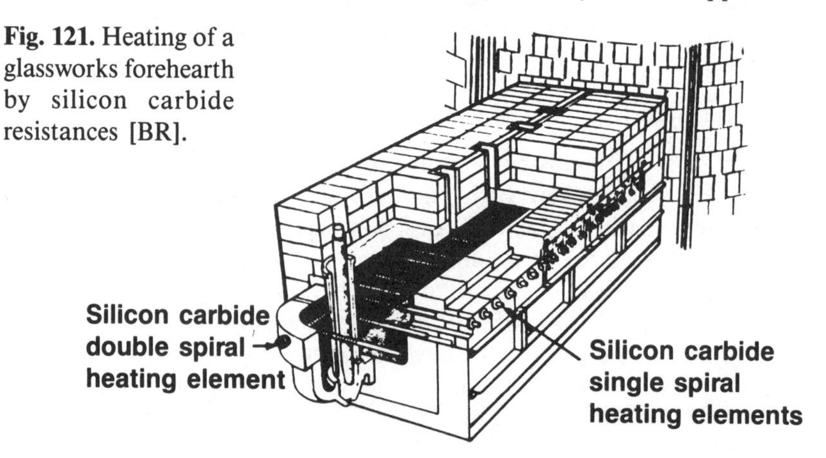


**Fig. 121.** Heating of a glassworks forehearth by silicon carbide resistances [BR].

### *7.3.3 Heat treatment of glass*

Heat treating of glass alters the properties obtained during its production. The most common treatments change mechanical characteristics but, with the development of special glasses, heat treatments are being used more to alter a set of mechanical, optical and thermal properties simultaneously. For example, air hardening, which consists of sudden cooling by blowing cold air onto both surfaces of a glass product previously taken to a temperature of 650°C, creates controlled stresses in the glass which enhance most of its qualities:

— mechanical, since for equal thickness, a hardened product is five or six times more resistant than an annealed product;

— thermal, since hardened glass resists thermal shock up to 300°C, while annealed glass breaks at a temperature of 40°C;

— safety, since in the event of breakage, hardened products break into small pieces which cannot cause deep wounds.

According to the type of glass and the temperature levels, heat transfer takes place simultaneously by forced convection and radiation, convection being the most widely used.

Numerous types of furnaces can be used for heat treatment of glassware and, under some conditions, an atmosphere is indispensable. The furnaces most widely used are of the chamber, conveyor or monorail type (with the latter type, products such as large windows are suspended vertically and continuously treated). Basically, specific consumption depends on temperature, the duration and type of treatment and the type of furnace, leading to a wide range of values.

### *7.3.4. Firing decorations*

Numerous glass products are decorated, in particular deluxe bottles and decorative objects; generally, the decoration process consists of enamelling or coating using special transfers. Most of the decorations require firing or drying at temperatures between 100 and 800°C. Resistance, chamber, conveyor, truck and monorail furnaces are widely used due to their simplicity, adequate energy efficiency and the quality these provide. The power for such furnaces, which are very similar to some ceramic firing or metal treatment furnaces, and their specific consumptions can easily be calculated using the method described at the beginning of this chapter.

## 7.4. Food industries

Resistance ovens are widely used in food industries. The most widespread applications are baking bread, pastry, biscuits, patés, pork products, cooked dishes and also toasting rolls, heating pots and boilers, drying salted products, fish smoking, etc. These operations are performed at low temperature, less than 500°C, and in most cases between 100 and 300°C. Heat transfer in ovens is either entirely by forced convection or by a mixed convection-radiation system. The radiation heat transfer method is dominant only in special cases such as grilling or browning, or heating certain pots [13], [16], [90].

### *7.4.1. Bread baking ovens*

Depending on the production rates required, bread is baked in discontinuous or continuous ovens (temperature approximately 250°C). Resistance ovens satisfy both amateur and industrial baking requirements. Ovens similar to those used in baking are also used to cook numerous other products: patés, pizzas, pastries, cooked dishes, etc. [AR], [AS], [AU], [AV], [AW].

#### *7.4.1.1. Discontinuous ovens*

Two types of ovens are used in this field: chamber ovens and truck ovens.

— *Chamber ovens (or fixed hearth ovens)*

Generally, these ovens consist of several independently heated, superimposed baking chambers. Two types of heating elements are used (*see* detailed description in "Enclosed electric resistance heating" chapter):

• stainless steel enclosed resistances, which are embedded in the hearth and either exposed (protected from mechanical shock by a metal grating) or encased;

• open thick wire resistances, fitted in appropriately shaped brick or steatite insulating parts (barrel mounted resistances) and systematically embedded in the hearth and roof.

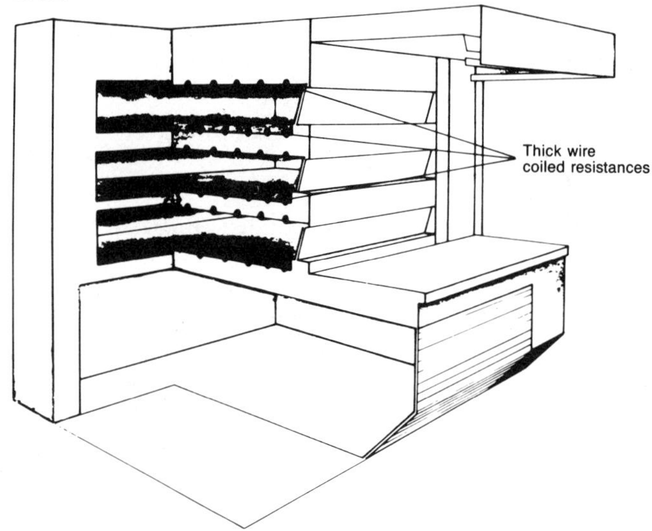


**Fig. 122.** Diagram of a chamber oven using coiled thick wire resistances in steatite supports [90].

Heat transfer comprises natural convection and radiation. In the past, accumulation ovens were generally used but have been more or less abandoned, since their very heavy structure called for heavy investment costs, relatively high energy consumption and lack of operational flexibility. In contrast, modern well-insulated directly heated ovens offer greater production flexibility, lower investment costs and power consumption (temperature build-up in such ovens and even part of the production can be made during off-peak rate hours; therefore, electrical energy prices remain highly competitive).

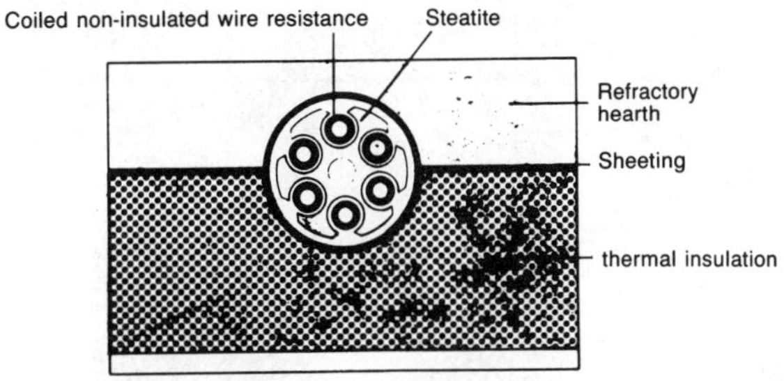

Coiled non-insulated wire resistance
Steatite
Refractory hearth
Sheeting
thermal insulation

**Fig. 123.** Cross-section of hearth showing resistances [90].

Protective grating
Encased resistances

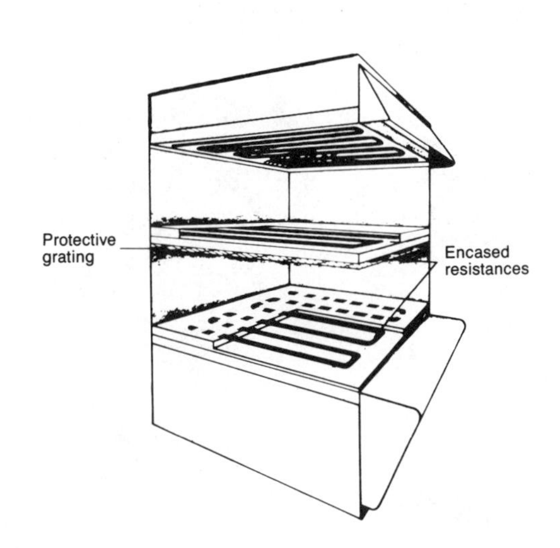

**Fig. 124.** Diagram of an encased waterproof resistance chamber oven [90].

Each baking chamber is equipped with an enclosed resistance-heated water vapor production unit. Loading and withdrawal are generally performed mechanically which improves working conditions. For example, a high capacity oven has four chambers representing a baking surface of 22 square meters and a power of 106 kW (including water vapor production), distributed over 144 encased resistances of 0.7 kW each (the remaining power is used for water vapor production); the temperature build-up time, which can be programmed automatically, is approximately 45 minutes. Baking time is 25 to 35 minutes. Figure 125 gives the characteristics of several chamber ovens.

The main advantage with this type of oven is that each chamber operates individually, each having its own regulation system, enabling different products to be cooked simultaneously. These can be used not only to bake bread, but other products such as patés, pizzas, quiches, etc.

Range of chamber ovens for bread baking

| Power (kW) | Baking surface ($m^2$) | Number of shelves | Floor area ($m^2$) |
|---|---|---|---|
| 34 | 7.50 | 2 | 7.1 |
| 50 | 11.25 | 3 | 7.1 |
| 67 | 15.00 | 4 | 7.1 |
| 82 | 16.75 | 3 | 10.1 |
| 109 | 22.50 | 4 | 10.1 |

**Fig. 125.**

Generally, specific consumption of electricity is around 0.55 to 0.70 kWh per quintal of flour converted into bread, according to operating conditions (0.5 to 0.6 kWh/kg of cooked bread).

— *Truck ovens*

The products to be baked are placed on a metal truck and introduced into a forced convection oven, generally heated by encased resistances.

The oven hearth can be fixed or rotary, the latter arrangement improving baking homogeneity. The water vapor required for baking is produced by an encased resistance-heated water vapor generator.

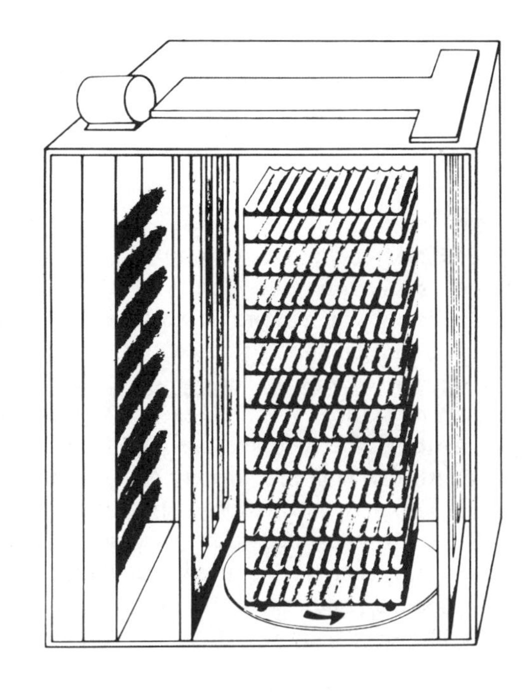

**Fig. 126.** Diagram of a rotary hearth trolley oven [90].

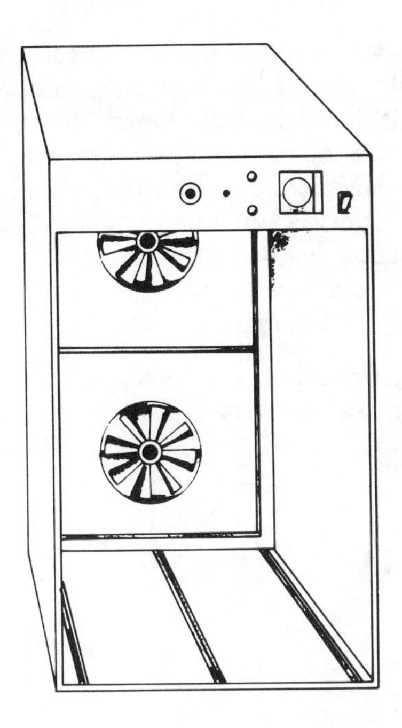

**Fig. 127.** Diagram of a fixed hearth trolley oven [90].

The primary interest in this type of oven is due to its low floor area and the ability to cook numerous products. Excellent temperature homogeneity and speed of heat transfer at low temperature provided by forced convection enables cooking of pastries, patés and meat products. Where high production rates are involved, it is possible to form a battery of several trolley ovens, increasing operational flexibility. For example, a 30 kW fixed hearth trolley oven (additional power of 3 kW for production of water vapor and ventilation) occupies a floor area of 1.75 square meters (oven height 2.5 m). The trolley has 14 shelves or levels capable of accepting 30 to 35 kg of baked bread, and the baking area is 7 square meters, enabling production of 65 to 70 kg of baked bread per hour (55 to 60 kg of flour converted into bread per hour); baking time at 250°C is approximately 25 minutes, while the oven warm-up time is 30 minutes. A high capacity 160 kW oven occupies a floor area of 8.4 square meters and can produce 400 kg of baked bread per hour.

Specific consumptions are similar to those obtained with chamber ovens, i.e., between 0.5 and 0.6 kWh/kg of baked bread.

*7.4.1.2. Continuous ovens*

These ovens are intended for high production rates of more than 300 kg per hour of baked bread. The products to be baked are placed on a metal conveyor belt, passing through a tunnel 10 to 30 m long. The conveyor speed can be varied according to the type of products being baked.

Heating is provided by coiled wire elements housed in steatite supports (barrel resistances) or encased resistances, located on either side of the conveyor. In the hearth, these encased resistances consist of tubular elements embedded in the refractory material or simply protected by a steel plate and, in roofs, consist of tubular resistances either exposed or enclosed in the refractory material, or of visible finned elements. Ovens of this type can also be fitted with atmosphere mixing fans or turbines. The temperature inside the tunnel is automatically regulated by thermostats, and auxiliary resistances produce water vapor. In the past, these ovens were often similar to fossil fuel ovens and of relatively heavy construction, but now they are very light in structure using only metal and light insulating materials, resulting in reduced specific energy consumption and high flexibility (*see* paragraph 7.4.2.).

Swing-tray ovens are sometimes used for bread baking; however, development of these has been very limited.

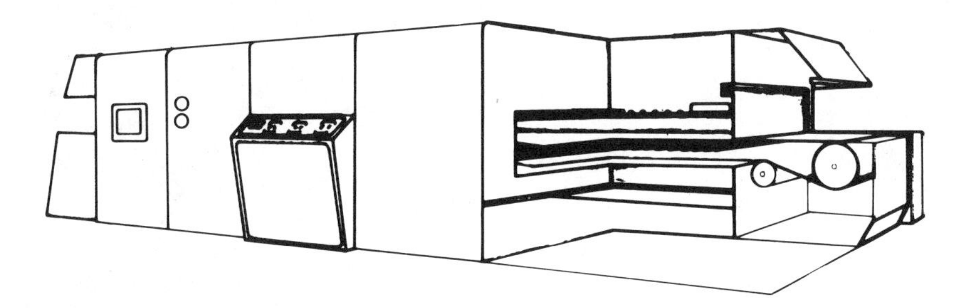

**Fig. 128.** Diagram of a continuous oven [90].

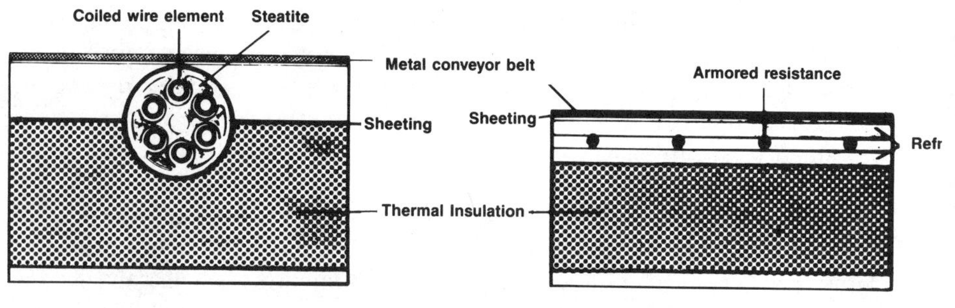


**Fig. 129.** Cross-section of a steatite-supported coiled resistance continuous oven [90].

**Fig. 130.** Cross-section of an encased resistance continuous oven hearth [90].

For example, Figure 131 provides the characteristics of a range of conveyor type tunnel ovens.

An oven of 290 kW produces 800 to 900 kg of baked bread per hour maximum (700 to 800 kg converted flour per hour). Specific energy consumption in continuous ovens is generally between 0.35 and 0.50 kWh/kg of baked bread (0.4 to 0.6 kWh/kg of converted flour).

Range of conveyor type bread baking continuous ovens (outside width 3 m).

| Power (kW) | Baking area ($m^2$) | Overall dimensions | |
|---|---|---|---|
| | | Length (m) | Height (m) |
| 106 | 15 | 10.30 | 1.40 |
| 126 | 20 | 12.80 | 1.40 |
| 146 | 25 | 15.30 | 1.40 |
| 166 | 30 | 17.80 | 1.40 |
| 128 | 7.85 | 7.85 | 1.95 |
| 168 | 10.52 | 10.52 | 1.95 |
| 208 | 12.66 | 12.66 | 1.95 |
| 248 | 15.33 | 15.33 | 1.95 |
| 288 | 17.46 | 17.46 | 1.95 |

**Fig. 131.**

However, specific consumption is highly sensitive to oven load and may increase substantially if the load differs greatly from that rated load. The main advantage of this type of oven is highly automated mass production with low energy consumption. However, high investment costs demand a high utilization factor.

### *7.4.1.3. Advantages of electric ovens in bakeries*

The main interest in resistance ovens in the baking industry is due to their space-saving quality, adequate energy efficiency, simple design, controllability, very low servicing requirements, operational flexibility (possibility of cooking numerous other products), automatic starting, improved working conditions, etc.

### *7.4.2. **Cooking ovens for preparation of meals and prepared foods***

The food industries are marketing more and more diversified products which require cooking. For amateur or semi-industrial products, cooking often takes place in forced convection electric ovens. The rapid industrialization of these products has, however, led to a more widespread use of mass production techniques. Continuous resistance ovens are the answer to this requirement. In addition to bread, which has already been described, some of the products produced in this manner are patés, roasts, pastries, pizzas, cooked dishes, braising and frying of meats and glazing of canned patés before closing.

Resistance heated conveyor ovens are highly suitable for these products. Forced convection systems further increase flexibility and a wide range of uses. Most modern ovens consist of:

— an all stainless steel casing;

— full thermal insulation of walls with mineral wool insulation;

— stainless steel conveyor belt (the conveyor belt texture varies as a function of the products to be conveyed) or a chain system for tray conveyance;

— juice recovery systems where such equipment is necessary;

— stainless steel encased resistances (tubes, finned resistances, etc.) arranged either in batteries (all forced convection heating) or over the load (mixed forced convection-radiation heating);

— ventilation devices;

— a control system, now commonly electronic.

Cooking time is generally varied by conveyor speed. An additional infrared radiation heating device can be used to enhance surface browing.

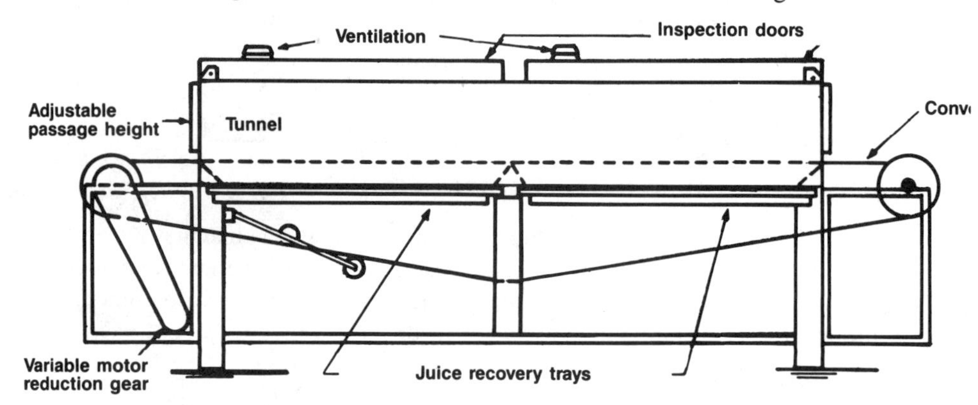


**Fig. 132.** Continuous tunnel oven for cooking dishes and prepared products (AS).

For example, Figure 133 gives the production capacity for a forced convection continuous oven.

Hourly production rate of a 36 kW forced convection oven

| Products | Temperature (°C) | Production rate (hourly) |
|---|---|---|
| Meat pies | 180-200 | 75 kg/h |
| Tarts | 250 | 400/600 per hour |
| Croissants | 220 | 600 per hour |
| Puff pastry | 180 | 600 per hour |
| Sausage rolls | 200 | 150 per hour |

**Fig. 133**

Numerous light-structured ovens of this type are modular, enabling adaptation to production capacity and keeping investment costs low. Figure 134 gives a range of ovens of this type (the oven described in figure 132 is at the "bottom of the range") with an adjustable passage height of 35 to 280 mm.

Range of forced convection continuous ovens for production of cooked dishes and prepared products (maximum temperature 400°C)

| Number of modules | 1 | 2 | 3 | 4 |
|---|---|---|---|---|
| Power (kW) | 36 | 72 | 108 | 144 |
| Total width (mm) | 1,300 | 1,300 | 1,300 | 1,300 |
| Width under conveyor (mm) | 650 | 650 | 650 | 650 |
| Total length (mm) | 3,600 | 6,250 | 8,900 | 11,550 |
| Length in tunnel (mm) | 2,600 | 5,250 | 7,900 | 10,550 |

**Fig. 134.**

These very simply designed ovens are the economic answer to many food industry requirements. When the oven fill factor is around nominal, efficiency is high. Specific consumptions vary widely with products but in most cases are between 0.4 and 0.6 kWh/kg of product treated. When calculating the power to be installed or the specific consumptions using the method given at the beginning of this chapter, remember to allow for latent heat of vaporized water eliminated during most treatments (drying, baking, etc.).

### *7.4.3. Other applications of resistance ovens in the food industry*

These applications are highly diversified and include chamber ovens, roasters, pots and boilers (these can also be heated by electrically produced steam or hot water), pancake machines (heating plates with built-in encased resistances; for example a 30 kW machine can produce 1,300 pancakes per hour), dryers and ovens for production of hams, sausages and similar products, biscuit baking ovens (where necessary with dielectric final heating), roll baking and grilling ovens, heating cylinders (drying of various products), snack and appetizer product ovens [16], [90], [AS], [BB].

## 7.5. Other industrial applications: laboratories

The metalworking, ceramic or glass product and food industries are the main users of resistance ovens and furnaces; equipment of this type is more and more frequently encountered in all industrial sectors and research laboratories where they are being used for everything from biological research to the study of materials from other planets. These furnaces are also being used as anti-pollution incinerators [57], [77], [83].

In particular, resistance ovens are used in numerous drying, baking and polymerization operations, or for applying films or coatings to a wide range of supports — textiles, paper, wood, plastics, etc. — or drying the products themselves.

Some examples of these applications are the plastic, wood and chemical/pharmaceutical product industries.

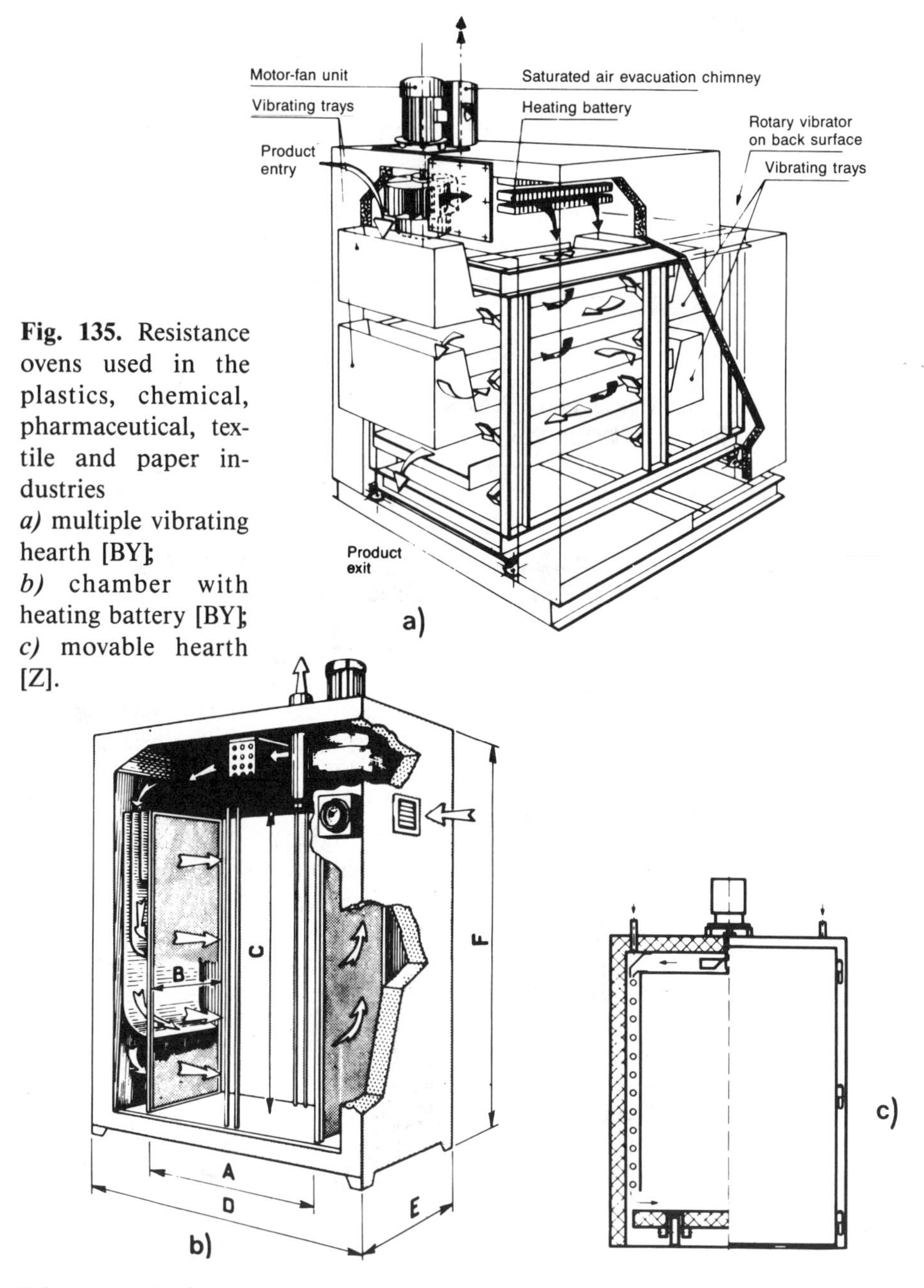


**Fig. 135.** Resistance ovens used in the plastics, chemical, pharmaceutical, textile and paper industries
*a)* multiple vibrating hearth [BY];
*b)* chamber with heating battery [BY];
*c)* movable hearth [Z].

### *7.5.1. Plastic forming industry*

Some products (powders, pellets, blocks, plates and sheets, etc.) are heated prior to conversion in forced convection ovens. Heating hoppers (used to dry and preheat products such as powders) directly mounted on injection moulding machines also exist. Discontinuous or continuous forced convection ovens are also used for polymerization and stabilization of a wide range of products. Generally, the temperature is between 120 and 250°C, but may be higher for special products.

Other resistance heated equipment is used for heat forming (heating trays), and welding and bending plastics (hot air generators), pre-heating blocks prior to shaping and firing decorations (hot air ovens). Numerous continuous or discontinuous forced convection ovens are also used for heating heat-shrinkable plastic films for packing a wide range of products before handling and shipment.

Range of resistance ovens

| Power (kW) | Heating chamber dimensions | | |
|---|---|---|---|
| | Width (mm) | Depth (mm) | Height (mm) |
| 6 | 400 | 600 | 600 |
| 8 | 600 | 600 | 800 |
| 12 | 700 | 700 | 1,000 |
| 17 | 800 | 900 | 1,000 |
| 22 | 800 | 1,000 | 1,200 |
| 26 | 800 | 1,000 | 1,500 |
| 30 | 1,000 | 1,000 | 1,500 |
| 34 | 1,000 | 1,200 | 1,500 |
| 40 | 1,000 | 1,200 | 1,800 |
| 46 | 1,200 | 1,200 | 1,800 |
| 56 | 1,200 | 1,500 | 1,800 |
| 74 | 1,200 | 1,800 | 2,000 |

**Fig. 136.**

These ovens are used in a wide range of industries — food stuffs, construction materials, metal parts, textile or paper articles, etc.

### *7.5.2. Chemical, pharmaceutical and biological industries*

Electric ovens are widely used in the chemical, pharmaceutical and biological industries for preparation of powders and pellets, drying medecinal plants, various sterilization operations, creation of microbe cultures, vacuum heating, isothermal holding, etc. [AX], [AY], [AZ], [BF], [BY].

### *7.5.3. Wood industry*

Mechanical ventilation chamber dryers are widely used in the wood industry; basically, they consist of (refer to Figure 137):

— one or two drying chambers fitted with carefully arranged baffles (1) ensuring correct air flow and homogeneous drying;

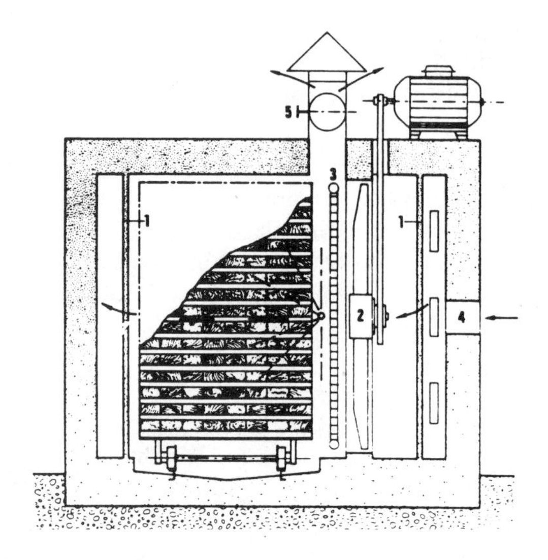


**Fig. 137.** Diagram of a chamber type resistance wood dryer [90].

— one or more large diameter helical fans (2) located inside the dryers and blowing air over the entire height of stacks, perpendicular to the wood grain;

— a water vapor or water spraying device which is optional and fulfills specific requirements: wetting before drying of closed pore dried woods, surface wetting of wood batches showing faults as a result of fast drying; a very damp atmosphere at the start of drying enables the dryer to be heated up rapidly;

— electric resistances (3). The large heating surface is distributed along the center of the dryer, over the available length and height;

— fresh air intake openings (4) and saturated air exhausts (5), which are generally manually controlled, vary the humidity;

— air temperature and humidity control systems.

In most cases, drying takes place during off-peak hours, the oven being simply ventilated outside of these periods. The power of the heating batteries, which generally consist of finned encased resistances, is about 4.5 $kW/m^3$ (ventilation power is about 0.2 to 0.3 $kW/m^3$). Wood dried in this manner must follow an accurate temperature and humidity cycle, and drying times vary greatly with the types of wood, and initial and final humidities. Specific energy consumption is highly variable and generally between 1 and 2 kWh/kg of extracted water, and between 120 and 300 $kWh/m^3$ of dried wood.

The main advantages of such dryers are low investment costs, limited energy consumption and low energy cost (use of off-peak hours), as well as operational simplicity. Over recent years, heat pump dryers have been developed and, for wood drying, are now competing with resistance dryers [90].

In addition to wood drying, resistance furnaces and ovens are used for drying and firing varnishes and coatings on wood products.

### 7.5.4. *Hot air and gas generators*

Numerous industries (chemical, plastics, foodstuffs, electrical construction, etc.) and laboratories require hot air in their manufacturing or testing processes. Numerous electric hot air dryers are available to satisfy a wide range of requirements both in terms of power, ranging from a few Watts to several thousand kilowatts, and in terms of temperature, ranging up to 1,000°C [4], [8], [56], [80], [L], [BE], [BN], [BZ], [BQ].

For temperatures less than 1,000°C, the heating elements consist of batteries of metallic resistances (for temperatures below 700°C often encased resistances, finned elements or others); conversely, for high temperatures, heating elements are non-metallic.

Silicon carbide elements, for example, enable heating of gases to more than 1,200°C. Other molybdenum wire resistance-equipped heaters enable heating of hydrogen to more than 800°C, with a range of 1 to 100.

These heaters consist of fixed or portable installations. Portable installations are used in welding, folding and shaping plastic sheets, shrinkage of plastic sheathing in the electrical and electronic industries, and shrinkage of plastic packaging, etc.

### 7.5.5. *Heating cylinders*

It is very easy to produce electric heating cylinders. The resistances are placed inside and radiate onto the cylinder. Electronic regulation provides adequate temperature accuracy. These cylinders are used for a very wide range of applications, including humidity profile rectifications on paper machines and drying and baking of thin film products in the food industry [BC], [BD].

## 8. BIBLIOGRAPHY

[1] W. TRINKS, *Les fours industriels*, Dunod, 1957.
[2] V. PASCHKIS, *Les fours électriques industriels*, Interscience Publisher New York, 1960, édition française, Dunod.
[3] W. H. MAC ADAMS, *Transmission de la chaleur*, Dunod, 1961.
[4] G. LEACH, *Developments in applying electro-heat to the large scale heating of air and process gases*, 5e Congrès international d'Électrothermie, Wiesbaden, 1963.
[5] W. BRUGGER et W. TAESLER, Le four tunnel électrique dans l'industrie céramique, *Revue BBC*, n° 10-11, 1964.
[6] G. STEINER, Bain de galvanisation à chauffage électrique pour pièces en fer, *Revue BBC*, n° 10-11, 1964.
[7] F. LAUSTER, *Manuel d'électrothermie industrielle*, Dunod, 1968.
[8] E. ASTORI, Emploi d'un réchauffeur d'air dans le domaine des recherches aérospatiales, *Revue BBC*, n° 3, mars 1968.
[9] R. STEINER, Une nouvelle usine de porcelaine d'Afrique du Sud équipée de fours tunnels, *Revue BBC*, n° 3, mars 1968.
[10] W. ROOTS, *Fundamentals of temperature control*, Academic Press, 1969.
[11] W. D. GILMOUR, *L'appareillage électronique industriel, équipements de commande et de régulation*, Eyrolles, 1969.
[12] R. GAUTHERET, *Les fours électriques à résistances*, SODEL, édition 1970.
[13] R. GAUTHERET, *Les résistances électriques de chauffage*, SODEL, édition 1970.
[14] W. HEILIGENSTAEDT, *Thermique appliquée aux fours industriels*, tomes 1 et 2, Dunod, 1971.

[15] G. TARDY, Fours de traitements thermiques, Éléments de calcul et de construction, chapitre V : Pertes thermiques, *Traitement thermique*, n° 57, 1971.
[16] ELECTRICAL TIMES, *Electroheat techniques for food processing*, n° 4, juillet 1972.
[17] P. RONGIER, *Programmateur régulateur de température pour fours à résistances*, B.I.S.T.-CEA, n° 172, juillet 1972.
[18] N. M. MUCHIN, *Neue Fragen bei der warmetechnischen Berechnung von Widerstandöfen*, Union internationale d'Électrothermie, Congrès international, Varsovie, 18-22 septembre 1972.
[19] R. S. PITT, *Reheating furnaces for steel slabs*, *Iron and Steel*, p. 542-546, octobre 1972.
[20] H. WESTEREN et A. BALAGUIER, Le développement de la cémentation à haute température, *Traitement thermique*, n° 72, 1973.
[21] H. GERBEAUX, Traitements thermiques en construction soudée, *Soudage et Techniques Connexes*, n° 10, vol. 27, octobre 1973.
[22] L. BROUTEAU, *Fours à chauffage électrique pour l'homogénéisation de billettes et le revenu de tôles d'alliages légers*, Comité français d'Électrothermie, 2 avril 1974.
[23] G. HERTAULT, *L'électrothermie au service des traitements thermiques de fils d'acier inoxydable*, comité français d'Électrothermie, 2 avril 1974.
[24] M. BOUREL *et al.*, *Équipements utilisés pour le traitement thermique des métaux ferreux et non ferreux en fonderie*, Comité français d'Électrothermie, 2 avril 1974.
[25] O. DRABEK, Die Wettbewerbslage des elektrischen Glasschmelzens in der BRD, *Elektrowärme International*, vol. 32, n° 4, 1974.
[26] INDUSTRIAL HEATING, *Rotary furnaces with $MoSi_2$ heating elements used in precision forging of turbine blades*, novembre 1974.
[27] R. THORSELIUS, *The use of electric furnace elements in the glass industry*, texte de conférence, documentation Bulten-Kanthal, 1974.
[28] J. A. POLISCHUK, Thyristorsteuerungen für elektrische Widerstandöfen und ihre energetischen Kennlinien, *Elektrowärme International*, n° 6, vol. 32, décembre 1974.
[29] J. YAN, Isolation thermique des fours, *Revue générale de Thermique*, n° 156, décembre 1974.
[30] H. SCHADLER, Rationeller Energieeinsatz für die industrielle Elektrowärme, *Elektrowärme International*, n° 2, vol. 33, avril 1975.
[31] R. MALGAT, *Cours d'Électrothermie*, École de Thermique, Paris, 1975.
[32] J. A. POLISCHUK et S. L. TREISON, Dynamische Kennlinien von Widerstandöfen und Auswahl einen Temperatureglers, *Elektrowärme International*, n° 4, vol. 33, octobre 1975.
[33] T. HUNT *et al.*, Energy conservation and environmental advantages of modern vacuum furnaces, *Industrial Heating*, novembre 1975.
[34] J. C. BOUCHON, *Récents développements industriels en électrothermie sous vide*, Comité français d'Électrothermie, 8-9 avril 1976.
[35] J. SCHMITT, *Progrès dans l'isolation réfractaire des fours à résistances*, Comité français d'Électrothermie, 8-9 avril 1976.
[36] G. LAVERGNE, *Les traitements thermiques en construction soudée*, Comité français d'Électrothermie, 8-9 avril 1976.
[37] U. KENTNER, Elektrische Ausrüstungen für warmbehandlungs Anlagen, *Elektrowärme International*, n° 2, vol. 34, avril 1976.
[38] M. MOUNIC, *Semi-conducteurs*, Foucher, Paris, 1976.
[39] F. BARBAS, Les éléments chauffants et la conception des fours électriques. *Matériaux et Techniques*, mai et juin-juillet 1976.
[40] CONGRÈS ÉLECTROTECHNIQUE MONDIAL, *Ensemble d'articles sur les fours à résistances*, Moscou, juin 1976.
[41] M. DE FRUYT et M. FOURNIER, Traitement en continu de boulonnerie diverse sous atmosphère contrôlée, *Traitement Thermique*, n° 107, août-septembre 1976.
[42] R. SCHWAB et G. WOLK, Erfassung von Ubergangsfunktionen an Industrieöfen

als Grunlage direkten Digitalsteuerung, *Elektrowärme International*, n° 5, vol. 34, octobre 1976.
[43] A. NIETO et F. PAUL, *Mesure des températures*, Éditions Radio, 1976.
[44] J. WHENTA, New electric heat treating furnace specifically designed to conserve energy, *Industrial Heating*, novembre 1976.
[45] O. N. PAYNE *et al.*, Four à résistance à haute température pour l'étirage des fibres à base de silice destinées aux communications optiques, *Ceramic Bulletin*, n° 2, 1976, Traduction EDF.
[46] G. THOMPSON, Ceramic fibres in furnace construction, *Metallurgia and Metal Forming*, décembre 1976.
[47] J. M. BLASKO et S. W. KENNEDY, Developments in vacuum sintering of powder metallurgy products, *Industrial Heating*, janvier 1977.
[48] G. RAMSELL, *Metal melting using resistance furnaces*, Congrès IEE, Londres, 8-9 mars 1977.
[49] G. MAGGETO, *Le thyristor : définition, protection, commandes*, Presses Universitaires de Bruxelles, Eyrolles, 1977.
[50] I. WELLS, Recent advances in the use of silicon carbide heating elements, *Metallurgia and Metal Forming*, mai 1977.
[51] A. IGIER et F. KEMEINER, Elektrisch beheizte Hochtemperaturöfen für Keramikteile der Mikroelecktronik, *Elektrowärme International*, n° 3, juin 1977.
[52] I. HOROWITZ, Arbeitswannen und Feeder-Vorherde mehrteiliger Elektro-Glasschmelzanlagen, *Elektrowärme International*, n° 3, juin 1977.
[53] M. ORFEUIL, *Le chauffage par résistances électriques*, École de thermique, Institut français de l'Énergie, 1977.
[54] K. KUHNKE et H. KUSCH, Betriebserfahrungen mit Thyristorofenschaltern in indirekt beheizten Widerstandsöfen, *Elektrowärme International*, n° 4, vol. 35, août 1977.
[55] DE FRUYT, Cuisson de produits céramiques en fours à multipassages en récupération, *Technique moderne*, n° 2-3, vol. 70, février 1978.
[56] INDUSTRIAL HEATING, *Aerospace technology applicable to industrial heating field* (*Electric Gas Heater*), mars 1978.
[57] INDUSTRIAL HEATING, *Lunar rock studies in specially designed furnace*, mars 1978.
[58] M. ORFEUIL et R. THOMASSIN, Les fours électriques à faible inertie thermique, *La Technique moderne*, avril 1978.
[59] K. FRANZ et H. MEINKE, Die Bestimmung von optimalen Ausmauerungen für Industrieöfen mit Hilfe eines Rechnenprograms, *Elektrowärme International*, n° 2, vol. 36, avril 1978.
[60] C. SOULET et J. C. BOUCHON, Installation de recuit sous vide moléculaire d'ébauches tubulaires en zirconium, *Bulletin d'Information Heurtey*, n° 68, 1978.
[61] Z. CHOJNACKI et J. HAUSER, Die Wärmeabgabe von Widerstandsheizwandeln in einem Bezugsystem, *Elektrowärme International*, n° 2, vol. 36, avril 1978.
[62] CONGRÈS DU COMITÉ FRANÇAIS D'ÉLECTROTHERMIE, Versailles 5-6 avril 1978.
**a)** M. D. FLOURY, *Four à passage de traitement thermique sous vide;*
**b)** R. KISSEL et R. WANG, *Amélioration des connaissances relatives aux conditions de fonctionnement des résistances moulées dans les fours continus de traitement;*
**c)** W. CZAKOWSKI et J. LOUARN, *Brasage en atmosphère réductrice halogénée à l'équilibre;*
**d)** R. SEJWACZ et M. DELEUZE, *Conception nouvelle d'un four à chauffage électrique isolé par matériaux fibreux et pouvant fonctionner jusqu'à* 1 000°C.
[63] J. GAULON, *Étude des différents modes de commande des thyristors*, Documentation EDF, 1978.
[64] M. MARTEUILH, *Résultats de mesures énergétiques sur des fours de cémentation gazeuse*, Comité français d'Électrohermie, 10-11 mai 1978.
[65] R. I. GRUBER, Modes for temperature control as dependent on process needs,

*Industrial Heating*, mai 1978.
[66] N. J. SHOR et M. WILSON, Indirect resistance heating for forging, *Metallurgia and Metal Forming*, juillet 1978.
[67] R. L. PEASLEE, Vacuum furnace brazing for repair of jet engine parts, *Industrial Heating*, août 1978.
[68] INDUSTRIAL HEATING, *Copper brazing of power steering reservoirs in specially designed continuous furnace*, septembre 1978.
[69] C. DEAN STARR, Destruction of 80 Ni/20 Cr heating elements by phosphorus, *Industrial Heating*, octobre 1978.
[70] M. ORFEUIL et M. BRIAND, *La fusion électrique des métaux non ferreux*, Comité Français d'Électrothermie, 12-13 octobre 1978.
[71] F. HOLDEN, Designing sintering furnaces and atmospheres for energy savings, *Industrial Heating*, janvier 1979.
[72] M. ORFEUIL et R. THOMASSIN, Le chauffage électrique des métaux avant formage, *Métaux Déformation*, n° 53, janvier 1979.
[73] T. BURAKOWSKI, *The emissivity of heating resistor alloys*, Documentation Union internationale d'Électrothermie, Paris 1979.
[74] P. LAPOUJADE, *Les produits réfractaires isolants*, Cours à l'École de Thermique, Institut français de l'Énergie, 1979.
[75] P. LAPOUJADE, *Les produits réfractaires industriels*, Cours à l'École de Thermique, Institut français de l'Énergie, 1979.
[76] L. HALLOT, *Les fours électriques de fusion et de maintien des métaux*, École Supérieure de Fonderie, 1979.
[77] INDUSTRIAL HEATING, *Efficient waste incinerators designed with electric heating and ceramic fiber insulation*, avril 1979.
[78] J. HEURTIN et J. GAULON, *Travaux de recherches*, Documentation EDF, 1979-1980.
[79] J. HEURTIN et J. GAULON, *Les fours à résistances et les économies d'énergie*, Documentation EDF, septembre 1979.
[80] INDUSTRIAL HEATING, *New « Porcupine » element makes heat gun safer and more efficient*, décembre 1979.
[81] M. BRIAND, L'électricité au service de la galvanisation à chaud, *La Technique Moderne*, n° 12, décembre 1979.
[82] J. GAULON et M. MAZOYER, *Possibilités techniques et technologies des fours à résistances*, Documentation EDF, janvier 1980.
[83] INDUSTRIAL HEATING, *New sulfur analysis system design with* $MoSi_2$ *heating elements in decomposition furnace*, mars 1980.
[84] J. E. FARKAS, Convenient method for power control of silicon carbide heating elements, *Industrial Heating*, mars 1980.
[85] M. D. FLOURY, *Fours sous vide pour le traitement de tubes de grande longueur*, Comité français d'Électrothermie, 6-7 mars 1980.
[86] M. VANDAMME, Exploitation d'un four à résistance en service aux aciéries d'Isbergues, Comité français d'Électrothermie, 6-7 mars 1980.
[87] A. ACCARY, *La métallurgie des poudres, les matériels électriques mis en œuvre*, Comité français d'Électrothermie, 6-7 mars 1980.
[88] M. COHEN, *Les résistances en bisiliciure de molybdène; évolution récente dans la transformation et la construction des fours*, Comité français d'Électrothermie, 6-7 mars 1980.
[89] J. L. MINGAUD et J. SOLA, Analyse et modélisation du bilan énergétique des fours de traitements thermiques continus dans les conditions d'exploitation industrielle, *Traitement thermique*, avril 1980.
[90] DOCUMENTATION EDF, 92000 Paris-La Défense.
[91] G. SÉGUIER, *L'électronique de puissance*, Dunod Technique, Paris.

**List of equipment manufacturers and suppliers mentioned in this chapter:**

[A] AUBE, 93100 Montreuil.
[B] AUBURTIN, 69007 Lyon.
[C] CFI, 93100 Montreuil.
[D] ECET (Meci), 94160 Saint-Mandé.
[E] INFRA-FOURS, 38000 Grenoble.
[F] HUNI, 92170 Vanves.
[G] MGR, 75018 Paris.
[H] IPSEN, 75013 Paris.
[I] RIPOCHE, 75014 Paris.
[J] CECF, 91160 Longjumeau.
[K] D.V.M., 38170 Seyssins.
[L] HEURTEY, 75823 Paris.
[M] FOURS ROISSAC, 69100 Villeurbanne.
[N] ELTI, 42400 Saint-Chamond.
[O] DELAGE, 42000 Saint-Étienne.
[P] ERSCEM, 95100 Argenteuil.
[Q] FOFUMI, 75009 Paris.
[R] SELAS, 92300 Levallois-Perret.
[S] STEIN-SURFACE, 91000 Évry.
[T] V.D.M.I., 75010 Paris.
[U] STEL, 91300 Massy.
[V] E.T.R., 92600 Asnières.
[W] NAGAT, 44550 Montoir-de-Bretagne.
[X] HUMERY, 37110 Château-Renault.
[Y] NABER (SOCOR), 75883 Paris.
[Z] BOREL, Peseux, Suisse.

[AA] FOURS ÉLECTRIQUES DE BESANÇON, 25000 Besançon.
[AB] SOLO, 93100 Montreuil.
[AC] CEM, 75008 Paris.
[AD] C.M.T.M., 75012 Paris.
[AE] COUDAMY, 87000 Limoges.
[AF] HERDIECKERHOFF, 77500 Chelles.
[AG] B.M.I., 38290 La Verpillière.
[AH] FERRO, 52100 Saint-Dizier.
[AI] HARTMANN, 90000 Belfort.
[AJ] MONTER, Novare, Italie.
[AK] DUPEUX, 75624 Paris.
[AL] PROTHERM, Velbert, RFA.
[AM] TABO, Kolbäck, Suède.
[AN] L.P.A.I. (La Physique Appliquée à l'Industrie), 95310 Saint-Ouen-l'Aumone.
[AO] GRAVITRON, Gosport Hants, Grande-Bretagne.
[AP] SFEAT, Milan, Italie.
[AQ] KOPO, Kone-Polya, Lippitie, Finlande.
[AR] BONGARD, 67810 Holtzheim.
[AS] THIRODE, 93700 Drancy.
[AT] TIBILETTI, 06800 Cagnes-sur-Mer.
[AU] PAVAILLER, 26500 Bourg-les-Valence.
[AV] GUYON, 74140 Douvaine.
[AW] MECATHERM, 67130 Schirmeck.
[AX] SAT, 73100 Aix-les-Bains.
[AY] ATA, 95230 Soisy-sous-Montmorency.
[AZ] ASET, 69800 Saint-Priest.
[BA] SLFI, 69100 Villeurbanne.

[**BB**] BALPC, 56320 Le Faouët.
[**BC**] SRRI, 75011 Paris.
[**BD**] STROMBERG, Helsinki, Finlande.
[**BE**] LEISTER, 75004 Paris.
[**BF**] TERMELEC, 75015 Paris.
[**BG**] METAL-IMPHY, 75008 Paris.
[**BH**] BULTHEN-KANTHAL, 92400 Courbevoie.
[**BI**] LA RÉSISTANCE RD, 05400 Veynes.
[**BJ**] BALLOFET (Plansee), 75016 Paris.
[**BK**] DRIVER-HARRIS, 78200 Mantes-la-Jolie.
[**BL**] SIGRI-OSI, 75000 Paris et Meitingen, RFA.
[**BM**] PYROX, CEA, 92260 Fontenay-aux-Roses.
[**BN**] LE MÉTAL DÉPLOYÉ, 92140 Clamart.
[**BO**] SEDIMMEC, 94200 Ivry.
[**BP**] BTU ENGINEERING, Chicago, USA.
[**BQ**] CARBORUNDUM, 92410 Ville-d'Avray.
[**BR**] MORGAN, 95500 Gonesse.
[**BS**] S.E.P.R., 92200 Neuilly.
[**BT**] LAFARGE, 92200 Neuilly.
[**BU**] JOHNS MANVILLE, 92500 Rueil-Malmaison.
[**BV**] CEC, 92120 Montrouge.
[**BW**] SAUDER INDUSTRIES, Glasgow, Grande-Bretagne.
[**BX**] JUNKER, 75009 Paris et Lammersdorf, RFA.
[**BY**] MANUTHERM (DAMOND), 91210 Draveil.
[**BZ**] THERMIDOR, 26100 Romans,
[**CA**] NORTON CERAMIC PRODUCTS, USA.
[**CB**] AJAX MAGNETHERMIC, Oxted, Grande-Bretagne.
[**CC**] LEYBOLD-HERAEUS-SOGEV, 91401 Orsay.

# Chapter 3

# Heating with encased electric resistance

## 1. PRINCIPLE OF HEATING WITH ENCASED RESISTANCES

Encased electric resistances (or "protected", "tubular", "coated" or "enveloped" resistances) are being used more and more, especially for non-furnace, high temperature apparatuses (more than 600°C) to heating fluids, liquids, gases and solids (molds, etc.). Heat transfer is generally by conduction, but sometimes by radiation.

## 2. CONSTITUTION OF AN ENCASED RESISTANCE

An encased metallic resistance consists of the following:
— an electric wire or ribbon resistance;
— an electrical insulator in which the resistance is mounted;
— an envelope, to protect the assembly mechanically and chemically.

In addition, when waterproof, these elements are known as shielded or armored resistances.

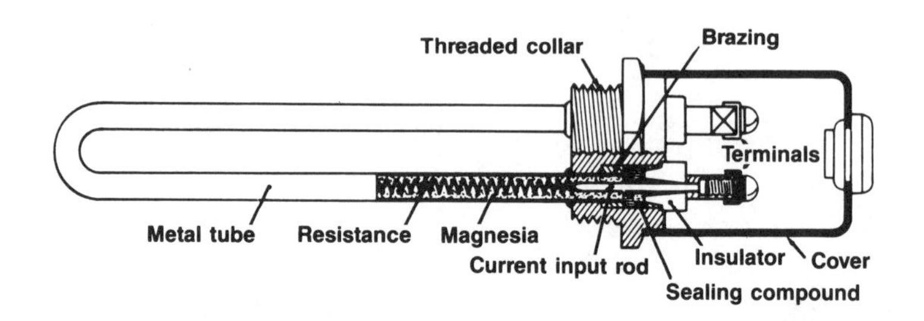


**Fig. 1.** Diagram of an encased electric resistance [17].

The insulating material must provide excellent electrical insulation at all temperatures and must offer adequate thermal conductivity so that the heat provided by the resistance can be transferred to the body to be heated, and prevent overheating of the resistance. Electrical insulation and thermal conductivity are properties difficult to find in the same material. The insulator must also offer adequate resistance to mechanical shocks and vibration. Magnesia, which is both an excellent electrical insulator and a good conductor of heat and also supports rather high maximum temperatures of 800 to 900°C, is often used as a coating material; other materials such as asbestos, mica, glass porcelain, steatite, cement, alumina and synthetic products are also used.

Operating temperatures of electrical insulators [17]

| Type of insulator | Usual operating temperatures of protected element (°C) | |
|---|---|---|
| | Current | Maximum |
| Asbestos | up to 250/350 | 400-600 |
| Mica (muscovite) | 300-400 | 500 |
| Mica (phlogopite) | 400-500 | 600 |
| Glass | 350 | 500 |
| Quartz | 600 | 800 |
| Powdered alumino-silicate products | 550 | 600 |
| Shaped refractory products | 800 | 1,000 |
| Cement | 800 | 800-900 |
| Powdered magnesia | 500-600 | 800-900 |

**Fig. 2.**

The envelope should offer excellent protection against corrosion, shock and vibration, etc., and, conversely, must not affect the environment. Generally, the envelope is metallic, but sometimes for low temperatures, plastic (polytetrafluoroethylene or polyvinyl chloride) or cement is used. The envelope may be composed of different materials, starting at the heating element, each playing a special role [6], [16].

The maximum operating temperature depends on the type of envelope and the electrical insulator, but equally on the product to be heated. Generally, temperature control is provided by thermostats.

Maximum temperature of casing in non-corrosive conditions [17].

| Type of casing | Maximum temperature of casing (°C) |
|---|---|
| Lead | 75 |
| Copper | 175/300 |
| Aluminum | 150/330 |
| Brass | 400 |
| Steels | 400/500 |
| Monel-stainless steels | 650 |
| Incoloy-inconel | 750/800 |
| Plastics | 100/200 |
| Mica | 400 |
| Special ceramics | 900/1,000 |

**Fig. 3.**

Generally, the resistance is a metallic alloy (iron-nickel-chromium, nickel-chromium or iron-chromium-aluminum) and, in most cases, consists of a wire, either coiled or straight. The resistive elements terminate in cold outlets; the current input terminals must be carefully constructed.

When selecting encased heating elements for a given application, one must ensure that:

— *the power density* (expressed in Watts per square centimeter, that is, the surface area of the outside casing) is low enough to prevent damage to the product touching the envelope, but also must be such that the heat transfer from the resistance to the product to be heated through the cladding material and the casing can take place normally, thus preventing overheating of the resistance;

— *the material forming* the sheath is capable of withstanding corrosion at operating temperature, and conversely does not pollute the environment.

Heating elements come in highly varied forms and have extremely diversified applications. They can be classified according to the type of insulator or casing, shape, maximum operating temperature, and type of application. The main types of elements are described below indicating their basic characteristics and standard applications.

## 3. DIFFERENT TYPES OF ENCASED RESISTANCES

There are numerous types of encased resistances.

### 3.1. Armored heating cartridges

Armored heating cartridges consist of a resistance wire, generally in nickel-chromium, wound around a ceramic core; the assembly is embedded in magnesia or highly compressed alumina and placed in a cylindrical metal envelope. The low thermal gradient between the heating wire and the outside tube enables high heat flux. The power density can reach 15 W/cm$^2$. Sometimes, special assemblies enable very high power densities of 20 to 40 W/cm$^2$ for pressurizing water and 500 W/cm$^2$ maximum for liquefying sodium.

Very long elements are sometimes known as heating rods, but their basic construction is the same.

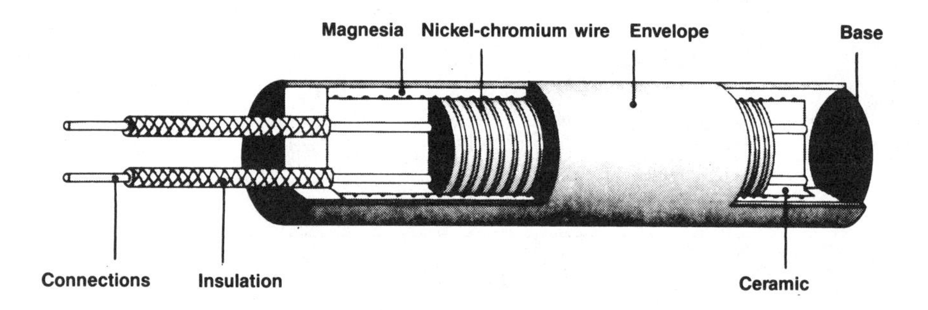


**Fig. 4.** Diagram of a heating cartridge [16], [D].

These elements are usually used to heat solids (moulds, platforms, core boxes, plastic welding tools, marking tools, heating bolts prior to tightening, etc.), but also to heat liquids.

Current unit powers vary from about 100 Watts to several kilowatts depending on specific power and length. By mounting these elements on flanges, it is possible to obtain equipment having a power of several hundred kilowatts, directly immersed in the liquids to be heated; single units can also be fitted in sleeves, to heat tanks [A], [B], [C], [D], [E], [F], [G], [R], [AD].

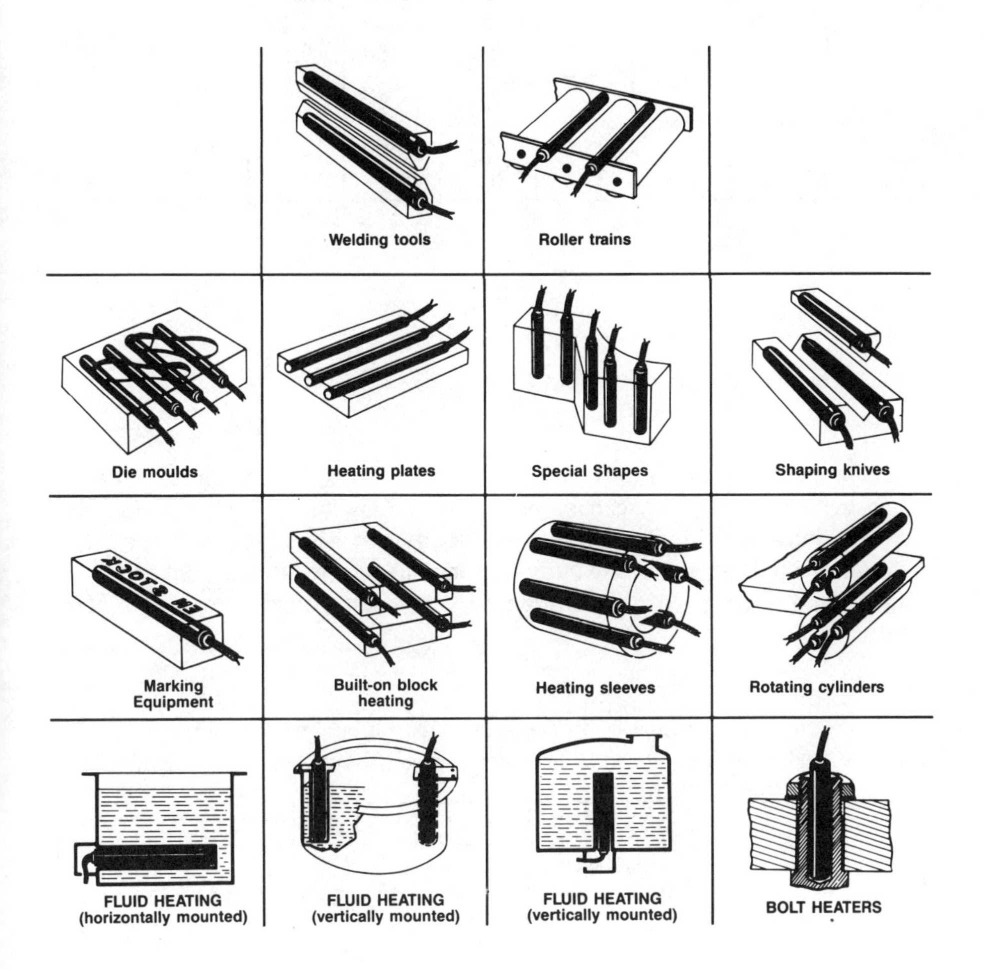


**Fig. 5.** Heating cartridge applications [A].

As an example, Figure 6 lists a wide range of heating cartridges applications. Generally, they are supplied from 220 V lines, but if required, can be supplied with other voltages.

Range of heating cartridges [A].

| Power density 3 W/cm² | | | Power density 7 W/cm² | | | Power density 16 W/cm² | | |
|---|---|---|---|---|---|---|---|---|
| Dia-meter (mm) | Length (mm) | Power (W) | Dia-meter (mm) | Length (mm) | Power (W) | Dia-meter (mm) | Length (mm) | Power (W) |
| 9.5 | 100 | 100 | 9.5 | 50 | 80 | 6.35 | 50 | 175 |
| 9.5 | 150 | 150 | 9.5 | 150 | 300 | 9.5 | 50 | 250 |
| 16.7 | 100 | 150 | 12.7 | 150 | 450 | 9.5 | 100 | 500 |
| 14.7 | 200 | 200 | 15.8 | 150 | 550 | 12.7 | 50 | 300 |
| 21.7 | 160 | 210 | 25.4 | 100 | 500 | 12.7 | 100 | 650 |
| 21.7 | 200 | 350 | 25.4 | 300 | 1,500 | 12.7 | 200 | 1,100 |

**Fig. 6.**

### 3.2. Armored tubular elements

These elements are cylindrical metal-envelope-armored resistances. Linear when manufactured, they can be bent into various forms. They are used for heating fluids (e.g. immersion heaters), solids and gases, and also form the heating element in some medium or long infrared emitters. Embedded in aluminum, bronze or cast iron during pouring, these elements are also used as heating collars.

Example of a range of linear armored tubular heating elements. Specific surface power density: 3.6 W/cm², specific linear power: 1 kW/m [B].

| Power (W) | 500 | 750 | 1,000 | 1,500 | 2,000 | 3,000 | 4,000 | 5,000 |
|---|---|---|---|---|---|---|---|---|
| Heating length (mm) | 510 | 740 | 910 | 1,390 | 1,830 | 2,850 | 3,830 | 4,830 |

**Fig. 7.**

Figure 7 gives a sampling of linear armored tubular elements. The specific powers in practice vary widely and depend on operating conditions and consequently on heat transfer. Thus, for a length of 5 m, the power produced by an element may exceed 50 kW, i.e., ten times that stated in the table [A], [B], [C], [D], [R], [AD].

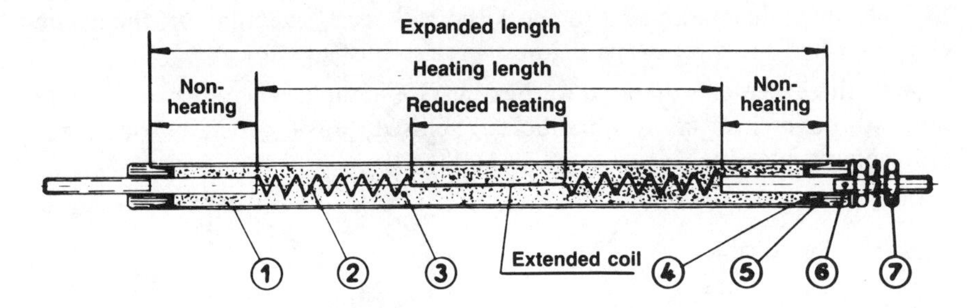


**Fig. 8.** Diagram of an armored tubular element: (1) metal tube; (2) coiled resistance, if necessary the heating part can be reduced in some parts of the wire; (3) cladding insulation; (4) and (5) seals; (6) connecting rod; (7) attachment and fittings [AD].

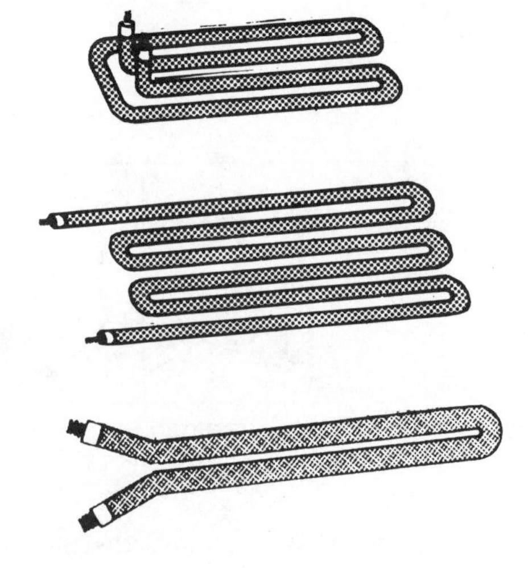

**Fig. 9.** Examples of tubular elements after shaping [AD].

## 3.3. Immersion heaters with armored resistances

Heating elements (*see* figure 1) are made from a spiral resistance in nickel-chrome or iron-chrome-aluminum alloys, insulated by magnesia or, more rarely, by alumina, and protected by a sheath generally of metal (mild steel, stainless steel, copper and copper alloys, lead, incoloy, monel, aluminum, titanium, polytetrafluorethylene coated steel, etc.).

These elements are used for heating fluids (baths, tanks, steam boilers or water boilers, circulation heaters, etc.). They are shaped according to application [A], [B], [C], [D], [E], [F], [G], [H], [I], [K], [L], [M], [O], [R], [AD].

### *3.3.1. Removable vertical or horizontal immersion heaters*

These heaters are designed for open tanks (it is not necessary to drill the tank and provide a seal). These elements are fitted on the wall or on bars, and can be used both in metallic and non-metallic tanks. The elements are equipped with cold ends long enough to ensure that the heated part remains immersed, but the only absolute guarantee is to check the bath level, switching off the heating element when the level drops below a pre-established value.

Also, these elements are used for heating closed tanks where they are inserted parts through inspection "manholes"; closed tanks generally operate at atmospheric pressure.

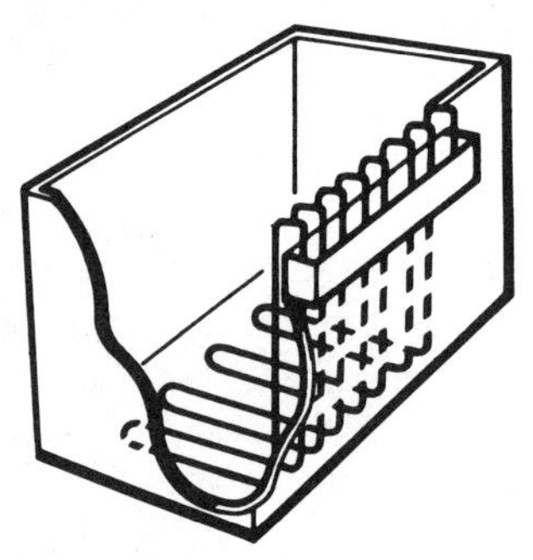

**Fig. 10.** Removable immersion heaters mounted on a wall [18].

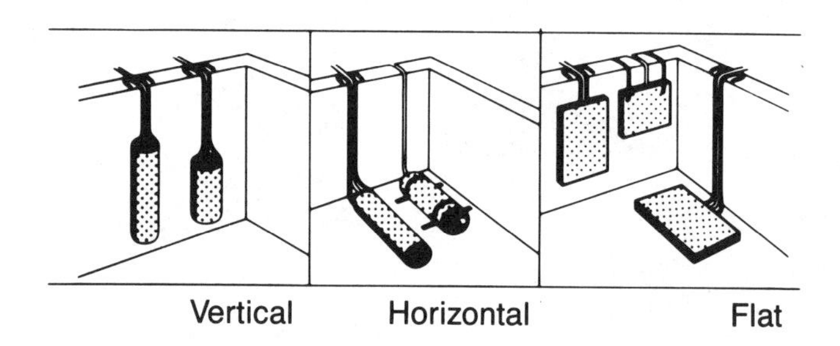


**Fig. 11.** Polytetrafluoroethylene-sheathed immersion heaters [K].

The connection box is located above or outside the tank. The most simple heaters are the single-phase type, with powers ranging from 500 W to 10 kW. The most powerful immersion heaters, with unit powers up to 60 kW approximately, are three-phase supplied, with the possibility of star-delta coupling.

Special power densities are generally between 1 and 10 W/cm$^2$, but special configurations can satisfy most requirements.

### *3.3.2. Screw-in or clamped immersion heaters*

These immersion heaters consist of heating elements brazed onto a support (threaded plug or flat flange) fitted with a seal. They are attached through the vertical walls of the tank or enclosure to be heated. Some heat transfers can also be screwed into the bottoms of tanks. In general, screw-in heaters are used for heating medium and small-sized tanks, while flange or clamped heaters are more appropriate for large capacity tanks. Both types are fully suitable for high temperatures and pressures, but are used only in cases where the tanks are easily emptied.

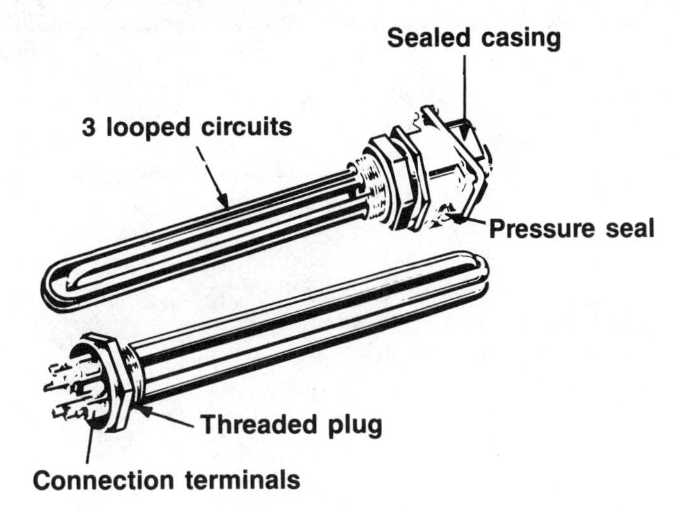


**Fig. 12.** Screw-in immersion heaters [B].

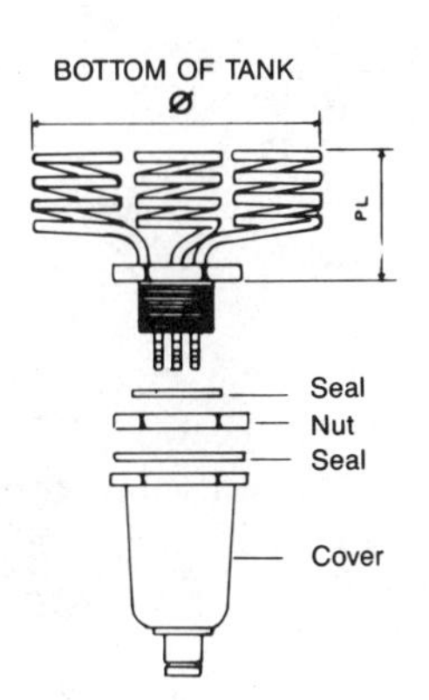


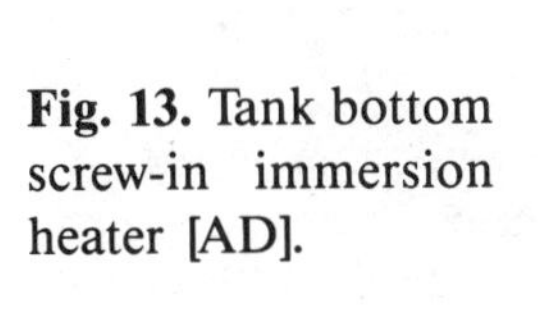

**Fig. 13.** Tank bottom screw-in immersion heater [AD].

**Fig. 14.** Flanged (clamped) immersion heater [AD].

### *3.3.3. Characteristics of armored resistance immersion heaters*

The power density and the type of casing varies with the fluids to be heated.

Immersion heater maximum specific power in Watts per square centimeter [17]

| Fluid to be heated | Maximum specific power ($/cm^2$) | | |
|---|---|---|---|
| | Copper | Steel | Stainless steel |
| Non-flowing water | 10 | | 10 |
| Flowing water | 15 | | 15 |
| Water heater | 8 | | 8 |
| Alkaline baths | | 6 | |
| Diluted acids | | | 2.5 |
| Phosphating baths | | | 1 to 4 |
| Highly fluid technical oils, up to: | | | |
| 50°C | | 4 | |
| 100°C | | 2.5 | |
| 250°C | | 2 | |
| 350°C | | 1.5 | |
| Comestible oils | | 5.5 | |
| Bitumen | | 1 | 1 |
| Lead baths | | | 4 |

**Fig. 15**

The values in Figure 15 above are given for reference only and, in the case of a new application, the choice of the immersion heater characteristics, type, casing and specific power must be thoroughly analysed with suppliers. To provide a safety margin and increase element service life, the specific powers used are often less than the maximum possible.

### 3.4. Barrel resistances

In the strict sense of the term, these heating elements are not encased resistances, but belong to this group due to their design and use.

With this type of resistance, the heating wire consists of spirals of nickel-chromium alloy, placed on refractory supports (in general steatite), equipped with grooves and known as barrels.

The usual surface power density values are between 1 and 4 $W/cm^2$, but for special applications, can reach 10 $W/cm^2$. These elements can be several meters in length. They are used for heating fluids and solids.

#### *3.4.1. Immersion heaters with barrel resistances*

For heating liquids, these resistances are used in two ways:

— either placed in horizontal metal casings, from which they can be easily removed without draining the tank;

— or inserted in a protective tube, for which any metallic or non-metallic material can be used as an envelope (silica, porcelain, graphite, polytetrafluoroethylene, etc.). Generally, this tube is mounted vertically in the bath, and forms a removable element immersion heater; the heating element can also be installed horizontally.

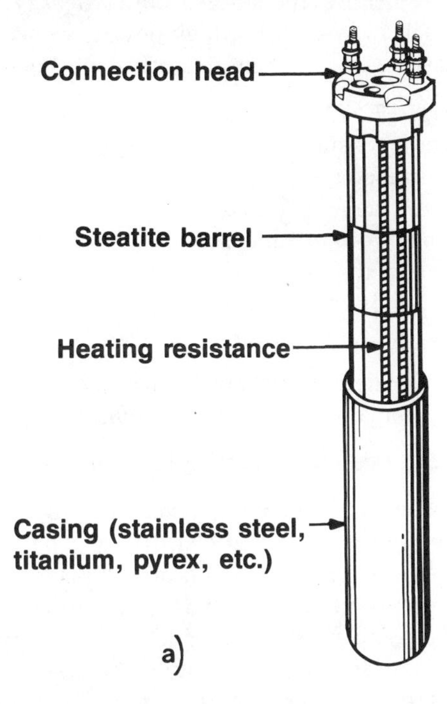


**Fig. 16.** Heating tanks using barrel elements:
*a)* barrel resistance in vertical protective tube [17];
*b)* element placed in horizontal casing [A];
*c)* element placed in vertical protective tube installed in tank [18].

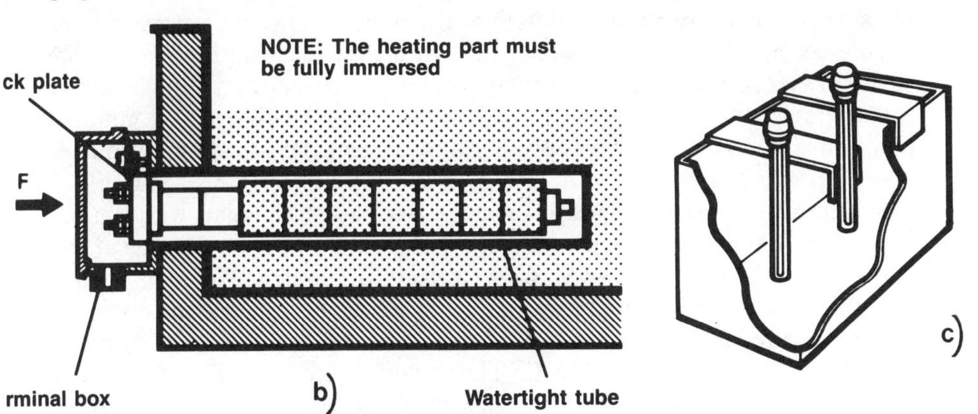


Transfer of heat to the outer tube is not as good as with armored resistances, and the external tube temperature can rarely exceed 300°C, which is sufficient for most liquids. The surface power density is therefore generally lower. Currently, the power of these elements ranges from 0.5 to 10 kW, and specific power from 1 to 4 W/cm$^2$.

### 3.4.2. *Heating solids with barrel resistances*

These resistances are also used for heating moulds, trays, etc., especially where it is necessary to have long heating elements with relatively high powers.

Generally, the surface power density is between 2 and 4 W/cm$^2$, but may reach 10 W/cm$^2$. For high power, connection is often three-phase. The heating cartridges sometimes enable much higher power densities to be reached.

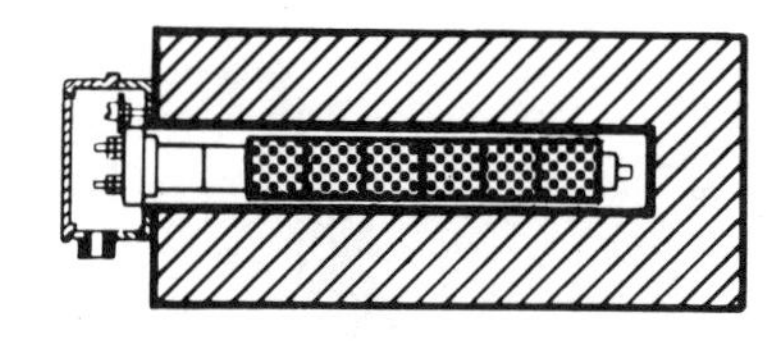

**Fig. 17.** Mould heating using barrel resistances; in practice, the resistance dimensions are much smaller than those of the body to be heated [A].

The design of some flat resistances with ceramic-protected heating elements is comparable to barrel resistances of this type.

## 3.5. Flat elements with metallic casings

Generally, these resistances are straight linear or circular; however, they can be shaped as desired [A], [B], [C], [D], [E], [S], [AD].

The nickel-chromium heating resistance consists either of a coiled wire wrapped several times around the element to ensure adequate distribution of heat over the complete surface, or a zig-zag cut ribbon.

A wide range of coating materials are used, magnesia, mica, steatite or refractory cement. Mica is a thermal insulator. The mica plate layer surrounding the resistance must therefore be thin for efficient heat transfer. Conversely, mica is an excellent electrical insulator with high dielectric strength enabling thin elements such as wire or ribbon and, if necessary to be supplied at high voltage (1,500 V and more). For example, for a total element thickness of 4 mm, the permissible surface power is 4 W/cm$^2$ for a casing temperature of approximately 400°C. However, they should not be used at temperatures over 300°C.

Steatite, when used as an encasing material, enables much higher temperatures than mica, 800 to 900°C approximately. The resistive wire is mounted in this material. The maximum specific power density is about 10 W/cm$^2$.

Refractory cements are used for encasing the resistance coil for elements whose operating temperature does not exceed 650°C. The usual operating temperatures are, however, lower: 400 to 450°C. The cement in contact with the heating resistance, whose temperature is much higher, may become brittle. However, special refractory cements enable higher temperatures. Specific powers are usually 1.5 to 3 W/cm$^2$.

The casing, which is applied to one or two surfaces, is generally metallic. These resistances can also be potted in aluminum. Some elements are armored and sealed; in others, the metallic casing is not sealed which must be allowed for in applications.

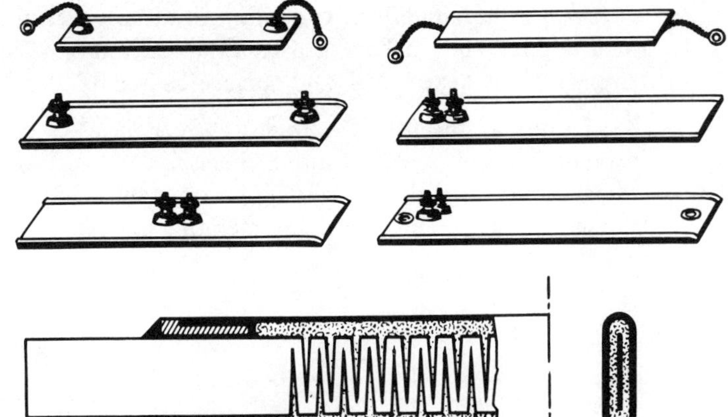

**Fig. 18.** Flat mica insulated resistances [17].

**Fig. 19.** Flat sawtooth shaped ribbon resistance [17].

Flat resistances are usually used for heating press plates, moulds and miscellaneous tools in industries such as rubber, plastics, wood, printing, metal transformation, etc. They are also used for external heating of tanks where immersion heaters cannot be used.

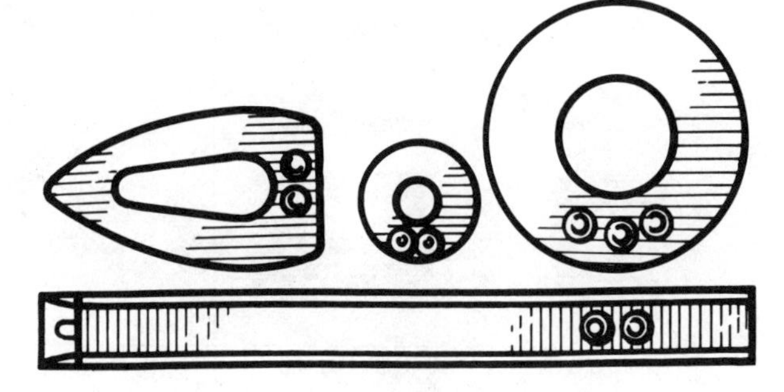

**Fig. 20.** Various flat resistances [S].

Other applications such as protection against humidity in control cabinets, protection of valves; pumps and strainers against freezing, heating simple ovens, etc. exist.

### 3.6. Heating collars

Heating collars are used to heat cylindrical shapes, to which they are attached by means of quick-release brackets or bolts fitted in lugs [A], [B], [C], [D], [E], [F], [R].

Numerous types exist:

— collars consisting of nickel-chromium wire or ribbon insulated by two mica plates; the casing is steel or brass and, in some elements can be used to seal the assembly. The maximum operating temperature is about 350°C and surface charge can reach 4 W/cm$^2$;

— sealed armored collars, consisting of an armored resistance brazed onto a thick copper plate behaving as a heat distributor. The maximum operating temperature is also 350°C and surface charge 4 W/cm$^2$;

— aluminum, bronze or refractory steel encased collars, used in outside installations or in explosive atmospheres, and consisting of an armored heating element embedded during casting in a metallic block; the supporting surface is then machined. The maximum operating temperature is 350°C for aluminum and 600°C for bronze;

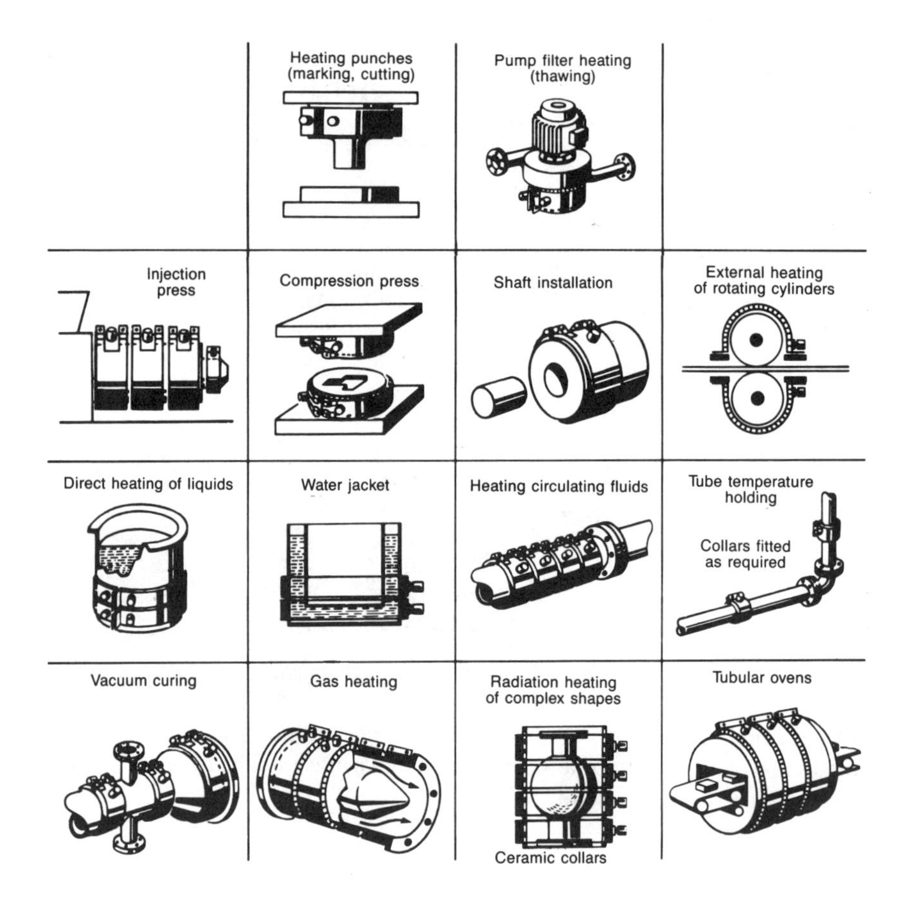


**Fig. 21.** Heating collar applications [A].

— articulated ceramic collars, in which the coiled resistance is mounted on steatite elements drilled with holes. Depending on temperature, the metallic casing is stainless steel (up to 700°C) or inconel (from 700°C); lagging is inserted between the casing and the steatite elements. Fabrication of the collar from basic elements renders the assembly highly flexible and enables it to conform to the part to be heated. Heat transfer is effected by conduction with direct contact between the ceramic elements and the part, or by radiation. The maximum operating temperature is 900°C approximately and maximum specific power 10 W/cm$^2$.

Most often collars are cylindrical but tapered, elliptical, flat cylindar, square, rectangular, with sharp or rounded edges and special shapes can be constructed as required.

Power ranges from about 100 Watts to several kilowatts; diameters range from 30 millimeters to several meters. For special shapes, one-third or half collars can be used.

Heating belts for barrels containing viscous products such as wax, grease, pomade, gelatin, organic acids, tars, etc. can be included in this category of heating element. With some heating belts, an armored resistance is, for example, brazed onto a coppered belt forming a heat distributor. Generally, unit power varies from 1 to 3 kW and surface power density is low, about 1 W/cm$^2$.

Heating collars are widely used in all industries.

### 3.7. Flexible elements in ceramic casings

Nickel-chromium alloy flexible heating elements consist of loops threaded through plates, with ceramic beads behaving as hinges, or coiled wires externally protected by sintered alumina insulating sleeves. Other alloys, such as iron-chromium-aluminum, can be used to reach very high temperatures. These elements are highly flexible, enabling them to be used on flat, cylindrical, tapered, spherical or complex surfaces. These resistances are assembled to form layers divided into circuits which are electrically connected in series or parallel. Numerous variants of such elements exist:

— "spiral wound" elements, consisting of a coiled wire encased in insulating sleeves (ceramic rings) of sintered alumina, offering high mechanical strength and resistance to thermal shocks. These encased resistances are quite flexible since the short sleeves are placed in series. For example, these resistances are used as heating elements in the hearths of flexible or temporary furnaces, and for preheating and annealing welded seams. The maximum operating temperature is 900°C. Unit powers are a few kilowatts. For example, the heating length of a 4 kW element is about 1.8 m. This type of element can be folded easily to form a U by removing one or two turns from the sleeves, thus placing both connections at the same end. It is then possible to place the element with the bare wire appropriately insulated in a refractory steel box; both arms are then very close, limiting the temperature to approximately 750°C and necessitating use in the horizontal position. These elements form heating levels or layers for internal and external parts requiring preheating or stress relief treatment (welding large parts such as reservoirs or spherical or cylindrical tanks, etc.);

— "cartridge belt" elements, which resemble a cartridge belt; these resistances are similar to those previously described, but consist of a series of rather short elements placed in parallel (often called fingers). The heating elements, forming waves, are housed in insulating sleeves, the latter being retained by a welded strip. The "fingers" form a "cartridge belt" which may be of different powers and lengths to suit the thermal requirements of the application. The maximum operating temperature is about 1,050°C, but special elements can reach 1,200°C. Basically, these elements are used for heat treating weld seams on pipes or conduits, but may also be used for heating all concave, convex or tapered surfaces.

**Fig. 22.** "Cartridge belt" element with stepped "fingers" [AF].

— Special elements with a maximum temperature of 1,200°C; these elements are produced from simple "fingers" assembled as required to obtain the desired form and thermal effect. They are widely used in brazing, welding, flanging, etc.

Where safety is of utmost concern, these elements are low-voltage supplied (24 V aboard ships, 48 V where accidental contact between the user and the assembly may occur — for example, on worksites); normal voltage (in general 220 V) is used where there are no particular safety problems.

These flexible ceramic-sheathed elements are mainly used in boilerwork and welding construction for preheating before welding and stress relief heat treatments after welding (*see* paragraph 5.8), but may be used for a wide range of applications since high temperatures can be reached.

### 3.8. Armored heating cables

The composition of armored heating cables varies widely from one manufacturer and application to another. They consist of a resistive core (nickel-chromium, copper-nickel, aluminum alloys, etc.) insulated by highly compressed magnesia inside a metallic sheath (copper, stainless steel, inconel, etc.). The cables are equipped with cold outlets having low resistivity, tough metals (in general copper is used). These cold outlets are fitted with terminations designed to ensure sealing and ease of connection to the supply circuit [A], [B], [C], [J].

The linear specific power generally ranges between 10 and 150 W/m, and the maximum temperature of the part to be heated can reach 700°C. Powers, lengths and supply voltages are diverse.

To adapt power density to the part to be heated, cables can be arranged in various ways: longitudinally, along the axis of a pipe, for example, or looped or spiralled, varying according to pitch. To improve heat transfer, cables are attached by ribbon, clamps, welding or a thermal cement. The assembly is insulated to limit thermal losses.

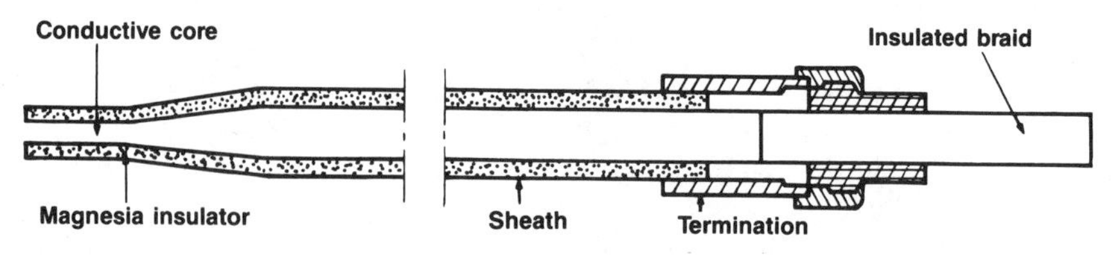


**Fig. 23.** Diagram of a heating cable [A].

Heating cables are used to heat flat or complex shaped metal surfaces (model plates, moulds) and to heat liquids (reservoirs, tanks, tubes, valves, etc.). Since they are completely sealed, these cables can also be used for immersion heating of baths.

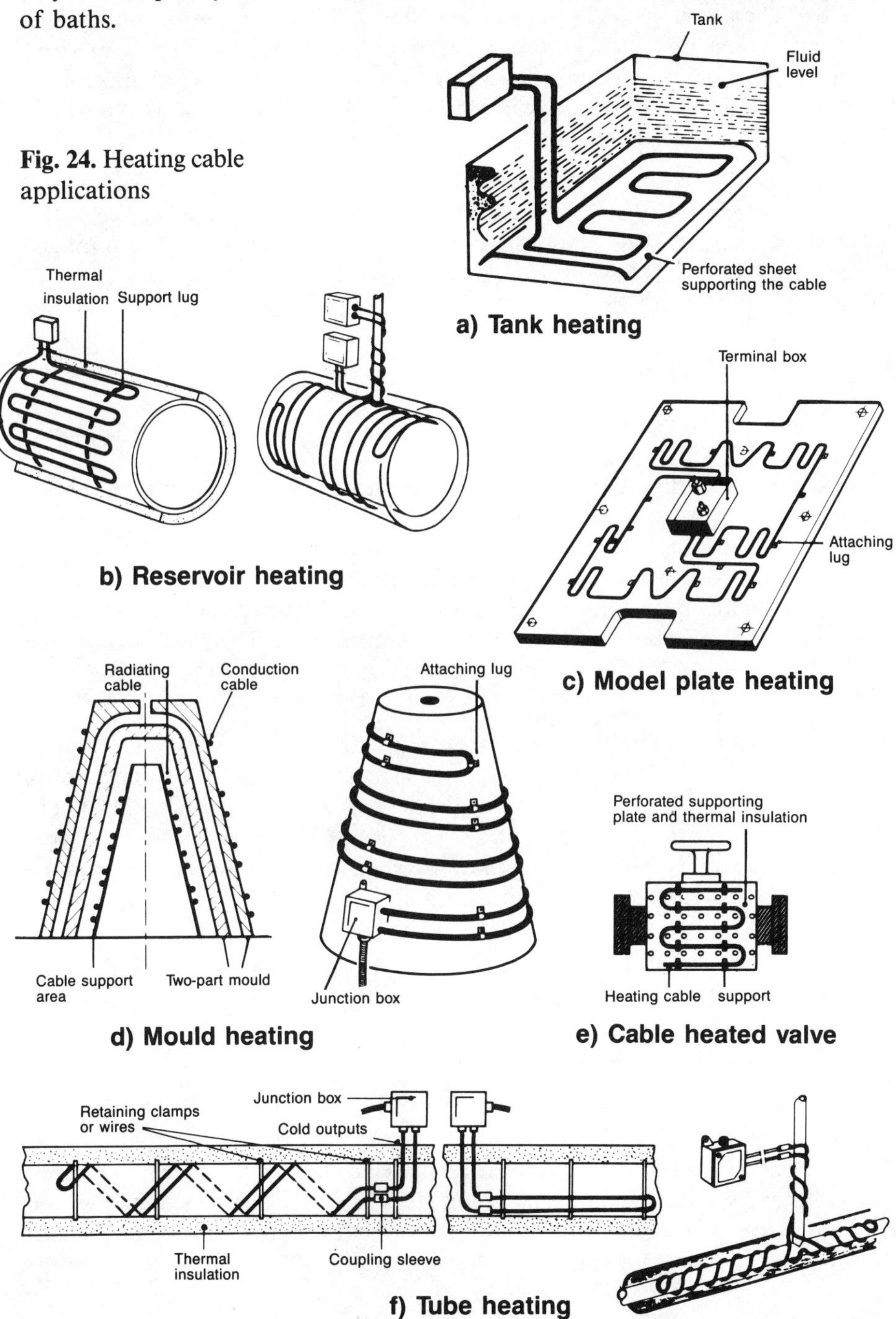


**Fig. 24.** Heating cable applications

Heating work or living quarters is another important application of cable heating. The heating cables are embedded in the floor concrete, which then acts as an energy accumulator during low rate hours when the price of electricity is low. The power generally varies from 50 to 200 Watts per square meter. The specific power of the cables used is, in most cases, between 30 and 40 W/m; the installed power per square meter of floor is adjusted by altering the pitch of the heating cable system (*see* chapter 5.9.). The cable is metal sheath is generally protected by a plastic coating, for example, PVC.

Heating cables embedded in concrete can also be used for de-icing footpaths, garage entries, etc. The power density in such cases is much higher, from 200 to 250 $W/m^2$.

## 3.9. Heating tapes

Fabrication of heating tapes involves many different techniques, but the terminations for these elements are always located at one end. Some tapes consist of two sheathed cables (forward and return of same resistive wire) attached to a flexible support. The resistance wire can be copper-nickel, nickel-chromium or constantan, insulated with PVC, polytetrafluoroethylene, silicon-impregnated glass, asbestos, quartz or magnesia. Sometimes, the tape is covered with glass or quartz fibre, copper, copper-nickel or stainless steel. The envelope covering the tape is made from PVC, stainless steel, copper, glass fibre or quartz.

The table in Figure 25 gives a manufacturer's range of tapes which can be waterproof or explosion-proof. Special assemblies enable very high power densities up to 800 W/m. Normally, these tapes are supplied with 220 V, but can be made for other voltages (24, 48, 110 and 380 V). When installed the tapes must not overlap, which would cause overheating and damage. Temperature control is provided by thermostats [1].

Other tapes consist of flat heating elements located between two layers of woven glass; the resulting envelope can be coated with rubber or plastic. As for the tapes described above, there must be no overlap; the temperature is also thermostatically controlled.

New types of self-regulating tapes have recently appeared on the market. The resistance consists of plastic material, radiated by electronic bombardment (*see* "Electron bombardment heating" chapter). The current is applied through two parallel conductors and flows through the resistive element, between the parallel conductors, forming a continuous heating circuit. The tape can be cut to the desired length.

Range of sheathed cable heating tapes (N)

| Heating element | Insulation element | Outside tape | Tube temperature (°C) | | Power density (W/m) |
|---|---|---|---|---|---|
| | | | Minimum | Maximum | |
| Copper-nickel | PVC | PVC | −20 | 50 | 5 to 15 |
| Copper-nickel | PVC | Copper braid | −20 | 60 | 5 to 16 |
| Copper-nickel | PTFE covered with silicon impregnated glass briad | Glass fibre | −20 | 200 | 15 to 50 |
| Copper-nickel | PTFE covered with silicon impregnated glass briad | Stainless steel braid | −200 | 200 | 15 to 50 |
| Nickel-chromium or copper-nickel | Silicon impregnated glass or asbestos | Glass fibre | −20 | 425 | 60 to 135 |
| Nickel alloy | Quartz braid | Quartz fibre | −20 | 800 | 135 to 210 |
| Constantan | Magnesia with copper-nickel sheath | Glass fibre | −20 | 350 | 15 to 90 |
| Constantan | Magnesia with copper-nickel sheath | None, wire separation varied by metal strips | −40 | 350 | 20 to 150 |
| Nickel-chromium and copper-nickel | Magnesia with stainless steel sheath | Glass fibre | −20 | 425 | 30 to 150 |
| Nickel-chromium and copper-nickel | Magnesia with stainless steel sheath | Stainless steel braid | −40 | 650 | 60 to 300 |

**Fig. 25**

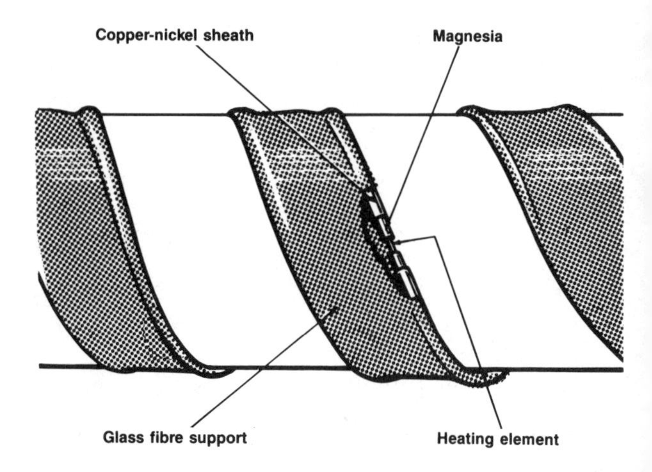


**Fig. 26.** Armored cable heating tape [N].

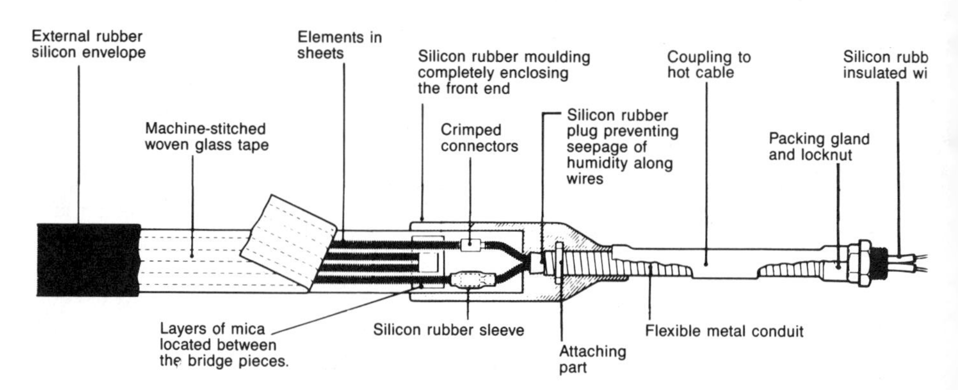


**Fig. 27.** Flat sheet-element heating tape

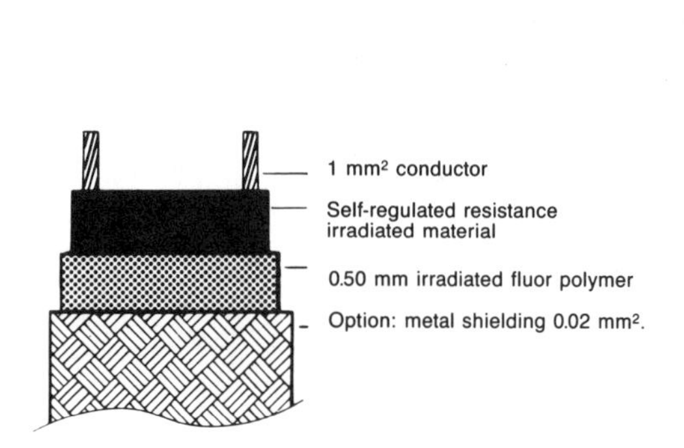


**Fig. 28.** Self-regulated resistance heating tape

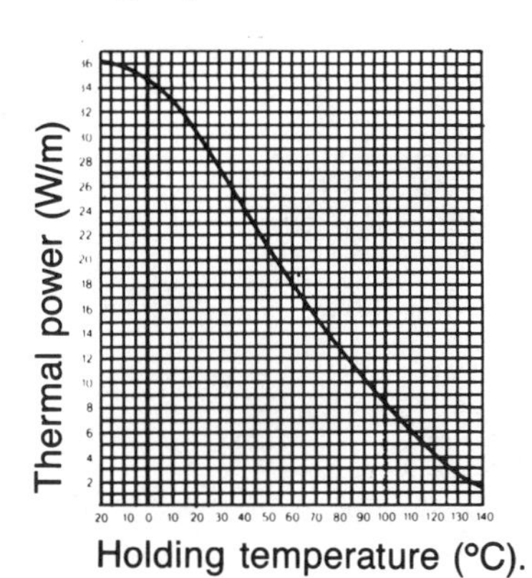


**Fig. 29.** Power variation as a function of temperature.

The nominal specific power varies between 10 and 33 $W/m^2$. The maximum continuous operating temperature is about 110°C, but can reach 150°C momentarily. This heating element self-regulates its power as a function of temperature changes. When the temperature decreases, the power increases and vice versa, the compensation taking place locally at each point of the tape. Self-regulation eliminates hot points, even at points where the tapes cross each other.

Due to their flexibility, heating tapes are used to heat tubes, valves, pumps and accessories. Generally, they provide temperature holding, de-icing and, in some cases, heating, since power densities are relatively low.

The products heated using this technique vary widely: chemicals, foodstuffs such as chocolate, sugar or oils, soap, low temperature molten metals, fuel, oil, waxes, greases, etc.

The tapes are placed longitudinally along or spirally around the parts to be heated. Fibreglass, plastic or metal attachments are sometimes required to ensure adequate heat transfer. Generally, the tape is coated with thermal insulation to reduce heat loss and installed powers.

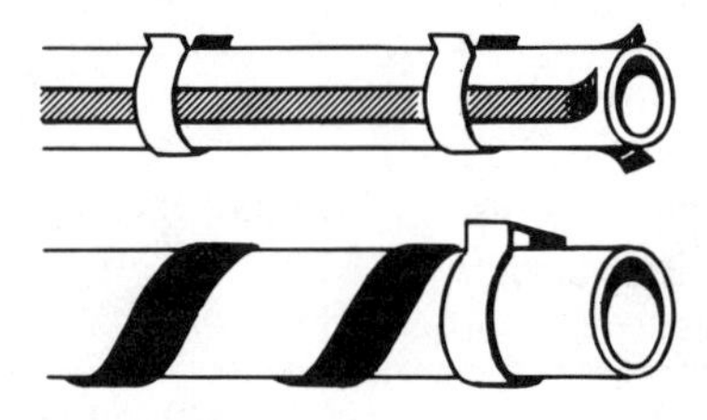

**Fig. 30.** Tube heating using heating tapes [17].

## 3.10. Heating panels and cloths

Heating panels are intended for the uniform heating of large surfaces. Heat transfer takes place by conduction, or, in some cases, by radiation. As a function of their application and type, such panels are sometimes known as heating coatings, jackets, sleeves, cloths or fabrics.

### *3.10.1. Thermal conduction panels*

These panels may be either flexible or rigid.

— *Flexible heating panels*

Flexible panels, which are also known as heating cloths or fabrics, consist of an encased resistance attached to a fabric or heat resistant metal mesh. The fabric is then used to distribute the heat. Panels can also consist of flat heating elements insulated with woven glass, with silicon impregnation. Also, woven asbestos heating cloths exist, with the asbestos warp forming the insulating part and the weft the heating part (however, this technique seems to be disappearing). The maximum operating temperature and power density depend on the coatings used. Figure 31 below gives the characteristics of some flexible heating panels.

Characteristics of flexible heating panels [N]

| Element sheath | Support | Maximum operating temperature (N.B. depending on power denzity) (°C) | Maximum power density (N.B. depending on working temperature ($kW/m^2$) | Physical characteristics | | | |
|---|---|---|---|---|---|---|---|
| | | | | Standard | Impermeable | Explosion-proof Area 1 and 2 | Mechanical protection |
| Silicon-impregnated glass thread | Glass fabric | 400 | 6.6 | • | | | |
| Quartz thread | Quartz fabric | 600 | 13 | • | | | |
| PTFE and silicon impregnated glass covering | Glass fabric | 170 | 2.4 | | • | | |
| PTFE with silicon impregnated glass fabric covering and stainless steel protective braiding | Glass fabric | 170 | 2.4 | | • | | • |
| Silicon rubber | Silicon rubber | 250 | 3 | | • | | |
| Copper-nickel | Glass fabric | 400 | 6.6 | | • | • | |
| Stainless steel | Glass fabric | 400 | 6.6 | | • | • | |
| Stainless steel | Stainless steel fabric | 600 | 13 | | • | • | • |

**Fig. 31.**

Flexible panels are produced in standard dimensions, but may be made to measure to fit different shapes. The most common attachment technique consists of attaching the panels by passing a lace through eyelets along the sides of the panels. The temperature can be accurately controlled by using surface-mounted thermostats. Several heating areas can, for example, be provided as a function of the level of a fluid [2].

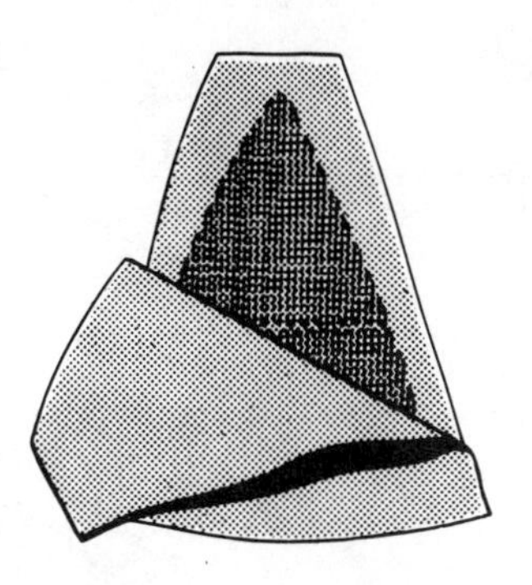

**Fig. 32.** Woven glass insulated nickel-chromium resistance flexible heating panel [17].

— *Rigid heating panels*

For example, some heating elements consist of a moulded resin and glass fibre panel, in which flat strip resistances are inserted. One of the sides is smooth, and a plastic foam-based thermal insulator is applied to the other, at least 25 mm thick. These panels can be produced in all shapes and dimensions. The power supply cable is connected to the elements through a waterproof packing gland moulded in the insulator. This type of panel can be exposed to weather or immersed in aqueous or organic solutions, since the polyester resin used in its manufacture is not affected by most chemical products and can withstand temperatures between — 30 and 100°C. The use of other resins satisfies most requirements, even if they involve highly corrosive products.

These panels have a maximum specific power of about 1 kW/m$^2$ and are usually supplied with single-phase 220 V.

### *3.10.2. Radiating panels*

The heating elements are placed at short distances (approximately 50 mm) from the walls to be heated and attached to a rigid panel. This heating surface often consists of a flexible heating panel. The back of the panel is thermally insulated and protected by a steel sheet.

The unit powers for panels of this type are highly variable, and generally between 1 and 10 kW, with maximum working temperatures of up to 600°C. Power densities range from 1 to 15 kW/m$^2$. Known as heating jackets, heating sleeves or heating coats, all these elements are used for heating recipients, barrels, pumps, valves, etc. The equipment is sometimes constructed from two or more separate parts, depending on the size and shape of the recipient to be heated. With the exception of low power equipment, the heating surface is generally divided into several areas which can be independently regulated.

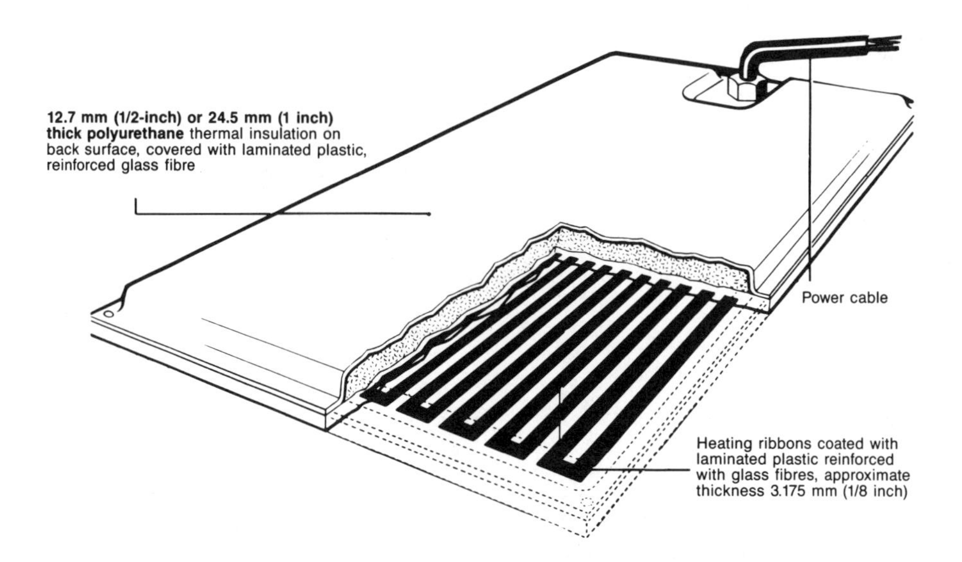


**Fig. 33.** Structure of a rigid heating panel

## 3.11. Finned resistances

These heating elements are used for heating gases. Heat transfer by convection from a solid to a gas takes place slowly, since the convection coefficient, unlike liquids, is low. Therefore, it is very important to increase the surface through which the heat transfer takes place; this is facilitated by the use of fins.

A finned resistance consists of a steel sheath-armored heating element, with cooling fins welded to the tube.

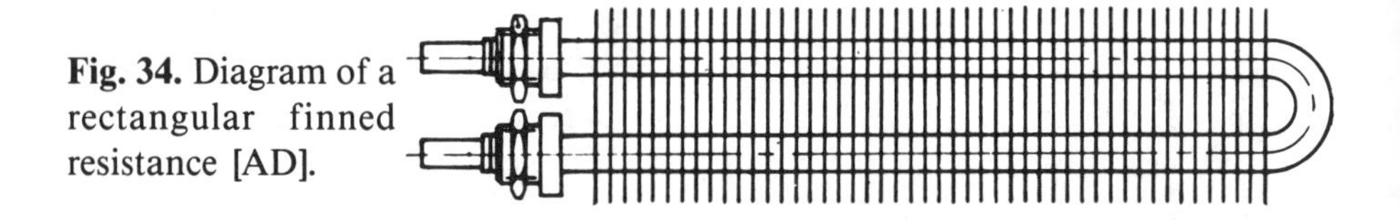

**Fig. 34.** Diagram of a rectangular finned resistance [AD].

While the rectangular shape shown in Figure 34 above is the most common, other designs are available: spiral fins, consisting of a fin helically wound around a cylindrical armored resistance, of practically any shape (straight, curved, folded, multiple looped, etc.) [B].

With forced convection, it is recommended that finned resistances be used in the vertical position, while with natural convection, they should be placed horizontally and on their edges for rectangular elements.

Range of rectangular finned resistances [B].

| Power (W) | 250 | 500 | 750 | 1,000 | 1,500 | 2,000 | 3,000 |
|---|---|---|---|---|---|---|---|
| Heating length (mm) | 150 | 300 | 420 | 520 | 770 | 1,020 | 1,520 |

**Fig. 35.**

Figure 35 gives a range of rectangular finned resistances 50 mm wide and supplied with 220 V. The overall length of the elements is slightly longer to enable connection to the mains. Finned elements available on the market vary greatly with power ranges, which are generally between 100 W and 10 kW. The mean specific power density, with reference to the supporting tube surface, generally varies between 2 and 5 W/cm$^2$, but some elements have a higher power density, up to 10 W/cm$^2$.

The limit utilization temperature is approximately 200°C for natural convection, but may be much higher with forced convection, from 500 to 600°C for some elements; in such cases, a device which switches off the electrical supply to the resistance in the case of a ventilation shutdown must be provided.

Finned armored resistances are used for natural or forced convection heating of buildings, and are often built into enclosed systems (heaters, radiators, etc.), and may also be directly mounted in ventilation ducts. Explosion-proof versions are available.

Also, these resistances are used as heating elements in low temperature furnaces and ovens, generally less than 150°C for natural convection and 600°C for forced convection.

## 3.12. Other encased heating elements

In the above discussion, we attempted to classify encased heating resistances according to major families, although the criteria used may have lead to overlaps or arbitrary divisions. However, many other encased elements exist, and the range is constantly expanding with the appearance of new materials and manufacturing methods; for example, with electron bombardment-radiated plastic heating elements. Many other examples could be mentioned.

### *3.12.1. Heating elements moulded in ceramic fibre refractories*

Heating elements of this type consist of iron-chromium-aluminum or nickel-chromium resistances which are vacuum moulded in a fibre refractory ceramic.

Generally, these elements consist of radiating panels, half-shells and cylindrical tubes, but can be practically any shape, since they are moulded.

The maximum power density is about 20 kW/m$^2$ of heating area (2 W/cm$^2$) with maximum heating element temperature between 1,100 and 1,200°C.

Their use is more comparable to furnace components than armored resistances, since they are most often used in small laboratory or industrial furnaces. Such recently manufactured elements could make use of other developments, such as radiation heating collars (AB), (AC). These elements offer the advantage of rapid implementation and supply their own insulation. Plastic films polymerized at high pressures provide adequate temperature homogeneity.

### *3.12.2. Printed heating elements*

The production of this type of element uses the techniques found in the printed circuit industry. These encased resistances consist of a chemically etched metal sheet inserted between two insulating plastic films. The plastic films are polymerized at high pressure to provide good mechanical and dielectric strength.

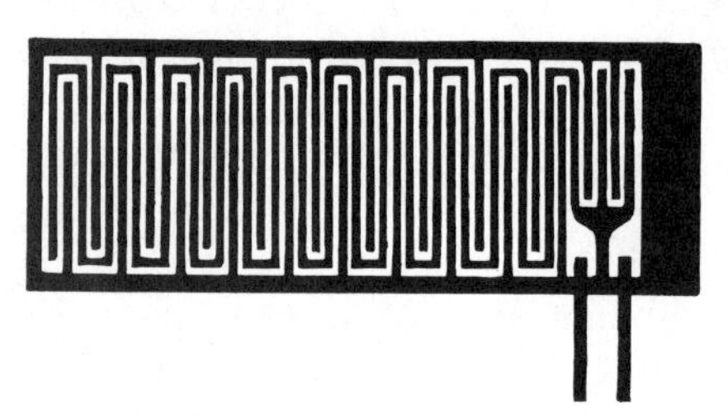

**Fig. 36.** Printed heating element [AC].

Elements of this type are very flexible, but rigid versions can be manufactured. The maximum operating temperature is about 250°C with protection against humidity being excellent.

Dimensions vary from 1 $cm^2$ to 300 $cm^2$ approximately. Power density, however, is low: less than 1 $W/cm^2$.

These elements are used for de-icing, as protection against humidity, in photcopying machines, for preheating tools, conduit anti-icing systems and in the production of small, low temperature ovens. Their lightness and low bulk make these elements suitable for aviation and electronics applications.

### *3.12.3. Heating cloths*

Flexible heating panels, which are in fact conventional heating cloths or fabrics, were described in paragraph 3.10. With these, the resistance consists of an encased element (e.g. cable) attached to the fabric or a metal grid, or textile glass insulated flat elements.

Other techniques have been developed in this field, which make use of new materials generally obtained from chemical research. In some cases, the heating element may be the cloth itself.

An example of this is a cloth in which the heating element consists of a textile glass support, coated with electrically conductive synthetic materials. Generally, this coating is a mixture of graphite, carbon black and resins. Power is supplied through two parallel electrodes (copper wires) located on each edge of the strip of cloth and throughout its complete length.

Sealing and electrical insulating of this heating surface and the electrodes are provided by a plastic film, for example polyester, deposited as a glaze on both surfaces of the cloth.

For 60 cm wide strips with a 50 cm heating element, the linear power varies from 75 to 180 W/m, i.e. a surface specific power of 0.015 to 0.036 $W/cm^2$.

These powers are low with a maximum continuous operating temperature of approximately 80°C [B].

Because of the low power density, this type of cloth is most often used to heat work areas or living areas through radiating ceilings, but can also be used to heat tanks or reservoirs where power is limited, and for de-icing, platform heating, etc. Some cloths of this type can withstand temperatures up to 150°C, increasing their range of applications.

## 4. ADVANTAGES AND LIMITATIONS OF ENCASED ELECTRIC RESISTANCES

Due to their diverse designs and applications, it is difficult to define the advantages of encased electrical resistances independently of application.

However, an important advantage is their diversity of shapes, types, materials, operating temperatures and power densities. This diversity enables an encased heating element to be found for most applications, if the requirements are carefully analysed and the rules concerning their use are respected.

Encased heating elements are, in such cases, highly reliable, require practically no maintenance and offer long service life; also, they provide excellent temperature control and uniform heating, which can be adapted to specific requirements.

Encased resistances also enable decentralized heating, in many cases where a centralized heating system with an intermediate thermal fluid is used (heating liquid reservoirs, moulds, etc.), a system which sometimes offers low thermal efficiencies due to distribution and regulating losses.

These resistances, which are easy to use and associated with thorough insulation and adequate regulation, often provide appreciable energy savings.

Also, the investments involved in the use of encased resistances are limited: sometimes less than those required with other techniques using electricity or fossil fuels.

There are few limits concerning the use of these resistances, except where very high temperatures and power densities are involved. Only a close examination of the application can enable a precise assessment of the advantages and limitations of this heating method.

## 5. INDUSTRIAL APPLICATIONS OF ENCASED ELECTRIC RESISTANCES. DESIGN ELEMENTS

Encased electric resistance applications vary widely and include all kinds of solids, liquids and gases, in all sectors of industry. These resistances are used in thermal production processes and for heating industrial or living areas. Therefore, it is impossible to describe all applications in detail.

In the following paragraphs, significant examples of applications which can be extrapolated to resolve other industrial heating problems, are described.

Generally, for a given application, it is easy to obtain at least an approximate estimate of the specific power to be installed in terms of heating elements. The specific power and type of application then enable a better estimate of the type of elements required and their characteristics, and therefore investment costs.

The price of heating elements is determined by type and dimensions, by the materials forming the casing, electrical insulation and the resistance, and also by power.

In particular, specific power is limited by the maximum temperature at which the element can operate, and therefore by:

— the maximum temperature of the material forming the resistance;

— the temperature at which the insulator will no longer fulfill its function;

— the maximum temperature which the envelope can withstand;

— the shape of the element and its cross-section.

The power density capability of encased resistances also depends on the characteristics of the environment, i.e., on heat transfer conditions:

— type of body to be heated;

— location of the heating element;

— thermal conductivity of the body to be heated;

— viscosity and rate of flow in fluid heating;

— boiling and melting point of the body to be heated;

— surface condition of the heating element;

— temperature difference between the body to be heated and the sheathed resistance.

To give only two examples, when it is necessary to heat aqueous solutions to a temperature below boiling point, the power density must not be too high, since the high temperature of the casing would then cause vaporization of the liquid coming in contact with it; therefore, it is important to provide an element area large enough to enable conductive heat transfer between the heating element and the fluid at a relatively low temperature difference, thus avoiding vaporization. Similarly, if a poor heat conductor such as bitumen is heated, the power density must be low (about 1 $W/cm^2$, or 8 $W/cm^2$ for water and 3 to 6 $W/cm^2$ for aqueous solutions): the heat of an overpowered resistance cannot be carried away, thus causing overheating, followed by destruction of the heating element.

Thus, determination of the specific power density of encased elements is a result not only of thermal calculations, but also of designer and user know-how; tests are often indispensable, at least where new applications are involved to confirm the conclusions of studies.

## 5.1. Using encased resistances for heating tanks and reservoirs

Encased resistances are often used to heat tanks, baths and reservoirs, etc., in numerous industries (surface treatment baths in metal transformation industries, tanks in chemical and foodstuff industries, bitumen tanks in the road industry, leather and skin transformation tanks, textile baths, etc.) [5], [7], [8], [14], [16], [17].

### 5.1.1. *Electrical techniques for heating tanks and reservoirs*

Electrical heating systems for tanks and reservoirs vary widely, as shown in the table of figure 37.

Four major systems, with numerous variants are, in fact, used:

#### 5.1.1.1. *Decentralized heating*

— *Immersed elements (immersion heaters)*

Immersion heaters can be arranged in two ways:

• astride the wall, which dispenses with drilling holes in the tank and fitting a seal;

• conversely, traversing the wall either through a seal (screw-in immersion heater) or flange (flanged immersion heater); immersion heaters of this type can also be fitted with removable heating elements placed in a metal tube or casing, to prevent draining of the tank if one of the elements becomes defective.

The casing for immersion heaters may be metallic (mild steel, stainless steel, copper, etc.) or non-metallic (silica, graphite, teflon, etc.).

Electrical tank and reservoir heating equipment

| | Centralized heating | Decentralized heating |
|---|---|---|
| Heating equipment | Electric resistance boiler<br>Electrode boiler | Immersed elements (immersion heaters)<br>Elements placed against walls (flexible heating elements, panels, etc.)<br>Water jacket<br>Circulation heater |
| Heated fluid | Hot water, steam, thermal fluid | Water (for water jacket, circulation heater) |

**Fig. 37.**

Immersion heaters must be selected carefully for each application to ensure that:

• the power density (expressed in Watts per square centimeter, the area considered being that of the external casing of the heating element through which the heat transfer takes place) is low enough to prevent damage to the solution being heated, but also to prevent overheating of the encased resistance.

• the material forming the casing is capable of withstanding the operating temperature and the corrosive effect of the bath to be heated.

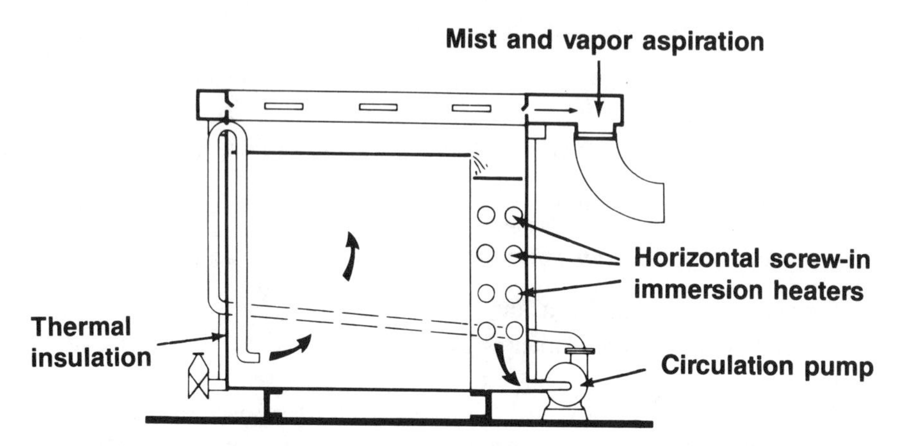


**Fig. 38.** Battery of horizontal immersion heaters placed in a fluid circulation tank [17].

Both points should be discussed in detail with the heating element suppliers.

In immersion heating, and where it is necessary for the heaters not to be located in the tank, the use of circulation heaters offers an interesting option.

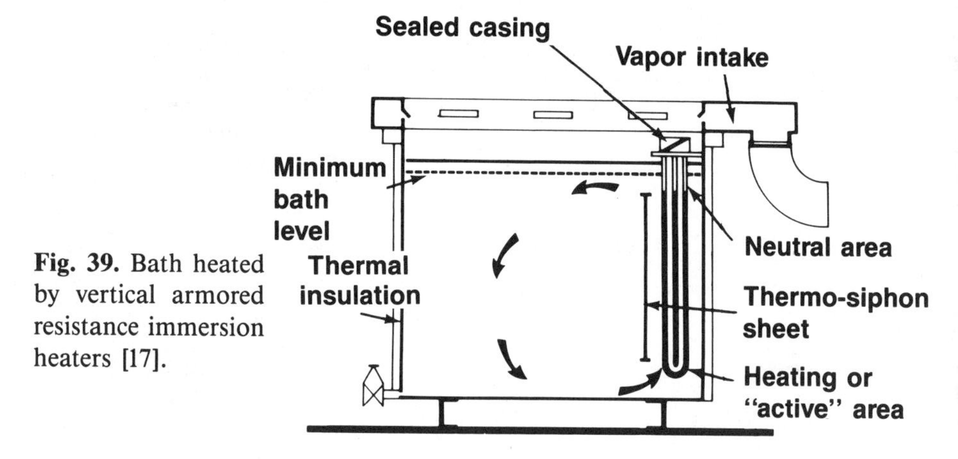


**Fig. 39.** Bath heated by vertical armored resistance immersion heaters [17].

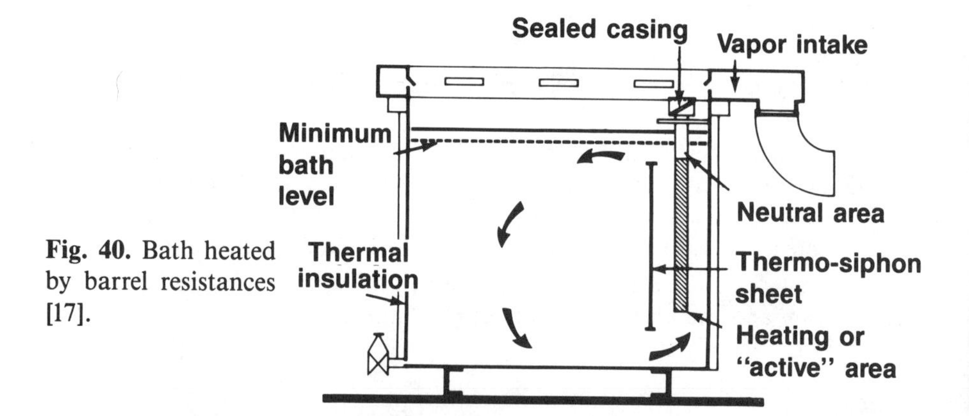


**Fig. 40.** Bath heated by barrel resistances [17].

— *Elements located on surface of tank*

When it becomes difficult to use immersion heaters (risk of very rapid contamination or damage to submerged parts, requirement for completely empty tank, or excessively corrosive liquid, etc.), it is possible to heat the bath using armored resistances clamped directly to the tank. In such cases, the tank must be correctly insulated if an acceptable level of efficiency is to be maintained.

The armored resistances used may be tubular, flat or annular. Metallic casing resistances can also be used for this application, as well as flexible, panel, strip, ribbon and jacket-type elements.

Externally clamped heating elements offer the advantage of not being subject to chemical corrosion, but in general are not used with PVC or rubber coated tanks, due to heat transfer problems (insulating effect of inside covering of tanks).

**Fig. 41.** Tank heating using horizontal resistance immersion heaters mounted on barrels and placed in casing [18].

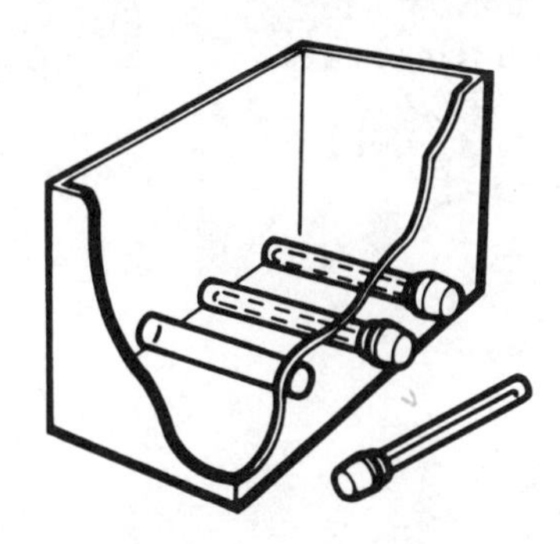

*5.1.1.2. Centralized heating*

Centralized electric heating must be used only when decentralized heating is impossible, since it is generally less efficient and economical.

Two main systems exist:

— *Water jacket tank (water bath)*. The water in which the treatment tank is immersed is heated by immersion heaters or circulation heaters. This can be considered as combined decentralized and centralized heating.

— *Hot water and steam electric boilers* (resistances or electrodes) or organic thermal fluid (resistance). The bath is then heated by an heat exchanger placed in the tank or outside the tank.

**Fig. 42.** Heating of tank by armored resistances placed against the wall [18].

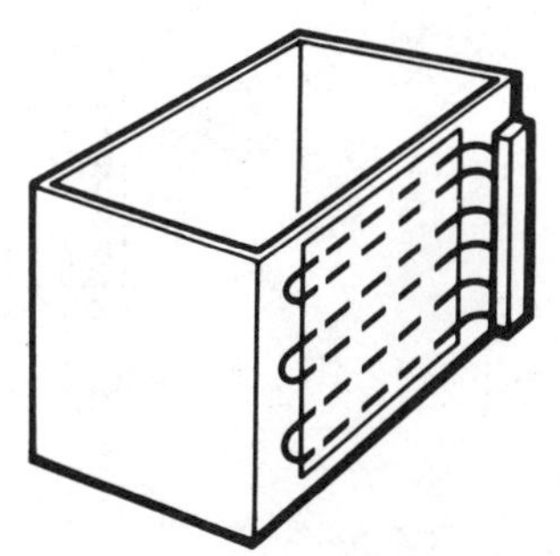

### *5.1.2. Equipment power*

Some tanks must be heated to treat immersed products (baths for preparation of metal surfaces prior to treatment, cleaning baths in the food industry, etc.); others contain products which must be heated before chemical or biological transformation. Finally, some tanks are used simply to store the products, keeping them at a temperature suitable for future transformation.

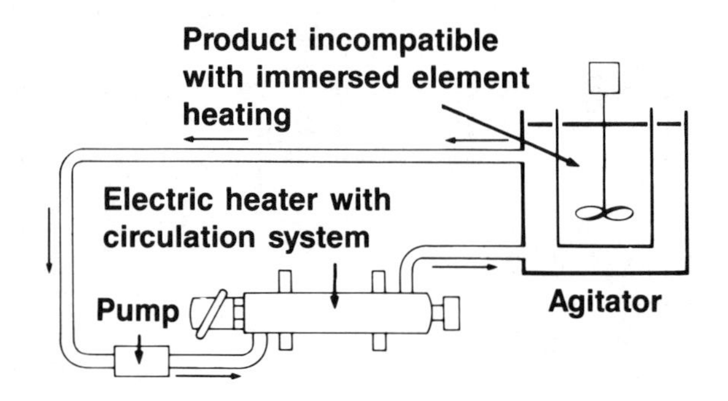


**Fig. 43.** Heating of tank through water jacket with circulation heater [17].

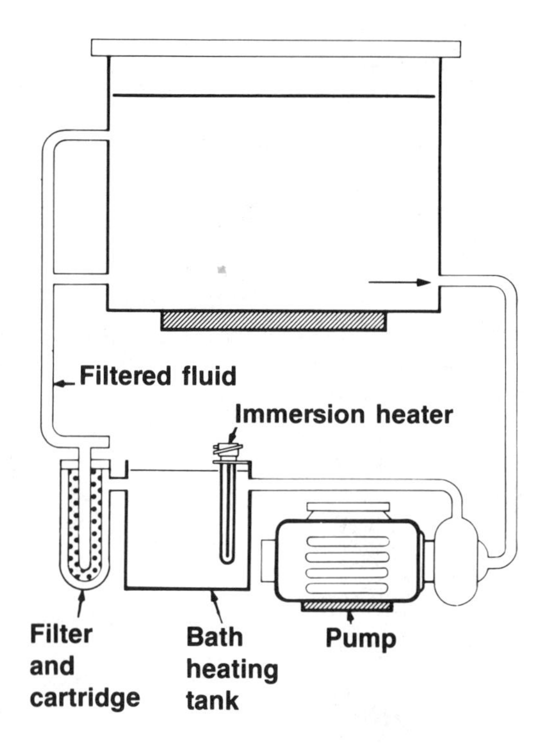


**Fig. 44.** Bath heated externally by immersion heater [17].

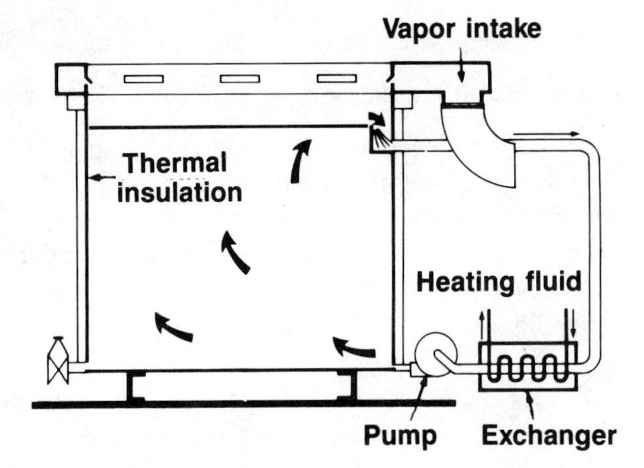


**Fig. 45.** Bath heated by steam or thermal fluid electric boiler [17].

In these three cases, thermal requirements are noticeably different, but the following design considerations will generally enable the installed power and the electrical energy consumption to be calculated.

*5.1.2.1. Thermal requirements for a tank in which products are treated by immersion*

The instantaneous thermal requirements (bounded by possible restrictions on fluid being heated) are represented by:

— the energy required to raise the temperature of the immersed products;

— tank thermal losses;

— heating fluid introduced to compensate for treatment temperature losses due to evaporation and treated-product movement:

$$\boxed{W_1 = m_1 c_1 (\theta_4 - \theta_1) + m_2 c_2 (\theta_3 - \theta_2) + D,}$$

$W_1$, power in Watts,
$m_1$, weight of product to be heated in kilograms per hour;
$c_1$, specific heat of products to be treated in Watthours per kilogram per degree Celsius;
$\theta_1$, initial temperature of products to be treated in degrees Celsius,
$m_2$, weight of liquid to be heated in kilograms per hour;
$c_2$, specific heat of liquid to be heated in Watthours per kilogram per degree Celsius;
$\theta_2$, initial temperature of liquid to be heated in degrees Celsius;
$\theta_3$, bath temperature in degrees Celsius;
$\theta_4$, final temperature of products on removal from bath in degrees Celsius, in general, $\theta_4 = \theta_3$
$D$, thermal losses through the walls and surface of the bath in Watts.

Addition of liquid to compensate for losses due to movement and evaporation is not always performed continuously, since it is often low. It is therefore advantageous to schedule this in periods when electricity is particularly cheap (off-peak rates); the electrical power required is proportionally reduced.

In practice, a safety factor of about 1.1 — 1.2 is used when evaluating power.

### *5.1.2.2. Power requirement for tank or reservoir during initial heating*

This power corresponds to the initial heating of the treatment bath and when the temperature of the products to be subjected to chemical or biological transformation has to be raised.

The installed power should be enable heating during time $t_m$, specified for the temperature rise of:

— the liquid;

— the tank;

and instantaneously, to compensate for losses due to:

— evaporation and radiation at the bath surface for open tanks;

— tank or reservoir walls.

#### *5.1.2.2.1. Simplified calculation of the power required during temperature rise*

The power to be installed $W_2$ is equal to the energy accumulated in the bath, between ambient temperature and the treatment temperature, divided by the temperature rise time, plus, as an initial approximation, 70% of the tank losses (walls and surface) at treatment temperature. This power is expressed by the equation:

$$\boxed{W_2 = \frac{m_2 c_2 (\theta_3 - \theta_2)}{t_m} + 0.7\,D}$$

$W_2$, power in Watts;

$m_2$, weight of liquid in tank or reservoir in kilograms;

$c_2$, specific heat of liquid in Watthours per kilogram per degree Celsius;

$\theta_3$, final temperature of the liquid in degrees Celsius;

$\theta_2$, initial temperature of the liquid in degrees Celsius;

$t_m$, temperature rise time in hours;

$D$, thermal losses through the walls and surface at final temperature 3, expressed in Watts.

This equation does not allow for the rise in temperature of the tank or reservoirs. The corresponding energy is generally low, and can be taken into account by adding the following term:

$$\frac{m_3 c_3 (\theta_3 - \theta_2)}{t_m}.$$

to the above equation.

If the liquid's final temperature is close to boiling point and the tank is open, it is better to use 0.8 $D$ to 0.9 $D$ instead of 0.7 $D$ for mean losses during temperature rise.

Generally, these approximations are sufficient for insulated tanks; in fact, power is estimated by this expression with an error margin of sometimes less than 10%. However, a more rigorous calculation can be performed, especially if it is desired to follow temperature change throughout the heating period.

*5.1.2.2.2. Temperature changes in the liquid during temperature raising, and corresponding power*

The change in temperature $\theta$ with time t is expressed by the equation:

$$\boxed{\theta=\theta_0+RW_2(1-e^{-t/RC}),}$$

if thermal losses D can be expressed as:

$$D=KS(\theta-\theta_0)=\frac{\theta-\theta_0}{R},$$

$K$ total heat transfer coefficient and $R$ total thermal resistance of the liquid-tank assembly; this expression is also only an approximation, since $K$ and $R$ generally depend on the temperature and are not constant
$\theta$, bath temperature at time $t$ in degrees Celsius;
$W_2$, heating element power in Watts;
$t$, time in hours;
$\theta_0$, bath, tank and ambient atmospheric temperature in degrees Celsius;
$R$, total equivalent thermal resistance between the bath and the environment expressed in degrees Celsius per Watt;
$C$, total thermal capacity of the liquid-tank assembly in Watthours per degrees Celsius; this is equal to the sum of all the elementary thermal capacities $m_i c_i$ (tank, liquid, insulator, etc.) where $m_i$ is the weight in kilograms and $c_i$ the specific heat in Watthours per kilogram per degree Celsius (for water, $c_i$ = 1.16 Wh/kg X °C).

The power required to obtain a temperature rise over time $t_m$ is then:

$$\boxed{W_2=\frac{\theta_{t_m}-\theta_0}{R(1-e^{-t_m/RC})}.}$$

If its value at temperature $\theta_{t_m}$ is used to estimate the total equivalent thermal resistance, the expression giving $W_2$ becomes:

$$\boxed{W_2=\frac{D}{1-e^{-t_m/RC}},}$$

$W_2$, heating power in Watts;
$D$, thermal losses;
$t_m$, duration of temperature increase in hours
$R$, equivalent total thermal resistance between the bath and the environment, expressed in degrees Celsius per Watt;
$C$, total thermal capacity in Watthours per degree Celsius.

This method leads to an "over" estimate of the power to be installed. Figure 46 gives the value of coefficient:

$$k=\frac{1}{1-e^{-t_m/RC}},$$

as a function of temperature rise time and the time constant, equal to the product *RC*.

### *5.1.2.3. Maintaining temperature in a tank or reservoir*

The power required to maintain the temperature is equal to the thermal losses of the tank or reservoir at storage temperature; for safety reasons, the power calculated in this manner is generally increased by 10 to 20%.

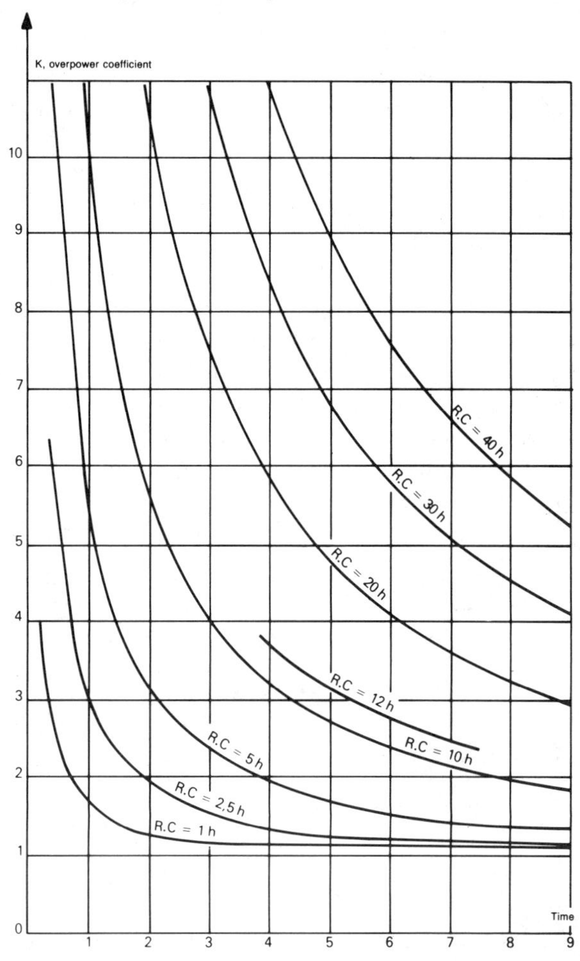


**Fig. 46.** Value of overpower coefficient *k* versus temperature rise time and time constant [17], [19].

When the tanks are correctly insulated, the corresponding powers are low.

### *5.1.2.4. Power to be installed in a tank or reservoir*

Each case requires special analysis. However, some guidelines can be given.

### *5.1.2.4.1. Duration of temperature rise for direct bath heating*

When the objective in heating a tank or reservoir is to take the liquid products to a temperature at which they are to be chemically or biologically transformed within a given time, a power equal to $W_2$ must be installed with a slight safety margin, where necessary.

If this time is short, the powers installed can be high. Particular attention must be paid to the surface power density of the elements ($W/cm^2$) so that an excessive power density does not cause va porization of the liquid at the heating element-liquid exchange surface, unless of course this is permitted.

Insulation of high powers increases investment costs. Therefore, an attempt should be made to ensure that, by careful organization of production, the temperature rise duration can be increased (tanks operating continuously, in which only the initial temperature rise is long).

### *5.1.2.4.2. Duration of temperature rise for a indirect bath heating*

Where heating of a fluid is intended to initially heat a bath which will then be used to treat pieces or products by immersion, the power required for a rapid temperature rise is generally much higher than that required during treatment.

In such cases, it is better to install a lower power, slightly higher than treatment power $W_l$ calculated above, and to provide for a longer temperature increase, for example automatically programmed during low rate hours. Only the initial temperature rise (after several days of interruption) may be long. If the precaution of covering open baths outside of production times is taken, the bath temperature will drop only very slightly and return to operating temperature will be rapid. The investment savings obtained by using this solution are often substantial, and important operational savings are also obtained by reduction of the power purchased from source.

### **5.1.2.5.** *Temperature rise of immersed products and bath cooling*

— *Temperature rise*

Generally, with the exception of large pieces which are poor heat conductors, temperature raising of immersed products is generally rapid. Changes in temperature, and therefore heating times (with the calculation method used, the part reaches the bath temperature after an infinite time only; to calculate the temperature rise time, it is considered that the temperature sought is reached when the temperature of the part is equal to a certain percentage of the bath temperature, e.g. 95 to 99%), can be estimated for thin parts of adequate thermal conductivity, by the expression:

$$\theta' = \theta'_0 + (\theta_3 - \theta'_0)(1 - e^{-(hs/mc)t}),$$

$\theta'$, temperature in degrees Celsius reached by part after time $t$;
$\theta'_0$, initial temperature of part in degrees Celsius,
$\theta_3$, bath temperature, considered as constant, in degrees Celsius;

$h$, surface transfer coefficient between part and fluid in Watts per meter squared per degree Celsius;

$s$, area of parts in contact with the liquid;

$m$, weight of parts immersed in bath in kilograms;

$c$, specific heat of parts in Watthours per kilogram per degree Celsius.

The value of $h$ W/m²/°C depends on numerous factors (shape and arrangement of parts, liquid viscosity, circulation rate, temperature difference between part and liquid, etc.), but, as an initial approximation, and for medium viscosity liquids such as oil, a value of 200 can be considered. If the liquid is in motion, these values can be multiplied by three or four.

For example, the temperature rise for metal parts of a few millimeters thickness is in most cases between 1 and 2 minutes.

For more rigorous calculations, it is necessary to have recourse to the conventional thermal laws concerning convection exchanges and heat propagation by conduction in a body.

— *Drop in bath temperature on introduction of parts*

When the temperature rise is rapid, and to calculate the temperature drop due to insertion of parts or products, it is possible to neglect the heat added by the heating elements during this short period, together with the tank thermal losses. Bath cooling $\Delta\theta_3$ can then be estimated by the expression:

$$\Delta\theta_3 = \frac{mc(\theta_3 - \theta_0')}{m_3 c_3},$$

$m_3$ and $c_3$ are bath weight and specific heat respectively.

The temperature drop is generally very low and, in many cases, less than 1°C.

— *Bath temperature drop in the event of prolonged heating interruption*

It is also of interest to measure the effect of prolonged element shutdown. Unlike the example given above, it is necessary to allow for bath thermal losses $Dt_1$ over the number of hours $t_1$. The temperature drop can be obtained from the relation:

$$\Delta\theta_3 = \frac{1}{m_3 c_3}[mc(\theta_3 - \theta_0') + Dt_1].$$

In some cases, the temperature drop may be lower. For instance, for a 70,000 l. aqueous solution bath treating 5,000 kg of steel parts per hour at a temperature of 50°C, the temperature drop, after switching off the heating elements for a period of two hours, is only 2 to 3°C. In such cases, and when high temperature accuracy is not required, the low temperature drop of the bath can be used to interrupt, or at least reduce heating during peak rate hours, offering a considerable energy saving without hindering production.

### 5.1.2.6. *Calculation elements*

In most cases, the data which follow enable calculation of the power to be installed by applying the methods described above.

#### 5.1.2.6.1. *Thermal losses at the free surface of an aqueous solution*

Generally, these losses are high and comprise losses due to evaporation, the highest, and losses due to radiation and convection.

Evaporation losses $D_l$ depend basically on temperature and the bath ventilation and agitation rate.

For an ambient temperature of 15°C, Figure 47 gives losses due to evaporation versus bath temperature. These thermal losses are expressed in Watts per square meter of bath. For an ambient temperature of betweeen 0 and 20°C, linear interpolation can be used to determine losses. If losses are $y_l$ for bath temperature $\theta$ and ambient temperature $\theta_1$, losses $y_2$ for ambient temperature $\theta_2$ will be:

$$y_2 = y_1 \frac{\theta - \theta_2}{\theta - \theta_1}.$$

The quantity of water evaporated can easily be calculated from vaporization losses as shown on Figure 47, if the latent heat of vaporization is known; at atmospheric pressure, latent heat of evaporation of water can be considered as 0.65 kWh/kg over the temperature ranges considered.

The evaporated solution must be replaced and reheated. However, this replacement can be carefully programmed (during off-peak hours).

Figure 48 gives convection and radiation losses at the surface of a bath versus temperature for an ambient temperature of 15°C. Within this temperature range, the curve representing losses can be considered as approximately linear. These losses are much lower than evaporation losses and, in the approximate calculation, it is not necessary to allow for bath agitation and ventilation which can increase losses considerably.

The total losses through the surface are high. A method of reducing these, for example, involves placing plastic balls (polypropylene, for example) on the bath surface, where compatible with production conditions. The gain in energy consumption can be 25 to 30%.

#### 5.1.2.6.2. *Thermal losses through walls of tank or reservoir*

Figure 49 gives losses for an ambient temperature of 15°C through the walls of a steel tank as a function of wall insulation (mineral wool). These losses, which are calculated for plane walls can be approximated for surfaces of different shapes. For more precise calculations, the fundamental thermal laws must be used.

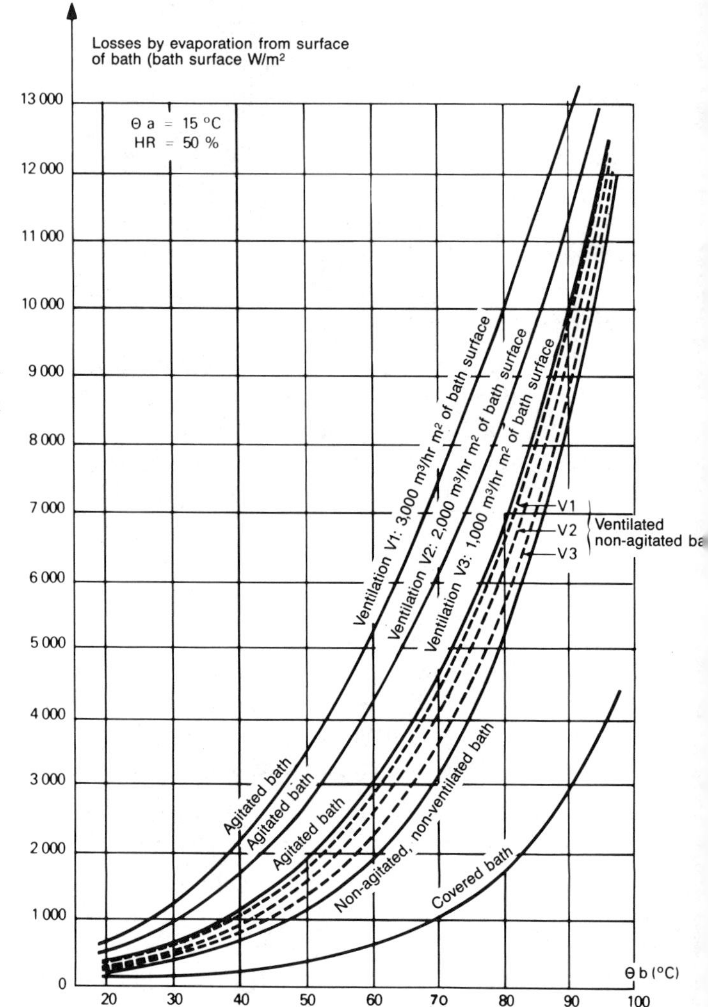


**Fig. 47.** Thermal losses due to evaporation at bath surface [17], [19].

For an ambient temperature of between 0 and 20°C, it is possible to calculate losses by linear interpolation as for surface losses.

However, some liquids are stored at much higher temperatures. The curve of Figure 50 gives losses through walls of a tank insulated by 5 cm of mineral wool with an outside temperature of 0°C.

If the outside temperature is other than 0°C, the thermal losses can be calculated, as an initial approximation, by linear interpolation.

These figures demonstrate the importance of correct insulation of tanks and reservoirs, which must never be less than 5 cm.

Generally, these data can be used to obtain a sufficiently accurate calculation of the power to be installed and the operational energy consumptions. More

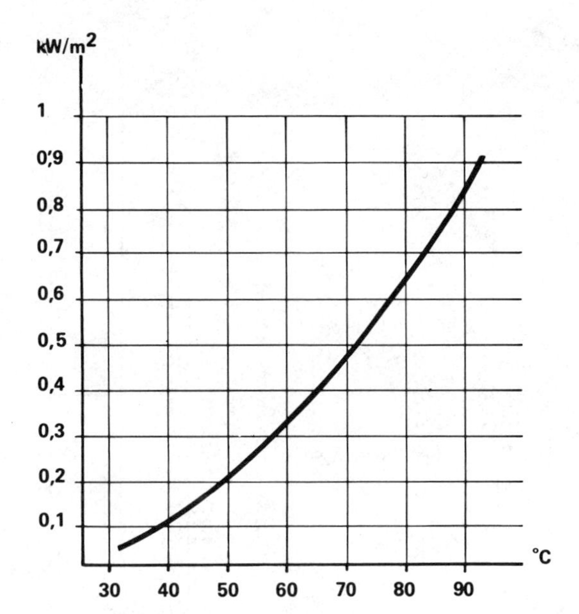


**Fig. 48.** Convection and radiation losses from the surface of an open bath [17].

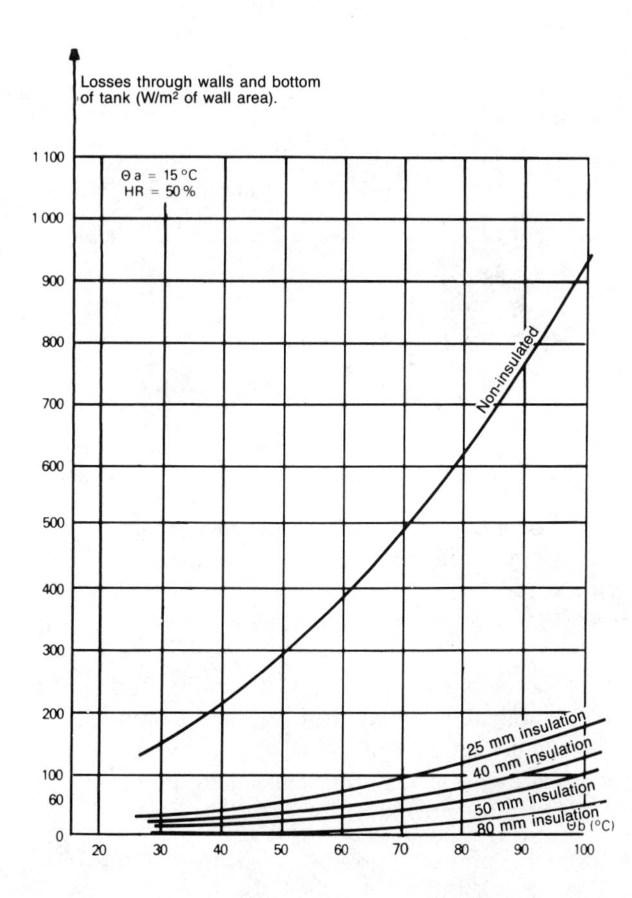


**Fig. 49.** Losses through walls of tank or reservoir [17], [19].

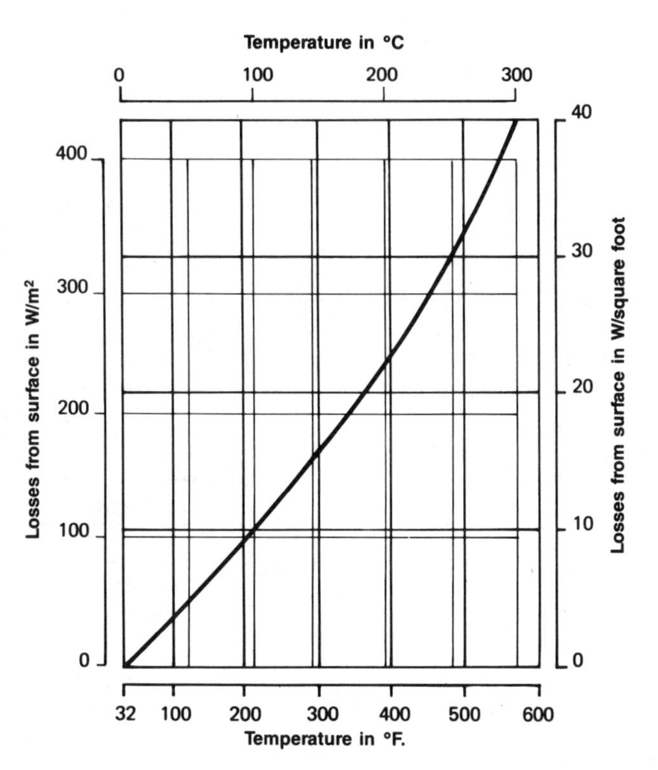


**Fig. 50.** Losses through walls of tank insulated by 5 cm of mineral wool with an outside temperature of 0°C [18].

| Thermal insulation (mm) | Correction factor |
|---|---|
| 0 | 20 |
| 25 | 1.9 |
| 50 | 1 |
| 75 | 0.7 |
| 100 | 0.54 |
| 150 | 0.36 |
| 200 | 0.27 |

**Fig. 51.** Correction factor for thermal losses for an insulation thickness other than 5 cm [18].

accurate calculations, allowing for transient modes can, however, be performed using the fundamental thermal laws.

## 5.2. Heating and maintaining tubes containing liquids

Essentially, tube heating is surface heating. Whenever it is technically possible in tank and reservoir heating, efforts are made to use immersed element heating, usually the most efficient thermally and economically. Conversely, with tube heating, external heating using encased elements is, in most cases, the simplest and most profitable method.

The industries involved vary widely — petroleum, chemical, food, road-building industries — including heating fuel and protection against freezing, which apply to all industries.

The most widely used heating elements are:

— tapes (*see* paragraph 3.9.);

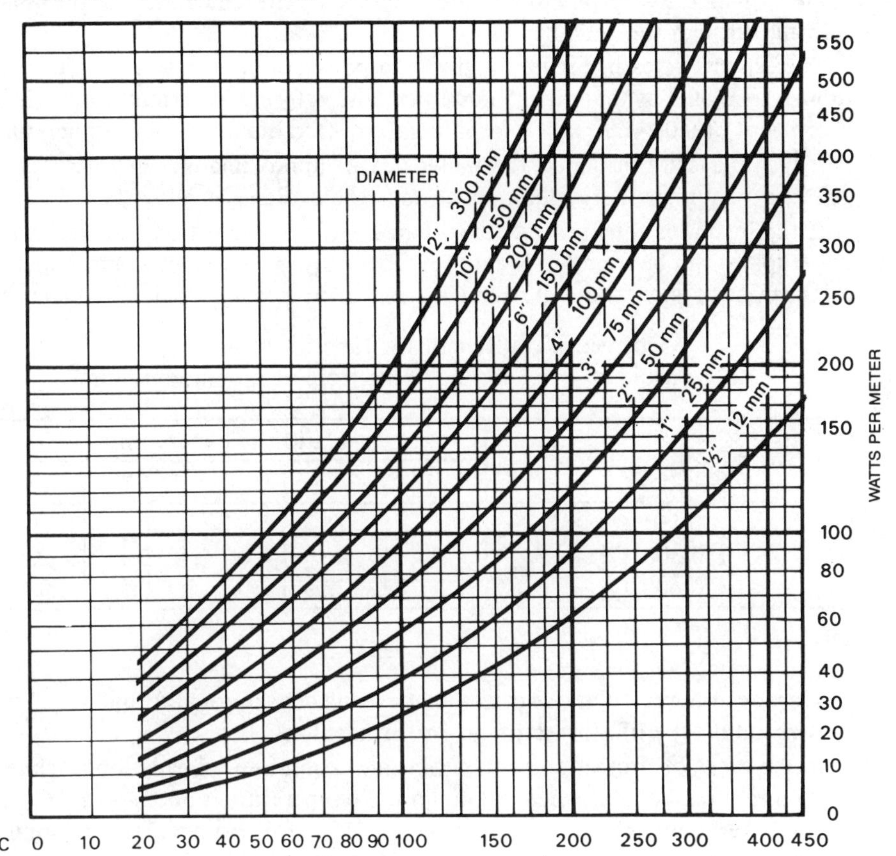


**Fig. 52.** Power required to maintain the temperature of a steel tube as a function of tube diameter (expressed in millimeters and inches) and the difference in temperature between the fluid flowing through the tube and the environment: thickness of insulator of thermal conductivity 0.06 W/m x °C equal to 25 mm [N].

— cables (*see* paragraph 3.8.);

— collars (*see* paragraph 3.6.).

In the first two cases, and as a function of the required power density and heating uniformity, the heating cable can be straight or helical. Numerous variants exist, suitable for most situations — explosive atmospheres, high or low temperatures, highly viscous products, grounding requirements, etc.

Armored heating elements can, however, be placed inside tubes which is equivalent to circulation heating. However, this type of equipment is not suitable for heating tubes of long lengths at constant temperature and low power density.

In such applications, it is more often necessary to maintain product temperature than to increase it. Circulation heaters and other systems are of interest when it is necessary to increase significantly the temperature of the liquid flowing through the tube.

The set of curves in Figure 52 gives the power to be installed per meter of tubing, insulated by material 25 mm thick, and offering a thermal conductivity coefficient of 0.06 W/m x °C (for mineral wool) to maintain tube temperature.

If the tube is not insulated, thermal losses are approximately ten times higher, giving a very high energy loss, which must be avoided at all costs.

If the insulator thickness is other than 25 mm, a correction coefficient for linear losses can be found in Figure 53. Obviously, optimum economy depends on the price of energy, and the trend is therefore to increase insulation thickness.

Corrective coefficient for tube linear losses;
*to be avoided, since too different from economic optimum

| Working temperature | Insulation thickness (mm) | | | | |
|---|---|---|---|---|---|
| | 25 | 37.5 | 50 | 75 | 100 |
| 0-150°C | 1 | 0.8 | 0.7 | 0.5 | * |
| 150-300°C | * | 0.8 | 0.7 | 0.5 | 0.4 |
| 300-600°C | * | * | 0.7 | 0.5 | 0.4 |

**Fig. 53.**

More accurate calculations can be made using conventional formulae (taking into account type of flow: laminar or turbulent, etc.).

In the event of protection against freezing, a minimum insulation thickness of 25 mm must be used when the minimum temperature is about — 10°C; for much lower temperatures, — 25 and — 35°C for example, the minimum insulation thicknesses must be 40 and 60 mm respectively.

When calculating power, a safety coefficient of 10 to 20% should be used, in particular for outside tubing, in which extreme atmospheric conditions may be encountered — very high winds, abnormal temperature drops, etc.

Electrical tracing systems for pipelines are being developped rapidly, since investment cost is relatively low, control excellent and energy efficiency high [1], [3], [4], [N], [AA].

## 5.3. Industrial electric water heaters

In numerous industries (food, textile, wood, leather dressing, tanners, etc.) large quantities of hot water are required for production. In such cases, it is profitable to produce and accumulate hot water in reservoirs or tanks at atmospheric pressure, using low electricity rates available during off-peak hours.

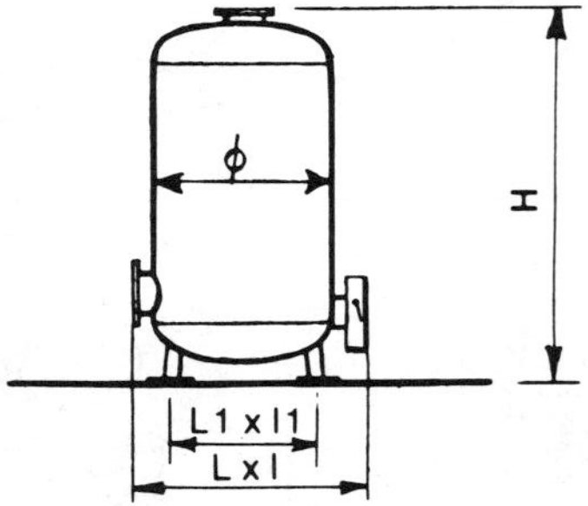

| Effective capacity in litres | Empty weight in kg | Power for heating to 80ºC in 8 hours in kW | ∅ in mm | L x l in mm | H in mm | L1 x l1 in mm |
|---|---|---|---|---|---|---|
| 5,000 | 1,200 | 72 | 1,500 | 2,300x1,500 | 3,650 | 1,350x1,350 |
| 10,000 | 1,850 | 120 | 1,900 | 2,700x1,900 | 4,400 | 1,700x1,700 |
| 15,000 | 2,200 | 192 | 1,900 | 2,700x1,900 | 6,200 | 1,700x1,700 |
| 20,000 | 2,700 | 240 | 2,500 | 3,300x2,500 | 5,050 | 2,250x2,250 |
| 25,000 | 3,150 | 312 | 2,500 | 3,500x2,500 | 6,100 | 2,250x2,250 |
| 30,000 | 3,500 | 360 | 2,500 | 3,500x2,500 | 7,150 | 2,250x2,250 |
| 40,000 | 4,000 | 480 | 3,000 | 4,000x2,500 | 6,700 | 2,700x2,700 |
| 50,000 | 4,650 | 600 | 3,000 | 4,000x2,500 | 8,130 | 2,700x2,700 |

**Fig. 54.** Range of vertical water heaters [AI].

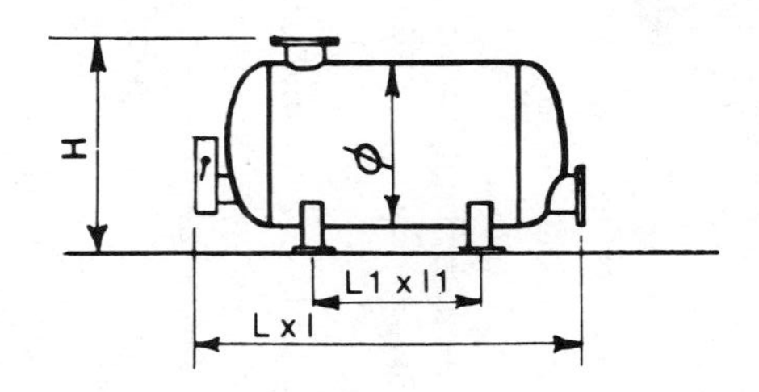


| Effective capacity in litres | Empty weight in kg | Power for heating to 80ºC in 8 hours in kW | ∅ in mm | L x l in mm | H in mm | L1 x l1 in mm |
|---|---|---|---|---|---|---|
| 5,000 | 1,200 | 72 | 1,500 | 4,150x1,500 | 2,200 | 2,300x1,350 |
| 10,000 | 1,850 | 120 | 1,900 | 4,900x1,900 | 2,600 | 3,070x1,700 |
| 15,000 | 2,200 | 192 | 1,900 | 6,700x1,900 | 2,600 | 4,870x1,700 |
| 20,000 | 2,700 | 240 | 2,500 | 5,550x2,500 | 3,200 | 3,750x2,250 |
| 25,000 | 3,150 | 312 | 2,500 | 6,600x2,500 | 3,200 | 4,800x2,250 |
| 30,000 | 3,500 | 360 | 2,500 | 7,650x2,500 | 3,200 | 5,850x2,250 |
| 40,000 | 4,000 | 480 | 3,000 | 7,190x3,000 | 3,700 | 5,440x2,700 |
| 50,000 | 4,650 | 600 | 3,000 | 8,630x3,000 | 3,700 | 6,880x2,700 |

**Fig. 55.** Range of horizontal water heaters [AI].

— *Typical installation*

The quantity of water required for one working day is stored in an enclosure, generally cylindrical and constructed from welded steel sheet. In most cases, contents vary between 5,000 and 100,000 litres, but much higher capacities are in service (150,000 litres in a concrete tank with an installed power of 1,100 kW). This tank is insulated by a mineral wool-type insulation at least 50 mm thick (75 to 100 mm is recommended) protected by a metal or plastic jacket.

The inside of the tank is protected against corrosion for hot water applications (paint, resins). The tanks are either horizontally or vertically mounted, and one or two manholes are provided for inspection and cleaning.

The water is heated by armored resistance immersion heaters but rarely by outside heating jackets since the power density is generally too low. Installed power is about 11 kW/m$^3$.

The maximum temperature of the water in these tanks at atmospheric pressure is about 80-85°C. If the hot water is used at a lower temperature, it is not necessary to raise it to these temperatures. Temperature monitoring is provided by one or more thermostats, which also protect the heating elements against overheating.

The water level is sensed by a pressure switch, and a solenoid valve enables automatic filling of the chamber at the time desired. System operation is controlled from a control cabinet [AH], [AI], [AJ], [AL].

— *Operation*

Filling takes place automatically at the end of the working day before the start of the off-peak rate hours. At the start of the off-peak hours, the heating system starts automatically; start of heating is often controlled by the power company's metering system to ensure that this takes place during off-peak hours. The installed power is designed to take the water to the desired temperature before the end of the off-peak period [AK], [AM], [AN].

During the working day, the hot water required is drawn without replacing it with cold water, providing all water required at the desired temperature from the enclosure without restarting the heating system; this also enables operation at atmospheric pressure.

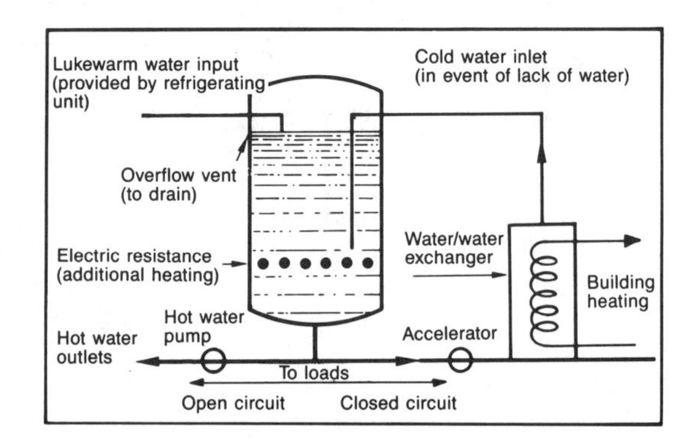


**Fig. 56.** Energy recovery on refrigerating unit: header tank with additional heating by immersion heaters; hot water used to heat buildings, provide hot water for sanitation and production processes (food, mechanical, plastics and textile industries).

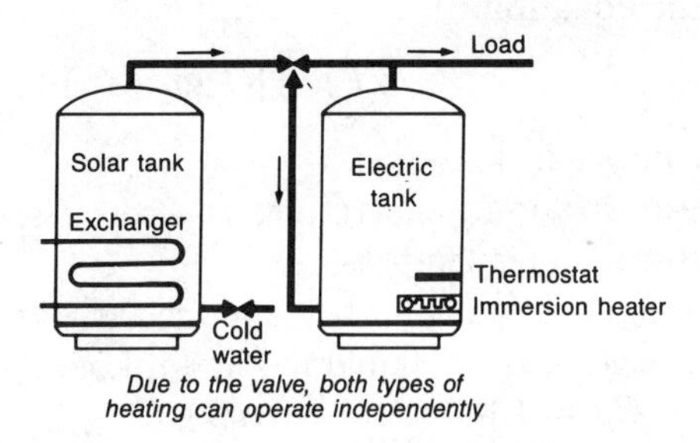


**Fig. 57.** Diagram of a combined solar/immersion water heater (independently operating tanks).

— *Range of industrial water heaters*

A very wide range of industrial water heaters exists. Figures 54 and 55 list a range of vertical and horizontal water heaters.

— *Advantages of industrial electric water heaters*

The electricity consumption varies with the initial temperature of the water; generally, this is about 85 kWh/m$^3$ for a temperature of 85°C.

The basic advantage of industrial electric water heaters is the cost per cubic meter of hot water. For investment costs comparable to those of a fuel-fired boiler, they often offer lower energy costs.

Also, these heaters are extremely simple, rugged tools requiring very little maintenance.

Water heaters can also be used in conjunction with energy recovery equipment (for example, installed on refrigerating machine condensers, waste waters and solar sources) preheat the water. The water is then taken to the final temperature during off-peak hours by immersion heaters.

## 5.4. Steam boilers heated by immersion heaters

The bottom part of steam generators features a bundle of armored resistances which are constantly kept immersed by the level control device, and the top part contains a vaporization chamber. The tanks must be adequately insulated to reduce heat losses but electric boilers are very compact and delivered ready for use. Designed for operation under pressure, these boilers must be submitted to stringent acceptance procedures.

The total heat (sensible heat and latent heat of vaporization) necessary to vaporize water at different temperatures and pressures varies only slightly (2,674 kJ/kg of steam, i.e., 0.744 kWh/kg at 1 bar, and 2,759 kJ/kg steam, i.e. 0.769 kWh/kg at 80 bars). With a power of 1 kW, it is therefore possible to produce 1.3 to 1.4 kg of steam per hour. The electric energy thermal conversion efficiency is excellent, being about 97%.

The power required to produce M kilograms of steam per hour is expressed by the equation:

$$Pw = [C_p(\theta_2 - \theta_1) + L] \times M \times 1.1$$

$Pw$, power in kilowatts,
$\theta_2$, vaporization temperature;
$\theta_1$, initial water temperature;
$M$, weight of water to be vaporized in kilograms;
$C_p$, specific heat of liquid in kilowatthours per kilogram per degree Celsius; for water, $C_p = 1.160 \times 10^{-3}$ kWh/kg°C;
$L$, latent heat of vaporization in kilowatthours per kilogram; this varies with vaporization temperature, and for water, can be expressed as $L = [703.5 - 0.806\theta] \times 10^{-3}$;
1.1, safety coefficient allowing for losses and voltage drops, etc. A higher coefficient can be used if overpower is required.

These boilers operate automatically. Pressure buildup is fast, ranging from 10 to 30 minutes and several heating modes are available to follow steam rate variations. Almost all qualities of water can be used with these boilers which require limited but regular maintenance.

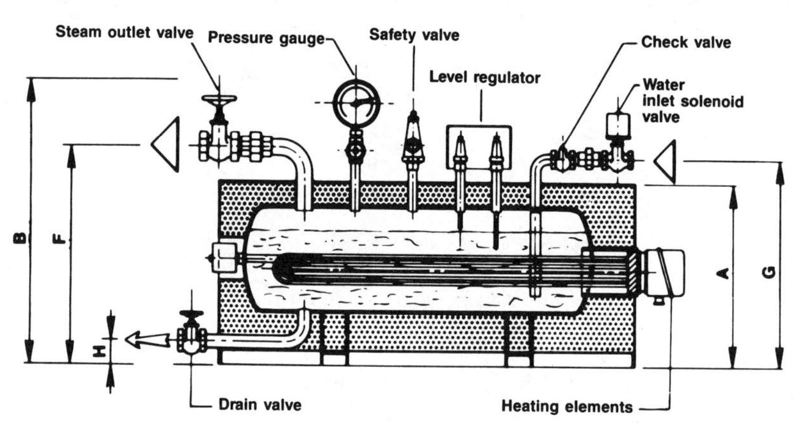


**Fig. 58.** Diagram of an immersion heater steam boiler [U].

Range of immersion heater steam boilers
The instantaneous specific consumption is 0.77 kWh/kg of steam [U].

| Power (kW) | 5 | 10 | 20 | 60 | 120 | 240 | 360 | 600 | 840 |
|---|---|---|---|---|---|---|---|---|---|
| Maximum rate (kg/h at 8 bars) | 6.5 | 13 | 26 | 78 | 156 | 312 | 468 | 780 | 1,092 |

The overall generator dimensions in meters (length, width, height) are 1 x 0.34 x 0.5 for the least powerful boiler, and 2.6 x 1.5 x 1.8 for the most powerful. Floor area can be further reduced by using vertical boilers.

**Fig. 59.**

These boilers, which are generally supplied with low voltage, are usually used to produce low pressure steam, at 10 bars maximum, in limited quantities of up to 1,000 kg per hour. Immersion heater boilers are also used to produce hot water or superheated water [T], [U], [V], [W], [X], [Y], [Z].

These highly compact boilers can be placed near user equipment. In fact, it is better to use electric boilers differently than fuel-fired boilers. While fuel-fired boilers are used in a central position with large steam distribution systems for reasons of convenience, electric boilers can easily be placed near user equipment thus preventing often high distribution losses.

The profitability of electrically produced steam depends, of course, on energy costs compared after the various conversion efficiencies (production, distribution, production cycle, etc.) have been taken into account. When this analysis is performed correctly, electrically produced steam is competitive in many cases (preferential use of off-peak rates, or use outside of peak hours; intense use during the months of April to October; weekend operation to relieve fuel-fired boilers employed much below their nominal power; auxiliary operation, for example when the feed system of equipment is located far from the main boilers requiring costly extension of the distribution network, many steam utilization points, calling for low power, or again, intermittent operation with the requirement to separate operation of equipment from the steam mains; decentralized supply to equipment, etc.). Such applications will increase in the future. In some countries where the price of electricity is highly competitive, electric boilers have often replaced fuel-fired boilers. In others, where electricity costs are low much of the year (generally 5 to 7 months, outside of the coldest winter months), electric boilers, in parallel with fuel-fired boilers, provide steam requirements during these periods.

For powers higher than about 1 kW, electrode boilers are rapidly becoming more attractive than immersion heater boilers.

When a factory or plant is considering using steam as a heating fluid for an industrial process, first of all a study must be conducted to seek whether or not it is possible to use a heating technique more efficient than steam (decentralized solutions such as immersion heaters for heating tanks containing liquid products or armored cartridges for moulds and trays), then, if a more efficient solution is not possible, to evaluate the attractiveness of electrically produced steam.

## 5.5. Circulation reheaters and electric hot water boilers

When a circulating liquid is to be heated, or if the heating element must not be placed directly in a tank, it is possible to use circulation re-heaters. These consist of a tubular body in which the liquid to be heated flows, and are equipped with screw-in or clamp-type immersion heaters. The steel body can withstand pressures of 15 to 40 bars. The system is insulated to reduce losses.

Equipment of this type is used to heat liquids up to approximately 350°C. Specific power is a function of the liquid and its flow rate. Protection against overheating is provided by a thermostat. As an example, the power density is about 2 $W/cm^2$ for fuels and thermal fluids, 4.5 $W/cm^2$ for water at a low flow rate, but may exceed 8 $W/cm^2$ for water at a high flow rate (*see* diagram, Figure 43).

Calculation of power W is very simple:

$$\boxed{W = mc(\theta_2 - \theta_1) + D,}$$

$m$, hourly weight of fluid to be heated in kilograms per hour;

$c$, specific heat in watthours per kilogram per degree Celsius;

$\theta_2 - \theta_1$, liquid temperature rise in degrees Celsius;

$D$, thermal losses in Watts.

These heaters can be mounted horizontally or vertically and arranged in series or parallel. The heating element power ranges from a few kilowatts to about a thousand kilowatts, and several heating rates are generally possible. For example, a 700 kW circulation reheater has 42 armored heating elements about 2.6 m long, with overall dimensions of 3.2 m, and diameter 0.5 m [B].

Equipment of this type is used for heating liquids such as water, oils, fuels and thermal fluids.

Hot water electric boilers belong to the circulation re-heater family.

## 5.6. Thermal fluid boilers

While remaining in the liquid phase and at atmospheric pressure, thermal fluids enable high temperatures of up to 350°C, while with steam or superheated water installations, it is difficult to exceed 200°C without complications and excessive costs. For even higher temperatures, it is necessary to use molten salts or metals, but with much more complex technology. The fluids most widely used are mixtures of hydrocarbons or other synthetic products.

These boilers, in fact, consist of a circulation re-heater with clamped or screw-in immersion heaters, a pump and a control and safety system. The surface power density is generally about 2 $W/cm^2$, but varies according to the fluid used. Between the initial temperature and operating temperature, the liquid expands by 25 to 30% thus necessitating an expansion chamber. These thermal fluid electric boilers are used in numerous industries for heating moulds or tanks but, as in the case of steam heating, should be used only if decentralized heating using immersion heaters and heating cartridges is impossible, since overall thermal efficiency is generally better.

## 5.7. Heating moulds, panels, and dies

In many industries — plastics, rubber, wood, foundries, forging and stamping, textiles, etc. — moulds, panels and dies are used to hot-form products. These tools may be heated for different purposes, for example to maintain an optimum temperature for the product concerned, to heat the materials to be formed, or to prevent thermal shock.

### 5.7.1. *Characteristics of encased resistances used for heating moulds, panels and dies*

Various electrical techniques are used to obtain this type of heating; the most attractive are those providing decentralized heating with encased elements. However, if it is difficult to use these elements, it is possible to use centralized or semi-centralized water or steam heating or electric generator thermal fluid heating. Efficiency is, however, lower than for decentralized systems.

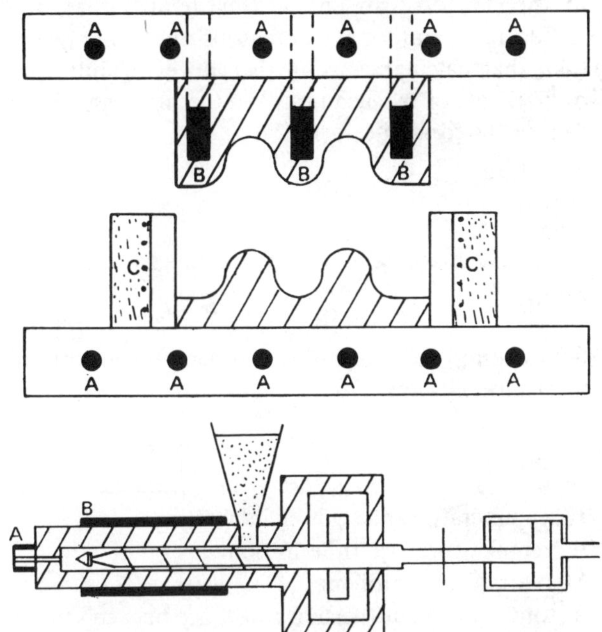


**Fig. 60.** Diagram of mould and panel heating using armored elements [18]:
A. Heating cartridges in panels
B. Heating cartridges in moulds
C. Heating cable with thermal insulation

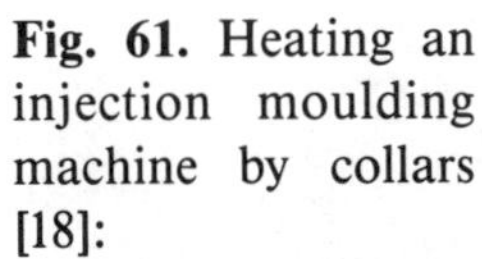

**Fig. 61.** Heating an injection moulding machine by collars [18]:
A. Nozzle heating
B. Material heating

Various heating elements can be used, but the most common are heating cartridges, flat resistances, collars and cables (*see* Figures 5 and 21).

### 5.7.2. *Calculation of power to be installed*

The power to be installed is based on many factors. In particular, it is necessary to distinguish the temperature rise before the start of a work station from the established production rate. Where a high tool heat-up rate is required, the power is generally higher than that required for production. Installing this power is not the best solution. It is, in fact, easy to program a temperature rise automatically outside of production times, and in particular during off-peak hours. The power to be installed is then that required for production.

#### 5.7.2.1. *Power requirements to reach a given temperature*

The power required to reach a given temperature is expressed by the equation:

$$W_1 = \frac{m_1 c_1 (\theta_2 - \theta_1)}{t_m} + 0.7\ D,$$

$W_l$, power in Watts;
$m_l$, weight of part (mould, die, etc.) to be heated, in kilograms;
$c_l$, specific heat of part to be heated in Watthours per kilogram per degree Celsius;
$\theta_1$, initial temperature in degrees Celsius;
$\theta_2$, final temperature in degrees Celsius;
$t_m$, temperature build-up time in hours;
$D$, thermal losses of part at final temperature, in Watts.

Generally, and for safety reasons, the power reserve is 10 to 20 percent higher than that calculated using the above formula. Factor 0.7 $D$ is generally an overestimate of the mean value of thermal losses during temperature rise. If losses can be stated in the form:

$$D = KS(\theta - \theta_0) = \frac{\theta - \theta_0}{R},$$

a more accurate value can be obtained from the expression:

$$\theta = \theta_0 + RW_1(1 - e^{-t/RC}),$$

representing the change in temperature $\theta$ with time $t$. The power is then determined by the equation:

$$\boxed{W_1 = \frac{D}{1 - e^{-t_m/RC}},}$$

$W_l$, temperature rise power in Watts;
$t_m$, temperature rise time in hours;
$D$, thermal losses of mould in Watts at final temperature,
$R$, total equivalent thermal resistance between the mould and the environment expressed in degrees Celsius per Watt (considered as equal to $(\theta_2 - \theta_0)/D$);
$C$, total thermal capacity in Watthours per degree Celsius ($C = m_l c_l$).

This expression is obviously similar to that given in paragraph 5.1.2.2. for heating a tank filled with liquid, and the values of coefficient:

$$k = \frac{1}{1 - e^{-t_m/RC}},$$

given for this can be used here, however, this is less accurate than for liquid heating when moulds operate at high temperature, since the thermal losses due to radiation become high, in particular if the mould is poorly insulated.

In fact, the initial approximation is generally sufficient, if power $W_1$ is approximately greater than $2D$. The losses are the sum of three terms obeying different laws.

— $D_l$, *dissipated power due to conduction* with:

$$D_1 = \lambda S\left(\frac{\theta_2 - \theta_3}{e}\right),$$

$D_l$, thermal losses in Watts;
$\lambda$, thermal conductivity in Watts per meter per degree Celsius;
$S$, exchange area or surface in square meters;

| Characteristics of sheathed resistance for heating of moulds and dies | Maximum temperature (°C) | Usual unit power (W) | Maximum power density (W/cm²) | Advantages | Limits |
|---|---|---|---|---|---|
| Barrel-wound resistance cartridges | 300 | 0.5-2.5 | 4 | Low cost | Limited temperature<br>Low power density<br>Requires drilling of hole |
| Compacted mineral insulating cartridges | 800 | 0.1-8 | 16 | High power density, highly reliable | Requires drilling of hole<br>Sometimes difficult to extract |
| Mineral-insulated heating cables | 800 | 0.4-5 | 6 | Long lengths<br>Easy to install<br>Adequate resistance to shocks | Can require use of low voltage power supply |
| Flat resistances insulation with:<br>• mica<br>• ceramic<br>• compacted magnesia | <br><br>300<br>800<br>600 | <br><br>0.1-3<br>0.1-4<br>0.1-4 | <br><br>4<br>10<br>6 | Low-cost but adequate heat distribution | May have to be fitted in grooves to improve heat transfer |
| Heating collars<br>• mica<br>• ceramic | <br>350<br>900 | <br>0.1-3<br>0.1-6 | <br>3<br>10 | Low cost<br>easy to install<br>adequate heat distribution | Mica elements somewhat fragile |
| Moulded collars or resistances | 400 to 700 | 0.5-4 | 3 to 5 | Very rugged<br>Adequate heat distribution<br>Cooling system can be provided<br>Suitable for corrosive and explosive atmospheres | High cost |

**Fig. 62.**

$\theta_2$, mould final temperature in degrees Celsius;
$\theta_3$, the temperature at the opposite end of the surfaces in contact in degrees Celsius;
$e$, thickness of wall in contact with the mould in meters.

Basically, these losses correspond to thermal losses through the mould, die or panel supports.

In fact, the initial approximation is generally sufficient, if power $W_1$ is approximalely greater than $2D$. The losses are the sum of three terms obeying different laws.

— $D_2$, *dissipated power due to convection*, with:

$$D_2 = 1.8\ \alpha_1\ S(\theta_2 - \theta_1)^{1.25},$$

$D_2$, losses in Watts;
$S$, exchange surface in square meters;
$\theta_2$, final temperature in degrees Celsius;
$\theta_1$, initial temperature (ambient temperature in degrees Celsius);
$\alpha_1$, coefficient equal to 1 for vertical walls; 1.4 for upward slanting horizontal walls, 0.7 for downward slanting horizontal walls;

— $D_3$, *dissipated power due to radiation*, with:

$$D_3 = \sigma\varepsilon\ S(T_2^4 - T_1^4),$$

$D_3$, losses in Watts;
$\sigma$, Stefan-Boltzmann constant, equal to $5.73 \times 10^{-8}$W/m$^2$ X K$^4$;
$S$, radiating surface in square meters;
$T_2$, final temperature in degrees Kelvin ($T_2 = \theta_2 + 273$);
$T_1$, initial temperature in degrees Kelvin ($T_1 = \theta_1 + 273$) (ambient temperature);
$\epsilon$, emissive power of part (0.8 approximately for oxidized steel, 0.3 for oxidized aluminum).

Losses due to convection and radiation, $D_2$ and $D_3$, are given in these expressions for non-insulated wall moulds. If the walls are insulated, just calculate the outside temperature of the surfaces allowing for the temperature drop in the insulation, and apply the above method to evaluate losses with sufficient accuracy.

Where high accuracy is not required, the estimated value of losses by convection and radiation can be obtained from the curves in Figures 63 and 64.

Figure 64 provides all the data needed to calculate the power required for a steel mould with only some parts insulated. Curve (1) gives the value of the specific heat of steel versus temperature; for an approximate calculation, a mean value over the temperature range is sufficient. Curve (2) gives the value of the outer wall temperature of an insulation around the mould (insulation of thermal conductivity 0.1 W/m$^2$/°C and 50 mm thick) versus mould temperature. Curve

(3) can be used to calculate thermal losses through the walls of a mould as a function of their respective temperature, whether insulated or not.

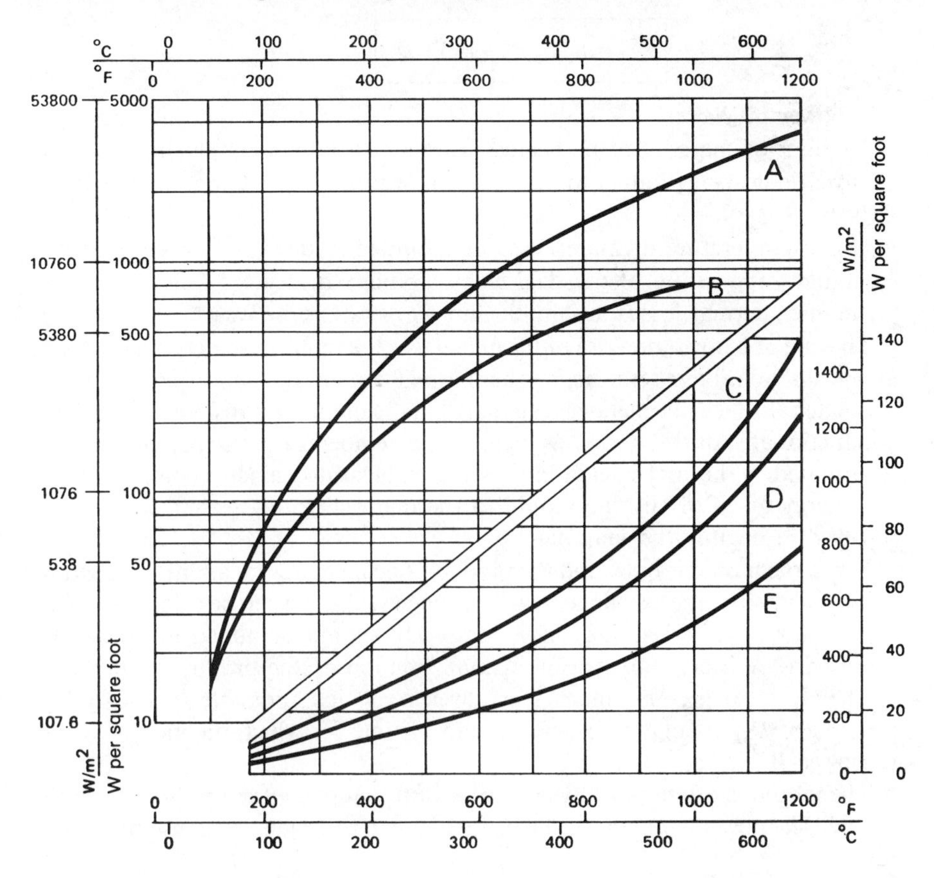


**Fig. 63.** Thermal losses through mould wall surfaces (W/m² and W/square foot with 1 W/square foot = 10.76 W/m²) versus mould temperature (°C and °F) [18].

A, oxidized steel moulds with non-insulated surface

B, oxidized aluminum moulds with non-insulated surface

C, D, E, moulds insulated by rock wool sheeting (thermal conductivity equal to 0.06 W/m/°C) 50, 100 and 150 mm thick respectively.

### *5.7.2.2. Power requirements for production*

This, of course, depends on the nature of the operations to be performed. For example, for casting moulds or stamping dies, heating during production is almost unnecessary since the liquid metal or parts at forging temperature keep the tool at an appropriate temperature, and it may even be necessary to cool it. Conversely, in industries such as plastic and rubber transformation, or the wood industry, a mould is often used to heat the material to be formed. In such cases, the power is expressed by the equation:

$$W_2 = m_2 c_2 (\theta_2' - \theta_1') + D,$$

$W_2$, power in Watts;
$m_2$, weight of material to be heated in mould in kilograms per hour;
$c_2$, specific heat of product to be heated in Watthours per kilogram per degree Celsius;
$\theta_2'$, final temperature of material to be moulded in degrees Celsius;
$\theta_1'$, initial temperature of material to be moulded in degrees Celsius;
$D$, mould thermal losses at equilibrium temperature, in Watts.

The rate of temperature rise of the material in the mould also depends basically, at a given mould temperature, on the thermal conductivity and specific heat of the material, since the energy accumulated in the mould is transferred by thermal conduction to the material as heat. This temperature rise is, in practice, determined by thermal calculations and tests. Heating elements compensate for the energy given off by the mould to the material to be heated, but are not involved in heating the material.

While the mould is open and the product is being removed, additional thermal losses occur since the exchange surface is increased (these can be calculated easily by the same method as $D$). These losses $D_M$ should be added to power $W_2$; however, they occur only during a small fraction of the production cycle.

Therefore, to prevent unnecessary overpower, it is possible to add only a fraction of $D_M$ calculated for the amount of time for which the mould is open to power $W_2$.

The mould is often partially cooled prior to removing the products from the mould. The power required for the temperature to rise again must also be allowed for.

If heating of the material causes it to melt, the latent heat of fusion $L$ must be taken into account, and a term $m_2L$ must be added to the power expression $W_2$.

It is recommended that moulds and other tools requiring heating be insulated, a practice still not widely encountered in industry, which causes high energy losses. Encased resistance heating, combined with careful thermal insulation and correct regulation, provides very high thermal efficiency.

## 5.8. Preheating and treatment of welded objects

An examination of welding production techniques shows that two special thermal operations are often performed:

— *Preheating before welding*

The two most common faults in a steel wall are due to cracks caused by hydrogen occlusion, and the appearance of hardened microstructures caused by high welding temperatures.

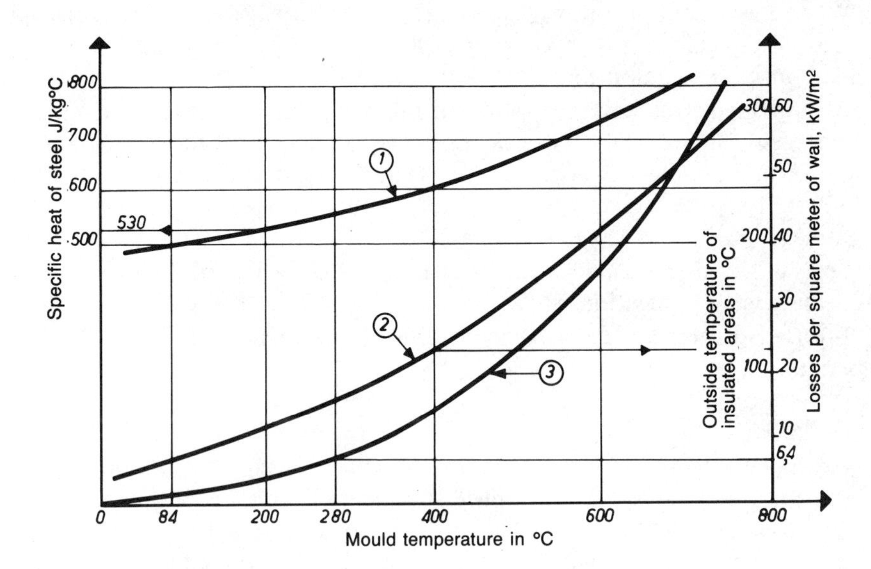


**Fig. 64.** Data required to determine power to be installed for heating a steel mould [15]
(1) Specific heat of steel in joules per kilogram per degree Celsius (1 J = 0.28 X $10^{-3}$Wh);
(2) Temperature of outside walls of insulation around the mould versus mould temperature (insulation thermal conductivity 0.1 W/m X °C, thickness 50 mm);
(3) losses due to radiation and natural convection of the walls, whether insulated or not, versus wall temperature.

To avoid these faults, which can cause cracking followed by weld bursting, it is necessary to heat the area adjacent to the weld to a temperature of between 100 and 300°C, depending on steel grade and the types of parts to be assembled, and to maintain this temperature throughout the welding operation. Some examples of preheating treatments are given below:

- non-carbon alloyed steel sheet is generally preheated to 100°C for contents of between 0.25 to 0.35% and between 200 and 300° for contents of more than 0.35%, while preheating is not often necesary for contents of less than 0.25%;
- carbon steel tubes are preheated above 200°C when their thickness exceeds 19 mm, while Cr-Mo tubes, whatever their thickness, require preheating to 200-250°C;
- stainless steel tubes should not be subjected to preheating, since it affects them adversely (risk of corrosion);

— *Stress relief treatments*

When a weld cools it contracts, causing stress on the edges of the two adjacent metal parts which can then cause distortion of the part and rupture during operation of the equipment.

Also, welded structures must undergo stress relief annealing at temperatures of between 600 and 800°C. The treatment cycle requires excellent temperature accuracy and homogeneous heating of the parts. Welded assemblies consisting of sheets less than 30 mm thick do not, in most cases, call for stress relief treatment.

Special treatments can be performed (hyperhardening of stainless steels, annealing, standardization, etc.).

Encased resistances are used in various forms to perform these thermal operations.

— *Encased resistances*

Heating elements are available in a wide range of shapes — ribbons, cords, plates, muffles, fingers, etc. — supplied at normal voltages, or for safety reasons, at low voltages (48, 24 V, etc.).

Resistances of this type are very adaptable and can be used both for preheating and heat treatment, in plants or on worksites, up to about 1,200°C. Once placed on the part to be treated, the heating elements are generally covered by a mineral thermal insulator to reduce heat loss.

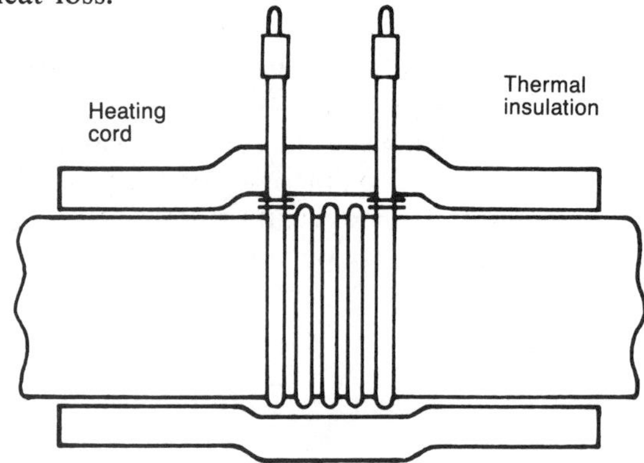


**Fig. 65.** Stress relief treatment of weld by heating cords [AG].

The service life of ceramic clad elements of the "finger" type is closely associated with the operating temperature; for preheating to 300°C, this should exceed 200 treatments. Conversely, at 1,200°C, it is difficult to exceed ten treatments, and at 1,050°C, about 20. If handling and operating requirements are respected, service life may be higher.

— *Temporary resistance ovens (furnaces)*

Temporary resistance ovens are particularly useful for treating large, complex-shaped structures. A temporary oven can be constructed around the metalwork structure.

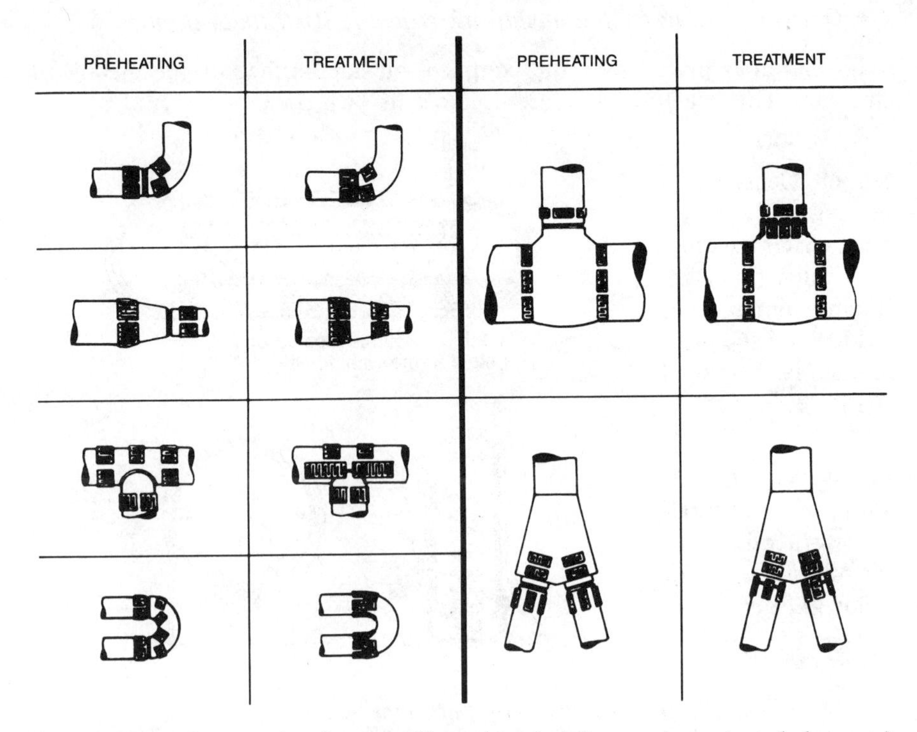


**Fig. 66.** Uses of encased resistances ("cartridge belt" ceramic protected elements) for preheating and heat treating metalwork assemblies [AF].

There are three types of temporary oven:

- *The temporary oven itself*

A true oven, generally constructed from prefabricated components, is built around all or part of the metalwork construction.

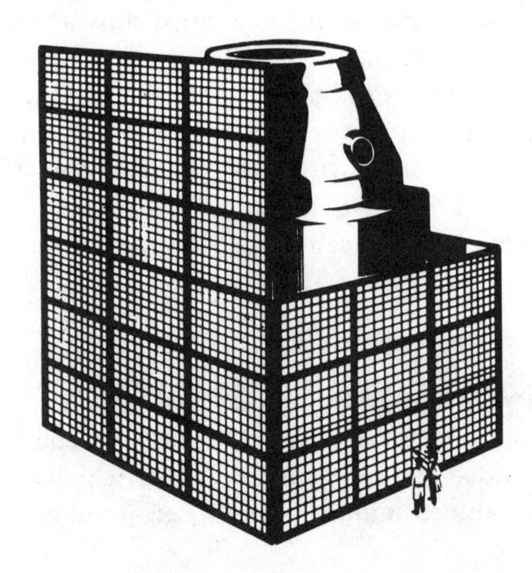

**Fig. 67.** Temporary oven constructed around a large metalwork reservoir [AG].

• *Ovens consisting of the metalwork structure itself (auto-oven)*:

Insulation is placed over the complete outside surface of the metalwork structure. The structure therefore becomes its own oven.

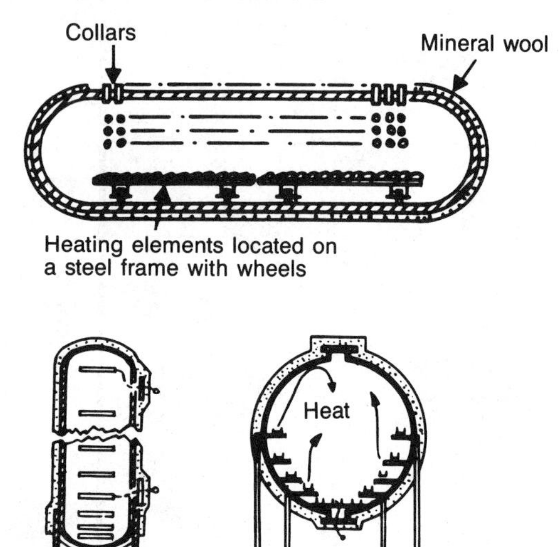


**Fig. 68.** General preheating of a boiler tank. After heating, each collar (there are several hundred) is welded and the insulation is removed [AF].

**Fig. 69.** "Ovens" consisting of the metalwork structure itself, for stress relief of welds [AF].

• *"Paired wall" or "insulated compartment" ovens*

This is a variant of the above method, but enables partial heating of the metalwork structure and is basically used to stress relief annular weld seams.

An insulated compartment containing heating elements is built inside the structure. The outside of the structure around the compartment is insulated.

In all three cases, heating is provided by "cartridge belt", "spiral" or "U-spiral" ceramic-encased resistances.

## 5.9. Electric heating through workshop floors

In most cases, a factory must invest relatively heavily in its electricity (transformer station and auxiliaries). Similarly, the electricity producer-distributor must invest in production and distribution equipment. Often, outside of production hours, such investments are not utilized. Accumulation electric heating, combined with careful insulation of the general area and adequate control, offers a particularly efficient and economic heating system. The heated floor, which is a decentralized accumulation heating process, is quite suitable for many workshops [17].

### *5.9.1. Heated floor construction*

Two very simple techniques are used to install a heated floor:

— For hard floors, a network of resistive cables attached to a welded framework is embedded in the poured concrete which is approximately 15 cm thick. The heat, produced mainly at night, accumulates and is then restored as required.

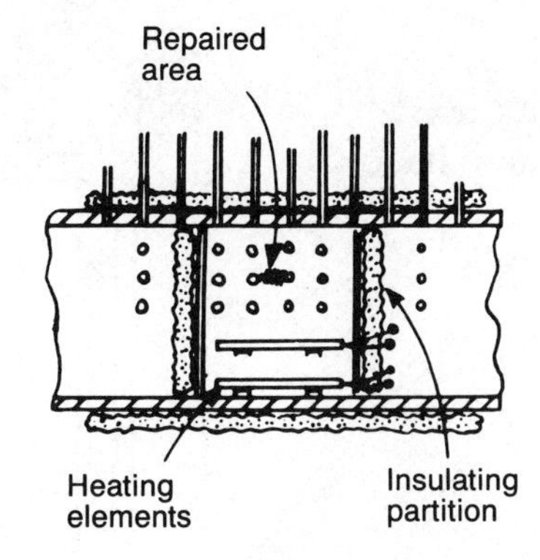


**Fig. 70.** Insulated compartment "ovens" for preheating and heating during welding and stress relief operations [AF].

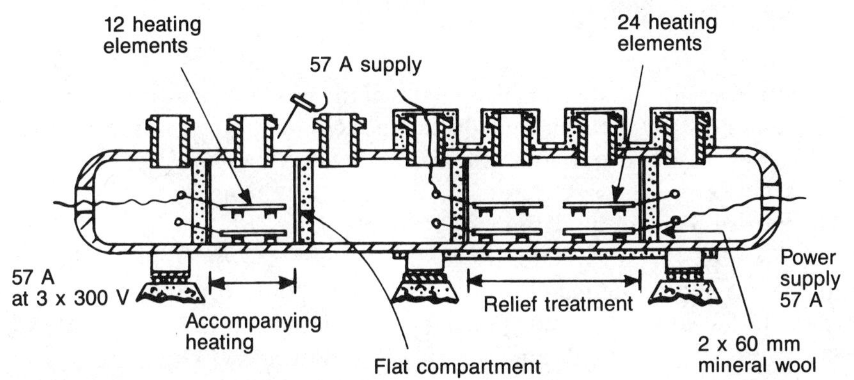


The paving is peripherally insulated by means of polystyrene panels, approximately 60 cm in height and 4 cm thick.

— For semi-rigid floors, the resistive cables are laid between two layers of cement-bound sand-gravel, covered with a semi-rigid coating which is made up of a hydrocarbonated covering with a hydraulic binder-based grout.

Three types of cables are used for such installations:

— a composite mineral-insulated cable, with a heated core embedded in an insulator (magnesia), protected by a double sheath of copper and plastic;

— aluminum cables, consisting of a circular, plastic-sheathed aluminum bar;

— a woven sheathed cable, consisting of a heating core covered with an insulator, a steel and copper metal braid and a plastic sheath.

The mineral insulator woven sheath cables are supplied with 220 or 380 V, while the aluminum bar is supplied with 48 V through a dry aluminum winding transformer [12], [13], [A], [J].

### *5.9.2. Power to be installed*

Depending on local environmental conditions and plant activities, two types of installation are possible:

— A high power installation, which is switched on during economic off-peak hours only. The heat is then restored as required.

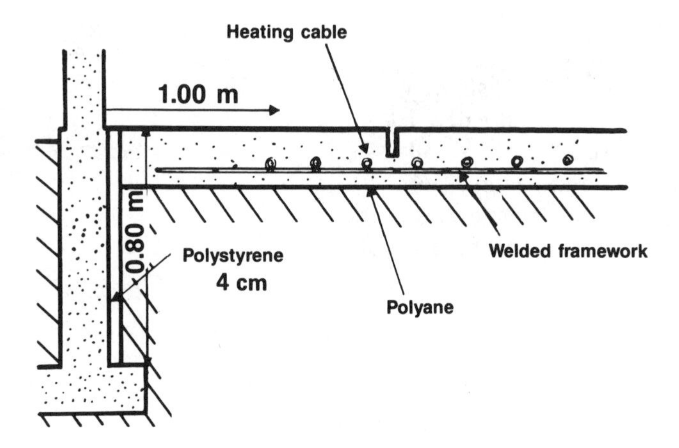


**Fig. 71.** Cross-section of a heated floor [17];

— A low power installation, which operates at off-peak hours on demand, with the option of restarting the heating during the coldest days to ensure maximum comfort. For reasons of economy, these short restarts always take place outside of peak hours.

The best system can be defined by a simple study.

The installed power is about 75 to 200 W/m², as a function of thermal requirements and inside temperatures. The linear power of the cables used is between 30 and 40 W/m. For adequate comfort, it is recommended not to exceed:
— 80 W/m² for 20°C (additional heating is almost always required);
— 150 W/m² for 18°C (additional heating is sometimes necessary in the coldest climates);
— 200 W/m² for 15°C (additional heating is almost never required).

### *5.9.3. Installation of machinery*

Heated floor heating does not in any way hinder installation or modifications to machinery in the workshops:

• low power machines can be placed on the floor without attachment: shock absorber pads prevent transmission of noise and vibrations. Shoes or plugs compatible with shallow anchoring can also be used.

• special machines or machines which require deep foundations are installed normally. The suppression of the heating network around these machines is of little importance, since such machines often give off more heat than the part of the floor neutralized.

### *5.9.4. Advantages of heated floors*

Heated floors are appropriate for large surfaces because:
— the effective volume is heated at floor level, where the personnel are located;
— even temperatures are provided throughout the area;
— overheating high areas is avoided, cutting heat losses through roofs (which account for 70% of the total heat losses);

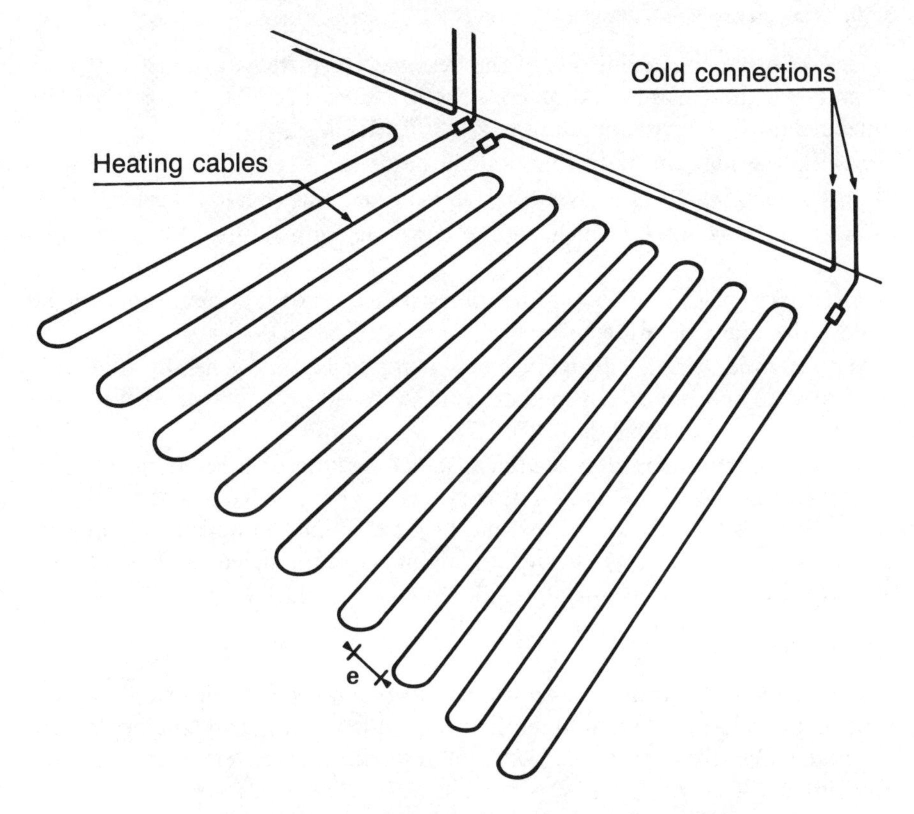


**Fig. 72.** Layout of cables in a heated floor [12].

— drafts and dust (particularly damaging in electronic assembly or precision work) are minimized.

Investments are kept low since heated floors:

— dispense with the requirement for a boiler and fuel storage installations and their ancillaries;

— forces the floor and technical premises from all constraints related to the presence of heating equipment and fluid lines.

Heated floors also offer reduced operating costs which may vary as a function of;

— the area of the building served;

— the type of areas (workshops, depots, etc.);

— the amount of thermal insulation;

— the local climatic conditions and the inside temperature desired.

Operation of a heated floor does not require supervision; this is provided automatically by means of sensors and thermostats. Servicing requirements are very low (checking contacts and regulation) and, in addition, energy is usually consumed at the most economical time of the day.

## 5.10. Accelerated hardening of concrete

Heat treating concrete is one of the best ways to increase productivity, both in prefabrication plants and at construction sites, and also to keep activity independent of climatic conditions [9], [10], [11].

Several thermal curing techniques can be used:

— lost steam, which only uses sensible heat and wastes latent heat;

— heat carrying fluids, which require constant temperatures due to thermal inertia;

— electricity, which combines armored resistance heating, accurate control and intense thermal insulation.

With electric thermal curing, primary energy savings are generally 40 to 75% compared to the lost steam method, and 20 to 50% compared to the boiler recycled thermal fluid method [9], [17].

The paragraphs below describe the different methods of concrete curing using armored resistances. Calculation of the power to be installed is made using the usual methods: calculation of thermal losses through the walls of the moulds or ovens, and the energy needed to maintain the concrete at the required temperature.

### *5.10.1. Heating elements*

Heating concrete through casing or a mould equipped with electric resistances is the most widely used method, both during on-site pouring and prefabrication. Economically, this heating method should be used whenever it can be reused more than 100 times.

Three types of heating elements are common:

— *Long elements (cables)*

The lengths can be up to 100 meters and more. The outside diameter of the casing is generally between 3 and 6 m. These can be shaped easily if the minimum bending radius is respected.

A copper-sheathed cable allows a maximum power of 40 watts per linear meter. A stainless steel cable can produce up to 150 to 200 W per linear meter.

Technically, this type of heating element should be used when a relatively limited power (for example, 1 kW/m$^2$ maximum) is to be distributed under the casing.

— *Rigid bar elements*

These elements do not exceed 7 meters in length. The casing diameters most widely used are 10 and 16 mm and always stainless steel. The maximum power is 3.6 W/cm$^2$ of casing surface which, for the 10 mm diameter, corresponds to a power of 1 kW per linear meter. By using a bending machine, these elements can be looped or made to form networks. But, for moulds, straight bars are often preferable.

A power per linear meter is chosen, which, as a function of the power to be installed per square meter, leads to a heating element pitch not exceeding 20 to 25 cm.

Economically, these elements should be used as soon as the power required reaches or exceeds 1 to 1.5 kW/m$^2$ [A], [B].

— *Finned elements*

These elements, which are constructed from a looped bar element and equipped with fins which diffuse the heat, are appropriate for oven heating. They are very simple to use and can be made from enamelled steel or stainless steel. The unit powers proposed by manufacturers range from 0.5 to 10 kW.

### *5.10.2. Heating techniques for casings, moulds and open areas*

— *Casings with power less than 1kW / m$^2$*

This type of casing is used for on-site pouring and allows vibration with probes as a rule. The relatively moderate power requirements leads to the use of long, low cost elements. These elements are arranged to form a network with a pitch of 5 to 15 cm placed against the walls of the casing and attached with welded straps or tabbed strips.

• Up to 0.5 to 0.6 kW/m$^2$, adequate distribution of the heating elements is provided by a cable producing 33 to 40 W per linear meter. If the number of times the casing is to be re-used is limited, the cable may be copper sheathed. For casings intended for use beyond one or two years, it is preferable to use stainless steel or copper-nickel encased heating elements.

• Beyond 0.5 to 0.6 kW/m$^2$, only copper-nickel or stainless steel elements provide sufficient power.

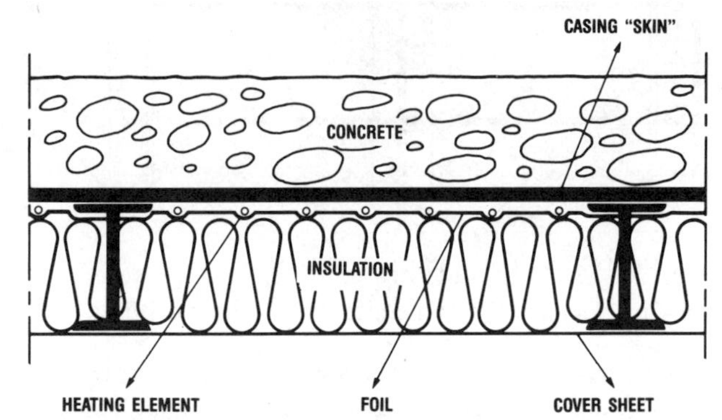


**Fig. 73.** Metal mould with long heating elements

A metal foil reflector is located under the heating elements. It prevents the lagging (for example, rock wool), intended to keep heat on the concrete side, from becoming mixed up with the concrete elements and the casing skin.

— *Casings with power exceeding 1 kW/m$^2$, or external vibration casings*

This is the typical prefabrication mould (for beams, facing panels, etc.) allowing several production cycles per day. The assembly method is as follows:

• long elements are attached by stainless steel brackets to a metal framework separated from the casing skin by 5 to 10 cm according to mould structure;

• where rigid bars are used, the best assembly method is shown in Figure 74: the structure of the mould itself, through which holes are drilled and in which the heating bars are inserted, is used. In general, the spacing of the stiffeners used to support the bars is adequate (1 every 50 to 70 cm). For 10 mm bars, the holes are drilled every 10.5 or 11 mm and carefully trimmed. It is necessary to prevent overhang at the ends since the bars would resonate and no longer resist vibrations. The bar is immobilized along its length by a crimped threaded sleeve (shown in the diagram) or by two stainless steel clamps located on either side of the stiffener near the center of the bar.

The length of the bars is calculated to allow for expansion (allow approximately 1 cm per meter of length). Also, the distances required to prevent parts under tension from touching, must be respected.

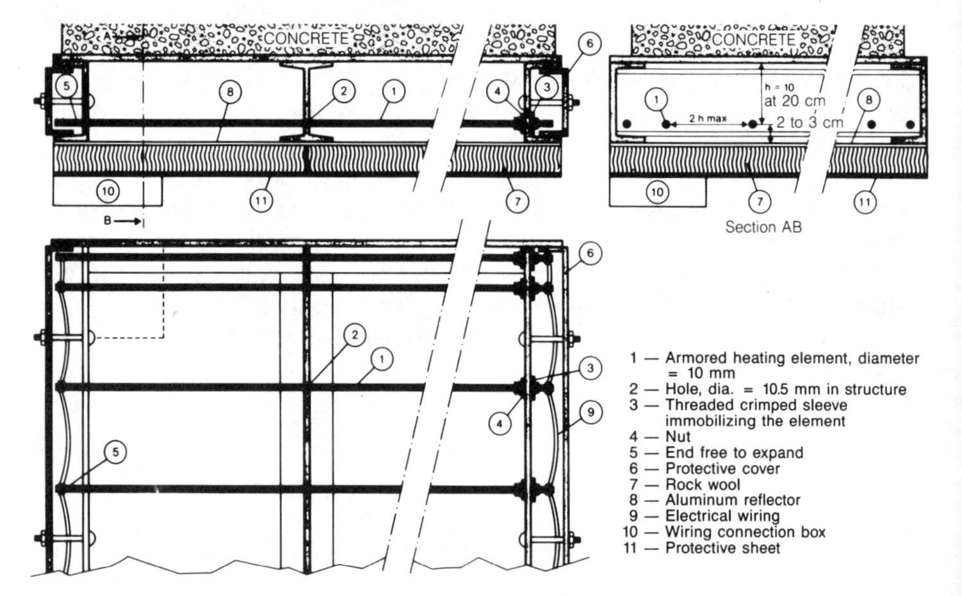


**Fig. 74.** Metal moulds heated by electric powered bar elements. Power 1 to 4 kW/m$^2$. Externally or needle-vibrated type [9].

This type of heating enables several kilowatts per square meter. The air trapped where the resistances are located behaves as a heat carrying fluid and provides adequate heat distribution. This principle enables high power density, and therefore low cost bars to be used, without creating hot points on the casing skin.

— *Moulds and areas*

Techniques similar to those described for casings can be used to heat the moulds and areas for the prefabrication of beams, struts, prepaving blocks and other prestressed products.

— *Heating covers*

During prefabrication, if it is necessary to harden a part thicker than twenty centimeters, it is best to heat it from the bottom and top, which considerably reduces the rotation cycle.

A heating cover is an insulated bell containing a reflector and a network of heating elements located 10 to 20 cm above the concrete. The assembly principle is the same as for the mould, except that the cover is not vibrated.

The electrical power is distributed between the mould and the cover in proportion to the quantities of concrete to be heated.

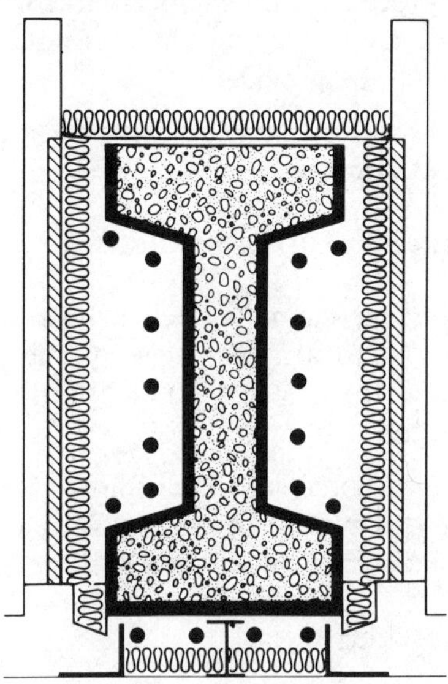

**Fig. 75.** Casing for accelerated hardening of beams [17].

If a cover is not used, it is necessary to cover the visible part with a tarpaulin. A simple tarpaulin suffices, or even a sheet of plastic, but an insulated tarpaulin is obviously preferable. The purpose is to retain the heat and to provide a humidity saturated atmosphere above the concrete, preventing unwanted drying. Moreover, at some sites, this provides protection from rain and other contaminants.

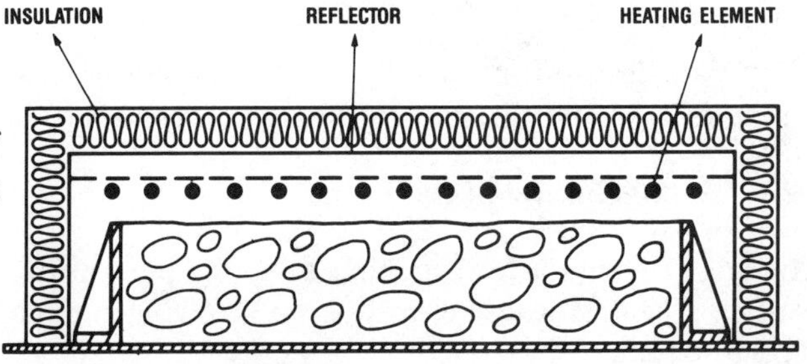


**Fig. 76.** Diagram of an electric heating cover [9].

### *5.10.3. Oven treatment of concrete blocks*

Various techniques are used for oven treatment of concrete blocks [10].

— *Heated floors*

These are the same type used for heating workshops, but the specific power is much higher, about 5 W/m$^2$. The combination of hot water mixing (accumulated tank contents heated during low rate hours) and heated floor is often the most economical solution for oven treatment, since most of the energy is consumed during off-peak hours.

— *Finned elements located at the bottom of ovens*

This process is suitable for ovens consisting of metal structures since drawing chimneys are formed naturally; if not, a spacing of about 10 cm between each pallet must be provided.

Stainless steel heating elements are recommended because of the hot, humidity-saturated atmosphere.

— *Linear elements, located under corners, and behaving as block pallet supports*

Here again, stainless steel elements are used. Small diameter, long armored elements (30 to 60 m) can be used. The power produced per linear meter of cable is about 100 to 150 W. All supporting corners are fitted, and excellent heat distribution is obtained.

It is also possible to use armored elements, in the form of 10 mm diameter tubes, 5 to 6 meters in length, which can produce 200 to 300 W per linear meter. Only one corner out of two need be equipped in this case.

## 5.11. Melting metals

Encased resistances are used to melt low melting point metals such as lead, tin or zinc. Although this is similar to heating fluids, the melting of metals must, in general, respond to heavier requirements than those encountered in heating the usual industrial liquids.

The advantage of this technique is basically that the energy is dissipated in the environment to be heated, ie no stack losses, and therefore high in energy conversion efficiency.

Two examples illustrate well the possibilities of encased resistances in this field: melting of lead and tin, and melting zinc in galvanization installations.

### *5.11.1. Melting lead and tin*

Figure 77 shows a lead melting oven equipped with barrel-type resistance immersion heaters whose elements can be changed rapidly, without having to empty the tank.

Immersion heaters facilitate the variation of installed power in different areas of the oven (melting area, maintaining and stabilization areas).

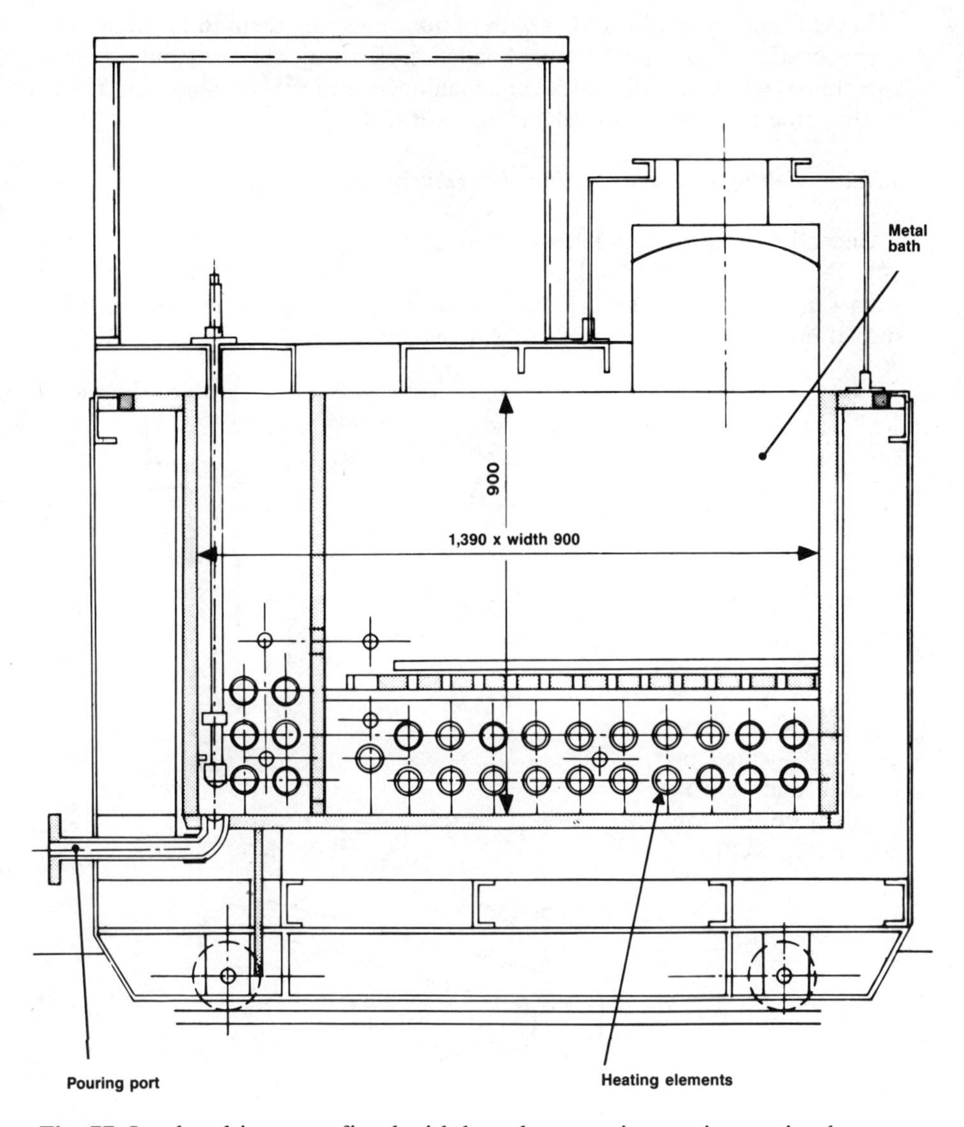


**Fig. 77.** Lead melting oven fitted with barrel-type resistance immersion heaters [17], [AO].

For example, an oven rated at 86 kW and 10 t capacity using 27 heating elements provides a maximum rate of 2,400 kg of lead per hour at 400°C.

Ovens of higher capacities and powers exist (one of the most powerfull being an oven of 20 t capacity, providing a maximum production of 10 t per hour for a specific consumption of 18 kWh/t). Solid metal feed to the oven as well as removal of molten metal can be fully automated, most often in fixed or tilting ovens.

These ovens are used together with options enabling them to be adapted to the application, for the production of batteries, electrical cables or moulded parts, patenting steel wires, refining lead and manufacture of rubber tubing, etc. Ovens of the same type are also used for tin bath melting.

### *5.11.2. Melting and maintaining zinc galvanization baths*

Generally, galvanization baths are heated by resistances which radiate either onto the mild steel metal tank or directly onto the bath when the tank is made from ceramic refractories (*see* "resistance furnaces"). New techniques, using special immersion heaters, are being developed.

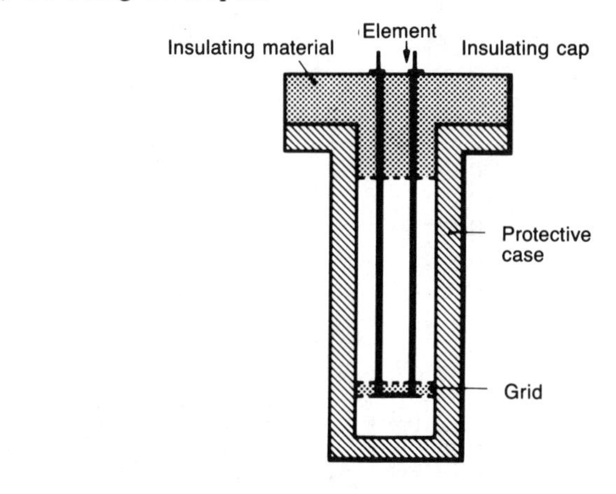


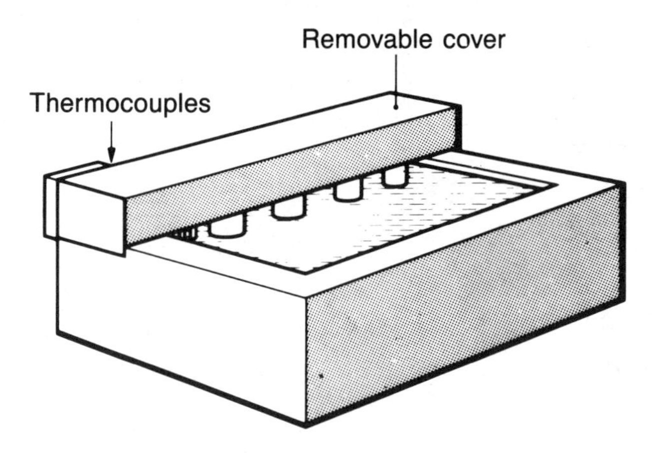


**Fig. 78.** Galvanization tank heated by carbide and silicon nitride immersion heater [17], [AP].

Molten zinc is a highly corrosive environment, therefore, corrosion-proof immersion heaters must be used. New materials using silicon nitride and carbide offer the required qualities. These immersion heaters consist of a silicon carbide heating element placed in a silicon carbide envelope and covered with silicon nitride, offering excellent protection. The unitary power of these elements is generally between 2 or 3 and 30 KW.

## 6. BIBLIOGRAPHY

[1] P. H. REIK, Chauffage électrique de surface, *Insulation* (*Londres*), n° 6, vol. 16, p. 289-290, novembre 1972.
[2] CH. SANDBERG, Design considerations for low watt density blanket and strip heaters, *IEEE Transcription on industry applications*, n° 1, vol. IA-11, p. 90-94, janvier 1975.
[3] A. J. CAHILL et C. J. ERICKSON, Self-controlled variable resistance heating system, *IEEE Transcriptions on industry applications*, n° 3, vol. 11, p. 314-318, mai 1975.
[4] J. MAC CALLUM, Heating elements for industrial process machines, *Electrical Review*, *n°* 24, vol. 197, p. 775-777, décembre 1975.
[5] ELECTROFINISH, Chauffage électrique de produits chimiques au moyen de réchauffeurs à immersion, *Galvano-Organo*, novembre 1975.
[6] D. MARTIGNON, How to select an electric immersion heater for the specific job, *Industrial Heating*, p. 10-15, mai 1976.
[7] TRAITEMENTS DE SURFACE, *Une installation d'oxydation anodique*, n° 157, octobre 1977.
[8] LABORELEC, *Le chauffage d'acide concentré*, Bruxelles, novembre 1977.
[9] J. CAUMETTE, Le durcissement du bêton par l'électricité, *Revue technique du Bâtiment et des Constructions industrielles*, n° 65, mars-avril 1978.
[10] J. CAUMETTE, L'étuvage électrique des blocs en béton, *Revue technique du Bâtiment et des Constructions industrielles*, mai-juin 1978.
[11] J. CAUMETTE, La mise hors gel des granulats, *Revue technique du Bâtiment et des Constructions industrielles*, janvier-février 1980.
[12] M. RECKEL, Chauffage électrique de locaux par plancher chauffant, Éléments pour cahier des charges, *Revue de l'Association des Ingénieurs de Chauffage et Ventilation de France*, novembre 1978.
[13] M. COUTELIER et M. RECKEL, Le chauffage électrique par le sol des locaux industriels, *Cahiers techniques du Bâtiment*, avril 1979.
[14] M. ORFEUIL et R. THOMASSIN, Le chauffage électrique des bains dans les traitements des surfaces des métaux, *La Technique moderne*, n° 5, vol. 71, mai 1979.
[15] J. GAULON et M. ORFEUIL *et al.*, Chauffage et maintien en température par résistances électriques blindées de moules métalliques, *Fonderie*, n° 391, vol. 34, juillet 1979.
[16] M. GAUSSIN, Le chauffage des liquides industriels par l'électricité, *Industrie Électricité Informations*, 1979.
[17] ÉLECTRICITÉ DE FRANCE, Documentation générale, 1980.
[18] ÉLECTRICITY COUNCIL, Documentation générale, Londres, 1980.
[19] LABORELEC, Documentation générale, Bruxelles, 1980.

**List of equipment manufacturers and suppliers mentioned in this chapter:**

[A] VULCANIC, 93330 Neuilly-sur-Marne.
[B] ETIREX, 02200 Soissons.
[C] MASSER, 91170 Viry-Châtillon.
[D] CETAL, 92320 Châtillon.

[**E**] ACEF, 75009 Paris.
[**F**] VULINCO, 75009 Paris.
[**G**] ELECTROTHERM, 69000 Lyon.
[**H**' CHIMIPLEX, 92250 La Garenne-Colombes.
[**I**] PARMILLEUX, 69003 Lyon.
[**J**] LES CABLES DE LYON, 92110 Clichy.
[**K**] PAMPUS, 95100 Argenteuil.
[**L**] TETATHERMIE, 27200 Vernon.
[**M**] ELECTROFINISH, 38600 Fontaine.
[**N**] ISOPAD, 75003 Paris.
[**O**] QUARTZ ET SILICE, 75008 Paris.
[**P**] HOTFOIL (Machinor), 75769 Paris.
[**Q**] RAYCHEM, 92250 La Garenne-Colombes.
[**R**] ELTRON (Comolet), 94300 Vincennes.
[**S**] BRAY-CHROMALOX, 92100 Boulogne.
[**T**] A.S.E.T., 69800 Saint-Priest.
[**U**] MANUTHERM (DAMOND), 91210 Draveil.
[**V**] Sté INDUSTRIELLE DE CREIL, 60100 Creil.
[**W**] LAGARDE, 26200 Montélimar.
[**X**] ALFA-LAVAL, 78340 Les Clayes-sous-Bois.
[**Y**] SULZER, 75000 Paris.
[**Z**] A.B.C., 93100 Montreuil.
[**AA**] COGETHERM, 93250 Villemonble.
[**AB**] S.E.P.R., 92200 Neuilly-sur-Seine.
[**AC**] BULTEN-KANTHAL, 92400 Courbevoie.
[**AD**] PLANCHER, 75011 Paris.
[**AF**] FOFUMI, 75009 Paris.
[**AG**] SELAS, 92300 Levallois-Perret.
[**AH**] COLLARD ET TROLLARD, 77100 Meaux.
[**AI**] DUFILS, 77000 Melun.
[**AJ**] DELROT, 81300 Graulhet.
[**AK**] S.C.I.M., 92400 Courbevoie.
[**AL**] Éts CHAROT, 89100 Sens.
[**AM**] Éts CHARRON, 44000 Nantes.
[**AN**] Éts DUCHEIN, 33000 Bordeaux.
[**AO**] SLFI, 69100 Villeurbanne.
[**AP**] MORGAN, 95500 Gonesse.

# Chapter 4

# Conduction heating

## Heating by Direct Current Flow Through Product to be Heated

### 1. PRINCIPLE OF CONDUCTION HEATING

In conduction heating, a current flow through the material heats it directly due to the Joule effect. This method of heating is also known as "direct heating by current flow" or, more simply, "direct resistance heating".

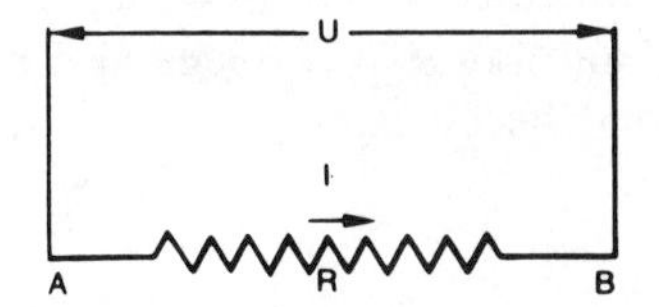


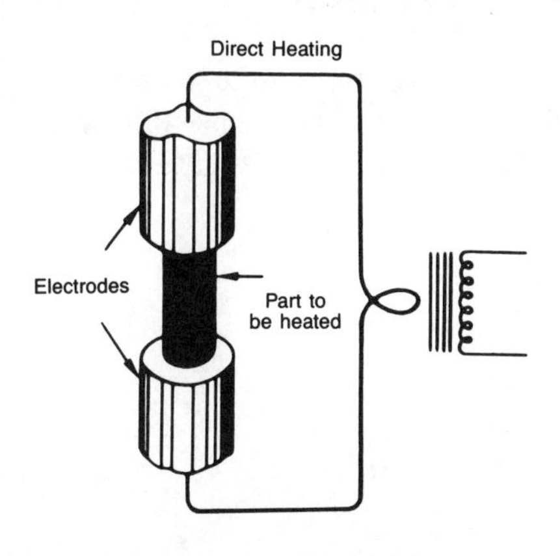


**Fig. 1.** Heating by direct flow of current through a part

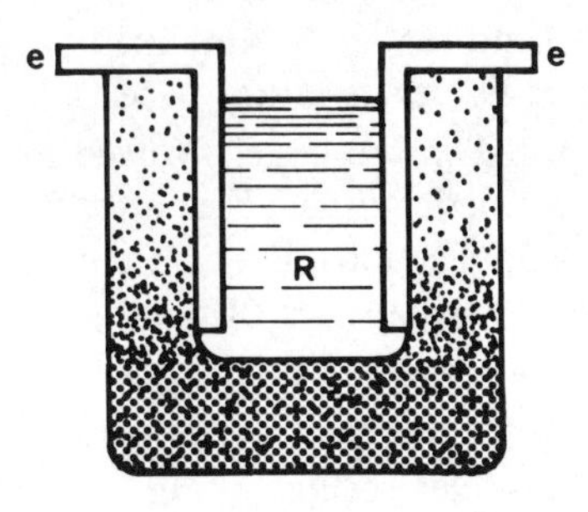


**Fig. 2.** Heating by direct flow of current through a mass:
e) electrodes;
R) mass heated by conduction

The dissipated power is given by Joule's law:

$$\boxed{P_w = RI^2 = UI = \frac{U^2}{R},}$$

$R$, resistance of body to be heated
$U$ and $I$, effective voltage and current.

Due to the very high currents generally involved, the power factor is not always equal to unity (the system reactance must be taken into account). In this case, the active power is expressed by:

$$P_w = RI^2 = R\frac{U^2}{Z} = R\frac{U^2}{R^2 + L^2\omega^2}.$$

**Fig. 3.** Conduction heating; equivalent diagram when self-induction cannot be neglected.

It can be easily demonstrated (by setting equal to zero the derivative of power $P_W$ with respect to $R$, with $L\omega$ considered constant) that the active power expression is maximum, for a given inductance, when $R = L\omega$, i.e. for a phase angle of 45° for which $\cos\varphi = 0.707$ and $\tan\varphi = 1$. Generally, active power when self-induction cannot be neglected becomes:

$$\boxed{P_w = UI\cos\varphi.}$$

A bank of capacitors can be used to compensate for the self-induction effect and render maximum dissipated power ($P = U^2/R$). However, in most cases, this device is not necessary.

The principle involved in this mode of heating is very simple. However, in practice difficulties are often encountered, which, up to now, have limited its applications. Extensive research is being performed to widen its application. The principal uses of conduction heating are:

— glass melting;
— heating of metals prior to forming;
— electroslagmelting of steel;
— heat treatment of metals;
— accelerated hardening of concrete;
— water heating;
— welding and brazing;
— preparation of ferrous alloys in reduction furnaces;
— electrochemical industries (preparation of silicon carbide and calcium carbide, etc.).

Although the last two areas use direct current flow through the charge, they are generally considered in submerged arc furnace applications, and are described in the chapter devoted to arc heating. Similarly, igneous electrolysis, used in the production of aluminum, can be compared to conduction heating; however, this is a very special field involving a unique technique based on electrolysis. In spite of the high energy consumptions (15,000 kWh/t of aluminum approximately), this method is mentioned for reference only.

## 2. CHARACTERISTICS OF CONDUCTION HEATING AND A TYPICAL INSTALLATION

Conduction heating applies to metallic or non-metallic materials, but the latter must be conductors of electricity.

The body or part of the body to be heated is placed between two electrodes or clamped between two connections, to which voltage is applied. Heat develops within the body directly.

The heat given off in the part or product to be heated is uniform if resistivity and cross-section are uniform.

However, many things can lead to temperature heterogeneity.

Slight irregularities in the density or texture of the material to be heated result in local variations of electrical resistivity, and therefore dissipated power. The area in contact with the walls of the furnace or with the atmosphere often heat more slowly, due to the combined effect of thermal losses, and, for bodies whose resistivity increases with temperature, the power given off is lower. However, in some cases, the tendency for the AC current to localize in peripheral areas (skin effect) tends to correct this minor overheating of the external parts of the body. The shape and position of the electrodes can also cause temperature differences in the material.

Therefore, theoretical calculations are not generally sufficient to determine the dimensions of entirely new equipment, and simply provide orders of magnitude. It is sometimes necessary to assume operating parameters and construct pilot installations to determine the final equipment.

The contacts of the current inlets on the body to be heated must be carefully designed. Due to the high current densities involved, these parts are generally massive; they must have very good electrical conductivity to reduce heating to a minimum. These electrodes are sometimes water cooled to increase service life.

In many cases, installations use low voltage (5 to 48 V), since the resistance offered by the products to be heated is often low; this low voltage power feed can also be imposed for safety reasons. However, high voltage equipment exists (up to approximately 30,000 V).

Generally, a conduction heating installation therefore consists of:

— current input electrodes, with their systems of connection on the product to be heated;

— an electrical power feed system, often equipped with a voltage step-down transformer;

— an insulated chamber or enclosure in which the product to be heated (furnace, tank, etc.) is placed: in this case, the furnace in the general sense of the word does not exist, the part to be heated being simply placed between the electrodes;
— load handling mechanisms;
— a control and regulation system.

## 3. ADVANTAGES AND LIMITS OF CONDUCTION HEATING

The main advantage in conduction heating is the heat transfer mode provided. Product heating is ensured by direct current flow, and the energy is given off within the material to be heated.

This process therefore offers very high electrical energy conversion efficiency, generally close to unity, and an overall industrial efficiency which is only slightly less, often between 70 and 95%.

The power densities applied to the products to be heated are high; therefore, the heating times are very short which tends to increase productivity and thermal efficiency. Similarly, the very low thermal inertia is a favorable factor in increasing thermal efficiency: this offers higher installation flexibility. Also, the power densities reached favor construction of highly compact installations.

The investment requirements with this technique are often relatively low, and, in many cases, may be less than with alternative processes.

Each application of this heating method offers its own advantages. However, conduction heating has inherent limitations. The contacts between the current inputs and the part to be heated impose permissible current limits, and must be very carefully designed. When a work piece is heated, it must offer a constant cross-section if heating is to be uniform. Also, in some cases, it may be necessary to operate at a frequency other than 50 Hz; this applies especially when heating is influenced by skin effect. A lot of research is being devoted to improving these techniques and to widening application since they can be very energy efficient.

## 4. CONDUCTION HEATING APPLICATIONS

Conduction heating applications are quite varied, and apply both to metallic and non-metallic bodies; some, such as welding, are well known, while others are more recent or are being developed.

Herein, the accent has been placed on applications concerning a wide range of users, although applications used by some enterprises in Europe only lead to very high energy consumption (electrochemicals, ferro-alloys, etc.). An example of the second type of application (production of graphite electrodes) is dealt with briefly.

### 4.1. Fabrication of graphite electrodes

Very large graphite electrodes are used in scrap iron melting electric arc furnaces.

These electrodes are produced as follows:
— petroleum cokes are crushed and sieved, then mixed with a binder consisting of coal pitch at about 160°C;

— the product is then extruded in a press at about 3,000 tons;

— the "raw" product is then heated slowly to carbonize the pitch, and again impregnated with pitch to fill cavities resulting from the distillation of volatile materials;

— finally, the conductive graphite coat is obtained by heating the baked products to 2,800°C.

**Fig. 4.** Production of graphite electrodes [44].

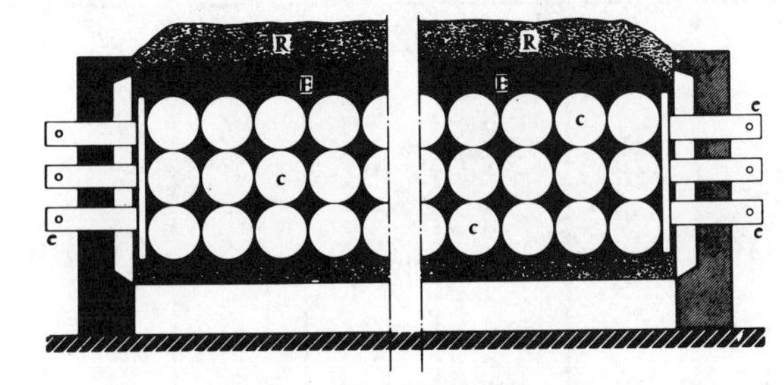


Cross-section of graphite coating furnace
C: Load; E: Carbonated packing; R: Refractory insulation; E: Current inputs

This high temperature is obtained by direct passage of the current through the mass consisting of the electrodes and grains of coke. The operation requires several days. The graphite electrodes are finally machined for use in arc furnaces. Furnace powers can be high, ranging up to several thousand kilowatts [44].

Similar techniques are used for firing other carbon-based products (in particular, complex materials using carbon fibres [43].

## 4.2. Glass melting

At normal ambient temperatures, glass is an excellent electrical insulator; it is also widely used as an insulator on power lines.

Conversely, at high temperature, especially in the liquid state, glass conducts electricity. Therefore, it can be heated by direct current flow.

Two applications are widely used throughout the glass industry:

— total electric melting;

— mixed electric-fuel melting, the electricity acting as supplementary energy (hence the name "booster").

### *4.2.1. Characteristics of melted glass*

Ordinary glass is a ternary compound $SiO_2$-$Na_2$(O-CaO) containing small portions of other elements; however, by altering the proportions of these three ingredients, or by replacing them with other materials, it is possible to obtain a wide variety of glass offering different properties.

Figures 5 and 6 give the composition and melting temperature of some of the most common glasses (they can also contain small proportions of different materials). The curves in Figure 7 show their resistivity versus temperature.

Percentage Composition of Some Widely Used Glasses (the numbers refer to the curves of figure 7 [44].

| Family | 1 | 2 | 3 | 4 | 5 | 6 | 7 | 8 |
|---|---|---|---|---|---|---|---|---|
| | Ordinary glass | | Glass Lamp | Neutral glass | Boro-silicate | Pyrex | | Lead crystal |
| $SiO_2$ | 70 | 72 | 67 | 64 | 61 | 79 | 81 | 56.37 |
| $B_2O_3$ | — | — | 2 | 8 | 25.5 | 14 | 12 | — |
| $Al_2O_3$ | 1.3 | 0.8 | — | 10 | 3.5 | 2 | 2 | 0.13 |
| CaO | 12.5 | 5.2 | 6 | 5 | — | — | — | — |
| MgO | 0.5 | 3.7 | — | — | — | — | — | — |
| $Na_2O$ | 14.7 | 15.5 | 11 | 7.5 | 9 | } 4.3 | } 3.9 | 5.4 |
| $K_20$ | — | 2.7 | 9.5 | — | — | | | 8.6 |
| PbO | — | — | — | — | — | — | — | 29.5 |
| BaO | — | — | — | 4 | — | — | — | — |

**Fig. 5.**

Melting Temperature of Some Glasses [21]

| Type of Glass | Melting point (°C) |
|---|---|
| Phosphate glass | 1,000 |
| Lead crystal | 1,100 |
| Boron glass | 1,130 |
| Barium borosilicated glass | 1,320 |
| Optical glass | 1,370 |
| Glazing glass | 1,410 |
| Translucent glass | 1,350 |
| Quartz | 1,700-1,800 |
| Enamels | 1,090-1,400 |

**Fig. 6.**

The resistivity of melted glass decreases rapidly with temperature, changing from approximately $10^{15}$ Ω X cm at normal temperatures to values of between 3 and 50 Ω X cm depending on the type of glass, around 1,300-1,400°C. Therefore, when designing a glass melting furnace, thermal runaway due to misadjustment must be taken into account, and unit operation assumptions carefully checked, if serious mistakes concerning the characteristics of the electrical equipment used with the furnace are to be avoided.

The resistivity of glass also varies with alkali content, since ions enable conduction through glass. For equal alkali content (sodium, potassium, lithium, etc.), electrical conductivity is inversely proportional to the volume of ions (schematically, the current flow takes place due to the mobility of ions in the silica framework).

**Fig. 7.** Change in resistivity in some glasses versus temperature [44]

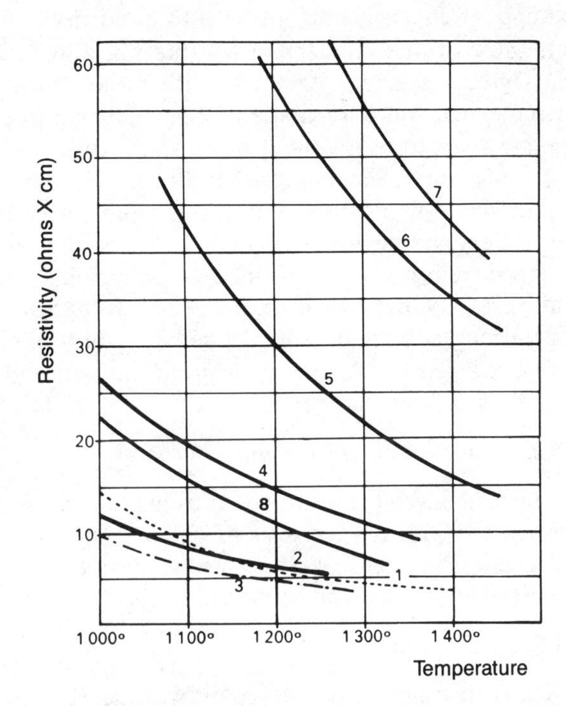


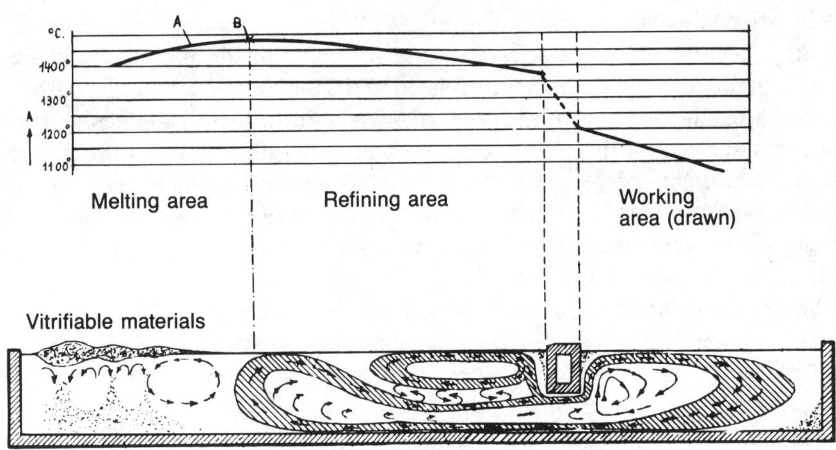


**Fig. 8.** Convection currents and materials in a glass melting furnace and temperature distribution [21].

It is therefore higher for sodium than for potassium glasses. However, conductivity is not related by a simple equation to the concentration of ions, or to their proportion in the composition of the glass, and only measurements can provide the resistivity value.

Since the conductivity of glass is of ionic origin, an AC current, generally at supply frequency, must be used to avoid risks of electrolysis.

Because of these characteristics, direct heating in a glass furnace is possible starting only at about 700°C, and the furnace must be started using another energy source, which is a major procedural handicap, since the furnaces can then operate several years without being shut down.

The viscosity of molten glass is also an important characteristic in furnace design, since this affects heat transfers within the bath, and in particular affects convection currents which build up in the bath. Glass melting furnaces can be designed from mathematical models combining the laws of thermicodynamics and electricity, but (when designing new furnaces) it is important to validate conclusions by work on a model and an industrial pilot installation [21], [30].

Figure 8 gives a diagram of material currents and temperature distribution in a glass melting furnace.

### *4.2.2. Conduction glass melting furnaces*

There are several types of conduction glass melting furnaces. Schematically, a furnace of this type consists of one or more intercommunicating refractory chambers. The electrical power feed is provided by electrodes passing through the refractory walls, and immersed in the bath (7).

#### *4.2.2.1. Electrodes*

The current input electrodes must withstand corrosion from the molten glass bath, offer adequate mechanical strength at high temperatures and have low resistivity; moreover, the electrodes must not contaminate the molten glass.

The quality of the glass is closely related to the buildup of convection currents in the bath. The position of the electrodes in the furnace is fundamental, since their arrangement with respect to each other and with respect to the furnace walls controls the energy given off and enables the best possible glass melting conditions to be obtained. The concentration of energy around the electrodes causes local overheating, resulting in an upward movement of the glass and convection currents in the bath.

Therefore, in addition to the main electrodes, electrodes located near the surface can be used to reduce the viscosity of the glass considerably in this area and ensure even mixing. Bubbles can then burst easily on the surface, and the glass becomes rapidly refined under favorable conditions.

Similarly, horizontal electrodes located on the vertical walls counterbalance the cooling effect of the walls, and reduce temperature differences between the center and the edges of the bath.

Electrode power supply circuits vary widely (closed triangle, Scott two-phase, etc.). The maximum supply voltage does not exceed 100 to 200 Volts, and is in general around 70 Volts.

Materials most widely used for electrodes are graphite and molybdenum. However, some furnaces producing special glasses are equipped with tin oxide or soft iron electrodes. In very special cases, platinum may be used [40], [44], [46].

**Fig. 9.** Current flow lines in molten glass [44].

Figure 9 shows the current lines around:
*a)* two horizontal electrodes located near the surface of the bath;
*b)* two vertical electrodes partially immersed in the molten glass bath;
*c)* cluster of electrodes.

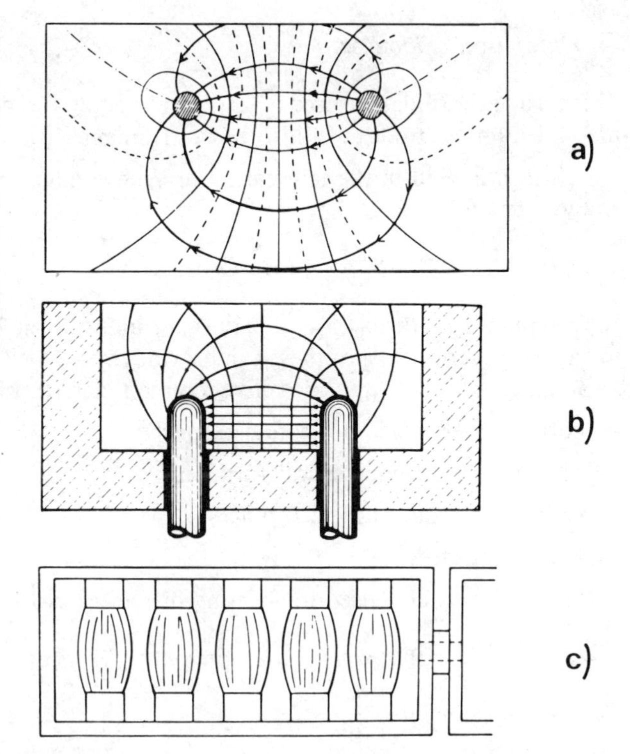


— *Graphite electrodes*

Graphite, which can be used in all types of furnaces, is used most often for the production of horizontal electrodes. Its high rigidity when hot and its density, slightly lower than that of glass, facilitate its use when the horizontal electrode must be deep in the bath.

Generally, electrodes are cylindrical, and range from 150 to 200 mm in diameter. The current density varies from 0.2 to 1 A/mm$^2$.

Hollow porous graphite electrodes offer the advantage of facilitating bath purification. The gas bubbles enter the porous mass of the electrodes due to hydrostatic pressure in the glass and are vented through an axial channel, instead of rising slowly to the surface of the bath.

— *Molybdenum electrodes*

Molybdenum can be used only for short, horizontal electrodes; at high temperatures, they have insufficient mechanical strength and buckle from pressure in the glass if long lengths are used. Conversely, this metal is widely used for the production of vertical electrodes which can be either solid or water cooled.

Electrodes are either flat or bar-shaped. Current density is high, about 3 A/mm$^2$, and dimensions are therefore low, about 30 to 50 mm diameter for cylindrical electrodes. Molybdenum oxidizes in air above 600°C. Therefore, the electrode outlets must be cooled [40].

— *Tin or iron oxide electrodes*

Tin oxide electrodes are used in lead glass melting furnaces (correctly designed molybdenum electrodes can also be used for melting crystal).

Soft iron is used in the fabrication of yellow glass, produced in a reducing environment.

### *4.2.2.2. Melting and booster furnaces*

The conduction furnaces used in the glass industry can be classified according to various criteria; most of these can provide overall melting, or simply booster or supplemental heating. The most common classification is as follows:

— horizontal-electrode horizontal furnaces;

— horizontal-electrode vertical furnaces;

— vertical-electrode vertical furnaces;

— special furnaces.

Conduction is also used for heating dies in glass fibre production.

#### *4.2.2.2.1. Horizontal-electrode horizontal furnaces*

— *Horizontal melting furnaces*

In design, these furnaces are similar to convection fossil furnaces. The electrodes located near the surface overheat the corresponding area and enhance fine glass production [5].

The furnace consists of a production area and a work area. The production area is divided into three parts. The first is used to preheat and melt the materials. The second completes melting and the third refines the glass. The current flows through the glass between each pair of electrodes. For example, these electrodes are closed-delta connected to the power transformer secondary. In the working area, the molten glass is cooled to the proper temperature and "drawing" takes place. Furnaces of this type are known as Borel or Romont-Saint-Gobain furnaces.

Graphite electrodes are used. In spite of the high bending strength of this material, a maximum electrode length exists, due to the hydrostatic property of the glass. This limits furnace width, bath surface and melting capacity.

These furnaces are used in all-electric melting and glazing glass production. The maximum daily capacity at present is about 120 tons per day for an installed power of about 6,000 kW. Specific consumption varies between 1.2 and 2 kWh/kg. Fossil fuel furnaces of the same type provide a production rate of 600 tons per day. Without technological improvements, electric furnaces of this type, for all-electric melting at very high daily production rates, compete poorly with fossil fuel furnaces because of the energy savings at present provided by the latter.

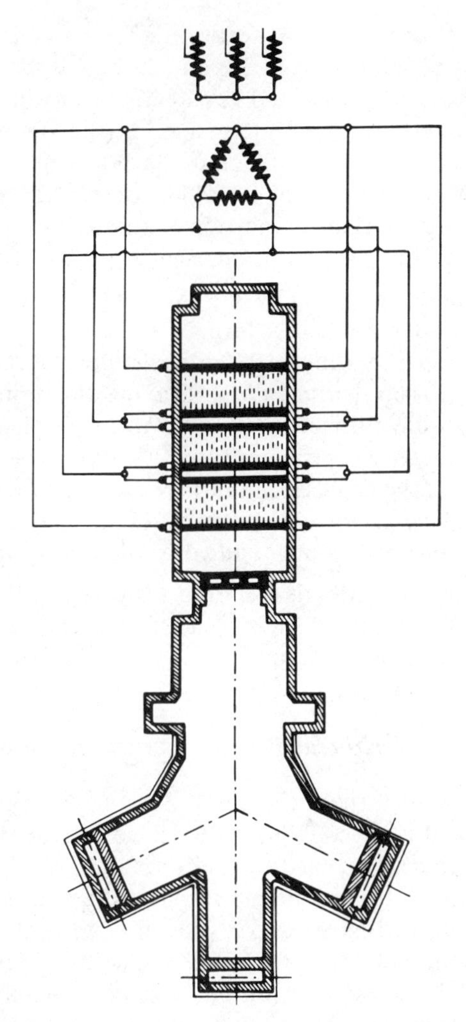

**Fig. 10.** Horizontal electrode furnace [5].

For lower production rates, down to approximately 130 tons per day, the choice between an electric furnace and a fossil fuel furnace for production of glazing glass or other glass products depends on the relative prices of the energies available after the respective efficiencies have been taken into account (that of the electric furnace is 1.3 to 2.5 times higher) and the quality desired. Moreover, investments are lower for electrical melting (20 to 30% less approximately). The development of the "float glass" process has led to significant changes in the furnaces used to produce flat glass.

— *Horizontal furnaces with booster heating*

Conduction heating can also boost the daily production rate of fossil fuel furnaces of a given capacity. This "booster" heating also offers special thermal characteristics, since the electrical energy conversion efficiency often exceeds 1. In reality, introduction of conduction heating, for which efficiency is slightly less than 1, as previously mentioned, creates convection currents in the bath. Heat transfer between the flames and the bath is then greatly improved, increasing total furnace efficiency, and apparently results in an electrical-energy-to-heat-conversion efficiency of more than 1 (the figures published give apparent efficiencies of between 80 and 200%). For a given furnace, fossil fuel consumption increases, as does that of electricity, but the molten glass rate increase is more than proportional. In addition, convection currents tend to homogenize the glass bath, improving quality and decreasing waste.

For example, boosting by 486 kW has enabled the production capacity of a 100-ton-per-day fossil fuel furnace to be increased to 136 tons per day, fuel consumption increasing by 0.77 tons per hour (8,900 thermal kW's) to 0.86 tons per hour (10,000 thermal kW's). The maximum installed booster power for a 200 ton-per-day furnace is about 4,000 kW. Depending on the type of furnace, electrical boosting contributes 10 to 40% of the energy required to produce glass. Figures published concerning furnaces equipped with a booster heating system, however, show a very wide range of results, and, in some cases, installation of a booster system has not resulted in the effect mentioned above.

In general, boosting is therefore a highly profitable and growing technique, since, for a given fossil fuel furnace size, it enables daily production rate to be increased, improve glass quality and allows for better adjustment of production to meet demand [14], [29].

#### *4.2.2.2.2. Horizontal-electrode vertical furnaces*

This type of furnace (Penberty, Sorg, etc.) makes use of vertical convection currents and is generally used for all electric melting. The various operations: loading, preheating, melting and refining take place in succesion. Loading is made through the top of the furnace; the bath surface, which is completely covered with vitrifiable material, does not overheat. The efficiency of the furnace is improved, and product losses are practically nils; the specific consumption, for normal operating conditions and rates of more than 10 tons per day, is between 0.9 and 1.2 kWh/kg. Covering the products greatly reduces emissions of noxious gas products and saves raw materials. For translucent glass production, fluorine losses are about 3%, while in a conventional furnace they may reach 50%; pollution and production costs are greatly reduced.

Convection currents are encountered only in the electrode area. Electrode arrangement prevents a continuous vertical flow of materials from occurring and confines convection currents to where they are useful for melting and refining the glass. Vertical temperature distribution enhances optimum refining of the molten glass. Generally, molybdenum electrodes are used, but tin oxide electrodes are sometimes used in crystal work furnaces. Electrodes are often located at two or more levels in the furnace, each level being independently supplied with power.

The refined glass passes through a siphon into the holding bath where it is cooled down to working temperature. The furnaces have one or more draw-off points.

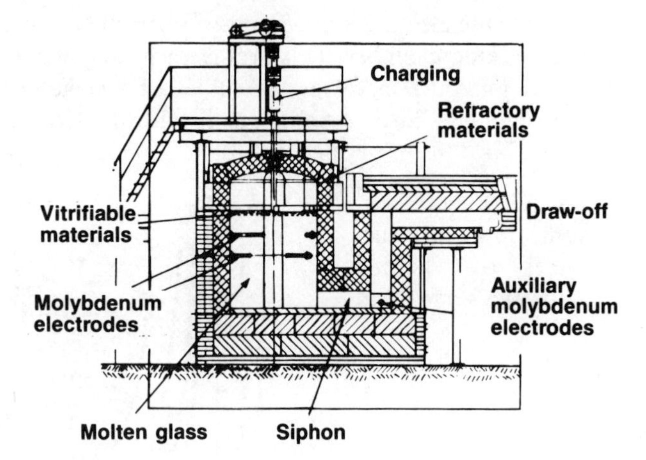


**Fig. 11.** Horizontal-electrode vertical furnace [21].

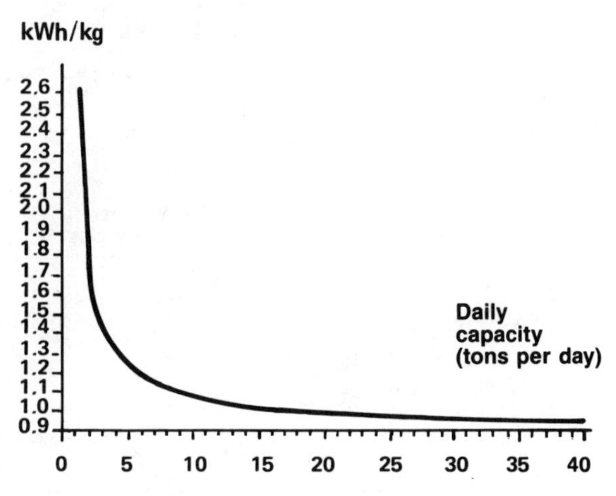


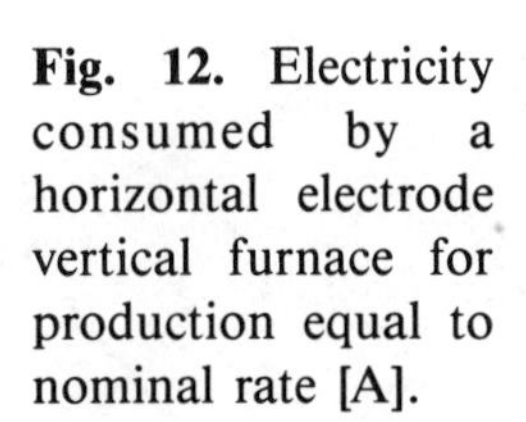

**Fig. 12.** Electricity consumed by a horizontal electrode vertical furnace for production equal to nominal rate [A].

These electric furnaces have many applications. Figure 15 shows some of their principal characteristics. The maximum production rate is about 100 to 150 tons per day.

The melting powers are equal to installed powers which cannot be continously sustained. The auxiliary power is used to heat the siphon and the drawing area. Specific consumption corresponds to nominal operation.

In the first furnace, changeover from normal to translucent glass requires 20 hours, and from a translucent to normal glass approximately 36 hours.

*4.2.2.2.3. Vertical-electrode vertical furnaces*

— *Melting furnaces*

Vertical electrode melting furnaces (Gell, Saint-Gobain, etc.) are similar to the vertical furnaces described above. The successive treatment operations take place in the vertical plane. The load is inserted through the top of the furnace, and refined glass passes through a siphon into the working bath.

The main electrodes are mounted vertically in the bottom part of the melting and refining chamber. This arrangement avoids the intense bending forces to which horizontal electrodes are subjected. Generally, the electrodes are flattened molybdenum. Their number varies with the size of the furnace.

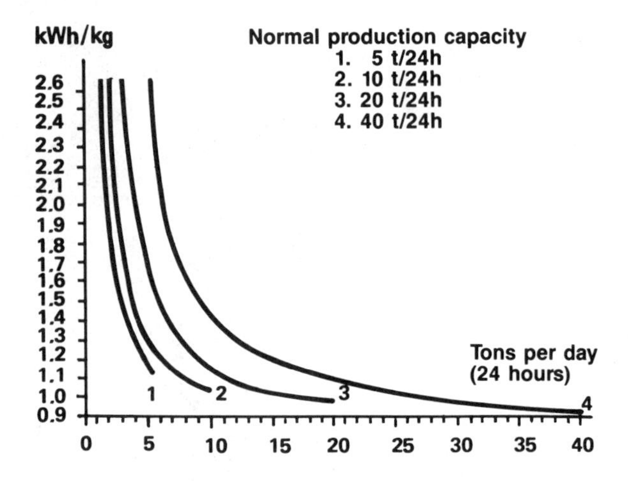


**Fig. 13.** Energy consumption for furnaces of various capacities versus load [A].

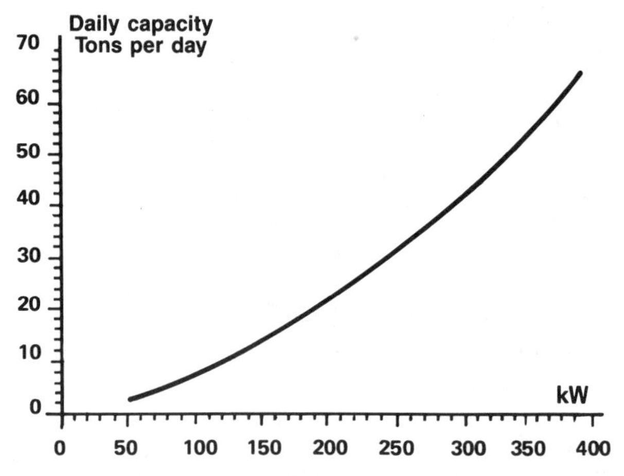


**Fig. 14.** Thermal losses of a range of vertical furnaces versus daily production capacity [A].

Various devices are used to vary furnace power (multitapping transformer, induction regulator, thyristors, etc.).

This type of furnace offers rather high specific production rates (tons per $m^2$ per day), of about 5 $t/m^2$ per day for high capacity furnaces, up to 200-250 tons per day and more [18], [29], [47].

Auxiliary electrodes are located at points in the furnace which satisfy a particular thermal requirement (bath homogenization, temperature holding prior to drawing, etc.)

Characteristics of some horizontal electrode vertical furnaces [A].

| Type of glass produced | Normal glass or translucent glass for small bottles (cosmetics) | Translucent glass for lighting equipment | Lead crystal 10-13% lead oxide | Boron glass for production of glass fibres |
|---|---|---|---|---|
| Production capacity (tons per day) | 14 | 3 | 30 | 40 |
| Bath area ($m^2$) | 2.85 | 1.2 | 10 | 11 |
| Specific productivity (tons per $m^2$ per day) | 4.9 | 2.5 | 3 | 3.6 |
| Melting power (kW) | 650 | 200 | 1,500 | 1,800 |
| Auxiliary power (kW) | 100 | 25 | 80 | 60 |
| Specific consumption (kWh/kg). | 1.1 | 1.25 | 0.9 | 0.85 |

**Fig. 15.**

Efficiency is high and specific consumption is between 0.8 and 1 kWh/kg of molten glass with gross raw materials. If supplied with glass scrap, specific consumption drops to 0.35-0.40 kWh/kg of molten glass.

This type of furnace may be used to produce a wide range of glasses — boron or lead glass, colored glass, high quality glass, bottle glass.

— *Booster heater furnaces*

In most fossil-fuelled furnaces, vertically-mounted electrodes provide booster heating. Both high and low capacity furnaces can be fitted with these; for example, bottle producing units of 300 to 400 tons per day [29].

In these, conduction heating is used both to increase melting capacity and to improve refining.

The conduction-heated areas, which are three-phase supplied, generally feature three or six electrodes.

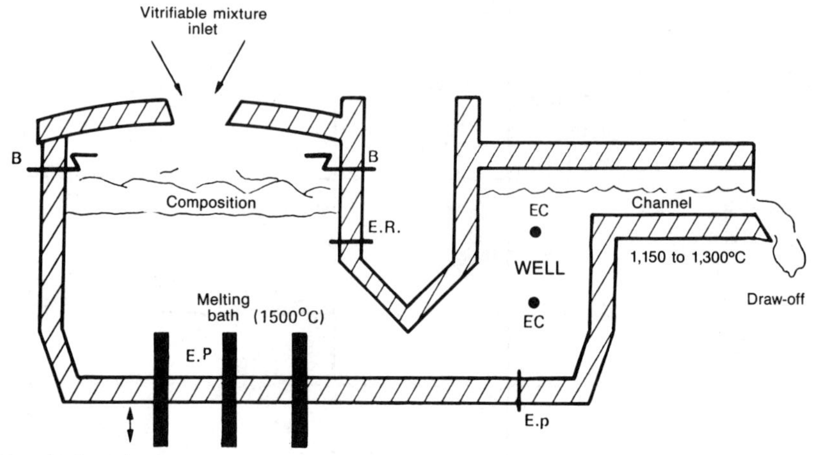


EP, main electrodes
EC, conditioning electrodes
ER, repulsion electrodes
Ep, bleed electrodes
B, fuel burners (used for starting)

**Fig. 16.** Vertical-electrode vertical furnace

Booster heating profitability is, in most cases, excellent, due to reasons similar to those mentioned in the paragraph devoted to horizontal electrode furnaces with booster heating.

*4.2.2.2.4. Special furnaces*

A wide variety of immersion-electrode glass melting electrical furnaces exist. These vary according to the shape, and electrode arrangement, feed method and number [36], [44], [C].

Figure 17 shows an example of a small electrode furnace used in glass-blowing work. This furnace is fitted with a booster fuel burner which, among other things, reduces power demand during peak hour rates.

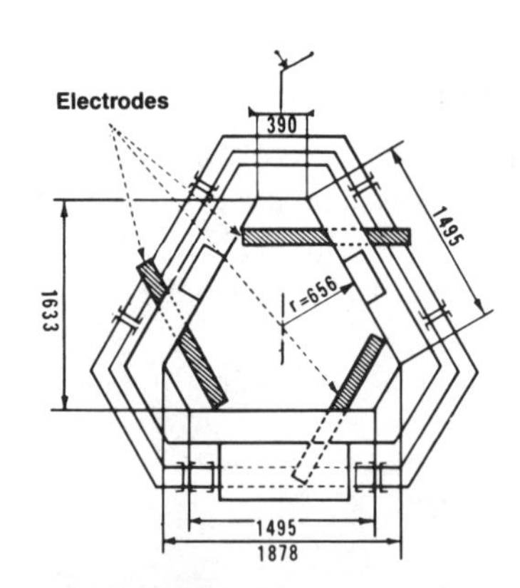


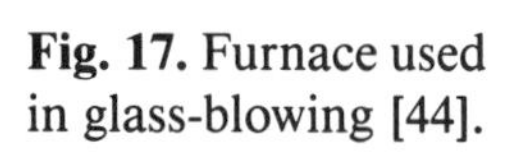
**Fig. 17.** Furnace used in glass-blowing [44].

The power of a furnace of this type is a few hundred kilowatts (300 kW, with the dimensions shown in millimeters in Figure 17). Each electrode is connected to one phase of the supply circuit.

Figure 18 shows another type of conduction furnace which was originally used for the production of microballs. This production involves the pulverization of glass at temperatures of between 1,400 and 1,700°C. Gradually, the use of this type of furnace has been extended to bottle production, glass firbres and special glasses (Leclerc de Bussy furnace)[C].

The electrodes are molybdenum and converge towards the center of ther furnace, and are offset by 120°. A part mounted on a vertical tube passing through the bottom of the furnace, and forming the orifice through which the glass flows, is located in the center of the furnace.

This is a cold wall-roof furnace. The furnace is loaded through the top, providing cold product coverage. The walls are made from red brass, cooled by a nest of copper tubes brazed directly onto the envelope through which water flows.

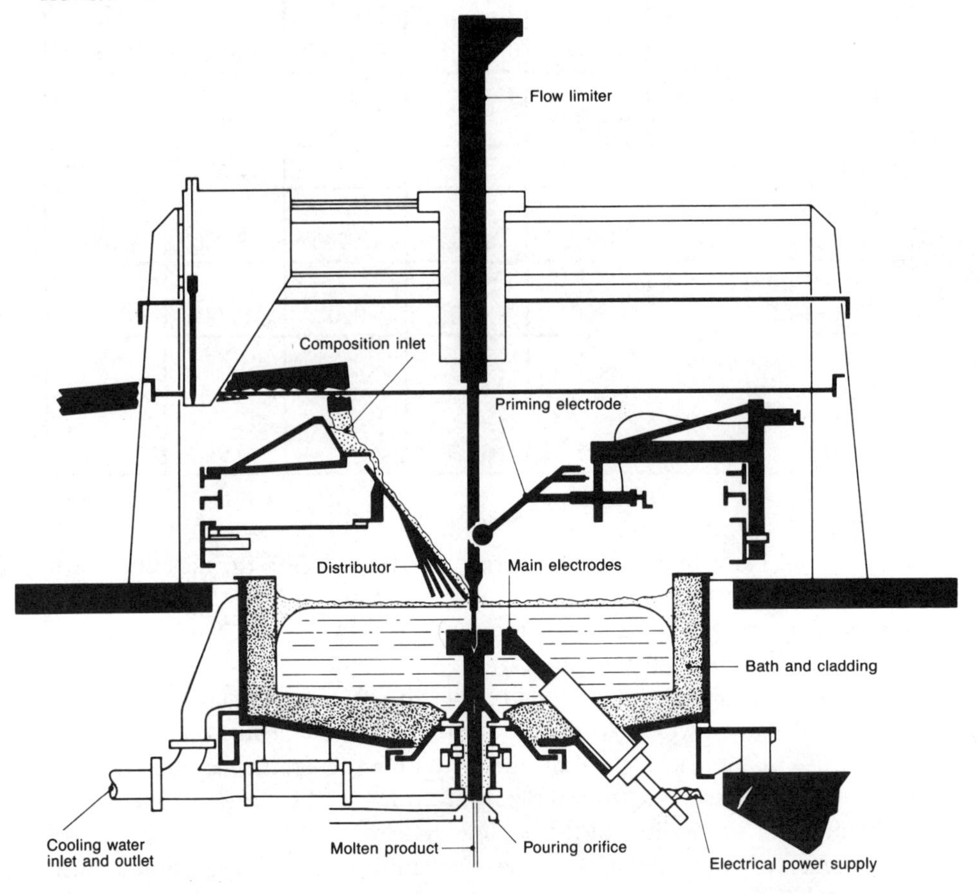


**Fig. 18.** Special glass furnace (Leclerc de Bussy furnace)[C].

The inside of the bath is lined with about ten centimeters of refractory material. In general, the diameter is about four times the height.

This furnace enables melting of glass products (sodium-calcium, borosilicate, alkali-free and special glasses, etc.) and also refractory products (alumina, silica, magnesia, lime, zirconia mixtures).

The capacity of these furnaces varies between 2 and 30 tons per day. The characteristics of a range of furnaces of this type are given in Figure 19. The rates and specific consumptions apply to the melting of sodium-calcium glasses. The specific consumption may be doubled for refractory products.

Compared to conventional furnaces, this type offers the following advantages:
— very high temperatures at center of furnace (2000°C), permitting the melting of glasses which are difficult to produce, and refractory products.

Characterisics of a range of glass furnaces
(Leclerc de Bussy type)[C]

| Daily production rate (t) | 1.5 to 3 | 4 to 9 | 6 to 15 | 9 to 22 | 13 to 35 |
|---|---|---|---|---|---|
| Furnace diameter (m) | 1.5 | 2 | 2.5 | 3 | 4 |
| Weight empty (kg) | 1,250 | 2,350 | 8,800 | 12,500 | 23,000 |
| Weight in production (kg) | 1,900 | 3,900 | 12,500 | 18,000 | 35,000 |
| Nominal power (kW) | 300 | 600 | 900 | 1,200 | 1,800 |
| Consumption per kilogram of molten glass (kWh) | 1.5 | 1.2 | 1 | 0.95 | 0.9 |

**Fig. 19.**

— cold wall bath, unaffected by thermal shock, enabling frequent shutdown and starting, together with high operational flexibility:

- starting time, furnace cold: 4 to 10 hours,
- shutdown time: a few minutes;

— high convection movement due to temperature gradient (2,000°C in center, 1,250°C on walls), which accelerates melting and homogenizes the bath;
— absence of unmelted material due to the high temperatures obtained and movement of glass away from the hottest area;
— recovery of energy lost in the cooling system;
— reduction of raw material losses due to volatilization, since the melting surface as a whole is covered;
— pollution free;

— working conditions improved due to silent operation;
— simplicity of operation, enabling the furnace to be controlled by non-skilled personnel;
— low bulk, for a unit of production capacity of 25 tons per day:
- furnace weight: 60 t,
- weight of conventional furnace: 600 t;

— low investment costs; for a 25 t per day unit, the investment cost is 30% less than that for a fossil fuel furnace [44], [C].

Small furnaces are also used for melting lead crystal. The design of such furnaces is similar to that of horizontal-electrode vertical furnaces, but their size is much smaller. The electrodes are often made from tin oxide ($SnO^2$) to

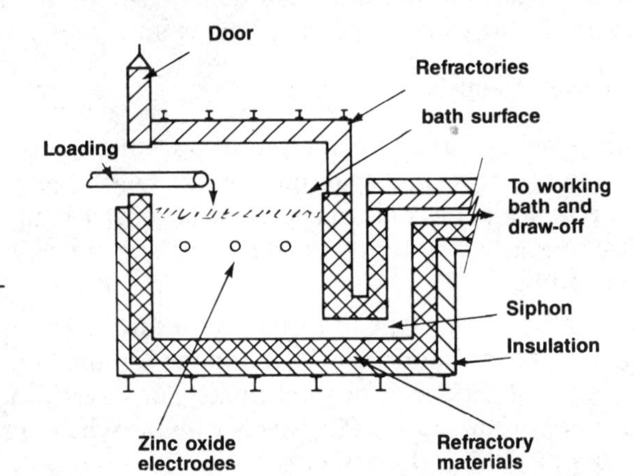


**Fig. 20.** Crystal melting furnace [22].

prevent any risk of reaction between molybdenum and lead oxide, and unwanted coloring of the crystal. Special molybdenum electrodes have been developed which, due to quick cooling, avoid contamination of the bath by molybdenum oxide [22], [26].

### *4.2.3. Advantages and disadvantages of glass melting by conduction*

The term glass in fact describes a vast family of products, and the advantage of using conduction heating can vary from one product to another. In general, this melting technique offers a number of advantages:
— direct transfer of heat into the molten glass bath and creation of convection currents within the bath, properties which enhance glass quality;
— reduction of losses of costly materials such as boron oxide or lead oxide;
— no toxic gases emitted during production of opaque glass (opal) containing components such as fluorine;
— production of colored glasses facilitated;
— operational flexibility, fast and frequent composition changes possible;
— highly compact installations;
— relatively low investment costs.

These advantages enhance the use of electricity for the production of special glasses and the creation of low or medium capacity units (less than 150 tons per day approximately). In such cases, electrical melting is very often highly profitable when compared to techniques involving the use of fossil fuels.

Conversely, for very high production rates, conduction heating has three main disadvantages. The first, technological, is the difficulty in obtaining high capacity electric furnaces (beyond 200-250 tons per day). The other two are economic: very high capacity furnaces, when the corresponding market is ensured, offer substantial savings; moreover, for large furnaces, fuel costs are, in spite of recent increases, still less than electricity costs for equivalent production.

However, for large furnaces, and also for smaller capacity furnaces, boosting has become highly profitable, since, for relatively low investment costs, it enables an increase in the capacity of installations and improves overall efficiency.

### 4.3. Melting enamels

Melting raw materials for the production of enamels can be provided by conduction furnaces similar to those used in glass melting. However, to make melting possible, the alkali content must be greater than 1%. If it is not, bath electrical conductivity is too low, and it is necessary to use conduction with other system heating.

The capacity of this type of furnace varies from 0.2 tons per day to 30 tons per day. Furnace operation may be continuous or discontinuous, and the single-phase (for small furnaces) or three-phase (for large furnaces) supply is used. The specific consumption varies widely with the type of product melted, which can be between 0.8 and 2.5 kWh/kg.

Basically, the advantage with conduction melting is that the fluorine, lead or boric oxide losses are very low, while efficiency and manufactured product quality is very high, due to the excellent homogeneity of the bath (A).

### 4.4. Electrode steam boilers

With these boilers, steam vaporization is obtained by direct current flow through the mass. The electric current is applied through electrodes immersed in the water, which itself forms the heating resistance.

The dissipated power, and therefore the steam flow, are a function of the surface area of the immersed electrodes, their arrangement, the power feed voltage and the resistivity of the water. Essentially, resistivity depends on the dissolved mineral and salt content. If an electrode boiler is to operate correctly, the water used must meet certain resistivity requirements. Distilled water or a condensate with low resistivities are appropriate for boilers supplied with medium voltage. Normal water, with a resistivity of between 500 and 1,500 ohm-cm at 100°C (at 15°C, resistivity varies from 1500 to 4,000 ohm-cm; at 200°C, this decreases by approximately 15% with respect to its value at 100°C), can be used in boilers supplied with low voltage or medium voltage boilers. In many cases, a closed water system with constant regulated resistivity is used [1], [2], [42].

Technically, these boilers can provide extremely varied steam rates, ranging from a few tens of kilograms per hour to almost 100 tons per hour, at pressures and powers ranging respectively from a few bars to 100 bars, and from a few tens of kilowatts to approximately 70,000 kilowatts.

### *4.4.1. Composition of an electrode boiler*

Schematically, these generators consist of a metallic chamber capable of withstanding working pressure, containing the water to be vaporized, and in which the metal electrodes connected to the power feed conductors are immersed. The electrodes consist of bars, cylinders or plates, either flat or curved, and generally made from steel or cast iron. The boiler body is insulated by thick covering (100 mm minimum of insulating material), making thermal losses almost negligible [2].

The boiler is supplied with three-phase AC at line frequency. DC would, in fact, break down the water into its components, hydrogen and oxygen, thus causing risks of explosion. The power supply voltages vary from 380 to 30,000 V.

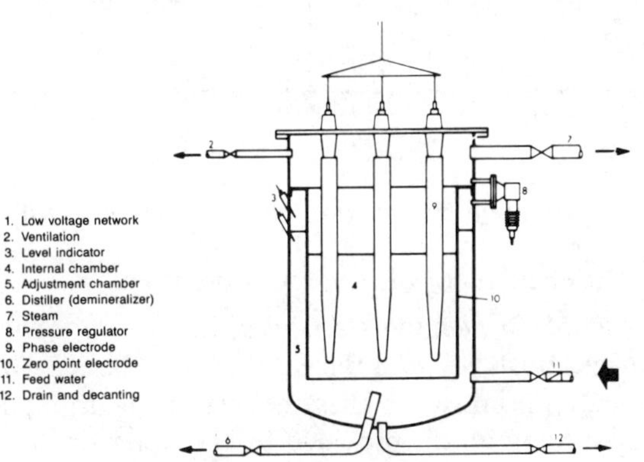


**Fig. 21.** Diagram of a low voltage electrode boiler with water-level power regulation [F].

Adjustment of the steam rate takes place basically by varying the resistance of the water to current flow, and therefore by modifying the cross-section or length of the fluid through which the electric current flows. The boiler power can therefore be regulated by increasing or decreasing the water level around the electrodes, or by altering the distance between an electrode or a counter-electrode. These boilers can operate only when water is present, and are protected against dry operation [42].

Ancillary devices — feed tank, pumps, regulation and control systems, safety equipment, bleeding devices, etc., complete the boiler itself.

### *4.4.2. Different types of electrode boilers*

Electrode boilers can be classified according to different criteria.

*a) High or low voltage power feed*

The power feed to electrode boilers varies between 380 and 30,000 V, and, as a function of voltage, are divided into:

— low voltage boilers, for which the voltage is less than 1,000 V;

— high voltage boilers, supplied at voltages of more than 1,000 V; with high voltage boilers, the resistivity must be kept low.

*b) Power level*

Powers range between a few kilowatts and (up to 70,000 kW approx.); in practice, voltage and power are not independent, and special combinations are possible.

Range of electrode boilers as a function of power and voltage

| Boiler category | Power range (kW) | Voltage range (V) |
|---|---|---|
| High power | 4,000 to 70,000 | 6,000 to 30,000 |
| Medium power | 1,000 to 4,000 | 1,000 to 6,000 |
| Low power | 15 to 1,000 | Low voltage or high voltage (beyond 100 kW) |

**Fig. 22.**

*c) Electrode system and electric current path*

Different types of electrodes are used:

— *simple immersion electrodes*; the electrodes are immersed in the water and current flows between the electrodes or the electrodes and a metal part forming a neutral point. With these boilers, the neutral point is often isolated from the boiler body and is provided by the water or the metal part (*see* paragraph *d*) below). These electrodes consist of steel or cast iron bars or plates. Most low power boilers, of less than 1,000 kW, operate at low voltage according to this principle, but some high voltage, high power boilers also use this principle [F], [O];

— *evaporation tube electrodes*; each electrode is fixed and surrounded by an insulating material vaporization tube which can be moved vertically and completely immersed in the boiler water, even in its highest position. Modification of the evaporation tube position with respect to the electrode alters the length of the current path between the electrode and the boiler walls ("grounded neutral" method; *see* paragraph *d)* below), and therefore alters the resistance of the water conductor and also absorbed power. For relatively low powers (up to 150 kW approximately per electrode, with maximum voltage of 8,000 V), vapor bubbles are vented through natural circulation inside the evaporation tube. For higher powers, it is necessary to provide a forced circulation device which draws away the bubbles when they form, since natural circulation is no longer sufficient to evacuate them when produced in very high numbers; if such a device were not fitted, the bubbles would adhere to the surface of the electrodes and form an insulator causing variations in current and a drop in power, and therefore in

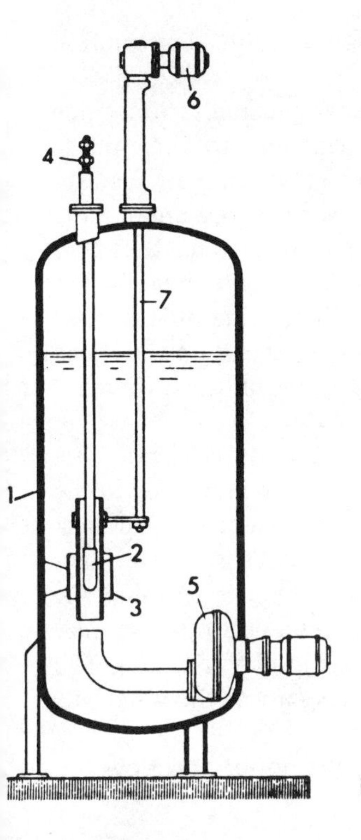


1. Boiler body
2. Electrode
3. Counter-electrode
4. Terminal
5. Circulation pump
6. Adjusting motor
7. Adjusting linkage

**Fig. 23.** Diagram of a counter-electrode, high voltage electrode boiler [G].

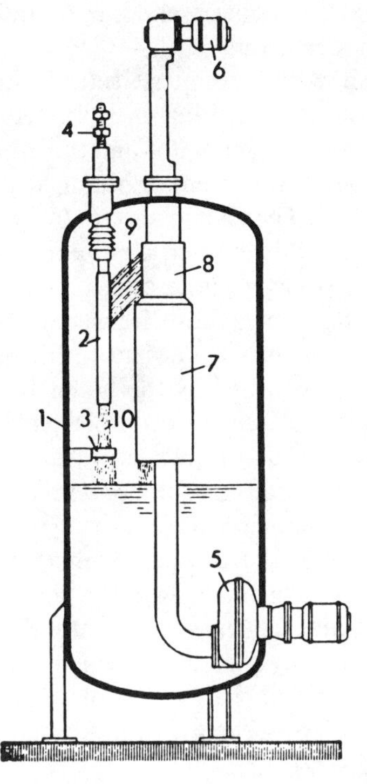


1. Boiler body
2. Electrode
3. Counter-electrode
4. Terminal
5. Circulation pump
6. Adjusting motor
7. Jet deflector
8. Tube and nozzles
9. Top jet group
10. Bottom jet group

**Fig. 24.** Diagram of a jet type electric boiler [G].

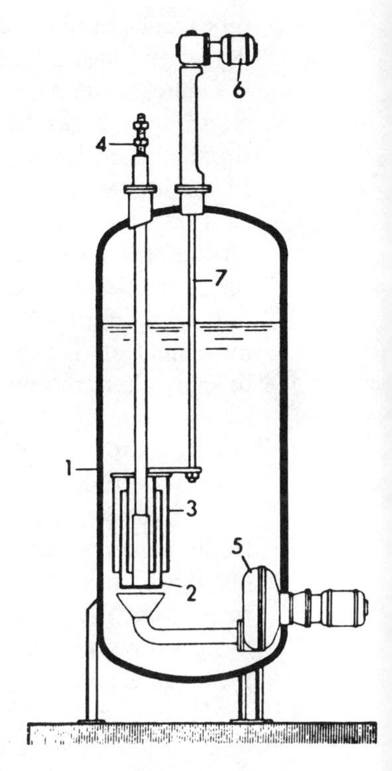


1. Boiler body
2. Electrode
3. Counter-electrode
4. Terminal
5. Circulation pump
6. Adjusting motor
7. Adjusting linkage

**Fig. 25.** Diagram of a counter-electrode, low voltage electrode boiler [G].

steam generation. Forced circulation is provided by a pump which discharges water picked up in the boiler under each electrode. These steam generators generally operate at high voltage [2];

— *electrodes fitted with counter-electrodes*; the current flows from one electrode to a concentric counter-electrode which is connected to the second conductor or the neutral point (current also flows between the electrode and the metallic boiler body, when this forms the grounded neutral). As before, a vaporization tube can be used to adjust the power, and incorporation of a forced circulation device provides a high steam rate.

Figures 23 and 25 show two counter-electrode steam generators (only one electrode is shown in the diagram, but these boilers generally are three-phase supplied); one is a high voltage generator and the other a low voltage generator. In the low voltage boiler, adjustment is obtained by varying the area of the electrodes facing each other. The counter-electrodes are attached to an adjustment linkage which can be raised or lowered by a motor. In the high voltage boiler, the electrodes and counter-electrodes are fixed, and an insulating tube, similar to the evaporation tube described above, is inserted between them; up and down movement of this tube, which is connected to the adjusting motor through a linkage, alters the power absorbed by the boiler [B], [G];

— *electrodes with multiple jets*; in multiple jet boilers, the electrodes are connected to each other through liquid jets, instead of being immersed in the boiler water. One of the electrodes is equipped with nozzles, each of which project a jet of water onto the counter-electrode. In this system, the electric current flows from the input terminal to the nozzles, then through the jets forming the resistance, to the counter-electrode and to the corresponding output terminal. In practice, the current and jets circuit is a little more complex (*Fig.* 24). The absorbed power is adjusted by a deflector which deflects part of the water jets from the electrodes directly to the bottom of the boiler.

Multiple jet boilers offer excellent flexibility, since the power and flow can be adjusted very rapidly [4], [G];

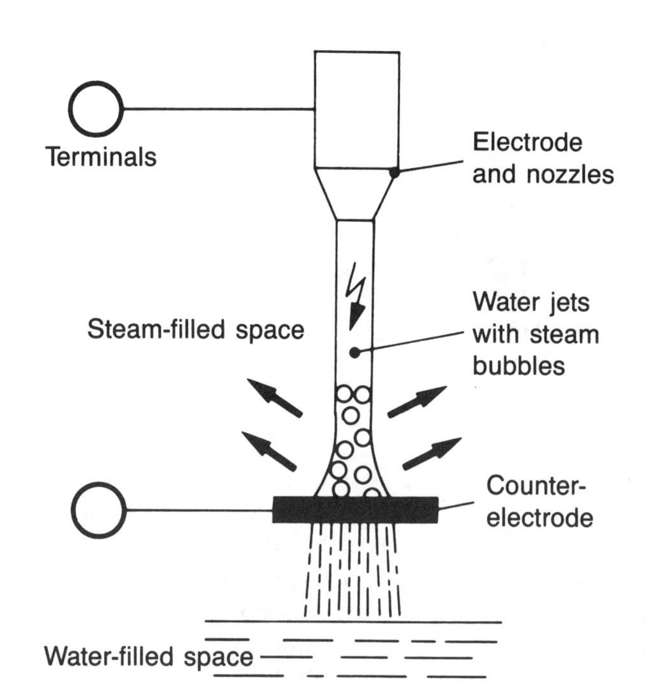


**Fig. 26.** Diagram showing operation of a jet boiler [G].

— *electrodes with directed current paths*: with this type of boiler, the electrodes are surrounded with insulating porcelain cylindrical separators. Radial current paths form around the electrodes.

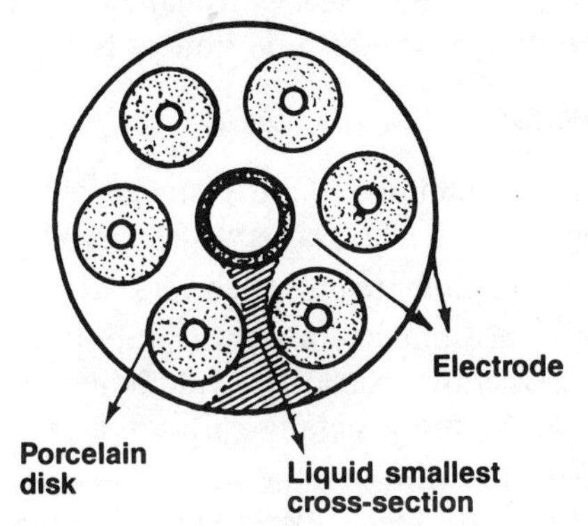


**Fig. 27.** Directed current path electric boiler [3].

The heat is not given off at the electrodes, which offer a wide cross-section to current flow at their contacts, but in the narrow area separating two porcelain cylinders. This prevents any formation of sparks or steam jackets at the electrodes. A large amount of the water helps convert electrical energy into heat energy, enabling compact boilers to be constructed [3], [Q].

The power is regulated by adjusting the water level, without using any internal mechanisms.

*d) Neutral mode*

Generally, two types of electrode boilers can be distinguished as a function of the neutral mode:

— *grounded neutral boilers*: the walls of the boiler or parts connected to these walls form the electric neutral, which is grounded through resistances and protective devices. The phase conductors are connected to the electrodes. Basically, this device is used with high voltage boilers;

— *isolated neutral boilers*: the neutral point is isolated from the boiler envelope. The neutral point either consists of an added metal part or is obtained by balancing appropriately arranged electrodes.

Isolation of the neutral point with respect to the boiler body is provided by the water or the metal part forming the neutral point. This system is used mostly with low voltage boilers, but also with some high voltage boilers [2].

The neutral mode must comply with power supply and electrical equipment safety standards and requirements.

*e) Power regulation system*

As a function of this criterion, electrode steam generators can be divided into a few major categories:

— boilers adjusted by varying the water level between fixed electrodes;

— boilers adjusted by varying the height of immersed electrodes;

— boilers adjusted by moving counter-electrode plates (or vaporization tubes) modifying the current paths [2].

In the future, thyristor power supplies which are more flexible, will most probably replace these electromechanical regulation systems and therefore simplify the design of electrode boilers [47].

*f) Conventional industrial boilers*

In practice, combining these systems, although all combinations are not possible, produces a wide variety of electrode boiler designs.

The most common steam generators are:

— for low or medium steam rates (up to approximately 1,300 kg per hour of steam, i.e. a power of 1,000 kW), steam boilers, three-phase supplied at the usual voltage of 380 V, and generally with a high resistance insulated neutral point.

The current flows through the water between the electrodes, which are vertical bars, or between the electrodes and a cylindrical tank; this metal tank, placed inside the boiler body and filled with water, then surrounds the electrodes and forms the neutral point (Figure 21). In general, power regulation is provided by varying the boiler water level.

As an example, Figure 28 gives the characteristics of a range of vertical electrode, low voltage-supplied boilers. The floor area occupied by the ancillary equipment, and in particular the feed tank, is approximately half that occupied by the boiler itself.

Range of low-voltage vertical electrode boilers [O].

| Power rating (kW) | 25 | 50 | 100 | 200 | 300 | 500 | 600 | 900 |
|---|---|---|---|---|---|---|---|---|
| Steam rate at 8 bars (kg/h) | 32 | 65 | 130 | 260 | 390 | 650 | 780 | 1,170 |
| Height (mm) | 1,585 | 1,495 | 1,495 | 1,685 | 1,685 | 1,685 | 1,685 | 1,685 |
| Width (mm) | 320 | 460 | 460 | 595 | 595 | 1,150 | 1,150 | 1,700 |
| Depth (mm) | 320 | 460 | 460 | 595 | 595 | 595 | 595 | 595 |

**Fig. 28.**

Counter-electrode low voltage electrode boilers also exist. These consist of a fixed electrode with a concentric counter-electrode, the vertical movement of which varies the absorbed electrical power (*Fig.* 25). Another type of boiler consists of three fixed-phase electrodes, connected to the cover from which they are insulated (the feed-through insulators behave as electrode supports). The counter-electrode, which is also insulated and subdivided into three elements, pivots (a device also used in hot water boilers). The electrodes and the counter-electrode are cylindrical, and, in the same manner as a variable capacitor, can be inserted gradually into the other by rotation, providing electrical power adjustment (*Fig.* 29).

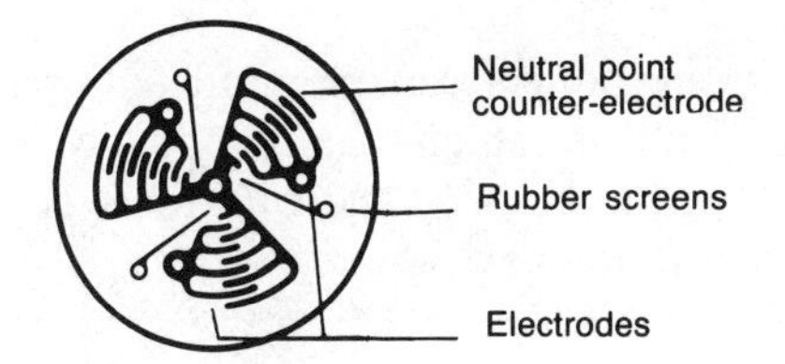


**Fig. 29.** Electrodes with rotary counter-electrode [F].

In most low voltage boilers, forced circulation of water under each electrode is not necessary;
— for high steam rates, in high voltage-supplied boilers, a forced water circulation device is required under each electrode. The three-phase supply may be either of the insulated neutral point high resistance type, or use direct grounding.

These can be counter-electrode boilers with evaporation tube movement regulation (*Fig.* 23) or multiple jet regulation (*Fig.* 24).

Another type of high voltage boiler consists of an external pressurized tank (boiler body), and an internal tank mounted directly on brackets and insulators inside the boiler body. The three-phase electrodes are immersed in the water contained in the inside tank; the latter, which also forms the electrical circuit neutral point, divides the boiler into two water reservoirs located one above the other (top tank with electrodes and bottom tank forming water reservoir). The circulating pump takes the water from the bottom tank and discharges it into the top tank through three pipes, opening onto each of the phase electrodes.

Figure 30 gives the characteristics of a range of high voltage electrode boilers. The dimensions are those of the boiler itself.

Range of high voltage electrode boilers [F]

| Voltage (kV) | 3 | 3 | 6 | 10 | 10 | 10 | 10 |
|---|---|---|---|---|---|---|---|
| Power (kW) | 5,000 | 8,000 | 12,000 | 10,000 | 25,000 | 40,000 | 70,000 |
| Steam rate (tons per hour) | 6.7 | 10.8 | 16.2 | 13.5 | 34 | 55 | 95 |
| Diameter (mm) | 1,650 | 1,950 | 2,350 | 2,650 | 2,850 | 3,050 | 3,550 |
| Height (mm) | 3,500 | 3,500 | 4,100 | 4,400 | 4,700 | 4,900 | 5,700 |

**Fig. 30.**

In particular, these figures show that high power electrode boilers can be very compact, due to the high power density, which, in numerous cases, is a significant advantage.

### *4.4.3. Efficiency and specific consumption*

Because of the steam generation method, the steam production efficiency of electrode boilers is excellent, about 95%. With 1 kWh, it is possible to produce 1.3 to 1.4 kg of steam.

### *4.4.4. Electrode boiler applications*

Technically, electrode boilers can meet all industrial steam requirements; however, they are best suited to high steam production rates, ranging from a few hundred kg of steam per hour up to almost 100 tons per hour. Armored resistance boilers are easier to use for low powers.

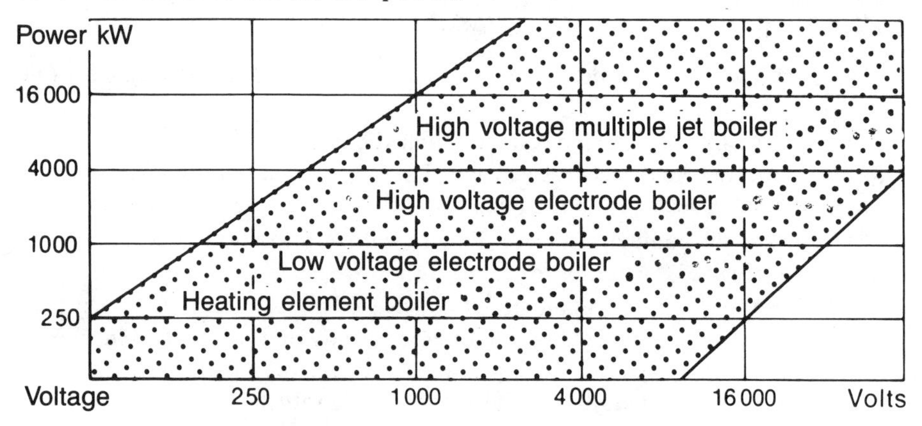


**Fig. 31.** Field of application of electric steam boilers [G].

The profitability of steam produced by electrode boilers is of course a function of the cost of energy compared after the various conversion efficiencies (production, distribution, regulation, reaction to variable loads, etc.) have been taken into account. When this analysis has been done correctly, and the possibilities of using electric boilers in ways other than fossil fuel boilers are fully explored, electrically produced steam may be shown to be competitive. In the part devoted to immersion-heated steam boilers, the chapter on "encased resistance heating" gives more information about the operation of electric boilers.

Electrode boilers are widely used in nuclear or fossil fuel power stations to provide "start-up" steam, maintain temperature or for auxiliary equipment (degassing, preparation of cooling agents, evaporation of boron, etc.).

These boilers can also be used in a wide range of industries — textiles, food, chemicals, paper, mechanical, etc.

### 4.5. Electrode water heaters

Electrode hot water heaters use the same principles as the steam generators described above, but are simpler in design. However, they operate at lower pressures, feature less equipment and require smaller power densities.

Generally, these are isolated neutral heaters, but may have a grounded neutral. As for steam generators, there are many types, the most common of which feature:

- concentric (*Fig.* 32) or rotary (*Fig.* 29) counter-electrodes;
- cylindrical tank forming the neutral point (*Fig.* 21).

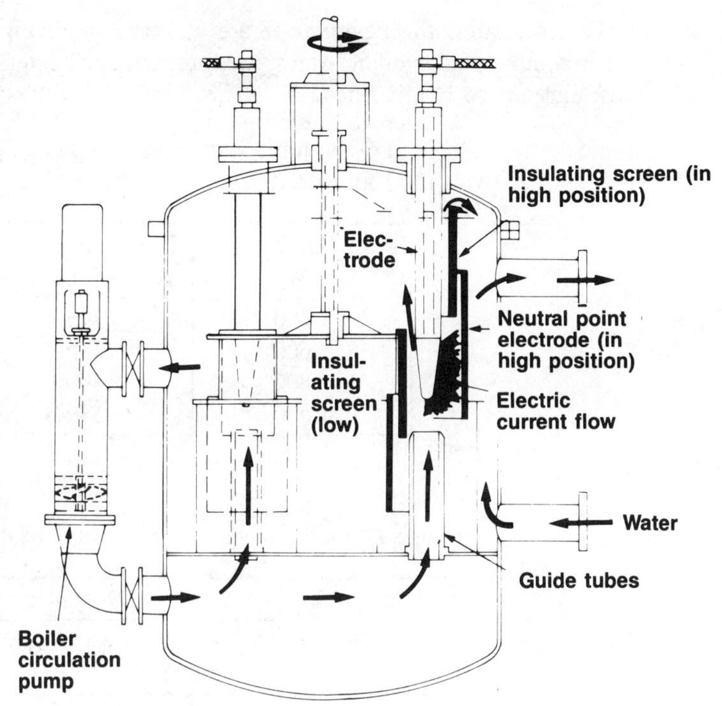


**Fig. 32.** Superheated electric water heater. The power is adjusted by movement of an insulating shield located between the electrode and the neutral point electrode [45].

The hot water is kept under pressure at a temperature less than its boiling point, defined approximately by the equation:

$$P_{\text{bars}} = \left(\frac{t°\text{C}}{100}\right)^4.$$

The pressure must always be maintained to prevent instantaneous vaporization. In comparison with steam, superheated water can be used profitably when the quantity of heat is low and the exchange surface is high, or transfer must be slow. These systems are cheaper than similar steam devices. There are practically no heat losses; the relative costs of the feed water and its processing are therefore low. Energy efficiency is also higher.

Temperature versus pressure for saturated steam

| Boiling temperature (°C) | Pressure (Bars) |
|---|---|
| 100 | 1 |
| 120 | 2 |
| 150 | 5 |
| 180 | 10 |
| 220 | 24 |
| 300 | 88 |

**Fig. 33.**

For low rates, the heaters operate at very low voltage (down to 1,000 kW approximately); for higher rates, medium voltage heaters enable very compact installations to be obtained.

Low voltage water heater range [F]
(width, 1,100 mm; depth, 750 mm; height, 1,750 mm)

| Power (kW) | Electrode length (mm) |
|---|---|
| 400 | 370 |
| 600 | 430 |
| 800 | 490 |
| 1,000 | 550 |

**Fig. 34.**

Range of medium voltage hot water heaters [F]

| Power (kW) | Water volume ($m^3$) | Overall dimensions | | |
|---|---|---|---|---|
| | | Width (mm) | Depth (mm) | Height (mm) |
| 3,000 | 2.6 | 2,500 | 1,920 | 3,600 |
| 4,000 | 2.8 | 2,500 | 1,920 | 3,900 |
| 6,000 | 4.7 | 2,700 | 2,120 | 4,400 |
| 8,000 | 5.0 | 2,700 | 2,120 | 4,600 |
| 10,000 | 6.5 | 2,900 | 2,320 | 5,100 |

**Fig. 35.**

Figures 34 and 35 provide typical characteristics of a range of low voltage boilers (380 V) with rotary counter-electrodes, producing superheated water at 120°C (pressure 6 bars), and those of a range of medium voltage heaters (10 kV) with concentric counter-electrodes, producing superheated water at 130°C (pressure 5 bars).

For example, these boilers are used to produce superheated water required for production processes. Immersion heaters are preferred for lower powers. Conversely, for high powers, electrode heaters enable compact equipment to be designed.

### 4.6. Electric steam accumulators

Storage of steam, or more accurately superheated water, may be advantageous when the steam demand occurs at a relatively low pressure. Steam accumulators can be used to:

— meet instantaneous steam demands, mainly when starting installations;

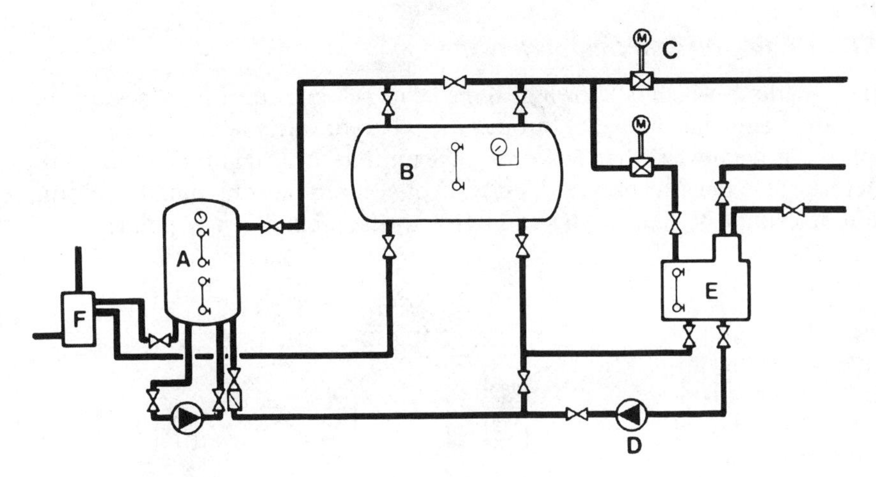


A. Electrode heater
B. Accumulator
C. Pressure reducing valve
D. Water feed pump
E. Feed tank
F. Pressure relief tank

**Fig. 36.** Electric steam accumulator installation diagram [F].

— equalize often fluctuating steam demands and, with a small boiler, enable variable requirements to be met;
— reduce or neutralize current demands during peak hours;
— produce and store steam during off-peak hours; the accumulator size is therefore determined by the steam capacity required over one day;
— release large quantities of steam over very short periods.

Energy is stored in the form of superheated water at high pressure, capable of producing steam by pressure relief. Production of super-heated water is provided by an electrode boiler. Accumulators can be large, up to approximately 100 $m^3$, and the storage pressure can reach 20 bars. It is therefore possible to maintain large quantities of low pressure steam (temperature from 100 to 160°C approximately). To profit from equipment of this type, where very high investment costs are involved, low cost electrical energy must be available [42].

## 4.7. Heating of tubes (tracing)

In some cases, it is possible to use the tubes themselves as an electric resistance for heating and maintaining the temperature of fluids, liquids or gases during transportation through metal tubes. Two techniques exist:
— heating by direct current flow in the tube carrying the fluid;
— heating by direct current flow through a "tracing" tube.

These processes can also be used to heat solids pumped through a tube, but as yet do not seem to have found industrial applications.

### *4.7.1. Heating of fluid transporting tube*

This method, which is known as impedance heating, consists of passing an AC current at mains frequency through the tube. For safety reasons, the voltage applied is generally less than 30 V. The tube must be divided into small sections. A return cable must be provided, and all connections must be shunted (if not, the tube cannot be welded and connected to the ground at any point).

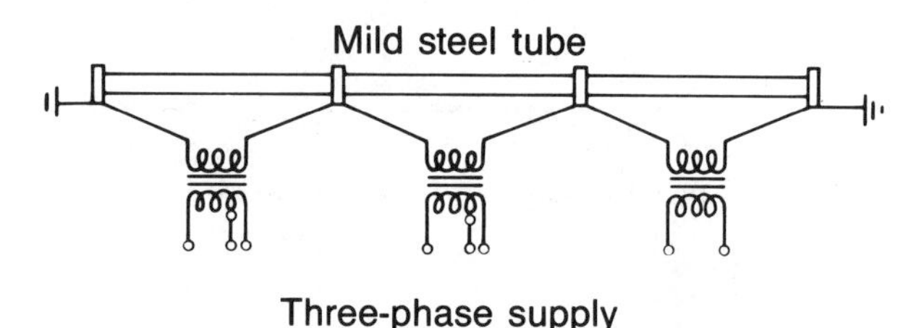


**Fig. 37.** Direct resistance heating applied to a tube divided into a number of sections, a multiple of three [45].

Generally, the powers reached vary between 40 and 1,000 W/m, and the voltage drop is between 0.3 and 1 V/m. If the return cable is laid close to the heating tube (if necessary against it), the power factor is high, of the order of 0.9. Electrical energy conversion efficiency is also high, almost 90%, and overall efficiency, if the tube is correctly insulated, remains close to this value.

This heating method is used to heat long conduits (contact resistance becomes negligible) in low mesh networks. However, this technique is prohibited for explosive or dangerous environments (contact problems). The maximum temperature is limited only by the softening temperature of the tube and the critical temperature of the fluid transported. This technique is used mainly to maintain temperatures, and more rarely for temperature buildup, in non-conductive liquids (petroleum and its derivates, chemical products, foodstuff liquids, etc.).

It can also be used to maintain temperature in hot gasses used in the chemical industry or for aerodynamic testing, etc.

This method of heating can also be used to warm up (at low temperature, some tens of degrees maximum) solid materials carried in the tube. The powers required are, however, higher than those currently encountered in fluid heating.

### *4.7.2. Heating using auxiliary tracing tubes*

In this method, which was developed in Japan, a tracing tube, welded to the line to be heated, is heated directly by current. The current is conducted by a copper conductor mounted coaxially in the tracing tube. This conductor is electrically insulated from the tube by a dielectric withstanding a high temperature (silicon rubber, high

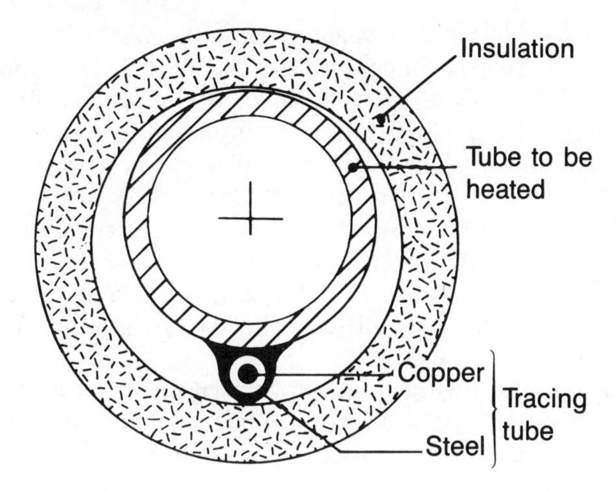


**Fig. 38.** Heating a pipeline by external tracing tube [46].

temperature PVC). The relation between the resistive tube and the power cable results in special properties. The current flows preferentially through the low impedance areas. In a coaxial system, the magnetic flux generated by currents flowing in the inner and outer conductors cause the current to flow on the interior of the outer conductor and on the exterior of the inner conductor. This results in skin effect (*see* paragraph 4.10.3.).

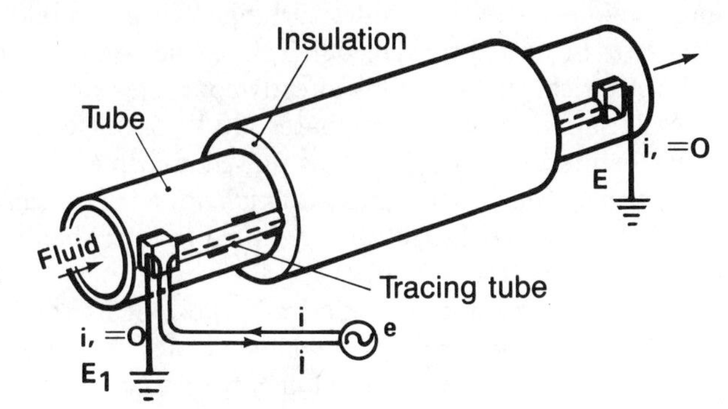


**Fig. 39.** Tracing of a pipeline [13].

With a ferromagnetic tube, this effect results in a total absence of current in the outside layers of the tube; this can therefore be welded to a large cross-section transportation tube, without causing current drain. In turn, the transport tube can be grounded at any point.

These special features explain the name of the process: "Skin Electric Current Thermo System" [12], [13], [25], [38].

This heating system offers the advantage of being supplied from a single point at a relatively high voltage. Therefore, there is no risk of arcing, and the process can be used in explosive or dangerous environments.

The voltage drop is about 0.3 to 0.7 V/m, and the linear power is 15 to 150 W/m. The system power factor reaches 0.9. The supply voltage is between 300 and 700 V/km. The maximum operating temperatures are 105 and 180°C respectively, with an electrical insulation consisting of high temperature PVC and silicon rubber. The outside diameter is between one-half inch (1.27 cm) and one and one-half inches (3.81 cm).

This process is used for tracing tubes carrying liquids of all types (pipelines, petrochemical, food industries, viscous products, etc.). For example with pipelines, the use of a voltage of 5 kV enables installation of a power supply point every 20 kilometers only. Efficiency of the system is much higher than that obtained with steam tracing.

This heating method can be used for applications other than tube tracing, and its range of application can be compared to that of armored resistance cables. A decision should only be made after doing a detailed analysis of investment and operation costs.

## 4.8. Heating concrete

The main reason for heating concrete is accelerated setting, thereby increasing productivity in the building and public works industries. Armored resistances are fully suitable for this application (see encased resistance heating and its applications), but a new process, based on conduction heating, is being developed. Three stainless steel electrodes, arranged in a triangle, are immersed in the concrete to be heated contained in a skip. The skip body forms the neutral point of the electrical circuit, and is directly connected to ground. The electrodes are generally supplied with three-phase, 380 V. Each electrode is connected to one phase. Regulation is provided by a pulsed multiple cycle thyristor power supply [47]. The installed power is about 200 kW for a bucket capacity of approximately 1.5 $m^3$. Approximately 1 kWh is required to raise the temperature of 1.5 $m^3$ of concrete by one degree [44].

This heating method is attractive because of its very high efficiency (better than 90%), high power density, which enables very rapid temperature rise, and the ease with which the injected energy can be adjusted. This process, which is currently used in some plants, is certain to become widely used in the concrete-using industries.

Another concrete conduction heating technique consists of placing the electrodes in a tube through which the concrete flows; however, this technique is in the experimental stage [44].

The concrete can also be heated by resistances embedded in the concrete itself during pouring. These resistances are considered disposable, and therefore must be very cheap. In some cases, the reinforcing rods armatures form the heating elements.

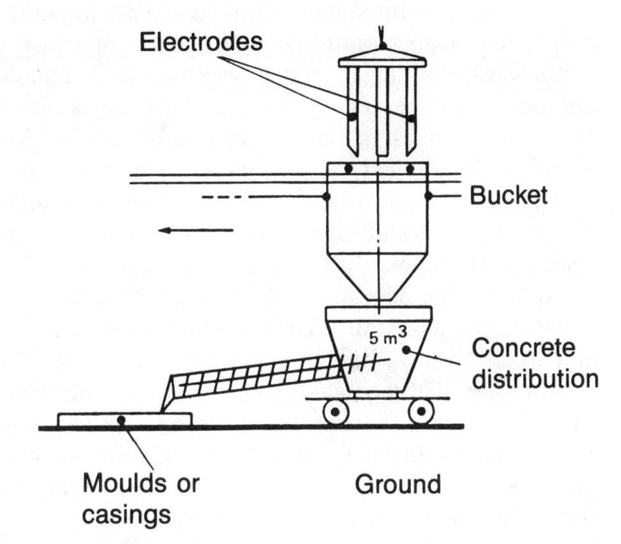


**Fig. 40.** Diagram of an immersion electrode bucket [44].

## 4.9. Heating metals before forming

Conduction heating is a simple technique for taking metals such as steels to their optimum plastic deformation temperature.

The part to be heated is short-circuited across the secondary of a single-phase transformer. The high current secondary therefore consists of the part to be heated and the transformer secondary winding. The current is applied to the part through contacts withstanding very high currents [8], [16], [32].

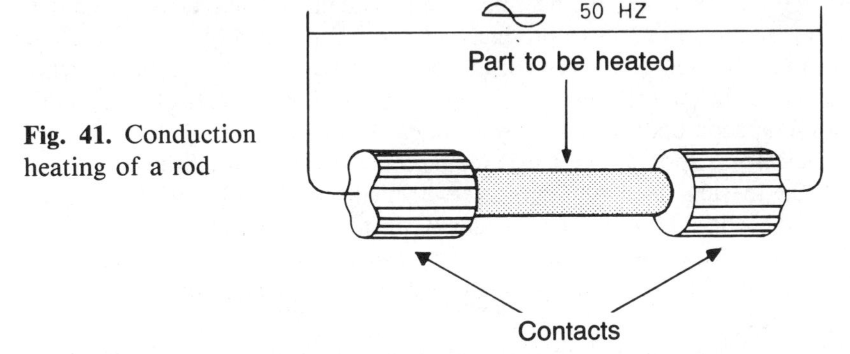


**Fig. 41.** Conduction heating of a rod

### *4.9.1. Conditions for the use of conduction heating*

The use of direct current flow for heating requires that the cross-section of the part to be heated be as constant as possible over its complete length; if otherwise, local overheating would occur which would be inversely proportional to heating time, since there would be no temperature equalization period due to thermal conduction.

Generally, to obtain high production rates and adequate efficiency, the ratio of length to diameter must be high, equal to or greater than 8. This phenomenon can be easily explained; the current density is limited by risks of overheating and rapid damage to contacts, and therefore cannot exceed a given value (4 to 10 kA for one contact). The maximum current flowing through the part is therefore determined by the surface area of the contacts. Now, at maximum current, the electrical power converted into heat increases with the resistance of the part. In addition, and for fixed current and cross-section, the higher the resistance the higher the power dissipated, resulting in higher production. Losses occur at the contacts and in the power cables, reinforcing the concept of a high length-to-diameter ratio. The longer the product to be heated, the lower the line and contact losses (and therefore the higher the efficiency).

Moreover, if the contacts were not a limiting factor, the current density, at least for ferromagnetic metals, must be kept below a certain threshold (about 15 A/mm$^2$) to prevent excessive temperature differences during the heating phase before reaching the Curie point. This is due to the very high skin effect (Kelvin effect) throughout the area in which the metal remains ferromagnetic (see description of skin effect in "Induction heating"). The resistance of the part to be heated is determined not only by its length and cross-section, but also by the resistivity of the material from which it is made. Conduction heating efficiency is proportional to the resistivity of the metal. Except for special cases (e.g. fine wire heating), it is not suitable for heating copper or aluminum alloys.

Therefore, conduction heating is appropriate for long products — billets, rods, bars, tubes — and flat products: sheeting, strips of ferrous metal.

### *4.9.2. Temperature distribution and choice of frequency*

Below the Curie point, AC heating produces an intense skin effect, such that the surface temperature of the products is higher than the core temperature. The temperature difference between the center of the part and the surface reaches 100 to 150°C during this phase for magnetic steels, but is practically negligible for non-magnetic bodies such as the standard stainless steels.

Conversely, as soon as the Curie temperature is exceeded (about 750°C for steels), the thermal gradient tends to reverse; current distribution through the part cross-section becomes practically uniform, and thermal losses due to radiation and convection from the surface cause cooling of the surface layer, so that the core temperature is finally equal to or greater than the surface temperature.

Figure 42 shows temperature distribution over the cross-section of a magnetic carbon steel billet (usual construction steel). With very short heating times, the temperature distribution is more or less uniform throughout the cross-section at the forming temperature. For longer heating times, the temperature in the center exceeds the surface temperature by 30 to 50°C.

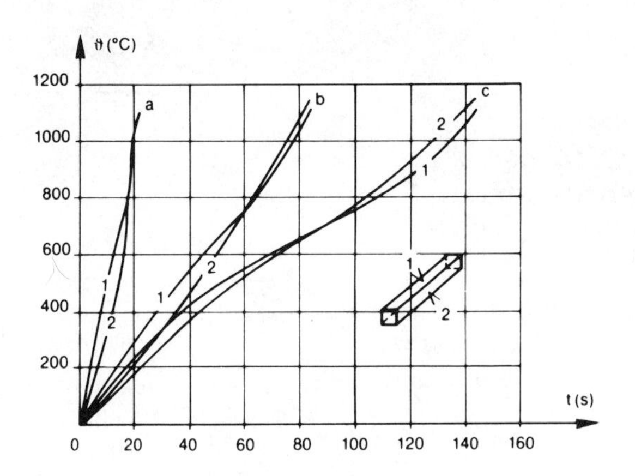


**Fig. 42.** Variation of surface temperature (1) and center temperature (2) of a carbon steel billet versus heating time [B].

Far from being drawbacks, skin effect and thermal gradient reversal are, on the contrary, favorable factors in heating products of a certain thickness such as billets, bars or rods. In fact, this temperature distribution ensures that:
— the core always contains a reserve of heat at end of heating which compensates for most of the energy lost by radiation, convection and conduction during transfer of the billet between the heater and the metal transforming equipment;
— since the temperature is higher at the core than at the surface, the products can be transferred as soon as the surface temperature is reached; oxidation and decarbonizing are then practically negligible.

AC heating at line frequency (50 Hz) is therefore widely used with these products, although DC is sometimes used to prevent variations in the power factor during heating, causing overrating of transformers (overrating of a transformer or installation of capacitors is, however, in most cases, less expensive than using an AC rectifier).

Conversely, for flat thin products such as sheets or strips, it is impossible to use AC with magnetic steels, due to the skin effect, and only DC offers even distribution of the temperature across the cross-section.

### *4.9.3. Typical conduction heater installation*

Schematically, a conduction heating installation consists of:
— a high voltage/low voltage transformer;
— a low voltage/very low voltage transformer;
— the heater itself consists of:
  • current input contacts,
  • the part loading, feed and ejection system;
— a water cooling system for the contacts and regulation equipment (these can also be air cooled);
— ancillary devices, control console, monitoring devices, etc.

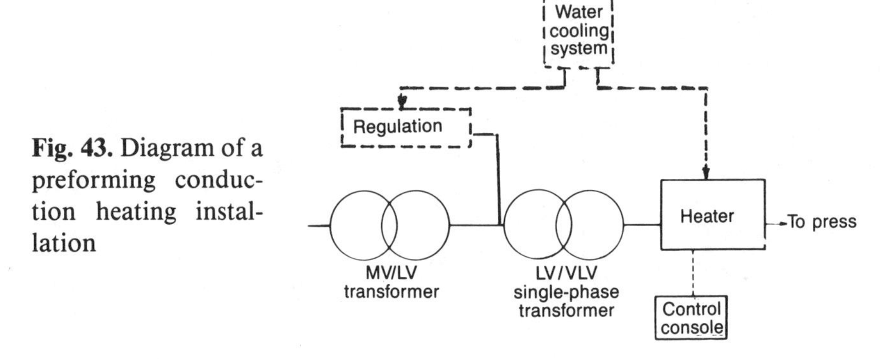


**Fig. 43.** Diagram of a preforming conduction heating installation

### *4.9.3.1. Current input contacts*

Contact-holder heads are used to hold the metal part and apply the current. These heads are installed on a mobile device enabling them to be adapted to the length of the heated product, to absorb length variations due to expansion during heating and, where necessary, to provide slight traction to prevent buckling for long products.

The current input contacts are the most delicate part of this heating process. In fact, the power of conduction heating equipment is determined by the current flowing through the part to be heated. This current is limited by the contacts. These must apply current to the part with a high safety margin, without local overheating of the metal or without causing metallurgical changes.

Each contact consists of a built-in copper element which can be changed very quickly. Contact designs vary with the type of heating, discontinuous or continuous, and the dimensions of the products heated.

With discontinuous heating, the contacts are arranged according to the temperature and pressure requirements at the extremities of the billets:
— if contact pressure is low, the contact area with the product to be heated is also low. The heat due to contact resistance cannot easily be dissipated through this low area, and risk of overheating limits power density;
— if the contact pressure is higher, the contact surface is greater; the heat given off in this area is lower, and the heat produced by the contact resistance can be carried off by the cooling water and the products being heated.

Heat resulting from contact resistances can be used to precisely set the temperature of the ends of a billet, by varying the pressure of the end contacts and altering the position of the lateral contacts with respect to the end of the billet.

Figure 44 shows various contact configurations used for discontinuous heating of products; these can also be clamped between current input clamps. The maximum current supported by a contact point is between 4 and 10 kA, depending on the contact material and the temperature requirements. For example, the configuration shown in figure 44 *e*) is used for a maximum cross-section of 10 m$^2$, with a current of 50 kA; the solution shown in figure 44 *f*) enables maximum currents of 100 and 130 kA to be applied to cross-sections of 50 and 200 $cm^2$ respectively. For larger cross-sections, it is possible to increase the number of contacts, although conduction heating is not widely used for large cross-sections, due to the required length-diameter ratio [15], [B], [K].

In continuous heating, the contacts generally consist of rotating rollers (*Fig.* 45). Contacts can also be provided by liquid metal baths (*Fig.* 59) [15].

To obtain adequate contact between the electrodes and products, the sections or billets must be carefully cut. The surfaces of parts in contact with the current inputs must be as clean as possible and free of rust and scale.

**Fig. 44.** Contact arrangement for stationary (discontinuous) heating [15].

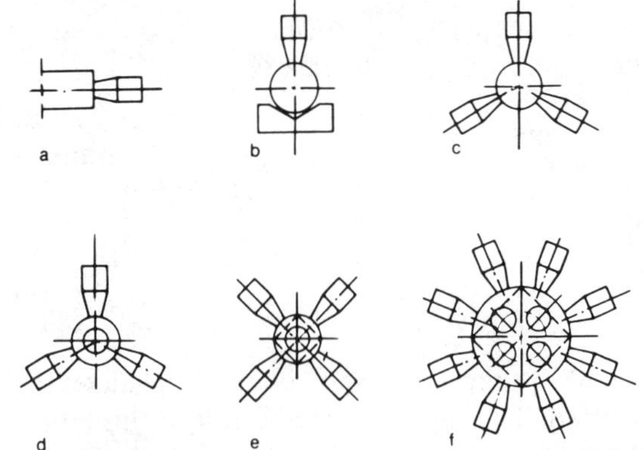


a) end contact only — b) lateral contacts only, two contact points — c) lateral contacts only, three contact points — d) lateral contacts and end contact, four contact points — e) lateral contacts and end contact, five contact points — f) lateral contacts and end contacts, 12 contact points.

**Fig. 45.** Contact arrangement for continuous heating

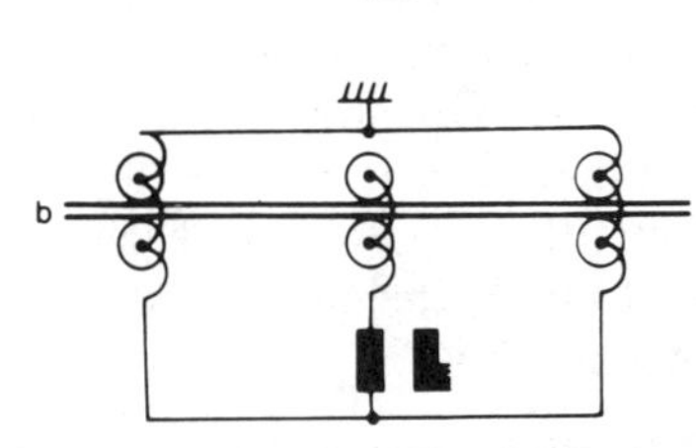


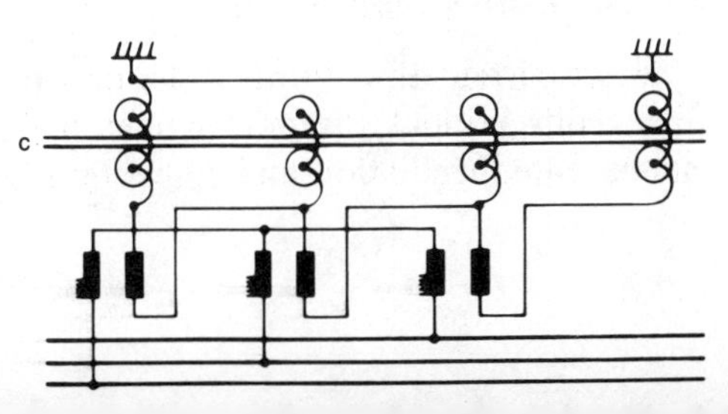


a) single-phase, one side to potential
b) single-phase, input and output on earth potential
c) three-phase, input and output at earth potential.

At the temperatures reached, thin layers of scale do not hinder current flow.

*4.9.3.2. Electrical power feed and regulation*

To ensure safety and prevent excessive currents which the contacts might not withstand, heating is made at low voltage, generally between 5 and 45 V. This very low voltage is pro- vided by a single-phase transformer.

It is possible to use current at line frequency, to rectify it or take it to a higher frequency. For long products, it is not generally necessary to use DC, and an AC line frequency supply is used in most cases. For flat products, DC must be used. Special cases exist in which high frequency AC is used.

**Fig. 46.** Power regulation

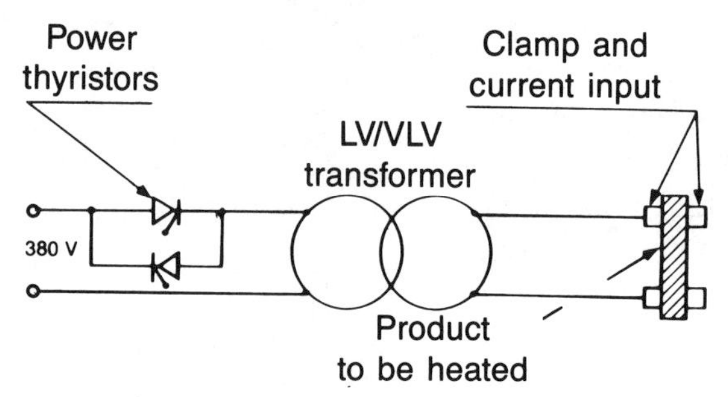


Various arrangements are used to regulate the power absorbed by the installation. The supply voltage can be adjusted according to product length, diameter or area to be heated. The current can also be adjusted using power thyristors in the primary circuit [47].

*4.9.3.3. Power factor and reactive energy compensation*

Very high currents must be taken from the transformer to the contacts. The connection must be such that its inductance is as low as possible (e.g. current input tubes parallel to the billet). The power factor, which, on starting, is about 0.3 to 0.4, increases during heating due to the increase in resistance, to reach values of 0.85 to 0.95 at end of heating.

For low power installations (a few kilowatts to several hundred kilowatts), compensation devices are not generally provided, but the supply transformers must then be overrated, since, on startup, the power factor of a well designed installation is about 0.4. Some devices enable this power factor to be improved from start- up. For larger installations, it is necessary to install a battery of compensation capacitors.

With rectified AC, the power factor is excellent throughout heating and varies from 0.9 to 0.95; therefore, whatever the power level, there is no requirement for compensation.

*4.9.3.4. Heating time and production*

Direct current flow through the part to be heated causes very fast heating, since conventional thermodynamic laws controlling heat transfer can be circumvented (radiation and convection heating).

Fig. 47. Power factor variation versus temperature for a 42 mm billet [B].

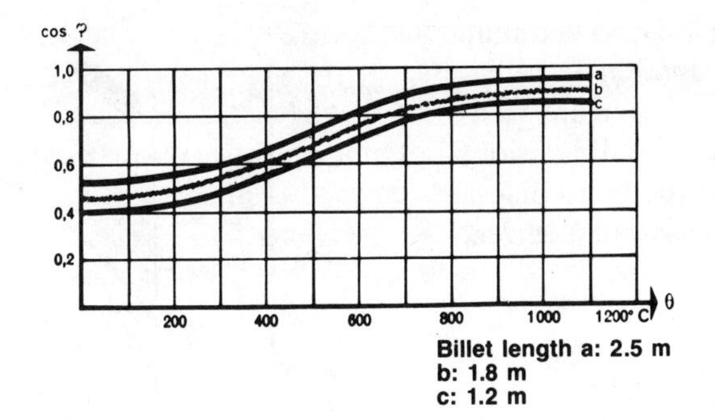


Fig. 48. Variations in nominal maximum power $N_s$ of the power transformer of a billet conduction heating installation as a function of cross-section $I_k$ and various billet lengths (valid for carbon steels. Contacts on all four surfaces and one front contact) [B].

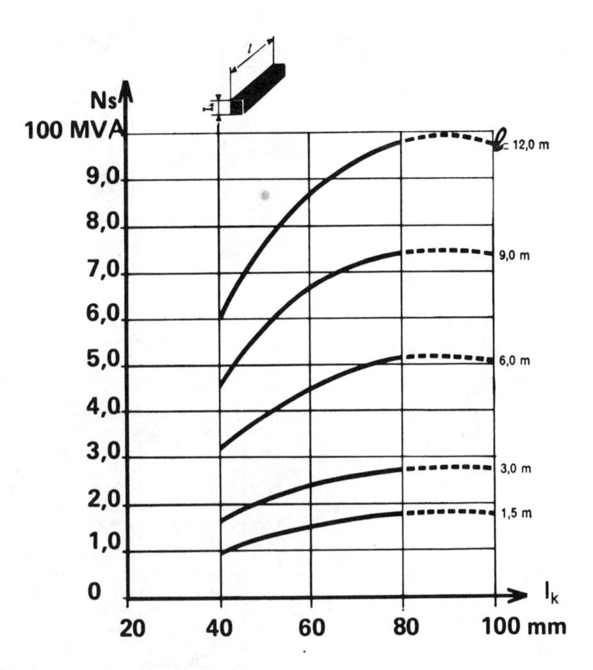


Maximum production is limited by the permissible current density in the product and the contacts, and also by product dimensions. Previous analyses have shown the attractiveness, both for productivity and efficiency, of an adequate length-to-diameter ratio; a high value of this ratio also has a favorable effect on the electrical power use factor. For a given current and product cross-section, this use factor increases, in fact, with product length, due to loading and unloading times, in discontinuous furnaces. Conversely, for continuous installations, which are rarely encountered in preforming heating, the use factor is normally equal to 1, and therefore has little effect on production capacity.

Figures 48 and 49 provide the production characteristics of high power installations.

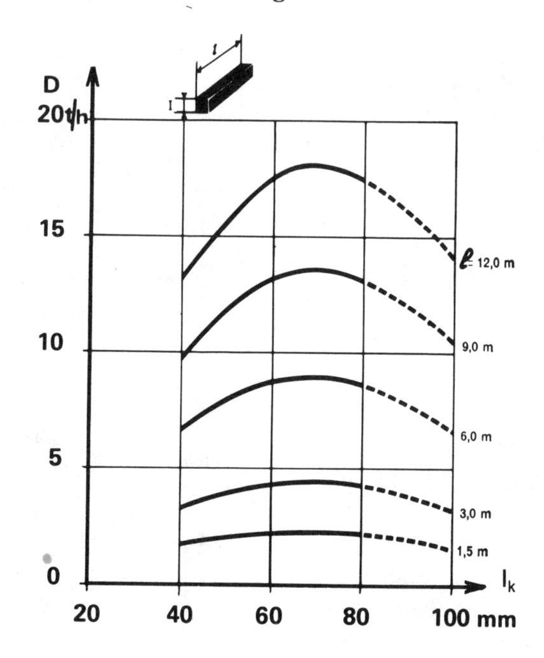


**Fig. 49.** Maximum production D versus lateral ($I_k$) and longitudinal dimensions of carbon steel billets, for maximum current [B].

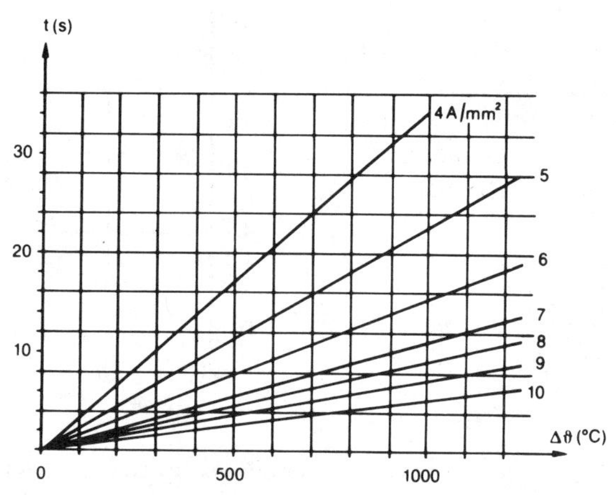


**Fig. 50.** Heating duration versus current density and temperature rise [B].

Basically, heating time depends on current density. Figure 50 shows the evolution of average temperature as a function of current density for a square cross-section billet of 42 x 42 mm. Overheating is very fast, since, with a current density of about 5 A/mm², the billet can be taken to 1,200°C in less than 30 seconds. In practice, normal heating times vary between about 10 seconds and two minutes.

Allowing for downtime due to handling of the parts, the power to be installed varies between 400 and 500 kW for a practical rate of 1 ton per hour. With hot forming, this rapid heating enables assembly line work. Due to its low and adjustable heating time, the conduction heater can easily be built into an automated work cycle.

The installation handles only one billet at a time (or two or three when several modules are mounted in parallel). In the event of a problem down-system from the heater, just cut power and no temperature holding problems arise. When the defect is repaired, a hot billet is available after a very short period.

Heating speed also leads to low oxidation and absence of surface decarbonizing, which are appreciable advantages, especially with high quality steels. Scale losses are less than 1%, and generally between 0.2 and 0.5% (the lowest percentages apply, of course, to relatively high cross-section products).

*4.9.3.5. Energy consumption and efficiency*

Conduction heater efficiency is excellent, about 80 to 90%, and specific energy consumption is lower than with any other heating process.

**Fig. 51.** Efficiency of a conduction heater versus billet length and diameter

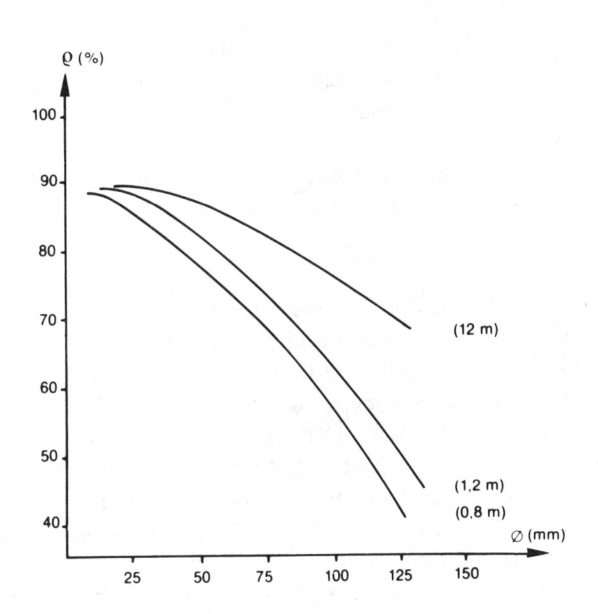


This process also provides substantial energy savings when compared to fossil fuel furnaces (for example, more than 50% primary energy saving in heating prior to stamping or forging). This result is obtained not only due to the thermal efficiency resulting from dissipating heat directly within the environment to be heated, but also due to the absence of off-load or starting losses, which therefore do not affect the overall specific consumption as they can in fossil fuel furnaces.

The mean specific consumption is about 300 kWh/t for steel heated to 1,250°C.

*4.9.3.6. Advantages and disadvantages of conduction heating*

Conduction heating offers numerous advantages:

— high thermal efficiency, up to 90%;
— very fast heating and operational flexibility;
— low scale losses, generally less than 1%, or even 0.5%;

— practically no surface decarbonizing;
— negligible metal grain enlargement;
— very low floor area;
— full automation capability;
— temperature distribution through the product favorable to transformation;
— relatively limited investment costs compared to other heating methods;
— capability of obtaining partial heating (at ends or on sections of bars).

In spite of these advantages, conduction heating has some disadvantages:

— it is appropriate only for high resistivity metals (basically ferrous metals) in the form of long products, therefore narrowing its application;
— construction of contacts is delicate, but problems seem to have been overcome; however, they must be regularly serviced, and only products whose ends are not oxidized can be heated;
— the installed power is used only during 60 to 75% of the total production cycle;
— the electrical power supply system is unbalanced; therefore, the phases must be balanced, either using several heaters in parallel, as is the case with numerous installations, or by connecting other equipment to the other phases.

However, these drawbacks are minimal and, in its specific field of use, conduction heating of metals before forming continues to develop rapidly.

#### *4.9.3.7. Comparison of conduction to induction heating*

Induction heating is widely used for heating metals prior to forming (see "Electromagnetic induction heating" chapter). Both methods, conduction and induction, can therefore compete for this application. Each case requires special analysis, but some general considerations can be provided. Conduction heating applications are much more restricted than with induction heating; in fact, this only applies to high resistivity metals (in general ferrous metals), offering sufficient length-to-diameter ratio (although present developments with conduction heaters may change this: *see* paragraph 1.9.4.1.1.) and a surface condition enabling current flow at the contacts.

Conversely, when these conditions are combined, the investment costs for a conduction heater are often lower (generally 20 to 40% less) and energy efficiency better. Conduction heating is therefore often preferred to induction heating. However, they become similar where high production rates, more than 10 tons per hour, are involved.

### *4.9.4. Preforming with conduction heating*

Conduction is used in numerous preforming heating applications, and especially prior to forging, upsetting, rolling and drawing [20], [33].

#### *4.9.4.1. Heating before forging*

##### *4.9.4.1.1. Heating steel bars or rods prior to forging*

Conduction heaters can be installed alongside most swaging hammers, presses and forging machines supplied with pieces of regular cross-section and 10 to 60 mm in diameter (or an equivalent square cross-section). Depending on the production rate to be ensured and the operational flexibility desired, one or more modules may be placed in parallel, each module forming an independent unit which can be synchronized with adjacent modules. These modules can be equipped with a gravity feed system and pneumatically controlled, interchangeable current input clips, which can be adjusted to the dimensions of the piece. Adaptation to pieces of different lengths and diameters requires less than 20 minutes and a few mechanical adjustments.

**Fig. 52.** Heating two parts in series [D].

The heating time and power levels are selected by the operator according to the series to be treated. Once adjusted, the module can be connected or disconnected immediately. Therefore, there is no waiting time, nor too many heated parts, which reduces the costs related to reheating and testing dies before starting a series. Also, the heater consumes only when used, providing very low overall specific consumptions with respect to the standard fossil fuel furnaces [D], [K], [L], [M].

Recent improvements have further enhanced the performance of this equipment and widened its field of application. Instead of treating pieces one by one, a new heater treats two (or more) pieces simultaneously, passing them over a series of heating stations; temperature rise takes place gradually at each heating station. This heating principle offers several advantages:

— the overall electrical resistance of the products to be heated and placed in series varies much less with time, since the temperature rise at each station causes a much lower variation in temperature than when the product is fully heated at one station;
— since the current is lower, it is possible to use a smaller transformer (the power factor is better), and risk of overheating at contacts is limited;
— parts with a smaller length-to-diameter ratio can be economically heated by conduction.

Transfer of parts is provided by a spar mechanism, and transfer time is very short (approximately 3 seconds), enabling a high production rate to be attained [D].

To illustrate the conductive possibilities in the heating before forging field, three modules, with an apparent power of 190 kVA each, operating at mains frequency (50 Hz) and heating only one piece at a time, have the following characteristics:

— each module is supplied by a single-phase, 190 kVA transformer, primary 380 V and 500 A, secondary 8.7 V and 20,000 A;
— piece length of between 150 and 600 mm and diameter varying from 12 to 50 mm;
— hourly production rate per heater of 450 pieces for a length of 150 mm and a diameter of 12.5 mm (approximately 200 kg per hour for the three heaters), 60 pieces for a length of 600 mm and a diameter of 38 mm (900 kg per hour approximately for the three heaters);
— heating temperature 1,250°C.

Synchronization of heaters is provided to obtain a regular rate in terms of level and duration. Two heating rates are used for heating each piece:
— start-off at high current, adjustable from 0 to 100% of the nominal current;
— end of heating at lower current, also adjustable.

The durations for both currents are controlled by timers.

The overall specific consumption, measured over a long period, is 300 kWh per ton of steel on average.

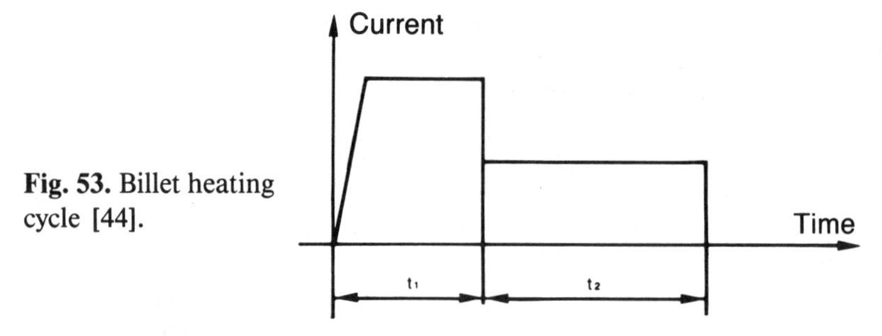


**Fig. 53.** Billet heating cycle [44].

*4.9.4.1.2. Upsetting on electrical machines*

Conduction heating is suited to heating before upsetting, which is a special forging application. If a force parallel to the axis of the bar is applied, when the forging temperature is reached, the heated part is crushed, enabling a blank of the part desired to be obtained. Figure 54 gives a diagram of an end-of-bar-mounted upsetting machine.

To limit buckling risks, the length of the part between the anvil and the vice must not exceed three times the diameter. In spite of the unfavorable conditions for this heating mode (low length-to-diameter ratio), efficiency remains high, since specific consumption is between 400 and 500 kWh/t of heated material around 1,100°C, compared to 270 kWh/t approximately under optimum conditions. This energy consumption is much lower than that obtained with a fossil fuel fur- nace heating the same products, and the gain is even higher than in the previous case, since, with a furnace, it is difficult to heat such a small piece with satisfactory overall efficiency [L].

It is also possible to upset in the middle of bars; in such cases, the current is applied through two vices, the inside separation of which must again be limited to three times the diameter to prevent buckling.

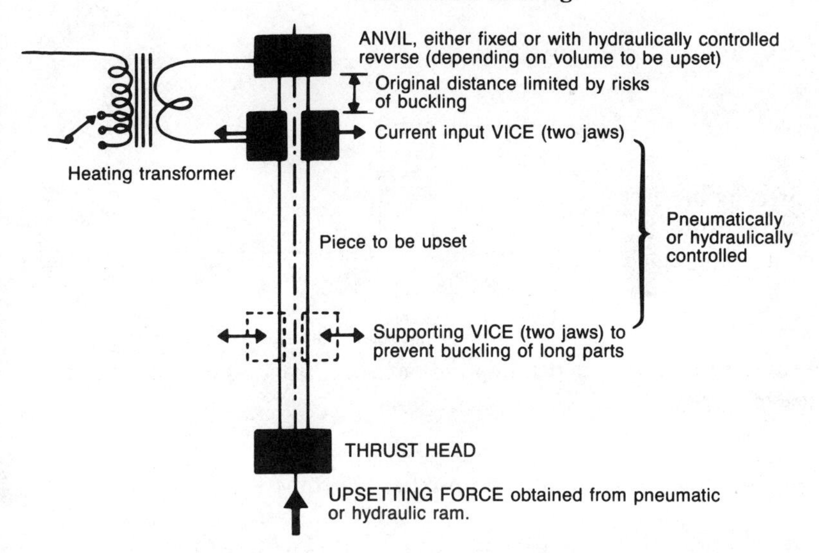


**Fig. 54.** Diagram of a flat anvil, end-of-bar vertical upsetting machine [L].

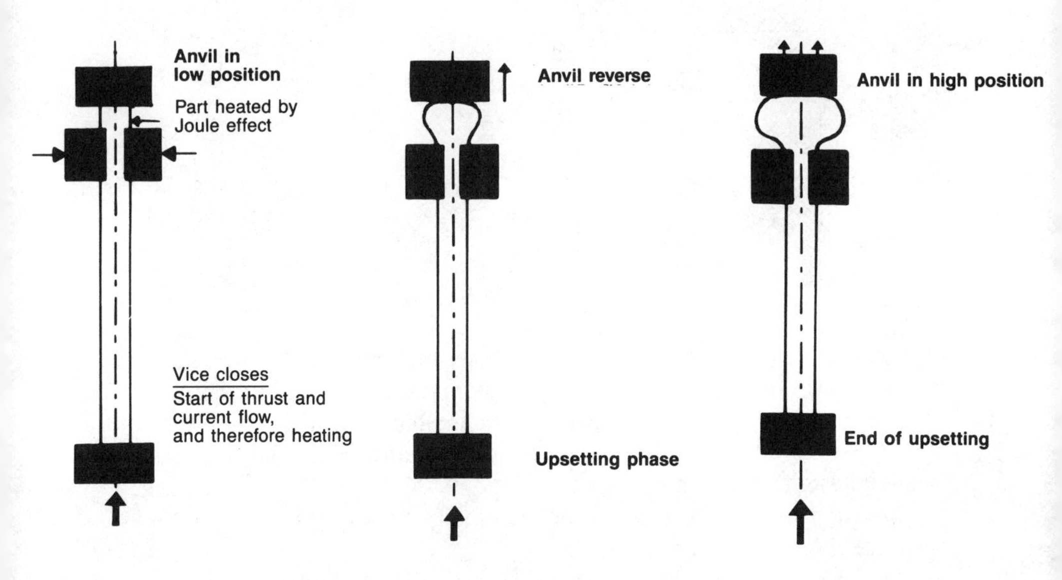


**Fig. 55.** Three successive stages in upsetting a valve blank [L].

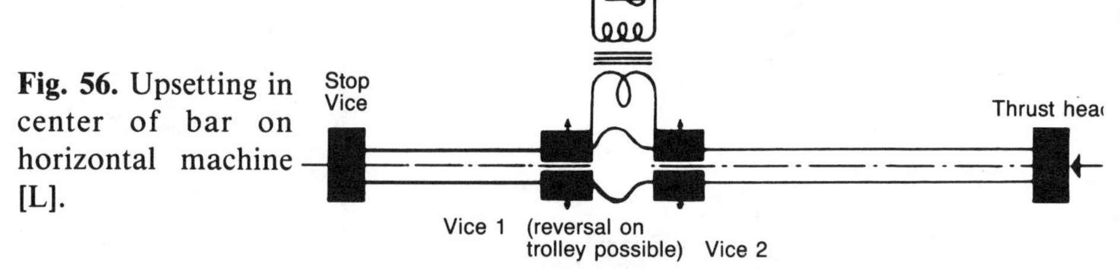


**Fig. 56.** Upsetting in center of bar on horizontal machine [L].

The upsetting force varies, depending on materials, from 50 to 500 N/mm², and the upsetting rate is 5 to 12 mm per second.

**Fig. 57.** Parts produced on an electric upsetting machine [L].

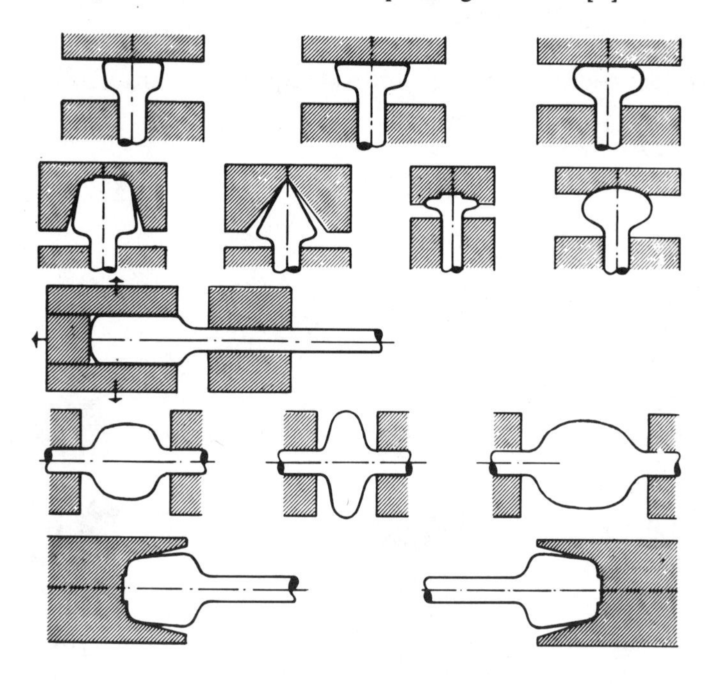

These heating systems are built into machine tools constructed around a vertical or horizontal frame; the hydraulic or pneumatic ram is located on one of the surfaces (sometimes the machine has two rams which are symmetrical with respect to the part to be upset). Both electrodes are connected to the secondary of a single-phase heating transformer.

Upsetting on electric machines offers many advantages:

— upsetting begins during heating;

— the upset blanks can be drop forged or swaged at the outlet from the upsetting machine;

— due to the accuracy obtained, upset blanks in dies do not require reworking operations, and simple grinding is sufficient in most cases;
— the grain structure is favorable to part quality;
— the temperature and the volume upset are constant and accurate which, after swaging, provides precise and regular parts, and therefore reduced runs after swaging;
— formation of scale is very low, increasing longevity of dies;
— the volume upset can have a wide range of shapes;
— only the useful part of the piece is heated, avoiding reworking of the rest of the part, and a high overall thermal efficiency is obtained;
— blank accuracy, absence of scale or decarbonizing and limitation of heating to the part deformed reduces machining to a minimum, and provides important savings in raw materials;
— working conditions for the operator are excellent (no radiated heat, noise, smoke or dust);
— energy savings are substantial;
— rectangular, hexagonal and square sections can be worked, and sometimes formed sections and various materials such as alloyed steels, refractories and even some copper alloys can be worked (if the resistivity is sufficient, as is the case for some brasses and bronzes; pure copper obviously cannot be heated using this method);
— handling parts is facilitated, and it is easy to design fully automatic machines;
— machine overall dimensions are minimal;
— worn parts (electrodes) are easily replaced at low cost.

Conversely, preparation of the parts to be heated (surface condition, end cutting) has a direct effect on the use of machines and production rates. (In particular, pieces must be obtained from descaled, straightened or drawn bars.) The end cut at the ram end must be as perpendicular as possible, especially for large diameters, so that the current density, at the contact with the anvil, is uniform. Chamfering may be of use, but for small diameters, correct shearing is generally adequate.

*4.9.4.2. Heating before rolling*

The equipment used with this type of heating is much larger than with heating before forging, but the basic principle remains the same.

Installations are intended for heating bars 40 to 180 mm in diameter, and up to 12 m long.

Generally, a heater consists of:
— an automatic bar feed and loading system;
— a movable contact-holder device, consisting of trolleys on which the contacts, connected by flexible cables to the low voltage/very low voltage supply transformer, are mounted;
— an automatic hot bar evacuation system;
— where necessary, a set of compensation capacitors.

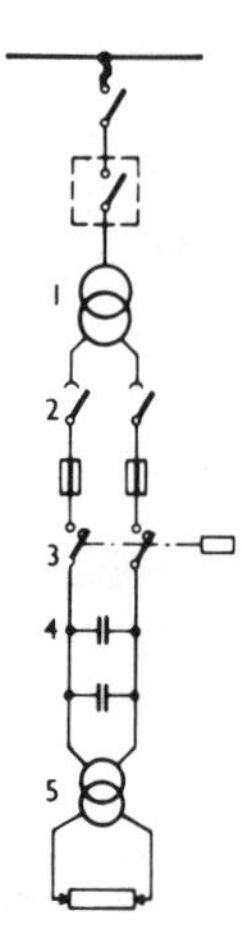

1- HV transformer
2- Isolating switch
3- Contactors
4- Capacitors
5- Heating transformer

**Fig. 58.** Diagram of a heating-before-rolling system [B].

In modern installations, the time interval between two heating cycles is about 10 seconds, which, for long products, provides an excellent electrical power use factor. Figures 48 and 49 give the characteristics of high capacity installations.

Basically, this type of equipment is used for discontinuous reheating of high quality steel bars and billets prior to rolling.

For example, these conduction heaters can provide:

— a maximum production rate of 1.25 tons per hour x m, i.e. 7.5 tons per hour approximately for 6 m round bars up to 60 mm in diameter;

— a maximum production rate of 2.3 tons/h x m, i.e. 18.5 t/h approximately for 8 m bars, for round products 110 mm in diameter.

Particularly profitable applications are reheating of billets produced by continuous pouring before being completely cooled (the specific consumption to take a billet to 1,200°C is only 100 kWh/t approximately, starting at 700°C), and heating during rolling.

Continuous operation installations also exist. The bars are end-to-end welded and heated by moving one or more modules (*see* Figure 45).

For example, a continuous machine consisting of three 550 kW heating modules (the voltage is adjustable over a wide range, but the voltages currently used are 40-45 V, which limits current) can heat approximately 6 tons of steel per hour to rolling temperature (1,100°C).

The maximum rate is 3 meters per second, and the mean rate 2 meters per second [39].

*4.9.4.3. Heating tubes before drawing*

Welded tubes undergo finishing operations in a reducer-drawing mill. This equipment enables tubes of many different diameters and thicknesses to be obtained from a single blank tube. Hot transformation, between 850° and 1,000°C, is used with temperature being a function of the final cross-section of the tube.

Conduction heating can be used to heat the tubes to rolling temperature.

For example, a heating line consists of three heating areas, each connected to one phase of the three-phase mains. Each area has its own single-phase transformer which provides impedance matching and isolation of the installation. Each transformer output is connected to a "heating cage" which, through contacts, allows the current to flow through the tube.

Depending on the cross-section of the tube blank, the separation between two "heating cages" for the same phase is adapted so that the nominal power of the corresponding transformer can be used. These cages heat the moving blank. The contacts consist of copper rollers whose base is immersed in a mercury bath.

A 4,000 kW installation can heat approximately 20 tons of tube lengths per hour, when approximately 6 meters long. Overall efficiency is very high and better than 95%. The installation is highly flexible, and its charge does not affect its efficiency.

*4.9.4.4. Other applications of metal heating before forming*

Conduction heating can be used for many other applications, with heaters comparable to those described above, but adapted to the special conditions of each application:

— heating prior to hot drawing for production of steel sections formed on hydraulic machines;
— heating rivets;
— heating bar ends prior to forging or stamping;
— heating the center of bars prior to forging or stamping;
— heating bars prior to bending (e.g. production of chains);
— heating prior to shearing (a 600 kW heater provides a production rate of about 6 tons per hour, if shearing is performed around 500°C).

**4.10. Metal heat and surface treatments**

Conduction heating is used in two different ways for heat treating metals. The product to be treated can be heated directly by the Joule effect; in this case, the installations are similar to those previously described for heating before forming, but are adapted to the special conditions required for the treatment involved. The product to be treated can be heated in an intermediate environment, which is itself heated by direct current flow; basically, equipment of this type consists of immersed electrode salt bath furnaces (*see* paragraph 4.11), but other equipment such as galvanizing furnaces is being developed (*see* paragraph 4.14) [A].

#### *4.10.1. Heat treatment of wires*

Conduction heating is used in continuous treatment of copper alloy wires (the diameter must be small so that the wire offers sufficient resistance) and more rarely for steel wires.

Figure 59 gives a diagram of a steel wire hardening installation.

Treatments most commonly provided by this technique are annealing of copper alloy wires (e.g. annealing prior to fabrication of telephone or electrical cables). Specific consumptions vary according to the temperature level required, but efficiency is always high and about 70 to 80%.

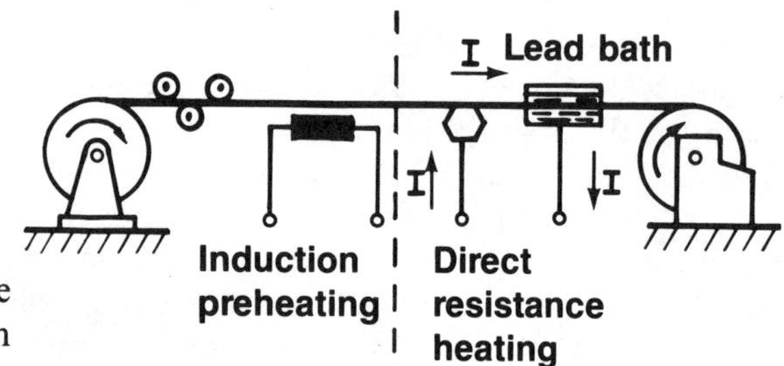


**Fig. 59.** Steel wire hardening installation

For steel, conduction heating can be used for annealing and heating prior to hardening or tempering. Difficulties encountered with current input contacts have, however, often led to the use of simpler techniques, such as resistance furnaces or induction heaters [17].

#### *4.10.2. Heat treatment of metal sheeting*

Conduction heating is widely used for continuous treatment of steel sheets. Sheets can also be heated up to austenitization temperature, then hardened. Brightening of tinned sheets, produced on electrolytic tinning lines, can also be ensured by remelting the surface layer of the tin by conduction. Techniques combining conduction and induction heating are often employed. An installation of this type, which offers higher productivity, is described in the "Electromagnetic induction heating" chapter, in paragraph 5.6.3. [17], [28].

#### *4.10.3. Heating before hardening by high frequency conduction*

Heating using high frequency conduction consists of heating a whole part or a portion of a part of metal using the Joule effect, and in applying a high frequency supplied by electrodes following a defined path.

With respect to conduction, in which the current used is at line frequency, high frequency provides better localization of the heat and enables high specific powers to be reached with very low contact losses. With respect to high frequency induction, conduction saves energy since the current is applied directly to the part without being affected by the electrical efficiency of the power transfer, i.e. inductor losses. Therefore, induction should be used only when conduction is not possible. High frequency conduction applications are more limited than for induction.

The theory of high frequency conduction heating is complex, and all the phenomena involved are not fully understood. Conversely, practical applications have already been mastered. Some simple explanations of the method in which high frequency currents act are available [27], [41], [P].

*4.10.3.1. Contact electrodes*

A contact can be compared to a resistor in parallel with an inductor and capacitor. The resulting complex admittance is expressed by the equation:

$$A = \frac{1}{R} + j\left(C\omega - \frac{1}{L\omega}\right).$$

The higher the frequency the more the inductive term becomes negligible and the capacitive term becomes dominant. If high pressure is also applied to the contacts, the capacitance, and therefore the admittance, are also increased. This elementary explanation shows that the higher the frequency the easier it is to apply current to a part through contact electrodes.

Electrodes, which must have low resistivity, are made from pure copper for static treatment of parts, and copper with a slight percentage of silver for sliding contacts intended for treatment of parts in motion. The electrodes are generally water cooled (sometimes air cooled), the energy to be dissipated being about 5 W/mm$^2$ of electrode in contact with the part to be heated. The shape of the contacts, shown in Figures 60 and 61, enhances heat dissipation so that the electrode need be cooled to within only a few centimeters of the contact point, optimizing electrode design.

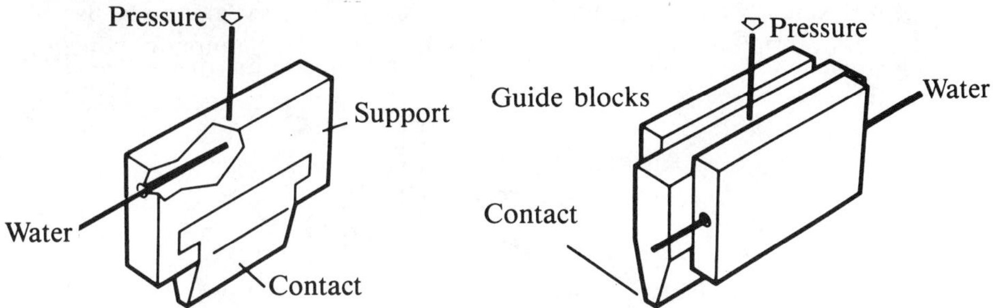


**Fig. 60.** Electrode for fixed part [41].

**Fig. 61.** Electrode for movable part [41].

*4.10.3.2. Localization of current and dissipated energy*

The position of the current input conductors with respect to the part determines the path followed by the currents through the part, and the dissipated power. AC currents take the path of least impedance and in the absence of electromagnetic constraints, distribute so that resistance $R$ is equal to self-inductance $L\omega$. At low frequencies, conductor dimensions are generally insufficient for the phenomenon to occur, and the resistance is much higher than the self-inductance.

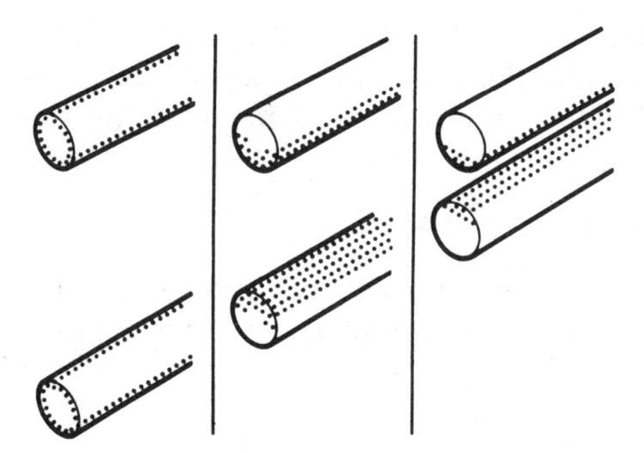

**Figure 62.** Current coupling at high frequencies [27].

At high frequency, resistance may equal self-inductance, and the current will tend to concentrate on the periphery of the conductor (skin effect, described in the "induction heating" chapter).

Moreover, the supply and return conductors, when close to each other, tend to couple due to the influence of the magnetic field of the two conductors, and the current flows only in a crescent-shaped section.

Condition $R = L\omega$ is obtained when the distance between the two conductors is equal to the thickness of the skin, and the energy given off by the Joule effect is then maximum.

If the return conductor consists of a plate, the proximity effect occurs in the same manner. The current in the plate concentrates on the surface of the supply conductor; this phenomenon will be more noticeable if the conductor is located near the plate (coupling is a function of the distance-to-skin thickness ratio).

To obtain localized overheating in a part, it is therefore necessary to apply current through two contacts located at opposite ends of the area to be heated and to concentrate it because of the proximity effect of the input conductors (Figure 63).

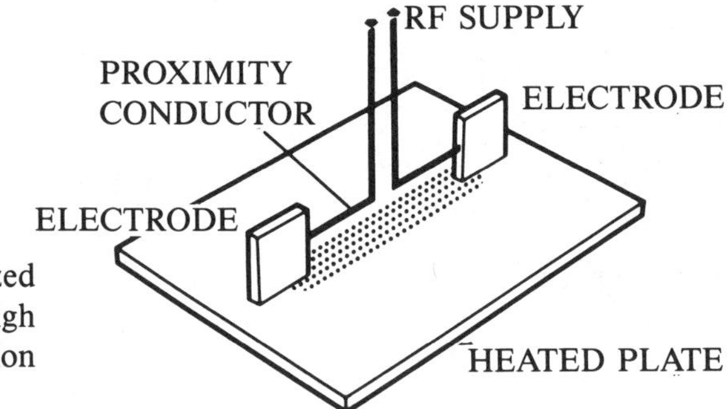


**Fig. 63.** Localized heating due to high frequency conduction [27].

*4.10.3.3. Heat treatment applications*

This technique can be used to heat any contour on a part, whatever its shape, as long as it lends itself to installation of contacts. This is used for hardening, annealing or tempering of localized areas on bulky parts or flat surfaces.

For example, high frequency conduction heating is used for static prehardening heating of automobile steering racks.

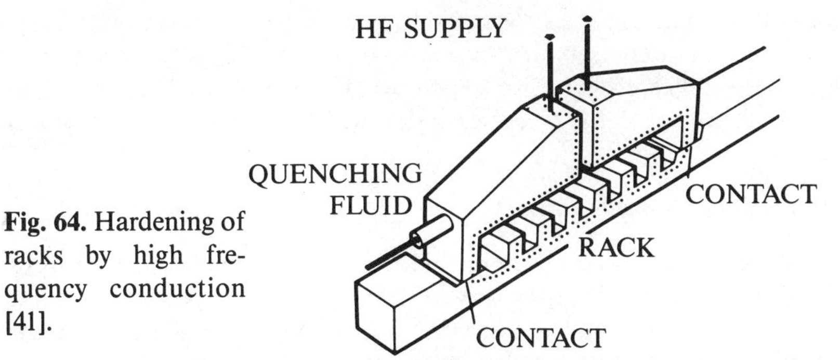


**Fig. 64.** Hardening of racks by high frequency conduction [41].

The high frequency current is applied to two ends of the teeth by contacts which cover the width of the rack. The contacts are connected to solid electrodes, the bottom surface of which is very close to the teeth to be heated. This surface contains holes to ensure hardening by sudden coolings of the teeth by means of a fluid after heating. Compressed air is used to drive out the remainder of the quenching fluid, in preparation for the next cycle.

For a tooth width of 20 mm, a length of 340 mm and a tooth root direct length of 180 mm, the power of the high frequency generator is 100 kW (measured at line) at a frequency of 250 kHz, and the heating time is 5 seconds.

The surface area of each contact is 20 $mm^2$. The current density at the contacts reaches 85 $A/mm^2$ (for sliding contacts, this does not normally exceed 20 $A/mm^2$) and the pressure applied to the contacts is 5 $kg/mm^2$. The high frequency current is about 1,700 A [P].

An overall efficiency of 50% is obtained, while induction heating for this type of application rarely exceeds 35 to 40% (with other techniques, efficiency is often much lower, since the mass of metal heated is more).

## 4.11. Electrode salt bath furnaces

Mixtures of melted salts (sodium and potassium chlorides, barium chlorides, fluorides, nitrates-nitrides, etc.) conduct electricity; therefore, they can be heated by direct current flow using electrodes arranged according to furnace design. Since the salts are conductive only in the molten state, these furnaces must be started by an auxiliary device (immersed resistances, burners, etc.). In comparison with externally heated or immersed resistance crucible furnaces, this type of furnace offers two major advantages:

— the specific power, expressed in kilowatts per square meter of bath, can be much higher, enabling much higher productivity; power densities 3 to 4 times higher than those obtained with externally heated crucible furnaces are common;

— high temperatures are reached without difficulty, and the service life of the crucibles is extended, since they are no longer submitted to a high temperature gradient. Crucibles as such need not be used, since the molten salt is in direct contact with the furnace chamber forming the crucible. With ceramic refractory brickwork, the maximum temperature is about 1,300°C, or even 1,650°C in special cases.

Refractory material crucibles are made from sillimanite bricks (a refractory with at least 60% alumina), bound by a special cement. The joints must be as thin as possible and carefully sealed to prevent salt infiltration. These crucibles are used for almost all salts, except cyanides, which call for the use of metal crucibles.

### *4.11.1. Immersion electrode furnaces*

In these furnaces, the electrodes are cylindrical or rectangular metal bars. They are vertically mounted parallel to each other, and generally grouped in pairs along the wall of the furnace. The low resistance offered by liquid salts and the use of electrodes close to each other and safety factors call for the use of low voltages, and therefore high currents; the supply voltage is generally equal to or less than 30 V. The electromagnetic forces then create intense mixing, which provides excellent bath homogeneity of the order of ± 2°C.

Range of immersion electrode salt bath furnaces
for heating steel prior to hardening [H]

| Power (kW) | Bath volume ($dm^3$) | Holding power at 850°C, cover open (kW) | Heating capacity at 850°C (kg/h) |
|---|---|---|---|
| 25 | 48 | 20 | 30 |
| 40 | 75 | 25 | 50 |
| 60 | 160 | 40 | 70 |
| 90 | 340 | 55 | 150 |
| 120 | 575 | 65 | 200 |

**Fig. 65**

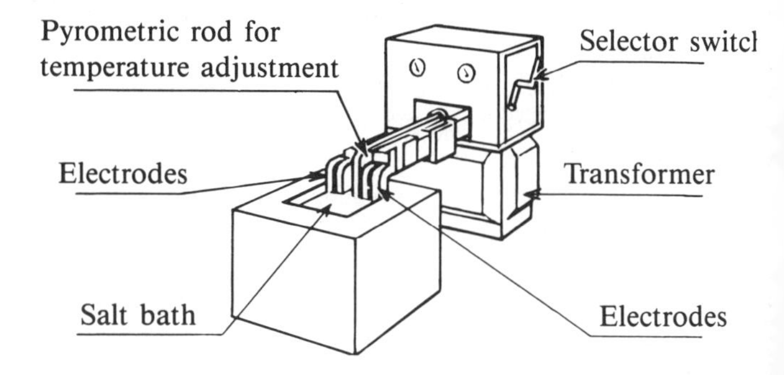


**Fig. 66.** Immersion electrode furnace [44].

There are variations of this system; one is to cover the current inputs at the electrodes with a refractory block. The entry of the electrodes into the bath is not at the salt surface, but below. This arrangement keeps the bath surface clear and reduces the air-salt corrosion area, increasing electrode service life. In addition, the bath's radiating surface is decreased, reducing thermal losses. These furnaces are similar in design to sub- mersed electrode furnaces described in the next section.

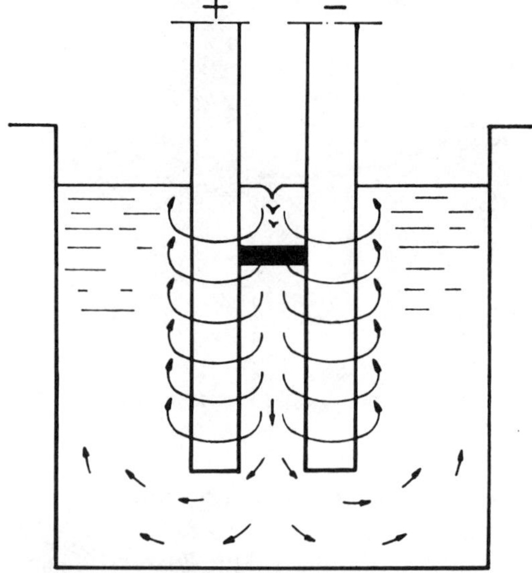


**Fig. 67.** Electromagnetic bath mixing [1].

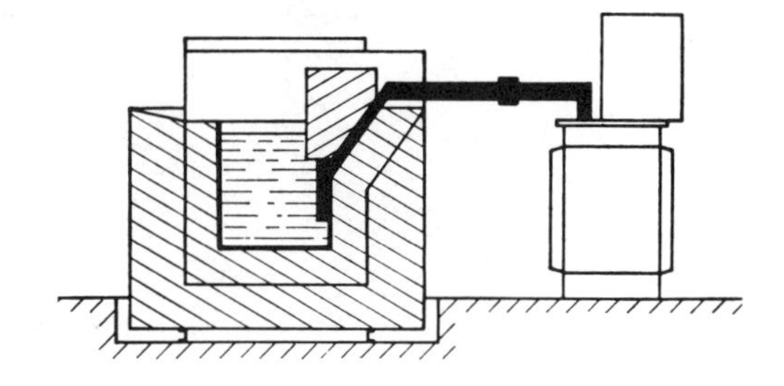

**Fig. 68.** Protected immersed electrode furnace

The electrodes can also be arranged symmetrically on the furnace wall. The current flows from one part to the other by traversing the complete bath. With three-phase furnaces, the bath is shaped like a hexagonal prism. This simple and cheap arrangement is widely used in small furnaces for treatment of high-speed steels, but current flows through the parts to be heated, with the risk of overheating. This fact often leads to the use of a parallel electrode arrangement.

In spite of their advantages — simplicity, high dissipated power, high operating temperatures, temperature homogeneity, etc. — immersed electrode furnaces have certain drawbacks:

— ease of current flow in the top layers of the salt, in particular when the electrodes are in parallel, creating hotter areas at the top than at the bottom;

— high potential differences (about 20 to 30 V) inside the bath, which can give rise to spurious current in the parts, and limit effective volume.

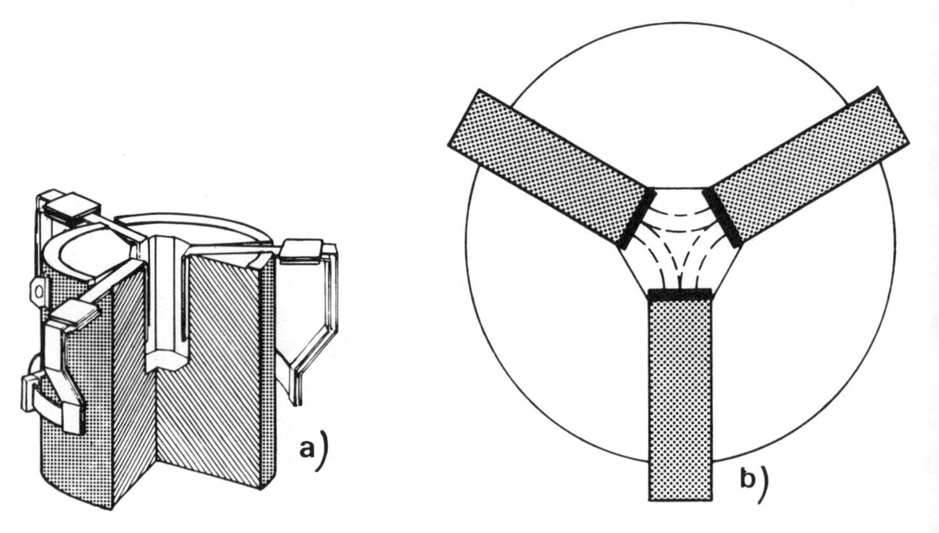


**Fig. 69.** Triangle-mounted immersion electrode salt bath furnaces:
*a)* Diagram of furnace [H]; *b)* current lines and electrodes [I].

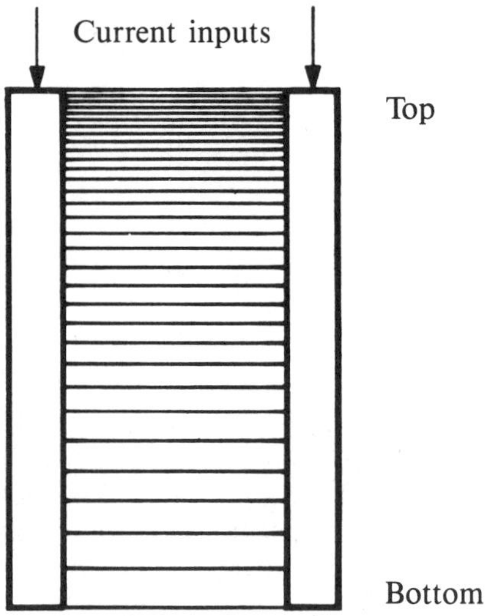

Current inputs

Top

Bottom

**Fig. 70.** Distribution of current densities between parallel electrodes [I].

— large volume occupied by electrodes, reducing working surface;

— loss of heat by conduction through large cross-section electrodes (the ends are sometimes water cooled);

— relatively short electrode service life (they tend to distort and are rapidly attacked where they are in contact with air and salt simultaneously);
— high formation of metal oxides, due to corrosion of electrodes, contaminating the bath.

However, in most cases, these drawbacks are minor.

### *4.11.2. Submerged electrode furnaces*

#### *4.11.2.1. Metallic submerged electrode furnaces*

Generally, the crucible is oblong shaped and is made from refractory brick, specially designed to withstand salt. There are two metallic electrodes (or a multiple of two) which are located opposite the base of the crucible, flush with the brickwork, over the complete length along one side, and over a height of approximately 60 mm. The current flows from one electrode to the other through the salt. The current lines are not parallel but opened out in proportion to electrode separation. The difficulty in determining the exact distribution of current lines renders calculation of the resistance of the part of the circuit consisting of the salt between the electrodes extremely difficult, and design of these furnaces is based on experience.

Also, very deep furnaces exist (several meters), in which the electrodes are located at different levels on the same wall.

The current inputs to these electrodes are made through the furnace masonry in the lower part of the bath. Salt leakage at this location is a threat, especially in very deep furnaces where the pressure is high. In fact, the masonry forms a very dense homogeneous block, containing no insulating material, and gives a steady decrease in temperature, enabling the salt to set rapidly inside the walls and thus provide excellent sealing.

This type of furnace offers several advantages:
— the electric current flows through the bottom of the crucible only; heating is therefore done only through the bottom, which promotes the formation of convection currents throughout the entire volume of the bath and provides adequate temperature homogeneity;
— if differences of potential remain in the bath, they limit the useful volume in terms of height but not area;
— the total surface area of the salt is used for treatment, and it is easier not to have conductors on the work surface;
— formation of metal oxides at the electrodes is practically nil;
— electrode service life is increased considerably by reducing corrosive areas which in turn decreases deformation; service life is between one and three years at 1,300°C (with conventional immersion electrode furnaces this rarely exceeds six months).

However, this advantage is counter-balanced by a major drawback. When the electrodes have to be changed, it is necessary to destroy and reconstruct the masonry completely.

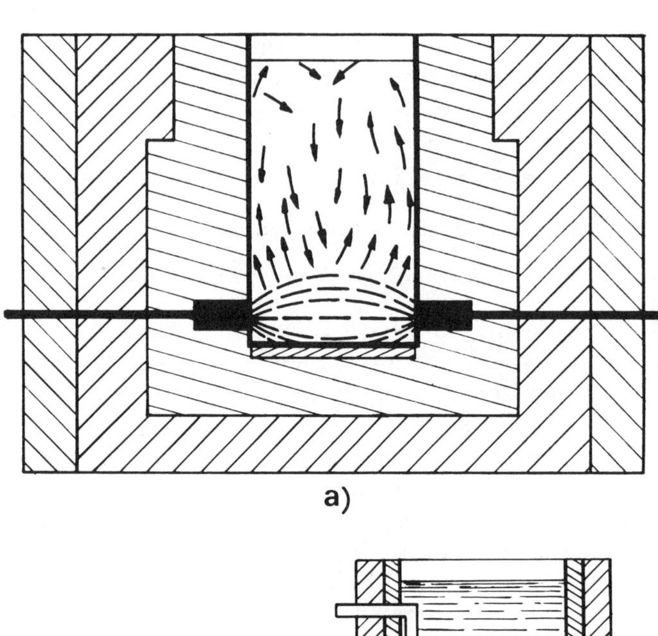


**Fig. 71.** Metallic submerged electrode furnaces
*a)* horizontal electrodes in bottom of furnace [I]
*b)* electrodes for deep furnace located on the same side of the wall [H].

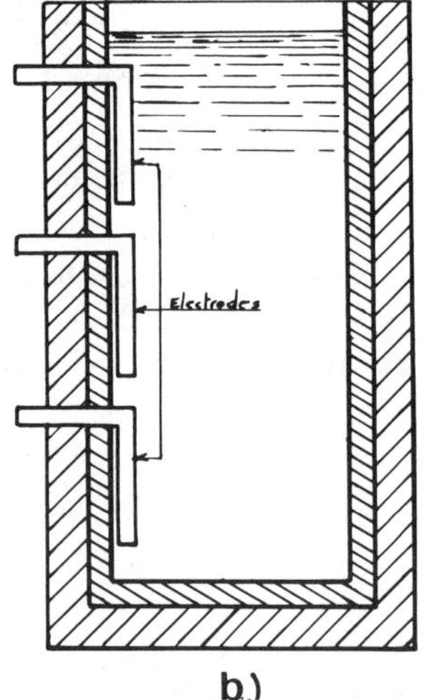


Moreover, if a large part is dropped to the bottom of the furnace and not withdrawn immediately, the electrodes may be short-circuited and damaged. The same incident can happen but more slowly if the slurry deposited at the bottom of the bath is not removed regularly. For these reasons, attempts have been made to use graphite electrodes and crossed-channel furnaces.

#### *4.11.2.2. Graphite submerged electrode furnaces*

Graphite electrode furnaces are identical to metallic electrode furnaces in construction and heating principles. The only difference is that metallic electrodes are replaced by cylindrical graphite rods, which can be pushed into the bath as they wear; electrode advance is controlled by screws, and spare rods can themselves be screwed into in-service rods when they become too short.

Range of submerged electrode salt bath furnaces (temperature 1,300°C). The maximum rate depends on the product heated [H].

| Power (kW) | Depth (mm) | Crucible volume ($dm^3$) |
|---|---|---|
| 25 | 450 | 15 |
| 60 | 800 | 60 |
| 70 | 800 | 70 |
| 100 | 1,520 | 185 |
| 120 | 1,920 | 300 |
| 120 | 960 | 170 |

**Fig. 72.**

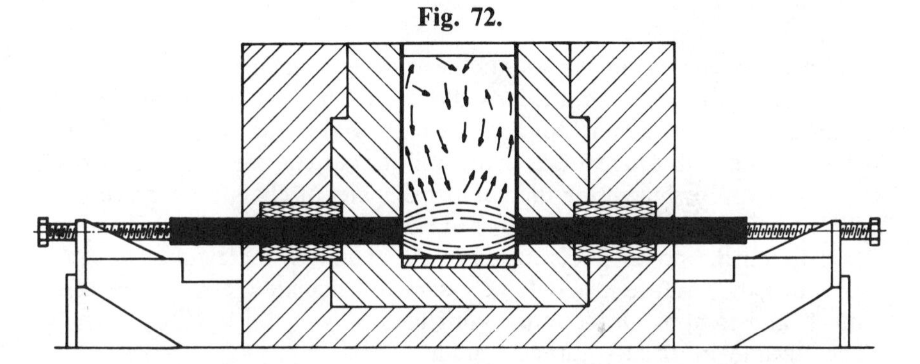

**Fig. 73.** Graphite submerged electrode furnace [I].

This system eliminates the necessity of rebuilding the masonry in metal electrode furnaces, and also permits very high temperatures. If appropriate refractories and salts are used, a temperature of approximately 1,650°C can be obtained with this type of furnace. A poorly maintained electrode advance system is, however, rather fragile, since the brickwork can become hollow at the electrodes allowing salt leakage. Electrode consumption is highly variable, from 25 mm per year in a 600°C brazing furnace, to 200 mm over ten days in a furnace operating at 1,300°C.

### *4.11.3. Crossed-channel electrode furnaces*

An examination of the various electrode furnaces shows that the electrodes themselves are the main cause of problems. Since the resistance of the salt is relatively low, these electrodes must permit high current flow, and are therefore large and difficult to house.

Cross-channel furnaces were developed to create a high resistance "salt" circuit. This resistive circuit is obtained by narrow salt channels connecting the main bath to small secondary wells, to which the current is applied by relatively thin rods due to the high supply voltage (80 to 110 V).

Fig. 74. Cross-channel furnaces

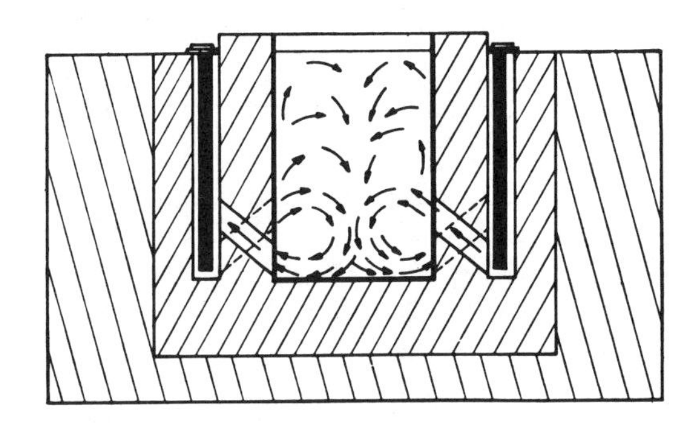

The channels connecting the wells to the working bath are inclined in opposite directions to form an "X" when viewed end-on. The heated salt can therefore rise through the well channel towards the bath, and in the other channel from the bath to the well. There are as many wells as current inlets. Parallel circuits can be obtained by increasing the number of pairs of channels per well (deep furnaces), or by increasing the number of wells (large surface area baths). Since practically all the power is dissipated in the channels, these furnaces can be designed easily, because it is only necessary to calculate the resistance of the channels and to apply Joules and Ohms laws.

The advantages of such furnaces are as follows:

— the current flowing through the working bath does not itself produce heat, and there are no unwanted potential differences in the bath. Potential differences reach 2 to 3 Volts only, and there is no risk of overheating parts due to direct current flow;

— temperature homogeneity is provided by convection currents;

— radiation losses in the wells used for the current inputs are almost negligible, since they are completely covered;

— conduction losses are very low, and the conductors do not have to be cooled;

— the conductors can be replaced easily.

However, these relatively complex and costly furnaces have seen only limited use, the most common being parallel immersion electrode furnaces; submerged electrode furnaces are used mostly for high powers and production capacities (powers of more than 100 kW).

### *4.11.4. Salt bath furnaces: applications and advantages*

Salt bath furnaces are used for the following operations:

— heat treatment of metals;

— surface treatment of metals;

— heating metals before forming;

— brazing (especially aluminum alloy assemblies);

— treatment of glass and plastics (much more rarely).

These furnaces are above all appreciated for the following advantages:
— fast heating (core heating times of parts 4 to 5 times shorter than gas atmosphere furnaces);
— uniform heating, minimizing part deformation;
— heating time easy to control and maintain;
— heating in an environment where corrosion can be controlled easily (e.g. elimination of oxidizing and decarbonizing action of combustion gases);
— ease of access;
— option of partial heating;
— high production rates, even for small compact furnaces;
— can be automated for high treatment capacities.

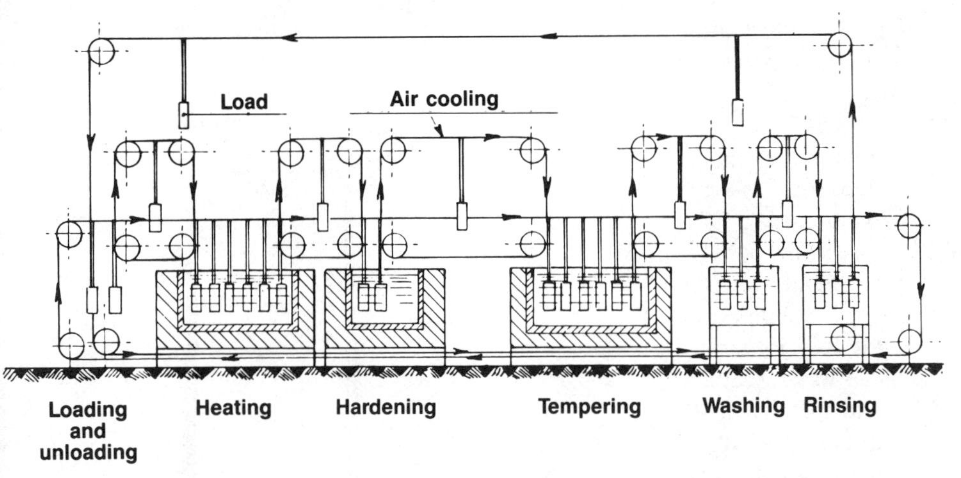


**Fig. 75.** Diagram of a fully mechanized installation [H].

The evolution of the use of salt baths is, however, controlled by several opposing factors.

The salts are pollutants (in particular, cyanides), and working conditions are often difficult; however, new non-polluting salts are beginning to become available.

Salts tend to corrode the parts immersed in them, and cling to their surfaces, in particular at rough points, calling for costly cleaning after an immersion in the baths. Vacuum furnaces are therefore being substituted for salt baths for heat treatments of precision parts and tools.

Operating costs, and in particular maintenance costs are rather high; energy costs are low, but, when it is difficult to cover the bath, losses due to radiation may be high (100 kw/m$^2$ at 900°C, 300 kw/m$^2$ at 1,300°C).

Other salt bath furnaces exist in which heating is provided by resistances radiating onto the crucible, or by armored resistance immersed in the molten salt. Performances, which are generally less than those of electrode furnaces, are, however, sufficient for numerous applications (see "Resistance furnace" chapter).

It is probable that salt baths, even if they lose some contracts to more sophisticated techniques (vacuum and induction furnaces, etc.), will continue to be of interest, in particular for simple treatment operations on parts of relatively low value, and in small and medium sized enterprises.

## 4.12. Remelting metals in electroslag

### *4.12.1. Principle*

This process, which is of Russian origin, consists of remelting a consumable metal electrode directly inside a copper water-cooled ingot mould.

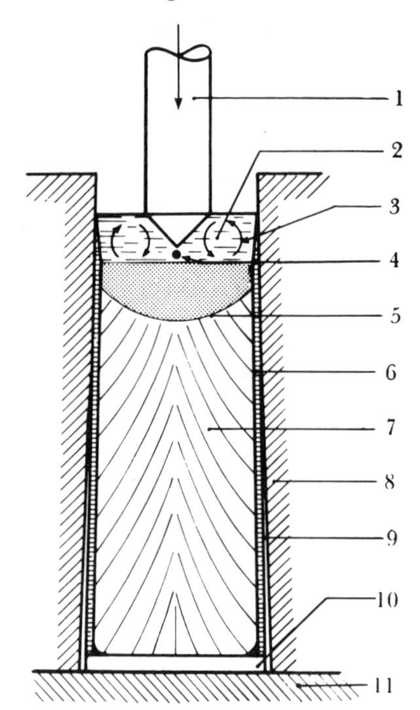


**Fig. 76.** Electroslag remelting principle [B].

1. electrode; 2. slag; 3. slag movement; 4. Liquid steel drop; 5. Liquid metal; 6. Slag crust; 7. Solid part of ingot; 8. Ingot mould inside wall; 9. Space due to shrinkage; 10. Priming plate; 11. Water cooled baseplate.

The electrode is brought into contact with the electroslag, in which heat is produced from single-phase AC current by the Joule effect. This heat maintains the slag in liquid form, and melts the electrode metal. This process is sometimes known as ESR (Electroslag Remelting or Refining). It is primarily of interest in high quality metal applications, and for its mechanical characteristics.

The baseplate (priming plate) and the electrode to be melted are connected to the secondary of a single-phase transformer. The electrode melts drop by drop. On passing through the slag, the drops are subjected to a highly active chemical effect, due to the large contact surface between the drops and the slag (desulphuration, outgassing). In addition, inclusions are decanted. The metal drops assemble and solidify very rapidly; the liquid bath is therefore shallow, and there are practically no segregations. Crystallization is directed along oblique or almost vertical lines, enabling the last inclusions to de- cant more easily [6], [9].

For starting, an arc is produced for a few moments between the electrode and a priming plate, consisting of a metal of similar grade to that of the metal to be remelted. The stable heating mode due to passage of current through the electroslag at high temperature then builds up rapidly. At the end of the operation, the melting rate is reduced very gradually.

The electrodes are produced using conventional processes, arc or induction furnaces (and, in some cases, vacuum induction furnaces).

### *4.12.2. Purpose of the slag*

The slag fulfills a multiple role:

— the slag bath protects the molten metal from all atmospheric effects. For steels and alloys containing highly oxidizable elements, this protection is reinforced by argon;

— a film of slag forms against the internal wall of the mould, slows down lateral heat transfers and causes progressive solidification from the bottom of the ingot. The surface condition of the ingots is excellent, and in general, do not require any surface preparation prior to hot transformation. After remelting, the difference in expansion coefficients enables easy removal, due to the slag film;

— the slag bath forms a thermal sink-head whose action, combined with slow cooling, provides a very dense ingot, free of separations and shrinkage cracks;

— the lateral slag film and the sink-head at the top of the ingot provide high metal purity; the sum of the drops at the bottom and head represent only approximately 5% of the weight of the metal, while there is practically no scaling or wear on the sides of the ingot;

— variation of the slag composition can be used to vary the metal refining conditions; also, the slag makes it possible to add carefully measured alloy elements during the operation;

— the slag film between the ingot and the mould not only provides a very smooth ingot skin, but also protects the mould; after remelting, the mould is clean without inclusions, greatly increasing its service life.

### *4.12.3. Comparison with consumable electrode vacuum arc furnaces*

The electrode melting process and gradual filling of the mould and formation of the ingot are similar to consumable electrode vacuum arc furnace remelting (see "Arc heating"). However, electroslag remelting differs in several ways:

— the heat required for melting is given off directly into the slag, and absence of arcs provides high operational stability;

— the use of electroslag provides high flexibility and offers many advantages as described in the slag analysis;

— the absence of vacuum pumps, sealed chambers and DC current generators required for the vacuum arc furnace remelting technique, reduces investment and operating costs.

Conversely, outgassing is often better (in particular, elimination of hydrogen) with vacuum arc furnaces. Some high capacity electroslag remelting furnaces require the use of very low frequencies, less than 10 Hz, due to the high currents reached (40 to 50 kA). The converters used are solid state thyristor generators [35], [45], [47].

### *4.12.4. Applications*

Installed powers vary from a few tens of kilowatts to several thousand, and ingot weights range from some tens of kilograms to approximately 200 tons. For example, a 1,000 kW furnace enables remelting of a maximum of 450 kg/h of metal (the rate varies bet- ween 100 and 45 kg/h as a function of ingot size and alloy grades) with an energy consumption of between 1200 and 1,700 kWh/t; ingot weight is between 150 and 2,500 kg. In general, this process is highly flexible as far as the shape and dimensions of ingots are concerned; they may be square, flattened, circular or octagonal [10], [11], [19], [23], [34].

The increase in metal quality, together with its special structure, greatly increases the transverse mechanical characteristics of the remelted metal. Forgability, in particular, is improved, due to the crystalline structure of the ingot, which permits heavy ingots used for hot forging to be produced [31], [37], [B], [R], [S].

There is a wide range of applications for electroslag remelting furnaces: steels for cold-rolling-mill cylinders, crankshafts, ball-bearings, stainless steels for the nuclear industry, dental or surgi- cal protheses, high-speed steels, superalloys for the aeronautics industry, gas turbines, the chemical industry, etc.

### *4.12.5. Process variation*

Electroslag remelting, as described above, is used basically for the production of high quality ingots from a consumable electrode. Since its development, variants have been developed either for applications of another type, production of moulded parts or welding, or to dispense with the constraint of using an ingot as an electrode.

#### *4.12.5.1. Continuous electroslag powder melting*

This process, known as CESPM (Continuous Electroslag Powder Melting), is not indicative of the principle involved. Instead of a solid electrode, this technique uses one or three continuous electrodes consisting of strips or wires; they are made from ordinary, non-alloy, and therefore cheap steel, and are melted in a layer of electroslag. The additives are introduced directly into the bath, mixed with iron dust or pellets.

However, the powder must be fed in at a regular rate and absorbed by the bath. This operation is performed by magnetically or mechanically accumulating a large quantity of powder on plain steel strip used to supply the current to the slag bath.

Since the powder offers a large surface to the slag, refining is much easier and faster. In practice, the very high current density in the electrode creates a strong magnetic field around the electrode, facilitating the introduction of the iron powder with additives, and mixing the bath. The powder is sprayed onto the magnetized strip which retains it and draws it into the bath.

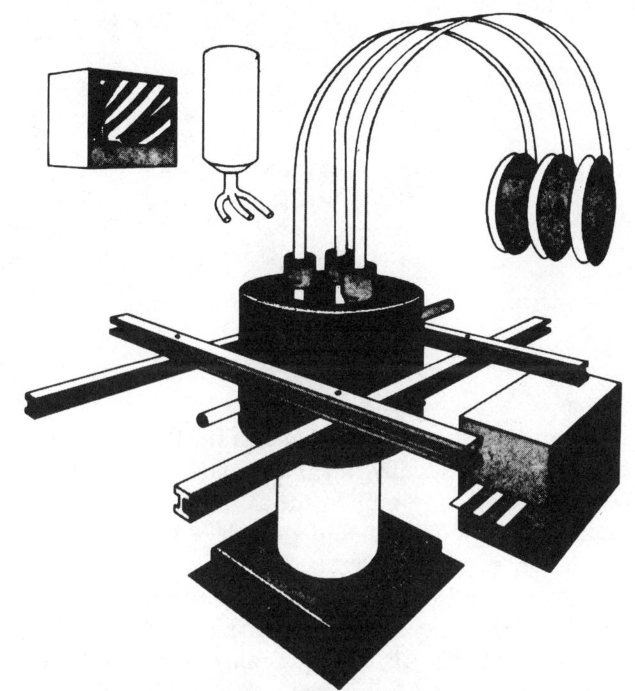

**Fig. 77.** Continuous electroslag powder melting [J].

A single or three-phase supply is used, depending on whether the system has one or three electrodes; the latter case is especially attractive since it balances the power supply [J].

This technique offers the advantage of obtaining a high percentage alloy ingot of any composition from ordinary steel, in the form of strips or wires. With respect to conventional electroslag remelting, this process also reduces specific energy consumption by at least 25%, increases the metal-slag contact surface and enables continuous operation and improves metal homogeneity.

The first industrial installation of this type has enabled ingots of a maximum weight of 50 tons in highly diverse alloys, stainless steels, high-speed steels, steels for rolling mill cylinders, cobalt steels, etc., to be produced. The process can also be used for recharging rolling mill cylinders or producing tubes.

*4.12.5.2. Production of castings using electroslag melting*

Another extension of electroslag melting is the production of castings (electroslag casting — ECS). This technique enables the metal to be melted and cast in a single operation.

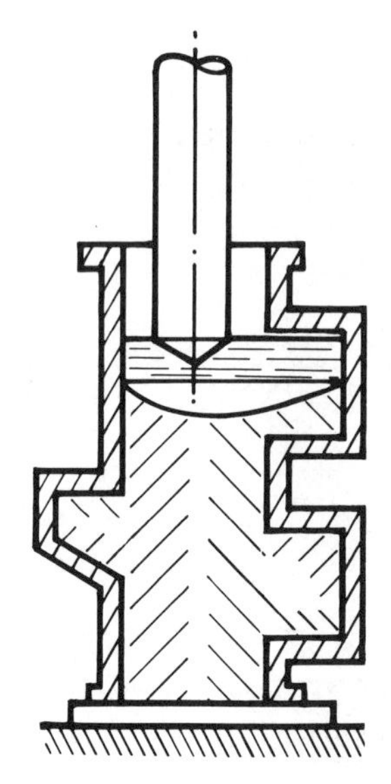

**Fig. 78.** Production of castings by electroslag melting [N].

The mould has the same shape as the part to be cast. The molten metal in the electroslag gradually fills the mould. Since it prevents any contact with air, the slag protects the molten metal from oxidation. The metal crystallization conditions within the mould provide a structure enhancing the mechanical qualities of the casting. The slag bath forms a thermal sink head. Due to refining provided by the slag, the quality of the metal obtained is very high.

This technique is used for producing a wide range of parts, both in terms of volume and shape. Also, a wide range of alloys are used.

Therefore, this process offers new possibilities in the production of castings having excellent characteristics. Components for nuclear power stations, rolling mills, furnaces, large dimension crankshafts, etc., are produced using this technique. Products of much smaller dimensions can also be cast, such as dental protheses or surgical instruments.

#### *4.12.5.3. Vertical welding in electroslag*

This welding technique uses localized remelting in electroslag. This was first of all applied to steel, and more recently to aluminum (electro-conductive flux welding).

This process is used with steel to make longitudinal joints on pressurized cylinders, or to assemble ships' hull sections. For aluminum alloys, it can, for example, be used for end-to-end joints on large conductive aluminum bars intended for electrochemical or electrothermal installations [24].

### 4.13. Melting in an electro-conductive crucible

Some electro-conductive materials such as graphite or silicon carbide are often used as a crucible for melting alloys or other metals. Generally, these crucibles are heated by resistances or burners providing heat transfer by radiation and, to a lesser degree, by convection. Induction heating is also used where high power densities or high temperatures are required.

Recent research has enabled the development of conduction furnaces in which the energy is dissipated directly into the crucible containing the product to be melted. Generally, the crucible is made from graphite. The furnace is supplied at low voltage through a single-phase transformer. The current input electrodes consist of a metal ring clamped against the top edge of the crucible and a metal plate in con- tact with the bottom. Regulation is provided by thyristors, variable chokes or any suitable technique. Tilting or fixed versions of such furnaces exist.

For example, the power required by a 15 litre capacity furnace (120 kg of bronze approximately) is about 40 to 50 kW. Furnaces of higher powers and capacities can be constructed.

These furnaces are suitable for melting copper alloys, but can also be used, not only for melting other metal alloys (e.g. light alloys), but also for melting non-metallic materials such as the complete glass product family.

The advantage of this technology is that it enables high power densities and temperatures and relatively limited investment and operating costs. This technology, which has reached indus- trial maturity, may find other important applications in foundries and other industries.

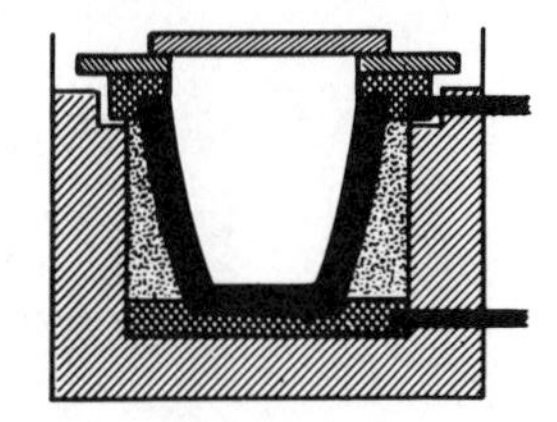

**Fig. 79.** Diagram of an electro-conductive crucible melting furnace

## 4.14. Conduction heating of galvanization baths

This galvanizing bath heating system, which was developed in Poland, uses a silicon carbide susceptor, immersed in molten zinc, as a resistive element. This susceptor takes the form of a crucible filled with molten zinc. The current is applied through graphite electrodes, one immersed in the zinc contained in the crucible, and the second in the bottom of the galvanizing tank. The silicon carbide crucible is heated directly. The electrodes are single-phase voltage supplied [44], [T].

This heating system offers high efficiency, from 75 to 90%, and enables high galvanizing temperatures to be reached (up to 600°C) and also greatly reduces formation of mattes and ashes. In industry, this is used for galvanizing during hardening of steel products, but also for melting and maintaining aluminum alloy baths.

## 4.15. Welding and brazing

Welding using direct current flow through the parts to be assembled has become a popular technique and has been the subject of many publications. Therefore, this paragraph simply outlines these

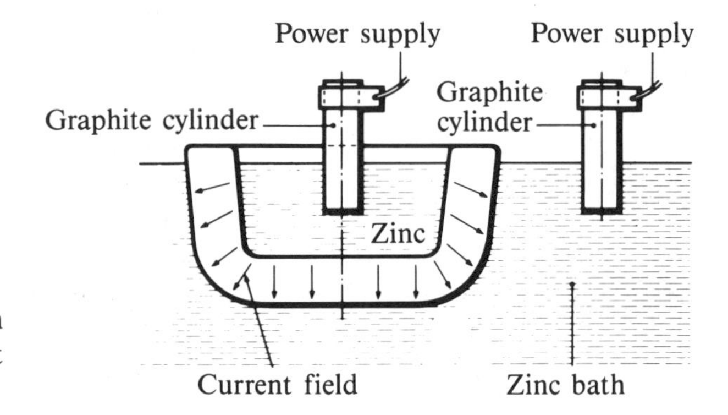


**Fig. 80.** Construction of a heating element [3].

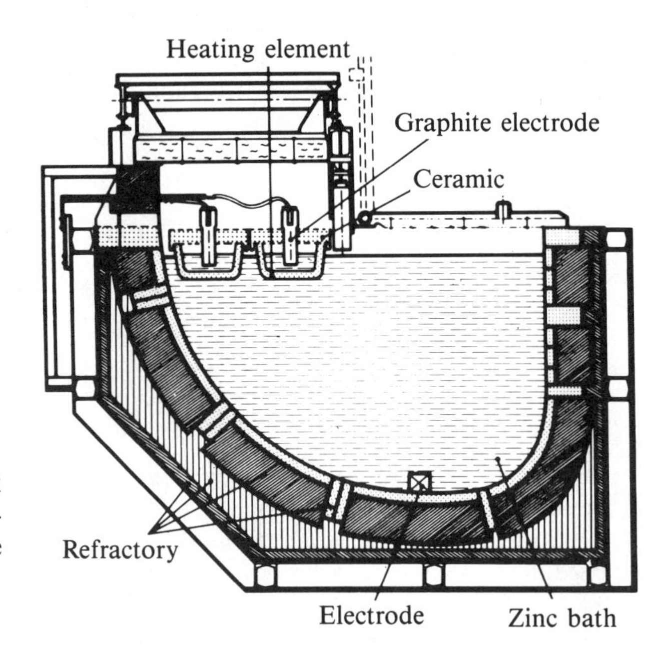


**Fig. 81.** Ceramic tank heated by direct current flow through the zinc bath [T].

principles. Two parts, clamped against each other, heat the area through which current flows, especially where the parts touch (contact resistance).

When the two touching surfaces are at welding temperature, high pressure alone will produce a weld. This principle has given rise to several techniques:
— butt welding, enabling the end-to-end assembly of two parts of different shapes or types;
— spot welding, which is widely used in the automobile industry;
— seam welding, enabling continuous treatment;
— boss welding, in which the section to be welded and the welding point are determined beforehand by bosses;

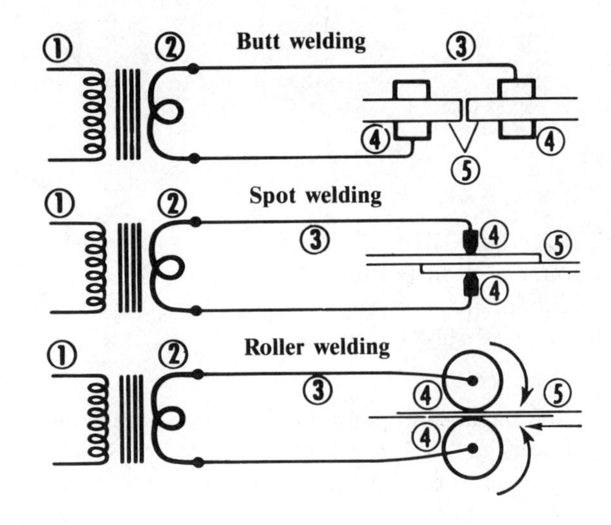


**Fig. 82.** Diagram of resistance welding machines

1. Welding transformer primary; 2. Welding transformer secondary; 3. Conductors applying welding current to the elements to be assembled; 4. Jaws, electrodes, rollers, through which the welding current flows at the same time as a certain pressure is applied to the elements to be assembled; 5. Elements to be assembled.

— capacitor discharge welding (pulse welding), in which a charge is pulse-injected into the parts to be welded.

Similar techniques are used for brazing metal parts. Conduction welding and brazing processes are used in many industries for welding steels, light alloys, copper alloys and special alloys.

Other welding processes using conduction heating have been developed for special requirements:

— electroslag or submerged arc welding (*see* paragraph 4.12.);

— high frequency conduction welding (*see* high frequency conduction heating, paragraph 4.10.3), [41], [P].

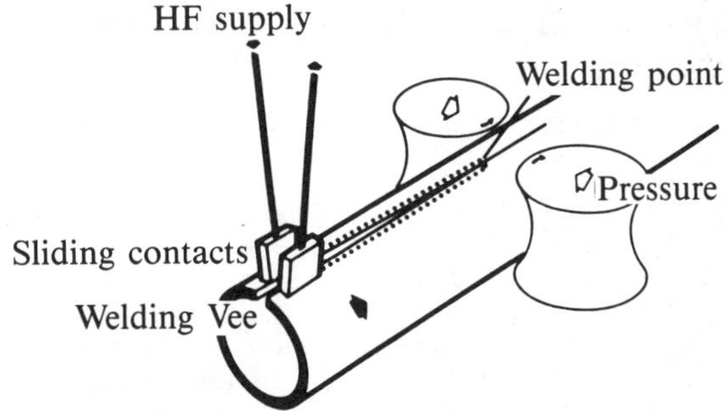


**Fig. 83.** Tube welding using continuous high frequency conduction [41].

In high frequency continuous conduction welding, the contacts are more or less rectangular and applied with moderate pressure. In order to prevent formation of sparks and to prolong contact life, the pressure should be a function of the surface condition of the sheeting. In continuous treatment, the current density at the contact varies between 15 and 20 A per $mm^2$ [41].

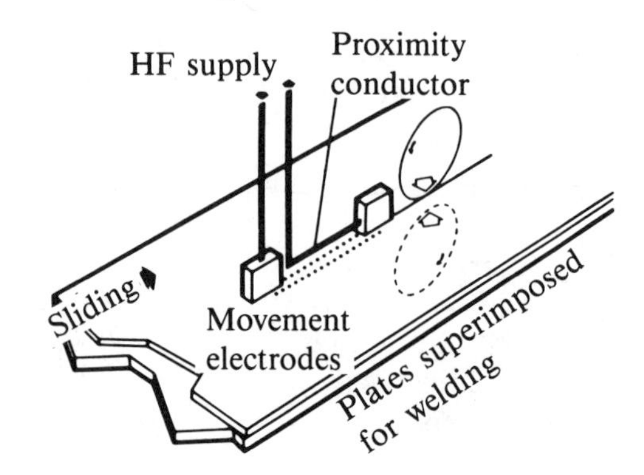


**Fig. 84.** High frequency continuous conduction welding of metal sheets [41].

Compared to induction heating, conduction offers a decrease in consumed energy for welding tubes which improves with tube diameter. For a 250 mm diameter tube, the energy gain may be as much as 40%. Conversely, the service life of the contacts is lower than that of the inductors, especially when the sheet has not been well cleaned.

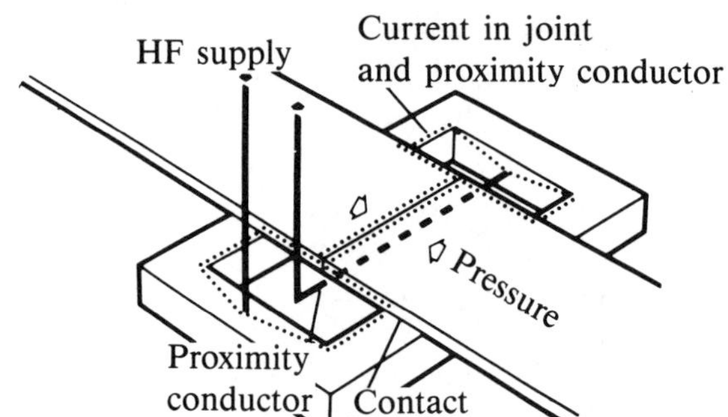


**Fig. 85.** Butt welding of sheets using high frequency conduction

Conduction can be used to weld different shapes (T or I, for example). Generally, the process is used for welding longitudinal, radial or spiral fins on tubes, concentric tubes and end-to-end or butt-welded plate and sheets. This process is also appropriate for stainless steel, aluminum and assemblies of these metals.

## 5. BIBLIOGRAPHY

[1] THE ENGINEER, *Chaudières électriques et accumulateurs de vapeur*, avril 1946 (même article, *L'Électricité*, mai 1946).
[2] R. MALGAT, Production de vapeur et d'eau chaude par les chaudières électriques à électrodes, *L'Électricité*, n° 118 et 119, juillet-août et septembre 1946.
[3] ALSTHOM, *Chaudières électriques à haute et basse tension*, Paris, 1948.
[4] P. PROFOS, Les chaudières électriques, *Revue technique Sulzer*, n° 1, 1949.
[5] E. BOREL, La fusion électrique du verre en Suisse, *Electroverre Romont*, Suisse, 1958.
[6] B. PATON *et al.*, Electroslagremelting of steel and alloys in water cooled moulds, *Iron and Steel*, Mars 1961.
[7] M. FORT, Some practical aspects of electric boosting, *Glass Technology*, vol. 5, n° 5, octobre 1964.

[8] J. SCHIFFARTH, Le chauffage direct par résistance appliqué au formage à chaud et au traitement thermique des métaux, *Revue BBC*, n° 10/11, 1964.
[9] B. PATON *et al.*, Refusion d'électrodes électroconsommables au sein d'une scorie électroconductrice, *Revue de Métallurgie*, février 1965.
[10] L. ANTOINE *et al.*, Résultats d'une exploitation industrielle de refusion sous laitier électroconducteur, *Revue de Métallurgie*, juillet-août 1966.
[11] N. HERMONT, Procédé de refusion sous laitier électroconducteur (ESR), *Revue BBC*, n° 3, mars 1968.
[12] LABORELEC, *Le chauffage des tuyaux par le système SECT*, Bruxelles, 1971.
[13] M. ANDO, Electric heating of pipelines with skin current, *Elektrowärme International*, vol. 29, n° 1, 1971.
[14] R. G. NEWTON, *Some new aspects of energy consumption in boosted glassmelting furnaces*, 9e Congrès International du Verre, Versailles, 2 octobre 1971.
[15] H. JURGENS, Widerstandserwärmung im Walzwerksbetrieb, *Elektrowärme International*, vol. 30, n° 1, 1972.
[16] E. J. DAVIS et A. L. BOWDEN, *Direct resistance heating of ferromagnetic billets*, Congrès Union internationale d'Électrothermie, Varsovie, 1972.
[17] J. KAZMIERCZAK et M. LOBODZINSKI, *Méthode numérique de calcul des systèmes de chauffage direct de fils et bandes en continu*, Congrès Union internationale d'Électrothermie, Varsovie, 1972.
[18] P. A. M. GELL, *Some further considerations on electric glass melting*, Congrès Union internationale d'Électrothermie, Varsovie, 1972.
[19] P. BONIS *et al.*, *Évolution des installations de refusion sous laitier électroconducteur et procédés nouveaux*, Congrès Union internationale d'Électrothermie, Varsovie, 1972.
[20] K. A. WALSHE et C. B. COOPER, *The application of direct electric heating (Metallurgy)*, Congrès IEE, Londres, 24 avril 1974.
[21] I. HOROWITZ, Grundlagen der Elektroglasschmelze, *Elektrowärme International*, vol. 32, n° 4, 1974.
[22] O. DRABEK, Die Wettbewerbslage des elektrischen Glasschmelzens in der BRD, *Elektrowärme International*, vol. 32, n° 4, 1974.
[23] G. LAMARQUE, *Application de l'électricité au traitement des métaux liquides : refusion sous laitier, dégazage avec réchauffage*, Comité français d'Électrothermie, 2-3 mai 1974.
[24] Y. DE BONY, *Le soudage vertical sous flux électroconducteur de l'aluminium*, Comité français d'Électrothermie, 2-3 mai 1974.
[25] P. J. HOLLIS, Skin effect current tracing warms fuel-oil pipeline, *Electrical Construction and Maintenance*, juin 1975.
[26] K. A. KOSTANYAN et M. AKHNAZAROV, Application de la fusion électrique au cristal au plomb, *Steklo I Keramika*, n° 4, avril 1976.
[27] A. TOURAINE, *Le chauffage par conduction haute fréquence; applications à la trempe et au soudage*, Comité français d'Électrothermie, 8-9 avril 1976.
[28] L. BECK et D. BISCARRAS, *Association du chauffage par conduction au chauffage par induction pour le brillantage des tôles étamées*, Comité français d'Électrothermie, 8-9 avril 1976.
[29] R. W. ROBERTS *et al.*, Thyristor controlled, 4300 kW power supply system for a mixed melter glass furnace, *IEEE Transcriptions on Industrial Applications*, n° 5, vol. IA-12, 1976 (USA).
[30] I. HOROWITZ, Mehrteilige Wannenofen System für die Electroschmelze des Glases, *Elektrowärme International*, vol. 34, juin 1976.
[31] W. J. MOLLOY et D. R. GREEN, Superalloys; development, production and application, *Metallurgia and Metal Forming*, juillet 1976.
[32] F. HEGEWALDT, Erwärmung im direkten Stromdurchgang, *Elektrowärme International*, vol. 34, n° 4, août 1976.
[33] A. ROHDE, Chauffage de barres par passage direct du courant, *Elektrowärme International*, vol. 34, n° 4. août 1976.
[34] S. B. LASDAY, Reliability of high temperature alloys enhanced by electroslag

remelting, *Industrial Heating*, septembre 1977 (un autre article d'un auteur anonyme sur la fusion ESR dans le même numéro).

[35] L. A. VOLOKHONSKY, *A study of the heat and electrical characteristics of electroslag remelting and vacuum arc furnaces intended for production of high quality steel ingots*, Congrès de l'Union internationale d'Électrothermie, Liège, 1976.

[36] I. HOROWITZ, Elektro-Glasschmelzwannen in Dreieckform, *Elektrowärme International*, vol. 36, n° 2, avril 1978.

[37] R. SCHLATTER, Direct current electroslag remelting of specialty alloys, *Industrial Heating*, octobre 1978.

[38] M. ANDO et J. E. GOODALL, *Une application du système de chauffage électrique SECT au pipe-line à longue distance*, Journées internationales d'Études sur le Chauffage par Induction (C.F.E.), 2-6 octobre 1978.

[39] ELECTRICAL REVIEW, *Resistance heater boosts steel strip production by* 30%, vol. 203, n° 14, octobre 1978.

[40] H. ENDRESS et W. MOMMERTZ, Energieverluste von wassergekühlten Elektrodenhalterungen in elektrischen Glas schmelzwannen, *Elektrowärme International*, n° 6, vol. 36, décembre 1978.

[41] A. TOURAINE, Le chauffage par conduction haute fréquence, *Revue générale d'Électricité*, n° 7/8, 1979.

[42] M. GAUSSIN, *Le chauffage des liquides industriels par l'électricité*, Documentation E.D.F., 1979.

[43] J. P. SLONINA, *Matériaux composites carbone-carbone, procédés d'élaboration, propriétés thermomécaniques*, Comité français d'Électrothermie, 6-7 mars 1980.

[44] DOCUMENTATION E.D.F., Paris-La Défense.

[45] DOCUMENTATION ELECTRICITY COUNCIL, Londres, G.B.

[46] DOCUMENTATION LABORELEC, Bruxelles, Belgique.

[47] G. SÉGUIER, *L'électronique de puissance*, Dunod Technique, Paris.

**List of equipment manufacturers and suppliers mentioned in this chapter:**

[A] SORG, Lohr, R.F.A.
[B] CEM, 75008 Paris, et BBC, Baden, Suisse.
[C] VERRERIES DU COURVAL, Le Courval, 80140 Sénarpont.
[D] M.G.R., 75018 Paris.
[F] ALFA-LAVAL, 78340 Les Clayes-sous-Bois.
[G] SULZER, 75300 Paris.
[H] INFRA-FOURS, 38000 Grenoble.
[I] ECET, 94160 Saint-Mandé.
[J] ELECTROTHERM, Bruxelles, Belgique.
[K] INDUCTOTHERM-SASI, 75840 Paris.
[L] LANGUEPIN, 93210 La Plaine-Saint-Denis.
[M] TOCCO-STEL, 91300 Massy.
[N] ENERGOMACHEXPORT, 75016 Paris et Moscou (URSS).
[O] MANUTHERM (DAMOND), 91210 Draveil.
[P] SINTRA-SEF, 92700 Colombes.
[Q] ALSTHOM, 75008 Paris.
[R] LEYBOLD HERAEUS SOGEV, 91400 Orsay.
[S] CREUSOT-LOIRE, 75008 Paris.
[T] BIPROMET, Varsovie, Pologne.

# Chapter 5

# Infrared Radiation Heating

## 1. INTRODUCTION

Infrared radiation heating, a special form of radiation heating and resistance heating, is an energy transmission method that uses electric resistances as radiation emitting sources.

Implementation, however, calls for special technologies, and its applications are often much different from those of conventional resistance furnaces. Accordingly, a special chapter has been devoted to infrared radiation heating.

The advantage of this heating method is in its high energy efficiency, which is due to several factors:

— it enables energy transfer from one body to another without requiring an intermediate support, and without noticeable absorption of the energy emitted by the environment separating the bodies; an intermediate fluid is required for convection heating;

— the thermal inertia of an infrared radiation heating system is generally very low, thus dispensing with the requirement for long heat-up or holding times;

— the energy radiated may be concentrated, focused, directed and reflected, in the same manner as light, which greatly increases its flexibility and adaptability;

— the power density can be very high, with the difference in temperature between the source and the body to be heated often high, thus leading to compact installations and high process speeds.

Conversely, it is difficult to treat complex shaped parts uniformly, since the energy received by each elementary surface of the body to be heated varies with the distance from the source, and angle of radiation incidence.

Infrared radiation heating therefore is best suited to industrial processes — drying, curing, polymerization, sterilization — in which products are in the form of strips or continuous layers, or again are repetitive in shape and size.

It should be noted that the effect of infrared radiation is purely thermal (identical to the thermal effect of the sun on human beings or nature), as opposed to other radiation such as ultraviolet, which may have photochemical effects. Therefore, infrared radiation is very safe.

## 2. FUNDAMENTAL LAWS OF INFRARED RADIATION

### 2.1. Spectral band

Infrared radiation is a form of electromagnetic radiation, with its spectrum (*Fig.* 1) located between the visible and micro waves, i.e. between wavelengths of 0.76 and 1,000 $\mu$m approximately. For industrial applications, the 0.76 — 10 $\mu$m range is usually divided into three bands:

- short infrared, from 0.76 to 2 $\mu$m
- medium infrared, from 2 to 4 $\mu$m
- long infrared, from 4 to 10 $\mu$m.

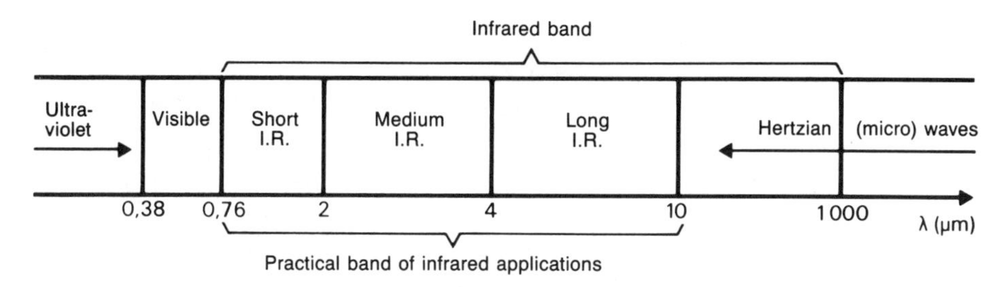


**Fig. 1.** Infrared radiation spectral band

These wavelengths correspond to very low photon energies, less than $4 \times 10^{-19}$ J, which are not likely to modify molecular structure, as occurs with higher energy radiation, such as ultraviolet, X and gamma.

### 2.2. Emission and absorption of radiation

All bodies taken to a temperature of more than absolute zero ($\theta$ = - 273°C) radiate and exchange energy with one another.

#### *2.2.1. Emission of radiation*

— *Emissivity (radiant emittance)*

The total emissivity M (radiance) at a point on the surface of a body is defined as the energy flux emitted per unit area:

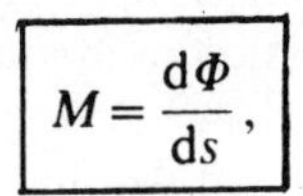

$$M = \frac{d\Phi}{ds},$$

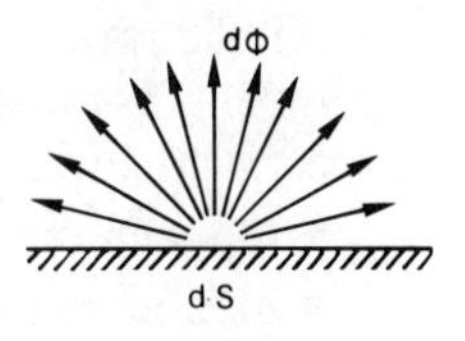


$M$ in Watts per square meter

**Fig. 2.** Total radiant emittance.

The total radiant emittance is an important element in comparing the power densities obtainable with various infrared sources; it is a function of the temperature and characteristics of the emitting surface [3], [14], [23].

— *Radiant intensity*

The total radiant intensity of a source in a given direction is the flux radiated per solid angle unit along this direction:

$$I = \frac{d\Phi}{d\Omega},$$

$I$ in Watts per steradian

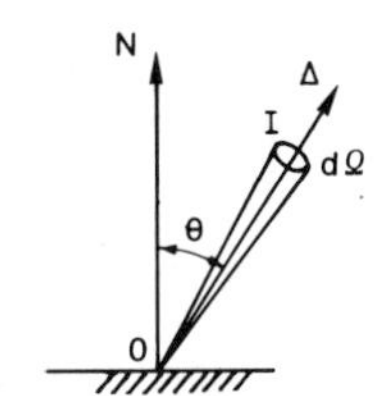


**Fig. 3.** Total radiant energy

— *Radiance*

The total radiance of a source in a given direction is defined as the intensity of the source along this direction, divided by the apparent area (projected surface) of this source in the same direction:

$$L_\Delta = \frac{d^2\Phi}{d\Omega\, ds \cos\alpha},$$

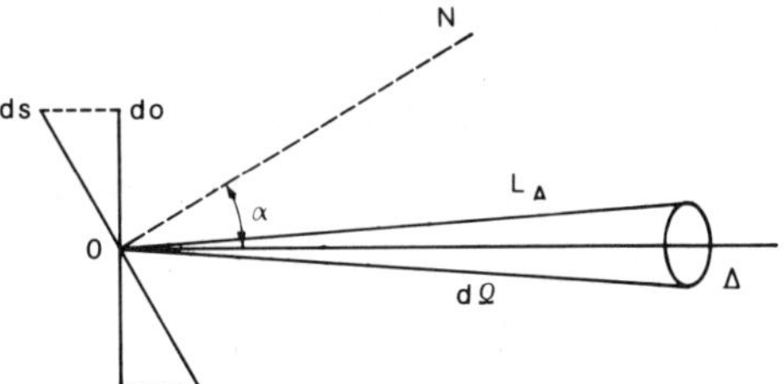


**Fig. 4.** Radiance.

$L$ in Watts per square meter per steradian.

This enables comparison of power radiated in a given direction by sources of different extents or orientations, with respect to this direction, together with the powers radiated by a given source in different directions.

— *Monochromatic emissivity, radiant intensity, and radiance*

All the variables previously mentioned concerning total radiation, (i.e. all the wavelengths radiated by the source) can be referenced to emission over a narrow spectral field centered on wavelength $\lambda$:

- monochromatic emissivity $M_\lambda$, in Watts per meter squared per meter;
- monochromatic radiant intensity $I_{\Delta\lambda}$ in Watts per steradian per meter;
- monochromatic radiance $L_{\Delta\lambda}$ in Watts per square meter per steradian per meter.

— *Lambert's law (or cosine law)*

Sources in which radial intensity is independent of direction obey Lambert's law, and are known as isotropic or diffuse emission sources. The radiant intensity in a direction $\Delta$ is then equal to the product of the intensity radiated in a direction perpendicular to this surface, times the cosine of the angle of this direction with the perpendicular:

$$L_g = L = \text{Constant},$$

$$I_\Delta = I_N \cos \chi.$$

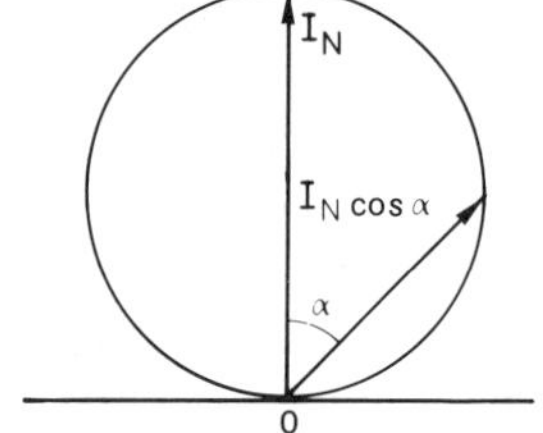


**Fig. 5.** Graphic representation of Lambert's law.

For diffuse emission sources, the total emissivity and radial intensity are then related by the equation:

$$M = \pi L.$$

These equations show the importance of the position of the product with respect to the emitters in infrared heating.

### *2.2.2. Absorption of radiation*

#### *2.2.2.1. Irradiance*

The irradiance of a surface is defined as the energy flux received per unit of receiving surface, from all directions from which radiation can be received:

$E$ in Watts per square meter.

$$\boxed{E = \frac{d\Phi}{ds},}$$

#### *2.2.2.2. Relationship between irradiance of the receiver and the radiance of the emitter*

When two surfaces exchange energy by radiation, the relation of irradiance and radiance are expressed by the equation:

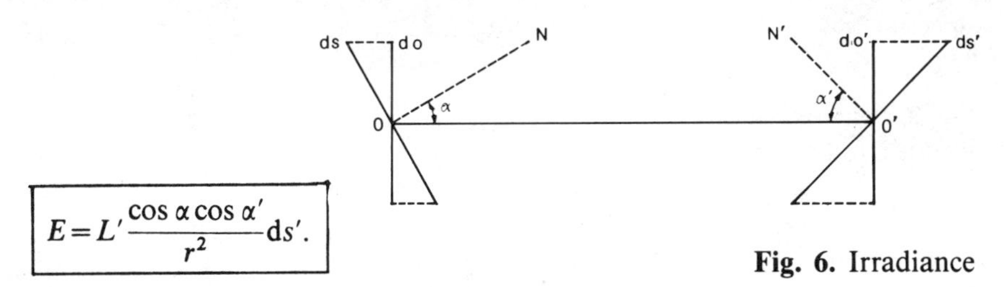


$$E = L' \frac{\cos\alpha \cos\alpha'}{r^2} ds'.$$

**Fig. 6.** Irradiance

As above, a monochromatic irradiance $E_\lambda$ is defined.

This formula highlights the importance, not only of the relative position of the emitters and the product to be heated, but also their distance.

*2.2.2.3. Absorption*

Emission of radiation is not, however, emission of heat. It is only when a body absorbs radiation that it is converted into heat. The overall absorption capability $\alpha$ (or absorption factor or absorptivity) at a point is the ratio between the flux absorbed and the incident flux. For all bodies present in nature, this ratio is less than 1, since a part of the radiation is reflected and, if the body is not opaque, transmitted.

Where $\varrho$ is the reflection factor (or reflectivity) and $\tau$ the transmission factor (or transmissivity), we obtain:

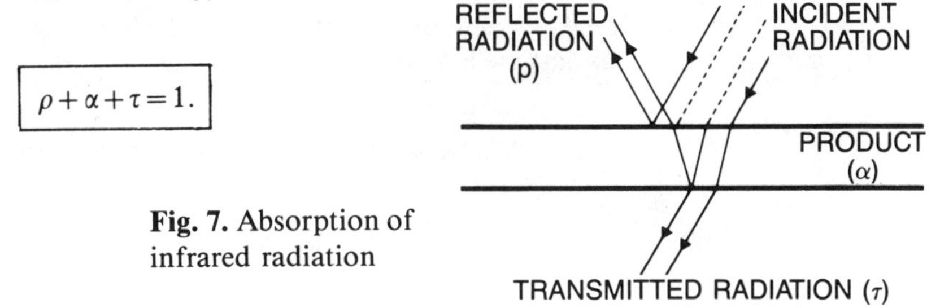


$$\rho + \alpha + \tau = 1.$$

**Fig. 7.** Absorption of infrared radiation

For a given body, the absorption factor varies with:

— spectral composition of the incident radiation and its direction;

— the characteristics of the body: surface condition, chemical nature, color, thickness and temperature.

Therefore, as before, it is necessary to define $\alpha_{\lambda\Delta}$, relative to a wavelength and a direction.

*Penetration depth*

Absorption of radiation by the body to be heated is gradual and takes place at a certain depth from the surface, according to a decreasing exponential (Bouguer's law). Radiation can penetrate several millimeters.

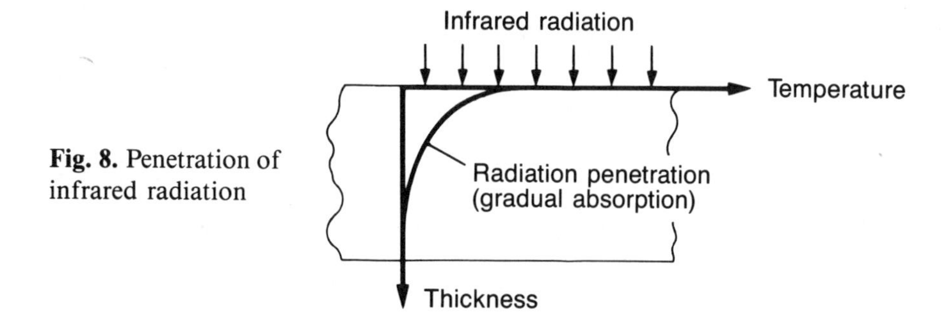


**Fig. 8.** Penetration of infrared radiation

The penetration depth depends in particular on the wavelength of the incident radiation; in general, this increases as wavelength decreases, and short infrared radiation is therefore more penetrating than the medium, which in turn is more penetrating than long infrared.

*Selective absorption*

Absorption of radiation is a selective phenomenon depending greatly on the incident wavelength. Therefore, ordinary glass and water, which are transparent to light radiation, absorb large quantities of infrared at wavelengths of 2.5 and 2.6 $\mu$m respectively. The absorption factor also varies with thickness. Thus, for water, the absorption factor becomes significant at a wavelength of 1.5 $\mu$m as soon as the water layer is thicker than 30 $\mu$m.

Figure 9 gives the absorption spectrum for a 10 $\mu$m film of water and a sheet of PVC 1 mm thick. Numerous data are available concerning absorption spectra.

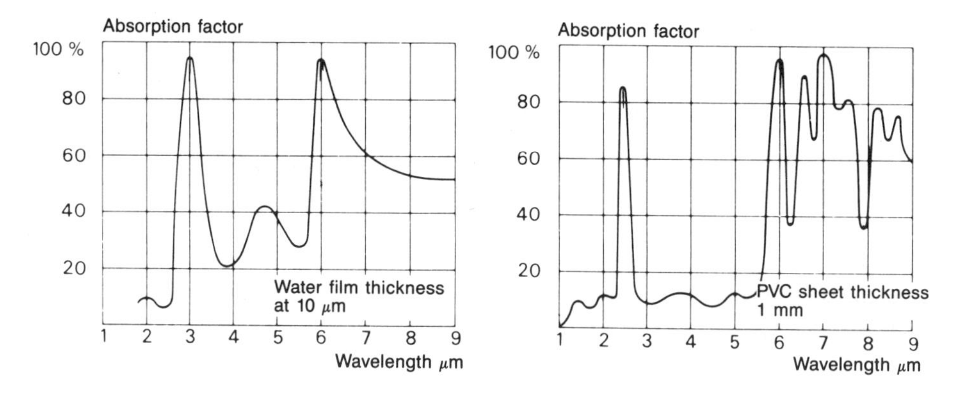


**Fig. 9.** Infrared radiation absorption curves

*This shows the benefit of adapting the range of wavelengths emitted by the source to the product to be treated.*

*Reflection*

Reflection depends in particular on the nature of the bodies, their surface condition and the incident radiation wavelength.

Some metals such as polished aluminum have a high reflection factor for infrared, about 0.9, and this reflection factor generally increases with wavelength; therefore, these metals will be used as reflectors in infrared radiation emitters.

### 2.3. Black body radiation

By definition, a black body is a body for which:

$$\boxed{\alpha = 1.}$$

The total absorption factor of a black body is therefore equal to 1, whatever the spectral composition of the incident radiation and the body temperature. Emission from a black body follows Lambert's law. This is a hypothetical body, an ideal emitter and absorbant, but some products such as carbon black and rough oxidized cast iron approach this.

#### *2.3.1. Total emissive power: Stefan-Boltzmann law*

Radiated power per unit area is proportional to the fourth power of the absolute temperature of the body surface:

$$\boxed{M = \sigma T^4,}$$

$M$, radiated power in Watts per square meter;

$T$, absolute temperature in degrees Kelvin (K);

$\sigma$, Stephan's constant, equal to $5.73 \times 10^{-8}$ W/m$^2$ $\times$ K$^4$.

*This law shows that very high power densities can be obtained with certain infrared electrical emitters, whose temperature can reach* 2,200°C. The value of radiated power versus temperature is given in Figure 10.

| $T$(K) | $0$(°C) | $M$(W/cm$^2$) |
|---|---|---|
| 300 | 27 | 0.046 |
| 500 | 227 | 0.357 |
| 700 | 427 | 1.370 |
| 1,000 | 727 | 5.710 |
| 1,500 | 1,227 | 28.900 |
| 2,000 | 1,727 | 91.000 |
| 3,000 | 2,727 | 462.000 |

**Fig. 10.** Radiated power versus emitter temperature

### *2.3.2. Spectral distribution of emitted energy: Planck's and Wien's laws*

Spectral distribution of radiated energy follows Planck's law, which, as a function of temperature and the wavelength considered, gives the monochromatic emissivity $M_\lambda$ of a black body for wavelength $\lambda$[3], [14], [23].

*Planck's law*

$$M_\lambda = \frac{C_1}{\lambda^5 (e^{C_2/\lambda T} - 1)},$$

$M_\lambda$, radiated power in watts per square meter per micron, corresponding to wavelength $\lambda$;
$C_1$, constant equal to $3.741 \times 10^8$ W $\times$ $\mu m^4/m^2$;
$C_2$, constant equal to 14,388 $\mu m \times K$;
$\lambda$, wavelength in microns.

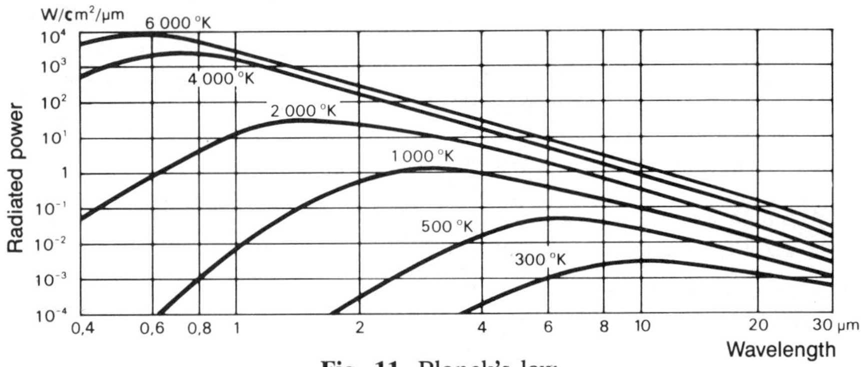


**Fig. 11.** Planck's law

The maximum monochromatic emissivity (emissive power referenced to wavelength) at absolute temperature $T$ corresponds to the wavelength $\lambda_m$, as determined by Wien's law:

$$\lambda_m . T = C,$$

$\lambda_m$, wavelength in microns ($\mu$m);
$T$, absolute temperature in degrees Kelvin;
$C$, constant, equal to 2,898.

*The hotter the body, the more it radiates in the short wavelengths.*

However, Wien's law is only an approximation, and can be correctly applied only for wavelengths between 0.3 and 10 $\mu$m, which is precisely the infrared radiation range used in industrial applications.

95% of the emitted energy is radiated within the wavelength limits of 0.5 and 5 $\mu_m$. An emission of 1% remains for wavelengths of less than 0.5 $\mu_m$m and 4% for wavelengths of more than 5 $\mu_m$.

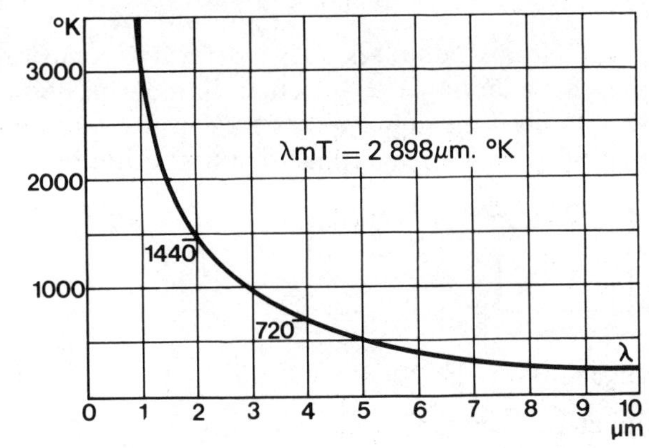


**Fig. 12.** Wien's law.

Spectral band corresponding to 95% of emitted energy

| Absolute temperatures ($TK$) | Celsius temperatures ($\theta$°C) | Maximum wavelengths ($\mu$) | Spectrum approximate limits ($\mu$) |
|---|---|---|---|
| 300 | 27 | 9.6 | 4.8-41 |
| 500 | 227 | 5.7 | 3-25 |
| 750 | 477 | 3.8 | 2-16 |
| 1,000 | 727 | 2.9 | 1.5-12 |
| 1,200 | 927 | 2.4 | 1.2-11 |
| 1,500 | 1,227 | 1.9 | 1-8 |
| 2,000 | 1,727 | 1.4 | 0.7-6 |
| 3,000 | 2,727 | 0.96 | 0.5-4 |

**Fig. 13**

For usual emitters whose temperatures vary between 400 and 2,200°C, almost all the emitted energy is located in the infrared.

## 2.4. Radiation from real bodies

### *2.4.1. Emissivity and dissipated power*

Radiation from a non-black body is defined by comparing it to that of a black body taken to the same temperature, introducing the emission factor $\epsilon$ (or emissivity); this is the ratio between the emissive power of the body considered and that of a black body at the same temperature.

The radiated energy is then:

$$\boxed{M = \varepsilon \sigma T^4,}$$

$M$, radiated power in Watts per square meter;

$T$, absolute temperature in degrees Kelvin;

$\sigma$, Stefan's constant, equal to $5.73 \times 10^{-8}$ W /m$^2$ X K$^4$;

$\epsilon$, emissivity.

Emissivity is always less than 1 and varies with the temperature and characteristics of the body. Figure 14 shows that emissivity varies widely with the bodies considered, but also for the same body, and in particular a metal, with surface condition. Moreover, as for emissive power, finer levels can be defined: directional monochromatic emissivity, monochromatic emissivity and directional emissivity. *In most industrial applications, is considered as a constant for a given temperature range and surface condition, and a given mean value.*

### *2.4.2. Relationship between absorption and emission: Kirchhoff's law*

Kirchhoff's law states that, for each wavelength and direction of propagation of radiation emitted from a surface, or incident to the latter, the directional monochromatic emissivities and absorptivities are equal:

$$\boxed{\varepsilon_{\Delta\lambda} = \alpha_{\Delta\lambda}.}$$

Conversely, this does not apply to the total emission and absorption factors. The total absorption factor $\alpha$ of a body at temperature $T$, with respect to radiation from a black body of the same temperature $T$, is equal to its total emission factor $\epsilon$, but the overall absorbant power of a black body with respect to radiation from another non-black body $B$ depends both on the type and temperature of body $B$.

In practice, and for each component of a radiative system, the absorption factor corresponding to the spectral band or the radiation temperature which it receives and the emission factor corresponding to its proper temperature are taken into consideration.

Total emission factors (emissivities)

| Ordinary metals | Body temperature (°C) | Coefficient of emissivity | Non-metallic materials | Body temperature (°C) | Emissivity coefficient |
|---|---|---|---|---|---|
| Steel or iron | | | Asbestos (blocks) | 30 | 0.9 |
| nickel-chromium | 20 | 0.06 | Wood | 30 | 0.9 |
| rusted | 20 | 0.68 | Brick | 30/1,000 | 0.9 |
| | | | Limestone | 30 | 0.9 |
| Aluminum | | | Leather | 30 | 0.7 |
| polished | 200 | 0.04 | Fabric | 30 | 0.9 |
| oxidized | 200/600 | 0.11/0.19 | Shellac | 30 | 0.7 |
| Copper | | | Red polished | | |
| fully polished | 100 | 0.03 | stoneware | 30 | 0.6 |
| slightly polished | 50/300 | 0.15 | Cork | 30 | 0.9 |
| rough or | | | Polished marble | 60/200 | 0.6 |
| oxidized | 100 | 0.75 | Mica | 30 | 0.75 |
| Tin | 30 | 0.12 | Paper | 30 | 0.9 |
| Cast iron | | | Plaster | 30 | 0.9 |
| non-oxidized | 200 | 0.40 | Porcelain | 30 | 0.9 |
| oxidized | 200 | 0.64 | Soot | 60/200 | 0.93 |
| rough | 200 | 0.85 | Bronze paint | 100 | 0.51 |
| rough-oxidized | 40/250 | 0.95 | Cream paint | 100 | 0.77 |
| Brass | | | Black paint | 100 | 0.84 |
| fully polished | 300 | 0.03 | Aluminum paint | 100 | 0.3/0.6 |
| mat | 50/350 | 0.20 | Glass | 30 | 0.9 |
| Mercury | 100 | 0.15 | | | |
| Nickel | | | | | |
| oxidized | 200 | 0.35/0.45 | | | |
| nickel-chromium | 50/1,000 | 0.65/0.75 | | | |
| Gold | | | | | |
| non-polished | 30 | 0.5 | | | |
| Platinum | | | | | |
| polished | 30 | 0.08 | | | |
| rolled | 30 | 0.10 | | | |
| Lead | | | | | |
| non-oxidized | 130 | 0.06 | | | |
| oxidized | 200 | 0.65 | | | |

**Fig. 14.**

## 2.5. Transmitted power

### *2.5.1. Calculation of transmitted power*

The power transmitted by infrared radiation emitters to products to be heated is a function of temperature, shape, relative position of surfaces, and emission coefficients.

The corresponding calculations, based on the laws of radiation described above, are, in most cases, surfaces of any shape, non-black bodies and extremely complex. In practice, however, actual problems can often be limited to a few special cases.

— *Transmission of heat between two black surfaces of any shape*

By definition, for both surfaces, emissivities $\epsilon_1$ and $\epsilon_2$ and $\alpha_1$ absorption factors and $\alpha_2$ are equal to 1, and all radiated energy received is absorbed.

Under these conditions, the heat flux emitted by the emitters to the body to be heated is expressed by:

$$\Phi = \sigma S_1 F_{12} (T_1^4 - T_2^4),$$

$\Phi$, transmitted heat flux;

$\sigma$, Stefan's constant;

$T_1$, emission temperature in degrees Kelvin;

$T_2$, receiving surface temperature in degrees Kelvin;

$F_{12}$, *surface angle factor* $S_2$ as seen from $S_1$;

$S_1$, emitter surface.

Angle factor $F_{12}$ is a dimensionless value, depending only on the shape and relative position of surfaces $S_1$ and $S_2$. For surfaces with simple shapes, common in infrared radiation applications, the calculation is easy.

— *Transmission of heat between two grey surfaces with high coefficients of emission*

A grey body is a body for which the emissivities relative to direction and wavelength are equal whether or not surface and wavelength are perpendicular to one another, and together equal total emissivity $\epsilon$.

If, in addition, emissivities $\epsilon_1$ and $\epsilon_2$ from the surfaces are high, a little less than 1, the approximate value of the transmitted flux, in which reflected flux is not taken into account, is:

$$\Phi = \sigma \varepsilon_1 \varepsilon_2 F_{12} S_1 (T_1^4 - T_2^4),$$

— *Transmission of heat between two flat grey surfaces*

In this case, allowing for reflected flux, it is easy to demonstrate that the heat flux transmitted is:

$$\Phi = \sigma S (T_1^4 - T_2^4) \frac{\varepsilon_1 \varepsilon_2}{\varepsilon_1 + \varepsilon_2 - \varepsilon_1 \varepsilon_2},$$

where $S$ is the area of two flat surfaces considered as very large with respect to the distance separating them.

Generally, this is written:

$$\varepsilon' = \frac{\varepsilon_1 \varepsilon_2}{\varepsilon_1 + \varepsilon_2 - \varepsilon_1 \varepsilon_2},$$

i.e.:

$$\frac{1}{\varepsilon'} = \frac{1}{\varepsilon_1} + \frac{1}{\varepsilon_2} - 1,$$

so that the above formula becomes:

$$\Phi = \varepsilon' \sigma S (T_1^4 - T_2^4),$$

if $\epsilon_1$ and $\epsilon_2$ are high, near 1, the formula given in the previous paragraph is obtained again.

### *2.5.2. Power density — comparison with convection heating.*

The power density transmitted by an emitter onto a product can be very high, up to 300 kW/m$^2$ (30 W/cm$^2$). However, in most cases, the power densities used are much lower.

Thus, for a source temperature of 800°C ($T_1$ = 1,073 K) and a receiver temperature of 100°C ($T_2$ = 373 K) and emissivities $\epsilon_1 = \epsilon_2 = 0.8$, the power transmitted is approximately 50 kW/m$^2$.

The same product, heated to 100°C by forced air convection at 450°C would receive only 8 kW/m$^2$ approximately for a speed of 10 meters per second and heat tranfer $h$ = 0.02 kW/m$^2$ X °C.

The power densities which can be exchanged are therefore much higher with infrared radiation, in which the transmitted power is proportional to ($T_1^4 - T_2^4$), than with convection where this power is proportional to ($T_1 - T_2$) or ($T_1 - T_2{}^{1.25}$). In addition, this energy is more easily absorbed by the products to be heated, since infrared radiation is absorbed over a considerable depth, while with convection, power is absorbed at the surface only and can only be transferred to the interior by conduction.

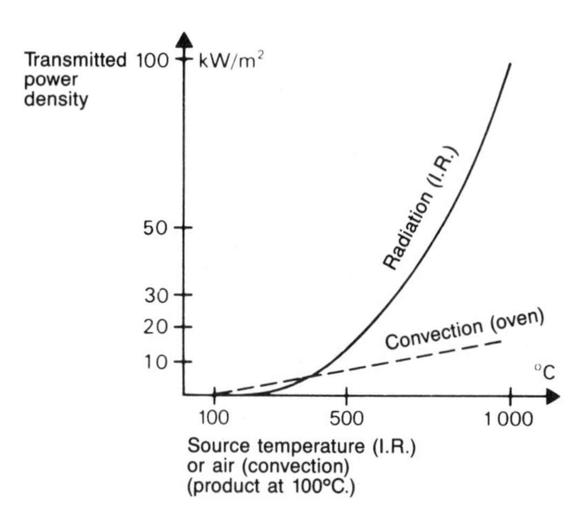


**Fig. 15.** Comparison of power densities obtained with infrared heating and convection heating

## 3. ELECTRIC EMITTERS OF INFRARED RADIATION

Emitter technologies correspond to each of the three infrared radiation spectrum bands (*see* paragraph 2.1.). The temperature of short infrared emitters is greater than 1,200°C, that of medium infrared emitters between 450 and 1,200°C and that of long infrared less than 450°C. The temperatures were calculated using Wien's law ($\lambda_m \times T = C$, *see* paragraph 2) using wavelengths limiting the three bands to maximum monochromatic emittance at the temperature desired. However, due to the shape of the curve giving the spectral distribution of the radiated energy versus temperature and wavelength (Planck's law, *see* paragraph 2), it is necessary for the following precautions to be respected to place the radiated energy within the desired band of wavelengths:

— for short and medium infrared radiation, the emission temperature must be much higher than that defining the upper limit of the band;

— for long infrared radiation, and so that most of the radiated energy is not of very long wavelength, at which it would not be absorbed by standard bodies, the emission temperature must be close to the limit of the lower emission band, or even in the upper part of the medium infrared band.

The various types of emitters all use the thermal effect of an electric current flowing through a resistive element (Joule effect).

### 3.1. Short infrared emitters

#### *3.1.1. Technology*

Short infrared emitters consist of vacuum tubes or lamps or, more often, inert gas lamps (argon, nitrogen) containing a tungsten filament heated to very high temperature (2,000 to 2,500°C).

The maximum monochromatic emittance is around 1.2 $\mu$m. Approximately 5% of the radiation is in the visible spectrum, which explains their bright yellow color.

Infrared lamps, which are very similar in design to lightbulbs (*Fig.* 16), have a glass envelope containing a silver or aluminum reflector. Unit power is low, generally 150, 250 or 375 Watts. The tungsten filament reaches 2,000°C corresponding to a maximum emission wavelength of about 1.4 $\mu$m.

**Fig. 16.** Infrared lamp

Infrared lamps consist of a quartz tube [1] filled with an inert gas, in which a spiral-wound tungsten filament, supported by disks, is placed and raised to a temperature of about 2,200°C (*Fig.* 17).

These tubes feature an internal gold-plated reflector or an external reflector. They are available in different effective lengths (0.2 to 1.5 m), and current unitary powers range from 0.5 to 7 kW. Tubes up to powers of 20 kW are available for special applications. The most common linear power is 50 W/linear cm of active tube.

For higher power densities, emitters, their bases and mountings can be air cooled or, in some cases, even water cooled.

Moreover, very high power density tubes exist in which the tungsten filament temperature reaches 2,700°C. To prevent filament evaporation which would cause blackening of the tube and diminish its energy efficiency and service life, a halogen, generally iodine, is added to the inert gas filling the tube. The evaporated tungsten reacts with the iodine to form tungsten iodide which is volatile. This iodide is taken by convection to the filament, where a high temperature breaks it down, and regenerates it. The linear power of this type of emitter can reach 300 W/cm of tube [20].

---

[1] Quartz is practically transparent to infrared radiation, since it absorbs only some 5% of infrared radiation. More than 50% of this energy is also re-emitted in the form of infrared radiation at a longer wavelength (with short wave infrared emitters, the tube reaches approximately 800°C, and about 400-500°C for the medium infrared emitters described below), the remainder is essentially exchanged due to the environment by convection. Quartz is only slightly sensitive to thermal shock (very low coefficient of expansion), offers adequate mechanical strength and is a poor conductor of heat. This material is widely used, therefore, in the manufacture of infrared emitters.

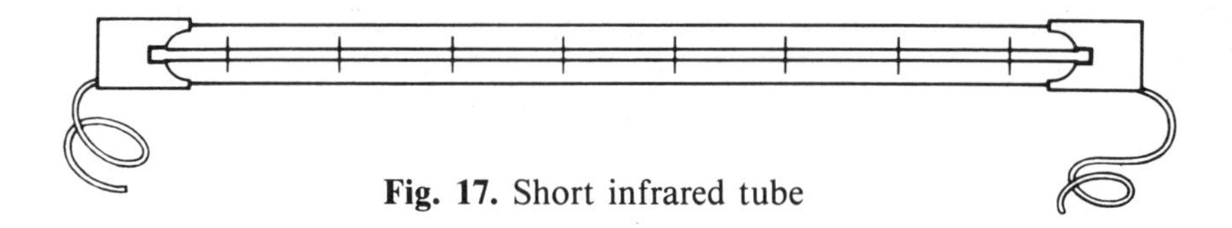

**Fig. 17.** Short infrared tube

Selection from a range of short infrared emitter tubes

| Heating length (mm) | Overall length (mm) | Voltage (V) | Power (W) |
|---|---|---|---|
| 140 | 241 | 110 | 500 |
| 272 | 368 | 220 | 1,000 |
| 410 | 508 | 380 | 2,000 |
| 700 | 798 | 380 | 3,000 |

**Fig. 18.**

### *3.1.2. Properties of short infrared emitters*

The low unit power of the lamps permits even radiation; conversely, the power density installed rarely exceeds 10 kW/m$^2$ of oven, while, with horizontal mounting, it can reach 25 kW/m$^2$.

The number of sources and connections is important, and although replacement of a lamp is a simple and rapid operation, they are gradually being replaced by quartz tubes much higher in unit power and less fragile.

Due to their linear structure, tubes are suitable for homogeneous continuous treatment of layered products. However, for normal supply voltages, tube lengths remain relatively low and do not enable widths of more than 2 meters to be treated without connections inside the furnace.

The most common powers are about 80 kW/m$^2$ of furnace, but it is possible with connection cooling devices and external reflectors to reach power densities of 200 to 300 kW/m$^2$ for horizontally-mounted tubes. Over relatively short periods, boosting emitter voltage enables power densities of 1,800 kW/m$^2$, to the detriment of emitter service life, normally around 5,000 hours; however, these possibilities are interesting for some research activities or testing of materials, since product temperatures can reach 1,700°C.

In general, short infrared sources offer a maximum monochromatic emittance of between 1 and 1.3 $\mu$m. They have very low thermal inertia (one second for the emitted power to be reduced by 90%), which enables fine adjustment and offers excellent safety in the event of production line shutdown.

The mean service life of such emitters is between 3,000 and 5,000 hours, but thermal and mechanical shocks should be avoided.

Penetration of radiation into the material to be heated is generally adequate and heating homogeneous throughout the material thickness, enabling high power densities to be used without excessive surface overheating.

Short infrared is therefore particularly good for heating thick products (a few millimeters). Since reflection is high for the short infrared, the furnace must be insulated to obtain adequate efficiency [2], [4], [6], [F], [G], [I], [L], [O], [Q], [T].

## 3.2. Medium infrared emitters

### *3.2.1. Technology*

Emitters consisting of the usual alloy resistances (nickel-chromium or iron-chromium-aluminum), generally operating between 700 and 1300°C belong to this range. They are mounted in tubes or silica or quartz panels, together with metal radiant tubes. Approximately 1% of the energy emitted by these emitters is in the visible range, giving them a light red color [28], [B], [D], [E], [H], [W].

Single or double, crystal or translucent silica tubes behave as a support for a resistive element, which, in most cases, is an iron-chromium-aluminum alloy heated to a temperature of 1,000 to 1,350°C. The tubes are often gold-plated at the back to reflect radiation from the filament. A wide range of useful wavelengths (0.2 to 3 $\mu$m) and powers (0.25 to 8 kW) exists for linear powers of approximately 30 W/cm of effective tube. For emitters of this type, cooling is not normally required. Tubes can be single or paired. The difference between these and short infrared tubes is that the incandescent wire does not, because of the alloy used and the temperature level, need to be protected from air, therefore simplifying fabrication.

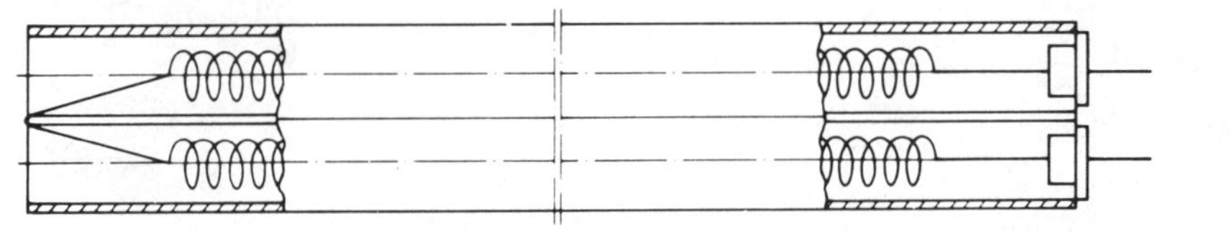

**Fig. 19.** Twin quartz tubes

Selection from a range of medium infrared emitters
consisting of transparent silica tubes without built-in reflectors

| Heating length (mm) | 800 | 1,000 | 1,200 | 1,500 | 1,800 | 2,500 |
|---|---|---|---|---|---|---|
| Power (W) | 2,650 | 3,300 | 4,000 | 5,000 | 6,000 | 7,600 |
| Voltage (V) | 220 | 220 | 380 | 380 | 380 | 380 |

**Fig. 20.**

*Silica or quartz radiant panels* use insulation and nickel-chromium or iron-chromium-aluminum resistances taken to a temperature of between 700 and 1,000°C (*Fig.* 21). Powers vary from 800 to 1,600 W for an effective area of 650 cm$^2$ (1.2 to 2.5 W/cm$^2$). However, high specific power quartz panels of up to 5 W/cm$^2$ exist.

**Fig. 21.** Radiant silica panel [H].

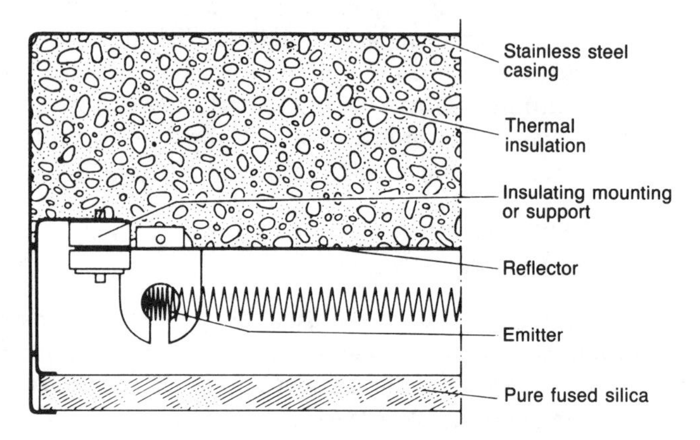


*Metallic radiant tubes* (*Fig.* 22) consist of a nickel-chromium resistance or even iron-chromium-aluminum resistances, embedded in magnesia (this material is both a good electrical insulator and a good conductor of heat) heating a sealed metal tube, generally in refractory stainless steel, by conduction. The radiation source is therefore not the filament but the metal sheet, which emits at a temperature of around 700°C (maximum 800°C). A large part of the energy radiated by these emitters is in the long infrared, and are sometimes classified in this category, since their operating temperature in balance with the product to be heated is often about 500°C.

**Fig. 22.** Radiant metallic tube

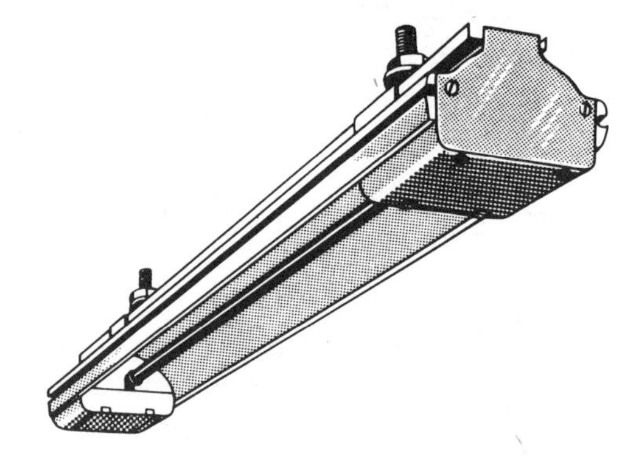

To increase efficiency, these elements are generally installed in a reflector. Lamp-shaped emitters of this type exist, in which the tube is spirally wound in one plane with a conical reflector concentrating the radiation.

These elements can also be shaped as required to fit the contour of the part to be heated.

The power of such elements varies from 0.8 to 8 kW for heating lengths of up to 4 m, corresponding to a linear power of approximately 20 W/cm of tube.

Selection from a range of medium infrared emitters consisting of metallic radiant tubes placed in an aluminum reflector

| Heating length (mm) | Overall length (mm) | Voltage (V) | Power ([1]) (W) | Power ([2]) (W) |
|---|---|---|---|---|
| 492 | 622 | 220 | 800 | 1,600 |
| 991 | 1,187 | 220 | 1,800 | 3,600 |
| 1,651 | 1,872 | 220 | 3,000 | 6,000 |
| 2,220 | 2,440 | 220 | 3,800 | 7,600 |
| 3,353 | 3,558 | 380 | 5,000 | 10,000 |
| 4,420 | 4,420 | 380 | 6,500 | 13,000 |

([1]) Single tube elements
([2]) Double tube elements

**Fig. 23.**

### *3.2.2. Properties of medium infrared emitters*

Quartz or silica tubes are used to obtain maximum power densities of about 80 kW/m$^2$ but, in order to guarantee high emitter service life, the most commonly used values are between 50 and 60 kW/m$^2$. Thermal inertia is relatively low, 30 seconds, for the emitted power to be reduced by 90%. Emission maximum is around 2.2 $\mu$m.

The mean service life is between 5,000 and 10,000 hours if mechanical shocks are avoided.

*Quartz or silica radiant panels* emit radiation with maximum monochromatic emittance between 2.5 and 3.5 $\mu$m, depending on the operating temperature. The power density is generally between 25 and 40 kW/m$^2$, but 60 kW/m$^2$ can be reached when the forced convection system is combined with infrared heating. Thermal inertia is higher than for tubes emitting in the medium infrared, being about 2 to 3 minutes.

These panels are very simple to use and, because of emission homogeneity, enable operation at very low distances from the products to be treated. They are also easy to clean and rugged, and withstand mechanical shocks well (in the event of damage to a layer of paper or fabric, for example) and also withstand thermal shock (water).

Metallic radiant tubes can radiate a power of 40 kW/m$^2$ in the oven onto the product, corresponding to a maximum monochromatic emittance wavelength of 3$\mu$m (but, depending on temperature, 30 to 70% of the energy emitted is in the long infrared). Thermal inertia of these elements is about the same as with quartz panels (2 minutes). They are particularly rugged and can be shaped as required to mate with the contours of products to be treated.

Medium infrared emitters provide power densities which are less than those of short infrared emitters, yet 4 to 5 times higher than with convection heating, part of the energy being transmitted by convection, 30 to 40%, which may suit some applications. Thermal inertia is higher, and fine regulation capabilities are therefore more limited, and special protection devices are required for fragile products in the event of a machine shutdown.

Conversely, these emitters are much more rugged. Radiation penetration is average and reflection low, which, in some cases, increases efficiency.

Also, they are often fully adapted to drying operations, since water offers a high absorption band at around 3$\mu$m, corresponding to maximum monochromatic emittance.

## 3.3. Long infrared emitters

### *3.3.1. Technology*

Basically, these sources consist of glass radiating panels, made electroconductive on the surface, and vitrified ceramic-covered panels heated to a temperature of between 300 and 600°C, or even 700°C. These sources do not radiate in the visible range, resulting in the name "dark emitters" [A], [J].

*Electroconductive-glass radiating panels* (*Fig.* 24) consist of a plate of hardened glass, the inside surface of which is coated with a thin layer of metal oxide acting like an electric resistance to heat the glass. The current is applied to two electrodes, placed along two opposite sides. An aluminum-plated sheet metal reflector and a glass wool insulator are located on the back surface.

The permissible surface temperature depends on the type of glass used, 80°C for ordinary glazing glass, 150°C for hard glass and 300 to 400°C for special glasses such as pyrex, which is the most widely used.

These elements give powers of 1,300 and 2,500 W for effective adjacent surfaces of 900 and 2,500 cm$^2$ respectively (1 to 1.5 W/cm$^2$). However, with pyrex emitting at 400°C, it is possible to obtain up to 2 W/cm$^2$. For the operating temperatures of these emitters, the emissivity coefficient of glass is 0.9 to 0.95, and therefore close to that of a black body.

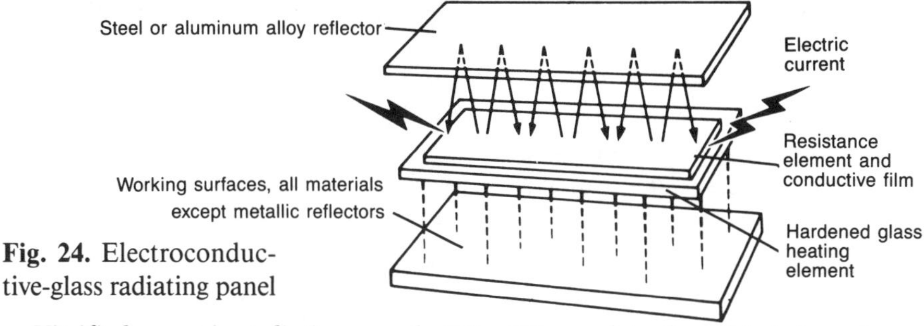


**Fig. 24.** Electroconductive-glass radiating panel

*Vitrified ceramic radiating panels* (*Fig.* 25) consist of a nickel-chromium resistance embedded in ceramic, coated with a special enamel. The maximum permissible surface temperature for these elements is around 700°C, but is normally between 400 and 600°C.

These emitters, which may be curved or flat, are available as rectangles or squares, with power ranges of 100 to 1,000 W for effective surface areas varying from 50 to 150 $cm^2$. Also circular lamp-like elements exist.

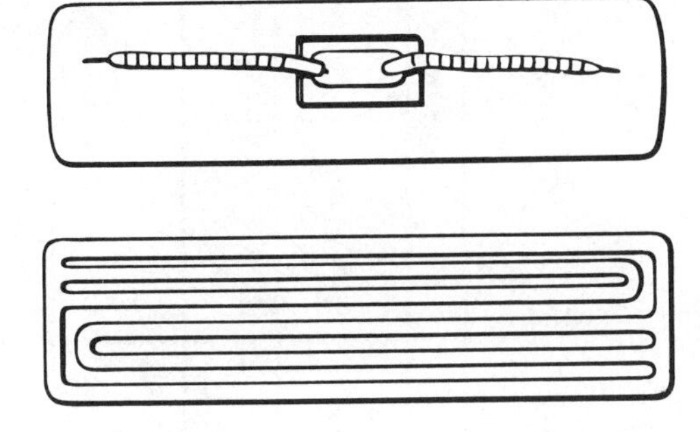

**Fig. 25.** Vitreous ceramic radiating emitter.

### *3.3.2. Properties of long infrared emitters*

*Electroconductive-glass radiating panels* emit radiation whose maximum monochromatic emittance wavelength is around 4.5 $\mu$m. They provide power densities of between 15 and 30 kW/$m^2$. Thermal inertia is high, approximately 4 minutes for power to be reduced by 90%. These emitters have service lives of several years, since almost half the energy is transferred by convection.

Depending on operating temperature (300 to 700°C), *vitreous ceramic radiating panels* offer maximum emission between 3 and 5$\mu$m. The emitted power is between 15 and 40 kW/$m^2$. Thermal inertia with these panels is slightly higher than with electroconductive-glass emitters, while the service life is similar.

Long infrared emitters offer relatively low power densities, but are 2 to 4 times higher than with convection heating, almost half of the energy being transferred by convection, which is good for some applications. Thermal inertia is very high, and for continuous products, precautions must be taken to prevent damage to the product in the event of a shutdown (lifting, transfer or rotation of panels).

Conversely, they are rugged, easily installed emitters, with adequate resistance to corrosion, and can be used for panels radiating mild and homogeneous heat. Long infrared radiation is generally well absorbed since numerous products have preferential absorption lines in the long infrared (see *Fig.* 9); this is surface absorption, and heat sensitive products must be protected against superficial burning by avoiding excessive power densities. These emitters are suitable for treating thin products.

## 3.4. Table of characteristics of infrared emitters

Figure 26 lists the main characteristics of infrared emitters in their most common applications.

Principal characteristics of infrared emitters

| Radiation range | Short IR | | Medium IR | | | Long IR | |
|---|---|---|---|---|---|---|---|
| Emitter types | Lamps | Quartz tubes | Quartz tubes | Silica radiating panels | Melallic radiating panels | Pyrex radiating panels | Ceramic elements |
| Operating temperature (°C) | 2,000 | 2,200 | 1,050 | 650 | 750 | 350 | 300 to 700 |
| Wavelength corresponding to maximum monochromatic emittance ($\mu$m) | 1.4 | 1.2 | 2.2 | 3.5 | 2.8 | 4.6 | 3 to 5 |
| Maximum installed power density ($kW/m^2$) | 10 | 300 | 70 | 25 | 40 | 15 | 40 |
| Thermal inertia corresponding to 90% reduction in emitted energy | 1 second | 1 second | 30 seconds | 2 minutes | 2 minutes | 5 minutes | 5 minutes |

**Fig. 26**

Principal charasteristics of infrared emitters

| Radiation range | Short IR | | Medium IR | | | Long IR | |
|---|---|---|---|---|---|---|---|
| Average service life (h) | 3,000 | 5,000 | 5,000 to 10,000 | Several years | Several years | Several years | Several years |
| Energy transferred by direct radiation [1] as percentage of emitted energy | 75 | 80 | 60 | 50 | 55 | 45 | 50 |
| Maximum product temperature [2] (°C) | 300 | 600 | 500 | 450 | 400 | 250 | 500 |
| Radiation reflected by products | High (requires insulation) | | Low | | | Low | |
| Radiation penetration | Adequate | | Average | | | Low | |
| Treatment quality | Homogeneous heating appropriate for thick products (a few millimeters) | | Relatively superficial heating Radiation well absorbed by wet products | | | Superficial heating appropriate for thin products | |

[1] This percentage varies with the importance accorded to forced convection.
[2] In normal operation

**Fig. 26.** (Continued)

## 4. AN INFRARED INSTALLATION

### 4.1. Reflectors

Different types of external reflectors (*Fig.* 27) can be used to increase the efficiency and adaptability of infrared emitters: flat, parabolic, elliptical, single or multiple reflectors. Thet enable concentration of radiation onto an area, a line or a point. Generally, reflectors are made from special aluminum or stainless steel, or, in special cases, ceramic. To prevent decreases in efficiency, they must be cleaned periodically. When the energy flux produced is high (for power densities greater than 80 kW/m$^2$), reflectors are either air- or water-cooled.

Emitters with built-in reflectors also exist. The backs of these reflectors are equipped with a metallic film, generally gold, not requiring maintenance.

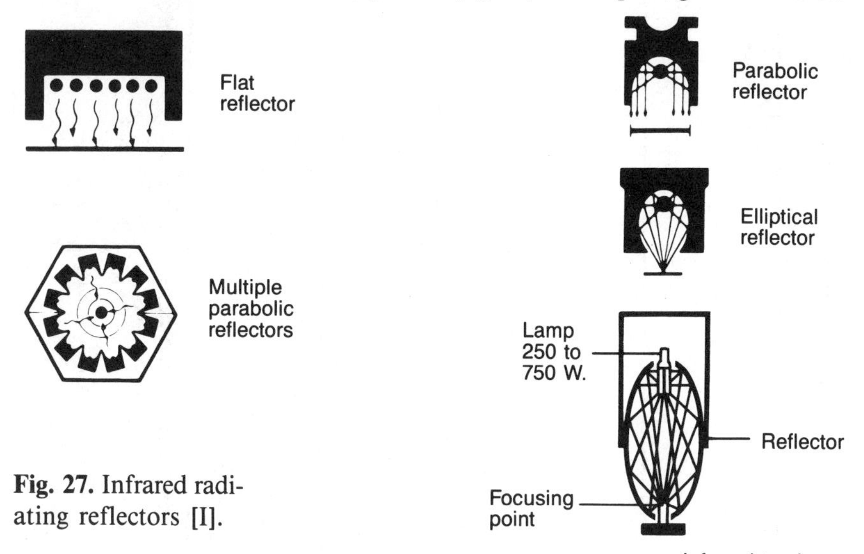


**Fig. 27.** Infrared radiating reflectors [I].

### 4.2. Radiating panels having multiple sources

The emitters can be arranged individually or built into radiating panels combining several sources. The radiating panels consist of lamps, quartz and metallic tubes, or ceramic elements (*Fig.* 28). Panels may be articulated to orient radiation onto the parts to be treated, and placed side by side to obtain large radiant heat surfaces (arches, hearths, walls). In some cases, emitters can also be bias-mounted with respect to product flow direction.

### 4.3. Infrared radiation furnaces

#### *4.3.1. Furnace types*

Depending on applications, the emitters are either open, which is rare since efficiency is low (*see* paragraph 4.3.2.), or contained in thermally insulated enclosures. The products treated are generally in the form of continuous belts or layers, or repetitive in shape and size; the furnaces most widely used are continuous tunnel furnaces (*Fig.* 29).

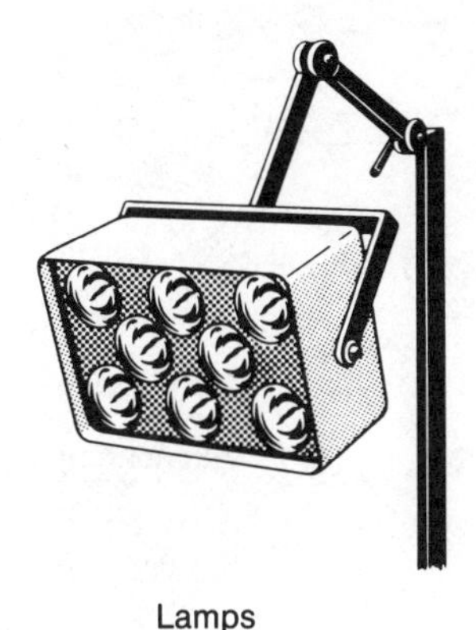

Lamps

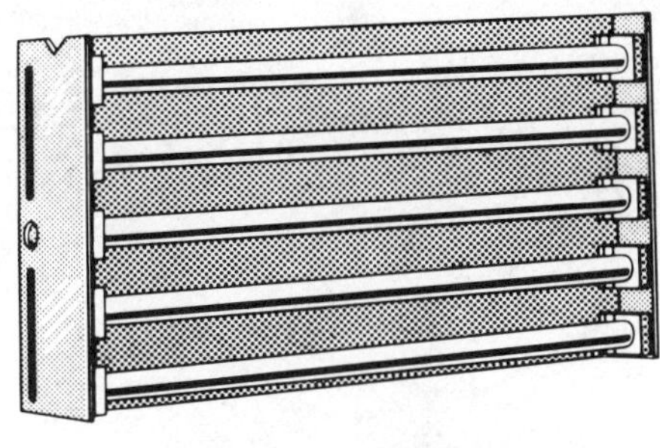

Quartz or metallic tubes

Ceramic elements

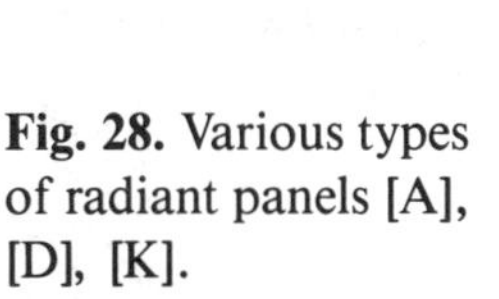

**Fig. 28.** Various types of radiant panels [A], [D], [K].

The products to be treated are conveyed on a roller, chain or rail conveyor (*Fig.* 29). In some cases, the product to be treated is the conveyor (textile, paper, plastic, polygraphic industries) in the form of a roller conveyor.

To obtain the required temperature-versus-time profile in the product (temperature rise, holding, cooling, etc.), the installed power density generally varies along the tunnel.

### *4.3.2. Insulation of infrared radiation furnaces*

The efficiency of heating systems consisting of simple side by side open air sources is low. This drawback applies both to the medium and long, as well as the short infrared, but for different reasons.

With short infrared, the energy transferred by convection is low and does not cause problems. Conversely, a large amount of radiation can be reflected by slightly absorbent bodies, and therefore lost. This defect can be partially corrected by fitting sheet aluminum reflecting inside walls (*Fig.* 30). However, this reflector will accumulate dirt and become less efficient, and part of the energy may be absorbed by it, and therefore lost.

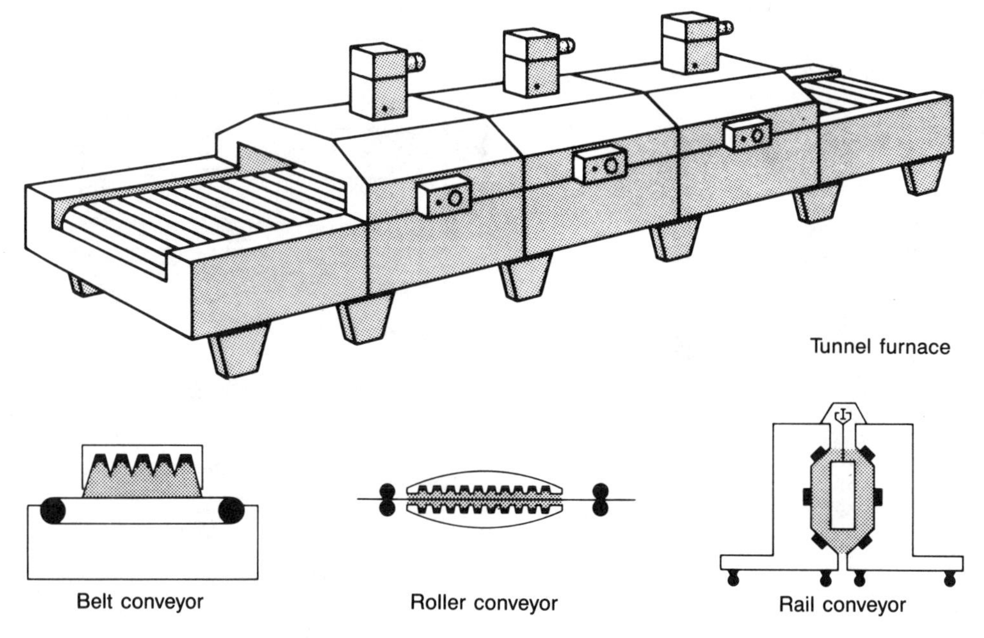


**Fig. 29.** Infrared radiation furnaces [28], [D].

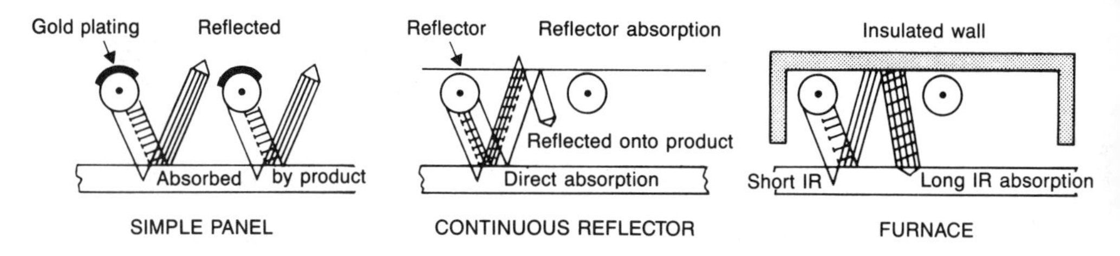


**Fig. 30.** Insulation of infrared radiation furnaces [22].

This is solved by using an insulated furnace (*Fig.* 30). The walls absorb the radiation reflected from the product. Since they are thermally insulated, the walls heat superficially and emit secondary radiation, mostly in the long infrared, onto the product. The effective radiation is therefore distributed according to the source emission spectrum and reflection from the walls. For a given furnace, power distribution between these two methods can vary from one product to another. This system enables short infrared, high power density, low inertia sources to be used, even for products which are only slightly absorbent in the short infrared.

It is possible to regulate the power transmitted directly in the short infrared. However, overall system inertia is increased, which may call for the use of safety devices in the event of conveyor shutdown. Moreover, if lamps are used, they may have to operate at very high ambient temperatures, 250 to 350°C, for which they are not always designed, calling for limitations in power density.

With *medium or long infrared*, radiation reflection is relatively limited and does not call for the use of special arrangements. Conversely, convection is high, and, here again, insulated furnaces must be used.

### *4.3.3. Infrared radiation furnace ventilation*

In many infrared radiation applications, vapors — water, solvents, etc. — with which the furnace atmosphere becomes charged, must be vented. In such cases, and for both elimination process and installation and personnel safety reasons (atmospheric saturation, radiation absorption and diffusion, risks of fire, toxicity, etc.), it is necessary to ventilate the furnace atmosphere.

However, in most cases, it is recommended that atmosphere renewal should be at least the minimum required to comply with hygiene and safety regulations and correct operation of the elimination process. Infrared radiation offers advantages in this respect, since the furnace atmosphere is heated only moderately, and losses due to ventilation are therefore relatively low. For example, in metal organic coating curing ovens operating at temperatures of between 150 and 200°C, the air temperature with short or medium infrared ovens is only 100°C approximately but between 340 and 450°C in convection ovens. The energy consumption required to ventilate solvent-charged air is therefore divided by a factor of more than 3.

Some of the energy lost in the extracted air can also be recovered either through an exchanger, or by channeling this air to the furnace entry to preheat incoming products.

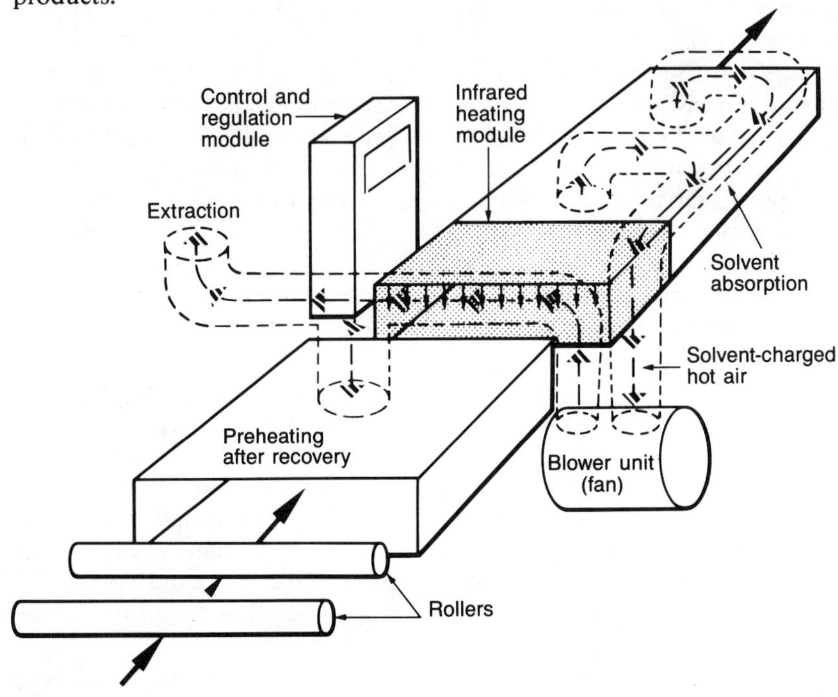


**Fig. 31.** Forced ventilation infrared furnace [I].

Fresh air can also be used for cooling emitters and reflectors in high power density furnaces and can supplement product heating by convection, which can cut energy losses.

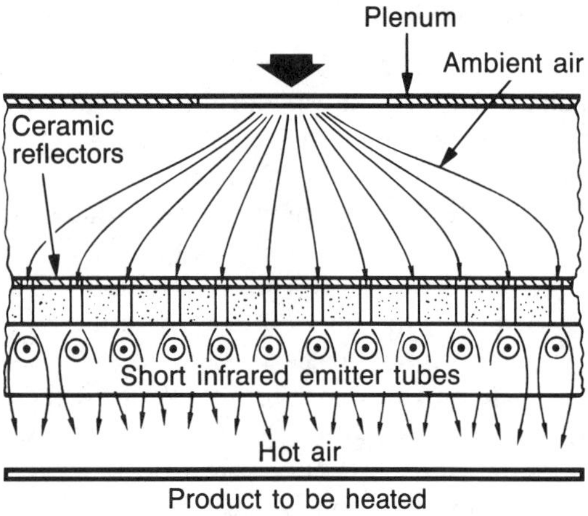

**Fig. 32.** Fresh-air-cooled short-infrared emitters with ceramic reflectors [I].

### *4.3.4. Regulation*

Measurement of the temperature of products on furnace conveyors is often delicate and costly. Measurement of the ambient air temperature in an infrared furnace would be of little use, and insertion of a probe in the infrared ray path would only provide indications of inherent absorption characteristics and temperature.

For reasons of facility and economy, temperature regulation methods using batteries of infrared emitter areas connected in series or parallel, or a combination of both, are used to obtain the required thermal profile previously determined by a combination of calculation and experimentation. With three-phase power, it is also possible to use a star-delta arrangement to provide two radiation levels, therefore minimizing starting constraints in lamp furnaces. With continuous furnaces, adjustment of speed and therefore product exposure times are supplementary temperature control methods. Also, in some cases, it is possible to vary the distance separating the products from the emitters, but this type of power adjustment is inefficient.

However, if it is absolutely necessary to constantly monitor temperature and radiated power, the regulation system is slaved to a product temperature measurement system consisting of a flexible thermocouple in contact with the product. When greater accuracy is required, which does not apply to most industrial processes, radiation pyrometers must be used; however, it is costly and is used essentially in laboratories. In such cases, the regulation system consists either of a dual emitter power supply system, or of activating the emitters for a variable percentage of cycle time only (cam system equipped electromagnetic relay).

The use of static pulsed multiple cycle regulation systems (thyristor regulation) is being developed. With such systems, regulation efficiency increases as source thermal inertia decreases.

For high thermal inertia emitters, cyclic power regulation is equivalent to lowering the source operating temperature, thus altering the emission spectrum and therefore product absorption. Since long-infrared emitters (where convection plays an important role and absorption is less sensitive to wavelength) are generally used, this phenomenon involves few drawbacks.

Short infrared does not cause problems *due to conveyor shutdown*, since thermal inertia is low: about one second. Conversely, for long and medium infrared emitters, separating devices (shielding system) must be provided for heat sensitive products.

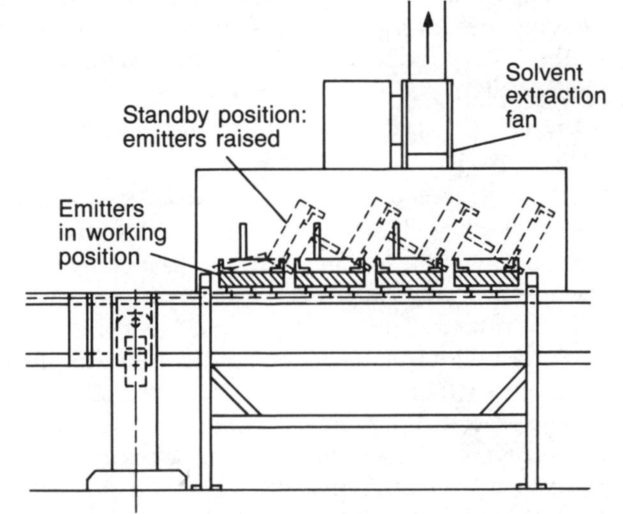


**Fig. 33.** Medium infrared radiation furnace (quartz panels) with shielding system [S].

Where the products are of different widths, manual controls can be used to switch off some of the emitters.

## 4.4. Selection of infrared equipment

Many factors go into making the right choice of infrared equipment for a given industrial process.

### *4.4.1. Selection of emitters*

Installation design begins with an examination of the wavelengths required (including secondary radiation) throughout the various product treatment phases.

Basically, the choice depends on the chemical and physical properties defining the absorption factor of the product (temperature, humidity, thickness, color, surface condition).

Therefore, it is generally desirable that the wavelengths emitted be located within the product's maximum absorption range.

For example, water has a maximum absorption baud at about 3$\mu$m (see *Fig.* 9). Medium infrared, which provides high power densities, and which emits in this range, is particularly suited for drying processes since part of the primary radiation and all the secondary radiation are located in the long infrared where water has a second maximum absorption band around 6$\mu$m.

Many organic coatings also have maximum absorption bands in the medium infrared, as well as in the short infrared.

Nevertheless, factors other than optimum use of the emission spectrum of source and absorption spectrum of the product to be heated control the choice of emitters:

— frequent variation in product color, simultaneous presence of products of different colors or the use of coatings with very different absorption characteristics (e.g. simultaneous baking of metallized or non-metallized paints) favor longer wavelength emitters (medium and long infrared), since its absorption is less selective for high than for short wavelengths; moreover, an important part of the energy is transferred by convection, offering better temperature homogeneity;

— irregularly shaped products are treated more homogeneously by long infrared where convection is substantial;

— very high power densities (rate increase for existing dryer, equipment thermal tests, etc.) are easily obtained with short infrared;

— sensitivity of the product treated to heat presents two very different options. The first, involving short infrared heating with moderate power density, provides better distribution of the energy applied to a product mass and reduces the risk of overheating in the event of sudden conveyor shutdown, due to adequate radiation penetration. The second, involving long infrared heating at low power densities and with a high rate of convection, guarantees mild homogeneous heating, but calls for a source shielding device in the event of conveyor shutdown. If very high, the load thermal diffusivity also affects the choice of solution, since the drawback with only surface radiation absorption (long infrared) will be minimized, and, if low, penetrating radiation (short infrared) will be more efficient;

— risk of destruction of emitters due to mechanical shock (treatment of solids, high-speed conveyors, etc.) or thermal shock (projection of fluids onto emitters) favors the use of more rugged emitters (long and medium infrared), but a well-designed furnace protects emitters;

— continuous treatment of variable width sheets requires easily dividable emitters, in which an abundance of radiation prevents energy wastage (short or medium infrared).

The factors to be taken into account are many, and any project analysis must begin with a thorough examination of these factors.

### *4.4.2. Power to be installed*

Knowing the hourly production rate and the successive temperatures required (preheating, heating, holding, etc.), the electrical powers required for each section of the furnace can be calculated. These powers are calculated from the mass of product to be treated, specific and latent heats of the product components, the heat of reaction, if transformations of a chemical nature occur (endothermic or exothermic), the ventilation rate and possible recoveries, thermal diffusivity of the load, exchanges due to environmental convection and thermal furnace losses.

Having chosen the type or types of emitters (emitters of different types can be used in successive areas of the furnace), it is possible to determine power density for the various parts of the furnace, and also, knowing the product width, it is possible to determine the length of each section and total overall furnace dimensions. It is then necessary to select the regulation, ventilation and safety devices.

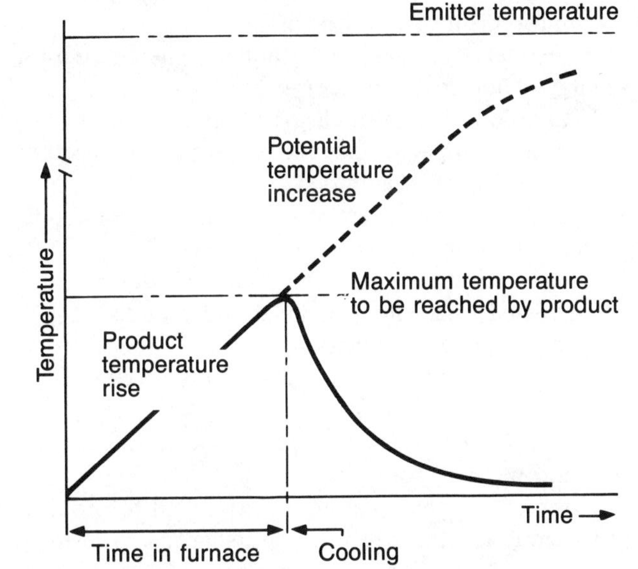


**Fig. 34.** Temperature rise in an infrared furnace

If calculations are necessary to determine an order of magnitude for the installed power and to ensure that the choices are suitable, one must also conduct tests on pilot installations, for both emitter choice and electrical power evaluation.

Numerous research centers with infrared equipment can provide this service (professional research centers, university laboratories, Electricité de France research centers, etc.), and most constructors can also conduct these tests. Also, several small simulators with interchangeable emitter panels exist which enable enterprises to perform in-field tests directly before having to do a more costly pilot.

## 5. ADVANTAGES AND DISADVANTAGES OF INFRARED RADIATION IN INDUSTRIAL PROCESSES

Infrared radiation in industry has the following advantages:
— direct transfer of thermal heat to product, according to the laws of optics;
— low thermal inertia and high temperature rise;
— the possibility of high power densities and fast treatment;

— heating homogeneity due to radiation penetration;
— performance of operations which are more difficult or impossible with other methods;
— easy addition to another heating process.

Wise use of these characterisics provides certain advantages over competing processes (forced convection furnaces, conduction heating cylinders):
— very good heat transfer accuracy;
— high productivity;
— reduction in overall furnace dimensions;
— improvement in product quality;
— investment cost gains and, under some conditions, operational cost gains (energy, labor, maintenance).

In addition, the use of electrical infrared as an energy source presents other advantages compared to fossil-fueled infrared heating:
— a wider range of emitters;
— adequate overall operation efficiency (often about 70%);
— regulation and adjustment flexibility;
— almost instantaneous startup and shutdown;
— increased safety for personnel and products;
— constructional simplicity;
— minimum maintenance requirements;
— improvement in power factor;
— no contamination or pollution by heating source;
— good working conditions.

However, infrared heating is best suited to products in layers or sheets, or sufficiently repetitive in form and shape; it is of interest only in applications where conventional resistance heating, which generally costs less, is difficult to implement or results in lower performance.

## 6. INDUSTRIAL APPLICATIONS OF INFRARED RADIATION

The use of electrically produced infrared radiation in industry is not new. Since its effect on materials is thermal, it is used in many applications such as drying, firing, heating, polymerization and sterilization. Infrared radiation is particularly suitable for treatment of thin film products which can be exposed over almost all their surface.

Industrial applications are extremely diverse. The list below gives just a few examples [1], [2], [4], [5], [7], [9], [13], [17], [22].

*Drying and firing, polymerization of coatings on various supports:*

— Paints and varnishes on metal, wood, glass and paper;
— coatings on leather and hides;
— dyes and primers on fabric;
— latex coverings on carpeting;
— PVC coatings on fabric;
— coatings and layers on paper;

— printing inks;
— teflon coatings on cooking utensils;
— silkscreen printing;
— gluing in shoemaking;
— manufacture of electronic components.

*Dehydration and partial drying:*

— papers, cardboards, textiles;
— ceramics;
— casting moulds and cores;
— metal parts after washing, rinsing and pickling;
— water paints and inks;
— tobacco leaves;
— plastic grains;
— bottles containing condensation;
— pharmaceutical products.

*Miscellaneous heating:*

— heating of plastics and glass prior to forming;
— heat treatment of metals;
— heat shrinkage for plastic packaging;
— enamel firing;
— welding and brazing;
— glass annealing;
— tin melting;
— preheating sheet prior to shot blasting;
— drying wood panels prior to coating;
— heating prior to assembly;
— cooking or grilling foodstuffs;
— pasteurization and sterilization of liquid foodstuffs;
— stabilization of packed foodstuffs;
— thawing aggregates;
— heating work stations;
— sterilization of pharmaceutical bottles.

Almost all industrial sectors use some application of infrared radiation.

## 6.1. Metal transforming, mechanical, electrical and electronic industries

There are many current or potential uses for infrared radiation in these fields.

### *6.1.1. Firing paint on metal supports*

Numerous metal products — automobiles, metal furniture, electrical convectors, toys, fire extinguishers, prefabricated building products, bottling, etc. — are coated with one or several layers of paint.

This provides a double function: protect products from corrosion while providing a more appealing look.

Whether the coating is liquid or powder, it must be baked at temperatures between 150 and 200°C to obtain its final qualities.

Infrared radiation, combined, in more complex cases, with convection, is very good for this operation, as shown by two examples chosen from the automobile industry and flat-product finishing industries [8], [11], [12], [16], [17], [21], [24], [25], [26], [29], [E], [M], [N], [Q], [T].

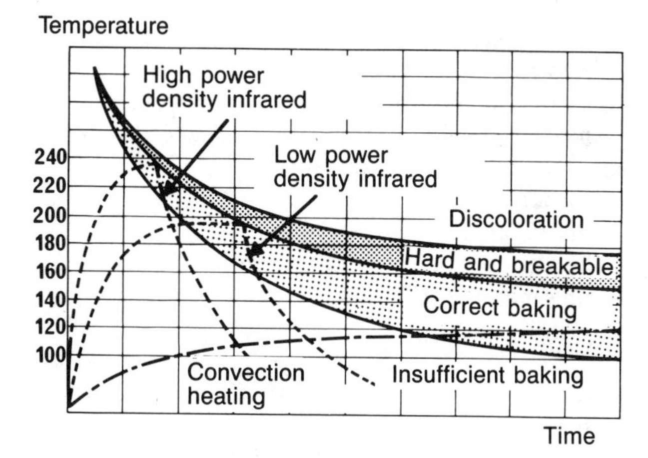


**Fig. 35.** Baking paint with infrared

In these applications, infrared radiation is attractive primarily because of:

— curing rate;

— compactness of equipment;

— almost immediate protection from dust and in-depth baking;

— energy efficiency;

— investment and operation costs generally lower than with hot air systems;

— absence of excessive air drafts (powder paints).

Conversely, it is difficult to bake coatings on complex shapes using infrared, although the thermal conductivity of the metal support favor even temperatures. This problem is resolved by using combination infrared/convection ovens.

Safety problems related to paints containing solvents must be carefully studied. It is generally recommended 60 to 100 m$^3$ of fresh air be introduced at 16°C into the oven for each liter of solvent evaporated. Finally, the use of infrared sometimes modifies paint structure.

### *6.1.1.1. Automobile industry*

Infrared radiation is used in the automobile industry for baking of primers and finishing paints and reworking of bodywork paint. Infrared baking does not only interest body builders. Numerous other automobile components (for example, wheel rims) can be painted, and these coatings can be baked using infrared radiation in ovens which are generally much more simple in design than those used for bodywork.

*Baking of primers and finishing paints*

Infrared radiation is excellent for protecting paints from dust. This operation consists of "seizing" the paint film immediately after application, so that its surface is stretched rapidly, fixed and hard enough to prevent incorporation of dust and formation of "grains".

The advantages of infrared radiation in these cases are:

— *rate* of temperature rise, which is two minutes for an ambient temperature of 150°C;

— *cleanliness*, since air circulation is reduced to the minimum required for evacuation of solvents;

— *smaller holding areas* due to rapid drying.

The thinness of the final coat of paint (20 to 50 $\mu$m) and penetration of radiation enhances energy transfer to the sheet metal which then transfers its heat to the paint coating. This process speeds the evaporation of solvents and, due to the sheet metal's thermal conductivity, temperature equalization. Installed power is about 20 to 100 kW/m of tunnel.

However, in more modern installations, the trend is to use combined infrared and convection tunnel furnaces, since direct radiation cannot reach all points of an automobile body. The infrared temperature rise section, in which convection is used only to evaporate solvents, is therefore followed by a convection — only section in which the temperature is equalized and baking takes place at a more or less constant temperature (the most widely used paints, acrylics or glyceropthalics, are baked at about 160°C, and baking time is highly sensitive to temperature: an increase of 5°C in baking temperature can decrease baking duration by a factor of 3 or 4).

In some baking tunnels, the infrared section has even been abandoned for primers only, in which the quality problem is not presented in the same manner. This trend is reversing, however, since infrared enables a rapid rise in temperature, therefore shortening of installations, increase in production rates and reduced energy consumption.

Short infrared, which enables excellent penetration and high power density, is the most widely used. Lamps are being abandoned for tubes which are easier to fit and less costly. Absorption of short infrared by paints is, however, color sensitive, as can be seen from the following color absorption factor mean values:

— black: 0.86 — red: 0.59
— blue: 0.62 — white: 0.48.

The importance of this phenomenon is lessened due to reflection of radiation by the oven walls and re-emission in the long infrared, in which absorption is only slightly sensitive to color.

Medium infrareds in crystal or metal tubes, whose use up to now has been limited, have also given adequate results.

Starting of the emitters is generally coupled to conveyor operation and ventilation without complex regulation; they are often combined so that power can be switched off as required, or operate at low power in the standby position [A], [12], [N], [P], [T].

*Retouch baking*

Retouch baking is provided by:

— emitting panels, generally of the lamp or quartz tube type, as described in paragraph 4.2, 20-30 cm from the parts to be baked; they can be adjusted so that only the re-worked part is treated (K), (H);

— tunnel furnaces, using short, medium or even long infrared, enable production line work. The three types of emitters can be used, since temperatures lower (about 100°C) than with paint baking are required, but longer duration than for polymerization. It is necessary to avoid damage to the internal coatings and accessories of finished products.

### *6.1.1.2. Flat-product finishing industry*

Prepainting of sheet (coil-coating) and, in general, finishing of flat products, are infrared radiation applications which, unlike North America, are little used in Europe [17], [21], [24], [29], [I], [Q].

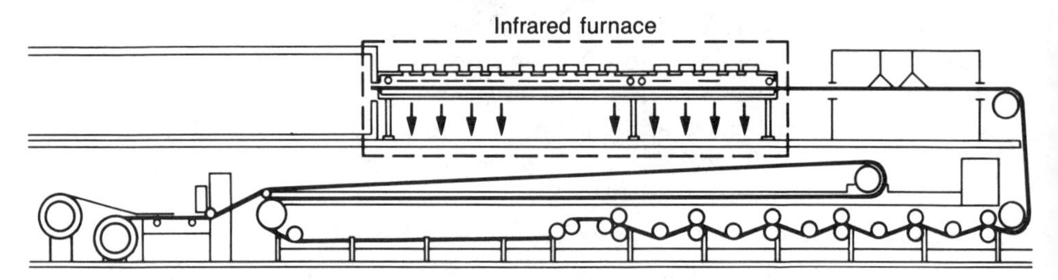


**Fig. 36.** Diagram of a prepainting line with air cooled infrared emitters [I].

Infrared radiation meets some of the requirements of this industry:

— power densities increase line speeds and reduce furnace lengths;

— heat losses by extraction of solvent-charged air are greatly reduced compared to forced convection furnaces.

According to an American study, the baking times with a high power density furnace combining quartz-tube-produced short infrared (300 kW/m$^2$) and forced convection (emitters and their ceramic reflectors cooled with fresh air, *see* paragraph 4.4.3.) are in fact very low [24].

The air flow varies from 2.8 to 5.6 m$^3$/min, for 10 dm$^2$ of reflector surface which is heated to a temperature of 100°C. The energy consumption would be 0.2 to 0.6 kWh/m$^2$ of product treated, and between 1.5 and 2.5 kWh/m$^2$ for the standard gas heated convection furnaces.

For example, in an American plant, a 4 m furnace with an effective heating width of 1.5 m is used to finish coatings of 0.2 mm on metal strip 1.25 m wide and 0.8 m thick, at a rate of 100 m/min. The polymerization time is about 2.5 seconds, and specific overall consumption is between 0.2 and 0.3 kWh/m$^2$ (*Fig.* 37).

### *6.1.2. Miscellaneous applications*

While paint baking is the most widely used application of infrared radiation in the metal transforming industries, there are many other applications:
— drying iron oxide (magnetite) for the manufacture of recording tapes.

Curing times for organic coatings on metal tapes in high power density infrared furnace

| Coating type | Base support (thickness in mm) | Layer ($\mu$m) | Baking time (s) |
|---|---|---|---|
| Single undercoat | Aluminum 0.63 | 5 dry | 2.5 |
| Water based primer | Aluminum 0.63 | 13 dry | 4 |
| Epoxy primer | Steel 0.63 | 5 dry | 4 |
| Final polyester coat | Steel 0.9 | 76 wet | 5 |
| Final coat silicone polyester | Steel 0.63 | 63 wet | 5 |
| Water-based acrylic, final coating | Steel 0.63 | 23 dry | 6.6 |
| Final vinyl coating | Steel 0.63 | 63 wet | 4 |
| Final solid polyester coating | Steel 0.63 | 25 dry | 12.5 |
| Electrocoat paint | Steel 0.16 | 8 dry | 1 |
| Epoxy (decorative) | Aluminum 0.51 | 33 wet silkscreen | 10 |
| Teflon | Steel 1.6 | 102 wet | 15 |

**Fig. 37.**

— polymerization of organic coatings on cooking utensils;
— drying parts prior to packing;
— heating parts prior to coating;
— curing of insulating products on electrical equipment;
— baking varnish;
— preheating parts prior to shot-blasting, welding, stamping;
— maintaining tool temperatures;
— heat treatments;
— drying parts after washing;
— heating glazing machines;
— preheating wires;
— preheating solder;

— heating casting moulds;
— baking sand cores;
— stamping die temperature control;
— polymerization and drying in the electronics industries.

## 6.2. Textile industries

Two methods of using infrared radiation are seen in this sector:
— as a complement to normal installations (train, hot flue, drum) to increase production or energy efficiency;
— for complete treatment of products.

### *6.2.1. Predrying fabrics after dying or sizing*

At the start of drying, the efficiency of most equipment is less than average. Efficiency and productivity can be increased by using infrared radiation.

Experience shows that medium infrared is best suited to this application, since it provides high efficiencies of up to 75% and good product quality. The difference with short infrared is, however, slight, and may be preferred if thermal degradation risks are high.

### *6.2.2. Heat treatments*

There are many forms of heat treatment: thermosetting reactive dyes, setting plastosoluble dyes on fibre polyester, thermosetting synthetic fabrics, polymerization of pigment setting resins, gelifying of coatings, setting latex foams to the backs of carpeting, drying sized layers of threads. In many cases, infrared can be used for the complete treatment with good energy efficiency and quality.

For example, a short infrared radiation oven, used for gelling PVC coatings on fabric, has replaced the forced convection oven, while reducing energy consumption by a factor of 5. Similarly, a medium 540 kW infrared furnace using metal radiant tubes can be used to thermoset latex foam on the back of carpeting at a temperature of 135°C at a rate of 7 meters per minute.

A thorough analysis, often involving prototype tests, is sometimes necessary to determine the best method [1], [4], [6], [10], [18], [22], [31].

## 6.3. Paper-cardboard (pasteboard) industries

In the paper industries, drying problems are encountered at all stages: fabrication, coating, printing, gluing, bonding, varnishing. This applies to both printing paper and cardboard, and packing paper and wallpaper [4], [15], [22], [28], [30], [31].

### *6.3.1. Paper-cardboard production*

While complete infrared radiation drying is not feasible, except for special papers produced in small quantities, it is an excellent alternative to the inherent problems of drying-cylinder heating in paper machines.

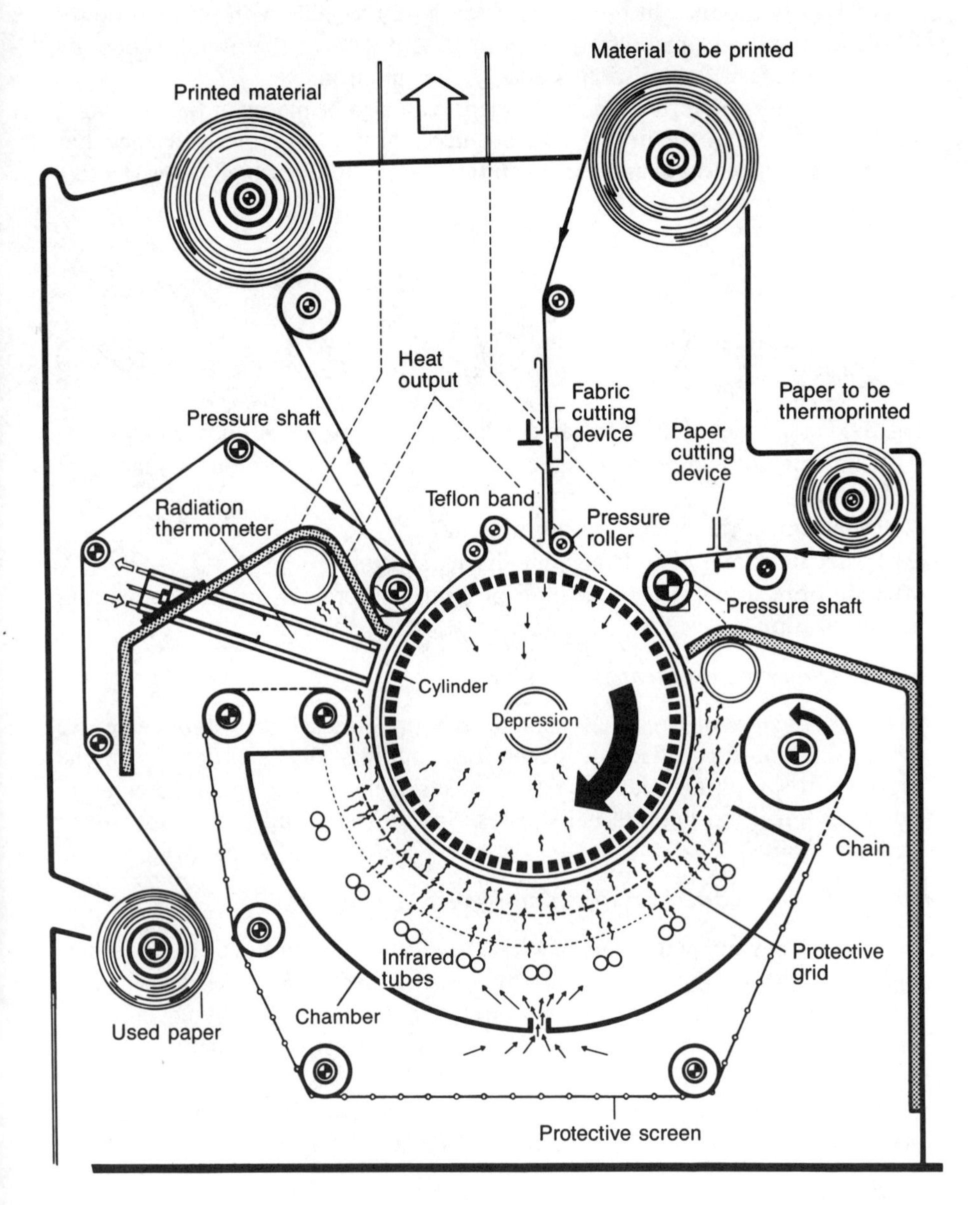


**Fig. 38.** Diagram of a textile thermo-printer machine with infrared heating [V].

### *6.3.1.1. Post-drying room drying to increase production*

Addition of an infrared compartment is an interesting solution to thermal conductivity problems encountered with dry pastes since the heavy paper is readily penetrated especially with short infrared.

The hourly production rate is increased by 20 to 40% with infrared power whose evaporation capability is between 15 and 20% of the overall capability of the dryer itself, and also cuts energy consumption.

Short or medium infrared compartments can also be placed at the beginning or in the middle of the drying line, not necessarily to improve efficiency, but to reduce machine length and increase productivity with their high power density.

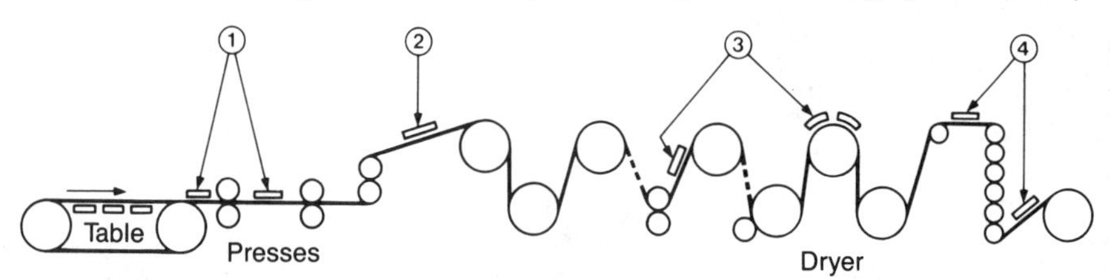


**Fig. 39.** Possibilities of using infrared radiation in the paper-cardboard industries:
(1) Reheating cold paste to decrease viscosity and to increase water removal at the presses
(2) Dryer entry: rapid temperature buildup
(3) In drying room, complementary drying stations
(4) Exit from drying room: heating prior to smoothing, dryness correction, polymerization . . . . . . . . . . . . . . . . . . . . . . . . . . . . . . . . . . . . . . . . . .

### *6.3.1.2. Obtaining excellent flatness*

The use of short infrared, traversing water vapor atmospheres without excessive absorption but with adequate penetration permits excellent flatness in the presence of saturating steam. It thus protects surface pores against premature closure, while opposing internal stresses. Subsequent straightening operations are thus avoided.

### *6.3.1.3. Dryness correction to obtain regular terminal profile*

Dielectric drying (high frequency) can be excellent at the end of drying, since high frequency energy is preferentially absorbed by water molecules.

However, it has been proposed that infrared radiation be used by slaving small lamps to a constant measurement of humidity. Selective illumination by the sources would correct humidity in small areas.

The rewetting operation would then be eliminated, with substantial energy savings. This process has yet to be applied in industry.

### *6.3.1.4. Other applications*

Other applications are also possible:
— preheating between wet presses to lower the viscosity coefficient of the water;
— predrying films;
— booster heating for glazing machines;
— drying thick, "rolled" cardboard (consumption 1 kWh/kg of evaporated water, drying time three times lower than with forced convection, adequate flatness without straightening);

— drying of swelling-texture paper (filter paper) in which steam cylinder compression is to be avoided during drying along with thermal shock on the cylinder (energy consumption 1 kWh/kg of evaporated water, drying two times faster than with forced convection);
— production of non-wovens on paper machine (compared to hot air, energy gains of 30 to 50%, three to four times faster heating in order to complete polymerization reaction after drying agglomerate products, six to ten times faster to obtain softening, self-bonding effects).

*6.3.2. Paper transformation*

Paper transformation operations are extremely diverse, but infrared ovens can be used for total or partial heating of products.

*6.3.2.1. Drying films and glazes*

The benefits of infrared increase with temperature, since hot air or steam temperatures (and therefore pressure) must also be high, complicating installations and accentuating radiation efficiency.

*6.3.2.2. Production of complex backings*

Drying backing by infrared consists of using several sheets, often combined with different materials (metal, plastic, etc.) enabling double or triple speeds to be obtained. Pre-drying some components can also be done with this technique.

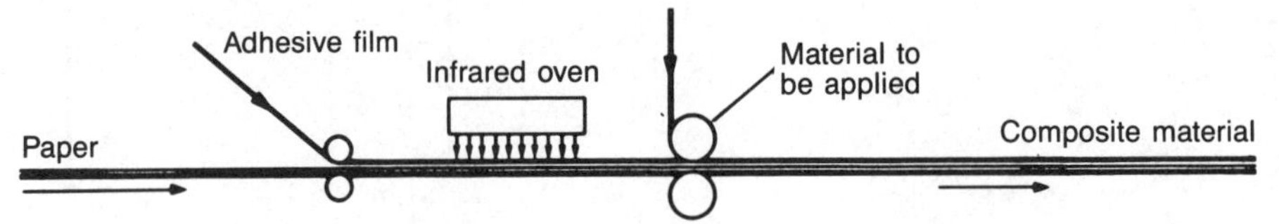


**Fig. 40.** Diagram of a complex back coating installation with adhesive film heating by infrared radiation [I].

*6.3.2.3. Other applications*

Infrared radiation can be used for other applications:
— polymerization of coatings;
— drying coatings on wallpapers.

**6.4. Graphics industries**

Infrared drying of inks is used mostly in offset printing and silkscreening [27], [30], [31].

Increases in printing speeds with modern presses cause problems in staining which infrared dryers have solved.

The dryer generally consists of a panel of short and sometimes medium infrared quartz tubes, producing high power densities and located just before the press stacker, to ensure rapid transfer of heat to sheets moving towards the outlet.

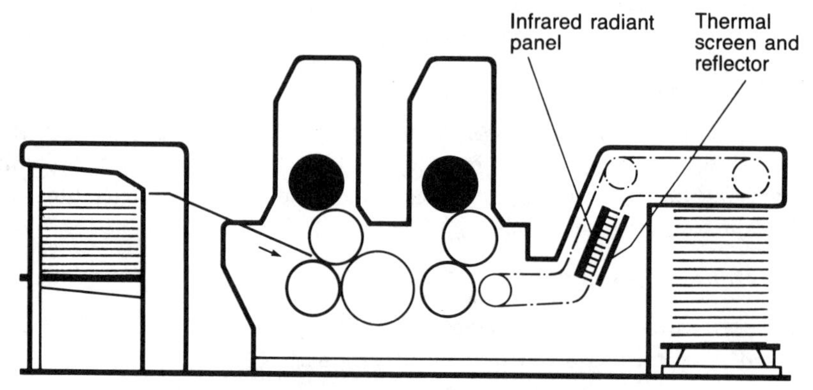


**Fig. 41.** Diagram of an infrared drying installation on a two-color offset press

An insulating screen is used as a thermal reflector to provide optimum energy efficiency. The tubes can be installed either in line with the sheets or transversely.

**Fig. 42.** Installed power as a function of paper width

| Width (m) | Power (kW) |
|---|---|
| .72 | 18 |
| 1.02 | 25 |
| 1.26 | 31 |
| 1.40 | 35 |
| 1.62 | 40 |

Infrared equipment power is, on average 30 kW for a printing rate of 6,000 sheets per hour but depends on paper width.

The use of infrared, which calls for inks which absorb radiation well and are slightly more expensive than ordinary inks, offers several advantages:

— the ink exposed to infrared radiation begins to dry before reaching the storage stack, which prevents the sheets from sticking to each other and blotting. Continuous drying for 15 to 30 minutes using the energy stored in the sheets (the temperature rise is 15 to 20°C when the sheet arrives at the stack). Drying time is therefore much shorter with thermal acceleration; otherwise it is several hours;

- the need for anti-smearing powder is reduced, or even eliminated, providing a cleaner and healthier atmosphere in the workshop;

— the presses can operate at optimum production speed, improving productivity;

— finally, quality is often improved due to powder brilliance and gloss.

In silkscreen printing, the supports can be dried rapidly in small, low power infrared tunnel ovens.

## 6.5. Food industries

As in other sectors, infrared radiation is used for a wide range of applications — cooking, pasteurization, sterilization, grilling, reheating, dehydration and preparation of packages. The examples given below illustrate the variety.

### *6.5.1. Roasting of hazelnuts*

Treatment lasts approximately half an hour, during which the nut is taken to a temperature of 320°C. An installed power of 180 kW enables production of six tons per day. The main advantages are the absence of product contamination, core heating, high efficiency, controllability and compactness of the installation [30].

### *6.5.2. Long storage treatment in industrial baking and pastry-making*

Some foodstuffs (breads, pastries, cooked foods, etc.), once cooked and packed under plastic film, can only be conserved after thermal sterilization which prevents growth of mildew during storage.

Short infrared radiation can be used to heat the product in complete safety, without absorption by the packaging. Treatment takes about 10 minutes, during which the core temperature for bread exceeds 80°C. An 80 kW oven can treat 400 kg of sandwich bread per hour. Specific consumption is about 0.25 kWh/kg of bread [30], [D].

### *6.5.3. Cookie and biscuit baking*

Infrared radiation can be used to bake cookies and biscuits. The specific consumption can be anywhere from 0.3 to 0.6 kWh/kg of final product [20].

### *6.5.4. Pasteurization and sterilization of liquid foodstuffs*

Infrared radiation can be used for heat treatment — pasteurization, sterilization, stabilization — of numerous liquid foodstuffs — milk, beer, wine, fruit juices, syrups, etc. [19].

The diagrams below show the general design of this type of installations.

Specific consumption varies according to treatment. For example, a 145 kW installation can produce 4,000 l/hr. of UHT sterilized milk [30], [U].

Similar equipment can also be used as hot water generators [U].

## 6.6. Leather industry

The various types of infrared emitters are being successfully used to polymerize coatings on skins and dry shoe soles [30], [H], [J], [R].

A long infrared oven, which was substituted for a fossil-fuel heated thermal fluid convection oven, has been used to reduce energy consumption by a factor of 5, which, with respect to primary energy, represents a saving of 50%.

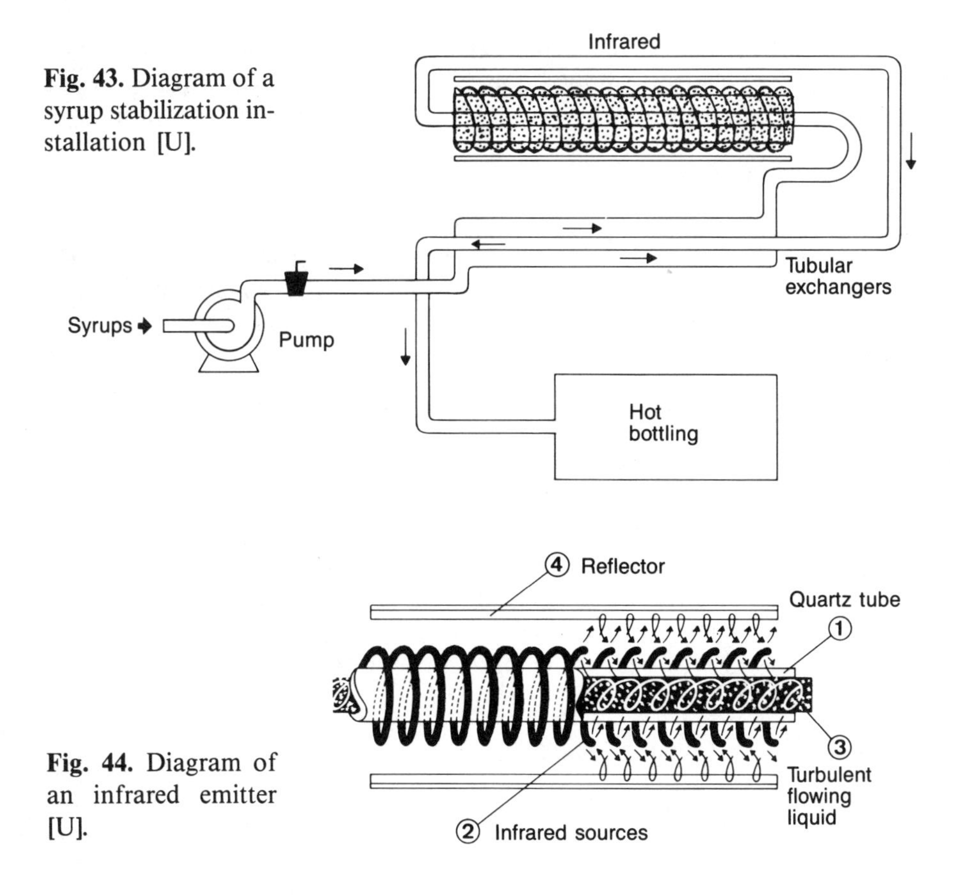


**Fig. 43.** Diagram of a syrup stabilization installation [U].

**Fig. 44.** Diagram of an infrared emitter [U].

### 6.7. Plastic and rubber transformation industries

Treatment of plastic materials using infrared radiation is widespread (heating prior to forming, drying, polyermization, etc.). Depending on the operations involved, temperatures vary between 80 and 180°C [6], [22], [30], [A], [O].

#### *6.7.1. Heat forming plastics*

Plastic sheets are reheated prior to bending, stamping or cutting, etc. Long infrared, and in particular ceramic emitters, are often used to provide mild homogeneous heating with limited power densities, about 15 kW/m$^2$ (medium or short infrared emitters can also be used if the power densities are comparable). Installed powers can, however, be high. A press intended for the forming of boats uses 1,600 ceramic emitters at 650 W, i.e. 1,040 kW, in two stations (preheating and heating) of a carousel which can form 15 tons of plaques per day.

Generally, infrared heating is preferred to hot air processes due to its better efficiency and adjustability, and to fossil fuel radiance for safety purposes. Electrical infrared can also be used for producing vacuums.

### *6.7.2. Polymerization of coatings*

Plastic coatings on various supports (textile, paper, etc.) are used to produce composite materials for a wide range of applications (clothing, tarpaulins, office equipment, leatherware, etc.). Infrared radiation can be profitably used for most polymerization and coating operations.

With a short infrared oven containing preheating, heating and holding sections, with a power of 115 kW, it is possible to produce 1,000 kg/hr. of product weighing a total of 900 g/m$^2$ (textile-plastic composite) with a thermal efficiency of around 65% (specific consumption of about 0.12 kWh/kg). By comparison, the overall thermal efficiency of hot air ovens performing the same operations is about 10%, the minimum specific consumption observed being 0.58 kWh/kg, ventilation not included. Primary energy economy with the infrared oven is 50%.

## 6.8. Construction materials industry

Infrared radiation has some applications in this sector: drying porcelain and ceramic, drying coatings, preheating prior to coating, anti-icing or warming of products [29], [30], [O].

### *6.8.1. Heating of asbestos-cement prior to coating*

Coloring of asbestos-cement calls for preheating before coating and drying. Preheating between 80 and 90°C of flat or corrugated panels can be obtained using medium infrared, in particular with metallic radiants, which are particularly rugged emitters of excellent efficiency, approximately 70% (specific consumption of about 0.20 kWh/m$^2$) and with adequate homogeneity.

The heat thus accumulated is restored after application for drying of the coating only requiring one infrared booster.

### *6.8.2. Curing of road surfaces*

This application is similar to those encountered in the plastic transformation industries. A insulated oven, fitted with short infrared lamps, radiating onto both surfaces of the product has enabled curing of a new type of PVC based, anti-skid road covering and tar-coated gravel coating with an efficiency of more than 60%.

In comparison, the efficiency of a fossil-fuel heated hot air oven did not exceed 15%, respective consumptions being 1.16 kWh/kg with forced convection and 0.2 kWh/kg with infrared.

## 6.9. Glass industry

There are several applications in this sector: firing paints and varnishes on mirrors, drying enamel, firing coatings, drying and production of glass-plastic composites (safety glass), preheating prior to press forming, preheating bottles, etc. [29], [30].

### *6.9.1. Drying and firing protective varnishes or paints on mirrors*

After silvering and coppering, the varnishes or paints deposited on the backs of mirrors can be baked by infrared radiation. Medium infrared, in the form of metal radiants, tubes or quartz panels, is frequently used.

### *6.9.2. Production of technical glasses*

Infrareds are used in several production stages, for example in the production of TV tubes, for drying slabs and cones after washing, preheating, drying of the various product coats (clay coatings, lacquers, graphite, iron oxide, etc.). In such products, the quality factor is fundamental.

## 6.10. Wood industries

Infrared radiation can be used for preheating panels prior to coating and drying-baking such coatings. When organic coatings are used, it is possible to use a solvent extraction system [29], [30].

The examples mentioned above provide an illustration of the capabilities of the infrared radiation industry, but are in no way exhaustive.

Whenever a thermal operation is required in a production line process handling products of relatively regular or repetitive shape, the ability of infrared radiation to perform this operation should be carefully examined.

## 7. BIBLIOGRAPHY


[1] M. DERIBÈRE, *Les applications pratiques des rayons infrarouges*, Éditions Dunod, 1953.

[2] R. GAUTHERET, *Le chauffage électrique par émetteurs de rayonnement infrarouge court*, Éditions Sodel, 1954.

[3] R. LOISON, *Chauffage industriel*, École nationale supérieure des Mines de Paris, 1956.

[4] M. LA TOISON, *Infrarouge et applications thermiques*, Éditions Eyrolles, 1964.

[5] E.D.F., *Le chauffage par rayonnement infrarouge*, AIE, 1964.

[6] G. SEURIN, *Le chauffage par rayonnement infrarouge appliqué à l'industrie textile*, extrait de *Rayonne, fibranne et fibre synthétique*, 1965.

[7] F. LAUSTER, *Manuel d'électrothermie industrielle*, Éditions Dunod, 1967.

[8] E.D.F., *Séchage de peintures sur carrosserie d'automobiles par rayonnement infrarouge*, AIE, 1968.

[9] R. MALGAT, *Cours d'électrothermie*, École de Thermique, 1970.

[10] J. CHABERT et P. VALLIER, *Optimisation de l'énergie infrarouge dans les traitements thermiques et le séchage des matériaux textiles*, Institut de Recherches des Textiles, juin 1973.

[11] M. PHILIBERT, Application et séchage des peintures et vernis, *L'industrie française*, n° 253, avril 1974.

[12] M. LEROY, Application des rayonnements infrarouges électriques pour le séchage de la peinture en construction automobile, *JIEI*, novembre 1974.

[13] KEGEL *et al.*, *Elektrowärme, Theorie und Praxis*, Verlag W. Girardet, Essen, RFA, 1974.

[14] J. GOSSE, *Rayonnement thermique*, Éditions scientifiques Riber, 1975.

[15] G. SEURIN, *Le chauffage par rayonnement infrarouge appliqué à l'industrie papetière*, exposé à l'École de Papeterie, 1975.

[16] M. CAMOU, Le problème de séchage-cuisson dans les chaînes de revêtement organique, *Galvano-Organo*, octobre 1976.

[17] R. C. GESLIN, *Quelques applications spécifiques du rayonnement infrarouge*, Colloque Comité français d'Électrothermie, avril 1977.

[18] J. CHABERT, *Le chauffage industriel par rayonnements infrarouges, haute fréquence et micro-ondes, principe de mises en œuvre*, Colloque Comité français d'Électrothermie, avril 1977.
[19] M. W. DE STOUTZ, *Stérilisation U.H.T. du lait par rayonnement infrarouge*, Colloque Comité français d'Électrothermie, mai 1977.
[20] LABORELEC, *Manuel d'électrothermie*, Bruxelles, juin 1977.
[21] R. GAUTHIER, *Cuisson de revêtement organique par infrarouge*, Colloque Comité français d'Électrothermie, octobre 1977.
[22] J. HEURTIN, J. P. METAIL et D. BIAU, *Utilisation des rayonnements ultraviolet et infrarouge pour les applications industrielles*, Conférence à la Société royale belge des Électriciens, octobre 1977.
[23] J. SCADURA *et al.*, *Initiation aux transferts thermiques*, Technique et Documentation, 1978.
[24] G. FERRY, *Séchage et cuisson par infrarouge à haute densité des revêtements sur produits plats*, Colloque Comité français d'Électrothermie, avril 1978.
[25] FECO CORP., Electric infrared oven designed for thermal efficiency in curing operations, *Industrial Heating*, juillet 1978.
[26] W. C. HANKINS, *Electric infrared heating*, Engineering, Grande-Bretagne, 1978.
[27] M. GAUSSIN, Le séchage des encres offset, Revue *l'Imprimerie nouvelle*, n° 270, mars 1979.
[28] E.D.F., *Le chauffage par rayonnement infrarouge dans les processus industriels*, Paris-La Défense, 1979.
[29] G. FERRY, *Utilisation des infrarouges haute densité pour les industries du parachèvement (métal, bois, papier, verre, céramique...)*, Colloque Comité français d'Électrothermie, mars 1980.
[30] Documentation EDF, 92080 Paris-La Défense, 1980.
[31] Documentation Electricity Council, Londres, 1980.

**List of equipment manufacturers and suppliers mentioned in this chapter:**

[A] ELSTEIN-TECHNOVA, 94400 Vitry-sur-Seine et RFA.
[B] ETIREX, 02200 Soissons.
[C] GTE SYLVANIA, 95380 Louvres et USA.
[D] INFRACOM et HERAEUS, 94800 Villejuif et Hanau, RFA.
[E] MASSER, 91170 Viry-Châtillon et Bruxelles, Belgique.
[F] MAZDA, 75008 Paris.
[G] PHILIPS, 75008 Paris et Eindhoven, Pays-Bas.
[H] QUARTZ et SILICE, 75008 Paris.
[I] MAIR et RESEARCH INC., 92100 Boulogne et Minneapolis, USA.
[J] SOVIREL, 92300 Levallois-Perret et Pittsburg, USA.
[K] TECALEMIT, 94150 Rungis.
[L] SETTON et WESTINGHOUSE, 92100 Boulogne et USA.
[M] A.E.T., 38240 Meylan.
[N] AIR INDUSTRIE, 92100 Courbevoie.
[O] SEPS, 92120 Montrouge.
[P] CARRIER, 92150 Suresnes.
[Q] SANEG, 92390 Villeneuve-la-Garenne.
[R] SCAAL, 38000 Grenoble.
[S] CASSO-SOLAR, Allendale, N.J., USA.
[T] INFRAROUGE SYSTÈM et INFRAROD TEKNIK, 69000 Lyon et Vaneborg, Suède.
[U] ACTINI-FRANCE, 74200 Thonon-les Bains.
[V] KANNEGIESSER, Vlotho, RFA.
[W] VULCANIC, 93330 Neuilly-sur-Marne.

# Chapter 6

# Electromagnetic Induction Heating

## 1. PRINCIPLE OF ELECTROMAGNETIC INDUCTION HEATING

Electric current flowing through an electrical conductor produces heat by the Joule effect. Heating is obtained either by applying a potential difference across the conductor — conventional resistance heating — or by placing the conductor in a variable magnetic field — electromagnetic induction heating.

**Fig. 1.** Induction heating

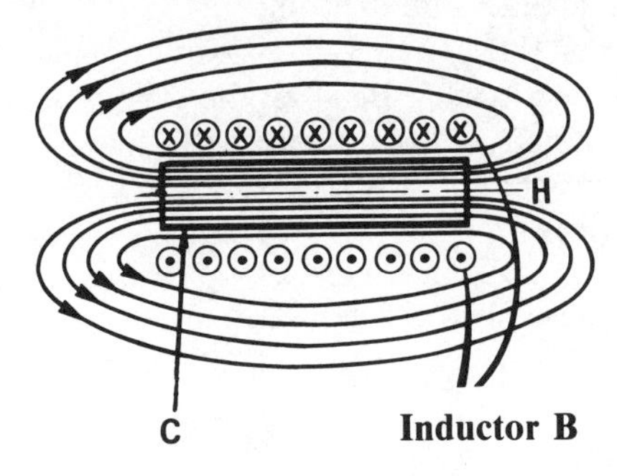


For example, if an AC potential difference V is applied to the terminals of solenoid B, the AC current at a frequency $f$ flowing through this coil creates a variable magnetic field H both inside the coil and around it. If an electrically conductive body C is inserted in the coil, the variation in magnetic field causes a variation in magnetic flux passing through this body and, according to Lenz's law, induces an EMF which causes eddy currents.

Eddy currents, or induced currents, are converted into heat due to the Joule effect in a heating body. Coil B forms the primary or inductor circuit, body

C the secondary circuit or load. Induction heating equipment is similar in principle to transformers, with or without iron magnetic circuits, the primary being supplied at line frequency (50 Hz) or higher.

However, it should be noted that any conductor through which a current flows creates a magnetic field, the intensity of which is proportional to current. *Induction heating applications therefore are not limited to the most common case where an object is placed inside a solenoid, but can use highly varied inductor configurations (flat inductors, linear inductors, tunnel inductors, etc.), and a wide range of relative part/inductor.* However, the advantage of the solenoid inductor is due to the fact that the magnetic field created by each turn is added to that created by the other turns, and that the total magnetic field, and therefore electromagnetic induction, is particularly intense inside the coil. Special devices, such as magnetic material field reinforcers, enable intense magnetic fields.

The induced emf is determined as follows:

$$E = -\frac{d\Phi}{dt},$$

$d\Phi$ being the variation in magnetic flux passing through the part to be heated during time $dt$.

The power converted into heat within the body is then $P = I^2R = E^2/R$. Resistor $R$ does not, however, have the value that it would have with a DC current flowing through it; this depends not only on the resistivity of the material, but also its magnetic permeability with respect to the frequency of the current flowing in the inductor.

When the substance to be heated consists of a magnetic material such as iron, cobalt and numerous steels, the thermal effect of the magnetic hysteresis is added to that of magnetic induction.

Added heat due to hysteresis is proportional to the surface area of the hysteresis cycle. The relationship between the energy developed by induced currents and magnetic hysteresis is expressed as follows:

$$\frac{P_i}{P_h} = af\,H_e^{0,4},$$

$P_i$, power developed by induced currents;
$P_h$, power due to hysteresis.
$H_e$, magnetic field
$f$, frequency
$a$, constant.

The power dissipated due to hysteresis is generally much lower (less than 10% in most cases) than that generated by induced currents; therefore, for an extreme case such as high hysteresis-tempered steels, the energy due to hysteresis reaches a maximum of 50% of the energy value given off by induced currents, and therefore represents one-third of the total power. Also, the energy due to hysteresis is not generated above the Curie point or magnetic transformation point since beyond this relative magnetic permeability becomes equal to 1 (ferromagnetic materials become paramagnetic at a temperature about 750°C for iron, 350°C

relative magnetic permeability may be very high). Induced currents themselves vary noticeably beyond the Curie point. It should be noted that magnetic induction heating involves three successive physical phenomena:
— electromagnetic transfer of energy from the coil to the body to be heated;
— conversion of electrical energy into heat in the load due to the Joule effect;
— heat transfer due to thermal conduction in the mass.

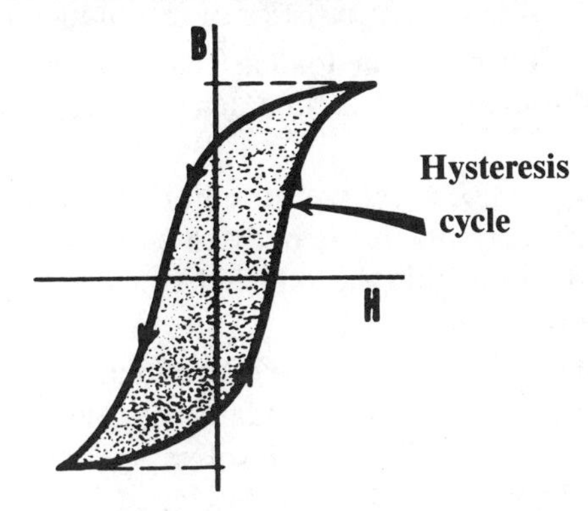


**Fig. 2.** Hysteresis cycle

Unlike conventional heating where, to heat a body, it is necessary to use a heat source having a higher temperature than the body, heat is applied to a much hotter body through the winding which remains relatively cold. The heat is produced in the part itself, without any material transmission agent; therefore, no contacts are required.

The very low thermal inertia in induction heating suppresses or very greatly reduces no load losses or losses while heating to temperature.

Also, electromagnetic induction heating can be used to obtain very high power densities, considerably reducing thermal losses.

These three properties — direct creation of heat inside the body to be heated, low thermal inertia and high power density — make induction heating highly energy efficient, and is often a source of substantial energy savings.

## 2. CHARACTERISTICS OF ELECTROMAGNETIC INDUCTION HEATING

For industrial applications, two aspects of induction heating merit attention:
— distribution of induced currents in the body to be heated;
— the power dissipated in this body.

These values in fact control the thermal effect of induction, but also control the conditions of its economic efficiency.

The characteristics of heat produced by induced current depend on numerous parameters, and in particular:
— the magnetic flux traversing the body to be heated, therefore:

• the type of material (relative magnetic permeability) and its condition (magnetic or non-magnetic body, effect of temperature);

• the inductor magnemotive force (ampere-turns per unit length, characteristics and saturation of the magnetic circuit);

• magnetic leakage (respective dimensions of the inductor and the part to be heated, coupling and characteristics of the magnetic circuit);

• frequency;

— electrical characteristics of the inductor and the load:

• resistivity of the load and the inductor at the temperatures involved;

• geometrical characteristics of the inductor and load;

• cross-section of the load affected by current flow, distribution of current density throughout this section, length of circuit through which induced current flows.

All these parameters are of major importance, since the penetration depth of induced currents in the body to be heated, the quantity of heat given off, its distribution within the load and efficiency of this method of heating depend on these parameters.

The energy dissipated in the load and distribution of current can be rigorously determined by using the fundamental laws of electromagnetism (Maxwell's equations), whose solution involves special functions; various analytic and numerical methods have been developed for this. These calculations, which are generally very complex, exceed the scope of this work, but can be found in several courses and fundamental works on electromagnetism [1], [2], [3], [5], [9], [26], [37], [40], [42], [112], [128], [137], [147*i*].

## 2.1. Penetration of induced currents

### *2.1.1. Definition of penetration depth*

The higher the frequency in the inductor, the greater the tendancy for the current flowing through the body to be heated to concentrate on its surface; the current density therefore decreases from the surface of the part to be heated towards its center. This phenomenon is known as skin effect or Kelvin effect. It can be shown (by simplifying integration of the Maxwell equations mentioned above) that the current density decreases exponentially from the peripheral of the part to be heated towards its center.

Current density is therefore of the form:

$$\boxed{i_x = i_0 \, e^{-x/d_0},}$$

$I_x$, current density at distance $x$ from periphery;
$I_0$, current density at load surface ($x = 0$);
$d_0$, constant, depending, in particular, on frequency (depth of penetration defined above).

Current $I$ flowing through the part, which is in the integral of the current density, is therefore equal to the area defined by the axes and the curve

representing the current density. Calculation of this area $O$, $i_0$, $D$, and therefore current, shows that this is equal to
$i_0 d_0$, that is to $O$, $i_0$, $A$, $B$:

$$I = i_0 d_0.$$

Areas $i_0CA$ and $BCD$ are equal, and the significance of $d_0$ clearly apparent; the total current flowing through the load is equal to the product of the current density at the surface $i_0$ times a certain depth $d_0$. Quantity $d_0$ is

**Fig. 3.** Current penetration depth.

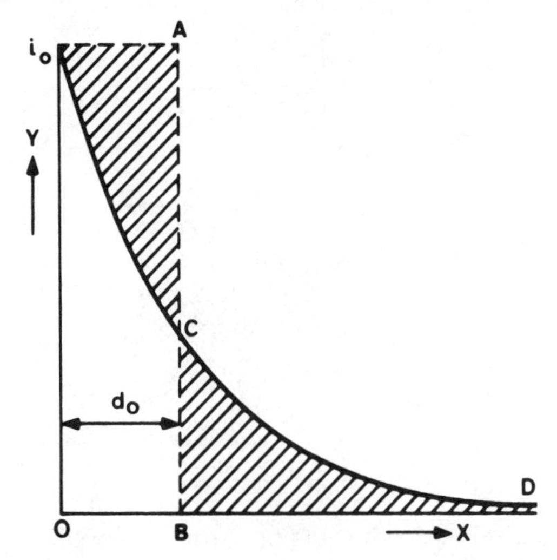


known as the current penetration depth or "skin thickness". Therefore, to facilitate analysis of the induction phenomenon, a non-uniformly distributed current in the load, the density of which decreases from the surface of the body to be heated, is replaced by a nonexistant equivalent current, uniformly distributed over an area of thickness $d_0$ from the surface whose density is equal to its surface value. This gives: $i(d_0) = i_0/e$, i.e. $i(d_0) = 0.368\ i_o$, and the current flowing through the layer of material extending from the peripheral to the penetration depth $d_o$ is:

$$I(d_0) = i_0 d_0 (1 - 1/e),$$

$$I(d_0) = 0.632\ I,$$

63% of the current is therefore concentrated in the skin depth thickness $d_0$. Now, according to Joule's law, the power given off is proportional to the square of the current density. The calculation then shows that the power given off in this surface layer of thickness $d_0$ is equal to:

$$P(d_0) = 0.865\ P.$$

*Therefore, the heating power is concentrated essentially in the skin thickness; this phenomenon controls most industrial applications of induction heating.*

**Fig. 4.** Penetration depth and frequency

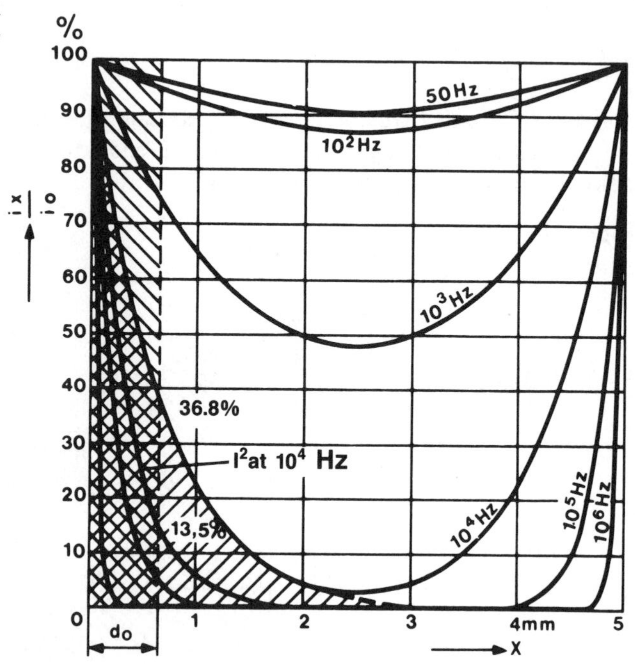


Penetration depth (obtained from Maxwell's equations) is given for a solid cylindrical load much longer than the inductor, and a diameter much larger than $d_0$, by the formula:

$$d_0 = \sqrt{\frac{\rho}{\pi \mu f}},$$

**Fig. 5.** Penetration depth and power dissipation

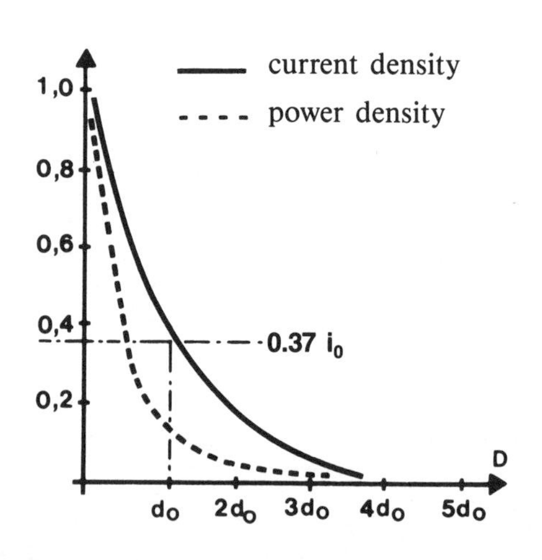

$d_0$, penetration depth in meters;
$\rho$, load resistivity in ohms-meters;
$f$, frequency of the current flowing through the inductor in Hz;
$\mu$, load magnetic permeability with $\mu = \mu_0, \mu_r$
$\mu_0$, magnetic permeability of vaccum, equal to $4\pi \times 10^{-7} H \times m^{-1}$,
$\mu_r$, load relative magnetic permeability.
i.e.:

$$d_0 = \frac{1}{2\pi}\sqrt{\frac{\rho . 10^7}{\mu_r f}},$$

or:

$$d_0 = 503.3\sqrt{\frac{\rho}{\mu_r f}}.$$

*Note:* in technical literature, the formula:

$$d_0 = \frac{1}{2\pi}\sqrt{\frac{\rho}{\mu_r f}};$$

is often found; this expression gives the value of $d_0$ in centimeters in the former C.G.S. unit system.

Penetration depth $d_o$ therefore depends on the resistivity of the body to be heated, its relative permeability and the frequency.

Approximate values of $d_0$ for various materials and frequencies is given in Figure 7 and the corresponding graph (*Fig.* 6).

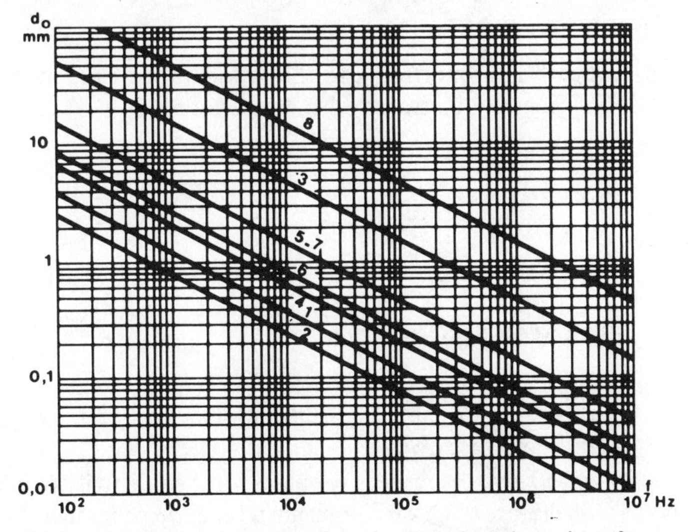


**Fig. 6.** Graph showing current penetration depth value $d_0$ against frequency for various materials and temperatures (see *Fig.* 7 for definition of materials) [4].

Value of $d_0$ for various materials at different temperatures [4]

| Materials | 1<br>Steel<br>20°C<br>$\mu_r=40$ | 2<br>Steel<br>20°C<br>$\mu_r=100$ | 3<br>Steel<br>800°C | 4<br>Copper<br>20°C | 5<br>Copper<br>800°C | 6<br>Alu-<br>minum<br>20°C | 7<br>Alu-<br>minum<br>500°C | 8<br>Graphite<br>20 and<br>1,300°C |
|---|---|---|---|---|---|---|---|---|
| $f$(Hz) | $d_0$ in millimeters | | | | | | | |
| 50 | 5.03 | 3.1 | 67.2 | 9.35 | 19.4 | 11.9 | 19.4 | 201 |
| 100 | 3.56 | 2.25 | 47.5 | 6.61 | 13.4 | 8.4 | 13.4 | 142 |
| $10^3$ | 1.424 | 0.71 | 14.6 | 2.09 | 4.26 | 2.66 | 4.26 | 45 |
| $10^4$ | 0.356 | 0.225 | 4.75 | 0.661 | 1.34 | 0.84 | 1.34 | 14.2 |
| $10^5$ | 0.112 | 0.071 | 1.46 | 0.209 | 0.426 | 0.266 | 0.426 | 4.5 |
| $10^6$ | 0.035 | 0.0225 | 0.475 | 0.066 | 0.134 | 0.084 | 0.134 | 1.42 |
| $10^7$ | 0.011 | 0.007 | 0.146 | 0.021 | 0.043 | 0.0266 | 0.043 | 0.45 |
| Graph number (*Fig. 6*) | 1 | 2 | 3 | 4 | 5 | 6 | 7 | 8 |

**Fig. 7.**

### *2.1.2. Effect of resistivity*

Penetration depth is proportional to the square root of the load resistivity. For metals, this generally increases with temperature:

$$\rho_\theta = \rho_0(1+\alpha\theta),$$

with:

$\alpha>0$ and $\theta$ in degrees Celsius.

Resistivity increases noticeably at melting point. For steels and cast irons, resistivity generally increases with the carbon content.

Figures 8, 9 and 10 give the respective values for the resistivity of some metals and alloys at temperature 0°C; variations are a function of temperature whereas for steel variations are a function of carbon content. For graphite, resistivity is more or less constant over the range 0-1,500°C.

| Body to be heated | Resistivity ($\mu\Omega$ - cm) |
|---|---|
| Silver | 1.47 |
| Copper | 1.60 |
| Aluminum | 2.56 |
| Zinc | 5.75 |
| Iron | 9.07 |
| Nickel | 12.32 |
| Tungsten | 10.92 |
| Molybdenum | 9.01 |
| Nickel-chromium (80/20) | 102-108 |
| Light alloys | 5-8 |
| Brasses | 5-10 |
| Zamak | 6-8 |
| Bronzes | 10-20 |
| Plain steels | 10-25 |
| Special steels | 40-80 |
| Cast iron | 60-100 |
| Pig iron | 30-35 |
| Ultra light alloys | 10-15 |
| Graphite | 1,000 |
| Liquid zirconium oxide | 100,000 |
| Liquid uranium oxide | 10,000 |

**Fig. 8.** Resistivity at 0°C (except $ZrO_2$ and $UO_2$) of some electrically conductive materials.

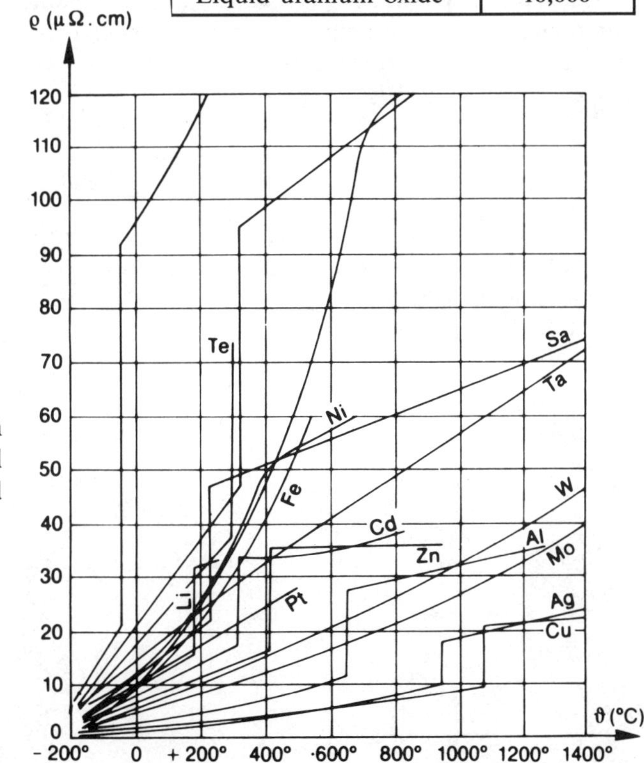


**Fig. 9.** Variation in resistivity with metal temperatures in solid and liquid states

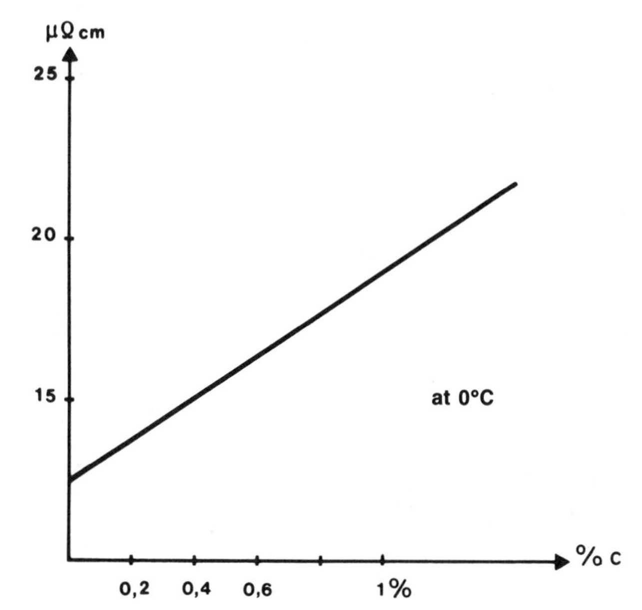


**Fig. 10.** Variation of the resistivity of steel with carbon content

In practice, and to facilitate calculations, a mean resistivity is used.

### *2.1.3. Effect of relative magnetic permeability*

The depth of penetration is inversely proportional to the square root of the relative magnetic permeability $\mu_r$. This is equal to 1 for non-magnetic materials such as aluminum alloys or copper, but reaches very high values for some ferromagnetic materials.

**Fig. 11.** Maximum magnetic permeability for some ferrous metals

| Metal | $\mu_r$ maximum |
|---|---|
| Pure iron | 14,000 |
| Cast steel | 3,500 |
| Annealed cast steel | 15,000 |
| Tempered steel | 100 |
| Cast irons | 300-900 |
| Mild steel | 2,000 |
| High permeability nickel alloys | Several hundred thousands |

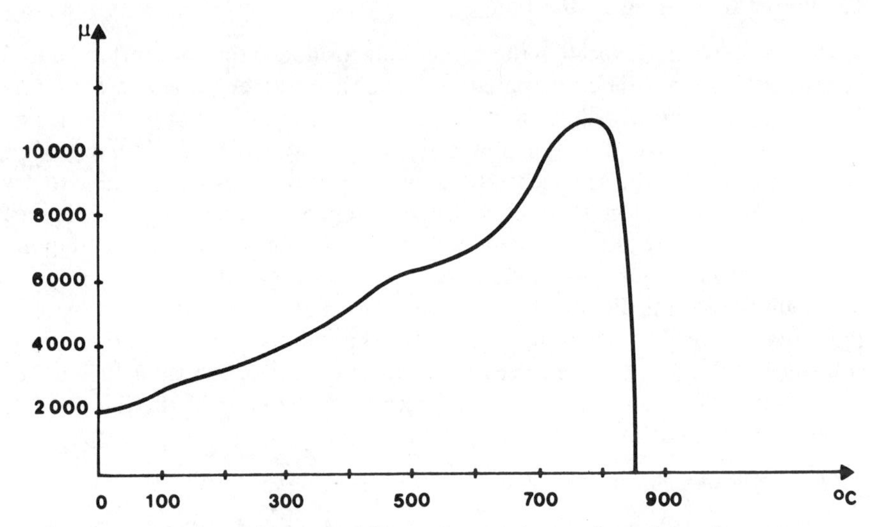


**Fig. 12.** Variation of permeability of a steel as a function of temperature

For these ferromagnetic metals, permeability varies widely, and it is difficult to provide orders of magnitude, since permeability is a function of temperature, alloy composition, impurities and physical condition (hardening, heat treatments, etc.), but also of the magnetic field to which the material is submitted and magnetic saturation of the circuit.

All ferrous alloys become paramagnetic beyond a certain temperature (Curie point), or their permeability decreases very suddenly to 1. For steels, this temperature is around 750°C.

The penetration depth is generally lower and skin effect more noticeable with a ferromagnetic body than with a paramagnetic body, at least below the Curie point.

### *2.1.4. Effect of frequency*

Penetration depth is inversely proportional to the square root of the frequency. While resistivity and magnetic permeability are characteristics of the body to be heated, the frequency can be chosen by the user; therefore, the latter gives of a method of controlling power dissipation within the body to be heated and therefore the amount of heat produced.

The frequencies used in induction heating are divided into four groups:

— low frequencies, for frequencies equal to or less than line frequency — 50 Hz in Europe, 60 Hz in the United States;

— medium frequency, 60 to 10,000 Hz;

— radio frequency, from 10,000 to 300,000 Hz;

— ultra radio frequency, beyond 300,000 Hz.

## 2.2. Power transferred to the body

The body, which is placed in the inductor, heats due to the Joule effect caused by induced currents. The dissipated power is therefore expressed as $P = I^2R$. Analysis of current distribution within a cross-section of a body to be heated has, however, shown that the power density is not uniform, but decreases exponentially from the surface to the center. Therefore, it is not possible to use the standard expression $R = \rho l/S$, where $S$ represent the area of passage of current $I$ and $l$ the length of the conductor as a resistance value without adapting it to the special conditions encountered in induction heating.

Calculation of the dissipated power therefore makes use of two of Maxwell's equations as already described. The idea of penetration depth, mentioned above, however enables very simple calculation, with an error of less than 10% of the power transferred to the part to be heated when the ratio of the diameter of the part to its penetration depth is high.

### *2.2.1. Simplified power calculation*

In the simplified calculation, the load current is considered to be completely concentrated in an area whose thickness is the penetration depth $d_0$, with a constant current density equal to the current density at the surface of the body to be heated (analysis of current distribution in a cross-section of a body to be heated, as described in paragraph 2.1.1., showed that current $I$ is equal to the product of the surface current density $i_0$ times the penetration depth $d_0$).

Therefore, induced currents flow through a conductor which can be equivalent to a cylinder. The conductor length is equal to the circumference of a circular cross-section of this cylinder, i.e. $\pi d$, $d$ being the diameter. The conductor cross-section is equal to $hd_0$, the product of the penetration depth $d_0$ (cylinder thickness) times cylinder height $h$. Therefore, the electrical resistance is approximately:

$$R_2 = \rho l_2 / S_2,$$

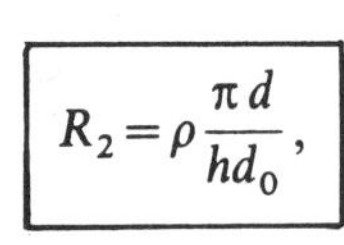

$$R_2 = \rho \frac{\pi d}{h d_0},$$

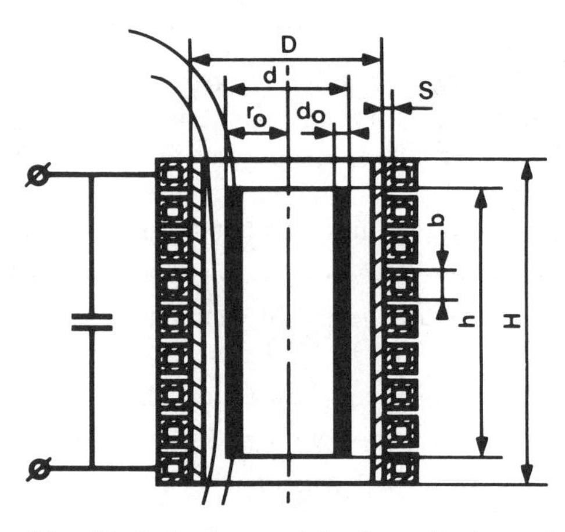


*Fig. 13.* Inductor and body to be heated

or again, by replacing $d_0$ by its value:

$$R_2 = 2\pi^2 \frac{d}{h} \sqrt{10^{-7}\,\rho\mu_r f}.$$

The energy dissipated in the body to be heated is then $P_w = I_2^2 R_2$. By replacing $R_2$ by the previous value and current $I_2$ by $n_1 I_1$ (the part to be heated is equivalent to the secondary of a single-turn short-circuited transformer secondary, the primary containing $n_1$ turns through which current $I_1$ flows); the dissipated power expression becomes:

$$P_w = 2\pi^2 \frac{d}{h} \sqrt{10^{-7}\,\rho\mu_r f}\,(n_1 I_1)^2,$$

introducing the rms magnetic field $H_e = n_1 I_1 / h$:

$$\boxed{P_w = 2\pi^2\, dh\, H_e^2 \sqrt{10^{-7}\,\rho\mu_r f},}$$

or again $P_w = 62.42 \times 10^{-4} dh H_e^2 \sqrt{\rho\mu_r f}$.

The linear power is:

$$P_{w/m} = 2\pi^2\, dH_e^2 \sqrt{10^{-7}\,\rho\mu_r f}.$$

The specific thermal power in Watts per square meter is equal to:

$$\boxed{P_{w/m^2} = 2\pi\, H_e^2 \sqrt{10^{-7}\,\rho\mu_r f},}$$

i.e.:

$$P_{w/m^2} = 2 \times 10^{-3}\, H_e^2 \sqrt{\rho\mu_r f}.$$

These expressions show that, to increase dissipated power, it is necessary:

— to increase the magnetic field, i.e., the number of ampere-turns $n_1 I_1$ of the inductor. However, this option is limited by the space available to fit $n_1$ inductor turns and by inductor current;

— to increase frequency which is limited for several reasons. The transferred power varies only with the square root of the frequency; moreover, the inductor impedance also increases with frequency, leading to a limitation of transmitted power. Losses in capacitor units, mountings, etc. increase with frequency and reduce efficiency. These high powers are reached at radio frequencies, therefore causing low penetration depths, and are useable only for surface heating; conversely, for deeper or core heating, it is necessary to use a deeper penetration depth and lower frequency, therefore limiting power.

The physical properties of the body to be heated also affect induced power. Thus, when heating ferromagnetic bodies, the induced power value for a given frequency and magnetic field, is much higher at a temperature below the Curie point than that applied above the Curie point; therefore, some induction heating systems use different frequencies below the Curie point where the line frequency is generally sufficient, and a higher frequency enabling a high power density above this point.

Resistivity is also a determining factor in transmitted power, which is generally higher for ferrous materials such as steel and cast iron than for non-ferrous metals such as copper and aluminum. The increase in resistivity with temperature, which is high for most standard metals, favors induction heating and, in particular, induction melting of metals since their resistivity increases at melting point (*see* paragraph 2.1.2.).

For ferromagnetic bodies, it is also necessary to allow for magnetic circuit saturation. When saturation is reached, the increase in magnetic field is followed only by a slight, then practically no increase in magnetic induction, which remains almost constant. An increase in the magnetic field then results in a decrease in heating system efficiency, but power density continues to increase.

### *2.2.2. Comparison of power densities with induction heating and other processes*

In spite of the limits previously mentioned, specific power (W/cm$^2$ of area), which can be transferred by induction, considerably exceeds that transferred in most conventional processes. Without difficulty, this specific power can be taken to values 1,000 times higher than those of a radiation furnace operating at 1,000°C, and easily reaches 50 times that of an oxyacetylene torch.

Moreover, with radio frequency surface heating (more than 10,000 Hz) of ferrous materials, powers of 3 kW/cm$^2$ are frequently used, while, for core heating, powers are more limited and of the order of a few tens to a few hundreds of watts per square centimeter, so as to limit the radial thermal gradient (where the limit power used is expressed by the experimental formula: $P_{w/cm^2} = 300/d$; $d$ is the body diameter in centimeters).

### *2.2.3. Parameters controlling the simplified power calculation: real transmitted power*

The expression of the power transmitted by induction, as determined above, was based on several hypotheses, uniform magnetic field, excellent inductor-load coupling, uniformly distributed current in surface layer equal to penetration depth, high diameter to penetration depth ratio, cylindrical part, etc. In practice, these conditions are not necesarily combined, and the simplified power calculation is notably altered.

#### *2.2.3.1 Effect of current distribution in the load, and penetration depth*

Current decreases exponentially from the periphery towards the center of the body to be heated, and the diameter to penetration depth ratio may vary considerably. Resolving of the Maxwell equations using Bessel functions shows that the calculated power expression remains valid on condition that a power transmission factor $F$, depending on the diameter to penetration depth ratio and part geometry, is assigned to the expression.

For a cylindrical part, the power and power density expression becomes:

$$P_w = 2\pi^2 dh H_e^2 \sqrt{10^{-7} \rho \mu_r f} F,$$

and:

$$P_{w/m^2} = 2\pi H_e^2 \sqrt{10^{-7} \rho \mu_r f} F.$$

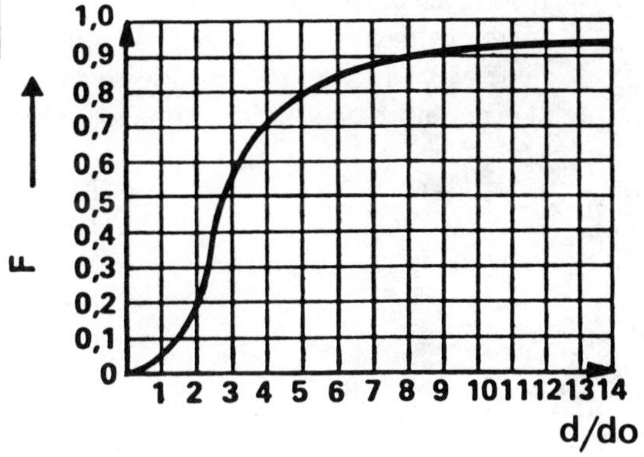

**Fig. 14.** Power transmission factor for cylindrical body

It is also possible to calculate the dissipated power density:

$$P_{w/m^3} = 8\pi H_e^2 \sqrt{10^{-7} \rho \mu_r f} \frac{F}{d}.$$

The power density is sometimes expressed as a function of factor $M$, defined by the equation $M = Fd_0/d$:

$$P_{w/m^3} = 4\pi H_e^2 \mu_0 \mu_r f M.$$

The curve in Figure 15 gives the value of factor $M$ versus body diameter to penetration depth ratio.

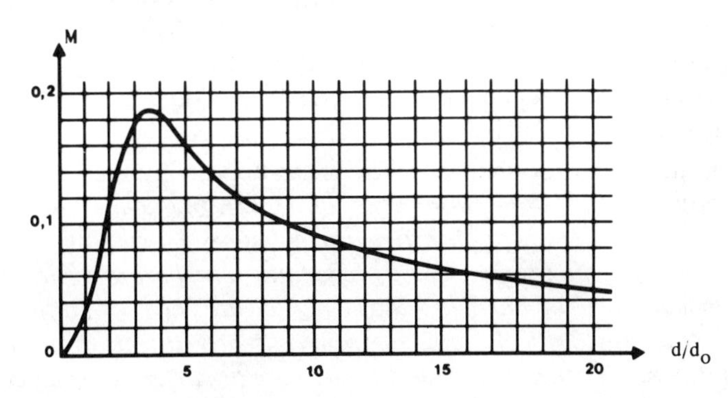


**Fig. 15.** Power density transmission factor M

The curve representing function $M$ passes through a maximum for a diameter to penetration depth ratio of the order of 3.5. Therefore, for a given frequency, a diameter value exists for which the power density is maximum. Conversely, for a given diameter, the power density is a function not only of factor $M$, but also of product $f \times M$, and it is better to use the expression giving power density as a function of power transmission factor $F$, since $F$ and $f$ increase simultaneously.

The curves giving power transmission factor $F$ can be plotted for various body and inductor shapes. For a square cross-section body, the cylindrical body curve applies accurately, if 1.1 times the side is used as an equivalent diameter. For an oblong part, placed within an inductor, Figure 16 gives the transmission factor value as a function of the ratio of the smallest dimension of the cross-section to penetration depth, the second side being considered as much longer [5], [100].

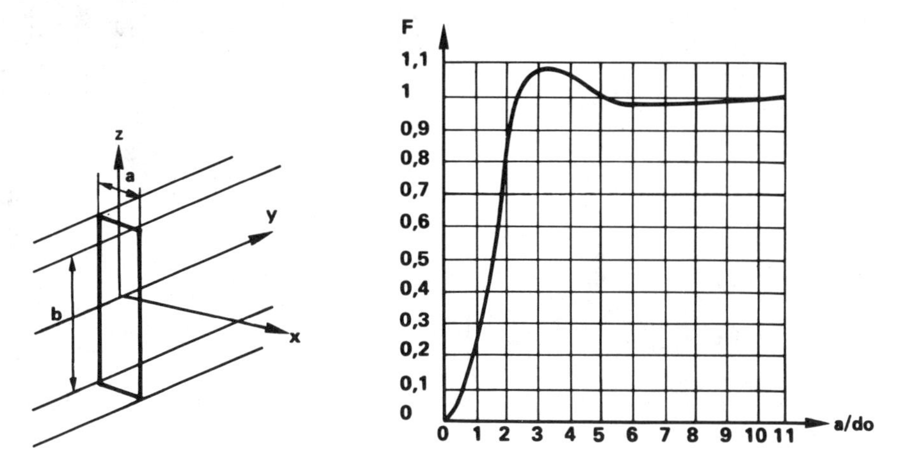


**Fig. 16.** Power transmission factor for rectangular shaped body, as a function of the ratio of the smallest dimension to peneteration depth.

The variation in power transmission factor shows that to obtain efficient heating with a cylindrical part, a diameter to penetration depth ratio of more than 3.5 must be used. Below this value, known as the critical diameter, the transmitted power and therefore efficiency drops rapidly. Moreover, an increase of this ratio beyond 8 or 9 does not alter transmitted power, and therefore provides no significant increase in efficiency.

Conclusions are similar for oblong bodies, as can be seen in the corresponding curve. *In general, a critical dimension exists for a given frequency below which the transmitted power, and therefore efficiency, collapses. Conversely, a critical frequency exists for a given dimension, depending on the properties of the material to be heated, below which heating is inefficient. The use of a frequency much higher than the critical frequency only slightly affects transmission efficiency, and other considerations; essentially core or surface heating control choice of frequency.*

### 2.2.3.2. *Effect of the relative dimensions of the inductor and the load*

While the lines of force are parallel to the center line of a long inductor coil, this does not apply to an inductor which is short with respect to its diameter. Moreover, end effects, corresponding to high dispersion of the field, exist in all cases.

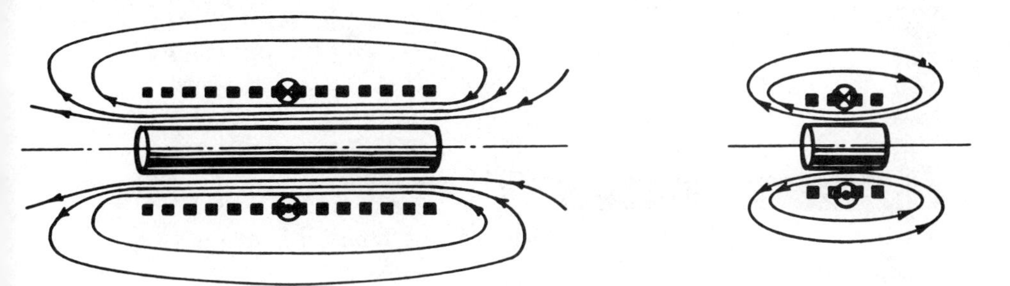

**Fig. 17.** Effect of body diameter-inductor length ratio and end effects

Both these factors reduce the strength of the magnetic field, and therefore the power transmitted to the body to be heated. In addition, this body does not totally fill the inside volume of the conductor, since, for simple shape considerations, its diameter must be less than that of the coil; this constitutes another magnetic dispersion factor.

The inductor-load coupling is therefore never perfect, and a second correction coefficient must be introduced into the simplified power expression. This coefficient, which is often known as coupling factor $C$, is always less than 1, and depends on the respective geometric characteristics of the bodies. Figure 18 gives the value of coupling factor $C$ for a cylindrical body. These curves show that if the inductor length is much longer than its diameter, the coupling factor is excellent, and around 1, even if the diameter of the body to be heated is much less than that of the inductor. Conversely, if the inductor is short and the part diameter much lower than that of the inductor, this coefficient can drop significantly.

Therefore, for a cylindrical body, the transmitted power becomes:

$$P_w = 2\pi^2 dh H_e^2 \sqrt{10^{-7} \rho \mu_r f}\, FC,$$

and:

$$P_{w/m^2} = 2\pi H_e^2 \sqrt{10^{-7} \rho \mu_r f}\, FC.$$

Comparable groups of curves can be plotted for bodies of different cross-sections; as before, a square body being considered as equivalent to a cylindrical body with a diameter equal to 1.1 times the side.

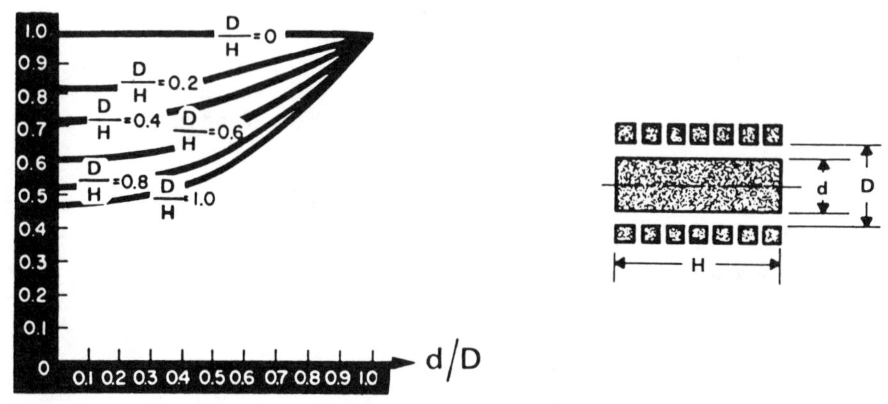


**Fig. 18.** Coupling factor for a cylindrical body

*2.2.3.3. Power transmission — hollow bodies*

Hollow bodies such as tubes can be induction heated. Distribution of the magnetic field, and therefore penetration depth, compared with the shape of the body still play a fundamental role.

If the hollow part of the body is a simple central bore, induced current distribution is practically the same as for a solid body whose calculated power expressions apply.

Conclusions concerning solid cylindrical loads are not, however, applicable to hollow cylindrical loads for which the penetration depth becomes significant compared to the thickness, or greater than the thickness.

Magnetic field $H_e$ in the hollow part, which is practically null when penetration depth $d_0$ is small with respect to the thickness $a$ of the cylinder, increases with penetration depth $d_0$. Induced currents will therefore develop throughout the thickness a of the body, and the current density on the inside surface of the cylinder increases with penetration depth. When the cylinder is very thin, the magnetic field becomes practically uniform in the hollow part; it is just as intense on the inside as on the outside of the body, and current density is constant within a given cross-section of the body.

Power transmission factor $Q$, which is similar to that defined for solid bodies, is a function of two ratios, that of the cylinder thickness to the penetration depth and that of cylinder thickness to cylinder diameter.

The transmitted power is therefore expressed as follows:

$$P_{u/m^2} = 2\pi H_e^2 \sqrt{10^{-7}\rho\mu_r f}\, Q,$$

The curves in Figure 20 give variation in factor $Q$ as a function of the two ratios previously defined; these are calculated using the Maxwell equations and Bessel functions [5].

If the cylinder thickness to diameter ratio $a/d$ is equal to 0.5, the tube is in fact a solid cylinder, and the curve representing factor $Q$ is the same as that of

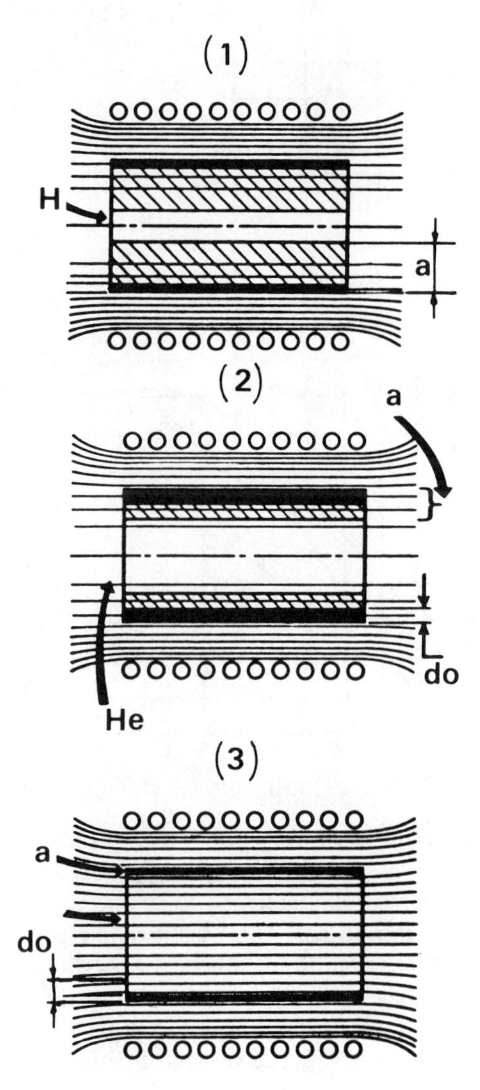


**Fig. 19.** Distribution of magnetic field in induction heating of hollow bodies
(1) hollow area forms a simple bore, penetration depth with respect to tube thickness is small, and the magnetic field at the center is null;
(2) penetration depth significant with respect to tube thickness — low magnetic field at center of tube
(3) tube thickness small with respect to penetration depth, strong magnetic field inside the tube.

factor $F$ for solid bodies. Function $F$ is therefore the limit of curve family $Q$ when $a/d$ is around 0.5.

If the tube is thin, (i.e. if tube thickness is relatively small with respect to its diameter), and if penetration depth is similar to or greater than the tickness, the current density is more or less unform across the tube cross-section. An excellent approximation of function $F$ is obtained if $a/d_0 \leq 1$ *and* $a/d \leq 0.1$, from the equation:

$$Q = \frac{(a/d_0)^3}{(a/d_0)^4 + 4\mu_r^2 (a/d)^2}\,.$$

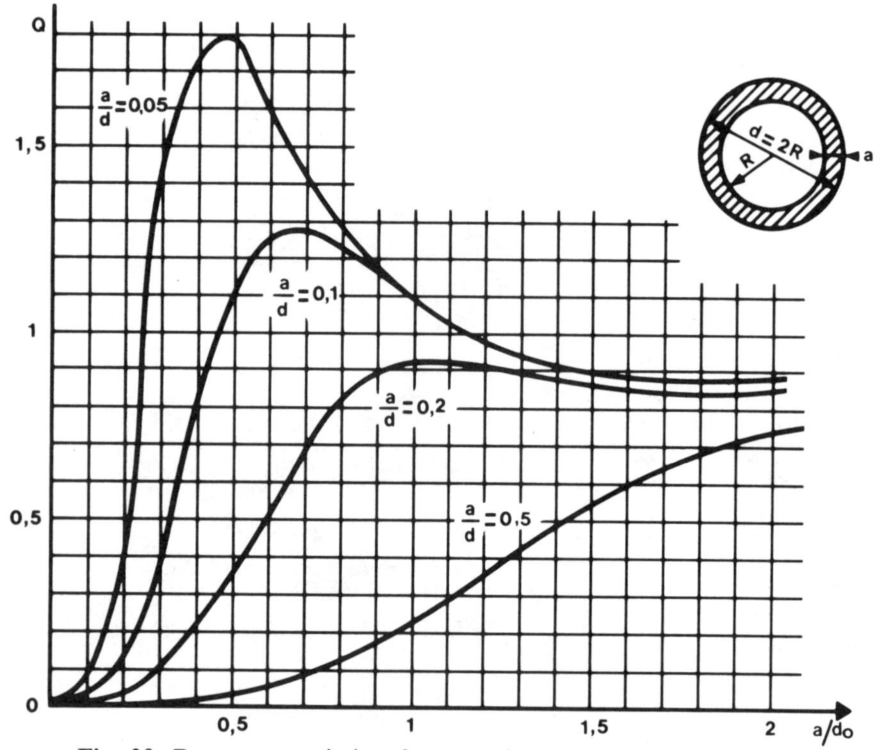


**Fig. 20.** Power transmission factor $Q$ for hollow body ($\mu_r = 1$).

Analysis of curves representing factor $Q$ show that the power transmitted to a hollow body is then, for certain appropriate values in body thickness to penetration depth and thickness to diameter ratios, much higher than that transmitted to a solid part of the same diameter. An optimum frequency value exists under these conditions, which makes efficiency maximum (this frequency corresponds to the value of $d_0$ for which function $Q$ passes through maximum — *see* electrical efficiency expression, paragraph 2.3., in which function $F$ is replaced by function $Q$). The value of this frequency is given by the equation:

$$f = \frac{2\rho\sqrt{3}}{ad\,\pi\,\mu_0}, \quad \text{or again} \quad f = \frac{10^7\rho\sqrt{3}}{2\pi^2 ad}.$$

Conversely, if the body diameter is known and the frequency, and therefore penetration depth are fixed, an optimum thickness exists which causes maximum energy dissipation in the charge, and is determined by the equation:

$$a = 2\mu_r \frac{d_0^2}{d}, \quad \text{in which} \quad Q(\max) = \frac{d}{4\mu_r d_0}.$$

For example, this approach can be used to determine the optimum thickness of a graphite crucible used to melt metal, or to behave as a susceptor in heating a product which does not conduct electricity. The crucible thickness is often much less than the penetration depth (the crucible is partially transparent to electromagnetic induction), and a residual magnetic field persists in the crucible charge, which will also be heated directly if the crucible content is conductive. Mixed heating can be used to reduce the crucible-charge thermal gradient and to increase transmitted power.

*2.2.3.4 Power transmission, general case*

Generally, it is difficult to calculate accurately the transmitted power for uneven shaped parts. However, some information can be provided.

Core heating of regular cross-section bodies, for example hexagonal, or ring gears can be adequately compared to that of a cylindrical body of diameter equal to the envelope circumference. However, it may be necessary to decrease power density to enable temperature to equalize due to the effect of heat conduction and to prevent hot points.

Similarly, for uniform heating of parts of regular cross-section, but the diameter of which varies with respect to a given inductor diameter, it is sometimes posible to vary power density, i.e. the number of ampere-turns per unit length (magnetic field) along the inductor.

For surface heating, the use of a suitable frequency often makes it possible to modulate the effective heating.

Therefore, in the case of a ring gear, and if frequency is very high, induced currents develop essentially on the surface of each tooth (fig. 21). Conversely,

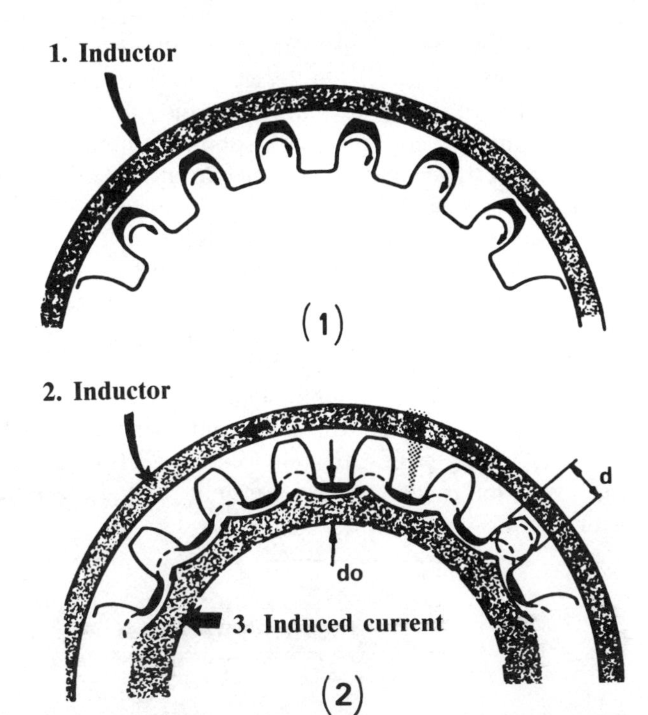


**Fig. 21.** Transmission of energy to a ring gear via induction: (1) radio frequency, low penetration depth: induced currents develop on the surface of the toothing; (2) lower frequency, penetration depth greater than tooth width d: induced currents build up in the hollow of each tooth and in the body of the ring gear.

if frequency is relatively low, tooth dimensions are such that no induced current can build up; in this case, it is the hollow separating each tooth which heats up. These properties are used in heating before case hardening.

*Therefore, in complex cases, as in the simpler cases previously examined, transmitted power is determined not only by calculation but also by the results of tests and experience.*

## 2.3. Electrical efficiency of induction heating

### *2.3.1. Calculation of electrical efficiency*

In induction heating, electrical efficiency is equal to the ratio of the effective energy induced in the part to be heated to the total energy involved:

$$\eta_e = \frac{P_w}{P_w + P_i}, \quad \text{or} \quad \eta_e = \frac{1}{1 + P_i/P_w},$$

$P_w$, energy induced in body to be heated;
$P_i$, energy dissipated in the inductor due to the Joule effect.

The energy dissipated in the inductor is equal to $I_l^2 R_l$, $R_l$ in which is the inductor resistance and $I_l$ the current flowing in the inductor. Therefore, the inductor must be constructed from a low resistivity metal such as very pure electrolytic copper, with very low resistivity, as is used in most inductors. However, as for the body to be heated, the current density is not always uniform in the inductor, and skin effect must be taken into account when determining inductor resistance. Two cases can occur:

— penetration depth in the conductor material is greater than the effective diameter of the inductor or of the same order of magnitude; the inductor resistance can then be calculated using the usual formula, giving its value as a function of resistivity, and the cross-section and length of the inductor. This case applies for the lower frequencies, and limited inductor cross-section, since the penetration depth for copper is about 9.5 mm at 50 Hz and 50°C. Therefore, to increase efficiency it is necessary to increase the inductor cross-section;

— penetration depth is much lower than the effective diameter of the inductor; the current then concentrates on the inside surface of the inductor opposite the body to be heated. The cross-section used for calculation of the resistance is equal to the product of the width of one turn times the penetration depth, the latter being defined using the same expression as that for the body to be heated. This situation is encountered at high and medium frequencies, and is similar for line frequency when inductors of sufficient cross-section are used. In such cases, it is unnecessary to increase inductor thickness to increase efficiency.

The efficiency expression below corresponds to the second case, which is the most common in industry. For a cylindrical body, electrical efficiency $n_e$ is then expressed as:

$$\eta_e = \frac{1}{1 + g(D/d)(H/h)(1/F)(1/C)\sqrt{\rho_1 \mu_1 / \rho_2 \mu_2}},$$

$D$, inside diameter of solenoid;

$d$, body diameter;

$F$, power transmission factor;

$C$, coupling factor;

$g$, constant depending on inductor geometry and tending towards 1 when the turns are very close to each other ($g \gtreqqless 1$);

$\rho_1$ and $\rho_2$, inductor and body resistivities;

$\mu_1$ and $\mu_2$, relative magnetic permeability of inductor and body;

$h$, body length;

$H$, inductor length (if $h \geqq H$, $h$ is considered as $= H$).

Therefore, electrical efficiency depends on numerous variables, but systematic respect of certain rules will provide high efficiency:

— use low resistivity inductors, generally consisting of very pure electrolytic copper, and limit inductor temperature which increases resistivity (the water or air cooling systems used with most inductors absorbs the energy dissipated in the inductor, and also aids high efficiency); the inductor magnetic permeability is then equal to 1, which also helps increase efficiency;

— provide adequate body-inductor coupling by reducing the difference between the diameter of the body and that of the inductor to a minimum, but also by using the complete length of the inductor to heat the body;

— use an adequate penetration depth ratio, at least greater than 3.5, to obtain a high power transmission factor;

— use a closely wound inductor, since factor $g$ is close to 1 if the inductor space factor is high (this factor is equal to the product of the number of turns times the width of one turn, referenced to the axial length of the inductor, and therefore represents the percentage of one turn of the inductor-enveloping cylinder occupied by the inductor metal; $g$ is the inverse of the space factor).

### 2.3.2. *Effect of relative magnetic permeability*

Relative magnetic permeability affects efficiency in two ways:

— the efficiency expression first of all shows that the higher the relative magnetic permeability of a body to be heated, the higher the efficiency;

— the power transmission factor $F$ is then larger, therefore the efficiency greater, when the ratio of the penetration depth to load diameter is smaller; the higher the permeability the lower the penetration depth. For a given frequency, it is therefore possible to heat bodies whose size is inversely proportional to their permeability, and vice versa for a body of fixed diameter; the frequencies to be used decrease as permeability increases.

#### 2.3.2.1. *Magnetic materials*

For high magnetic permeability materials, product $\varrho_2\mu_2$ is very high, and electrical efficiency can exceed 90%. Therefore, for steel in which the relative magnetic permeability and mean resistivity over the temperature range considered, are respectively equal to 100 and 60 $\times$ $10^{-8}$ Ω.m, the maximum electrical efficiency is 98%, taking only the effective quantities $\sqrt{\rho_1\mu_1/\rho_2\mu_2}$ and $F$ into account.

Beyond the Curie point, relative magnetic permeability becomes equal to 1 and efficiency decreases. However, it remains excellent provided the resistivity of the body to be heated is high and power transmission factor $F$ is high, and the frequency $f$ has been chosen for conditions prevailing beyond the Curie point. Therefore, for heating to a temperature of about 1,200°C, maximum electrical efficiency will still be 88%, taking the effect of quantities $\sqrt{\rho_1\mu_1/\rho_2\mu_2}$ and $F$ only into account (mean $\varrho_2$ is then considered as equal to 100 $\times$ $10^{-8}$ Ω.m).

The curves in Figure 22 show that efficiency is maintained at a high value for a relatively high resistivity body, such as steel, when the final temperature exceeds the Curie point.

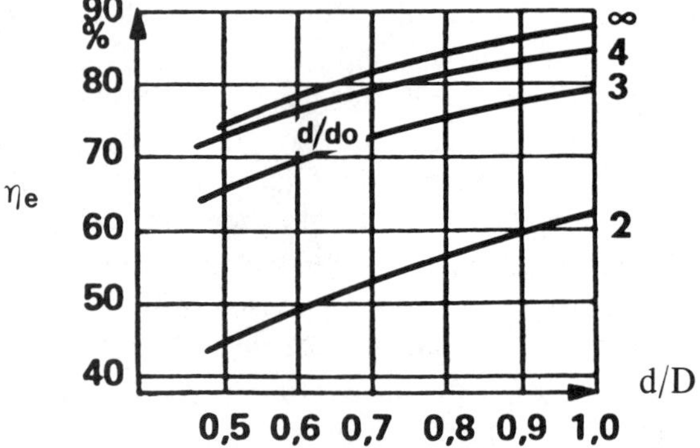


**Fig. 22.** Electrical efficiency for steel, induction heated to a temperature beyond the Curie point.

However, ferromagnetic bodies are a special case. When the magnetic circuit is saturated, and if the magnetic field continues to increase rapidly (equal to $n_l I_l / H$) resulting in increased losses due to the Joule effect in the inductor, the magnetic permeability begins to decreases (at saturation $\mu_2$, approximately expressed as $\mu_2 = B_s / \mu_0 H_e$, where $B_s$ is magnetic induction at saturation and whose value is about 1.8 T, and $\mu_2$ therefore decreases with $H_e$). Efficiency can then drop, and power density should be limited to avoid this phenomenon.

*2.3.2.2. Non-magnetic materials*

For non-magnetic materials, electrical efficiency is often lower than for ferromagnetic bodies. For example, for copper of the same resistivity as the inductor, the maximum electrical efficiency at the start of heating is 50%. Since resistivity increases with temperature, efficiency then increases with temperature. This maximum efficiency is, however, much higher, between 60 and 75%, for copper alloys like brasses and bronzes in which resistivity varies between 5 and 20 $\times$ $10^{-8}$ Ω.m. For aluminum and its alloys, maximum electrical efficiency is between 55 and 70% for resistivities of between 2.6 and 10 $\times$ $10^{-8}$ Ω.m. Efficiency is also much higher when these alloys are in the molten state, since resistivity is then much higher.

However, electrical efficiency remains very high for high resistivity, non-magnetic bodies. For example, graphite, whose resistivity is of the order of 1,000 $\times$ $10^{-8}$ Ω.m, maximum electrical efficiency is 96%.

### 2.3.3. *Effect of resistivity*

The paragraph above shows that the increase in resistivity, like that of relative magnetic permeability, has a beneficial effect on electrical efficiency. Unlike magnetic permeability where an increase contributes to decreasing penetration depth and therefore improves the power transmission factor and efficiency, an increase in resistivity leads to an increase in penetration depth. Therefore, the frequency must be high enough if the penetration depth to diameter ratio is to remain low (ultimately, when resistivity is very high as for bodies which do not conduct electricity, it becomes almost impossible to heat by induction).

Nevertheless, electrical efficiency is directly proportional to resistivity, and induction heating offers better efficiency for materials such as ferrous alloys, graphite and silicon carbide than for copper and aluminum alloys.

### 2.3.4. *Effect of frequency*

In most cases, frequency does not directly affect efficiency, although it is of fundamental importance due to the diameter-penetration depth ratio, which controls the power transmission factor. Therefore, frequency is all the more important in that, unlike magnetic permeability and resistivity, it is a variable which the user can control. Therefore, frequency adjustment enables high transmission factors to be obtained.

With cylindrical parts, as soon as the diameter-penetration depth ratio exceeds 4, the power transmission factor is more than 0.7 (*see* curve, Figure 14), and the maximum electrical efficiency, as for bodies of high resistivity with respect to copper, such as steel, whether magnetic or not, remains higher than 80%. For low resistivity, non-magnetic bodies such as copper and aluminum alloys, it is better to remain well above the previous value of this ratio, between 5 and 10, to increase an already relatively low efficiency (if core heating is used, these bodies offer excellent thermal conductivity, and this is not necessarily a drawback).

When the diameter to penetration depth ratio is between 2 and 4, efficiency remains high, between 70 and 80% for magnetic bodies heated below the Curie point, but decreases noticeably, about 50% above the Curie point, and becomes unacceptable, of the order of 20%, for low resistivity, non-ferromagnetic bodies.

Below a diameter to penetration depth ratio of 2, efficiency becomes very low, whatever the nature of the body to be heated.

In general, to obtain high efficiency, close to the possible maximum, attempts should be made to choose a frequency such that the diameter-penetration depth ratio is greater than 4.

Similar conclusions can be reached for non-cylindrical parts in which electrical efficiency, for equal resistivity and magnetic permeability, reaches higher values, as shown in the power transmission factor analysis (*see* curves, Figures 16 and 20).

### 2.3.5. *Effect of inductor geometry*

To obtain high efficiency, the load-inductor coupling must be as close as possible. However, practical values limit coupling (electrical insulation on inductor turns, mechanical and thermal protection, etc.). The complete length of the inductor should be used to heat the body. Where possible, to minimize end effects, the inductor length must be large with respect to diameter.

In some cases, attempts are made to improve coupling by using field concentrators (magnetic cores and sheets, etc.) or transverse flux inductors; the latter enable continuous heating at line frequency (50 Hz) of low resistivity flat products such as aluminum strip with an efficiency of about 90%, which would be impossible with conventional inductors [25], [147*d*], [148].

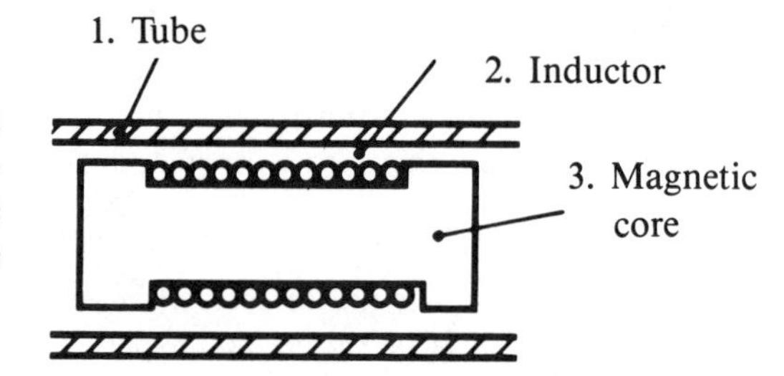


**Fig. 23.** Induction heating of a tube using a laminated core inductor inside the body

### *2.3.6. Electrical efficiency: sensitivity analysis*

The above analysis shows that many parameters control inductor electrical efficiency. However, the sensitivity of efficiency to these parameters should be taken into account to make full use of the operational latitudes available.

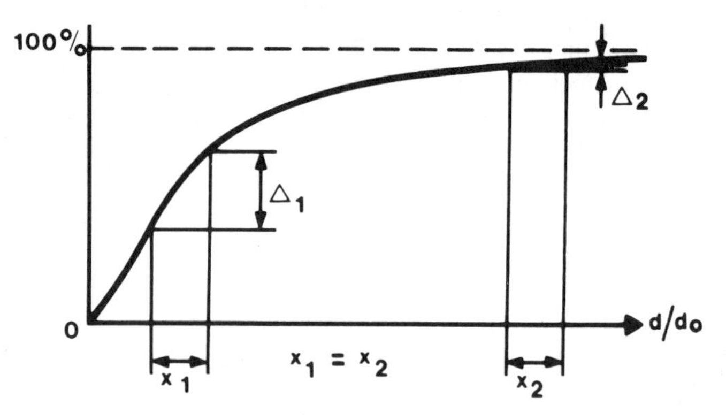


**Fig. 24.** Sensitivity of efficiency to a variation in diameter to penetration depth ratio (this curve was plotted from the curve providing the power transmission factor *F*, but is not identical to it).

An examination of the curve giving the power transmission factor versus diameter to penetration depth ratio fully illustrates the importance of sensitivity. If we consider an area in which this curve is practically horizontal, an increase in frequency has almost no beneficial effect on the power transmission factor, and therefore on efficiency. Conversely, on the steep portion of this curve, the same increase in frequency will cause an important increase in power transmission factor, but not of efficiency, since, if the body is ferromagnetic, efficiency may already be very high, which of course cannot exceed 1.

For ferromagnetic bodies below the Curie point, the quantity $\sqrt{\varrho_1\mu_1/\varrho_2\mu_2}$ of the efficiency expression is still low, less than 0.1, corresponding to a maximum efficiency of more than 90%, offering a wide choice in diameter to penetration depth ratio, and therefore frequency and inductor geometry. For example, efficiency will remain satisfactory, even if the load-inductor coupling is relatively loose; for a coupling factor of about 0.5, efficiency is still better than 80%. This low sensitivity of efficiency to coupling may be good for applications where the area to be heated must be kept clear to perform other operations (preheating or stress relief annealing of tubes during welding operations). However, in all cases, the highest efficiency compatible with operational requirements should be sought.

For non-magnetic bodies or magnetic heated beyond the Curie point, this freedom of choice is restricted. If resistivity is high a wide degree of freedom exists for coupling, but the choice of frequency is restricted since the diameter to penetration depth ratio must be adequate if an important decrease in efficiency is to be avoided. Conversely, where body resistivity is low, optimum values of the various parameters must be used to maintain an acceptable efficiency.

Finally, electrical efficiency is generally between 75 and 95% for ferromagnetic materials below the Curie point, and for high resistivity non-ferromagnetic materials such as graphite.

For correctly coupled, medium resistivity, non-ferromagnetic materials (between 10 and 100 $\times$ $10^{-8}$ $\Omega$.m, efficiency is between 50 and 75%.

Conversely, efficiency is lower, between 35 and 55%, for non-ferromagnetic bodies whose resistivity is less than 10 $\times$ $10^{-8}$ $\Omega$.m, such as copper, aluminum and some of their alloys (in the liquid state, resistivity is much higher, between 20 and 30 $\times$ $10^{-8}$ $\Omega$.m for these alloys, and efficiency increases noticeably). Therefore, the utmost attention must be devoted to the design of the inductor and to the choice of frequency.

## 2.4. Overall efficiency and energy efficiency

Overall efficiency is a function of electrical efficiency, but also depends on heat losses in the body to be heated due to radiation and convection. Losses due to compensation capacitors, supports, power supplies, etc., must be added to these losses. The overall efficiency is generally 5 to 15% less than that of electrical efficiency, since the power densities obtainable with induction heating are high and heating times therefore very short. However, an accurate appreciation of total efficiency can only be obtained in terms of each application.

When compared to other heating methods, overall efficiency is excellent and often enables substantial energy savings.

## 2.5. Power factor

Leakage flux is high in most induction heating installations, and depends on the specific application and the spatial relationship between the coil and the body to be heated. The reactive power increases as the difference between the inside diameter and the outside diameter of the body to be heated increases.

Generally, the power factor varies between 0.05 and 0.6, depending on load, dimensions and frequency. In all cases, reactive power must be compensated by insertion of capacitors which take the power factor to around 1 [60], [78], [111*a*].

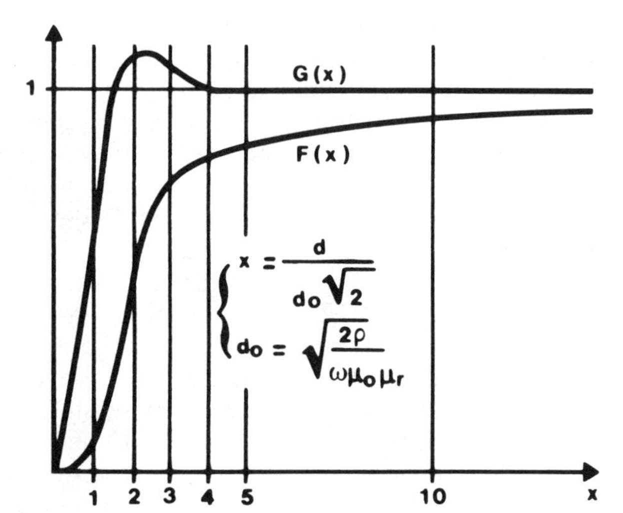


**Fig. 25.** Active power transmission factor $F$ and load reactive power coefficient $G$.

The power factor is equal to:

$$\cos \varphi = \frac{P_w + P_i}{\sqrt{Q_t^2 + (P_w + P_i)^2}},$$

$P_w$, active power dissipated in the load;
$P_i$, active power dissipated in the inductor;
$Q_t$, total reactive power.
For a cylindrical body, $Q_t$ can be calculated from the reactive power in the air gap $Q_E$ equal to $L_E I_l^2$, i.e.:

$$Q_E = \frac{\mu_0 \pi^2 (D^2 - d^2) f N^2 I_1^2}{2H},$$

and in the body to be heated $Q_w$, defined by:

$$Q_w = \frac{\pi d \rho G N^2 I_1^2}{h d_0},$$

$G$ being a factor similar to power transmission factor $F$.

### 2.6. Current and heat penetration

Current penetration $d_0$ must not be confused with heat penetration. The current penetration depth is simply a value used to obtain an approximation of the transmitted power dissipation area. Heat penetration depends on current penetration depth, transmitted power, the thermal diffusivity of the material and its emissivity.

## 3. INDUCTION HEATING EQUIPMENT

Basically, induction heating system consists of:
— one or more heating inductors;

— a power source, equipped with a battery of compensation capacitors and, where necessary, a medium or radio frequency generator;
— a system for cooling the inductor and other components (capacitors, generators, etc.);
— a load handling system;
— installation monitoring and control devices.

Such equipment varies with the application involved, and is described in further detail in the corresponding chapters. However, these types of equipment have common features.

## 3.1. Inductors and matching to the current source

Generally, to keep Joule effect losses to a minimum, inductors made of very pure electrolytic copper are used, and in most cases are water- and, rarely, air-cooled [2], [8], [9], [21], [25], [26], [59], [76], [85], [86].

Thermal efficiency depends largely on inductor design. Faced with the wide variety of part shapes heated by induction, it is therefore necessary to adapt inductors to the thermal effect desired.

### *3.1.1. Matching of the inductor to the source*

Matching of the inductor to the electrical source supplying it is generally provided by transformers and capacitors.

Where it is possible to connect capacitors in parallel, or more rarely, in series with the inductor, matching of the inductor system to the source is made through a transformer, or directly via the generator.

In other cases, common in heat treatment such as brazing and welding, compensation is made by capacitors at the generator itself.

Electrically, the equivalent diagram of the inductor is simple and consists of:
— a load inductance $L$, which is that of the inductor coupled to the part or body to be heated;
— a resistance $R$, equivalent to the primary, connected in series with $L$.

**Fig. 26.** Inductor-load equivalent diagram.

At a given frequency $f$, the inductive load offers a quality factor or overvoltage factor defined by the ratio $Q = L\omega/R$, where $\omega = 2\pi f$.

In most practical applications, $Q$ is located between extreme values $Q_{min}$ and $Q_{max}$, such that:
— $Q_{min}$ is greater than 3-4 for inductors closely coupled to a magnetic load (iron or steel) or a highly resistive load (graphite);
— $Q_{max}$ is less than 50-100 for inductors loosely coupled to non-magnetic materials or good conductors (copper, aluminum, etc.).

At medium frequencies from 1 to 20 kHz, the mean value is around $Q = 10$; with radio frequency heating, this factor is generally between 15 and 20.

The value of inductance $L$ depends basically on the shape of the inductor. They vary considerably depending on whether the inductors are intended for very localized heating — single turns, loops (with low inductances — some 1/100 $\mu H$ or 1/10 $\mu H$), or elongated inductors for overall heating, consisting of numerous turns, the self-induction factor being commonly in the tens of $\mu H$. Therefore, when the same sources are used with different inductors, it is necessary to adapt the inductor to the source over a rather wide range.

For reasons of safety and to prevent arcing in heat treatment or brazing applications, efforts are made to operate at rather low voltages, of the order of some tens of volts. The inductor is then adapted to the source through a transformer:

— a magnetic circuit transformer for medium frequency heating;
— an air-core transformer for radio frequency heating.

The values of the $Q$ factor and inductance $L$ are used to calculate the transformation ratio.

In other applications in which high impedance inductors are used in environments which do not give rise to insulation or electrical leakage problems (melting, heating prior to forming, core annealing, etc.), the inductor can be driven directly by the source. However, a transformer may be used to vary dissipated power (for example, in low frequency crucible induction furnaces) [109].

### *3.1.2. Different types of inductors*

The most simple and widely known inductor is the solenoid type already described. This is widely used for heating for forging, melting and heat treatment of metals. This type of inductor is suitable for field, power and efficiency calculations. Moreover, efficiency is often the main factor when selecting a solenoid inductor, especially in overall heating applications where energy costs are important.

Induction heating applications would be limited if the solenoid were the only configuration used. Conversely, due to development of inductors of various shapes, induction heating has satisfied numerous heating requirements such as heat treatment, brazing and welding.

Inductor shapes are highly varied. In many cases, magnetic field concentrators are used to improve inductor efficiency or modify the heating effect in some areas.

For example, in a linear inductor, the copper conductors, are laid parallel to instead of enveloping the body. The thermal efficiency of an inductor of this type is poor, due to high magnetic leakage. It is therefore necessary to limit or eliminate such leakage to obtain high thermal efficiency. The solution is to use magnetic field concentrators which consist of laminated magnetic sheets 5/100 to 20/100 mm thick. They cover the copper conductors of the inductor, and thus concentrate almost all the current on the inductor surface facing the part. For inductor-part air gaps of 1 to 2.5 mm, efficiencies of 50 to 75% are obtained.

The specific power applied to the part is between 1 and 3 kW/cm$^2$, close to that obtained with enveloping inductors [109], [S], [V].

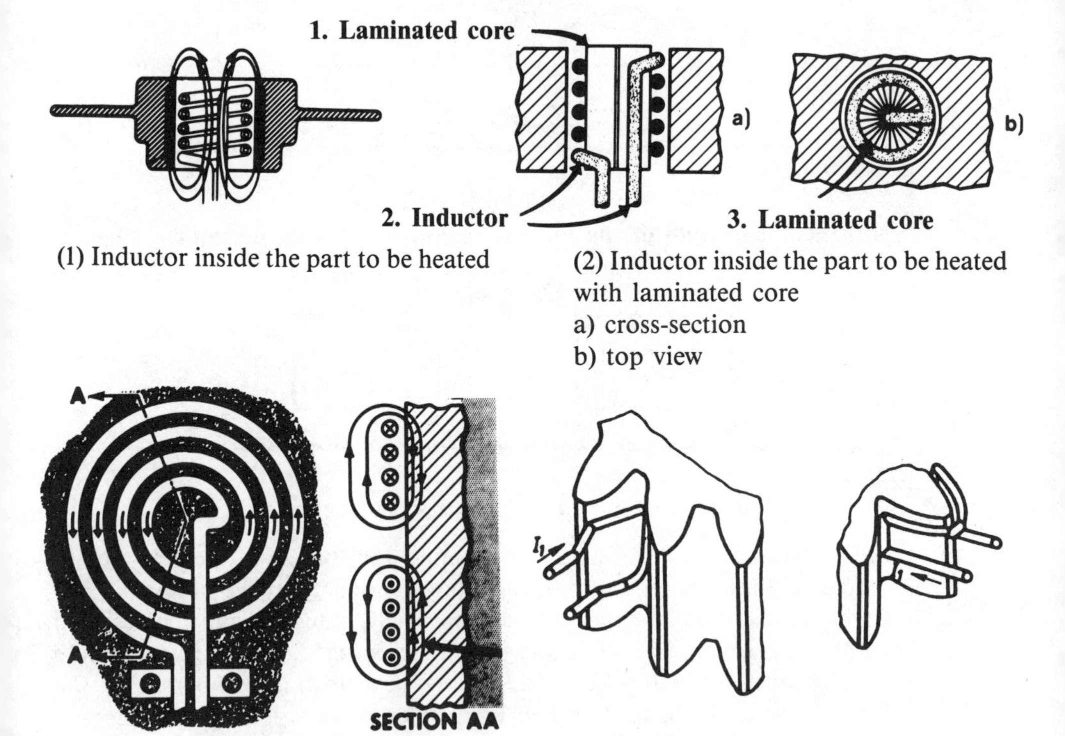


(1) Inductor inside the part to be heated

(2) Inductor inside the part to be heated with laminated core
a) cross-section
b) top view

(3) Flat inductor and part heating

(4) Inductors for heating gear teeth.

**Fig. 27.** Different types of inductors

Linear inductor technology, which involves the use of magnetic circuits, is used over the medium frequency range from 1 to 10 kHz or even 20 kHz. However, use of very fine magnetic sheets — 2/100 to 5/100 mm — enables them to be used up to radio frequencies of 100 to 250 kHz.

In addition to the efficiency aspect, the use of magnetic field concentrators enables very accurate adjustment of the thermal effect on a clearly determined area of the part by selective positioning of the inductor magnetic sheets.

To treat a revolving part with this type of inductor, the part must be rotated in front of the inductor (or, if necessary, the inductor must be rotated around the part, which is more difficult) to obtain uniform heating.

Due to the high magnetic fields and currents involved in induction heating, the electromagnetic forces resulting from interaction of the magnetic fields and the inductor and load currents can have good and sometimes bad consequences.

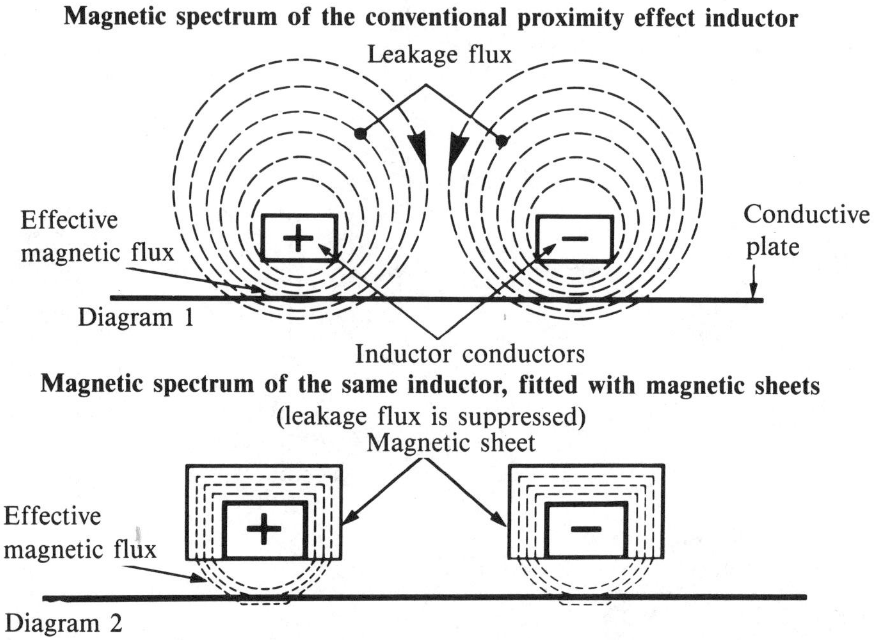

**Magnetic spectrum of the conventional proximity effect inductor**

Leakage flux

Effective magnetic flux

Conductive plate

Diagram 1

Inductor conductors

**Magnetic spectrum of the same inductor, fitted with magnetic sheets**

(leakage flux is suppressed)

Magnetic sheet

Effective magnetic flux

Diagram 2

**Fig. 28.** Linear inductors with magnetic sheets [S].

In induction melting furnaces, for example, the stirring in the bath due to the effect of these forces improves metallurgical quality of the products, only if it remains below a certain threshold. Conversely, inductors must be designed to withstand mechanical effects caused by the electromagnetic forces. For core heating products (reheating prior to forming, heat treatment of bars, etc.), the inductor is often embedded in concrete which protects it from the thermal radiation of the part; in crucible melting furnaces, the inductor is maintained by wedge and stiffener systems. Various solutions are used in heat treatment applications such as brazing, or the manufacture of semiconductors; copper conductors offering high mechanical inertia (thick walls) or lined with copper plating to increase stiffness (thin wall inductors), inductor assembled by copper welding and rigorous mechanical attachment, etc.

The electromagnetic forces are subject to Laplace's laws. Interaction between two parallel linear conductors is $F = 2(I_1 I_2/d) \times 10^{-7}\ N/cm$, where $I_1$ and $I_2$ are the currents flowing in the two conductors in Amperes, and $d$ the distance between the two conductors in centimeters. Moreover, an interaction exists between field and current in materials where induced currents develop, which, in cylindrical bodies, results in a radial centripetal force or pressure and an axial force. In solid structures, the radial force $F_r$ is inoperative. Axial force $F_z$ can cause levitation phenomena. In liquid baths, both types of force cause electromagnetic stirring which is low at radio frequencies, significant at medium and very high at low frequencies.

The various types of inductor, their special features and the effect of their electromagnetic forces are described in detail in the paragraphs devoted to the different applications.

### 3.2. Frequency converters

From the line frequency, frequency converters supply current at a usually higher frequency to the heating inductor. Four types of generators are used:

— rotary converters;
— solid state thyristor generators;
— vacuum tube generators;
— saturated-core transformer static generators.

#### *3.2.1. Rotary converters*

Rotary converters are used as a medium frequency source for frequencies of 500 to 10,000 Hz [113].

Essentially, a rotary converter unit consists of a three-phase induction motor, supplied from 50 Hz line, driving an alternator which provides the current at the desired frequency.

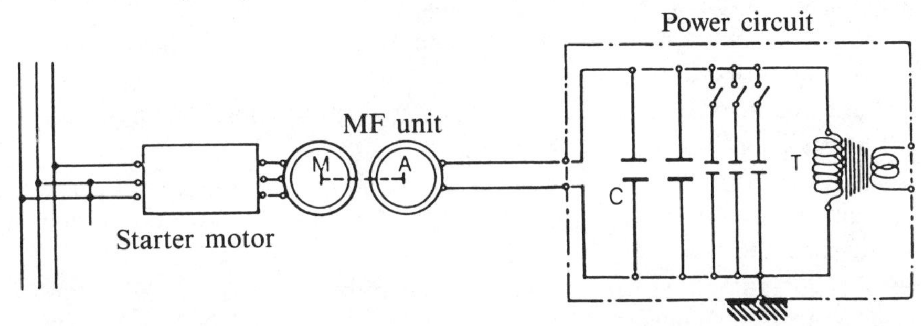


**Fig. 29.** Diagram of a rotary converter. The load circuit is located beyond transformer T and is not shown. This arrangement, in which power factor compensation takes place at the input to the power transformer, is used for low powers (heat treatment, etc.); for higher powers (e.g. in melting) compensation is made at the load circuit.

This alternator supplies the circuit consisting of the inductor and the compensation capacitors. In general, these capacitors consist of a fixed battery of tuning capacitors and capacitors which can be switched, which are automatically placed in service in modern installations, so as to provide resonance $LC\omega^2 = 1$.

The voltages produced by alternators are often higher than those required by the inductors. In this case, matching transformers are used. The excitor enables gradual and simple on-load adjustment of the alternator output voltage, and therefore converter power.

The high currents flowing through the coil and the capacitors imply that paths must be as short as possible. For noise reasons, converters are often located in premises separated from the remainder of the installation; the capacitor batteries, which are generally located in separate premises for high powers, must be as near as possible to the work station. In some heaters intended for heating of metals prior to forming, the capacitors are located under the inductor.

Except for low powers, the converter units at present used are, in most cases, very silent, water-cooled vertical shaft monoblock systems, with rotation speeds of 3,000 rpm. The biggest converters at present constructed have a power of 1,250 kW vertically placed, and 1,700 kW horizontally placed.

Rotary converters efficiencies are of the order of 85% over the frequency range 1-3 kHz, and 75 to 80% over the frequency range 3-10 kHz.

For a given frequency, unit efficiency decreases slightly as unit power decreases and vice versa; for a given unit power, efficiency drops slightly as frequency increases.

Also, when used at a small part of its rated power, the efficiency of rotary units decreases.

### *3.2.2. Medium frequency thyristor solid state generators*

Medium frequency thyristor solid state generators provide frequencies ranging from a few hundred Hertz to approximately 10,000 Hz.

With this type of generator, the 50 Hz line AC is first rectified; the DC obtained is then converted into AC.

Basically, an induction heating thyristor static generator installation consists of:

— a voltage step-down transformer;

— a Graetz bridge thyristor rectifier, with a smoothing inductance; the converter output voltage, which is a function of the DC voltage, can be adjusted by controlling the rectifier;

— a single-phase thyristor converter which converts the DC flowing through it into medium frequency AC;

— the load circuit, consisting of the furnace coil or the heater and the battery of compensation capacitors.

Thyristors are water or air cooled, each technique having its advantages and drawbacks.

Together with the inductor and the capacitors, the load circuit forms an oscillator circuit which tends to oscillate spontaneously at its resonant frequency. The converter operates automatically at load circuit resonant frequency if thyristor triggering is slaved to this frequency. Therefore, over a certain frequency range, operating frequency is adjustable.

The furnace circuit can be a series or parallel oscillator circuit. With a series circuit, the thyristors are calibrated to provide the total current required by the furnace. In a parallel circuit, the thyristors must maintain the total voltage across the terminals of the load circuit. As long as the voltage across the furnace terminals is less than 1,200 V approximately, a parallel converter is used since this is the most simple to design and the thyristors are more easily protected. For higher voltages, series converters are often used since the thyristors are then subjected to a much lower voltage than that across the furnace terminals. The load circuit is supplied either directly or through a transformer.

On-load power is adjustable by varying the rectified voltage, i.e., the rectifier thyristor phase angle which can be accurately adjusted. The power performance

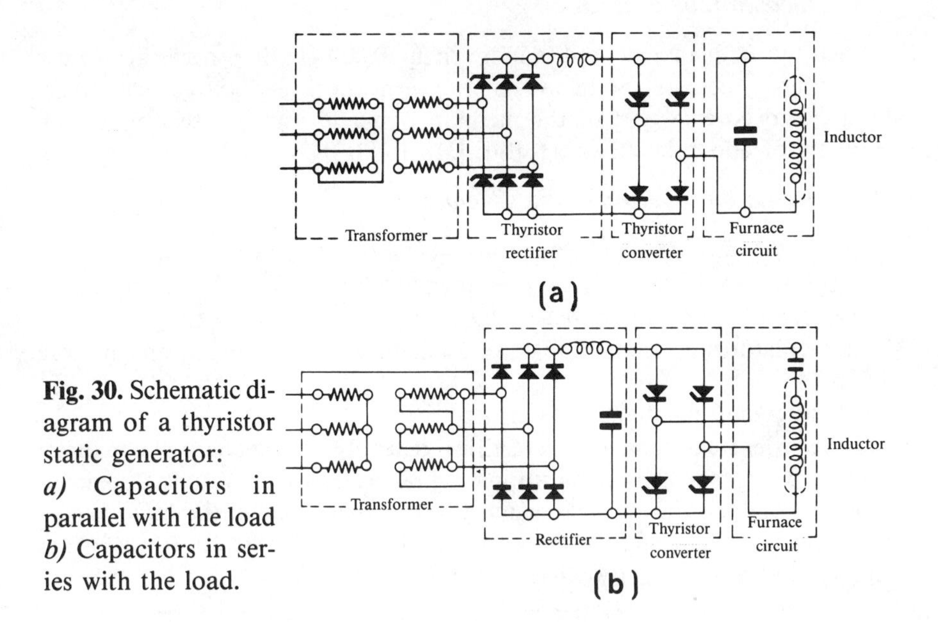


**Fig. 30.** Schematic diagram of a thyristor static generator: *a)* Capacitors in parallel with the load *b)* Capacitors in series with the load.

of a converter depends on frequency. If the frequency is much lower than the rated frequency, the converter cannot provide full power; if higher, the converter cuts out as soon as it reaches the maximum frequency for which it is designed. The frequency range is generally 25 to 30% of the rated frequency. At present, thyristors enable frequencies of the order of 10,000 Hz. This upper limit can, however, be exceeded with special circuits such as sequential converters in which the thyristor operating frequency is only one-third of the output frequency. Devices of this type are also used to design two-frequency converters operating at rated power over two frequency ranges within a ratio of 3 (for example, 3.3 and 10 kHz); this concept may be of interest in heating magnetic bodies before and after the Curie point, although two-frequency converters are rarely encountered in practice.

At present, converter maximum power is about 2,000 kW (500 to 3,000 Hz), but it is always possible to increase power by connecting several converters in parallel.

Thyristor generator efficiency is high for frequencies of between 500 and 3,000 Hz and can reach 90% and 95% for frequencies of between 3,000 and 10,000 Hz. The off-load absorbed power of thyristor generators is less than 2% of the rated power and can start almost instantaneously. Therefore, overall efficiency is excellent.

Recently, the reliability of thyristor generators has increased greatly due to progress in power electronics and very simple repair techniques that have been developed with automatic fault-finding and plug-in modules [10], [12], [39], [41], [46], [62], [65 *c,d,f*], [81], [91], [106], [108], [113], [151].

### 3.2.3. *Vacuum tube generators*

When the frequency desired is more than 10,000 Hz, it is necessary to use vacuum tube generators. Some vacuum tube generators (periodic generators) are also used to cover a part of the medium frequency range with a minimum frequency of about 4,000 Hz [1], [14], [31], [108], [113], [136].

#### *3.2.3.1. Design of a vacuum tube generator*

Basically, a vacuum tube generator consists of the following items:
— a rectifier, producing the power required at 5,000 to 15,000 VDC, depending on the type of oscillator triodes;
— an oscillator stage, consisting of an oscillator circuit excited by one or more triodes;
— a load-circuit matching transformer.

The rectifier stage consists of a rectifier, generally a three-phase silicon diode bridge, and a voltage step-up transformer supplying the triode. Various techniques are used to adjust the voltage, mercury vapor thyratrons in the high voltage rectifier, autotransformer or magnetic amplifier in the step-up transformer primary, and more recently, thyristor controllers.

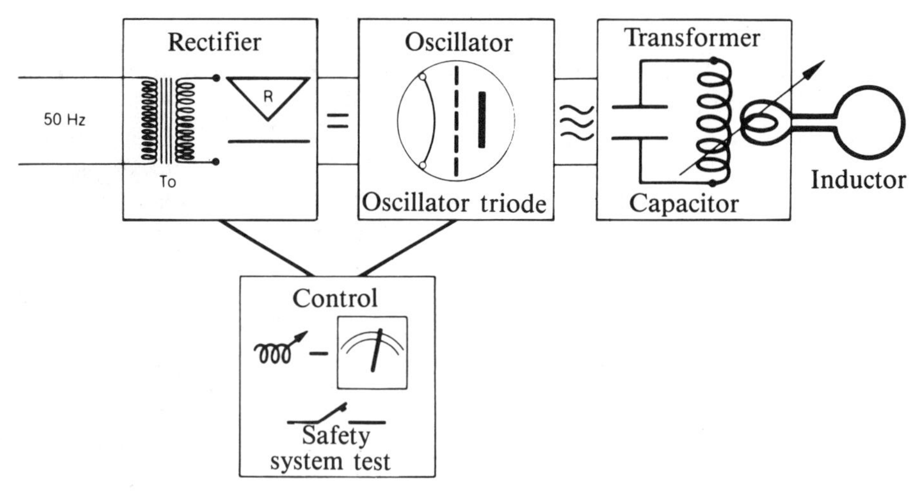


**Fig. 31.** Diagram of a radio frequency vacuum tube generator [4].

The rectified voltage maintains the oscillator circuit, comprising the inductor and the capacitors associated with it, in oscillation. The oscillator circuit is designed to accept the very high reactive energy passing through it.

The inductor often operates at a much lower voltage than that of the tube. In such cases, a matching transformer is required. Generally, and because of the radio frequencies, air transformers are used whose transformation ratio can be adjusted either by an autotransformer configuration, or, more frequently, by a secondary winding which moves with respect to the primary. Matching transformers used in practice most often consist of one turn of a copper tube in the secondary in series with the heating inductor, and six to twenty primary turns coupled to the oscillator circuit. Matching transformers must be used when the inductor has only one or two turns transmitting a high specific power, as in surface heat treatment or brazing. Such devices enable the same generator to heat various parts with inductors of different configurations.

The matching transformer is located either between the inductor and the tank capacitors, or between the capacitors and the tube. This type of radio frequency generator is known as an aperiodic generator.

The power can be adjusted by varying the rectified voltage with the controller. Since the radio frequency circuit is a self-triggered oscillator, it operates at any frequency, and automatically provides a purely resistive load across the anode-cathode terminals of the tube provided the excitation voltage is correct. For any load offering a self-induction factor $L$ and quality factor $Q = L\omega/R$ (*see* paragraph 3.1.1), the frequency is automatically such that $f = 1/2\pi\sqrt{LC}$ and the anode resistance is equal to $2\pi Q\sqrt{L/C}$. To obtain full power, the inductor must be matched to the capacitor. Regulations concerning radio frequency generators (radio frequencies) are such that one standard frequency is gradually being adopted [108], [136].

The efficiency of radio frequency generators is about 70 to 90% for frequencies between 10 and 100 kHz, and slightly lower, 55 to 70%, for radio frequencies. Starting is almost instantaneous, and varies from 15 to 120 seconds, depending on power, which corresponds to the vacuum tube warmup time. The off-load power is about 3% of full load. Tube service life is generally between 7,000 and 10,000 hours. At present, there are industrial radio frequency generators with powers between 1 and 1,200 kW.

### *3.2.3.2. Aperiodic generators*

Aperiodic generators offer several advantages over conventional radio frequency generators in which the capacitors are an integral part of the generator. In a conventional generator, the active power required to heat the load placed in the inductor is tapped off the generator's oscillator circuit and applied to the inductor through a matching transformer and an electrical line connecting the secondary of this transformer to the inductor.

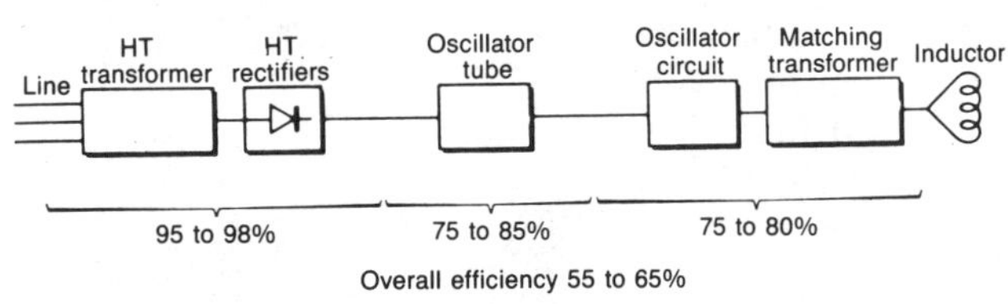


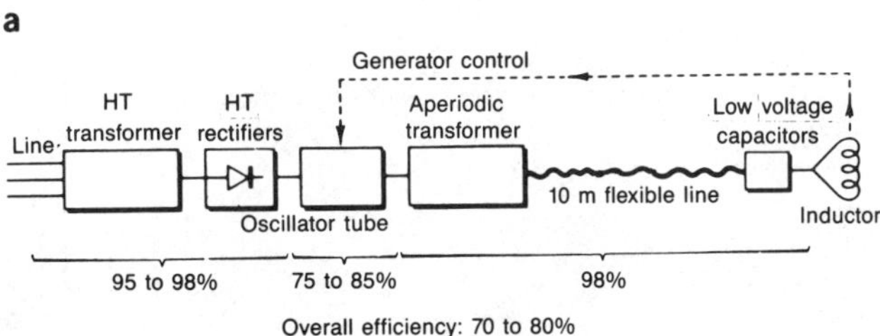


**Fig. 32.** Vacuum tube generators [E]:
*a)* conventional oscillator
*b)* aperiodic generator

The reactive power involved is very high, 10 to 20 times higher than active power. Apparent power $S (S = \sqrt{P_\omega^2 + P_r^2})$ is essentially due to the reactive power and the current $I (I = S/U)$ is very high, since voltage $U$, for safety reasons in applications such as surface heat treatment or brazing, must be low. This technology therefore precludes the use of long lines between the generator and the inductor due to the high current flowing in the inductor. The generator must be located immediately beside the inductor.

Conversely, an aperiodic generator does not have a separate oscillator circuit. There is no inherent operating frequency, and the oscillator is completely controlled by the load circuit comprising the inductor and a battery of low voltage capacitors which form an oscillator circuit tuned to the heating frequency [35], [65*k*], [110].

The load circuit is connected to the vacuum tube through a special, high efficiency matching transformer. Therefore, the generator only fulfills its basic role, i.e. provision of active power $P_\omega$ — only the active power $P_\omega$ is carried in the transmission line and the apparent power $S$ is limited to $P_\omega$.

Therefore, it is no longer necessary to place the generator near the inductor which greatly increases system flexibility. Efficiency is increased 5 to 10% approximately. The frequency range for vacuum tube generators has been increased to include 4,000-30,000 Hz previously inaccessible to conventional vaccum tube generators.

### *3.2.4. Saturable core transformer static generator*

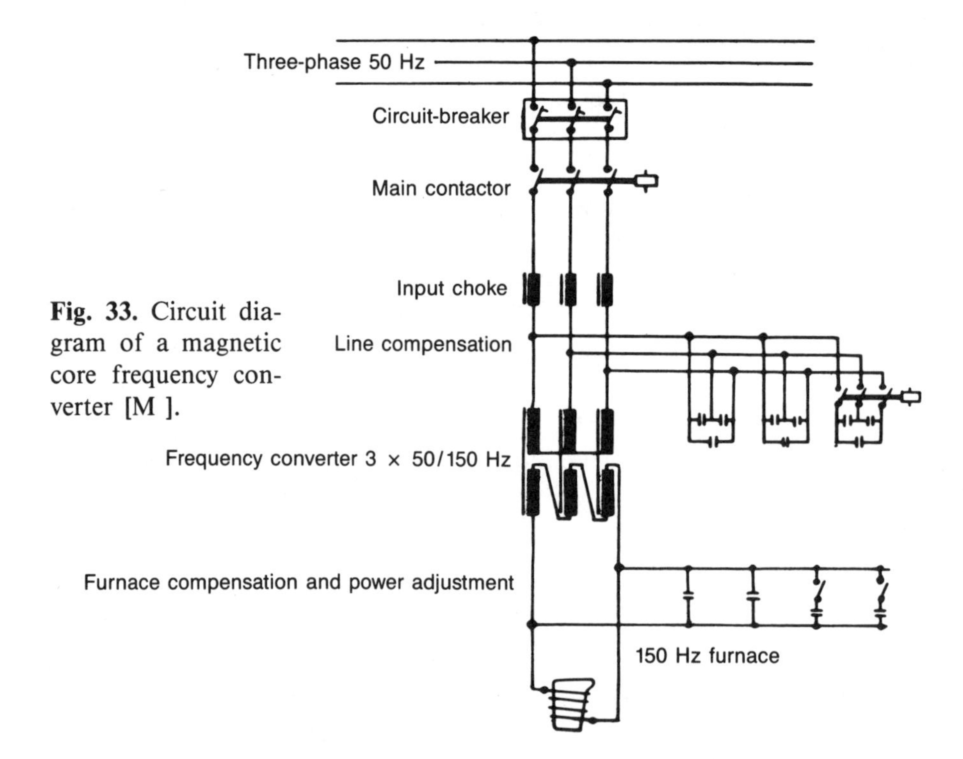


**Fig. 33.** Circuit diagram of a magnetic core frequency converter [M ].

These generators are used to produce frequencies equal to 3, 5 and 9 times the line frequency (in Europe, 150, 250 and 450 Hz). Basically, these generators are used to supply crucible induction furnaces and furnaces for heating of metals prior to formming [135], [L], [M].

A frequency multiplier static converter uses low loss, oriented crystal lamination core, single-phase transformers, operating well below saturation.

The special primary and secondary winding arrangement (the identical primary coils of the single-phase transformers are star-coupled, with the secondary windings in series, for a frequency trebler) and harmonics due to saturation of the magnetic cores enable a single-phase voltage, the frequency of which is equal to 3, 5 and 9 times the mains frequency, to be induced in the secondary.

The chokes located at the converter input limit harmonic feedback to a rate acceptable by the electricity distributor.

In magnetic converters, saturable core transformers can be considered as valves which allow current to flow when their core is saturated, and block current when the cores are not saturated. In thyristor generators, this magnetization characteristic has practically replaced electronics for opening and closing semiconductor switches.

### *3.2.5. Comparison of frequency converter characteristics*

#### *3.2.5.1. Comparative fields of use of frequency converters*

Vacuum tube generators are primarily used for the production of radio frequencies. Their uses are highly diversified: surface heat treatment, brazing, welding, fabrication of semiconductors, bonding metals, bonding metals with other materials, low quantity melting of metal or non-metallic bodies using a sensor, etc.

The field of application of saturable-core transformer-type frequency multipliers is basically limited to crucible-type induction melting furnaces and to furnaces for heating metals prior to forming. They can compete with other medium frequency generators, rotary converters or thyristor generators where the frequencies desired are close to those provided by multipliers. Basically, economic considerations, and sometimes frequency flexibility dictate selection which can be difficult when choosing between static thyristor generators and rotary converters that share a wide frequency range.

#### *3.2.5.2. Comparative characteristics of medium frequency thyristor generators and rotary converters*

Rotary converters are extremely rugged, simple and common failures are very easy to locate. Thyristor-type static generators, however, offer many advantages over rotary converters.

— *Automatic frequency adaptation*

During heating, magnetic permeability, resistivity and sometimes coupling vary. Therefore, furnace inductance is not constant.

In alternator supplies, where the rate of rotation and therefore frequency are fixed, it is necessary to provide a battery of capacitors whose value can be adjusted during heating to maintain a high power factor and therefore adequate efficiency. Batteries of relay-switched capacitors must then be installed. With static converters, the frequency is adjusted when the furnace impedance varies to maintain resonance; the maximum power is constantly transmitted, increasing productivity.

— *Excellent performance*

Static converter efficiency is much higher at full load than at partial load. At low load, the advantage of converters becomes very clear. Now, during down times, or with low loads which is frequent, the inertia of rotary generators (starting times of several minutes) prevents these being shut down during down times; therefore, these continue to rotate and absorb a significant amount of energy (bearing friction, cooling, etc.).

**Fig. 34.** Compared efficiencies of a thyristor static generator *(a)* and a rotary converter unit *(b)*.

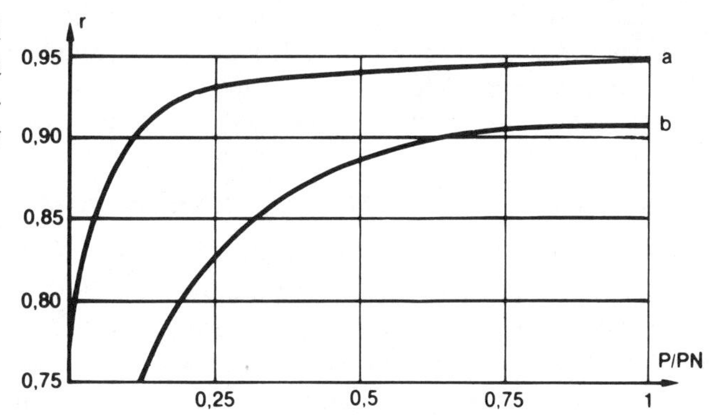


Conversely, since it starts quasi-instantaneously, the converter can be shut down. In terms of total efficiency, the advantage of the static generator over the rotary converter is even more notable. Analysis performed on industrial installations shows that energy savings of between 10 and 20% have been obtained by using static converters.

— *Light support structures*

The high power in rotary converters necessitates heavily anchored foundations, while cabinet-mounted static generators only require special cable runs or passages. Static generators are lighter and less bulky than rotary units.

— *Absence of noise*

Noise generated by rotary converters often requires them to be isolated, thereby increasing cable lengths and therefore losses. Thyristors, however, are very quiet and can be located near the furnace.

— *Water saving*

Even when water cooled, static converters have very low water requirements due to their high efficiency.

— *Ease of control and regulation*

Control of thyristor static generators is easily automated (automatic adjustment of the converter to the inherent frequency of the load oscillator circuit, no handling of balancing capacitors).

— *Low maintenance requirements*

The maintenance required by static generators is low since there are no moving parts, but rotary units require servicing of bearings, ball-bearings and slip-rings. If protected, the service life of thyristors is long.

Due to these numerous operational advantages, thyristor static generators are now being substituted for rotary converter units since they cost the same, or even less.

## 3.3. Induction heating installation cooling systems

### *3.3.1. Cooling equipment*

Each of the components of an induction heating installation loses power through heat conversion which must be transfered to heat sinks with high exchange coefficients. Water is therefore the most common cooling fluid especially for the inductor. Air cooling is sometimes preferred for simplicity and reliability. In many cases when the power transferred is low or efficiency high and coil losses limited, capacitors and thyristor static generators and inductors are air cooled (e.g. channel melting furnaces).

The internal circuitry of the cooling equipment follows sinuous, inaccessible paths, with low cross-section. They cannot accept ordinary water, which is full of impurities, causes scale and can even cause electrolytic corrosion due to the voltages applied.

Therefore, water offering the qualities required by the cooling equipment must be used. Water cooling systems can be classified as follows:

— single passage system;

— open cooling tower;

— cooling tower with heat transfer;

— aerorefrigerant cooling;

— *Single passage cooling system (open-circuit)*

The water flows through the heat transfer or through the equipment to be cooled itself and is then discharged. For a given flow rate, heat removal depends entirely on the increase in water temperature.

For a mean increase of 10°C, heat transfer is 10 kcal (or 42 kJ approximately) per kilogram of water. This cooling method is feasible only for low powers or where water is very cheap.

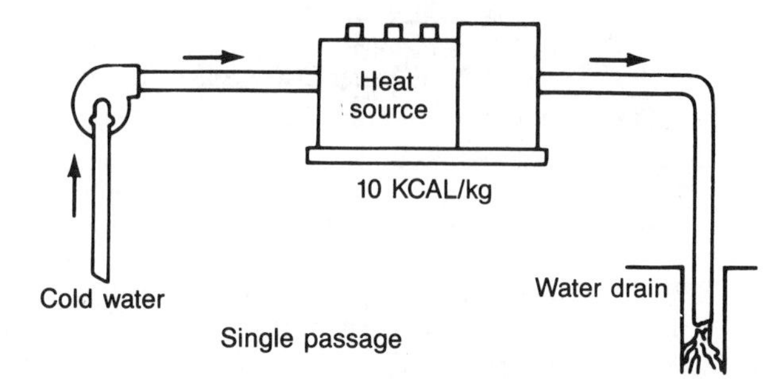


**Fig. 35.** Single passage cooling system

— *Open cooling tower*

Open cooling towers use the evaporation cooling principle. The water to be cooled is in contact with the moving air; the impurities in the air are drawn off by the water.

Open cooling towers save 95% of the water required for a single passage system, but do not eliminate scale and corrosion problems. In fact, these problems are aggravated since evaporation of the water increases the impurity content. Moreover, since water is highly aerated in open towers the danger of corrosion due to oxygenation is greater.

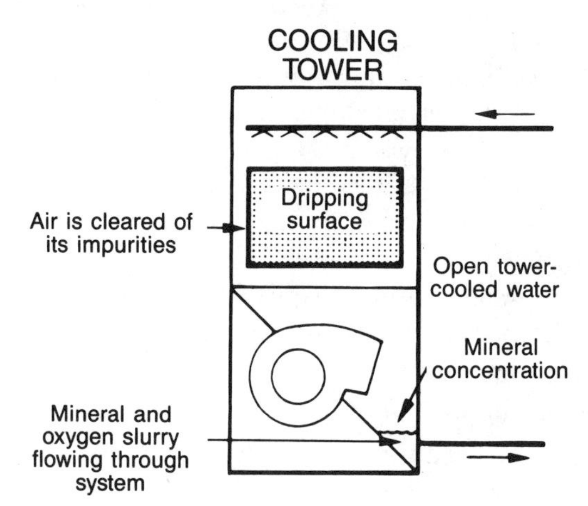


**Fig. 36.** Open cooling tower

— *Cooling tower with heat exchanger*

To benefit from the efficiency of evaporation cooling and from the maintenance advantages of closed circuit cooling, two-component installations exist: a cooling tower and separate heat exchanger. The cooling tower water (open-circuit) flows from one side of the exchanger, and the fluid to be cooled (closed-circuit) flows through the other side of the exchanger.

Most high power installations are equipped with cooling systems that supply excellent quality for water inductor cooling efficiency.

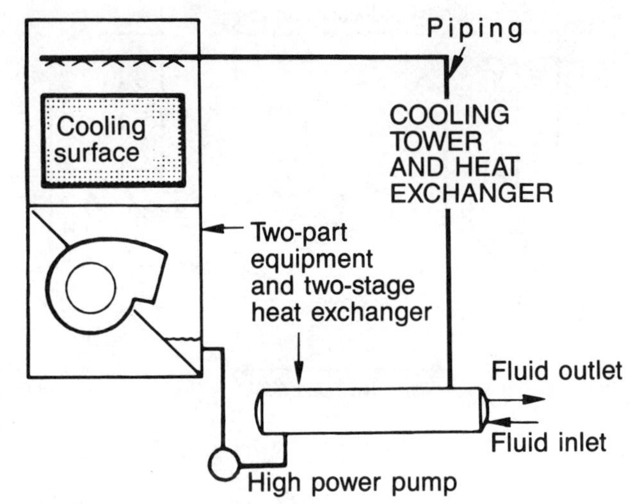


**Fig. 37.** Cooling tower with heat exchanger

— *Air-to-air cooling tower*

Air coolers are closed-circuit coolers, but do not offer the energy savings of evaporative cooling. Generally, air coolers (air-to-air coolers) consist of a heat transfer with a finned tube stack and several fans or blowers. Efficiency is determined by sensible heat transfer.

With an air-cooled system, it is possible to obtain cooling of up to 8 to 10°C of dry bulb-measured ambient temperature.

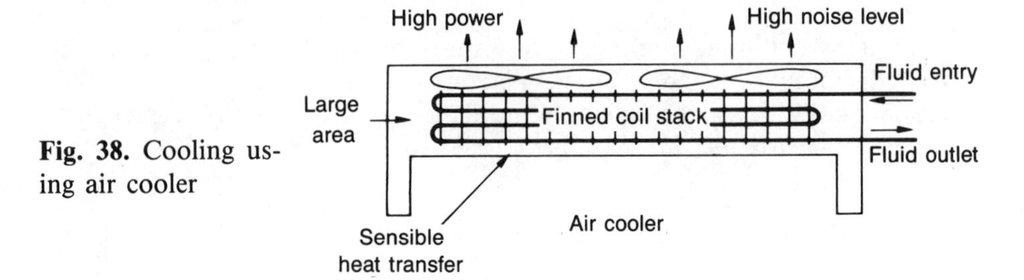


**Fig. 38.** Cooling using air cooler

### *3.3.2. Energy recovery on induction heating installations*

Although efficiencies obtained with induction heating installations are often very high, certain energy losses can be recovered to further improve the energy efficiency.

Figure 39 gives an example of the thermal balance of a low frequency induction crucible.

The heat given off by the transformers, capacitor batteries, cables, and, where applicable, generators is of only secondary importance; in winter, it can provide basic heating for small premises, in particular when air conditioning equipment is used.

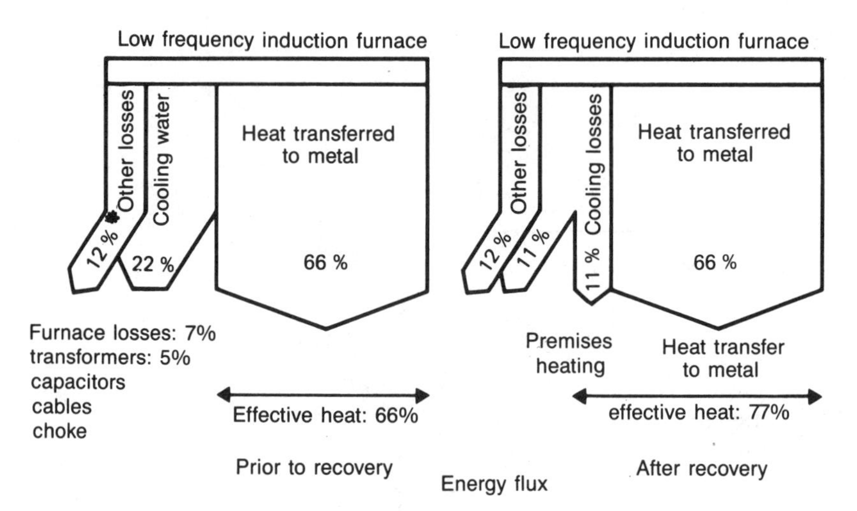


**Fig. 39.** Thermal balance of low frequency crucible induction melting furnace [148].

The heat given off by the inductor cooling circuit represents most of the losses and is generally vented from the plant. This field is rich in recovery opportunities [66], [111*c*], [114], [127].

Two techniques can be used to recover the energy contained in inductor cooling water: heat pump and exchanger systems.

From the energy viewpoint, the heat pump is a high performance process, but, due to its relatively high investment costs, exchanger systems are often preferred.

The recovery installation must not disturb the manufacturing process, and the inductor must be kept at low temperature so as not to damage the electrical insulation. However, the cooling water temperature must be high enough to prevent condensation of atmospheric water vapor.

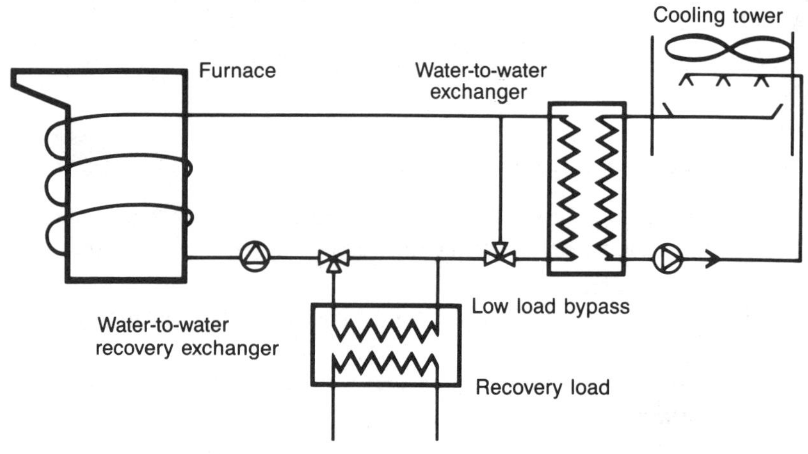


**Fig. 40.** Diagram of an induction furnace recovery installation [148].

For most induction heating equipment, the cooling water inductor inlet temperature must not be less than 30°C and the outlet temperature must be 60 to 70°C if the furnace is to operate correctly.

Around 60°C, it is therefore possible to recover energy for multiple applications. At the output from the recovery exchanger, water can easily be taken to a temperature of 45 to 55°C. The uses to be preferred are those having the longest operating times and which offer the best possible payback;

— technical hot water;
— sanitary hot water (showers);
— office and workshop heating.

However, the layout of the installations and investments to be made may reverse this order of priority.

When recovered energy is used to heat premises and produce sanitary hot water, investment cost recovery occurs in one to three years.

## 4. ADVANTAGES AND DISADVANTAGES OF INDUCTION HEATING IN INDUSTRIAL PROCESSES

Whatever the industrial application, induction heating offers several advantages:

— fast heating because of high power densities;
— -precise localization of the thermal effect due to adequate design of the inductor and the effect of frequency on current distribution in the body to be heated;
— very low thermal inertia, induction heating system response being practically instantaneous in most applications;
— possibility of heating to very high temperatures with efficiency practically independent of temperature;
— ease of automation of equipment;
— reproducibility of operations performed;
— heating efficiency often very high;
— no pollution due to heat source;
— very good working conditions.

Other advantages are highlighted by an examination of each application.

Conversely, induction heating systems sometimes require high investment costs. To limit capital expenditure projects should be examined carefully.

## 5. INDUSTRIAL APPLICATIONS OF INDUCTION HEATING

Induction heating is widely used in industry; applications are not simply limited to heating conductive materials such as metal alloys or graphite, since by using a susceptor, for example a graphite crucible, it is possible to heat products which do not conduct electricity (glasses, ceramics, etc.).

The following paragraphs describe the principal applications of induction heating in industry, emphasizing those where the installed powers are highest or most diversified. Such applications can also be widely used to analyse other applications [8], [21], [52], [55], [65*g*], [74], [80], [87], [89], [128], [129], [147*g*].

## 5.1. Choice of frequency and industrial applications of induction heating

The study of induction heating characteristics has shown that frequency plays a major role, due to its influence on skin effect and transferred power. In particular, this analysis showed that:
— *to obtain very localized surface heating* (and very rapid to prevent heat migration) and for heating — even core heating — small parts, radio frequencies should be used. In particular, this applies to case hardening, continuous welding of tubes, hot machining, total or local preheating of small parts, brazing, etc.;
— *for in-depth and regular heating of large parts,* medium or low frequencies (line frequencies) provide a higher heating rate and better homogeneity than conventional processes. Therefore, these frequencies are to be preferred for heating billets prior to forming, annealing large weld seams, melting furnaces and treating large heavy parts.

## 5.2. Melting metals

Induction metal melting furnaces are being widely developed. In general, they consist of:
— crucible induction furnaces;
— channel induction furnaces.

### *5.2.1. Crucible induction furnaces*

#### *5.2.1.1. Characteristics of crucible induction melting furnaces*

##### *5.2.1.1.1. Overall composition*

Crucible induction furnaces (magnetic coreless) consist of a water-cooled induction coil surrounding a ramming mixture refractory lining, forming a crucible in which the metallic mass to be melted is located. In some furnaces, this ramming mixture crucible is replaced by a graphite or metal crucible. To provide the required rigidity, it is installed in a metal carcass. For tilting furnaces, tilt generally takes place round an axis passing through the pouring spout, enabling the molten metal to be poured easily. Where furnace capacity is in the hundreds of kilograms, tilting is provided by hydraulic rams. The furnace must be fitted with covers to prevent radiation losses, which, at the temperatures involved, may be considerable (400 kW/m$^2$ for iron at 1,500°C) [16], [24], [28], [82], [97], [130].

##### *5.2.1.1.2. Inductor*

Generally, the inductor is constructed from helically wound, very pure, low resistivity electrolytic copper tubing through which the cooling water flows (the cooling water is a source from which energy can be recovered by a system of exchangers or heat pumps). To facilitate cooling, the coil is often divided into several parallel circuits. Water must be specially treated to prevent scale forming in the circuits. The successive, almost touching turns are insulated from each other.

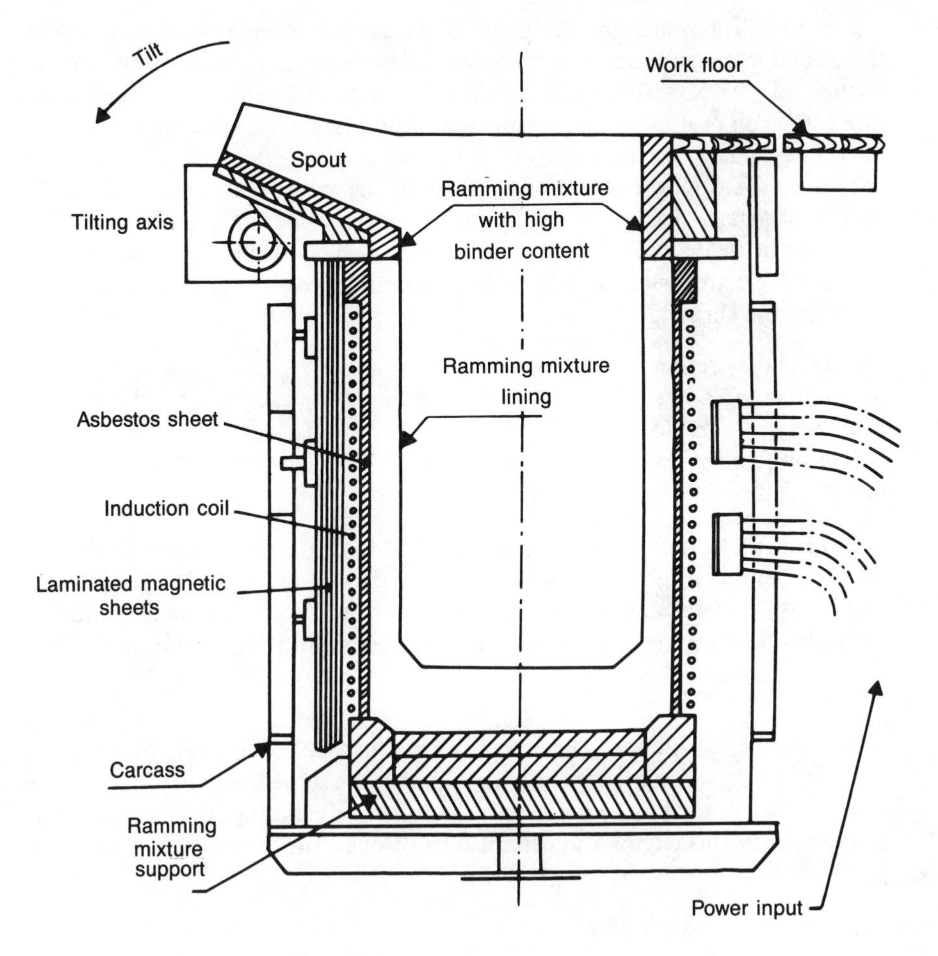


**Fig. 41.** Diagram of a crucible induction furnace

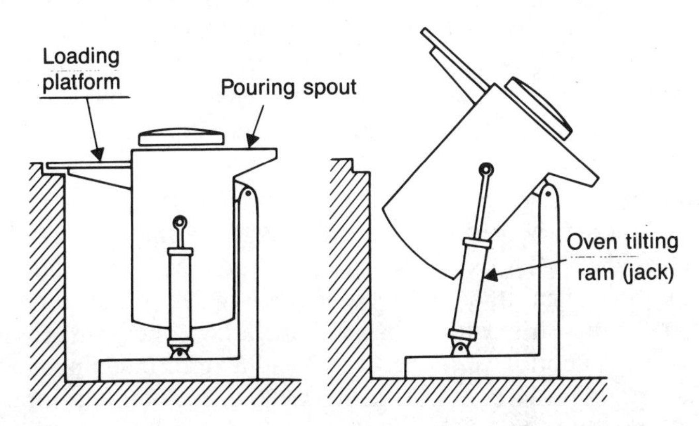


**Fig. 42.** General view of a low frequency induction furnace

This electrical insulation often consists of successive ribbons: the first in teflon, the second in micanite and sometimes a third in treated asbestos, the whole being coated with varnish. Wedges are inserted to maintain separation between the turns. The coil profile is studied to obtain the highest possible electrical efficiency; the thickest part faces the inside of the furnace for skin effect reasons, and water cooled on the opposite side. For medium frequency furnaces, the coil may be more symmetrical, since penetration depth in the copper inductor is lower. The coil must be held solidly, since it is subjected to high mechanical forces due to forces of electromagnetic origin and ramming mixture thrust (expansion, metallostatic thrust).

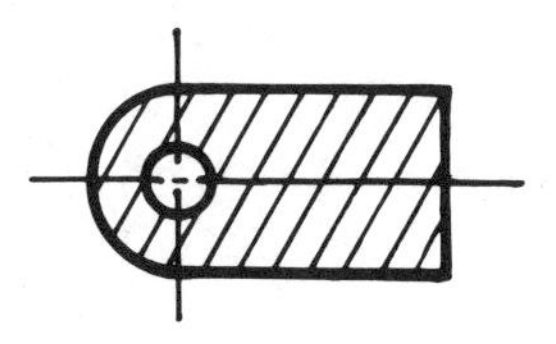

**Fig. 43.** Cross-section of one turn of a low frequency, crucible induction furnace inductor (full scale).

Electrical leak detection processes are used in most large capacity furnaces. To prevent molten metal reaching the coil if the lining bursts, "cogemicanite" and asbestos plates are inserted between the coil and the ramming to protect the coil and reduce thermal losses [66], [85], [96], [98], [111*c*], [114], [127], [130].

*5.2.1.1.3. Magnetic shields*

To reduce magnetic leakage and prevent the furnace carcass from overheating, magnetic shields are placed around the coil; they are sometimes known as furnace yokes or frames (ends), consisting of silicon laminate plating. The magnetic flux is enclosed by this circuit and cannot pass through the metal furnace structure [130].

*5.2.1.1.4. Refractory lining*

The type of ramming mixture depends on the alloys melted. For cast iron, an extra-silicious ramming mixture is used. In furnaces intended for melting steels, a basic or mixed refractory is often used, and sometimes an alumina-based, neutral refractory. For aluminum alloys, dry dead-burned fire clay-based mixtures (aluminous based ramming mixtures) have replaced alumino-silicate ramming mixtures. Also, for copper alloys, neutral or alumina-based acid ramming mixtures are used.

If, in terms of electrical efficiency, it is preferable to use a thin lining, it is better, for refractory behavior and safety reasons, that greater thicknesses be used. The curve in Figure 44, based on the experience of a large automobile group, is an excellent trade-off between these two opposing requirements. The lining service life varies with metals and operating conditions. For example, for cast iron, consumption varies between 2 and 6 kg of ramming mixture per ton cast for melting furnaces (reworked every 200 to 300 pourings for cast iron, but every 100 to 150 pourings for steels), and between 0.5 and 2 kg for holding furnaces. Service life is much longer when the working temperature is low: approximately three years for aluminum alloys, if the lining is partially repaired regularly.

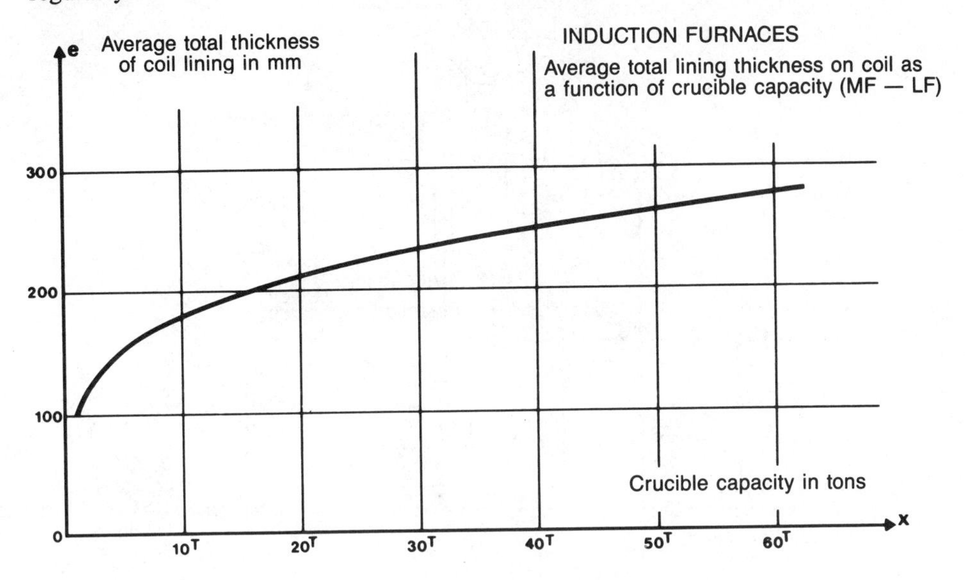


**Fig. 44.** Lining thickness against capacity [130].

The lining is installed by ramming the mixture between the protective plates of the coil and a metallic form of diameter equal to the inside diameter of the crucible, and of the same shape as the latter. After sintering the ramming mixture, the metal framework is recovered [120], [121], [122], [124], [130], [141].

*5.2.1.1.5. Electrical power supply*

Crucible induction furnaces can be divided into several types, as a function of frequency:

— crucible induction furnaces operating at line frequency (50 Hz in Europe);

— medium frequency crucible induction furnaces, or radio frequency furnaces for heating at a higher frequency.

The electrical power supply and some properties are different for both types (stirring, starting, etc.).

For medium frequency crucible furnaces, the electrical installation designed depends on the type of generator used (*see* paragraph 3.2.).

*5.2.1.1.6. Capacitor battery*

Leakage flux is high, and the power factor is therefore low; the use of a radio frequency further decreases the power factor. Therefore, this is generally of the order of 0.1 to 0.2, and it is therefore indispensable to install a battery of capacitors to return the power factor to an acceptable value. These capacitors

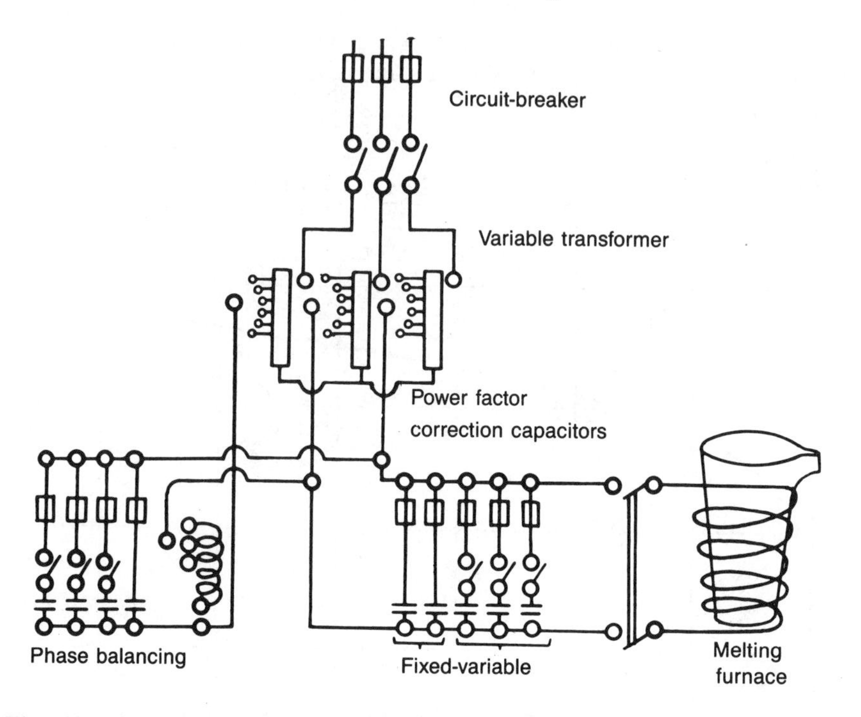


**Fig. 45.** Circuit diagram of a low frequency crucible induction furnace.

are air or water cooled. A simple formula providing an adequate approximation has been proposed to determine this reactive power:

$$\frac{Q_w}{P_w}=1.2\,\frac{e+(d_{01}+d_{02})/2}{(d_{01}+d_{02})/2},$$

$Q_\omega$, reactive power in kilovoltamperes;
$P_\omega$, active power in load in kilowatts,
$e$, thickness of refractory lining in meters;
$d_{01}$, penetration depth in coil in meters;
$d_{02}$, peneteration depth in load in meters [130].

*5.2.1.1.7. Bath stirring*

Interaction of the magnetic field and electrical currents create electromagnetic forces (governed by the three Laplace laws) applied from outside the crucible towards the inside. This results in strictive tendencies in the mass to be heated, which has no effect on a solid mass, but, where the metal is in a molten state, results in elevation of center of the bath and stirring of the molten metal. Height h of this dome is equal to:

$$h=3.22\times10^{-5}\,\frac{P_s}{m}\sqrt{\frac{1}{\rho f}},$$

$P_s$, specific power in Watts per square meter;
$m$, specific mass in kg per cubic meter;
$\varrho$, metal resistivity in ohms-meters;
$f$, frequency in hertz.

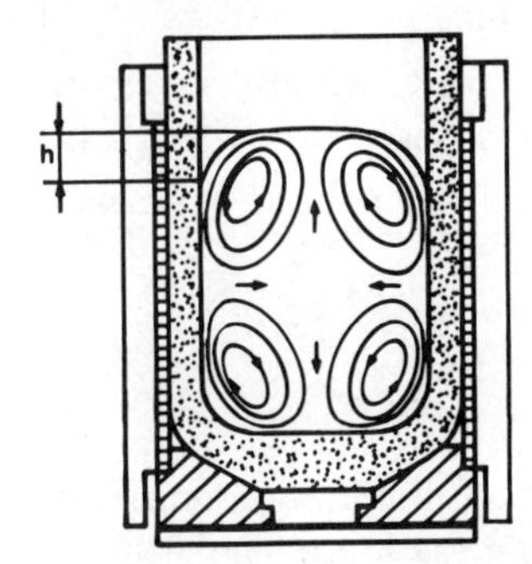


**Fig. 46.** Stirring and elevation of center of bath [130].

This height can reach 20 to 30 cm for furnaces operating at line frequency.

For equal specific power, stirring is inversely proportional to frequency, if the metal is not very dense and a good conductor (by replacing this specific power in dome height expression $h$ by value $P_s = 2 \times 10^{-3} H_e^2\sqrt{\varrho\mu f}$, it will appear that the dome height depends neither on frequency nor or resisitivity, and is expressed by $h = 6.44 \times 10^{-8} H_e^2/m$; therefore, frequency and resistivity cancel each other for a given specific power).

Conversely, for a given dome height value, and therefore stirring, the higher the frequency the higher the specific power, resulting in faster melting.

The penetration depth for induced currents remains the same for a given frequency whatever the furnace capacity; the metal mass in which stresses develop is much higher in a small diameter than in a large diameter furnace. Therefore, at a given frequency, stirring is inversely proportional to furnace capacity.

If the presence of a dome on the bath surface has few drawbacks (only if, however, the height is not such that the metal breaks the slag on the top of the bath isolating the air), this does not apply as far as the effect of the metal on the refractories is concerned. This metal flows along the walls of the crucible, and is in continuous friction with the refractory materials. If specific power is too high, this may result in rather rapid wear of the refractories. Conversely, stirring should be sufficient for the metal produced to be homogeneous, and for loads and additives to be rapidly absorbed by the bath.

Therefore, for a given furnace diameter and frequency, the electromagnetic forces define a maximum installed power which must not be exceeded if excessive stirring is to be avoided. The specific power of furnaces operating at line frequency generally varies between 220 and 400 kW/t of capacity, while, for medium frequency furnaces, it is between 500 and 1,200 kW/t of capacity. To prevent excessive stirring, and instead of limiting specific power, it is also possible to place the inductors over a part of the bath height only. Medium frequencies can be used to obtain furnaces with a capacity of a few kilograms, and radio frequency for very low capacity furnaces. However, stirring is not the only parameter *used in the definition of furnace diameter and frequency, since starting with a cold load also greatly influences the selection (see* paragraph 5.2.1.1.8) [32].

While the stirring phenomenon imposes some limits in induction crucible furnace design, at least for furnaces operating at line frequency, it is also one of the major advantages with this type of furnace, since:

— it enables the use of small dimension scrap and waste, without excessive oxydation losses, since they are rapidly drawn into the bath;

— additives are easily and rapidly absorbed by the molten bath with high efficiency;

— bath composition is uniform.

### *5.2.1.1.8. Cold load starting and bottom of bath*

To start a crucible induction furnace with a cold load, the induced currents must be capable of developing in the element forming the load. The element diameter-penetration depth ratio must therefore be greater than a minimum value, which, as seen from the power transmission study, is of the order of 4. When starting with a cold load consisting of fine scrap, juxtaposed with very small contacts (pointed and limited, due to the presence of oxide, dirt or grease), it is not possible to consider this mixture as a single metal mass. The load therefore consists of as many inducting circuits as there are separate pieces. Therefore, a minimum load element dimension exists for each frequency and alloy.

This limit disappears when the bath is liquid and alternate loadings and withdrawals are used, and bath bottom equal to at least 30% of the furnace capacity is conserved. Line frequency furnaces must be supplied on cold starting with large scrap (or by a solid block representing 25 to 30% of the furnace capacity obtained by ingoting on emptying the furnace). For subsequent loads, and when the "alternate loads and withdrawl" method is used together with a bath bottom equal to at least 30% of the furnace capacity, there are no restrictions on element dimensions. With medium frequency furnaces, cold load starting is generally made using elements of standard dimensions, and the bath bottom is of less interest.

Outside of production periods, the advantage of retaining a liquid bath bottom with a low frequency crucible furnace depends on numerous factors such as the number of work stations, furnace capacity and the alloys produced. With a medium frequency crucible furnace, there is no need to maintain a bath bottom under these conditions.

### *5.2.1.1.9. Vacuum or controlled atmosphere crucible induction furnaces*

Low, medium or radio frequency crucible induction furnaces facilitate vacuum or controlled atmosphere melting. Vacuum induction furnaces are divided into two families: surface vacuum and integral vacuum furnaces [11], [13], [36], [67], [68], [90].

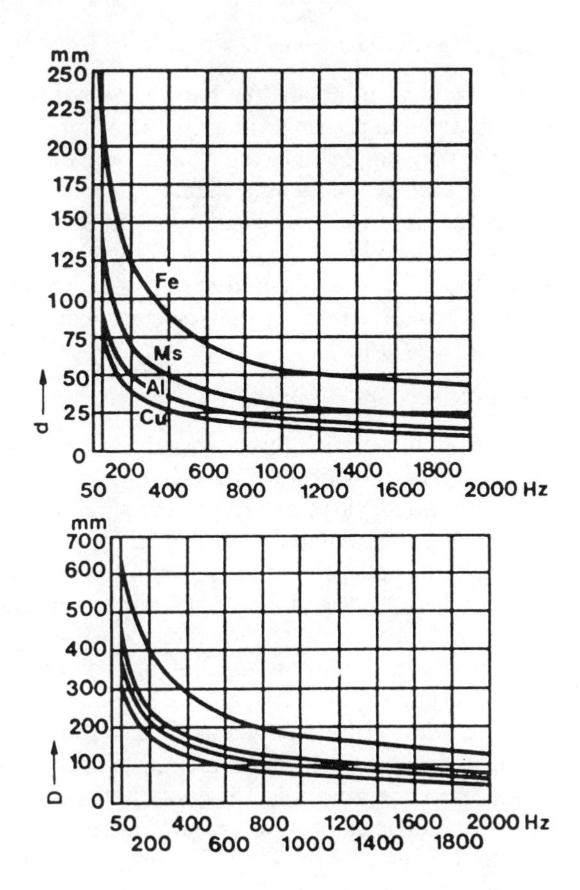


**Fig. 47.** Minimum dimension of parts forming the load against frequency for cold load starting

**Fig. 48.** Minimum diameter of crucible against frequency and metal (from top to bottom, cast iron, brass, aluminium and copper).

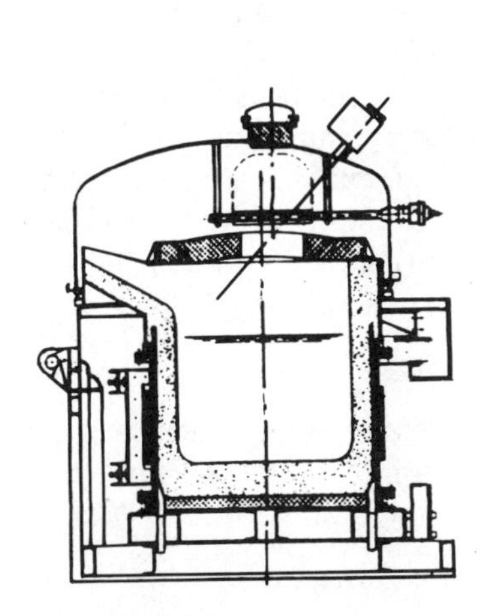

**Fig. 49.** Surface vacuum crucible induction furnace [M].

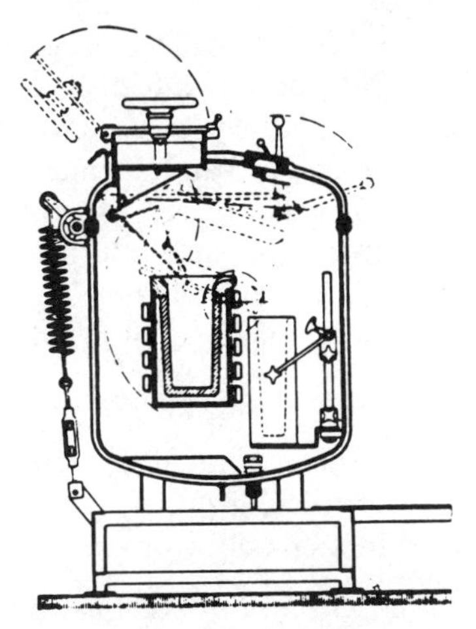

**Fig. 50.** Integral vacuum crucible induction furnace [M].

— *Surface vacuum furnaces*

A low vacuum, of about $10^{-3}$ bars, is created at the surface of the bath only. Therefore, vacuum pouring with this type of furnace is difficult. Furnace design, however, is very simple, since the inductor and the current inputs and water inlets are located outside the vacuum chamber. It is also easy to produce high capacity furnaces, and investment costs are relatively low. Specific power (kW/t capacity) is of the same order of magnitude as for other crucible induction furnaces.

Basically, this furnace is used in steel-making and steel foundries to produce high quality steels (refractory stainless steels, etc.) with a very low carbon content (less than 0.03%) [M].

— *Integral vacuum furnaces*

The overall installation (crucible, inductor, pouring ladle) is placed in a large sealed chamber, in which a high vacuum, of the order of $10^{-8}$ to $10^{-9}$ bars is created. Pouring is done at vacuum pressure into a mould or pouring ladle. A refractory material crucible is often used for pouring instead of a water-cooled copper pouring ladle, since this is a simpler solution. Conversely, there is a risk of reaction between the molten metal and the crucible wall. These furnaces are often divided into several sealed compartments — one for the furnace, one for the pouring ladle, one for introduction of additives, etc.; this facilitates operation, reduces the time required to reach vacuum pressure and therefore increases the use factor, while reducing vacuum creation costs.

Compared with a vacuum arc or electron beam furnace, the vacuum induction furnace offers the advantage of easily holding the load in the liquid state as long as metallurgical operations require. Degassing and deoxidation are therefore complete. It is easier to add alloy elements to the bath at the moment and in the sequence desired.

These furnaces are above all used for the production of iron and nickel alloys, with high oxygen affinity additives such as aluminum, titanium, boron, cobalt and hafnium. In particular, they are used to produce superalloys (alloys which can be used at high mechanical and thermal stresses, at temperatures from 600 to 1,100°C). When the metal produced is used to make forgings, it is also remelted in a consumable electrode vacuum arc furnace, an electron beam furnace or an electroslag remelting furnace, to obtain optimum structure for forging. These furnaces are also used to melt special metals such as uranium [B], [AB].

— *Atmosphere or pressure crucible induction furnaces*

Crucible induction furnaces are also used for melting in an atmosphere or under pressure. As for vacuum furnaces, two solutions can be used: place the furnace as a whole in a chamber, or only protect the bath surface.

*5.2.1.2. Fields of application of crucible induction furnaces*

Crucible induction furnaces are used for melting numerous metals in casting foundries, for fabrication of partial-products in non-ferrous metals, refining non-ferrous metals, steel-making, non-ferrous metal metallurgy, etc. Numerous examples are described in the paragraphs which follow; in particular, this applies to casting foundries. The characteristics of the equipment used in other fields are often comparable, and generally undergo minor modifications.

In equipment performances, the melting rate does not correspond to any downtime (cleaning, pouring) and is an intrinsic characteristic of the furnace. Conversely, the maximum hourly rate allows for each of the operations indispensable to correct melting, practical hourly rate or mean integrated rate, and all programmed shutdown times (conveyor shutdown, casting waiting times, casting or melting incidents, etc.). The last two variables are related to the activity of the industry involved. When choosing a furnace, these factors must be carefully analysed, since the furnace power practical use factor is often between 60 and 80% [B], [C], [F], [G], [H], [I], [J], [K], [L], [M], [N], [Q], [U], [W], [AB].

*5.2.1.2.1. Steel foundries*

Crucible induction furnaces are often encountered in steel foundries; however, the choice of an induction furnace should be the subject of a thorough analysis, since, in general, these cannot meet all the requirements of this industry [102], [104], [119].

Crucible induction furnaces are, in fact, unsuitable for metallurgical operations, and in particular slag refining, since, due to their principle, only the metal part of the bath is heated.

The starting conditions with low frequency furnaces make frequent alloy changes difficult. The limited power density leads to relatively long melting times, which limits productivity and may adversely affect the service life of the refractories, which are kept at high temperature for long periods. Therefore, this type of furnace is little used, except for the production of steels intended for casting or large items. *Conversely, the medium frequency induction furnace forms a remelting device fully adapted to the production of special steels in limited quantities with frequent alloy changes,* since its melting time is very high and cold load starting is very easy [69].

Example of a range of high power density, medium frequency crucible induction furnace for melting of steel at 1650°C (melting speed starting with hot furnace)

| Capacity (kg) | 45 | 90 | 135 | 170 | 150 | 250 | 350 | 450 | 650 | 1400 | 2600 |
|---|---|---|---|---|---|---|---|---|---|---|---|
| Power (kW) | 75 | 125 | 175 | 225 | 200 | 300 | 400 | 500 | 700 | 1200 | 2100 |
| Melting speed (kg / h) | 100 | 185 | 270 | 375 | 315 | 530 | 725 | 900 | 1270 | 2180 | 3800 |
| Frequency (Hz) | 3000 | | | | 1000 | | | | | 500 | |

Fig. 51

At medium frequencies, specific power is high, of the order of 1000 kW/t capacity for small furnaces and 700 kW/t for low frequency furnaces.

The rated specific consumption is about 700 kWh/t for very low capacity furnaces and 600 kWh/t for larger furnaces, from a crucible content of 200 kg approximately, but mean consumption is generally around 850 and 750 kWh/t respectively for the more modern furnaces using thyristor solid state frequency converters.

*5.2.1.2.2. Cast iron foundries*

Low frequency crucible induction furnaces are becoming more widely used for mass production of malleable spheroidal graphite grey iron. Conversely, for more limited production rates, or for production of special cast irons, a medium frequency crucible induction furnace should be used because of its flexibility [44], [49], [57], [119].

Range of low frequency crucible induction furnaces for melting cast iron at 1,500°C; melting speed with bath bottom starting at 2/3 of furnace capacity

| Furnace capacity (t) | 1 | 2 | 3 | 4 | 6 | 7.5 |
|---|---|---|---|---|---|---|
| Power (kW) | 350 | 600 | 750 | 1,250 | 1,800 | 2,000 |
| Melting rate (t/hr) | 0.5 | 0.95 | 1.2 | 2.2 | 3.3 | 3.6 |

| Furnace capacity (t) | 10 | 15 | 20 | 25 | 30 | 40 |
|---|---|---|---|---|---|---|
| Power (kW) | 3,000 | 4,500 | 6,000 | 7,000 | 9,000 | 10,000 |
| Melting rate (t/hr) | 5.5 | 8.4 | 11.3 | 13.2 | 17.1 | 19 |

**Fig. 52.**

There are numerous reasons for this development:

— possibility of using a high proportion of relatively low cost steel waste, which is then recarbonized;

— very low oxidation losses, both for the metal load, even finely divided, and for alloy additives, due to bath stirring;

— decrease in ingoting, due to the possibility of returning hot metal to the bath;

— easy correction of analysis and temperature;

— high constant energy efficiency, which is a much appreciated characteristic, particularly with superheating;

— no pollution of metal due to sulphur, as with other melting methods, which is a most important point in the production of spheroidal graphite cast iron;
— ease of obtaining of carbon content cast irons intended for the production of malleable cast iron;
— excellent metallurgical quality of the alloys produced (homogeneity, composition precision and regularity, no metal pollution, accurate temperature control);
— improvement in working conditions and reduction of pollution.

Conversely, equipment of this type involves rather heavy investment costs and must therefore be thoroughly studied [28], [84], [97], [98], [118], [130].

The capacity and power of such furnaces are constantly increasing. Units of a capacity of 60 t and powers of 21,000 kW produce up to 35 tons per hour.

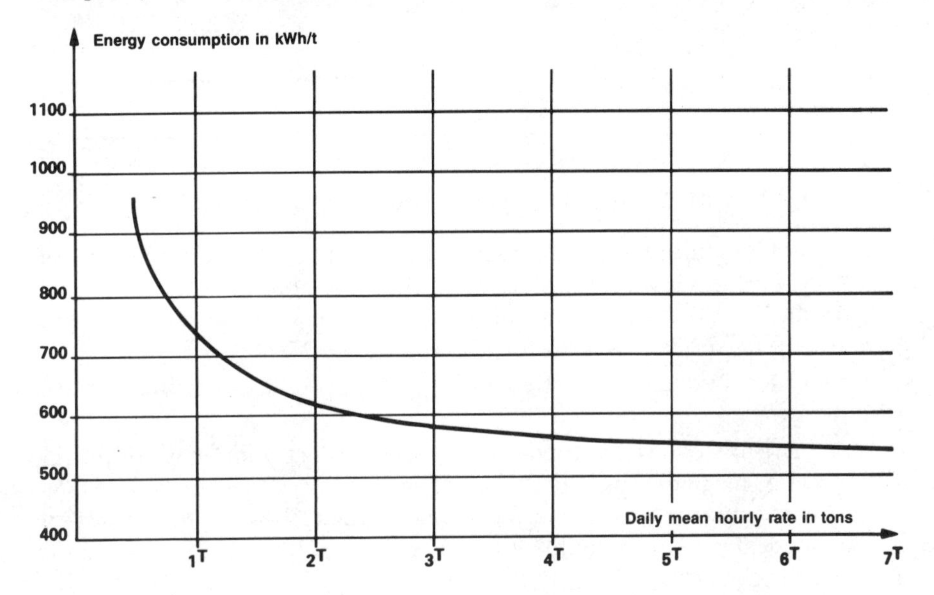


**Fig. 53.** Low frequency crucible induction furnace, capacity 25 t, power 5,500 kW: energy consumption versus mean hourly rate for production of synthetic grey cast iron, load and withdrawal 3 t [130].

For low frequency furnaces, specific power is about 250 to 400 kW/t of capacity. The rated specific consumption is around 500 kWh/t at 1,500°C, but in practice, mean consumption varies between 550 and 750 kWh/t as a function of foundry activities, number of stations worked, pouring temperature, furnace capacity and method of operation (alternate loading and withdrawal or complete batches), specific power, etc.

Low frequency crucible furnaces are also used together with other melting methods — arc or cupola furnace; they also behave as buffer furnaces between melting and casting (permanent availability of molten metal, return of liquid metal without ingoting, etc.), or grading furnaces using addition of additives or homogenization effect; both functions can, of course, be complementary.

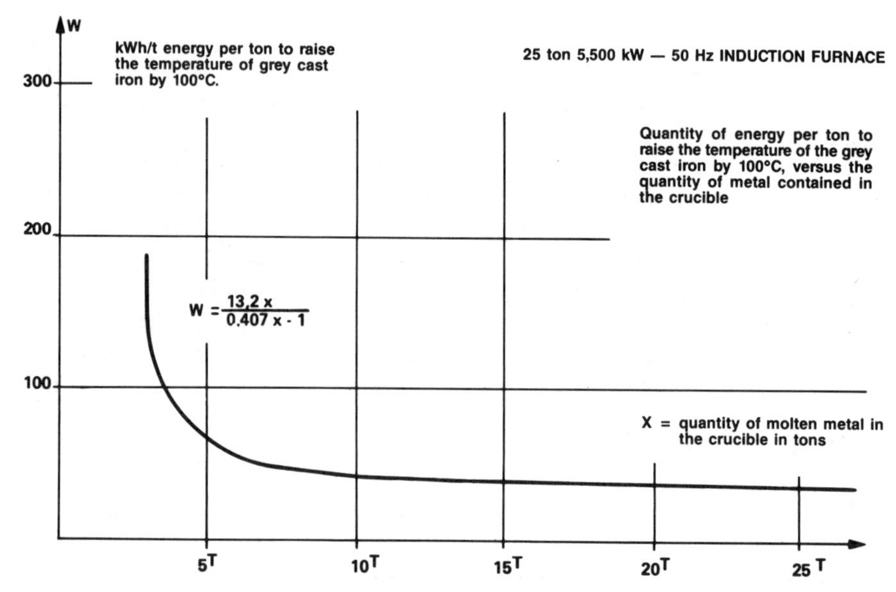


**Fig. 54.** Energy consumption for superheating [130].

These holding furnaces are also used to superheat metal. The specific power of holding furnaces is generally much lower than that of melting furnaces, from 100 to 200 kW/t of capacity, depending on requirements. Some manufacturers have developed special furnaces for this application.

For medium frequency induction furnaces, the specific power is about 700 to 1,200 kW/t of capacity.

With thyristor solid state generators, the specific consumption is now of the same order of magnitude as that of low frequency furnaces; higher melting rates in medium frequency furnaces in fact result in a reduction of thermal losses, which practically compensate the very limited energy loss occurring when the frequency is changed.

Specific consumption is about 600 to 750 kWh/t, depending on utilization conditions [69].

#### *5.2.1.2.3. Light alloy melting (aluminum and magnesium)*

Low frequency crucible induction furnaces are widely used for melting aluminum alloys in pressure moulding foundries, which often produce only a single alloy (sometimes two or three, but in such cases, several furnaces are used) in large quantities, generally more than 700 kilograms per hour.

This furnace then replaces numerous fossil fuel furnaces and can be used to feed the holding furnaces at the casting stations.

The rate is high and bath stirring rapidly absorbs the loads, including scrap with minimum oxidation, and provides a composition and thermally homogeneous metal. Crucible service life is high at 2 to 3 years, and metal gasing is practically null.

Example of a range of high power density, medium frequency crucible induction furnaces for melting cast iron at 1,500°C (melting rate, hot furnace start)

| Capacity (kg) | 50 | 150 | 200 | 300 | |
|---|---|---|---|---|---|
| Power (kW) | 125 | 175 | 225 | 300 | |
| Melting rate (kg/h) | 205 | 300 | 405 | 590 | |
| Frequency (Hz) | 3,000 | | | | |

| Capacity (kg) | 180 | 350 | 500 | 700 | 950 |
|---|---|---|---|---|---|
| Power (kW) | 200 | 350 | 500 | 700 | 900 |
| Melting rate (kg/h) | 345 | 680 | 1,000 | 1,400 | 1,810 |
| Frequency (Hz) | 1,000 | | | | |

| Capacity (kg) | 1,400 | 2,100 | 2,600 | | |
|---|---|---|---|---|---|
| Power (kW) | 1,200 | 1,800 | 2,100 | | |
| Melting rate (kg/h) | 2,400 | 3,600 | 4,100 | | |
| Frequency (Hz) | 500 | | | | |

Fig. 55.

Range of low frequency crucible induction furnaces for melting aluminum alloys at 700°C

| Capacity (t) | 0.8 | 1 | 1.5 | 2.3 | 3.5 | 4.5 | 5.5 | 7.5 | 10 | 15 |
|---|---|---|---|---|---|---|---|---|---|---|
| Power (kW) | 240 | 280 | 400 | 650 | 900 | 1,200 | 1,450 | 1,900 | 2,100 | 2,200 |
| Melting rate (kg/h) | 440 | 520 | 770 | 1,300 | 1,800 | 2,400 | 2,900 | 3,900 | 4,350 | 5,500 |

Fig. 56.

Loading can be easily mechanized and operation is highly flexible. These furnaces are also used in steel work and sand foundry work where rates per alloy are high but occur more rarely.

With aluminum casting, furnace power utilization is high, generally of about 80 to 85%. Therefore, the actual rate is about 80% of the maximum rate. Specific consumption, in most cases, is between 500 and 550 kWh/t poured, and reaches 600 kWh/t if the furnace is maintained over the weekend, which is not absolutely necessary, but possible [17], [28], [79], [111*e*], [114], [131], [142].

Medium frequency crucible furnaces are beginning to be used in aluminum casting where hourly production rates are relatively low or work times limited (less than 2 of 8 hours shifts), since installations are more compact and cold load starting is facilitated. These furnaces can therefore be used in pressure, shell and sand casting.

Range of medium frequency induction furnaces for melting of aluminium at 700°C

| Capacity (kg) | 75 | 125 | 175 | 225 | 300 | 350 | 500 | 600 | 700 |
|---|---|---|---|---|---|---|---|---|---|
| Power (kW) | 75 | 125 | 175 | 225 | 250 | 300 | 400 | 500 | 600 |
| Melting rate (kg/h) | 110 | 205 | 300 | 405 | 475 | 590 | 815 | 1,000 | 1,200 |

**Fig. 57.**

Foundries producing large quantities of magnesium alloys also frequently use this type of furnace, but, in such cases, it is necessary to use a steel crucible instead of a refractory lining to prevent the metal reacting with the refractory material. If the crucible is thick, 50 to 60 mm, induced currents develop almost exclusively in the steel crucible (for steel at 800°C, penetration depth is about 60 mm). The load is heated indirectly. The crucible weight and thermal inertia are high. If it is desired to reduce these values, thinner crucibles, thickness 8 to 12 mm thick, are used. Heating is, usually direct but sometimes indirect. However, efficiency is slightly lower in the second case, since the resistivity of magnesium is low [79].

*5.2.1.2.4. Copper alloy foundries*

Low frequency crucible induction furnaces are used mostly in copper alloy foundries for mass production with one alloy, or production of large parts, such as ships' propellers. In the first case, specific power is high, of the order of 200 to 300 kW/t of capacity, while in the second, it is much more limited, from 30 to 100 kW/t of capacity.

Stirring produces a homogeneous metal even if it contains elements of highly different densities. Oxidation losses are low whatever the load composition, and gasing is reduced to a minimum. However, channel induction furnaces compete well with low frequency crucible furnaces, especially for relatively low melting point alloys such as brass [79], [99], [131].

Range of low frequency crucible induction furnaces for melting copper alloys

| Capacity (kg) | Power | | Melting speed | |
|---|---|---|---|---|
| | 60/40 brass (kW) | Copper (kW) | 60/40 brass at 1,000°C (kg/h) | Copper 1,200°C (kg/h) |
| 600 | 175 | 150 | 565 | 365 |
| 900 | 230 | 200 | 810 | 460 |
| 1,200 | 290 | 250 | 1,050 | 590 |
| 1,800 | 400 | 350 | 1,450 | 870 |
| 2,400 | 550 | 480 | 2,100 | 1,250 |
| 3,600 | 750 | 650 | 2,900 | 1,750 |
| 5,000 | 1,050 | 900 | 4,150 | 2,450 |
| 6,000 | 1,150 | 1,000 | 4,600 | 2,750 |
| 7,500 | 1,400 | 1,200 | 5,700 | 3,300 |
| 10,000 | 1,800 | 1,550 | 7,350 | 4,350 |

**Fig. 58.**

Specific consumption varies noticeably with alloy pouring temperature. Between 400 and 500 kWh/t cast of pure copper low resistivity metal is poured at around 1,200°C, dropping to 300 to 350 kWh/t for brass poured at 1,000°C. Specific consumption for bronzes is between that of brass and copper.

Medium frequency induction furnaces are widely used in small and medium-sized copper alloy foundries, and in particular for metals poured at high temperature (copper, high resistance brasses, bronzes, aluminum bronzes, copper-nickel, copper-nickel-zinc alloys, etc.). The crucible is either in graphite or consists of a ramming mixture lining. Starting flexibility with these furnaces provides them with a definite advantage in foundries producing a wide range of sand cast alloys: various types of removable graphite furnaces (*lift-coil* or *push-out* furnaces, removable crucible tilting furnaces, etc.) are used in such cases, since these facilitate alloy changes, and specialization of crucibles according to alloy families prevents metal pollution by unwanted additives. Aluminum bronze foundries also use this type of furnace since, in most cases, these produce their alloys themselves from their elementary components, and bath stirring provides adequate homogeneity.

In graphite crucible furnaces, the dissipated energy is distributed equally between the metal to be melted and the graphite crucible, which then transfers the energy it has absorbed to the load by thermal conduction. The crucible is also used as a pouring ladle.

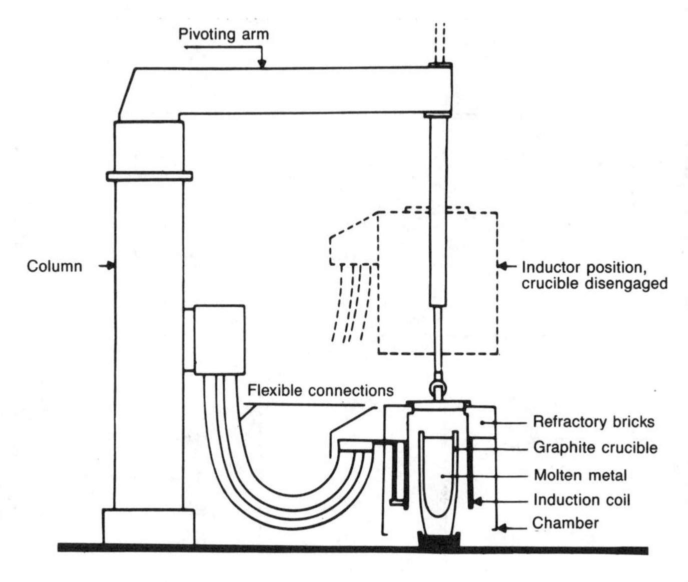


**Fig. 59.** Mobile crucible and inductor or lift-coil furnace [B].

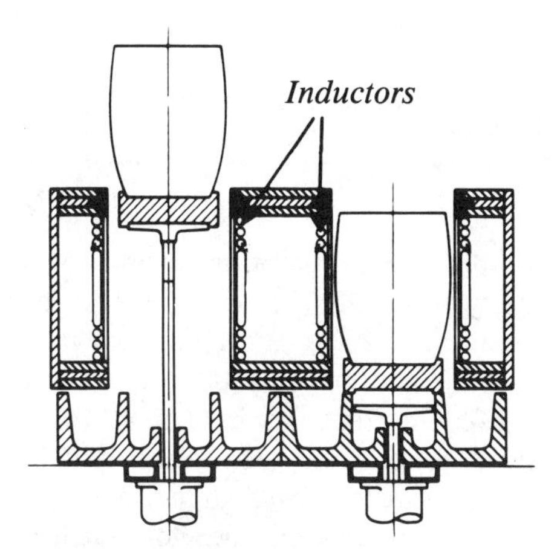


**Fig. 60.** Mobile crucible fixed-inductor or push-out, push-up crucible [L].

The power densities obtained with medium frequency crucible induction furnaces are high, and melting is very rapid with copper alloys.

The practical specific consumption is between 0.5 and 0.6 kWh/kg poured, and can reach 0.7 kWh/kg for high temperature castings (nickel alloys poured around 1,350°C).

For bronzes, melting specific consumption with refractory-lined furnaces generally varies between 350 and 500 kWh/t, depending on furnace capacity, melting speed, pouring temperature and the duration of cleaning, loading and pouring operations.

Frequency tripler furnaces, which can be considered as intermediate between the usual low and medium frequency furnaces, are also sometimes used.

Melting time of bronze load in graphite mobile crucible,
medium frequency induction furnace, with furnace hot
(pouring temperature: 1,175°C);
with furnace cold, add 10 to 20% to melting times

| Capacity (kg) | Power | | | | | |
|---|---|---|---|---|---|---|
| | 50/60 kW | 100 kW | 150 kW | 125 kW | 175 kW | 250 kW |
| | 2,500/3,000 Hz | | | 900/1,200 Hz | | |
| 50 | 26/21 | | | | | |
| 65 | 33/27 | | | | | |
| 80 | 39/31 | 18 | | | | |
| 95 | 45/36 | 21 | | | | |
| 110 | | 24 | 17 | 18 | | |
| 125 | | 28 | 19 | 22 | | |
| 140 | | 32 | 22 | 25 | | |
| 170 | | 40 | 27 | 32 | 23 | |
| 200 | | | 35 | 43 | 30 | 21 |

**Fig. 61.**

Melting time in minutes for bronze load at 1,175°C
in refractory-lined, medium frequency induction furnace
(operating frequency 900/1,200 Hz)

| Capacity (kg) | Power (kW) | | | | | | |
|---|---|---|---|---|---|---|---|
| | 25 | 175 | 250 | 375 | 500 | 750 | 1,000 |
| 150 | 24 | | | | | | |
| 250 | | 26 | | | | | |
| 300 | | 32 | 20 | | | | |
| 500 | | 52 | 33 | 22 | | | |
| 750 | | | 50 | 33 | 25 | | |
| 1,000 | | | 66 | 44 | 33 | 22 | |
| 1,500 | | | | 66 | 50 | 33 | 25 |
| 2,000 | | | | | 66 | 44 | 33 |

**Fig. 62.**

Frequency tripler crucible induction furnace range
for melting bronze at 1,175°C

| Power (kW) | Usual capacity (kg) | Melting speed (kg/hr) |
|---|---|---|
| 125 | 255 - 300 | 365 |
| 200 | 300 - 450 | 550 |
| 250 | 300 - 750 | 730 |
| 300 | 500 - 1,000 | 910 |
| 350 | 750 - 1,500 | 1,000 |
| 400 | 750 - 2,000 | 1,275 |
| 500 | 1,000 - 2,000 | 1,600 |
| 600 | 1,500 - 3,500 | 1,900 |
| 750 | 2,000 - 4,000 | 2,200 |
| 1,000 | 2,500 - 6,000 | 3,000 |
| 1,250 | 3,000 - 6,000 | 3,850 |
| 1,500 | 3,000 - 7,500 | 4,450 |
| 2,000 | 4,000 - 10,000 | 5,900 |
| 2,500 | 5,000 - 10,000 | 7,400 |
| 3,000 | 4,000 - 20,000 | 8,900 |

**Fig. 63.**

For bronzes, the mean specific consumption during melting is between 400 and 500 kWh/t.

*5.2.1.2.5. Other crucible induction furnace applications in metal melting*

Low and medium frequency crucible induction furnaces are used for numerous other applications:

— *Melting zinc alloys in casting foundries*

This application is valid for high production rates of 1.5 to 2 tons per hour only. The mean specific consumption is between 100 and 150 kWh/t poured [79].

— *Melting zinc alloys for production of partially-finished products*

Since the quantities melted are generally high, the crucible induction furnace is well suited for this purpose. The specific consumption is more or less the same as above (100 to 150 kWh/t poured).

— *Remelting zinc cathodes*

The low frequency crucible furnace is suitable for this application, but for investment costs and capacity reasons, the channel furnace is preferred.

— *Melting and distillation of zinc for production of zinc powder*

The low frequency induction furnace is suitable for this application, since both operations can be performed in the same furnace.

— *Melting iron and zinc mattes*

Mattes, galvanization plant wastes, consist essentially of iron and zinc; the zinc, a low melting point expensive metal, can be recovered by melting in an induction crucible furnace. Practical specific consumption is between 150 and 200 kWh/t of mattes.

— *Copper alloys for production of half-finished products (bars, strips, tubes, wires, panels, etc.).*

Low or medium frequency furnaces similar to the refractory-lined tilting furnaces used in casting alloys; in many cases, a channel induction furnace is used for temperature holding and pouring.

— *Melting copper alloys in refineries*

Recovered copper-based alloys are melted in crucible induction furnaces and channel furnaces to produce graded ingots. The furnaces are similar to those used in casting.

— *Melting aluminum alloys in refineries*

The furnaces are the same as those used for casting [17].

— *Melting aluminum alloys for production of half-finished products*

Low frequency crucible induction furnaces are generally used [17].

— *Holding aluminum in aluminum production plants*

Low frequency induction furnaces are used to adjust the composition of the various alloys produced [17].

— *Melting nickel alloys*

The thermal characteristics of iron and nickel are very similar, and medium frequency crucible induction furnaces are suitable for melting of nickel and its alloys. In practice, specific consumption ranges from 700 to 800 kWh/t poured. The alloys produced are nickel- copper (70/30 and 75/25), monel, inconel, nickel-chromium alloys used in the production of electrical resistances and superalloys. Vacuum melting is often used [22], [67].

— *Melting precious metals*

Gold, silver and platinum are always melted in low quantities; therefore, medium or radio frequency crucible induction furnaces are used so that as little metal as possible is lost, the load being placed in a graphite crucible. Platinum must also be vacuum melted.

— *Melting and holding steel*

Induction furnaces are used to produce high alloy steels and ferrous alloy melting (ferrosilicon and ferrochromium). In association with other melting methods, these furnaces are also used as maintining furnaces [58], [147 *k*].

— *Melting superalloys prior to casting*

Medium or radio frequency induction furnaces are used in the production of superalloys (basically nickel, cobalt and chromium alloys), at temperatures of 1,150 to 1,600°C. These alloys, which are widely used in the manufacture of aircraft engines, are then poured using the lost wax and centrifuging technique. Vacuum melting can be used [111 *d*].

— *Levitation melting*

Melting highly reagent metals in crucibles sometimes causes unwanted chemical reactions with the crucible material, which contaminates the load. To prevent this, it is possible to melt small quantities of metal in an induction furnace in which the load is levitated by electromagnetic forces. However, this technique is not widely used in industry.

### *5.2.2. Channel induction furnaces*

#### *5.2.2.1. Channel induction furnace characteristics*

Basically, a channel furnace consists of:
— a molten metal tank, which connects with both ends of the channel; this reservoir or crucible, which is refractory lined, contains most of the metal;

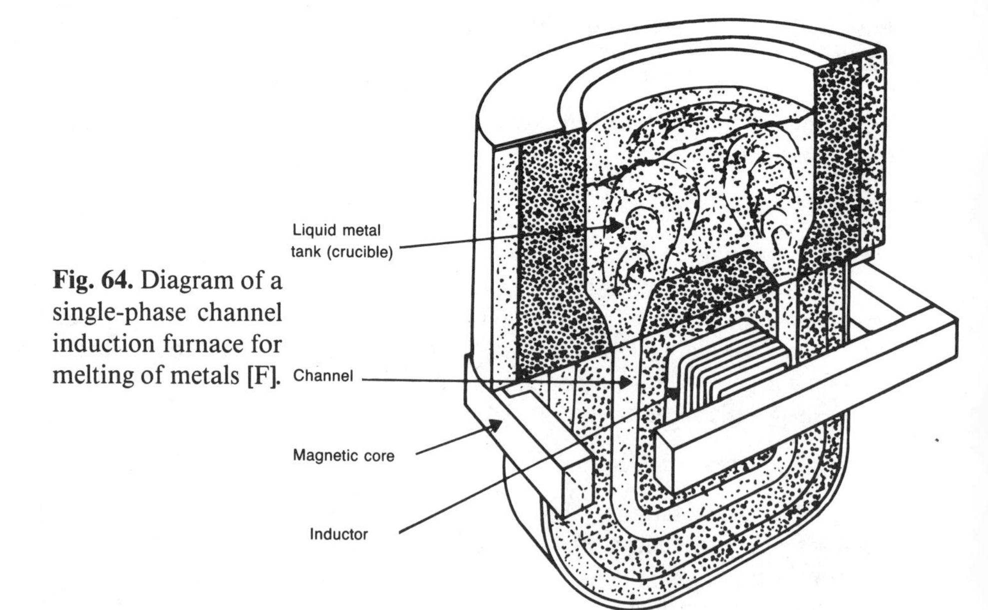


**Fig. 64.** Diagram of a single-phase channel induction furnace for melting of metals [F].

— a channel filled with molten metal, generally in the form of a loop, both ends of which open into the bottom of the chamber;
— an induction coil wound around a laminated steel core, surrounded by the channel.

Electrically, the channel furnace is equivalent to a transformer with a closed magnetic circuit in which the channel forms the secondary.

*Principal characteristics*

The induced currents cause the metal to melt due to the Joule effect. Metal flow in the channel is obtained by convection and the electromagnetic effect, thus enabling transfer of heat from the channel to the crucible. This migration of metal between the channel and the crucible causes light stirring (much less than in crucible furnaces), and is more noticeable below than on the surface, where agitation is weak.

To start these furnaces, it is first necessary to fill them with molten metal, then apply power to the inductors; the molten metal heats in the channel and begins to circulate. A furnace of this type can operate only if the secondary is closed, and on condition that the metal is not allowed to "freeze" in the channel.

When not in use, these furnaces must be kept at low power. It is also possible to empty the furnaces, but this adversely affects the service life of the refractory lining.

In the metal vein, which can be analagous to an infinite number of parallel conductors, the currents attract, causing contraction. In some cases, this contraction or pinching effect can cause a break in the liquid vein. Since the channel is generally covered over by a large quantity of molten metal, the pressure applied by the bath tends to cancel the pinch or contraction effects.

The specific power is limited to prevent the pinch phenomenon from occurring and to prevent excessive overheating of the metal in the channel. Specific power is 700 kW maximum per channel for cast iron, and 400 kW for copper alloys. Channel furnaces are therefore not very rapid, and, to increase the power of a channel furnace, several inductors, and therefore several channels, are required. Consequently, they are better for holding purposes than for melting.

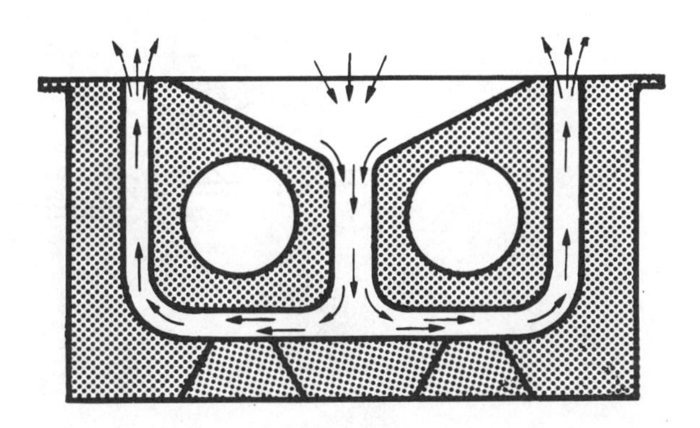

**Fig. 65.** Diagram of a high power density channel furnace [C].

However, new channel furnaces are being developed which, due to a special channel shape, are capable of transmitting much higher specific power, of the order of 3.000 kW maximum per channel. Superheating of the metal in the channel is limited to approximately 30°C (metal flow rate in the channel is high, approximately 300 tons per hour for a 2,500 kW inductor).

### *5.2.2.2. Various types of channel furnaces*

A wide range of channel furnaces exists, and, depending on the position and shape of the channel, are classified as follows:
— inclined and vertical channel furnaces;
— horizontal channel furnaces;
— open channel furnaces (now more or less abandoned).

A furnace can have one, two or three channels. All operate at line frequency (50 Hz) and may be single-phase, two-phase or three-phase. In most cases, they are tilting furnaces; fixed furnaces are used for low production capacities, or, if necessary, as temperature holding furnaces for metal previously melted in another furnace.

— *Inclined-channel furnaces*

At present, this type is more widely used than channel furnaces. It consists of a cylindrical reservoir of liquid metal, rotated around the horizontal axis.

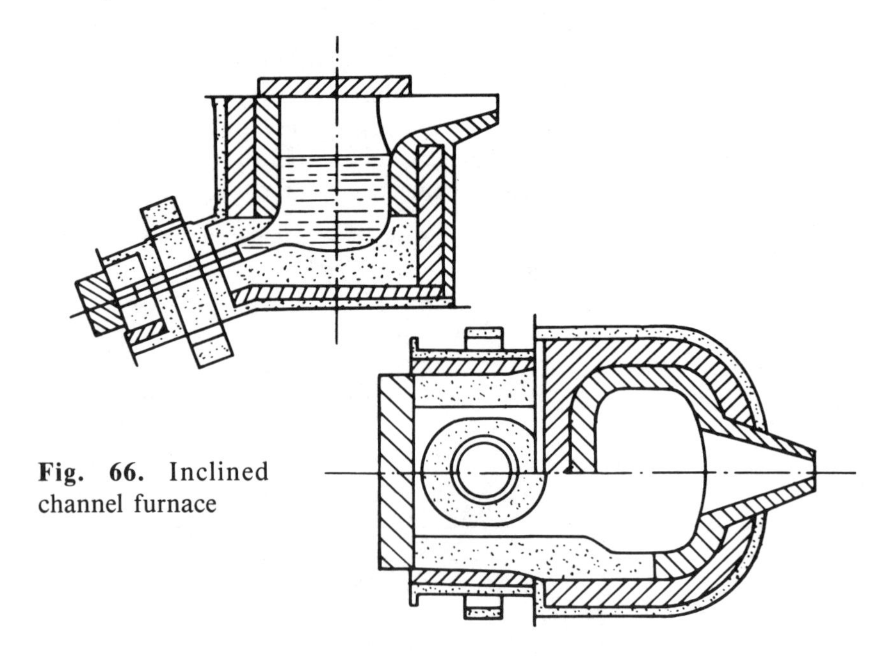

**Fig. 66.** Inclined channel furnace

— *Vertical channel furnaces*

These furnaces operate according to the principle described above. However, in some furnaces, the channel is removable, which facilitates reworking and, where necessary, means that a part of the furnace can be kept in reserve. Linear channel furnaces are easier to clean.

— *Rotating or drum furnaces*

Generally, these are large, relatively high power furnaces and are used in high production rate foundries.

With these, the furnace itself is inserted in a cylindrical drum or casing, mounted on rollers. The bottom part of these furnaces in which the channels are fitted, and which also supports the coil and its magnetic core, is removable. This part is generally known as the "inductor". Furnaces of this type sometimes have two or three inductors, as shown in figure 70 (they are then known as inclined or oblique channel furnaces).

When the melting operation is completed, pouring is performed by rotating the cylinder or drum on its rollers. This causes the molten metal to move towards the outside through the pouring hole located on one side of the cylinder, or at the end of the furnace.

This construction method enables rapid and easy replacement of an inductor without having to shut down the furnace or to empty it completely.

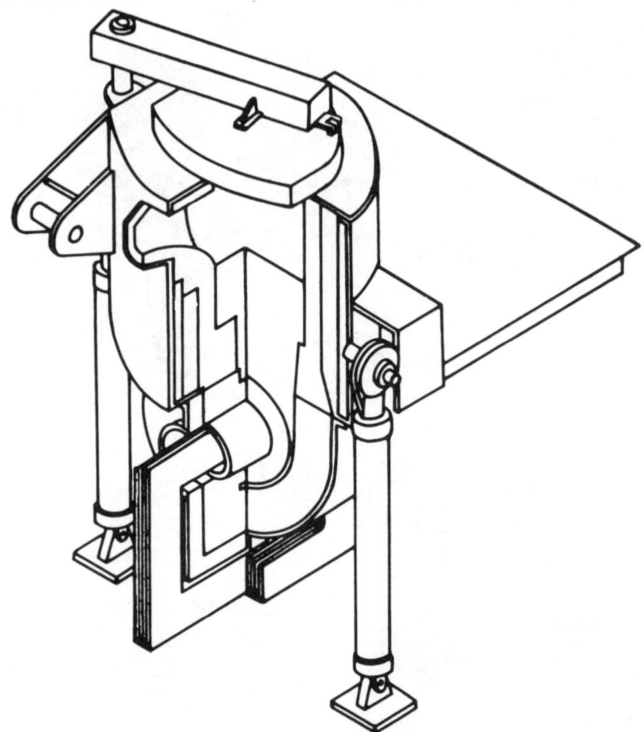

**Fig. 67.** Vertical channel furnace [L].

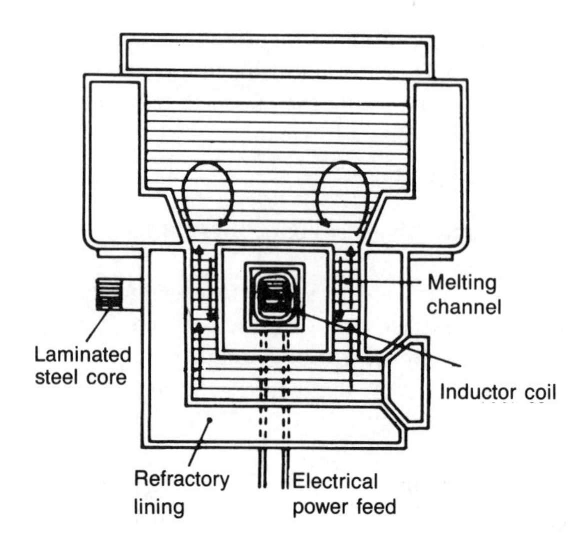


**Fig. 68.** Linear vertical channel furnace [AE].

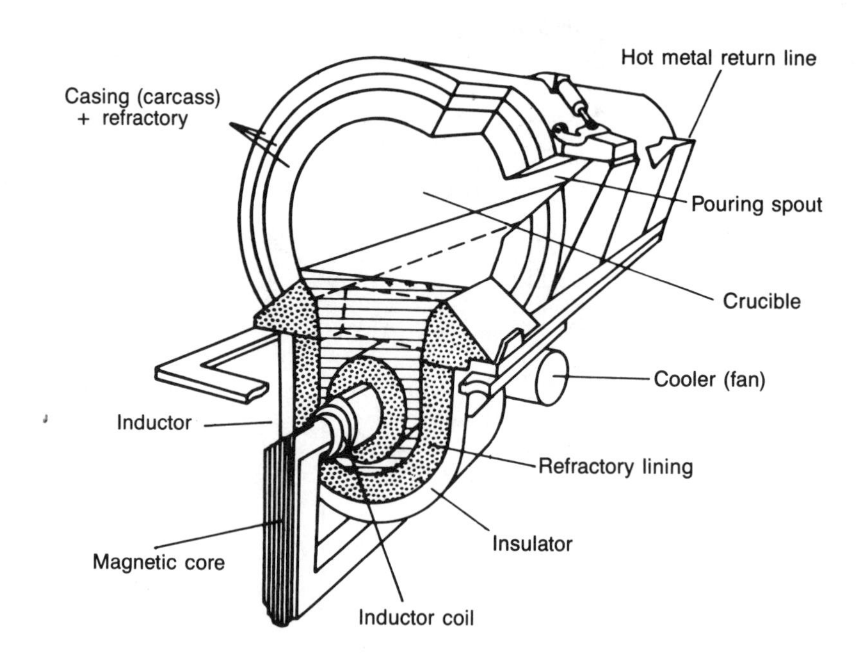


**Fig. 69.** Drum-type channel induction furnace [L].

This method also enables the design of inductors better adapted to the operations to be performed, and higher powers to be dissipated, considerably facilitating "superheating" operations. At present, the capacity of such furnaces is around 200 tons.

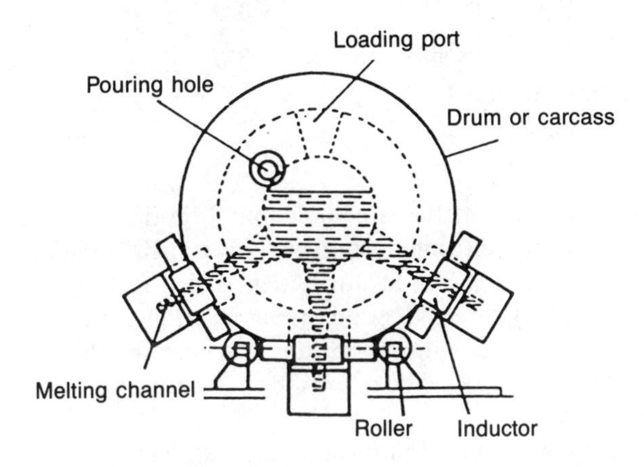


**Fig. 70.** Multi-inductor channel furnace.

— *Horizontal channel furnaces*

a) Single-chamber, horizontal channel furnaces

With these furnaces, the channel is horizontal and is easier to clean than the fixed vertical channel type.

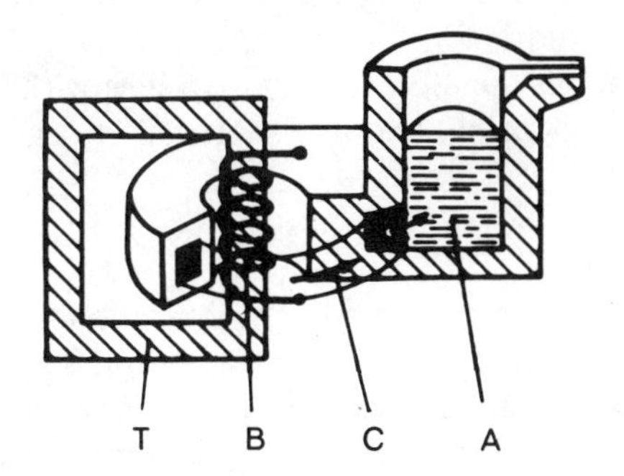


**Fig. 71.** Horizontal channel induction furnace:
A, molten metal
B, inductor
C, channel
T, magnetic core

The inclined channel furnace is preferred to this type of furnace.

b) Double tank or two-chamber horizontal channel furnaces

These furnaces operate according to the principle described above, but have two chambers. The channels connecting these chambers are linear and slightly inclined. Channel inclination is greater in light alloy furnaces than in copper alloy furnaces.

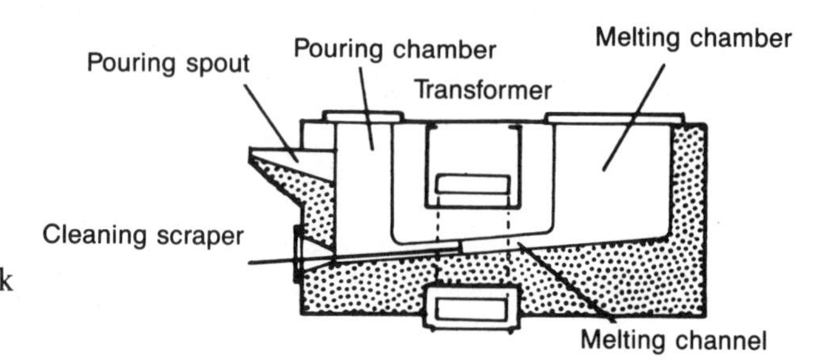


**Fig. 72.** Double tank induction furnace

The products to be melted are loaded into the first chamber, and the molten metal drawn from the second chamber. This principle implies that the metal in the second chamber is free of impurities, which remain in the first chamber. Moreover, in this chamber "drawing off" at bath temperature, which is unaffected by cold loads, remains almost constant. Furnaces of this type are designed so that the channels can easily be cleaned by "scraping" to prevent obstruction due to deposited impurities.

Two-tank channel induction furnaces are restarted easily with solidified metal, on condition that the metal forms a closed loop, so that the induced currents can develop, and on condition that starting is very gradual to prevent thermal shocks affecting the refractory materials.

*5.2.2.3.* Operation

*Composition of loads*

Due to the permanent presence of a bath bottom in channel furnaces, materials of any shape can be loaded. However, fine and light products tend to agglomerate and form "sponges" which stick to the wall or float in the bath and are very difficult to withdraw.

In melting, it is therefore necessary to start with compounded materials, such as scrap from pouring, and it is preferable to shot-blast these materials so as to avoid cleaning operations.

*Loading*

With channel furnaces, loading is very gradual, and load supervision, which is generally performed by the furnace operator, is as easy as with crucible furnaces although the loading orifices are slightly smaller. However, with side loading orifices, operations are more complicated than with channel furnaces.

*Lining — maintenance*

To line channel furnaces, complex shapes must be produced; therefore, wet ramming mixtures, calling for very careful insertion and gradual drying to low temperature are used.

One or two thicknesses of asbestos are first placed against the metal work to compensate for expansion, followed by a layer of insulating bricks, one or more layers of refractory bricks and finally, in contact with the metal, either bricks or a high alumina content ramming mixture (95 to 98%).

Generally, the inductor is lined with a high magnesia content mixture. Due to the shape of the channels, it is rather difficult to check the condition of the lining in operation. Therefore, the channels must be carefully reworked.

The quantity of refractory materials used for lining a channel furnace is high; for a furnace of total capacity of 32 t, with 12 t of bath bottom, 40 tons of refractory materials, including 8 tons of brick, are required, and the 800 kVA inductor requires 1,500 kg of "concrete". The service life of the lining varies according to application; for a cast iron holding furnace, this is about one year and a half for the chamber, and 3 to 9 months for the inductors.

For non-ferrous alloys, service lives are much longer. For brass melting furnaces, service life exceeds 3,000 operations (4 years for twin-tank furnaces).

For drum furnaces, installation of the new lining and drying is a long process. On average, from demolition up to putting back into service, one and one-half months shutdown is required. Therefore, reserve chambers must be kept available. For smaller furnaces, this operation generally requires a few days [120], [121], [123], [130].

*Control and supervision*

Channel furnaces, which must be maintained in operation with a permanent bath bottom during weekends, have special requirement: the presence of an operator during holding periods outside of production times to remedy any operating problems.

While indispensable with large furnaces, supervision is sometimes superfluous for low capacity furnaces.

Control itself does not give rise to any special problems, except in cast iron melting, for which only a narrow safety margin exists for superheating the metal.

Moreover, when used in duplex with another furnace, it is necessary to avoid contamination by dirt, since cleaning operations are very difficult.

### *5.2.2.4.* Advantages and disadvantages of channel furnaces

*Advantages of the channel furnace*

*Production of high quality metal:*

— adequate homogeneity of the alloys produced, due to uniform temperature throughout the furnace and light stirring;
— regular cast product quality;
— high precision metal composition;
— clean, dust-free pouring.

*Moderate investment costs:*

— relatively low cost; — simplified electrical equipment (relatively small capacitor battery);
— good power factor: 0.5 to 0.7.

*Improved working conditions:*

— clean, noiseless furnaces which do not give off odors;
— very low quantities of smoke and dust produced.

*Principle disadvantages of channel furnaces*

*Costly, delicate refractory materials* (for ferrous products):
— linings are very difficult to install, and it is almost impossible to inspect the channel linings without destroying the channel;
— since the metal is heated in the channel and the heat is then convection-transferred to the remainder of the furnace, the metal temperature in this area is higher than 100 to 150°C; therefore, the refractories are more stressed and must be carefully chosen;
— due to the temperature difference of 100 to 150°C, alumina refractories behave satisfactorily only if the metal temperature is kept below 1,500-1,550°C.

*Lack of flexibility:*

— continuous operation is required, and the bath bottom must be maintained (15 to 20% of furnace total capacity);
— restart is delicate (needs external addition of liquid metal);
— impossible to change alloy quickly;
— melting times relatively long due to limited specific power.

### *5.2.2.5. Field of application of channel induction furnaces*

Channel induction furnace specific powers are relatively low. Therefore, they are used for holding and superheating metals; however, in some industries, they are high performance tools, especially in the non-ferrous metal industries [93], [130], [C], [F], [G], [H], [J], [L], [M], [T], [U], [AE].

#### *5.2.2.5.1. Holding temperature of steel and cast iron in ferrous metallurgy*

Since steels are normally produced at temperatures higher than 1,600°C, the 100 to 150°C superheating of metal in the channels, as used in melting furnaces, leads to an excessive temperature for correct behavior of the refractory materials. Therefore, channel furnaces are not used for steel melting.

Channel furnaces are, however, used in ferrous metallurgy as holding furnaces, in duplex with other melting equipment [45], [47], [54], [93], [117], [146].

Very large channel furnaces (in the U.S.A., the largest have a capacity of 1,500 tons and an installed power of 15,000 kW) are also used for holding and superheating iron from blast furnaces prior to refining in the oxygen converter.

#### *5.2.2.5.2. Melting and holding the melt in iron casting*

Channel induction furnaces are often used as holding furnaces in large foundries, in parallel with crucible or cupola furnaces. In such cases, the channel furnaces behave as a buffer between the melting and casting chambers, in which molten metal is permanently available, enabling homogenization of the metal, limiting composition variations and, where necessary, enabling the composition to be adjusted. Overall energy consumptions are somewhat higher than those that might be expected from the adequate energy efficiency of these furnaces (approximately 80%), since the furnaces must operate permanently. Generally, they are used to superheat the metal with respect to the temperature obtained in the melting furnace [33].

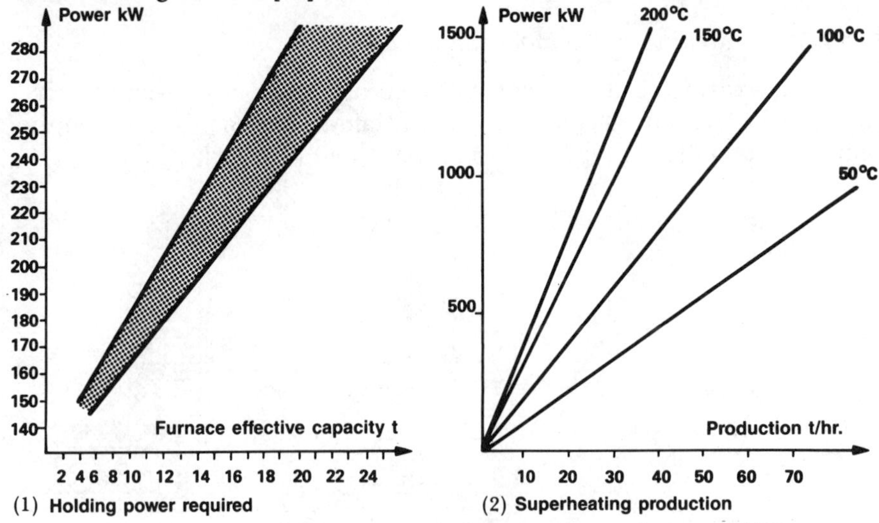


(1) **Holding power required** (2) **Superheating production**

**Fig. 73.** Channel furnace for holding and superheating [149].

Low frequency crucible induction furnaces, which are more flexible, are successfully competing with channel induction furnaces for temperature holding applications.

Sometimes channel furnaces are used for melting cast iron, in particular in foundries in which large parts are produced. Melting is preferably performed at night to benefit from low electricity costs [48], [C].

Range of high power density channel induction furnaces used for melting cast iron at 1,500°C [C]

| Furnace capacity (t) | | Melting Speed (t/h) | | | | | | |
|---|---|---|---|---|---|---|---|---|
| | | Power (kW) | | | | | | |
| Useable | Maximum | 400 | 540 | 750 | 1,100 | 1,300 | 1,500 | 2,000 |
| 7.5 | 10 | 0.54 | 0.90 | 1.45 | 2.36 | | | |
| 10 | 17 | 0.54 | 0.90 | 1.43 | 2.33 | 2.81 | | |
| 15 | 22 | | 0.88 | 1.42 | 2.32 | 2.32 | 2.81 | 3.36 |
| 20 | 27 | | | 1.36 | 2.27 | 2.72 | 3.27 | |
| 25 | 33 | | | 1.32 | 2.22 | 2.72 | 2.72 | 3.25 |
| 30 | 40 | | | 1.25 | 2.14 | 2.66 | 3.18 | |
| 40 | 50 | | | | 2.12 | 2.63 | 3.14 | 4.15 |

**Fig. 74**

High power density channel induction furnaces developing 3,000 kW per inductor and a melting rate of 6.3 tons per hour are now available.

In some channel furnaces, pouring is made through gas pressure, generally air on the surface of the bath, the metal being poured through a siphon [73], [75], [147 *I*], [J], [M].

*5.2.2.5.3. Melting and holding copper alloys*

Since it is easy to obtain the limited power densities required, channel furnaces are widely used for melting of copper alloys. Depending on company requirements, different types of furnaces are used [77], [99], [131].

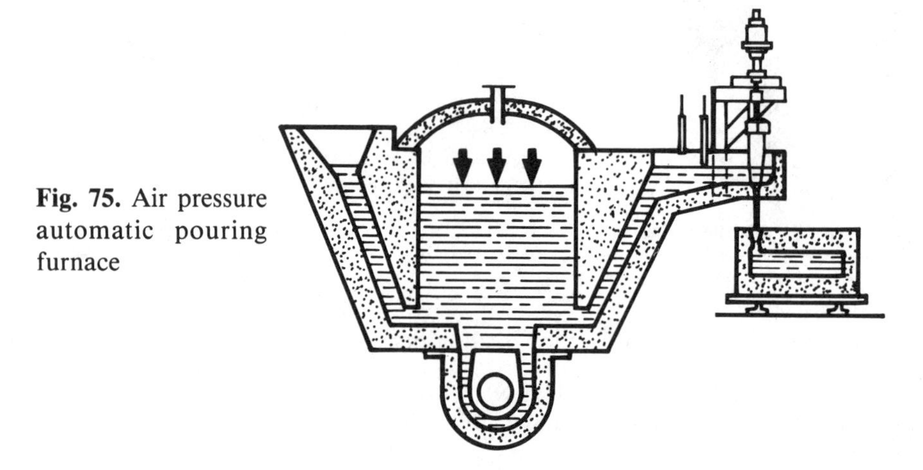

**Fig. 75.** Air pressure automatic pouring furnace

— *Melting copper alloys in double tank channel furnaces*

Range of double tank induction furnaces

| Power (kW) | Capacity (kg) | Brass hourly rate (kg/h) | Bronze hourly rate (kg/h) |
|---|---|---|---|
| 40 | 800 | 120 | 100 |
| 50 | 800 | 160 | 120 |
| 65 | 800 | 220 | 160 |
| 80 | 800 | 300 | 200 |
| 100 | 1,200 | 400 | 300 |
| 120 | 1,200 | 500 | 350 |
| 150 | 1,500 | 650 | 500 |

**Fig. 76.**

When melting, specific consumption varies with furnace size. Of the order of 280-300 kWh/t for the smallest, this drops to 230 kWh/t for large brass melting furnaces; similarly, for bronze, specific consumption is between 300 and 400 kWh/t.

For holding purposes, and with draw-off tanks open, holding power requirements vary from 20 to 40 kW; with covered tanks, this is from 12 to 25 kW.

These furnaces are most widely used in shell or pressure casting foundries.

— *Melting copper alloys in vertical channel furnace*

During melting, specific consumption is of the order of 330 kWh/t for copper, 240 kWh/t for 70/30 brass and 220 kWh/t for 60/40 brass.

Range of vertical channel induction furnaces

| Capacity (kg) | Maximum Power (kW) | Hourly production rate | |
|---|---|---|---|
| | | Copper (kg/hr) | 70/30 brass (kg/hr) |
| 250 | 60 | 180 | 250 |
| 500 | 120 | 360 | 500 |
| 750 | 180 | 550 | 750 |
| 1,000 | 240 | 750 | 1,000 |
| 2,000 | 480 | 1,500 | 2,000 |
| 4,000 | 600 | 1,800 | 2,550 |

**Fig. 77.**

Induction furnaces, which are fully suitable for mass production, are used mostly for the fabrication of half-finished products and in refineries.

In all cases, evaluation of overall specific consumptions must allow for holding periods. For example, for brass, mean consumption will be of the order of 290 kWh/t for single-station operation, but only 220 kWh/t for continuous operation.

— *Holding copper alloys*

Channel furnaces are often used as holding and pouring furnaces, in conjunction with low or medium frequency crucible induction furnaces.

*5.2.2.5.4. Melting and holding aluminum alloys*

In aluminum melting using channel induction furnaces, clogging of the channel is a handicap. Appropriate design of the channel, with limited superheating of the metal in the channel and high circulation rates, have limited this phenomenon; the channel furnace is therefore suitable for this application. Crucible induction furnaces for high capacities and resistance furnaces for low capacities are, however, preferable to the channel furnace [17], [131], [C].

For melting, specific consumption is of the order of 550 kWh/t for small furnaces, dropping to 430 kWh/t for high capacity furnaces.

Range of channel induction furnaces for melting aluminum at 750°C

| Capacity (kg) | 200 | 400 | 600 | 1,000 | 2,000 | 4,000 | 6,000 |
|---|---|---|---|---|---|---|---|
| Power (kW) | 60 | 110 | 150 | 300 | 500 | 720 | 1,080 |
| Melting rate (kg/hr) | 100 | 200 | 300 | 700 | 1,150 | 1,650 | 2,500 |

**Fig. 78.**

Conversely, channel clogging in holding furnaces, in which the power density, and therefore superheating of the metal, are very limited, is low. Tilting or fixed furnaces are used, draw-off being made through gas pressure, mostly air, from the bath surface.

Range of channel furnaces for holding of aluminum alloys at 750°C

| Capacity (t) | 4 | 6 | 9 | 12 | 15 | 25 | 30 | 40 |
|---|---|---|---|---|---|---|---|---|
| Power (kW) | 60 | 75 | 100 | 150 | 200 | 250 | 270 | 300 |
| Holding mean absorbed power (kW) | 27 | 35 | 47 | 56 | 65 | 90 | 105 | 135 |

**Fig. 79.**

The installed powers, which are much higher than the absorbed powers, can be used to increase the metal temperature if required. Moreover, lower powers may be installed, so the figures given are approximate.

For example, tests performed on a 14 t, 77 kW furnace have demonstrated that the absorbed holding power is only 45 kW on average [111 *e*] at 470°C.

#### *5.2.2.5.5. Melting and holding zinc alloys*

Channel induction furnaces are used in numerous zinc foundries. Mainly, they are used for remelting cathodes, or as pouring furnaces.

Furnace capacity may be high, more than 100 t for cathode melting, and can feature up to four inductors with a power of 400 to 500 kW each.

Melting specific consumption is generally between 90 and 120 kWh/t. Holding power is very low, for example about 100 kW for a 60 t furnace.

## 5.3. Heating metals prior to forging

Heating metals considerably facilitates forging; metals are frequently heated before performing this operation, although cold forming techniques are also used. Heating is apparently very simple, since it consists of taking the partially finished metal products, most often of simple geometrical shapes — billets, bars, tubes, etc. — to a given temperature, which is uniform throughout the part, over as short a time as possible.

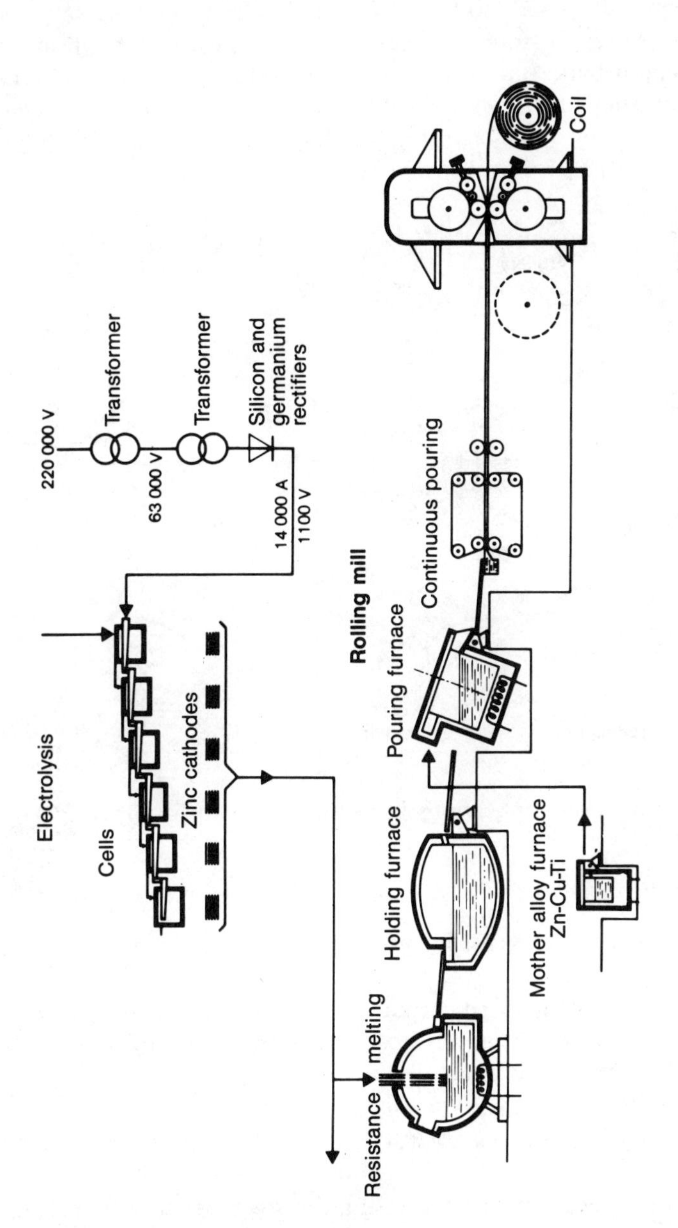


**Fig. 80.** Diagram of a zinc and flat partial-product production installation using resistance melting and channel induction pouring furnaces [148].

Once this temperature has been reached, the metal is hot formed using an appropriate technique — rolling, stamping, die-stamping, forging, drawing, extrusion, rolling, drilling, flow turning, thermal expansion, etc. The temperature required for such operations is generally between 1,100-1,300°C for steels, 750-900°C for copper alloys and 450-550°C for aluminum alloys. However, lower temperatures can be used for some applications such as heating tubes prior to re-rolling or rolling of bars (900-1,000°C for steels). Mostly, heating concerns all the product, but sometimes only a small part of the metal has to be heated, for example during forging of bar ends [15], [18], [19], [52], [53], [87], [115], [134].

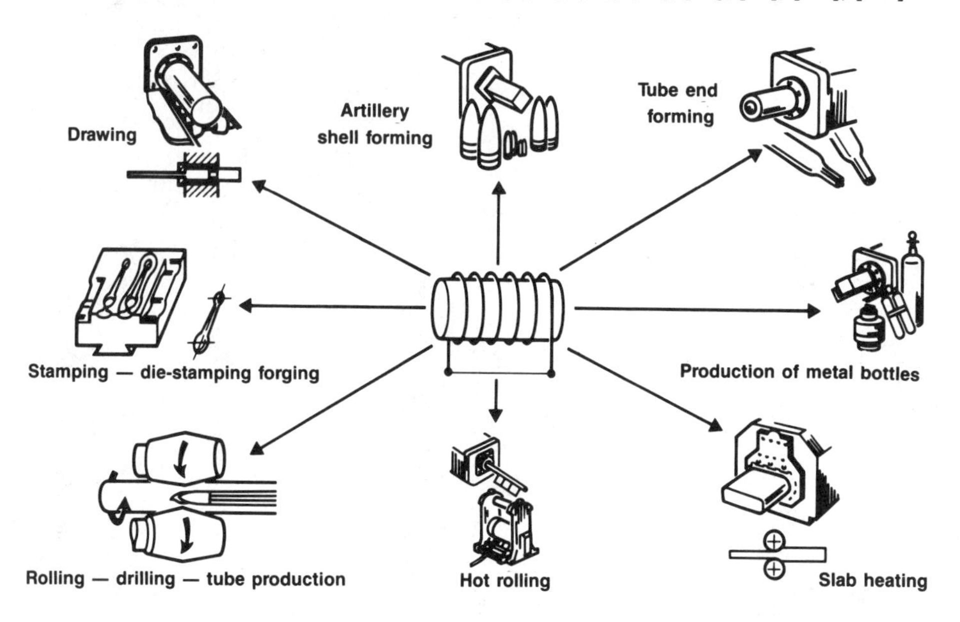


**Fig. 81.** Induction applications in heating prior to forming

### *5.3.1. Choice of frequencies*

Preforming heating operations are intended to through heat a metal product homogeneously.

Therefore, penetration of induced currents into the part must be as deep as possible and the frequency used relatively low, since penetration depth is inversely proportional to the square root of the frequency.

Now, the higher the penetration depth to diameter ratio, the lower the efficiency. Conversely, a higher frequency will be needed if a high energy efficiency is required.

Moreover, to prevent excessive temperature gradients between the external and internal layers of the part, power densities must be limited, since very high values would cause superficial overheating and eventually risk of melting, heterogeneous heating and energy losses due to surface radiation. For example, for steels, the maximum permissible power density is calculated from the experimental formula below, in which diameter d is expressed in centimeters:

$$P = \frac{300}{d}\ \mathrm{W/cm^2}.$$

If this value is exceeded, unwanted surface oxidation phenomena would occur, which can be avoided by carefully employed induction heating, unlike fossil fuel furnaces [88].

Heat transfer inside the part becomes important, and thermal conduction ensures homogeneous heating.

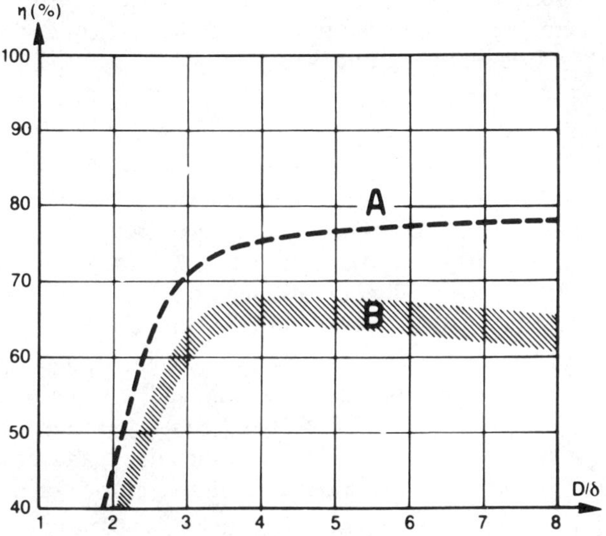

**Fig. 82.** Induction heating efficiency for steel billets:
curve A, inductor electrical efficiency
curve B, overall efficiency.

Therefore, selection of the frequency is a tradeoff between competing requirements, and the frequency chosen is the lowest in keeping with acceptable efficiency and adequate heating speed (the transmitted power density being proportional to the square root of the frequency).

The optimum trade-off can be determined experimentally, or by modelling the phenomenon combining the heat transfer laws and those controlling dissipation of the power induced in the part.

For example, for steel heated to 1,200°C, the curve in figure 82 shows that overall efficiency is not only affected by inductor optimum efficiency, but also by:
— losses at connections, which are proportional to the square of the current;
— losses due to compensation capacitors, which are proportional to reactive power;
— heat losses due to radiation and convection;
— losses in slippage guides, which depend on the length of the inductor, and therefore the linear power density.

Optimum efficiency is obtained for a diameter-to-penetration depth ratio (if a square cross-section product is treated, the diameter is defined as 1.1 times the side of the square) of the order of 4, while that of a single-diameter-product inductor is obtained for a value of 7 to 10 (*see* paragraph 2); however, the curve shows that, insofar as a range of products and not a single diameter product is treated, it is better to exceed the optimum value of this ratio at which efficiency is only slightly sensitive to variation, than to remain below it, where efficiency decreases very rapidly. The frequency to be used can then be estimated from this ratio.

### *5.3.1.1. Magnetic materials*

For magnetic materials, induced current penetration depth is inversely proportional to the square root of the relative magnetic permeability of the body. Therefore, it is possible to use much lower frequencies at temperatures below the Curie point than above.

For very large installations, in which high investment costs are justified, it may be advantageous to use two frequencies to obtain very high energy efficiency in cases where the final temperature is higher than the Curie point.

As long as the temperature is below the Curie point (around 750°C) and the metal remains ferromagnetic, the lower frequency will be used, generally line frequency, dispensing with the requirement for a frequency converter. The energy applied to the product in this phase represents 50% approximately of the total energy required to take the product to 1,200°C.

As soon as the metal becomes non-magnetic, the higher frequency is applied. In this manner, optimum heating and efficiency are obtained over both temperature ranges. These two-frequency installations are rarely used in installations other than those treating large diameter products, more than 100 mm, with high production rates of several tons or tens of tons per hour.

For smaller installations, it is often more economic to use a single frequency, the highest, to obtain adequate efficiency beyond the Curie point (in general, medium frequency). Over the possible frequency range, the bottom limit is retained so as to obtain the best heating conditions below the Curie point.

In most cases, the same heater is used to heat parts of different cross-sections. The frequency used will be the optimum frequency for parts whose dimensions represent most of the annual specified production, insofar as this frequency is not a major drawback for parts of other dimensions. Therefore, one frequency can be used to treat a complete range of parts with a rather wide variety of dimensions (diameter or cross-section) as shown in figure 83 below concerning heating steel beyond the Curie point (equivalent diameter for square cross-section products, is previously defined, as 1.1 of the cross-section).

The line frequency of 50 Hz only enables large cross-section products to be heated economically; for efficiency to be acceptable, the diameter must be more than 200 mm.

Medium frequency heating (rotary generators or thyristors) is fully suited to most of the cases encounter in industry.

When products to be heated are of small diameter (e.g. wires), much higher frequencies are used and therefore require vacuum tube generators.

Choice of frequency as a function of part diameter for heating to temperatures beyond the Curie point (steel, resistivity 100 $\mu\Omega$.cm).

| Optimum diameter (mm) | Economic diameter range (mm) | Frequency (Hz) | |
|---|---|---|---|
| 300 | 200 and beyond | LF | 50 |
| 90 | 60 to 210 | MF | 500 |
| 70 | 40 to 150 | | 1,000 |
| 45 | 30 to 100 | | 2,000 |
| 35 | 25 to 80 | | 3,000 |
| 30 | 20 to 60 | | 4,000 |
| 20 | 16 to 45 | | 10,000 |
| 3 | 2.5 to 8 | HF | 450,000 |
| 2 | 1.4 to 6 | | 1,000,000 |

**Fig. 83.**

For intermediate diameter between the high and medium frequency ranges (up to 10,000 Hz), aperiodic generators, providing frequencies of some thousands of hertz to some tens of thousands of hertz are used.

Tables such as above can be drawn up for metals other than steel. Only the effective resistivity and, where applicable, magnetic permability, notably change the economic range.

For heating below the Curie point, it is difficult to provide economic ranges as above. Variations in relative magnetic permeability can be very high and have a determining effect on the frequency used. The figures given in figure 84 cannot be considered as reference values, and each alloy must be carefully examined.

Choice of frequency as a function of part diameter for carbon steel ($\mu_r$ = 100) for heating below the Currie point

| Optimum dimension (mm) | Economic diameter range (mm) | Frequency (Hz) | |
|---|---|---|---|
| 30 | 25 and above | LF | 50 |
| 10 | 8 to 40 | MF | 1,000 |
| 4 | 3 to 20 | | 3,000 |
| 2 | 1.5 to 4 | | 10,000 |

**Fig. 84.**

From a diameter of 25-30 mm, the line frequency (50 Hz) can be used systematically.

In most cases, it is not necessary to use radio frequency, even for small diameter products. Wires can often be heated at medium frequency.

In the final analysis, the choice of frequency is such that:

— a range of frequency values exists which enables heating with excellent efficiency (and vice versa for a given frequency value), and for given product diameters and electrical and magnetic characteristics;

— below these frequency values, heating takes place with lower efficiency and this must be avoided;

— beyond these frequency values, the decrease in efficiency is slight (inductor efficiency only increases, since the penetration depth-to-diameter ratio drops; conversely, losses due to radiation and convection increase, and the time taken for heat to diffuse to the core increases);

— because of the penetration depth values below and beyond the Curie point, optimum frequencies are much lower for heating up to 750°C than for higher temperatures.

*5.3.1.2. Non-magnetic materials*

Beyond the Curie point, non-magnetic materials are equivalent to magnetic materials and consist essentially of non-ferrous metals such as aluminum and copper alloys [95].

The operating frequency is essentially related to the resistivity of the metal and to the dimensions of the products to be heated. Copper at 850°C and aluminum at 500°C have more or less the same resistivities, and are combined in figure 85 below, which, for a given frequency, can be heated by induction with adequate efficiency.

Choice of frequency as a function of diameter for non-ferrous alloys

<table>
<tr><th colspan="2">Aluminum and alloys (500°C)<br>Copper (850°C)</th><th colspan="2">Brass</th><th colspan="2" rowspan="2">Frequency (Hz)</th></tr>
<tr><th>Optimum Diameter (mm)</th><th>Economic Diameter Range (mm)</th><th>Optimum Diameter (mm)</th><th>Economic Diameter Range (mm)</th></tr>
<tr><td>90</td><td>60 and above</td><td>180</td><td>110 and above</td><td>LF</td><td>50</td></tr>
<tr><td>30<br>20<br>15<br>6</td><td>20 to 250<br>15 to 230<br>10 to 200<br>4 to 50</td><td>60<br>40<br>30<br>12</td><td>40 to 250<br>30 to 230<br>20 to 200<br>9 to 100</td><td>MF</td><td>500<br>1,000<br>2,000<br>10,000</td></tr>
<tr><td>1<br>0.6</td><td>0.75 to 10<br>0.4 to 8</td><td>2<br>1.2</td><td>1.3 to 10<br>0.8 to 8</td><td>HF</td><td>450,000<br>1,000,000</td></tr>
</table>

**Fig. 85.**

Whether for heating prior to extrusion or prior to rolling, re-heated non-ferrous alloys are often of large diameters. Therefore, the line frequency, 50 Hz, is used in most cases.

Other alloys such as magnesium or zirconium can also be induction heated. Extrusion of magnesium alloys requires a temperature of between 280 and 450°C, depending on the type of alloy, the complexity of the shape to be obtained and the drawing ratio. Reheating can be made in normal atmospheres, in open air, while all other processes (flame or resistance furnaces) involve the risk of hot points, and therefore require a protective atmosphere. For zirconium, the drawing temperature is between 600 and 800°C, but the alloy must be protected against superficial oxidation. Several techniques have been developed to protect the metal, for example preliminary copper plating of the billet to be heated; in this case, the advantage of heating by induction is the heating speed, which prevents any risk of oxidation. In most cases, line frequency is used for heating prior to drawing magnesium and zirconium alloys, since the billet diameter is generally large, and therefore the diameter-to-penetration depth ratio is sufficient to provide high efficiency at this frequency [65 H2].

### *5.3.2. Heating time*

High power densities must be accepted for rapid heating. But, as the considerations concerning choice of frequency have shown, induced currents develop on the part surface, and the heat produced reaches the interior by conduction; therefore, if heating time is too short, the temperature may not be sufficiently homogeneous in the cross-section. Therefore, heating times may be as short if thermal diffusivity is high. Therefore, heating will be faster for

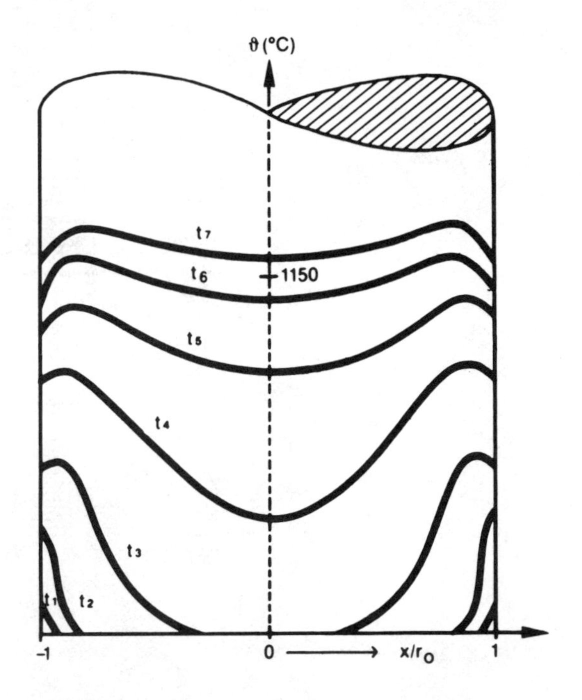


**Fig. 86.** Temperature distribution within a plane containing the axis of a steel cylinder, after various heating periods : $t_1 < t_2 < \dots < t_7$.

for non-ferrous metals such as copper and aluminum alloys, whose mean thermal conductivities over the heating range are about 400 W/m × °C for copper, 100 W/m × °C for brass and 200 W/m × °C for aluminum respectively.

### 5.3.2.1. *Ferrous metals*

As can be seen from Figure 86, temperature distribution varies greatly with heating time. Therefore, the heating time and power are chosen so that, at the end of the heating period, the difference in temperature between the core of the metal and its periphery is low. For steel, the generally accepted difference between the center line and the surface of the part to be heated is 6 to 8% (for example, 75°C if the final temperature reached in the surface layer of the metal is 1,150°C).

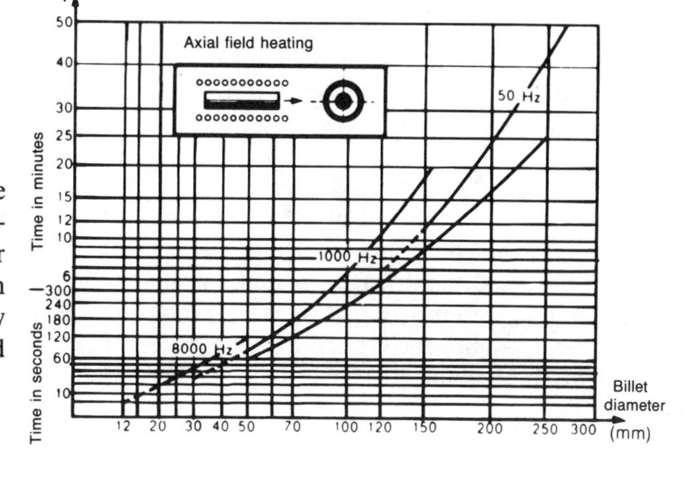


**Fig. 87.** Heating time using a uniform axial field (The lower curve is the minimum time arrived at by varying field strength) [B].

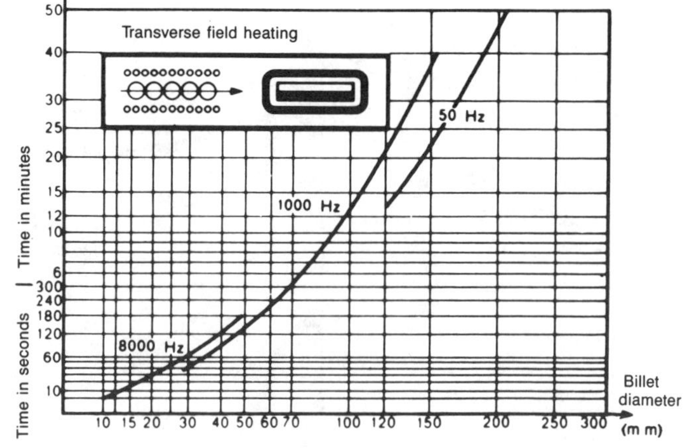


**Fig. 88.** Heating time using a transverse field [B].

To calculate the approximate core heating time of a steel billet for a peripheral temperature of 1,200°C and a temperature difference of 40°C within the billet, the following empirical formula may be used:

$$t = 910\, d^2,$$

$t$, heating time in minutes;
$d$, diameter of the part to be heated in meters.

Figures 87 and 88 provide the heating times for a cylindrical steel body as a function of diameter, to reach a surface temperature of 1,200°C with a temperature difference of the order of 70°C within the part.

### 5.3.2.2. *Non-ferrous metals*

Generally, the line frequency is used. Figure 89 provides the usual heating times for various alloys.

Non-ferrous alloy billet heating time

| | Diameter (mm) | Seconds |
|---|---|---|
| Aluminum to 500°C | 100<br>200<br>300 | 30<br>120<br>300 |
| Brass (70/30) to 750°C | 100<br>200<br>300 | 90<br>300<br>720 |
| Copper to 800°C | 100<br>200<br>300 | 25<br>100<br>210 |

**Fig. 89.**

### 5.3.3. *Efficiency and specific consumptions*

Overall efficiency, which is defined as the ratio of energy theoretically required to take a product to a determined temperature to the energy effectively expended, is of the order of 50% for low resistivity metals such as copper and aluminum alloys, but, for steel, reaches 70%. When the steel part diameter is less than 200 mm, these must be heated at medium frequency, which, using a solid state generator, reduces efficiency to 60-65%.

Induction heating efficiency for forging applications

| | Steel at 1,200°C | Aluminum at 550°C | Copper at 800°C |
|---|---|---|---|
| Theoretical energy requirements (kWh/t) | 240 | 145 | 95 |
| Specific consumptions (kWh/t) | 330 to 600 | 250 to 400 | 190 to 300 |
| Practical efficiency (%) | 40 to 73 | 36 to 56 | 32 to 50 |

**Fig. 90.**

However, for comparative economic analyses, it is better to use mean consumptions. Figure 90 compares the theoretical energy required to take products to the correct temperature and the measured specific consumptions on normally operating industrial installations.

For steel, the most currently encountered values in heating stamping billets to 1,200°C is between 450 and 550 kWh/t. Conversely, in steel-making, re-heating slabs from ambient temperature to 1,200°C requires only 300 kWh/t approximately.

Figure 91 gives the variation and specific consumption against diameter, for heating at 50 Hz, for aluminum.

Heating efficiency of aluminum

| Diameter (mm) | 150 | 205 | 255 | 305 |
|---|---|---|---|---|
| Consumption (kWh/t) | 275 | 260 | 255 | 250 |

**Fig. 91.**

For copper alloys, specific consumption varies highly with the alloys treated, as they differ greatly in resistivity.

In heating prior to drawing, specific consumption is about 250 kWh/t for magnesium alloys, and 150 kWh/t for zirconium alloys.

This difference in specific consumptions is explained by numerous factors such as product dimensions, part-inductor coupling, frequency, heating rate, thermal insulation, inductor construction technique, frequency converter used, operating conditions, etc.

Figure 92 shows the effect of diameter and frequency on specific consumption for heating steel bars to 1,250°C.

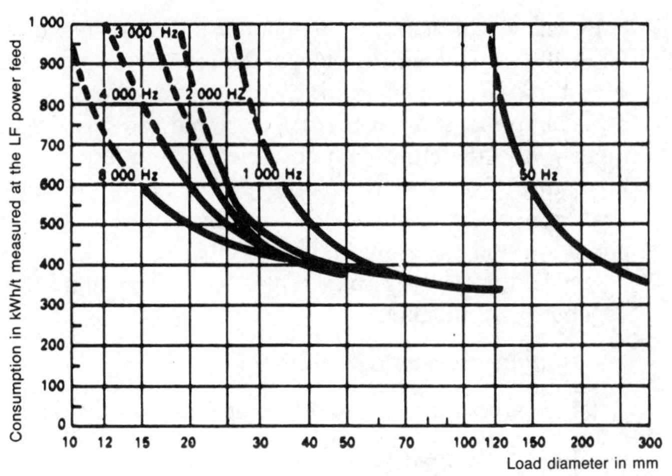


**Fig. 92.** Electrical energy consumption required for induction heating of steel bars to 1,250°C [B].

When it is necessary to compare induction and flame heating consumptions, it is necessary to use overall specific consumption for both cases. For example, the overall consumption observed for heating steel billets is 2,300 thermal kWh on the average in the flame furnaces at present use in this profession. Therefore, the primary energy saving obtained is of the order of 30%.

### *5.3.4. Advantages and limits of induction heating in through heating prior to forging*

#### *5.3.4.1. Advantage of induction re-heating*

Induction heating is being more and more widely used in industry for numerous reasons [50], [51], [63], [65 j], [71], [72], [125], [134], [138].

— *High productivity*

In many cases, the introduction of an induction heater in a hot forming line results in a noticeable increase in overall productivity. Increases of 20 to 30% are frequently encountered in enterprises equipped with induction heaters. This productivity increase is basically due to the regular rate of the induction heater, which imposes its working rhythm on the line as a whole, by automatically providing a regular flow of products at the correct temperature. The response speed of induction heaters producing the first hot products reduces tool heating times after each tool change. Moreover, due to the increase in their service life, tools are changed less often.

— *Heating speed*

Depending on product cross-section, a few seconds to some tens of minutes are sufficient to reach the required temperature, versus some tens of minutes to a few hours for fossil fuel furnaces.

— *Reduction of oxidation losses*

Basically, scale loses are attributable to heating rate over the temperature range 800-1,300°C. For ferrous metals, the gain in metal weight is generally between 1 and 3%. This gain is lower for copper alloys, and almost negligible for aluminum alloys. With the present trend towards the transformation of high quality, and therefore high cost metals, and also the necessity of saving raw materials, this saving is becoming more and more important.

Moreover, the scale reduction obtained provides parts offering better surface condition and therefore higher quality.

For alloys such as magnesium and zirconium, this oxidation reduction capability is fundamental.

— *Adequate temperature accuracy*

Temperature accuracy and homogeneity is adequate.

— *Absence of decarburization*

The decarburization of ferrous metals is almost zero, which improves the characteristics of the parts and results in better quality. Less decarburization also means less machining to remove the surface layer of affected metal, and therefore savings in materials, energy and machining operations.

— *High quality of parts*

With limited oxidation and decarburization, homogeneity and temperature accuracy, holding the fine grain metallurgical structure and core ductility enable high quality parts to be produced.

— *Possibility of heating a clearly defined area of the part.*

This advantage further increases the energy and economy performance of the process and its flexibility.

— *Atmosphere heating capability*

Where almost total elimination of oxidation (special steels, steels intended for precision forging, titanium, zirconium, etc.) is desired, an induction heater can easily be equipped with a device creating a protective atmosphere.

— *Increase in tool service life*

For example, the service life of forging dies is considerably increased. Estimates vary, but increases of 15 to 30% are often mentioned by users.

Moreover, the gain for tools is not simply limited to forming machines. The increase in service life of tools used for subsequent machining can be considerable and, in some cases, tripled.

— *Low energy consumption*

The heaters are started almost instantaneously, off-load losses are negligible, at least with static converters — no requirement to maintain temperature in the event of production shutdown, repairs, tool changes, pauses, etc., and the heating system can be restarted without delay. Only the part to be formed is heated. Energy efficiency is therefore excellent, and the method of heating in many cases offers large primary energy savings. Energy costs are often much lower with induction heating than with fossil fuel heating.

— *Automatic operation*

An induction heater can be built easily into an automated production line. Mechanization of a forming line can also be made step by step; this evolution is considerably assisted by induction heating. For example, an initial stage could be a system for the automatic extraction and transfer of heated products to forming machines. At a later stage, an automatic loading system can be put into service and, if series production requirements justify it, the line can be totally automated.

— *Smaller size*

For a given hourly production rate, an induction heating installation generally requires less workshop space than a fossil fuel furnace.

— *Improvement in working and environmental conditions*

Over the years, this aspect has become more and more important. Induction heating significantly contributes to the effort to improve working conditions, either by dispensing with particularly onerous tasks such as withdrawal of loads, or by creating a better environment by significantly decreasing radiated heat and eliminating smoke and noise due to heating equipment. Moreover, induction heating is a clean process, which fully satisfies increased requirements concerning atmospheric pollution.

### *5.3.4.2. Limitations of induction heating*

These limitations are purely technical and inherent in the principle itself, or economic under present or foreseeable conditions.

The first category essentially concerns the requirement of heating products of regular shapes and cross-sections. Moreover, this situation is the most frequently encountered in industry. However, when it is necessary to provide intermediate heating for a part during forging, induction heating is rarely suitable; conversely, this is a highly efficient method for intermediate re-heating of half-finished products during rolling.

Economically, induction heating encounters limits in terms of investment, and therefore the possibility of amortizing investments by sufficient production, the necessity of treating large series of identical or similar cross-sections to prevent frequent replacement of inductors, and the availability of sufficiently qualified personnel to ensure normal maintenance of the heaters.

Therefore, as in all investment choices, a technical-economic balance, allowing for technical, economic, financial and social aspects, is used to determine the most suitable equipment.

### 5.3.5. *Composition of an induction heating system for heating before forming*

Generally, heating equipment is installed in a production line. For example, figure 93 shows the composition of a stamping line for steel slabs [A], [B], [C], [D], [E], [F], [M], [N], [R], [S], [W].

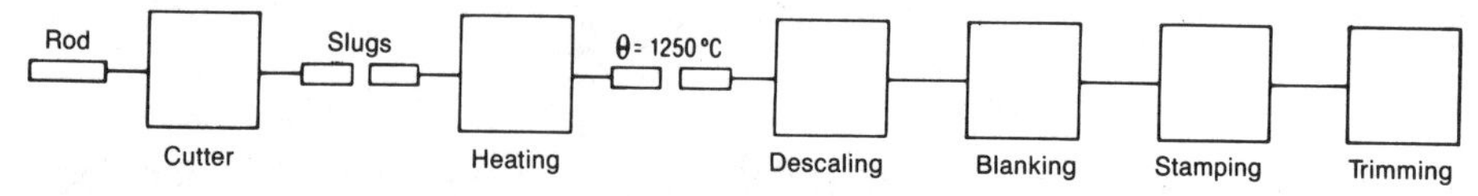


**Fig. 93.** Steel slug stamping line

Basically, induction heating preforming equipment consists of:
— an inductor and its power supply;
— a load handling system — loading, feed, ejection;
— an inductor cooling system, capacitors and a generator, if a frequency other than the line frequency is used;
— ancillary equipment: control console, monitoring equipment, capacitors.

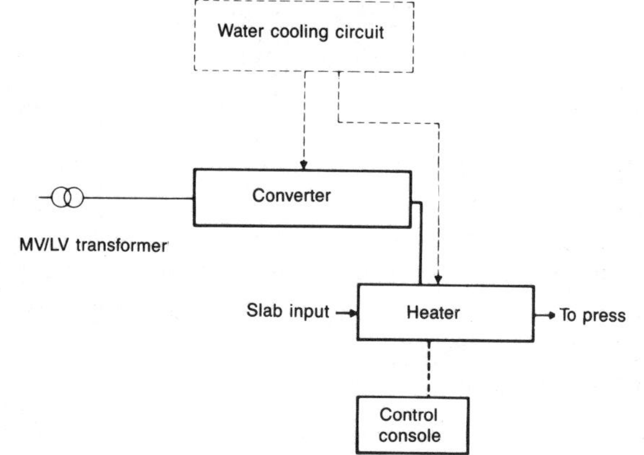


**Fig. 94.** Diagram of an installation for induction heating for forging

Induction heating is fully suitable for high production rates; very fast heating, loading, feed, ejection and transfer are easily automated.

When it becomes necessary to operate at a frequency other than the line 50 Hz, the frequency required is provided by one or more rotary converter units or solid state thyristor units, for medium frequencies up to 10,000 Hz, while vacuum tubes are used for higher frequencies.

The part to be heated — slab, billet, bar, etc. — passes through one or more inductors. At the end of the path, the part, which is at forging temperature, is generally automatically carried to the forming system. When changing production series, it may be necessary to replace the inductors to adapt to part cross-sections.

In the event of a line incident, only a few parts are in process; heating can be interrupted immediately, and only some parts have to be withdrawn.

Preforming induction heating equipment differs in the following ways:
— the systems for handling parts inside the heater: roller, ram, chain or other feed systems;

— their adaptation to heated parts — slugs, billets, bars, slabs, tubes, etc. — and the type of heating — complete or partial;
— the design of the inductor and the transport of the parts inside the latter — longitudinal or transverse, etc. [15], [18], [50], [51], [52], [55], [65 *i*], [92], [103], [116], [126], [139], [144], [145], [147 *c*], [151].

### *5.3.5.1. Different types of induction heaters*

#### *5.3.5.1.1. Total heating*

— *Pusher systems*

This type of heater is particularly suitable for relatively small slugs and billets. Loading may be manual, from an intermediate magazine (oblique fixed or vibrating chute) or an automatic magazine, such as a wormscrew conveyor, which accumulates billets inserted randomly, and places them in the desired position for presentation to the heater, or for longer billets, by a chain conveyor which takes the billets one by one, or from a shelf-type magazine.

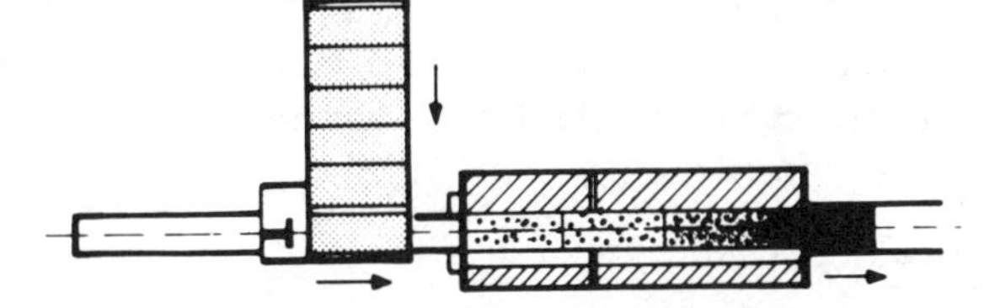

**Fig. 95.** Intermittent pusher [B].

The products advance one after another through the coil, pushed by a piston actuated by an electric motor, a pneumatic system or a continually moving chain. Products are guided through the inductor by non-magnetic, wear resistant water-cooled steel runners.

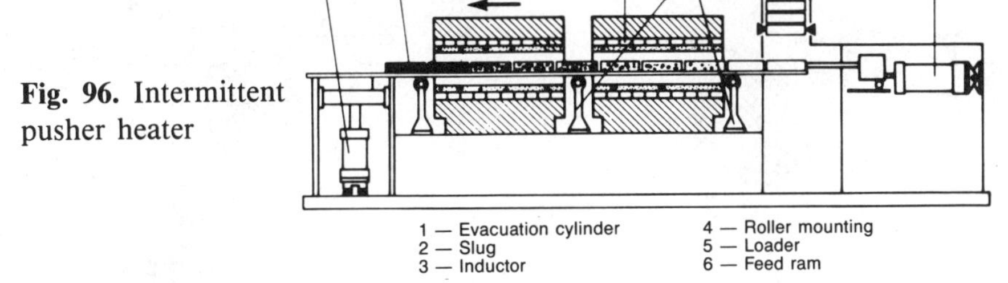


**Fig. 96.** Intermittent pusher heater

Most of the time with intermittent systems, the part is immobile in the coil and moves only when a new part is inserted.

With continuous pusher systems, the column of billets moves constantly through the inductor.

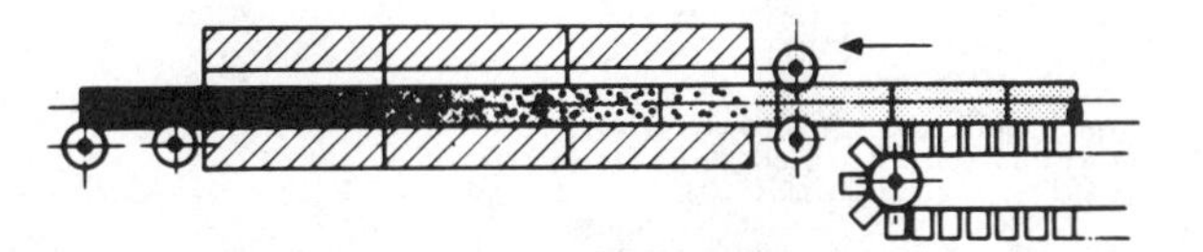

**Fig. 97.** Continuous pusher system [B].

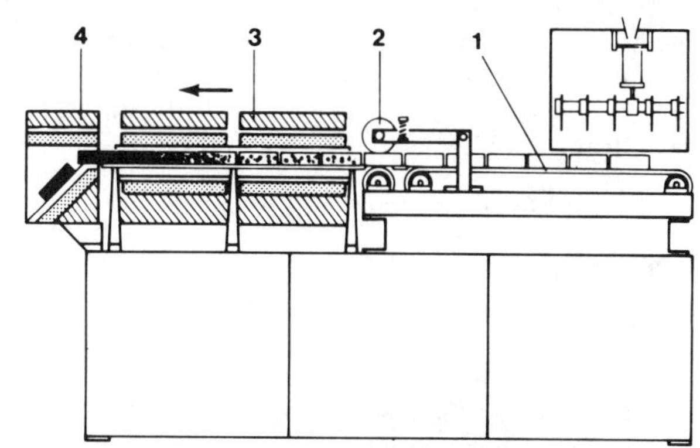


**Fig. 98.** Continuous pusher system heater

1. Conveyor chain
2. Pusher roller
3. Heating inductor
4. Temperature holding inductor

In most cases, parts are simply ejected by tilting at the end of the inductor (short billets). The parts drop into a chute, where they are picked up by manual devices or pneumatic pincers. Rollers are generally used to extract longer parts.

With these feed systems, and in particular at high temperatures, the parts sometimes tend to become welded to each other. Very simple systems are used to remedy this drawback on extraction (for example, when the extraction orifice is swept by a metal arm; if the billet has not been correctly extracted, the shock uncouples it from the next billet).

— *"Walking beam" system*

This system is mostly used for handling large parts, in general from lengths of 500 mm for high rate, fully automated installations. Ejection and transfer are generally provided by pincers.

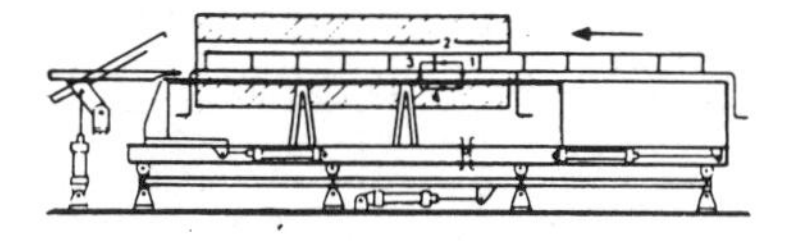

**Fig. 99.** "Walking beam" heater (B)

1. Mobile rails lift the slug
2. Slug advance
3. Slug deposited on the fixed rails
4. Mobile rails reverse

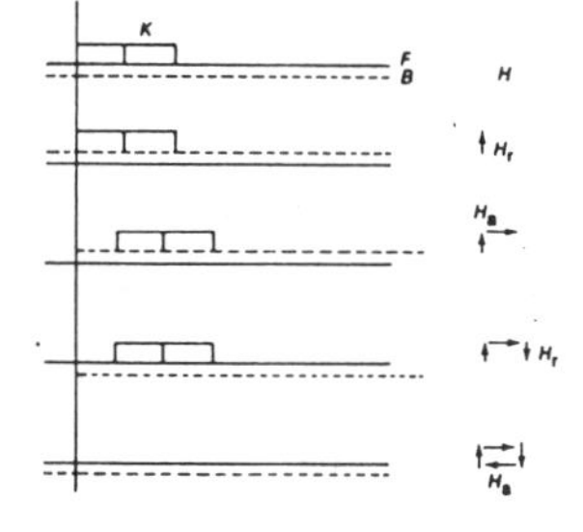


**Fig. 100.** "Walking beam" system movement [F].

The conveyor device consists of a pair of fixed rails F installed in the inductor, and forming part of the heater, and a pair of mobile rails B

$H$ = movement $H_r$ = radial movement
$H_a$ = axial movement $K$ = part to be heated

Although more complex, this system reduces the risk of sticking, overlapping, lifting and jamming. The movement is easily matched to automatic press operation.

— *Bar continuous heating system*

This type of heater is mostly used for automatic forging of bars on horizontal presses. Generally, production rate is high, up to 10 tons per hour, and handling is fully automatic.

The bars pass through the inductor continuously. After ejection, the bars can be cut individually in the hot forming machine, forged in a horizontal forging machine or formed by winding round a former (for example, production of large springs).

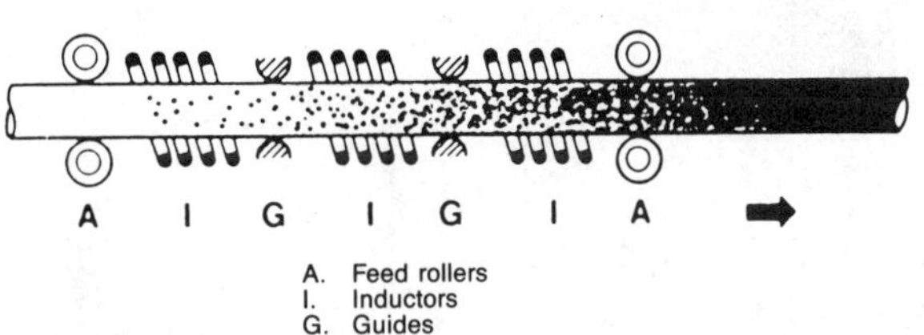


**Fig. 101.** Bar continuous heating system [B].

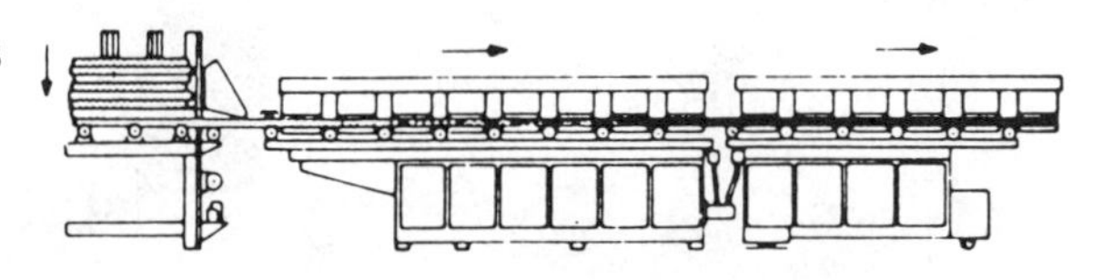

**Fig. 102.** Continuous bar heating [B].

— *Special heaters*

For parts of complex geometry or dimensions, where the diameter is greater than the length, special heaters have been constructed. For example, the inductors may be vertically mounted, ejection being made through the top onto a slope, with loading through the bottom. Special heaters with specially designed inductors are used for reheating slabs in steel making.

Multiple inductor heaters mounted in parallel are also used for high production capacities, for example, re-heating of billets prior to rolling [52], [55], [56], [65 *h*], [83].

*5.3.5.1.2. Partial heating*

— *Bar end heating*

Movement may be semi-automatic, with manual loading and unloading; a temperature control system or heating time control system causes automatic ejection prior to withdrawing; this can be fully automatic from loading to withdrawal. Various types of inductors and handling systems are used.

• *Field axial with respect to the part and tranverse movement*

Insertion of the parts in the inductor, their translation and extraction are controlled by three rams or cylinders. Movement is transverse with respect to the inductor, and the field is axial with respect to the part.

This configuration is used for heating the ends of square or flattened bars or cylindrical rods, with maximum production rates of about 1,200 parts per hour, part diameters of 10 to 80 mm and lengths of 30 to 800 mm; these heaters can, of course, be adapted for parts of different characteristics.

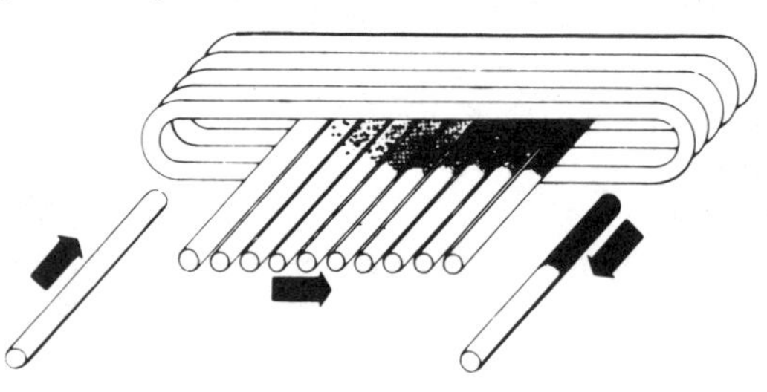

**Fig. 103.** Rod and bar end heating, axial field, transverse movement [B].

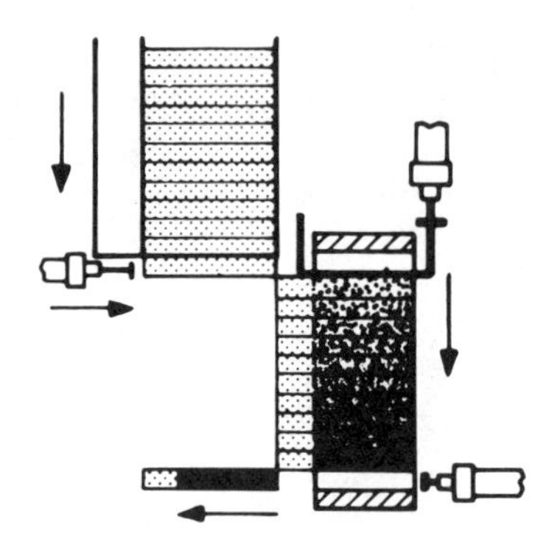

**Fig. 104.** Axial field transverse movement heater

• *Open or semi-open channel inductors*

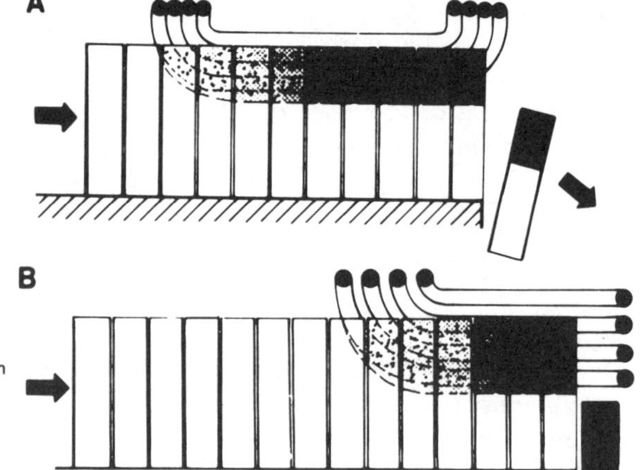


**Fig. 105.** Heating principle with channel inductor and lateral conveyance [B].

A. Inductor open at both ends
B. Semi-open inductor, axial extraction

The parts are pushed by electro-pneumatic mechanisms or driven by a device moving under the inductor. This type of heater is suitable for heating long parts at high production rates of up to 3,600 parts per hour for a diameter range of 10 to 80 mm and heated length of up to 120 mm.

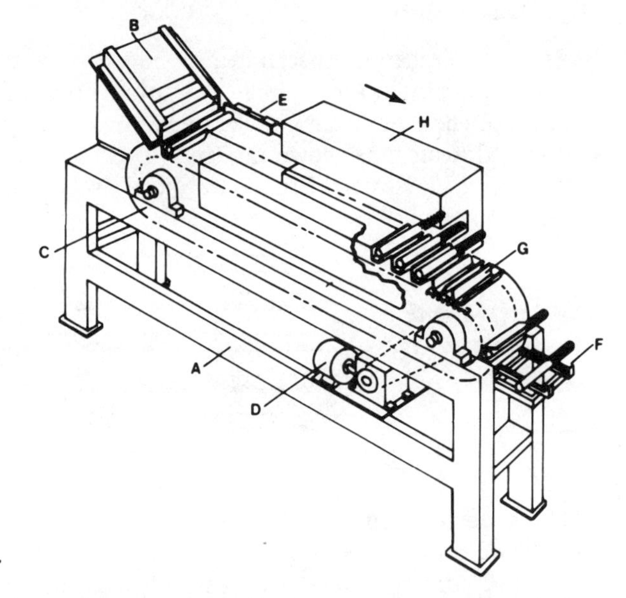


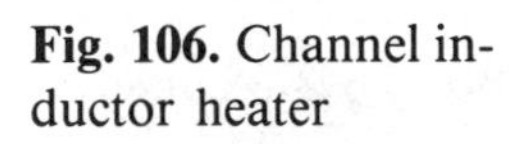
**Fig. 106.** Channel inductor heater

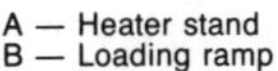
A — Heater stand
B — Loading ramp
C — Conveyor
D — Conveyor control
E — Bar gauge
F — Bar withdrawal
G — Billet conveyor
H — U-shaped inductors

• *Multi-inductor heater*

Axially heated multi-inductor heaters are used before stamping bolts, or before rolling threads on anchor bolts. There is only one part per inductor.

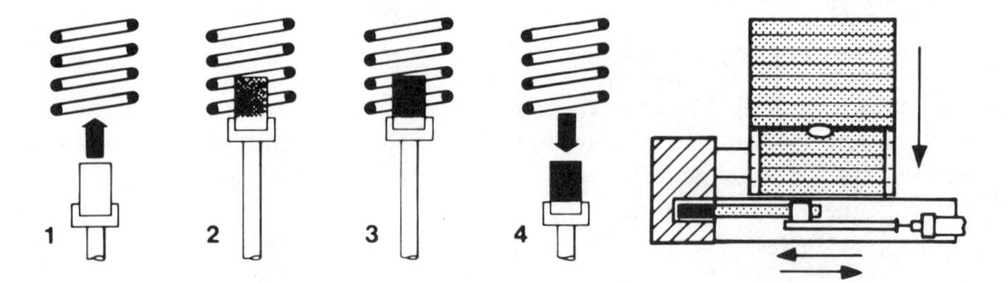


**Fig. 107.** Axial heating vertical multi-inductor heater [B].

**Fig. 108.** Automatic axial insertion and extraction [B].

The inductors may be vertical or horizontal. In some cases, this type of heater is used for total heating of parts (e.g. valve forming). Production rates vary from 100 to 1,800 parts per hour as a function of the number of heating heads.

— *Partial heating of bars*

It is possible to continuously heat parts of bars which are then cut and formed. The bars are then fed intermittently through a series of inductors.

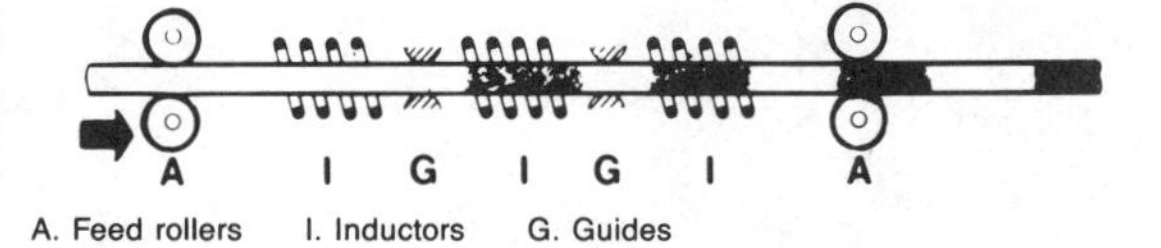


**Fig. 109.** Localized heating of bar with regular "pulsed" feed [B].

— *Special heaters*

As for total heating, special heaters intended to resolve particular problems can be custom built. For example, two principles are used in heating tube ends:
• the tube ends are heated in channel inductors (a principle identical to that of bar end heaters as shown in figure 106);
• the ends of a cluster of tubes are successively inserted in moving inductors before being transferred to a press [20].

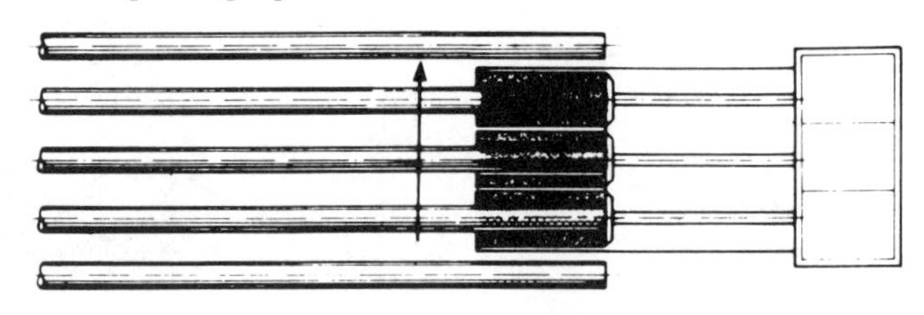

**Fig. 110.** Mobile inductor tube end heating system

### *5.3.5.2. Inductors*

The inductor consists of a very high electrical conductivity, very pure electrolytic copper tube of round or square cross-section formed according to the desired contour and cooled by water under pressure. If the diameter of the product to be heated is less than 75% of the rated diameter of an inductor, this has to be changed, since the coupling will be insufficient causing an unacceptable drop in efficiency.

— *Composition of an inductor*

Figures 111 and 112 show the composition of an inductor and various construction options. Particular attention should be paid to the insulation between turns and to the refractory material coating protecting the turns against the hot billets moving through the inductor.

1. electrical connection
2. water-cooled coil
3. water-cooled guide tube
4. refractory lining
5. insulating housing
6. connection for water coolant

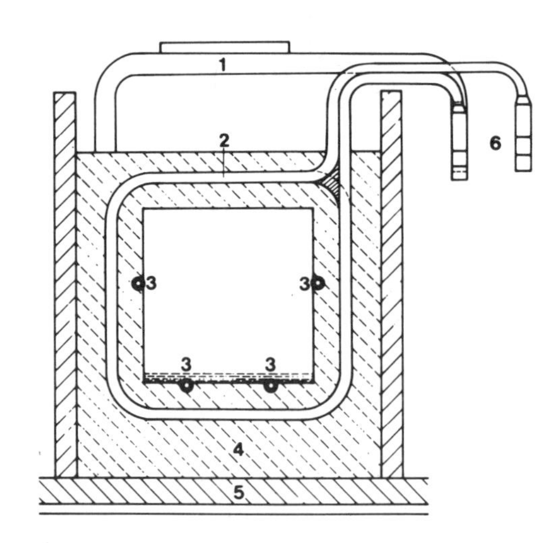


**Fig. 111.** Cutaway view of an induction coil

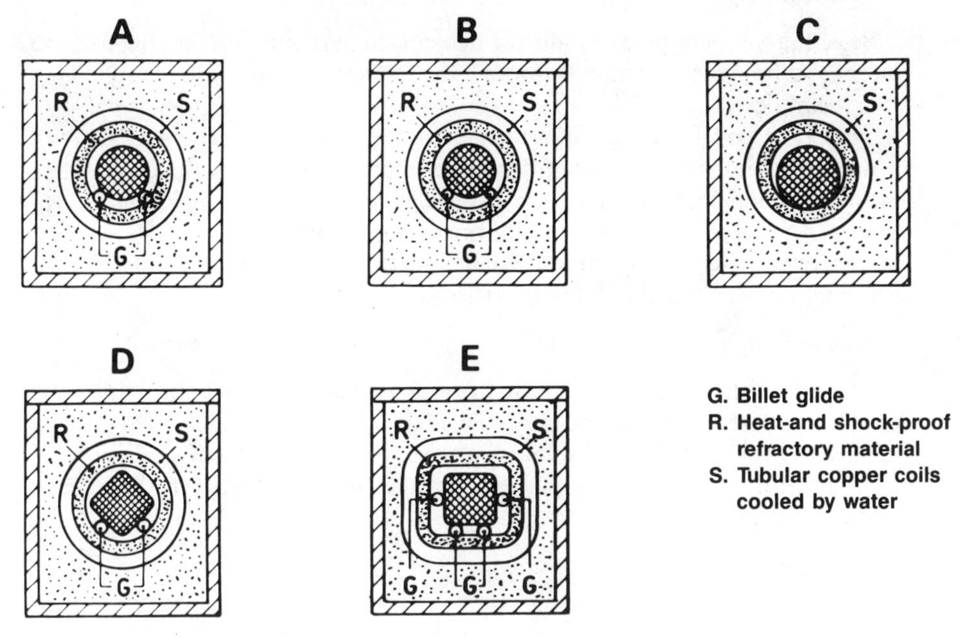


**Fig. 112.** Cross-section of an inductor and guides for the parts.

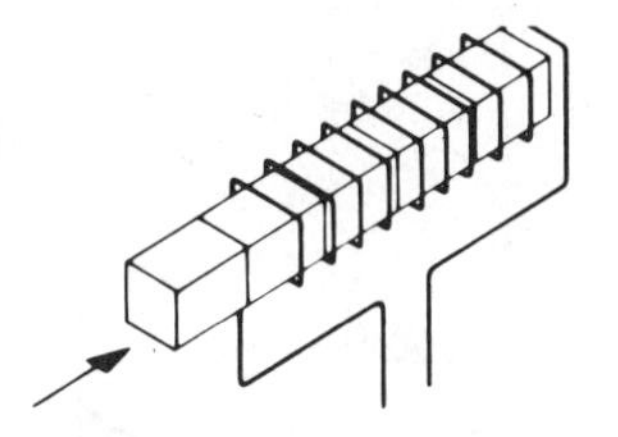

**Fig. 113.** Coil for axial movement

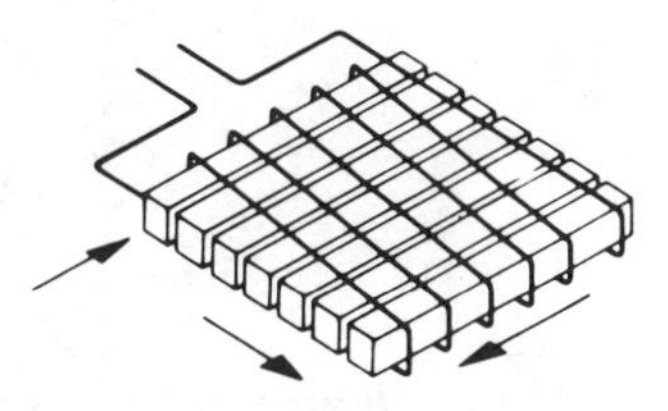

**Fig. 114.** Coil with longitudinal input and output, and transverse displacement during heating.

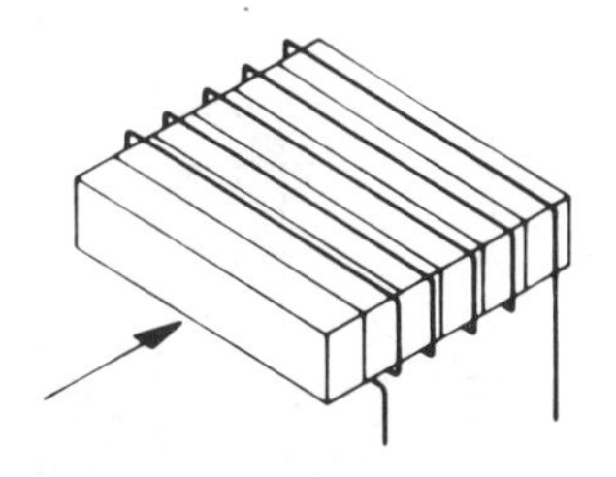

**Fig. 115.** Coil for transverse movement

**Fig. 116.** Tunnel coil

— *Inductor types*

Selecting the right type of inductor depends in particular on the shape of the pieces and how they are moved through the inductor as shown in the following four examples:

- axial field coil for in-line movement, the most common arrangement;
- axial field coil with a system for transverse movement where the billet enters the inductor longitudinally, moves transversely then exits in the longitudinal direction; yield is higher, but the mechanism is more complex;
- transverse field coil, suitable for billets of limited length;
- tunnel field coil, suitable for short slugs.

— *Heating area and inductor length*

The inductor length depends, for the various diameters of billets, upon the production rate required. The curves in Figure 117 show inductor length as a function of the diameter and frequency necessary to heat 1 kg of steel per hour to a temperature of 1,200°C with a temperature difference between the surface and core of about 5%.

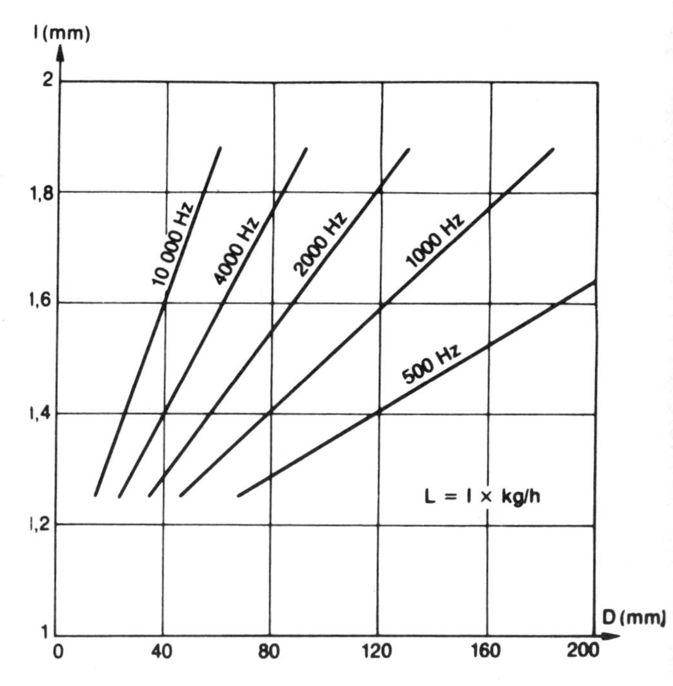


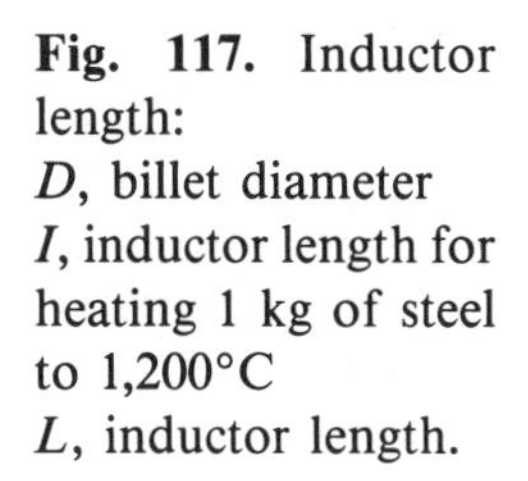

**Fig. 117.** Inductor length:
*D*, billet diameter
*I*, inductor length for heating 1 kg of steel to 1,200°C
*L*, inductor length.

In practice, the heater consists of several heating sections, for example with the heater intended to take ferrous products to a temperature beyond the Curie point:

- preheating area, to 400°C approximately, with widely spaced turns;
- heating area, to Curie point, 750°C approximately, with tighter wound turns;
- area corresponding to final heating with very tightly wound turns to reduce losses due to radiation and formation of scale to a minimum.

— *Extraction of heated products*

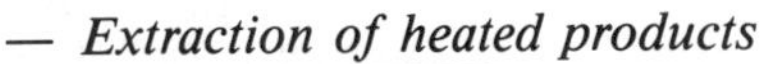

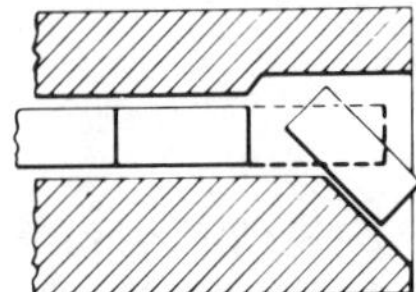

**Fig. 118.** Gravity extraction

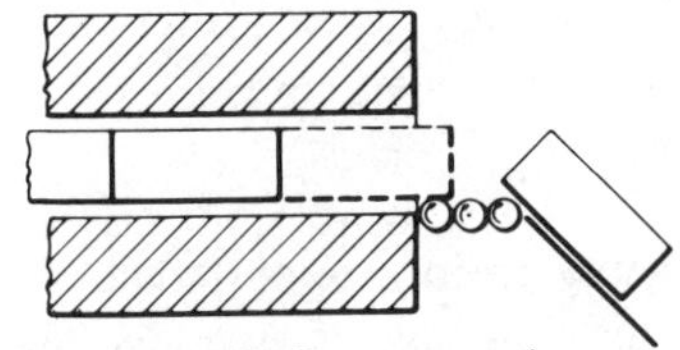

**Fig. 119.** Roller extraction

Various extraction systems are used. For short billets, gravity extraction is often used, while for longer billets, roller or pincer extractors are used.

— *Replacing the inductor*

Once the frequency has been fixed, an inductor is designed for a product of rated diameter, thus providing optimum efficiency. Products of smaller cross-sections may be heated but with lower efficiency. Therefore, it is better to have several inductors if the production run involves very different cross-sections. For example, if products of 60 to 120 mm diameter are reheated, three inductors can be used to cover diameter ranges 60-80, 80-100 and 100-120 mm.

Also, for long products, the internal dimensions of the inductor must also be chosen as a function of part dimensions, and also as a function of risks of deformation, so as to prevent jamming.

Inductors are designed for rapid removal and installation. Replacement of an inductor generally requires 20 to 40 minutes, which underlines the advantage of using sufficiently long working periods for products of similar cross-sections.

**5.3.5.3.** *Controlled atmosphere heating*

Formation of scale is much lower with induction heating than with fossil fuel furnaces. However, where total elimination of scale is required (high quality steels used for precision stamping, or highly oxidizable metals such as zirconium, etc.), it is easy to work under protective atmospheres, using simple devices such as gas jets at the inductor inlet and outlet. Nitrogen, which is a neutral, non-toxic protective gas, is fully suitable for this application. In some cases, it is possible to keep the products as a whole in a neutral atmosphere.

**5.3.5.4.** *Power factor*

A battery of capacitors is connected in parallel with the coil to correct the power factor; in small heaters, they are often mounted directly under the coil, inside the welded frame.

#### 5.3.5.5. *Generators*

Different types of medium frequency, or more rarely high or radio frequency generators are used. Rotary converters, thyristor solid state converters, aperiodic generators or vacuum tube generators are used, depending on requirements (*see* paragraph 3.2.).

#### 5.3.5.6. *Cooling systems*

Most cooling systems can be used, the most current being cooling towers and refrigerating units.

### 5.4. Brazing

Brazing is an assembly technique consisting of joining two or more parts together by means of an added material, which is melted. In the joint area, the parts must be heated to the added metal melting point. Induction heating is widely used in brazing. This enables accurate localization of the thermal effect required to melt the added metal. Heating is very fast, and oxidation or changes in structure or chemical composition of the parts are reduced to a minimum. Usually with added metals, a distinction is made between those having a melting point of less than 500°C, such as tin and lead-cadmium alloys (soft brazing) and those whose melting point is high, such as copper and its alloys (brass, bronze) or silver and its alloys (hard brazing).

To obtain good quality brazing, the relative positions of the inductor and the parts must be studied carefully so that it is the area to be brazed which heats, due to the effect of induced currents, and not the added metal which is taken to melting temperature by thermal conduction. The coil must therefore be located so that the added metal cannot reach its melting point before the surfaces to be brazed have reached the brazing temperature.

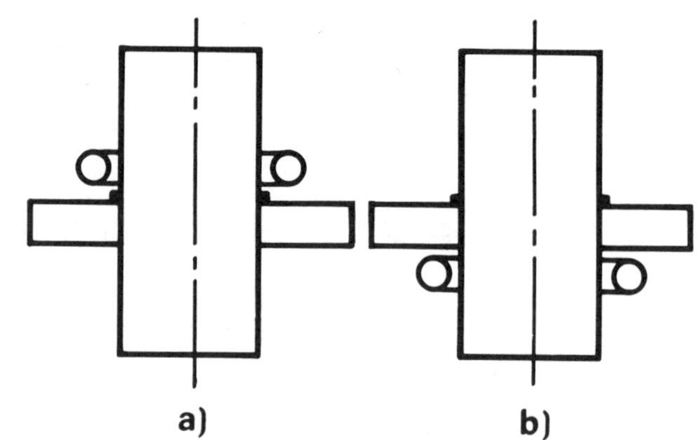


**Fig. 120.** Induction brazing:
*a)* incorrect inductor arrangement, the added metal melting point is reached too fast;
*b)* correct inductor arrangement; heating by conduction through the parts.

The type and position of the inductor must also be such that the added metal can fill the joint between the parts through gravity. Brazing can also be performed in a controlled atmosphere.

To prevent oxidation, the added metal is generally coated with a protective flux, except when working in controlled atmospheres.

Fig. 121. Brazing by induction

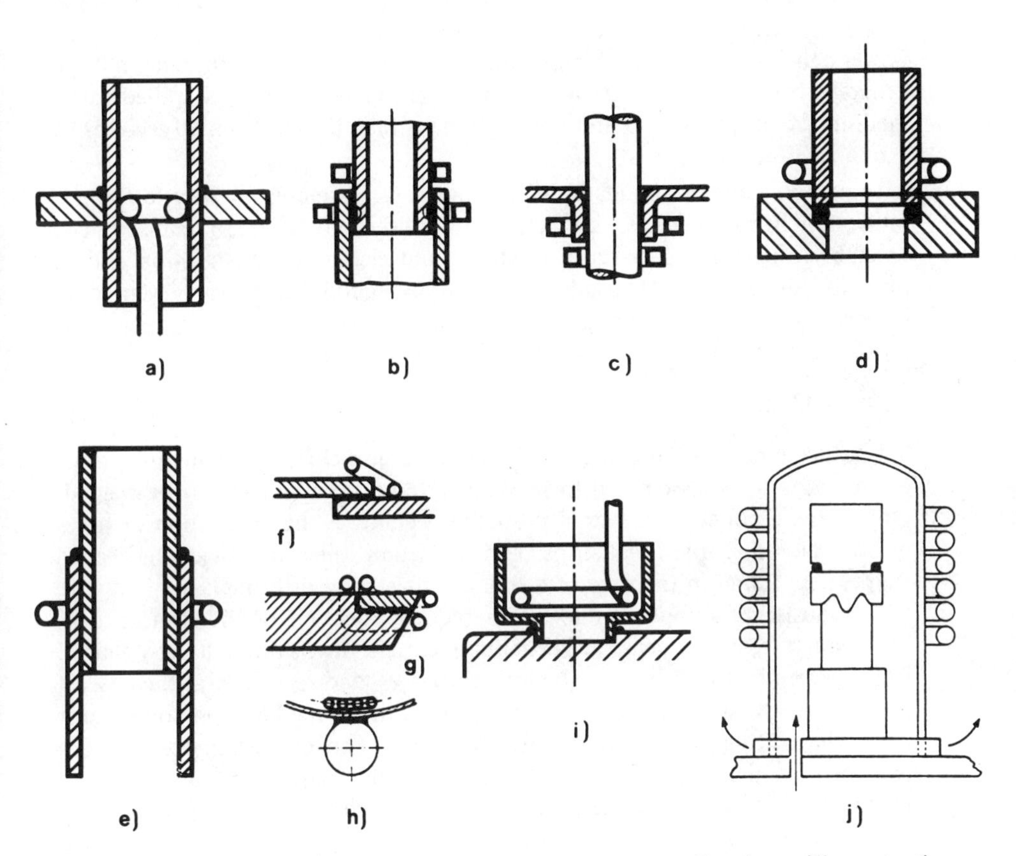


Brazing with protective atmosphere (brazing with vacuum is also possible)

When the weight or thermal conductivity of the parts to be brazed differ greatly, or if one part is magnetic and the other non-magnetic, it is also necessary to take precautions with the design and position of the inductor so that both parts reach brazing temperature simultaneously. After heating which varies according to the type of brazing, there must be a cooling period.

If these conditions are fulfilled, induction heating offers excellent brazing quality, while providing important energy savings. In general, brazing operations call for the use of radio frequency, low power generators, of 0.5 kW to some tens of kW. In most cases, treatment times vary from 1 to some tens of seconds. Power densities can be very high, a few kilowatts per square centimeter, but are generally between 50 and 500 W per square centimeter. With thick walled parts, medium frequency is used and heating time is then longer, from 10 seconds to 3 minutes approximately.

Induction brazing applications are numerous: assembly of bicycle frames, fabrication of cutting tools with metal carbide disks, production of tennis rackets, assembly of numerous parts in electrical and electronic construction and in precision engineering. The advantages of induction heating are even more important when the components to be assembled are small, or operations repetitive, since automation is facilitated.

## 5.5. Welding

Localized induction heating enables temperatures close to the melting points of the parts to be assembled to be obtained and permits parts to be welded. One of the main applications in induction welding is the production of tubes or other closed profiles. Flat strips are deformed by rollers to provide the desired shape, (e.g. round in the case of a tube). This is moved through the inductor, which consists of a single turn for radio frequency heating, or one or more turns for medium frequency heating. The induced current travels through the V-shaped slot. Current density is very high along the edges, and especially at the base of the Vee, which favors welding, and the temperature at the junction point is just below the melting point of the metal to be welded. The tube then passes between two rollers which makes a pressure weld. With the frequencies currently used in this process, 100 to 500 kHz, the depth to which the metal is softened is 1 mm on each side; it depends on the feed rate, and decreases as this rate increases. The weld quality increases as the soft area decreases, i.e., as speed increases. Below 10 to 15 m/minute, induction welding becomes difficult. Transition from the cold area is relatively sudden, enabling high pressures to be applied to the welding rollers without deforming the tube, and producing a small welding bead or seam, from which all impurities or oxide residue has been removed [6], [101], [147 *m*], [B], [F], [P].

It is also possible to use a linear inductor supplied at medium frequency to produce induced currents parallel to the weld seam. Contrary to the case described above, there are no spurious induced currents but the power density is lower, and the process is not widely used in welding (conversely, it is used in annealing heat treatment).

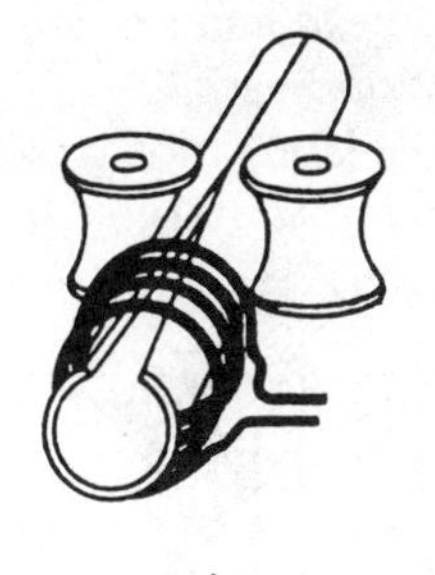

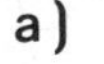


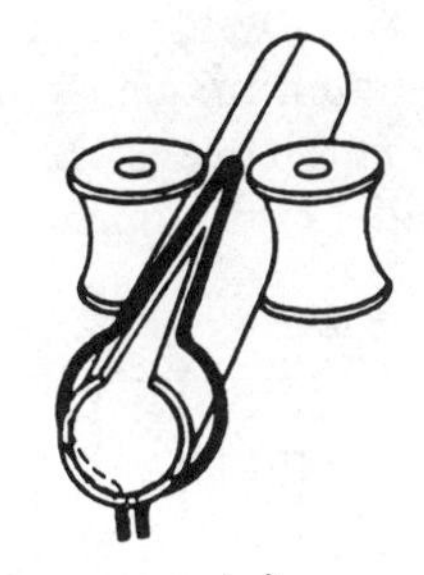


**Fig. 122.** Induction welding of tubes [F]:
a) *multi-turn inductor*
b) *single-turn inductor*

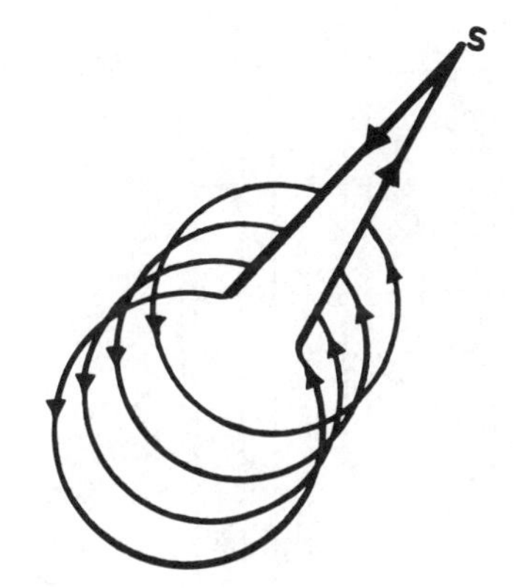


**Fig. 123.** Paths followed by the useful induced current in the tube blank. The current concentrates along the edges causing them to heat (S: welding point) [23].

**Fig. 124.** Paths followed by radio frequency spurious current. This current is suppressed by using a magnetic core [23].

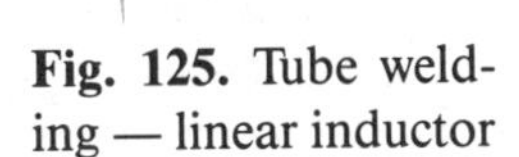

**Fig. 125.** Tube welding — linear inductor

To almost fully eliminate spurious induced currents which do not flow through the weld, and therefore represent a loss, a ferromagnetic core is inserted into the blank which considerably increases the leakage current circuit impedance.

To reduce Joule effect losses in the lines through which very high currents flow, the radio frequency generator is located near the welding location. The radio frequency power is applied through a transformer whose output voltage is some tens of volts.

After welding, the weld bead and sometimes the entire tube are annealed by medium frequency induction heating.

This process is used in the fabrication of steel, stainless steel, aluminum, brass and copper tubes. Feed rates of 100 to 150 m/min. are common.

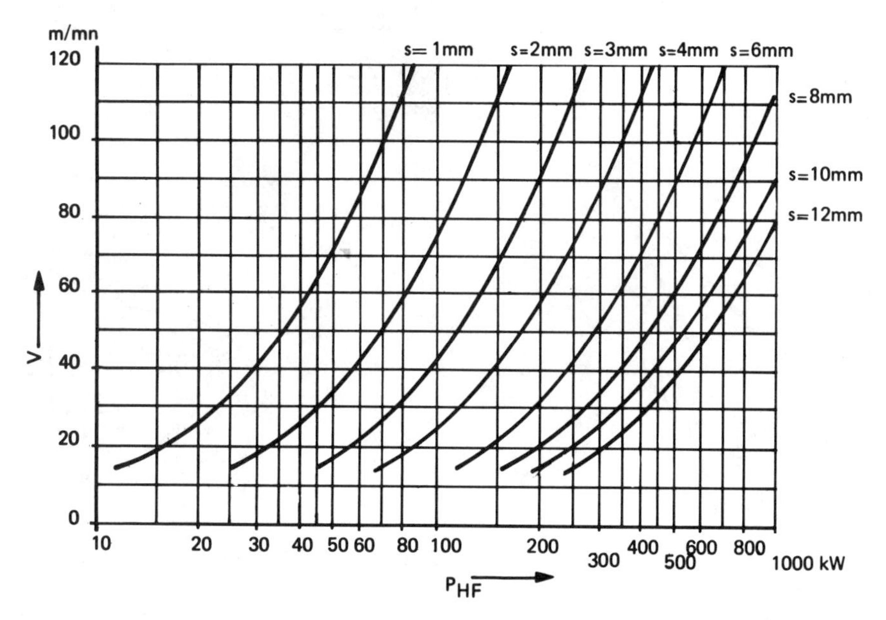


**Fig. 126.** Optimum feed speeds for the production of steel tubes: s, tube thickness

For steel, figure 126 gives the optimum production speed as a function of power across the radio frequency generator and tube thickness. For aluminum and brass, the optimum rate is 30% higher; for stainless steel and copper it is 65% of the optimum for steel.

The advantages of radio frequency welding are high production rates, weld quality and regularity, the possibility of welding non-pickled sheet and low power consumption. However, resistance and submerged arc welding or welding under inert gases (usually argon) competes with this process, especially in the production of helically welded tubes.

## 5.6. Heat treatments and surface treatments of metals

Induction heating can be used both for surface heat treatment and through treatment of metals. While surface treatment of steels is the most widely known application of induction heating, numerous other treatments can be performed using this technique: localized annealing, through hardening, continuous hardening of bars, continuous annealing of wire, some tempering operations, etc.

There are at least two reasons why this method of heating is suitable for heat treatment:

— the thermal effect can be concentrated where desired without heating other areas of the part treated;

— heating time is very short with respect to other heating processes.

This fast and selective heating, combined with efficient energy transfer, makes induction heating saves a significant amount of energy in heat treatment.

### 5.6.1. *Surface hardening of steels*

Only steel containing more than 0.3% carbon can be quench-hardened. Two methods are used to harden the steel surface:

— either heat the surface to be hardened only, then quench-harden it;

— or use a non-hardenable steel, recarburize it superficially (case hardening, carbonitriding in furnaces — either resistance or fossil fuel in a controlled atmosphere — or in salt baths), then quench-harden it.

The first method presupposes the concentration of high power around the periphery of the part to be surface hardened; heating time must be short enough to prevent the overall temperature of the part from rising due to thermal conduction.

High or medium frequency induction heating, which provides high specific powers of the order of 1.5 to 5 kW/cm$^2$ and produces a marked skin effect, is therefore particularly suited to surface hardening of steels.

This application has been developed considerably in the automobile, agricultural machinery and general mechanical engineering industries. Numerous contractors also offer their services in this field [27], [29], [43], [110], [132].

The main reason for this success is economic. In fact, this process enables less costly steels to be used and prevents distortion of parts which is frequent in conventional treatments, and therefore dispenses with the requirement for straightening operations.

While surface hardening using induction heating calls for high production runs to amortize the construction of an inductor (in general, a few hundred for the most current parts), it is possible to construct inductors which adapt to the most complex shapes, while very precisely localizing the area hardened.

Some of the most common induction hardening applications in industry are:

— crankpin and crankshaft bearing surfaces;

— gearing, in which the profile of the hardened layer must generally mate with the teeth contour, camshafts, center pins, etc.;

— tank track shoes or tractor tracks;

— rolling mill cylinders;

— valve stems, tappets;

— perforating shell heads;

— saw blades, shears;

— machine tool benches.

Induction enables more precise local treatments than any other method, especially, flame heating. In fact, the shape of the inductor, the power applied, heating time adjustment and the quenchant flow rate enable the area heated to be defined accurately, and provide the required mechanical characteristics with excellent fidelity [65 b, d], [*70*], [*101*], [*105*], [*107*], [*109*], [*140*].

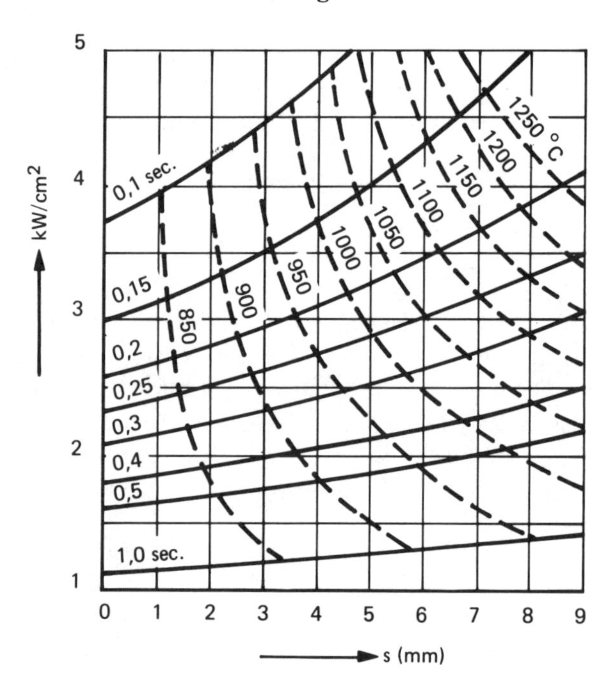


**Fig. 127.** Relationship between hardened layer depth(s), temperature, power density and heating time (frequency 100 kHz).

Induction surface hardening offers the following advantages:
— ease of automation and integration in production lines;
— low energy consumption.

Energy consumption varies from 0.5 to 1.5 kWh/kg of heated metal for surface hardening at 900°C; the weight of metal heated is often 1/3 to 1/20 of the total weight of a part. Electrical energy consumption is therefore very low, and primary energy consumption much lower than that of a fossil fuel furnace or a flame-hardening installation (this is generally decreased by a factor of 2 to 10);
— hardening located as desired;
— excellent reproducibility;
— cleanliness;
— high speed and productivity;
— no surface oxidation or contamination of the metal.

The graph of figure 127 shows the relationship between the depth of the hardened layer, heating temperature, power density and heating time required to obtain adequate results.

Moreover, the frequencies used are related to the desired treatment depths; the table in figure 128 below gives the frequencies generally used.

The optimum frequency can be estimated by calculation using the expression for the penetration depth of induced currents, considering that, at start of heating, the steel is magnetic, and then between the Curie point and the hardening temperature becomes non-magnetic. At high power densities, heating time is very short; conduction heating of the part is therefore reduced, especially when heating is followed by rapid cooling such as in quenching. The value of the

Surface hardening of steel
Selection of frequency for a Rockwell hardness after hardening of 50 HRC minimum.

| Hardening depth | Diameter of the Part (mm) | Frequency | | | |
|---|---|---|---|---|---|
| | | 1,000 Hz | 3,000 Hz | 10,000 Hz | 100,000 Hz |
| 0.3 to 1.2 mm | 6 to 25 | | | | 1 |
| 1.2 to 2.5 mm | 11 to 56 | | | 2 | 1 |
| | 16 to 25 | | | 1 | 1 |
| | 25 to 50 | | 2 | 1 | 2 |
| | 50 | 2 | 1 | 1 | |
| 2.5 to 5 mm | 19 to 50 | | 1 | 1 | 2 |
| | 50 to 100 | 1 | 1 | 2 | |
| | 100 | 1 | 2 | 3 | |

1, best; 2, satisfactory; 3, acceptable

**Fig. 128.**

penetration depth is therefore fixed to a value slightly lower than that of the hardening depth, and the frequency is calculated from this value using the formula $f = (10^7/4\pi^2)(\rho/\mu_r d_0^2)$. However, for this frequency, it is necessary to ensure that the diameter penetration depth ratio is sufficient, for which $f \geq 2.5(\rho/\mu_r)(10^7/d^2)$ Hz and $d/d_0 \geq 10$ is retained. If this condition is satisfied, efficiency is high, which is generally the case, since the heating sought is highly superficial [88].

Hardening depth depends not only on frequency, but also, as can be seen from figure 129, on power density.

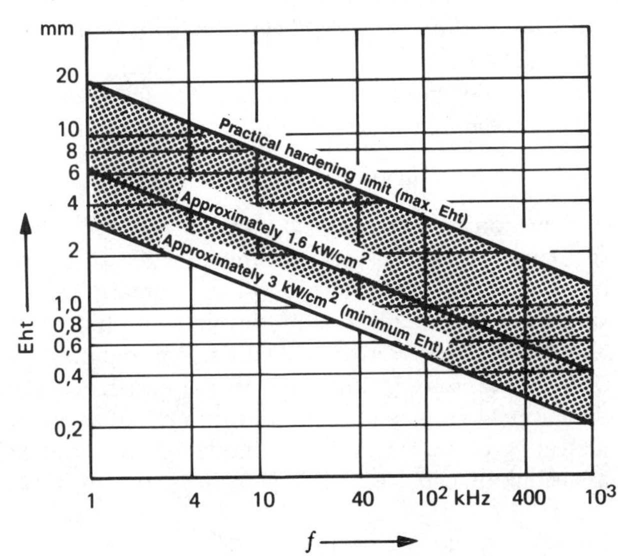


**Fig. 129.** Relationship between hardening depth Eht and frequency, and specific power [23].

The generators for medium frequencies in general are rotary units with thyristor generators becoming more common and vacuum tube generators for radio frequencies.

Inductor shapes vary widely. The inductors can be equipped with a spray system enabling immediate hardening of the part after heating, which is particularly appreciated in treatment of heavy parts such as rolling mill cylinders or continuous treatment of mechanical parts, which can then be easily automated. It is also possible to produce inductors equipped with protective gas feed systems. Linear inductors with magnetic field concentrators (*see* paragraph 3.12) are often used for surface hardening [109], [133], [143], [147 *h*].

While guided by calculations, development of inductors also depends on experience. When the parts treated vary greatly, the same generator can be used on condition that the inductor and its load is matched to the generator by a transformer.

With solenoid inductors, four hardening modes can be used depending on the position and relative movements of the part to be hardened and the inductor:
— *fixed hardening:* the part and inductor are fixed with respect to each other. The quenchant is projected onto the part at end of heating, or the part is dropped into the quenchant. In particular, this process is used for small parts or tools;
— *rotation hardening:* symmetrical parts rotate while being heated in a coil which surrounds them to prevent irregular overheating. Quenching is provided by sprays and the hardening thickness is uniform. This technique is used for bearing seats, spherical heads, sleeves, wheels, etc.;
— *continuous movement hardening:* the part or inductor is moved horizontally or vertically but parallel. Only that area of the part under the active portion of the inductor is heated. The quenchant spray immediately follows the inductor and projects the quenchant against the surface of the part at an angle such that the inductor does not become wet. The main range of application is hardening of lathe benches, saws, knives, slides, shafts, etc.;
— *continuous rotation hardening:* this technique combines both the above processes and provides excellent results, in particular for hardening of shafts and axles.

Due to the wide range of applications involved, surface hardening is one of the most common applications of induction heating [B], [E], [F], [N], [O], [P], [S].

### 5.6.2. *Through heat treatments*

Induction heating can also be used to perform heat treatments throughout the cross section. These treatments consist of continuous annealing of longitudinally welded tube welds, total annealing of tube blanks prior to hot sinking in sinking mills, annealing in protective gases of stainless steel tubes, annealing of butt welded tube beads, annealing and heating prior to hardening of bars, etc. However, induction annealing can also be used for parts of smaller dimensions and weights, annealing the ends of small springs, after cold working, annealing or heating prior to hardening of small hollow items, continuous annealing of wires, etc. [25], [94], [101], [111 *a*], [147 *d, e, h, j*].

**Fig. 130.** Inductors used for surface heat treatments:
*b), d), e)*, linear inductors;
*g)*, linear inductor with magnetic field reinforcing strips;
*f)*, linear inductor with rotation of the part to be treated.

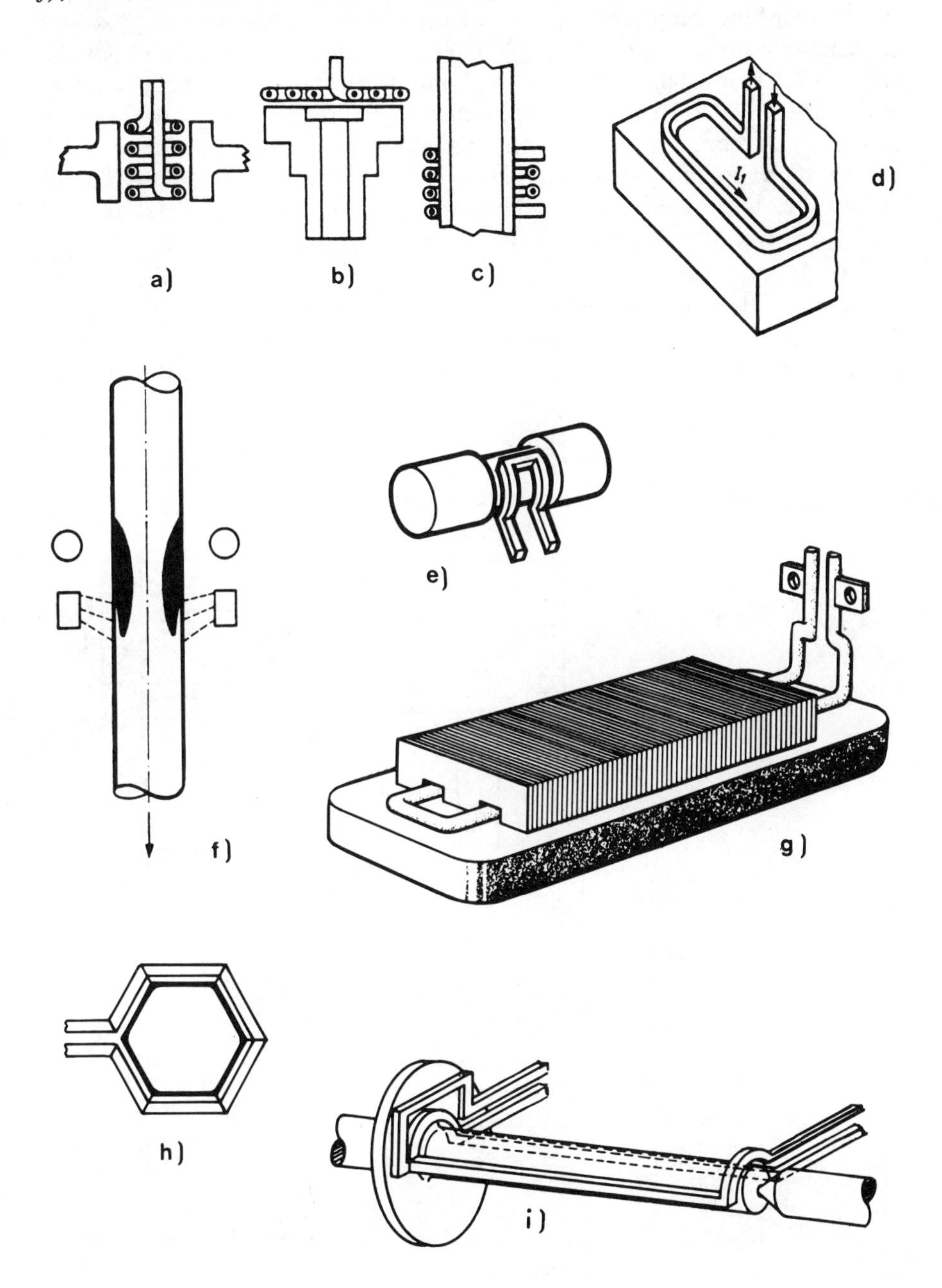

For annealing tube weld seams, a multi-turn inductor, arranged around the welded joint or a linear inductor with a field concentrator similar to that described for welding (*see* paragraph 5.5) is used. The operating frequency is generally between 50 and 10,000 Hz, and depends on the thickness of the tube and the working temperature (effect of tube thickness-to-diameter ratio and diameter-to-penetration depth ratio). The powers involved can be high, of the order of 1,000 to 2,000 kW for 10 to 12 mm thick tubes with treatment rates of the order of 20 to 30 m/min. The specific power used is generally of the order of 1 to 20 kW/cm$^2$. For example, on large pipeline tubes, a linear inductor 1.5 m long can be used to anneal longitudinal weld beads at 1,050°C over a heating width of 15 to 20 mm. The specific power is 1.5 kW/cm$^2$ at 3 kHz, and the inductor power 350 kW, for a treatment speed of 15 to 20 m/min.

Induction heating is also used for annealing and through hardening of bars. The situation is similar when heating, although the temperature is lower than that to re-heat before forming, as described above (paragraph 5.3); the equipment and calculations methods used are similar [B], [E], [F], [N], [O], [P], [S].

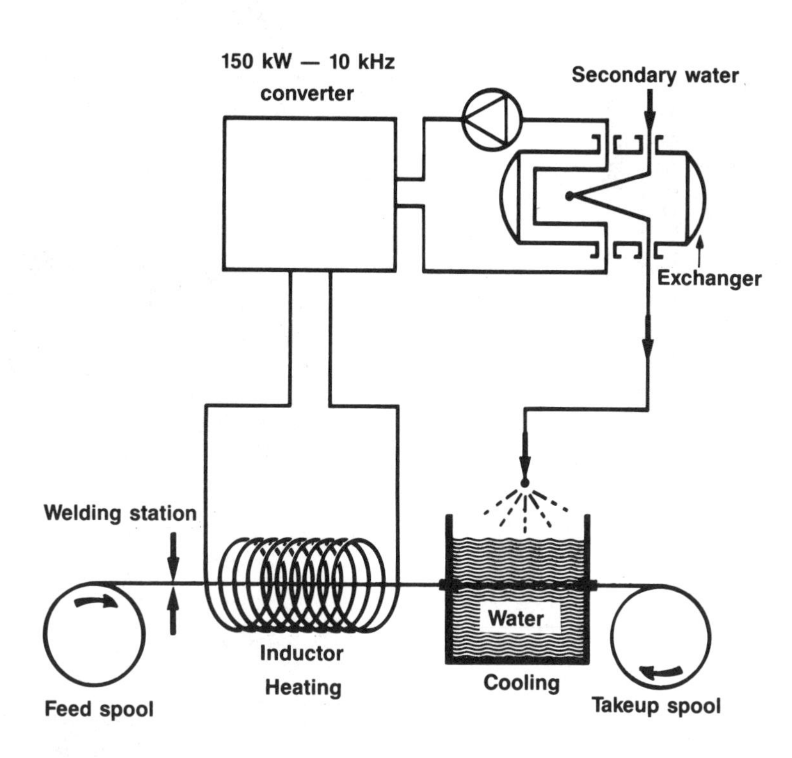


**Fig. 131.** Diagram of a heat treatment installation in which steel wires are heated to 400°C [S].

### 5.6.3. *Surface treatments*

Induction heating is not widely used in surface treatment. However, in special cases, it is used to bake paints (baking of paints on strips, drums, etc.), to heat wires prior to plasticization and for re-melting tin, either alone or as a backup to direct current heating on tinning lines [25], [65 *e*], [147 *d*], [B], [E].

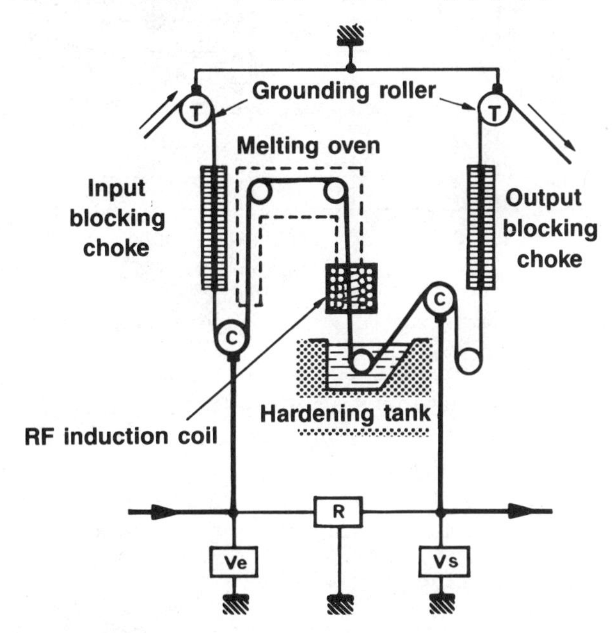


**Fig. 132.** Tinning line, with tin re-melting using induction-conduction heating [65 e].

## 5.7. Fabrication of semiconductors

In the semiconductor industry, operations using induction heating include the preparation of high purity polycrystalline material (zone melting refining), the growth of monocrystals (floating zone drawing), epitaxial layer growth, depositing of thin conductive films (connectors) and insulators (structure passivation) by evaporation and spraying [38], [65 *a*], [147 *a, b, f*], [E], [S], [X], [AB].

Induction heating is used in these applications for various reasons:

— cleanliness, flexibility, utility;

— high power density;

— rigorous localization of energy release and precise adjustment of applied power;

— ease of work in controlled atmosphere (inert, reactive, high or low pressure, vacuum).

In induction heating applied to semiconductors, special characteristics such as their very high resistivity, which decreases with temperature, must be taken into account. At ambient temperature, resistivity increases with purity. In the liquid state, resistivity drops noticeably ($10^{-3}\Omega \times$ cm, i.e. the resistivity of graphite, ten to twenty times that of the standard metals).

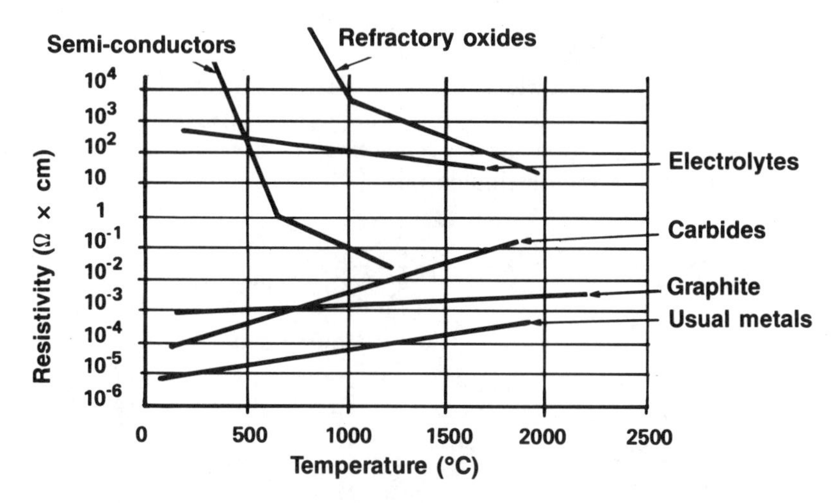


**Fig. 133.** Resistivity of semiconductors as a function of temperature [S].

The penetration depth of induced currents is therefore very high and working frequencies are also high, of the order of 0.5 to 3 MHz. Susceptor heating is also used, enabling work at lower frequencies. In most cases, the use of a susceptor during starting is also necessary, since the penetration depth at ordinary temperatures is of the order of 0.5 m at a frequency of 1 MHz, while, for a material in the molten state, it is only 0.5 cm at this frequency.

### *5.7.1. Purification by zone refining*

This technique is partly based on the important difference of solubility of the impurities in a material in the solid and liquid states, and also on the possibility of running a narrow homogeneous melted zone along a solid ingot. Induction heating is ideal for this operation.

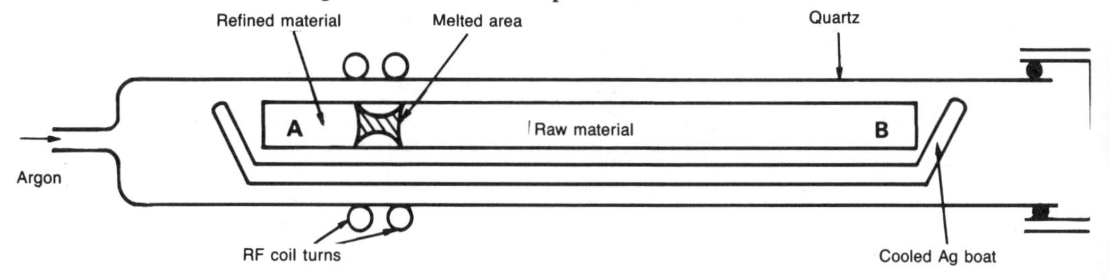


**Fig. 134.** Purification by zone refining [65 *a*].

The basic method consists of forming a melted zone of as uniform and stable length as possible, on a water-cooled silver boat, so that selective migration of impurities can take place constantly through the solid liquid interface during zone advance. The molten material is maintained by through induction heating.

The ingot is submitted to about ten successive purifications with the inductor always moving in the same direction. This technique is used especially for recovery of silicon (drawing residue). Operating frequencies are of the order of 400 to

500 kHz. This is also used for floating zone refining of other materials such as germanium.

### *5.7.2. Growth of monocrystals*

The fabrication methods for semi conductor monocrystals (silicon, germanium, etc.) are numerous (Czochralski, Kyropoulus, Bridgman, zone melting processes, etc.). Induction heating competes with electric resistance heating in these technologies. For vertical growth drawing with zone melting, only induction heating is used.

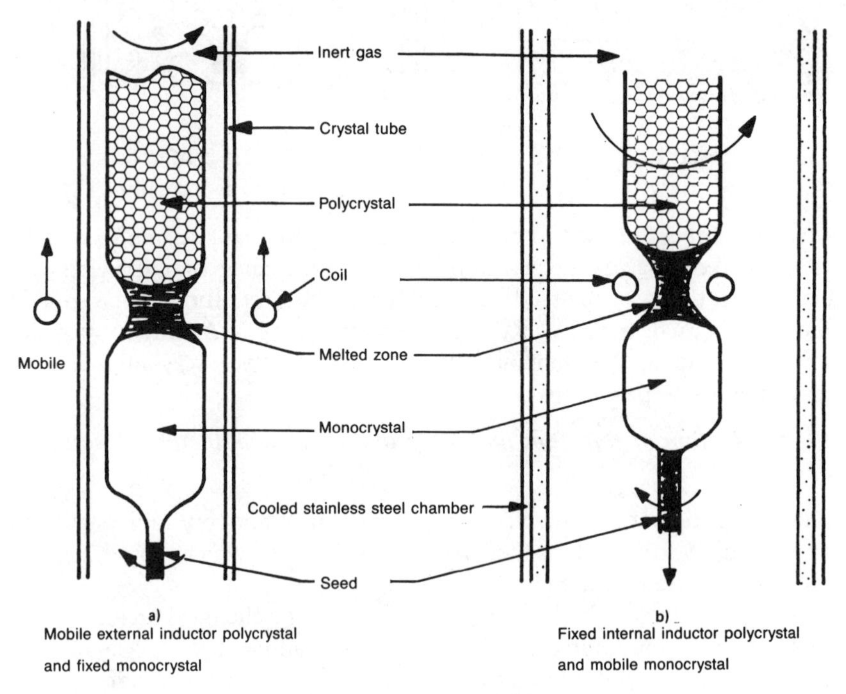


**Fig. 135.** Coaxial drawing [65 *a*].

In this technology, the polycrystalline material placed in a sealed chamber with inert gas is melted locally. The liquid drop which is started by an appropriately oriented seed breeds the crystal. The molten product is kept in balance by its surface tension and by levitation forces. There are several variants, depending on whether the polycrystal-crystal assembly describes an upward movement with respect to the material, and whether the coil is located outside the chamber (generally quartz) or inside (water-cooled chamber). The diameter of the polycrystal may be less than, greater than or equal to that of the monocrystal, and drawing may be coaxial or off-center. Heating is started by a graphite susceptor.

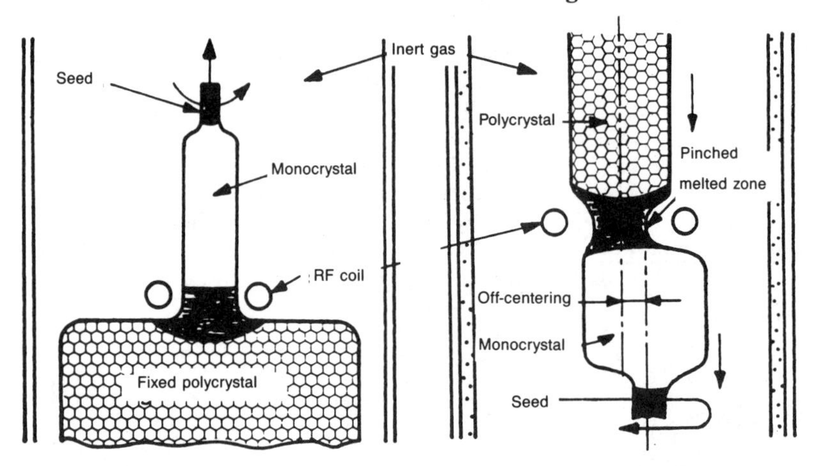


**Fig. 136.** Pedestal drawing [65 *a*].

**Fig. 137.** Off-centered "pinched" ZF drawing

These techniques are used to produce small or large diameter monocrystals. For the growth of large diameter monocrystals, more than 50 mm, drawing with gradual off-centering of the seed during drawing and restriction of the zone at the inductor is used; a 20 kW RF generator (operating at 2.5 MHz) provides a translation rate of 2.5 mm/min. This method enables very high quality monocrystals to be produced.

### *5.7.3. Production of thin film, monocrystalline semiconducting structures: epitaxy*

Epitaxy is an operation which consists in growing a monocrystalline layer on an appropriately oriented monocrystal or substrate, which continues the crystalline structure of the material.

For example, silicon-on-silicon epitaxy is obtained by chemical breakdown of silicon hydro-halogenides in the vapor phase (for example dichlorosilane). One of the main interests of the epitaxial growth process is the production of homogeneous doping films, of controlled thickness and high crystalline perfection at relatively low temperatures (1,000-1,200°C).

More recently, the epitaxy technology principle has been extended to depositing thin films of insulating materials (silica, nitrides, alumina) and also conductive materials (in particular refractory metals). This semi-conductor structure fabrication process also uses chemical breakdown reactions during the vapor phase.

Epitaxial technology is mainly based on radio frequency induction heating, although techniques using infrared radiation are also being developed.

Basically, the essential elements of an epitaxy reactor consist of:

— a reaction chamber, generally of quartz, in which reactive gases circulate (hydro-halogenide, carrying gas);

- a conductive material susceptor (silicon carbide-coated graphite, for example) placed inside the chamber, and on which the substrata are deposited.

The susceptor is taken to the required temperature by an inductor located inside the chamber. Induction heating using susceptors is fully adapted to this technology, since it provides the power densities required for epitaxial reaction, but also prevents any contamination of products and offers high energy efficiency. Various epitaxy reactor configurations are currently: — horizontal reaction chamber, circular or rectangular cross-section, fitted with a fixed, flat susceptor inclined along the chamber center line;
— vertical reaction chamber, of circular cross-section, with a cylindrical or truncated cone-shaped vertical susceptor, the center line of which is aligned with that of the chamber, and which rotates around this axis;
— bell-shaped vertical reaction chamber with a rotating flat horizontal susceptor.

The radio frequency generators used have powers of some tens of kilowatts and operate at frequencies of the order of 500 kHz. For example, energy consumption is of the order of 3 Wh per micron deposited and per centimeter of substrate surface, and the energy cost is less than 0.5% of the cost of the operation.

### *5.7.4. Vacuum depositing of thin film materials*

Vacuum deposits of thin film materials for contacts and for protection of junctions on semi conducting structures use different technologies, both in terms of the nature of the materials involved (metals, alloys, insulators such as oxides, nitrides or glasses), and by the physical, chemical and electrical properties they possess before and after they are deposited. Therefore, heating methods enabling sublimation of materials are various — direct Joule effect (refractory metal crucible containing the material to be deposited, heated through current flow), indirect Joule effect, (crucible radiation heated through high temperature resistances), electronic beam beam and radio frequency induction.

Some applications of radio frequency induction heating are the evaporation of aluminum in boron nitride crucibles, with a graphite susceptor for contacts, or lead-in tantalum crucibles for soldering connections. Equipment of this type uses radio frequency generators of the order of 5 kW, operating at frequencies of approximately 0.2 MHz. For aluminum and lead, energy consumptions are 0.08 and 0.02 kWh respectively per micron deposited and per square centimeter of substrate.

Spraying techniques are now being added to these evaporation techniques. This consists of bombarding a target with positive ions accelerated to form a gas plasma. In such cases, very high frequency is used (between 10 and 30 MHz, but the industrial frequencies of 13.5 and 27 MHz are generally preferred for telecommunication restrictions) to heat the cathode, and this application is, in the strict sense of the word, another induction heating application. For example, this technique permits vaporization of quartz to protect planar structure junctions.

In the electronic component production field, induction heating is widely used because of its special characteristics — high power density, ease of heating non-metallic bodies to high temperatures by means of a susceptor, cleanliness. These applications are developing rapidly.

## 5.8. Bonding metals and non-metallic bodies

Bonding is becoming a more common industrial assembly method. Bonding applies both to assembly of different metals and to assembly of metals and non-metallic bodies (for example, metal to glass). Induction heating, generally at radio frequency, is used to accelerate hardening of the cement joint.

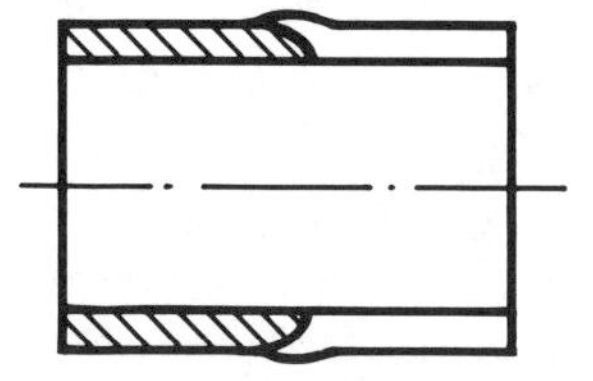

**Fig. 138.** Metal to glass bonding

This operation can be performed easily in a protective atmosphere.

## 5.9. Heating non-conductive bodies: application to liquid and gas state materials

Only electrically conducting bodies can be heated directly by induction. However, it is possible to heat such bodies indirectly using a conductive susceptor. The material most widely used for this purpose is graphite but other metals can also be employed.

Induction heating applications using susceptors are numerous, melting refractory materials, glasses, quartz, sintering carbides, semi conductor fabrication, heating liquids, heating gasses during chemical synthesis [30], [64], [65 *g*].

With steel or graphite susceptors, electrical efficiency is very high, between 80 and 95%.

For example, induction heating can be used to heat food products prior to extrusion (production of special types of expanded snacks, dietary products, textured proteins, confectionary products, etc.) [80], [AD].

Induction heated cylinders are used to dry very high quality paper or special textiles but other applications can also be envisaged [148], [Y].

Induction heated metal tanks also exist [M], [AA], [AC].

In many cases, induction heating often competes with resistance heating for applications of this type. However, it provides much higher power densities, and its thermal inertia is very low which justifies its use in numerous industrial applications.

For heating tanks, the power density can exceed 55 kW/m$^2$, but is generally limited, due to the thermal conductivity of the liquid to be heated. The table in figure 140 below gives a range of line frequency induction heated tanks. Much bigger tanks can be heated with higher installed powers (e.g. 22,500 litres with a power of 1,400 kW).

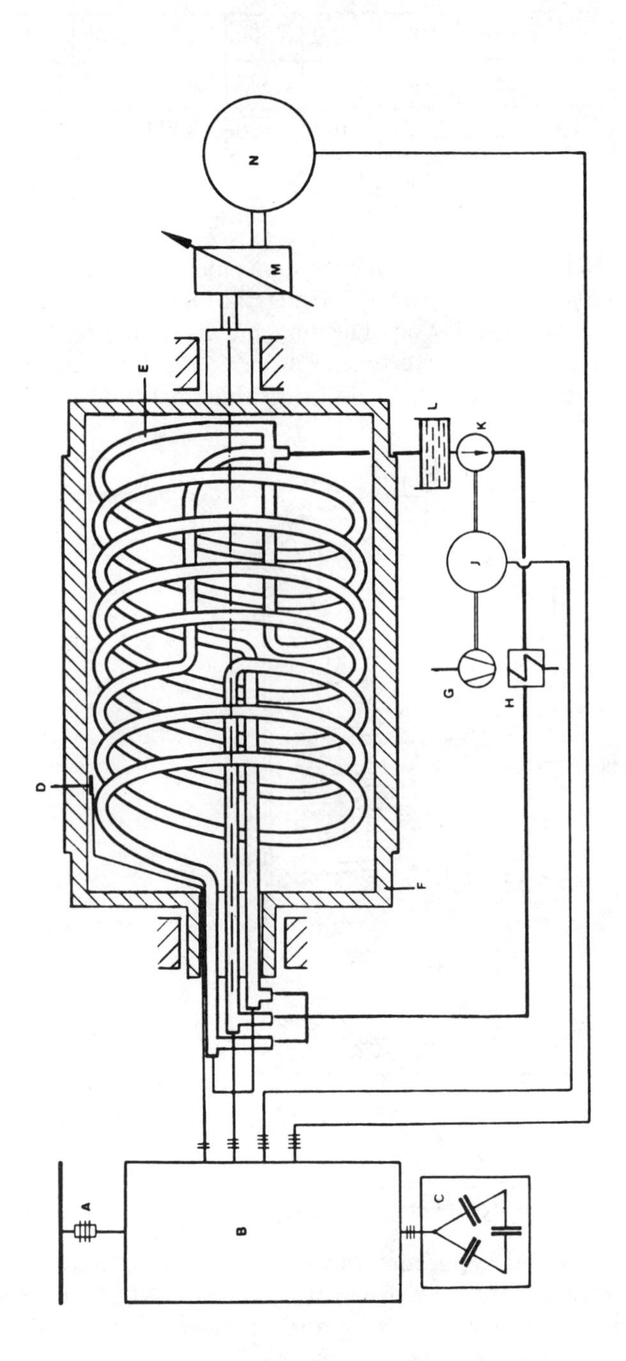


A, AC line; B, cabinet; C, capacitors; E, heating coil; F, dryer cylinder; G, coolant fan; H, coolant; J, motor for G and K; K, circulation pump; L, balancing reservoir; M, mechanical controller; N, cylinder motor

**Fig. 139.** Induction heated cylinder [Y].

Tank heating by induction (temperature: 300°C) [AA]

| Capacity (l) | 10 | 50 | 100 | 250 | 500 | 1,100 | 2,200 | 3,000 |
|---|---|---|---|---|---|---|---|---|
| Diameter (mm) | 380 | 610 | 760 | 1,000 | 1,375 | 1,830 | 2,300 | 2,590 |
| Power (kW) | 10 | 30 | 50 | 100 | 150 | 300 | 550 | 675 |

**Fig. 140.**

This type of tank is used primarily for heating chemical products such as resins, or more rarely, food products (e.g. palm oil). The interest of the process resides not only in high power density enabling rapid heating, but also in the ease with which a relatively high temperature, often between 250 and 650°C, is maintained, which is difficult to obtain with indirect steam or even thermal fluid heating [30], [64], [M], [AA], [AC].

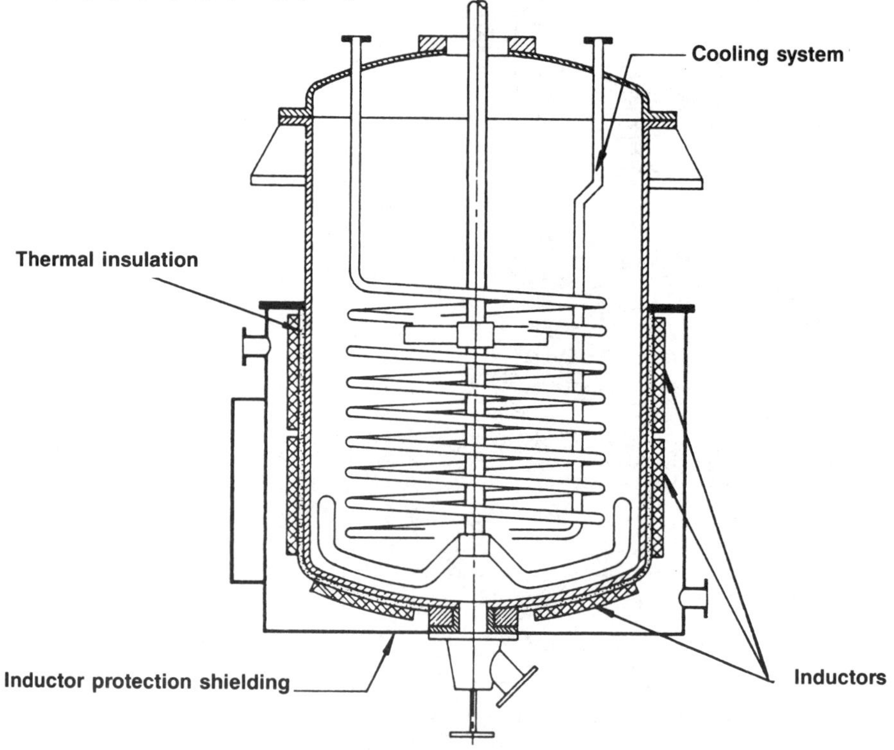


**Fig. 141.** Induction-heated chemical reactor or tank [AC].

To conclude, induction heating applications are extremely diverse, and apply both to metals and non-metallic bodies. Whenever one or more of the following conditions is desired, high power density, rapid heating, fast treatment rate, high or medium temperature, thermal effect localization, low thermal inertia, energy efficiency, etc., the possibilities of induction heating should be carefully examined.

## 6. BIBLIOGRAPHY

[1] S. DUPERRIER, *Pratique du chauffage électronique*, Paris-Chiron, 1956.
[2] P. G. SIMPSON, *Induction heating, coil and system design*, MacGraw Hill, New York, 1960.
[3] A. BUSSON, *Cours d'électricité industrielle*, Riber, 1963.
[4] P. SIMEON, Électrothermie et haute fréquence, *Techniques CEM*, n° 59, 1964.
[5] J. PARENT, Les principes généraux du chauffage par induction, *Bulletin scientifique de l'A.I.M.*, n° 1, 1966.
[6] E. RUNTE, Le soudage de tubes au défilé par chauffage inductif haute fréquence, *Revue BBC*, n° 3, mars 1968.
[7] F. LAUSTER, *Manuel d'électrothermie*, Dunod, 1968.
[8] M. G. LOZINSKI, *Industrial applications of induction heating*, Pergamon Press, Novembre 1968.
[9] A. F. LEATHERMAN et D. E. STUTZ, Basic induction heating principles, *Metal Treating*, juin 1970.
[10] F. BRICHANT, *L'ondistor*, Dunod, 1970.
[11] G. BAZANTE, Utilisation du chauffage par induction et des techniques du vide pour la fusion et le traitement des métaux, *La Métallurgie*, décembre 1971.
[12] B. DE MIRAMON, Application du convertisseur statique au four à induction moyenne fréquence, *Revue générale de Thermique*, n° 2, février 1972.
[13] STEEL TIMES, *New breed of induction furnaces*, n° 4, avril 1972.
[14] H. LINDT, Générateurs HF modernes de petite puissance pour chauffage par induction, *Revue BBC*, mai 1972.
[15] W. FUCHS, Installation de chauffage par induction de billettes devant être forgées, *Revue BBC*, mai 1972.
[16] M. RIETHMANN, Installations modernes de fusion pour fonderies, *Revue BBC*, juin 1972.
[17] H. P. SCHAUB, Fours à induction pour fusion et maintien en température dans l'industrie de l'aluminium, *Revue BBC*, juin 1972.
[18] W. ANNEN, Le réchauffage par induction de billettes dans les laminoirs à petits fers et à fils, *Revue BBC*, juin 1972.
[19] E. RUNTE, Chauffe par induction dans les lignes à tubes, *Revue BBC*, juin 1972.
[20] INDUSTRIAL HEATING, *Induction heating system for rapid production of truck axles with minimum scaling*, juin 1972.
[21] UNION INTERNATIONALE D'ÉLECTROTHERMIE, Congrès de Varsovie, ensemble de communications, 1972.
[22] METALLURGIA AND METAL FORMING, *L'électricité dans la production des alliages de nickel*, n° 12, 1972.
[23] COMITÉ SUISSE D'ÉLECTROTHERMIE, *Le chauffage électrique des métaux par induction*, rapport n° 16, 1972.
[24] W. SCHOTT, L'emploi de l'ordinateur pour le dimensionnement des fours à induction à creuset, *Revue BBC*, n° 10-11, octobre 1972.
[25] ELECTRICAL REVIEW, *Inducteur à flux transversal pour un chauffage à rendement élevé*, 26 janvier 1973.
[26] A. GUILBERT, *Circuits magnétiques à flux alternatif; transformateurs, théorie, fonctionnement et calcul*, Masson éditeur, 1973.
[27] K. H. ANDRÉ *et al.*, Trempe de l'acier par impulsions, *Traitement thermique*, n° 72, 1973.
[28] COLLOQUE DU COMITÉ FRANÇAIS D'ÉLECTROTHERMIE, *Le four à induction en fonderie, ensemble de communications*, Versailles, 5 avril 1973.
[29] H. LAPLANCHE, La trempe superficielle par chauffage inductif, *Traitement thermique*, n° 76, juin 1973.
[30] R. F. ANGEL, The place of standard frequency induction heating in the technology of electric surface heating, *Elektrowärme International*, Vol. 31, n° 3, juin 1973.
[31] A. L. TOURAINE, Évolution des générateurs haute fréquence pour le chauffage par induction, *Ingénieurs et Techniciens*, juillet 1973.

[32] J. D. LAVERS *et al.*, Current distribution, forces and circulation in the coreless furnace, *IEEE Transactions on Industry Applications*, n° 4, vol. 9, août 1973.
[33] W. JAGT, Metallurgische und betriebliche Erfarhungwerte des Induktions-Rinnenofens für Gusseisen, *Elektrowärme International*, n° 5, vol. 31, octobre 1973.
[34] M. TAMA, Development of channel type induction furnaces, *Elektrowärme International*, n° 5, vol. 31, octobre 1973.
[35] A. JAKOUBOVITCH, Les générateurs apériodiques en moyenne fréquence, *Ingénieurs et Techniciens*, novembre 1973.
[36] M. COLOMBIE *et al.*, Aciers ferritiques à haute teneur en carbone renfermant du molybdène, *Matériaux et Techniques*, n° 3, mars 1974.
[37] J. D. LAVERS et P. P. BIRINGER, An improved method of calculating the power generated in an inductively heated load, *IEEE Transcriptions on Industry Applications*, n° 2, vol. I A-10, mars 1974.
[38] M. E. MACNEAR, Techniques for doping semi conductor materials, *Solid State Technology*, 1974.
[39] Y. SUNDBERG, Convertisseurs de fréquence pour la fusion et le réchauffage par induction, *ASEA Revue*, n° 2, vol. 46, 1974.
[40] R. STOLL, *The analysis of eddy currents*, Oxford-Clarendon Press, 1974.
[41] EL BEDWEIHI *et al.*, SCR control of power for induction melting, *IEEE Transcriptions on Industry Applications*, n° 4, vol. 10, septembre 1974.
[42] C. A. TUDBURY, Electromagnetics in induction heating, *IEEE Transactions on magnetics*, Congrès IEEE, Toronto, n° 3, vol. 10, septembre 1974.
[43] F. REINKE, Automatic tool changing and work cycle on a new semi-automatic machine for the induction hardening of crankshafts, *Elektrowärme International* n° 5, octobre 1974.
[44] F. NEUMANN, Gegenüberstellung der Schmelzverfahren für die Güsseisenherstellung, *Elektrowärme International*, n° 5, octobre 1974.
[45] R. PERIE, Élaboration d'acier au four à induction à canal à l'Usine Sacilor d'Hagondange, *Revue de Métallurgie*, n° 10, octobre 1974.
[46] G. CAUSSIN et R. CHAUPRADE, Alimentation statique de fours moyenne fréquence, *Revue Jeumont-Schneider*, n° 19, décembre 1974.
[47] C. SCHIVDASANI, Fours à induction à canaux pour maintien en température, surchauffe et accumulation d'acier, *Revue BBC*, n° 1, vol. 62, janvier 1975.
[48] M. A. AHMAD, L'utilisation des fours à induction à canaux dans les fonderies de fonte, *Revue BBC*, n° 1, vol. 62, janvier 1975.
[39] H. P. SCHAUB, Conception d'une installation de fusion à induction pour une grande fonderie, *Revue BBC*, n° 1, vol. 62, janvier 1975.
[50] W. ANNEN, Chauffage par induction de lingots et barres dans les forges à grande puissance, *Revue BBC*, n° 1, vol. 62, janvier 1975.
[51] W. ANNEN, Le réchauffage par induction augmente la qualité et la production dans les laminoirs, *Revue BBC*, n° 1, vol. 62, janvier 1975.
[52] M. DEHOVE, L'expansion thermique, *Ingénieurs et Techniciens*, n° 93, mars 1975.
[53] J. COLE, Chauffage par induction pour la forge, *Formage et traitements des métaux*, n° 64, 1975.
[54] J. ANTOINE et F. SAUVAGE, La métallurgie sous vide dans un four à canal d'aciérie, *Revue de Métallurgie*, n° 6, juin 1975.
[55] E. JAHN, Erfahrungen bei der Herstellung von Rohrbögen auf der Induktionbiegenmaschine, *VGB Kraftwerktechnik*, n° 6, vol. 55, juin 1975.
[56] J. F. PUPIER, *Le réchauffage par induction dans l'aciérie MacLouth*, Cessid, juillet 1975.
[57] P. BONIS, Domaines d'utilisation des fours à arc et à induction, *Revue générale d'Électricité*, n° 7/8, vol. 84, août 1975.
[58] E. DÖTSCH, F. HEGEWALDT, Zum einsatz des Induktionstiegelofens für die Stahlerzeugung, *Elektrowärme International*, n° 6, décembre 1975.
[59] S. LUPI *et al.*, A method of calculating the induction heating parameters of sectionalized inductors, *Elektrowärme International*, n° 6, décembre 1975.

[60] W. PINKOFSKY, Kondensatoren für Anlagen zur induktiven Erwarmung, Rückblick und Ausblick, *Elektrowärme International*, n° 6, décembre 1975.
[61] E. HOROSZKO, Asymmetry in induction channel furnaces, *Elektrowärme International*, n° 6, décembre 1975.
[62] B. DE MIRAMON, Application des onduleurs à moyenne fréquence au chauffage et à la fusion par induction, *Revue générale d'Électricité*, n° 2, vol. 85, février 1976.
[63] R. COURDILLE, Laminage et forgeage à chaud. Énergies utilisées dans le réchauffage de l'acier, *Revue technique de l'Apave*, n° 193, vol. 57, mars 1976.
[64] ELECTRICAL REVIEW, *Tank heating by induction aids coating process reliability*, vol. 198, n° 10, mars 1976.
[65] COMITÉ FRANÇAIS D'ÉLECTROTHERMIE, Colloque « Chauffage par induction », Versailles, avril 1976 (ensemble de communications) :
*a*) J. C. BOUCHAUD, *Quelques aspects de l'utilisation de l'électrothermie inductive dans l'industrie des semi-conducteurs.*
*b*) R. CORBIER, *Régulation des équipements de traitement thermique par induction.*
*c*) M. F. BRICHANT, *Les onduleurs moyenne fréquence.*
*d*) J. REBOUX, *Évolution récente des traitements thermiques en haute et moyenne fréquence.*
*e*) L. BECK et D. BISCARAS, *Association du chauffage par conduction au chauffage par induction pour le brillantage des tôles étamées.*
*f*) X. HENRY, *Nouveaux développements des contacteurs statiques dans les lignes de traitement en continu.*
*g*) E. GRAY, *Chauffage par induction pour le frittage sous charge du carbure de bore et autres matériaux réfractaires.*
*h*) C. SOULET, *Réchauffage par induction de billettes de métaux non ferreux avant extrusion.*
*i*) M. MEINEN, *Le chauffage par induction; chauffage partiel avant formage.*
*j*) F. SCHLUCK, *Coût de fonctionnement dans une forge pour l'industrie automobile utilisant des groupes tournants et des convertisseurs statiques.*
*k*) A. JAKOUBOVITCH, *Les générateurs apériodiques.*
[66] R. KRAUS, Rückgewinnung von Verlustwärme bei Induktionsofenanlagen, *Revue BBC*, vol. 58, mai 1976.
[67] W. J. MOLLOY, Development, production and applications of superalloys, *Metallurgia and Metal Forming*, juillet 1976.
[68] METALLURGIA AND METAL FORMING, *From vacuum melting to finished forging*, juillet 1979.
[69] U. KETNER, Betrieberfahrungen mit Schwingkreisumrichtern für MF-Induktionsschmelzanlagen, *Elektrowärme International*, n° 4, août 1976.
[70] P. BRAISCH, Zur Technologie der Induktions-Härtung von Dieselmotorteilen, *Elektrowärme International*, n° 4, août 1976.
[71] N. GRULKE, Schmiede Block Erwärmungsanlagen in Kompaktausführung, *Elektrowärme International*, n° 4, août 1976.
[72] E. STANGL, Induktive Erwärmung von Blöckchen, Stangen und Rohren zum warmformen, *Elektrowärme International*, n° 4, août 1976.
[73] H. G. DOMRES, Druckgasbetätigte Vergiesseinrichtung, *Elektrowärme International*, n° 5, octobre 1976.
[74] CONGRÈS DE L'UNION INTERNATIONALE D'ÉLECTROTHERMIE, Liège, octobre 1976 (ensemble de communications).
[75] A. ARCHENHOLTZ, Holding of nodular iron in channel induction furnaces, *ASEA Journal*, n° 4, vol. 49, 1976.
[76] P. BRADDON, Importance of coil design in electric induction furnaces, *Metallurgia and Metal Forming*, novembre 1976.
[77] G. PHILLIPS, Fusion des alliages de cuivre dans des fours à induction à canaux, *Hommes et Fonderie*, novembre 1976.
[78] W. PINKOFSKY, Neue Aspekte bei der Projektierung von Ofenkondensatoren für Induktionsanlagen, *Elektrowärme International*, n° 6, décembre 1976.

[79] E. CALAMARI, Le four à induction à creuset dans les fonderies de métaux non ferreux, *Journal du Four Électrique*, n° 10, décembre 1976.
[80] J. F. DE LA GUÉRIVIÈRE, Snacking - Techniques de cuisson-extrusion, *Revue des fabricants de produits à base de sucres et de farines*, décembre 1976.
[81] G. F. BOBART, Solid state controls and miniaturization serve industrial heating industry, *Industrial Heating*, janvier 1977.
[82] G. TEVAN et S. NAGY, Optimierung der Abnessungen induktiver Tiegelschmelzofen, *Elektrowärme International*, n° 1, vol. 35, février 1977.
[83] H. DELLA CASA *et al.*, Le réchauffage électrique des produits avant laminage, *Cessid*, mars 1977.
[84] E. KOLBE *et al.*, Möglichkeiten der Analyse des Betriebsverhaltens von Induktiontiegelofen mit Hilfe von Modellen, *Elektrowärme International*, avril 1977.
[85] P. SCHERG, Zur Auslegung von Induktionsofenspulen, *Elektrowärme International*, avril 1977.
[86] S. LUPI *et al.*, The calculations of inductors with periodical fields, *Elektrowärme International*, avril 1977.
[87] J. C. MEMBRE, Un procédé nouveau, l'expansion thermique, *Formage et traitement des métaux*, n° 83, 1977.
[88] LABORELEC, *Manuel d'électrothermie*, Bruxelles, juin 1977.
[89] F. PIETERMAAT, Le chauffage par induction; applications actuelles et futures, *Bulletin du Comité Belge d'Électrothermie et d'Électrochimie*, n° 45, 1977.
[90] F. HOECHTL, Induktiv beheizte Vakuumöfen für Aluminiumschmelzen, *Elektrowärme International*, n° 3, juin 1977.
[91] H. G. MATTHES, Der statische Frequenz–Umrichter zum Einsatz in der industriellen Elektrowärme, *Elektrowärme International*, n° 3, juin 1977.
[92] E. RICHTER et G. THIEME, Anlagen für die induktive Stangenerwärmung, *Elektrowärme International*, n° 4, août 1977.
[93] J. ANTOINE et P. BONIS, Channel induction furnaces for holding, superheating and storing steel, *Elektrowärme International*, n° 4, août 1977.
[94] K. PFEIFFER, Metallveredelung im Hoch-und Mittelfrequenzfeld, *Elektrowärme International*, n° 4, août 1977.
[95] ELECTRICAL REVIEW, *Induction heating for aluminium extrusion*, vol. 201, n° 21, novembre 1977.
[96] L. FRANÇOIS *et al.*, Bobines de fours à induction, *Techniques CEM*, n° 100, décembre 1977.
[97] F. ARELMANN, Einsatz von Induktionsofen, *Elektrowärme International*, n° 6, vol. 35, décembre 1977.
[98] D. COMSA et J. PAUTZ, Uber den zusammenhang von Leistungs-grösser und geometrischen Abmessungen von Induktionstiegelofen, *Elektrowärme International*, n° 6, vol. 35, décembre 1977.
[99] W. BUCHEN, Elektro-schmelzöfen für Kupfergusslegierungen, *Elektrowärme International*, n° 6, vol. 35, décembre 1977.
[100] A. VON STARCK, Ein Verfahren zur Berechnung der induktiven Erwärmung metallischer Werkstücke mit rechteckigen Querschnitt in Induktionsspulen, *Elektrowärme International*, n° 6, vol. 35, décembre 1977.
[101] K. SCHAUFLER, Installations moyenne fréquence stationnaires et mobiles pour le chauffage par induction, *Revue BBC*, n° 2, vol. 65, février 1978.
[102] H. METZWER, Installations de fusion à induction à fréquence industrielle pour fonderie de cylindres de laminoirs, *Revue BBC*, n° 2, vol. 65, février 1978.
[103] Z. TASEVSKY, Installation de chauffage par induction pour la déformation à chaud de lingots et de barres, *Revue BBC*, n° 2, vol. 65, février 1978.
[104] F. HEGENWALDT, Schmelzen von Stahl in Induktionstiegelöfen, *Elektrowärme International*, n° 1, vol. 36, février 1978.
[105] M. R. EL HAIK *et al.*, Le chauffage par induction pour le traitement thermique des métaux, *Traitement Thermique*, mars 1978.

[106] S. N. OKEKE, Application of thyristor inverters to induction heating and melting, *Electronic Power*, n° 3, vol. 24, mars 1978.
[107] G. TARDY, Les traitements thermiques par induction, *La Technique Moderne*, n° 4, vol. 70, avril 1978.
[108] A. TOURAINE, Les générateurs de courant pour le traitement thermique par induction, *Traitement Thermique*, n° 123, mars 1978.
[109] J. REBOUX, Les inducteurs pour le chauffage par induction, en particulier en traitement thermique, *Traitement Thermique*, n° 123, mars 1978.
[110] A. JAKOUBOVITCH et M. TAMALET, Application du générateur apériodique à la trempe de contour, *Traitement Thermique*, n° 123, mars 1978.
[111] CONGRÈS DU COMITÉ FRANÇAIS D'ÉLECTROTHERMIE, Versailles, 6-7 avril 1978.
*a*) J. P. MÉTAIL, *Induction moyenne fréquence appliquée au chauffage de fil; étude des paramètres de rendement énergétique.*
*b*) B. T. LINDBERG, *Four à induction de maintien et coulée automatique.*
*c*) M. BAU, G. ULMER *et al.*, *Récupération de chaleur sur les eaux de refroidissement des fours à induction en fonderie d'alliages ferreux pour le chauffage d'ateliers.*
*d*) C. MONTHUY, *Chauffage par induction et coulée centrifuge de pièces en superalliages.*
*e*) R. POCHARD, *Gestion économique d'un ensemble production-distribution d'alliage d'aluminium dans une fonderie sous pression.*
[112] E. HOROSKO, Induction heating of rotating bodies, *Elektrowärme International*, n° 2, vol. 36, avril 1978.
[113] H. G. MATTHES et E. MAULER, Stromversorgungseinrichtungen für Induktionserwärmungsanlagen, *Elektrowärme International*, n° 2, vol. 36, avril 1978.
[114] G. ULMER, Fonderie, réduction des consommations spécifiques, *Revue générale de Thermique*, n° 197, mai 1978.
[115] A. DARQUE, Le chauffage par induction, *Métaux Déformation*, n° 48, mai-juin 1978.
[116] D. F. DUFF, Induction heating for automated forming operations, *Industrial Heating*, juin 1978.
[117] R. J. DOWSING, Channel furnace superheat can help cut steelmaking costs, *Metal and Materials*, juin 1978.
[118] F. OSTLER, Wirtschalftliche Aspekte für gusseisen-, warmhalte- und giessöfen mit induktiver Beheizung, *Elektrowärme International*, n° 3, vol. 36, juin 1978.
[119] E. DOTSCH, Stahlerzeugung in Netzfrequenz-Induktionstiegelofen und Bedeutung für die gusseisener Zeugung, *Elektrowärme International*, n° 3, vol. 36, juin 1978.
[120] H. FLESSA *et al.*, Neue zustellungs Methoden für Induktionstiegelofen und Induktions rinnenöfen für stahl und Grauguss, *Elektrowärme International*, n° 3, vol. 36, août 1978.
[121] K. H. GORSLER, Ablege und Instandhaltung der keramischen Ofenausklein-dung, *Elektrowärme International*, n° 4, vol. 36, août 1978.
[122] K. E. GRANITZKI et L. H. HALLOT, Erfahrungen mit der Dauersehablone für Induktionstiegelöfen verschiedener Grösse und Metall-Legierungen, *Elektrowärme International*, n° 4, vol. 36, août 1978.
[123] H. G. FELDHUS, Neuerung beider Zustellung von Induktionsrinnenöfen, *Elektrowärme International*, n° 4, vol. 36, août 1978.
[124] W. FRERKING, Neüste Verfahrenstechnik für das Zustellen und Sintern von Induktionstiegelöfen, *Elektrowärme International*, n° 4, vol. 36, août 1978.
[125] S. E. ROGERS et R. H. OGLESBY, Energy conservation in the drop forge, *Metallurgia and Metal Forming*, septembre 1978.
[126] P. GUEZ, Travail à chaud des métaux par chauffage inductif, *Métaux Déformation*, n° 50, septembre-octobre 1978.

[127] K. AHN, Wärmerückgewinnung bei Induktionstiegelöfenanlagen, *Elektrowärme International*, n° 5, vol. 36, octobre 1978.
[128] COMITÉ BELGE D'ÉLECTROTHERMIE ET D'ÉLECTROCHIMIE, Communications des Journées Internationales d'Études sur le chauffage et la fusion par induction, Liège, octobre 1978.
[129] F. HOROSKO, Induction heating of railway switches, *Elektrowärme International*, n° 6, vol. 36, décembre 1978.
[130] L. HALLOT, *Les fours électriques de fusion et de maintien des métaux*, École supérieure de Fonderie, 1979.
[131] M. ORFEUIL et M. BRIAND, *La fusion électrique des métaux non ferreux*, document EDF, 1979.
[132] J. PAGEL, Les aciers pour trempe après chauffage superficiel par induction, *Traitement Thermique*, n° 131, janvier 1979.
[133] M. TAMALET, Le traitement de surface par induction, *Traitement Thermique*, n° 131, janvier 1979.
[134] M. ORFEUIL et R. THOMASSIN, Le chauffage électrique des métaux avant formage, *Métaux Déformation*, n° 53, janvier 1979.
[135] D. SCHLUCKERBIER, Weiterentwicklung der magnetischen Frequenzumformung auf 450 Hz, *Elektrowärme International*, n° 1, vol. 37, février 1979.
[136] K. HENSS et H. LINN, Die Einführung der Transistortechnik bei Industrie Hochfrequenz-generatoren kleiner Leistung, *Elektrowärme International*, n° 1, vol. 37, février 1979.
[137] J. REBOUX, Techniques du chauffage par induction et ses applications, *Journal du Four Électrique*, n° 4 et 5, 1979.
[138] R. COURDILLE, Réchauffage des métaux avant formage, *APAVE*, n° 205 et 206, 1979.
[139] H. JURGENS, Anwendung der induktiven erwärmung in der schmiede, *Elektrowärme International*, n° 1, vol. 37, février 1979.
[140] R. EPPINGER, Induktives Härten von Gusswerkstoffen, *Elektrowärme International*, n° 1, vol. 37, février 1979.
[141] K. E. KOLLENBERG, Der Einsatz von trockenen Korundstampfmassen in Induktionsöfen zum schmelzen und warmhalten von Leichtmetall, *Elektrowärme International*, n° 1, vol. 37, février 1979.
[142] P. FOISY et J. GUILLERMET, Choix d'une installation de fusion pour alliages d'aluminium, *Hommes et Fonderie*, n° 99, 1979.
[143] J. WOLF, Progressive induction hardening of gears submerged in quenchant, *Industrial Heating*, août 1979.
[144] INDUSTRIAL HEATING, *Automated induction heating heats various size billets for forging operations*, septembre 1979.
[145] R. SIGFRIDSON, Chauffeuse à induction pour le traitement de tubes en continu, *Journal du Four Électrique*, n° 8, octobre 1979.
[146] INDUSTRIAL HEATING, *Superheating hot metal with channel induction furnaces for QBOP and BOP steelmaking*, octobre 1979.
[147] CONGRÈS DU COMITE FRANÇAIS D'ELECTROTHERMIE, Versailles, 6-7 mars 1980 :

*a*) J. M. DESVIGNES et H. LE GALL, *Épitaxie en phase liquide des fibres minces pour mémoire à bulles magnétiques.*

*b*) J. C. BOUCHON, *Machine de tirage de monocristaux de grenat de gallium gadolinium.*

*c*) R. LENTZ, *Fours cloche à induction basse fréquence pour le préchauffage des lingots.*

*d*) J. P. METAIL *et al.*, *Le chauffage de produits minces par inducteurs à flux transverse.*

*e*) M. TAMALET et P. LAILLER, *Le recuit en continu de tubes de cuivre.*

*f*) B. SARRETTE, *Electrothermie et microprocesseurs, applications actuelles et perspectives.*

*g*) J. P. SLONINA, *Matériaux composites carbone-carbone, procédés d'élaboration.*

*h*) M. G. ANTIER, *Utilisation du microprocesseur dans les traitements thermiques par induction en série et en continu.*

*i*) P. CREMER *et al.*, *Recherches en cours dans le domaine des fours à induction.*

*j*) J. REBOUX et M. DALLET, *Une installation moderne de traitement par induction en ligne de barres d'acier.*

*k*) C. LECHEVALLIER *et al.*, *Procédé de chauffage par induction de l'acier liquide en poche : essais à l'échelle de 5 t.*

*l*) F. OSTLER, *La coulée automatique de la fonte par induction.*

*m*) S. BARBINI, *Application du chauffage par induction haute fréquence pour la soudure sur chantier de « trains de pipes » servant au transport de gaz et d'hydrocarbures et les possibilités de raccordement souterrain hyperbare.*

[**148**] DOCUMENTATION EDF, 92080 Paris La Défense.
[**149**] DOCUMENTATION ELECTRICITY COUNCIL, Londres, G. B.
[**150**] DOCUMENTATION RWE, Essen, RFA.
[**151**] G. SEGUIER, *L'électronique de puissance*, Dunod Technique, Paris.

**List of equipment manufacturers and suppliers mentioned in this chapter:**

[A] ALSTHOM, 75008 Paris.
[B] ACEC-ELPHIAC, 75019 Paris.
[C] AJAX MAGNETHERMIC, Oxted, G. B.
[D] BIRWELCO, 92290 Châtenay-Malabry.
[E] CELES, 68610 Lautenbach.
[F] CEM-BBC, 75008 Paris.
[G] ELIN-UNION, Vienne, Autriche.
[H] FOMET, 77610 Fontenay-Trésigny.
[I] CRECENZI, 75017 Paris et Turin, Italie.
[J] ASEA, 95340 Person.
[K] E.M.A., 75010 Paris.
[L] INDUCTOTHERM-SASI, 75840 Paris.
[M] JUNKER, 75009 Paris.
[N] RADYNE, 75009 Paris.
[O] MHM, 78630 Orgeval.
[P] SINTRA-SEF, 92700 Colombes.
[Q] SIEMENS, 93200 Saint-Denis.
[R] STEIN-SURFACE, 91130 Ris-Orangis.
[S] TOCCO-STEL, 91300 Massy.
[T] RUSS ELEKTROOFEN, 75009 Paris et Cologne, RFA.
[U] CALAMARI, Milan, Italie.
[V] POLYPENCO (FERROTRON), 91330 Yerres.
[W] SIDERGIE-AEG, 75016 Paris.
[X] HUTTINGER, 67230 Boofzheim.
[Y] ESCHER WYSS, 75000 Paris.
[Z] CGEE ALSTHOM, 90000 Belfort.
[AA] FOFUMI (HYGROTHERM), 75009 Paris et Manchester, G. B.
[AB] LEYBOLD-HERAEUS-SOGEV, 91400 Orsay.
[AC] CHELTENHAM INDUCTION HEATING, Cheltenham, G. B.
[AD] CREUSOT-LOIRE, 75008 Paris et 42700 Firminy.
[AE] BIRLEC, Aldridge, Walsall, G. B.

# Chapter 7

# Dielectric Hystereris Heating

- **High frequency dielectric heating**
- **UHF dielectric heating (microwave heating)**

## 1. PRINCIPLE OF DIELECTRIC HYSTERESIS HEATING

A dielectric, i.e. a material which is an electrical insulator, becomes polarized when placed in an electric field, for example between the electrodes of a capacitor. If the electric field is alternating, successive distortion of the molecules causes heating. This thermal effect is known as dielectric hysteresis heating, dielectric loss heating or simply dielectric heating. However, the term dielectric loss heating is a misnomer since these "dielectric losses", an expression inherited from electrical engineering, which tries to reduce such losses as much as possible, in fact represents the useful effect of this technique. The heat dissipation mechanism is extremely complex, but can be explained by the movement of electric charges due to the electric field within a given atom (electronic polarization) and at the limits between two heterogenic environments (ionic polarization). Thus, due to the effect of an electric field, an atom or molecule with negative charges (electrons) and positive charges (protons) tends to distort, since the charges are attracted by electrodes of opposite polarity. When electrode polarity is inverted, the charges of the atom or molecule are drawn in the opposite direction; these successive changes of direction cause heating.

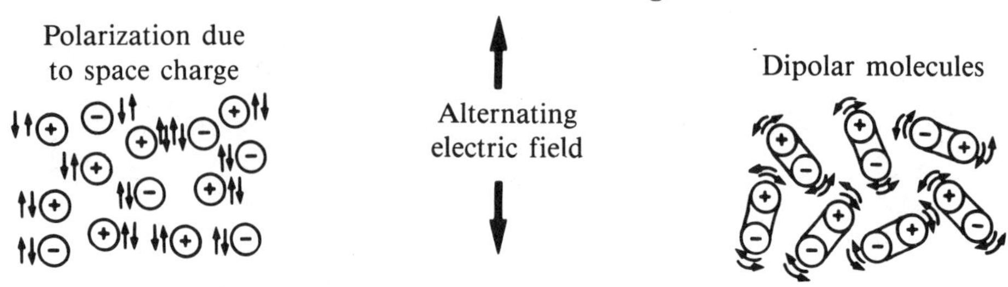


**Fig. 1.** Schematic representation of dielectric hysteresis heating.

Heating is particularly high for molecules which, in the absence of an electric field, form a dipole; i.e. where the positive charge center of mass is different from that of the negative charge. When a dipole of this type is exposed to an electric field, it tends to align with the field. When the electric field changes direction, the dipoles are attracted in the opposite direction and the molecules tend to rotate. Therefore, a large part of the field is dissipated in inter-molecular friction, which results in heat being given off.

The higher the frequency of the electric field, the more intense the friction and the higher the heat. Generally, a distinction is made between radio frequency dielectric hysteresis heating, in which the frequency is between 10 and 300 MHz (radio wave band) and UHF or microwave heating, which uses the 300-30,000 MHz range (frequency and wavelength are related by the equation $\lambda f = V$, where $V$ is the speed of propagation of electromagnetic waves in a vacuum; i.e. $\lambda f = 3 \times 10^8$ m/s). This difference between microwave and radio frequency is not only due to the difference in frequency ranges, but also to their respective characteristics, which, in numerous areas, are not identical (generator, applicator, waveguide, heat penetration, etc.)

Microwave heating is described in paragraph 6, and radio frequency heating is described in the first part, devoted to dielectric heating [1], [50].

The heat is produced directly and solely in the mass of the material to be heated. It is this property which provides the essential interest of dielectric hysteresis heating. In fact, electrical insulators are often bad conductors of heat. With conventional energy transfer methods such as radiation and convection of a heat source on the surface of a body and mass heating by thermal conduction, the surface-core temperature gradient of the part to be heated is often high, and heating rate and therefore productivity low.

Other aspects of dielectric heating can increase its attractiveness, such as high power density and selective heating, etc. However, with respect to installed power, the cost of dielectric heating equipment is high.

The use of this heating method must be carefully studied, and the possibilities of using other, less "costly" techniques such as resistance or infrared heating should be analysed first.

## 2. CHARACTERISTICS OF RADIO FREQUENCY DIELECTRIC HEATING

The characteristics of dielectric heating depend essentially on the nature of the material to be heated and the applicator.

## 2.1. Simplified calculation of the power dissipated in a homogeneous load

In a homogeneous oblong load, placed between the two flat parallel plates of a capacitor to which a radio frequency AC potential difference is applied, it is easy to express the power dissipated in the body to be heated.

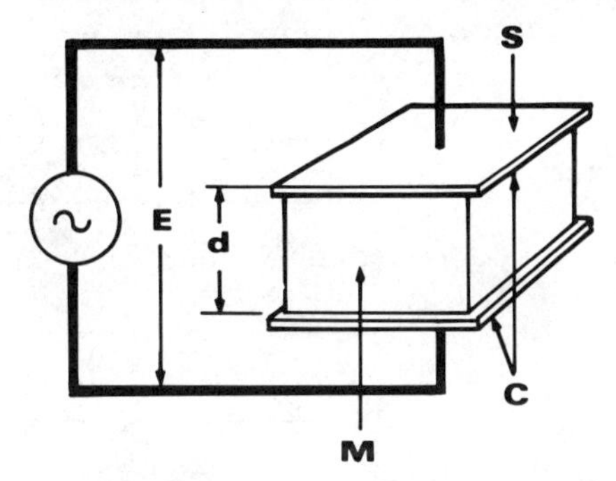


**Fig. 2.** Dielectric heating diagram

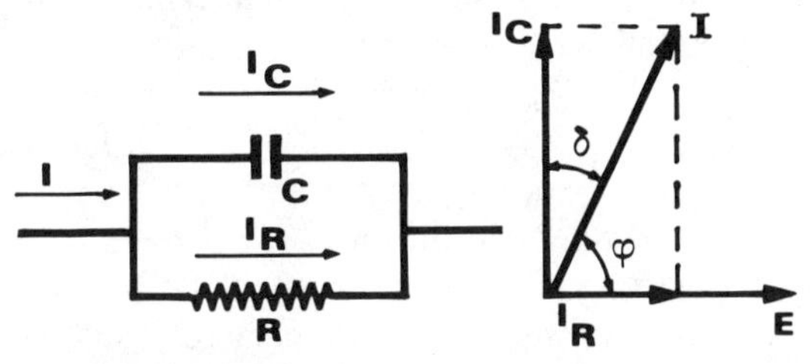


**Fig. 3.** Dielectric heating: equivalent circuit diagram

In a perfect capacitor, the power absorbed between the plates is zero. The current has a phase angle of $\pi/2$ with respect to the voltage. Conversely, if a dielectric is inserted between the plates, heat is given off, and the capacitor can then be replaced, from the electrical point of view, by a perfect capacitance C connected in parallel with resistance R. The active current flowing through the resistance is in phase with the radio frequency voltage applied, while the reactive current is phase-shifted by $\pi/2$ with respect to the voltage. The resulting current is phase-shifted by angle $\varphi$ with respect to the voltage. In dielectric heating, instead of current phase-shift angle $\varphi$ with respect to voltage, we use its complement, angle $\delta(\varphi + \delta = \pi/2)$, which is known as the "loss angle" (here again, as previously stated, this "loss angle" is, in fact, the useful heating effect) [1], [22], [89].

The active power applied to the body to be heated is equal to:

$$P_w = EI \sin \delta = EI_c \operatorname{tg} \delta$$

with:

$$I_c = \omega EC = 2\pi fEC,$$

i.e.:

$$P_w = 2\pi fE^2C \operatorname{tg} \delta.$$

For a flat capacitor, the capacitance is expressed as:

$$C = \epsilon_0\epsilon_r\frac{S}{d}$$

$C$, capacitance in farads;
$S$, plate surface in square meters;
$d$, distance between plates in meters;
$\epsilon_r$, relative dielectric constant (relative permittivity);
$\epsilon_0$, dielectric constant of vacuum, considered as equal to $1/(36\pi \times 10^9)$F/m.

The power given off is therefore:

$$P_w = 2\pi f E^2 \frac{S}{d} \varepsilon_0 \varepsilon_r \operatorname{tg} \delta.$$

The density of power per unit volume, when replacing voltage by the electric field $V$ equal to $E/d$, is:

$$P_{w/m^3} = 2\pi f V^2 \varepsilon_0 \varepsilon_r \operatorname{tg} \delta.$$

By replacing the constants by their value, the calculated power density, in Watts per cubic centimeters, becomes:

$$P_{w/cm^3} = 5.56 \times 10^{-13} f V^2 \varepsilon_r \operatorname{tg} \delta,$$

where $V$ = Volts per centimeter.

The product $\epsilon_r \tan\delta$ is known as the "loss factor"; it would be more suitable to call this the "power transmission factor" since it in fact characterizes the capability of a product to be heated by dielectric hysteresis. To provide a simple interpretation for this value, it can be considered that $\tan\delta$ represents the resistance of the material to the movement of the molecules forming it (comparable to "viscosity" at molecular level) and that $\epsilon_r$represents the polarizability of the material.

The power density is proportional to frequency, to the electric field and to the loss factor. Current values for this power density are between 0.5 and 5 W/cm$^3$.

### *2.1.1. Effect of frequency and authorized frequencies*

Increasing the frequency increases the power density in the material to be heated. However, since the frequencies used are within the radio frequency range, and therefore must conform with telecommunications regulations, the frequencies which can be used in France are 13.56 MHz ± 0.05%, 27.12 MHz ± 0.05% and 40.68 MHz ± 0.05%, but these frequencies, which are specified in international agreements, may vary from one country to another.

When the frequency increases, it becomes more difficult to develop high power sources. Also, the use of large electrodes gives rise to standing wave problems, and therefore non-uniform heating.

However, these limitations have practically no effect on dielectric heating industrial applications, since the power densities obtained are already very high compared with other heating methods.

### *2.1.2. Effect of electric field*

To obtain high power densities, the electric field must be as high as possible. However, increasing the voltage increases the risks of breakdown. The breakdown voltage of dry air is around 3 kV/mm and, in general, dielectric heating equipment operates with electric field values of between 80 and 300 V/mm, for safety reasons; the most frequently encountered values are between 80 and 160 V/mm. To prevent discharge, the inter-electrode voltages rarely exceed 15,000 V.

As in the case of frequency, these limits have no practical importance, since the power densities reached generally suffice for industrial applications.

### *2.1.3. Effect of "loss factor"*

This factor is a characteristic of the material to be heated, but is not a constant. Both its components $\epsilon_r$ and tan $\delta$ depend on frequency, temperature, moisture and also on the production conditions of the product to be heated. If the "loss factor" is too low, heating takes place slowly, and it becomes difficult to reach the desired temperature due to heat losses. Conversely, if the "loss factor" is too high, current leakage takes place through the material, which is in reality a bad electrical insulator; it is then not possible to use high voltages. Therefore, for dielectric hysteresis heating to be successful, the "loss factor" must, although this is not absolute, satisfy the condition:

$$0.01 < \epsilon_r \operatorname{tg} \delta < 1.$$

If the "loss factor" increases with temperature, power is concentrated at the hottest points, which may cause local overheating. The power density must then be reduced, and the temperature allowed to equalize by conduction. It is also possible to apply power to the load in pulses, rather than continuously. Temperature equalization then takes place between pulses.

Figure 4 gives the relative permittivities and "loss factors" of numerous bodies at ambient temperature. When it is desired to use dielectric hysteresis heating in industrial practice, it is important that these values be obtained from the material manufacturer, and that tests be performed in collaboration with them and specialized equipment suppliers and specially equipped research laboratories (textile research center, EDF, etc.), since the values of Figure 4 are simply intended as an illustration [1], [6], [27], [71], [93 *b*].

These values show that there are major differences between loss factors of materials, and therefore between their capability of being heated by dielectric hysteresis. Water, which has one dipolar molecule in the absence of an electric field, absorbs the energy of a radio frequency or microwave electric field very easily. This property is very interesting in the numerous drying applications encountered in industries.

## 2.2. Heterogeneous load heating

The dissipated power is calculated considering that the load was homogeneous and fills all the space between the electrodes. In practice, the material to be heated is always more or less heterogeneous, and often only fills part of the inter-electrode space.

It is easy to determine dissipated power when materials which are considered as homogeneous are placed in series or parallel between the electrodes.

### *2.2.1. Series connection*

In numerous industrial processes, several materials have to be heated together (manufacture of complex materials, bonding, etc.). Moreover, even if there is only one material to be heated, there is often a layer of air between this material and one of the electrodes.

Relative dielectric constant and loss factor

| Material | Dielectric constant $\epsilon_r$ | | Loss factor $\epsilon_r$ tg $\delta$ | |
|---|---|---|---|---|
| | at 10 MHz | at 2,450 MHz | at 10 MHz | at 2,450 MHz |
| Water: | | | | |
| pure ice at −12°C | 3.7 | 3.2 | 0.07 | 0.003 |
| pure water at 25°C | 78.0 | 76.0 | 0.36 | 12 |
| salt water | 80.0 | 75.5 | 100 | 18 |
| Wet wood (perpendicular to grain) | 2.6 | 2.1 | 0.1 | 0.07 |
| Dry wood | 2 | 1.9 | 0.04 | 0.01 |
| Cellulose acetate (rayon) | 6 | 6.1 | 0.07 | 0.09 |
| Melamine | 5.5 | 4.2 | 0.23 | 0.22 |
| Bakelite | 4.3 | 3.7 | 0.18 | 0.15 |
| Polyamid (nylon) | 3.2 | 3.0 | 0.09 | 0.04 |
| Polyester | 4.0 | 4.0 | 0.04 | 0.04 |
| Polyethylene (Polythene) | 2.25 | 2.25 | 0.0004 | 0.001 |
| Polystyrene | 2.35 | 2.55 | 0.0005 | 0.0005 |
| Polytetrafluorethylene (PTFE) | 2.1 | 2.1 | 0.0003 | 0.0003 |
| Polyvinyl chloride | 3.7 | 2.9 | 0.04 | 0.1 |
| Perspex | 2.7 | 2.6 | 0.027 | 0.015 |
| Paper | 3.5 | 3.5 | 0.4 | 0.4 |
| Wool (humidity 20%) | 1.2 | — | 0.01 | — |
| Cotton (humidity 7%) | 1.5 | — | 0.03 | — |
| Pyrex glass | 4.84 | 4.82 | 0.015 | 0.026 |
| Rubber | 2.5 | 2.5 | 0.08 | 0.03 |
| Water-based glue | 5 | — | 0.25 | — |
| Kidneys | 4.5 | 2.5 | 4.2 | 0.18 |
| Steak | 50 | 40 | — | 12 |

**Fig. 4.**

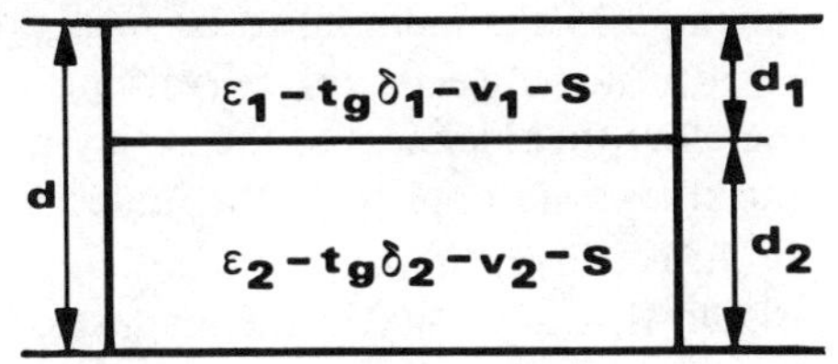


**Fig. 5.** Series connection

Voltage $E$ between the electrodes is broken down into two voltages, $E_1$ and $E_2$, and the electric field in each material is respectively:

$$V_1 = \frac{E_1}{d_1} \quad \text{and} \quad V_2 = \frac{E_2}{d_2}$$

Since both materials are connected in series, voltages $E_1$ and $E_2$ are related by the equation $E_1\omega C_1 = E_2\omega C_2$, i.e.:

$$\boxed{\frac{E_1}{E_2} = \frac{\varepsilon_2 d_1}{\varepsilon_1 d_2}} \quad \text{or also} \quad \boxed{\frac{V_1}{V_2} = \frac{\varepsilon_2}{\varepsilon_1}}.$$

Much information can be obtained from these equations.

If a layer of air with a relative dielectric constant of the order of 1 exists between the material to be heated and one electrode, the electric field intensity in this material decreases when the thickness of the layer of air increases, and is therefore always lower than in the absence of an air layer. If indices 1 and 2 apply respectively to the air layer and to the material to be heated, the ratio between the field intensity $V_2$ in the dielectric, separated from one electrode by an air layer $d_1$, and the field intensity $V_{20}$ in the absence of an air layer, is given by the expression:

$$\boxed{\frac{V_2}{V_{20}} = \frac{1}{1 + \varepsilon_2 (d_1/d_2)}}.$$

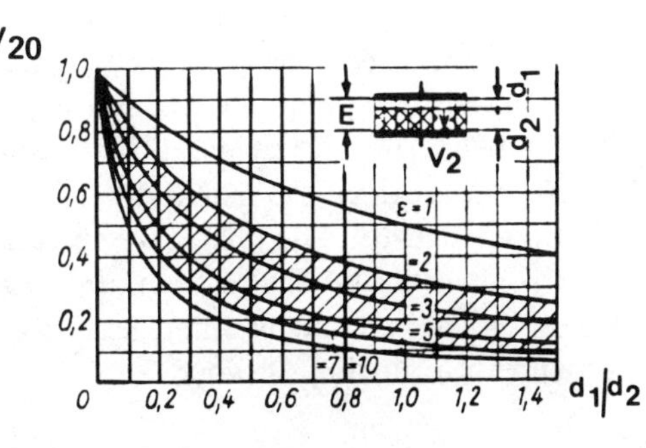


**Fig. 6.** Effect of air layer $d_1$ on electric field intensity $V_2$, in a dielectric of thickness $d_2$ and dielectric constant $\epsilon_2$.

This expression shows that:

— the power dissipated in the material to be heated can be modified, for a constant voltage between the electrodes, by varying the distance between the electrodes, and therefore the thickness of the air layer, which provides a simple method of varying the power applied to the material;

— the higher the dielectric constant of a material, the higher the increase in the proportion of the field intensity in the air layer;

— the thicker the material to be heated, the higher the field proportion inside this material, and the more efficient the heating.

It is also possible to compare the dissipated power densities in materials connected in series:

$$\boxed{\frac{P_1}{P_2} = \frac{\varepsilon_2}{\varepsilon_1} \frac{\operatorname{tg} \delta_1}{\operatorname{tg} \delta_2}.}$$

If it is considered that all other variables (loss angle, density, specific heat, etc.) are equal, it is therefore the material with the highest dielectric constant which will heat slowest. Therefore, series connection is used if the dielectric constant of the material to be heated is low.

In fact, loss angle $\delta$ is also highly important, and the power dissipated in the air is practically zero when an air layer and the dielectric to be heated are in series. The only effect of the air layer, as shown in the previous analysis, is to reduce the electric field and therefore the power density in the material to be heated. This air layer is, however, useful in many cases since it facilitates elimination of water vapor in drying operations, or enhances uniform heating if the load surface is not flat.

Generally, the behavior of materials connected in series with respect to dielectric heating can be forecast only if the relative dielectric constants and loss angles are known.

### *2.2.2. Parallel connection*

When two materials are connected in parallel, the electric field in each is modified by the adjacent material. As an initial approximation, the mutual effect of two materials can be ignored, and it is possible to consider that the power expressed as described in paragraph 2.1. above is applied separately to each.

Therefore, it is always the material with the highest "loss factor" that has the highest power density. The relative temperature rise rates, however, depend on the relative density and specific heat of both materials.

This differential heating is widely used in dielectric heating applications, and in particular for glueing of woods.

### *2.2.3. General case of a heterogeneous load*

When, from the dielectric characteristic viewpoint, the material heated is highly heterogeneous, it is generally the component offering the highest "loss factor" which is the source of the most intense heat. Thus, for wet bodies, water which has a high loss factor heats first, facilitating drying of the materials, not only by enabling rapid elimination of water, but also by preventing overheating of the material which often enhances product quality.

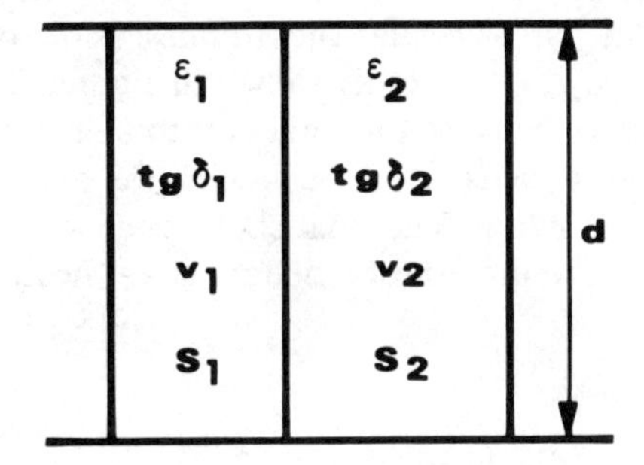


**Fig. 7.** Parallel connection

Moreover, as drying progresses, the water content drops and the absorbed power decreases automatically, preventing any risk of overheating. This selectivity and self-regulating capacity encountered in dielectric hysteresis heating, are profitably used in numerous industrial applications.

## 2.3. Power calculations — General case

The simplified calculation performed above is based on several hypotheses — homogeneity of the heated material or its various components, homogeneity of the capacitor electrostatic field, absence of standing waves, regularly shaped products, electrodes comparable to the plates of a capacitor, etc.

In practice, these conditions are more or less satisfied. The electric field of a flat capacitor is distorted around the plates. Therefore, the electrodes must extend beyond the materials to be treated by a minimum width $x$, estimated as:

$$x = \frac{d}{2\pi} l_n \frac{135}{p},$$

in cases where the body to be heated occupies all the distance between the electrodes, and in which:

$x$, in millimeters;
$d$, distance between electrodes in millimeters;
$p$, permissible percentage tolerance in temperature across load.

Moreover, the electrodes used can differ from the flat plates of a capacitor; the product to be heated is often irregularly shaped and composition is heterogeneous.

Thus, in order to heat a material with non-parallel surfaces, as shown in figure 8 a), and if parallel electrodes are used, the electric field is much higher in the thickest part of the material as shown by the series connection analysis; heating is highly irregular. If the configuration shown in figure 8 b) is used, the electric field is much lower in the thick part of the material than in the thin, and heating is as irregular as in the previous case. To obtain regular heating, the electrode configuration shown in figure 8 c) should be used; the inclination of the electrode can be easily determined using the relation established above *($V1/V2 = \epsilon_2$)* concerning electric fields in two bodies in series. If the body to be heated has a dielectric constant $\epsilon_2$, and to obtain the same electric field in the material as a whole, and therefore the same temperature rise, the thickness of the air layer at the thinnest side of the material must be equal to $(d_1 - d_2)/\epsilon_2$. Also, as shown in figures 8 d) and e), it is possible to leave an air space between the total part to be heated and the electrode but, in this case, the power density will be much lower and thermal equalization will be obtained by conduction. The final choice is generally a function of the conditions under which the material or materials are heated.

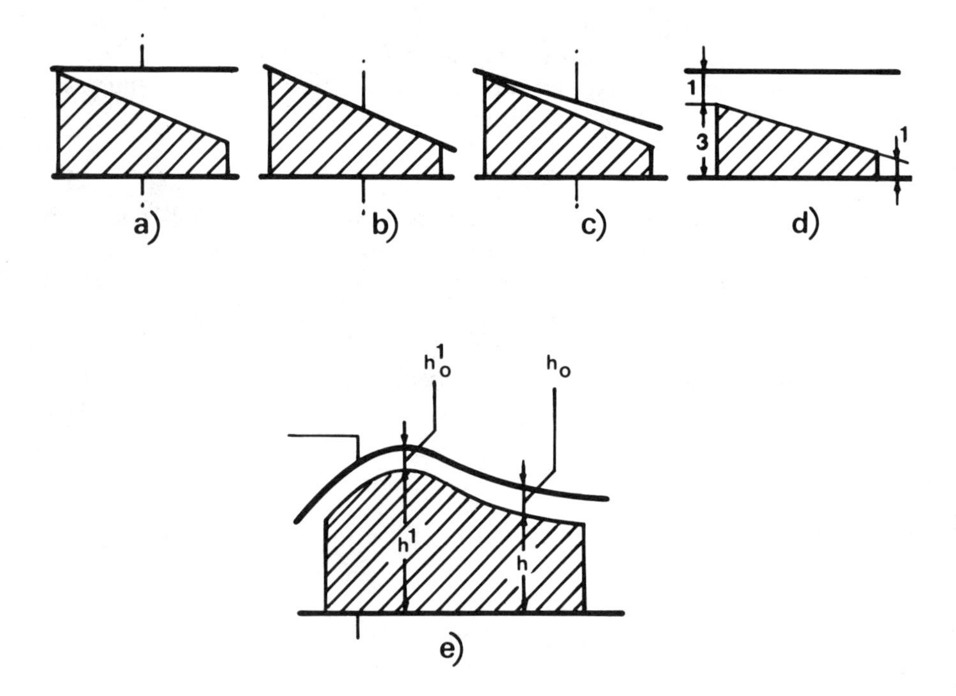


**Fig. 8** Influence of the electrode configuration.

Generally, the load is placed between the electrodes, but it is also possible to use the leakage electric field of cylindrical electrodes parallel to the load (*see* paragraph 3.1); in particular, this applies to thin items (sheets, strips, etc.) or to obtain purely superficial heating. However, efficiency is lower with this electrode configuration.

Also, long electrodes can create a heterogeneous field, due to the attenuation of the voltage and creation of standing waves. This phenomenon occurs when the electrode length approaches or exceeds one-quarter of the length (in meters) defined by equation $\lambda = 300/(f\sqrt{\epsilon_r})$. This drawback is overcome by the use of tuning inductances.

In spite of these difficulties, it is generally possible to calculate the power density using the fundamental laws of electromagnetism, and especially Maxwell's equations. The calculations are highly complex and exceed the framework of this introduction to dielectric hysteresis heating.

The power density equation is:

$$\frac{dP_w}{dm} = 2\pi \varepsilon_0 f \varepsilon_r \operatorname{tg}\delta V^2,$$

$P_w$, dissipated power;

$m$, part or body volume;

$f$, frequency;

$\epsilon_0$, dielectric constant of vacuum;

$\epsilon_r$, relative dielectric constant;

$\tan \delta$, loss angle tangent;

$V$, electric field.

This expression is identical to the simplified formula previously established, but applies only to an infinitesimally small volume of the load. To obtain total dissipated power, this equation must be integrated for the total volume of the load. In numerous cases, the simplified calculation sometimes gives an adequate approximation, and the corresponding formula can be used to calculate the power given off in the load and heating time, the final adjustment being made after a series of tests.

## 2.4. Penetration depth

Although the simplified calculation for the power applied to the part to be heated does not explicitly refer to this, dielectric hysteresis radio frequency heating in fact results in the absorption of electromagnetic radiation by the material to be heated.

Now, when an electromagnetic wave comes into contact with the dielectric, part of it is reflected by this material, and the other part, which is usually higher, penetrates the material. The energy of this wave gradually attenuates, and is converted into heat within the material. Radiation is absorbed according to a decreasing exponential law, such that:

$V_x = V_0 e^{-\alpha x}$;

$P_x = P_0 e^{-\alpha x}$;

$V_x$, electric field at distance $x$ from surface;

$P_x$, power density at distance $x$ from surface;

$V_0$ and $P_0$ are the values at the surface.

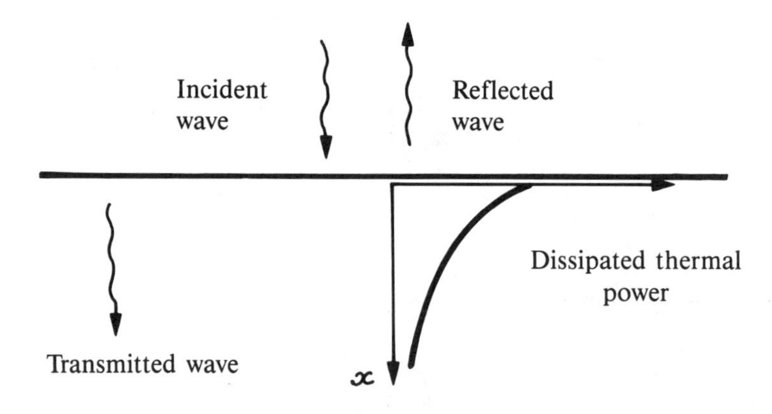


**Fig. 9.** Penetration depth [93 *a*]

The attenuation constant $\alpha$ depends on the dielectric properties of the product and the incident radiation wavelength, and is equal to (Von Hippel's equation):

$$\alpha = \frac{2\pi}{\lambda_0}\left[\frac{\varepsilon_r(\sqrt{1+\mathrm{tg}^2\,\delta}-1)}{2}\right]^{1/2},$$

with $\lambda_0 = c/f$, $c$ speed of propagation of wave in vacuum, equal to $3 \times 10^8$ m/s.

By convention, the wave penetration depth into the material is the depth at which the incident power is reduced by $1/e$, i.e. approximately 37% of its initial value (this definition is not identical to that of penetration depth in electrically conducting materials as described in the electromagnetic induction heating chapter; the peneteration depth for this method of heating was the distance from the surface at which the current density was reduced by $1/2\epsilon$). Therefore, penetration depth is equal to $1/(2\alpha)$, i.e.:

$$d = \frac{\lambda_0}{4\pi}\left[\frac{\varepsilon_r(\sqrt{1+\text{tg}^2\,\delta}-1)}{2}\right]^{-1/2}.$$

In most cases, tan $\delta$ is low, much less than 1, and the penetration depth formula can be simplified:

$$d = \frac{\lambda_0}{2\pi\sqrt{\varepsilon_r}\,\text{tg}\,\delta},$$

or again, by introducing frequency:

$$d = \frac{c}{2\pi f\sqrt{\varepsilon_r}\,\text{tg}\,\delta}, \quad \text{or} \quad d = \frac{4.77 \times 10^7}{f\sqrt{\varepsilon_r}\,\text{tg}\,\delta} \quad \text{in meters.}$$

Penetration depth is inversely proportional to frequency, to the tangent of the loss angle and the square root of the relative permittivity of the material to be heated. The latter two values are characteristic of the material, and the penetration depth is lower for bodies offering adequate dielectric heating aptitude. The only variable available to modify penetration depth is therefore frequency, but the choice is limited to those frequencies authorized by telecommunications regulations.

With radio frequency dielectric heating, penetration depth is high, generally greater than 1 m and may reach several tens of meters. Therefore, the idea of penetration depth is of limited importance for most applications, and the heat given off in the mass of the material to be heated can be considered as homogeneous insofar as the material itself is homogeneous.

With microwave dielectric heating, penetration depth is much lower since the frequency is generally 100 to 300 times higher and is therefore an important factor which must be carefully examined.

These conditions have led to the use of dielectric hysteresis, radio frequency heating for applications in which the materials to be heated were very thick. However, radio frequency dielectric heating processes are not limited exclusively to "mass" treatment of materials. By using applicators consisting of divided field electrodes, adapted to the geometry of "sheet" materials, it is also possible to heat such products efficiently.

### 2.5. Heating times

If the power density per unit volume is known, it is possible to estimate the time required for a heating operation:

$t$, temperature rise time;
$C_p$, specific heat of the material to be heated;
$u$, density of the material to be heated;
$\Delta\theta$, temperature increase.

$$t = \frac{C_p u \Delta\theta}{2\pi \varepsilon_0 f \varepsilon_r \operatorname{tg} \delta V^2},$$

This expression does not include the latent heat required to vaporize a product. If the latent heat of vaporization is known, it is possible to calculate the energy to be applied once the boiling point has been reached.

This calculation also neglects heat losses due to convection, radiation and conduction. However, since heating time is very short, these losses are very low during temperature rise, unless, of course, the final temperature is high.

## 2.6. Comparison of power densities between radio frequency dielectric heating and other processes

With dielectric heating, power densities can be very high, of the order of 0.3 to 5 W/cm$^3$, i.e. 300 to 5,000 kW/m$^3$. For flat products, this often varies between 30 and 100 kW/m$^2$. These values are simply representative, however, and depend on numerous factors such as the dielectric characteristics of the materials ($\epsilon_r$, tan $\delta$, dielectric strength, etc.) and the working voltage which determines the transmissible power density, and also on the conditions under which product heating takes place, as these can impose limits on temperature rise time (for example excessively fast drying can distort some products).

Moreover, the heat is generated within the product itself, thus avoiding the time taken to transfer heat by conduction, which may be very long with poor conductors of heat.

Therefore, treatment times are often considerably reduced, for example in some cases by a ratio of 5 to 10 to one.

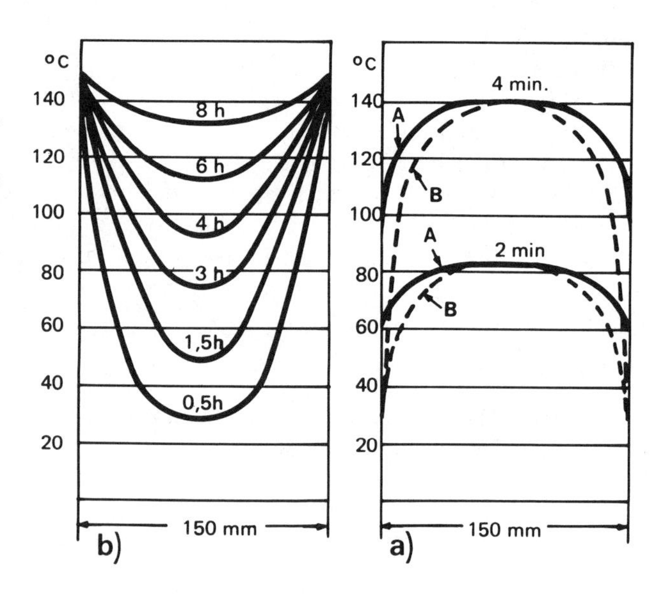


**Fig. 10.** Heating rate: *a)* with radio frequency dielectric heating, power density 300 kW/m$^3$; *b)* in conduction heating using heating panels [98]; *A)* with thermal insulation; *B)* without thermal insulation.

### 2.7. Power factor

In radio frequency dielectric heating, the reactive current is much higher than the active current, as can be seen from the values of tan $\delta$, which are often very low, and in most cases between 0.01 and 0.1 with important exceptions such as water, wet products or some glues or cements, for which tan $\delta$ is between 0.5 and 2. The load impedance is sometimes simply a function of its ohmic resistance, and capacitance has practically no appreciable effect on impedance at radio frequency. This property can be easily explained by examining the equivalent circuit diagram of a dielectric heating system. The capacitance of the perfect capacitor in parallel with the resistance is much higher than the value of the latter, and therefore the resulting impedance is basically a function of the resistance.

Therefore, the power factor is low and must be corrected by using an induction coil.

### 2.8. Efficiency and specific consumptions

There is a wide range of possible applications for radio frequency heating, and it is difficult to provide general specific consumptions. Efficiency in radio frequency energy generation is between 55 and 70% and overall efficiency is very similar, of the order of 50-60% (*see* paragraph 3.2.4.1. and all paragraphs devoted to process applications).

The power consumed from the line is therefore 1.5 to 2 times the power dissipated in the load. The power of dielectric heating generators is often given in radio frequency kilowatts, and therefore conversion efficiency must be taken into account when calculating the power absorbed from the line.

## 3. RADIO FREQUENCY DIELECTRIC HEATING EQUIPMENT

Basically, a radio frequency dielectric heating installation consists of a radio frequency current generator, an adaptor or matching device between the generator and the load, the applicator itself and its system of electrodes and the ancillary control and handling devices.

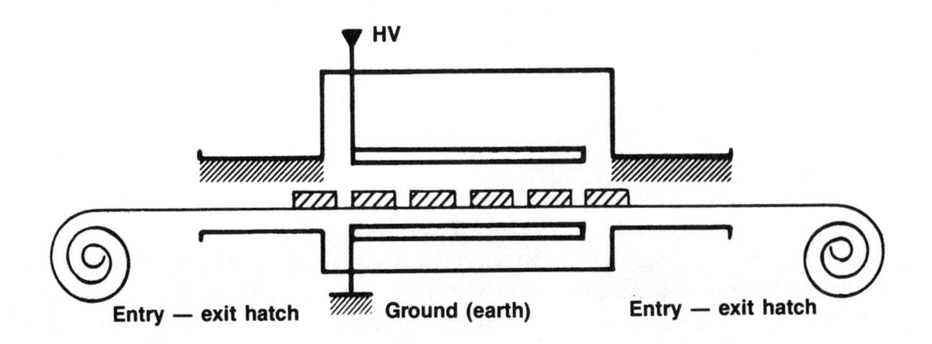


**Fig. 11.** Continuous dielectric heating using flat electrodes [89].

### 3.1. Applicators

Various types of applicators are used as a function of the nature and shape of the product, and depending on whether the production process is continuous or discontinuous. The electrodes can be separate or consist of the plates of a flat capacitor or capacitor of another shape (concentric tubes or spheres, etc.).

#### *3.1.1. Flat electrodes*

The product to be treated is placed between the plates of a flat capacitor, or move either as a layer or mass between the plates. The first arrangement, in which the product may be in contact with the electrodes or not, is used mostly for discontinuous operations (plastic welding, wood gluing, etc.), while the second enables continuous production.

#### *3.1.2. Garland electrodes*

The product to be heated generally moves in sheets or belts between the divided electrodes, tubular or in bars, mounted vertical to the direction of movement of the belt. The electrodes are staggered on either side of the belt and connected in parallel on each side of the belt.

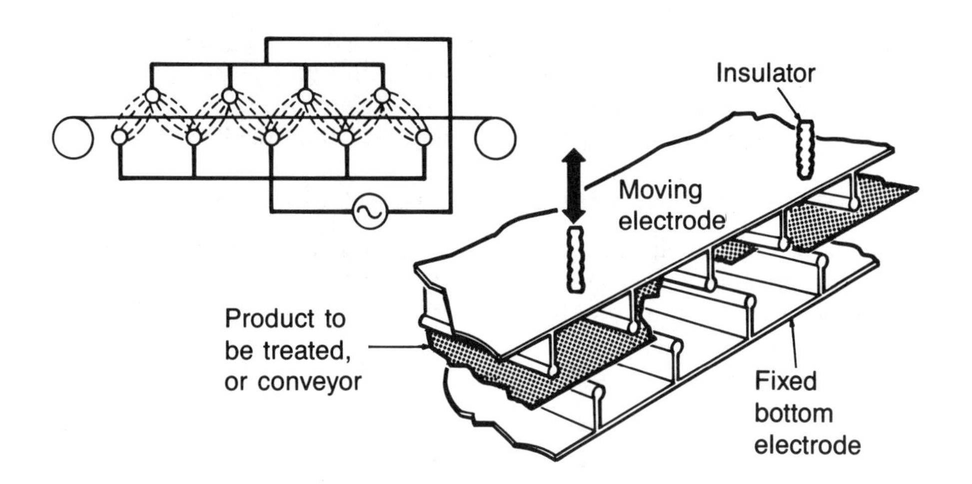


**Fig. 12.** Garland electrodes [97].

Garland (or staggered) electrodes are used to produce intense fields in thin belts, since the electric field is not perpendicular to the product, as with plate electrodes, but directed along the direction of movement of the belt. Therefore, it is possible to transfer high powers of the order of 30 to 100 kW/m$^2$ to the moving belt. The separation between the electrodes is often adjustable, so that the transmitted power can be easily modified.

### *3.1.3. "Strayfield" electrodes*

This system consists of a set of electrodes (tubes, rods or belts) located in the same plane and parallel to the plane of the material to be heated. Two successive electrodes are of opposite polarities.

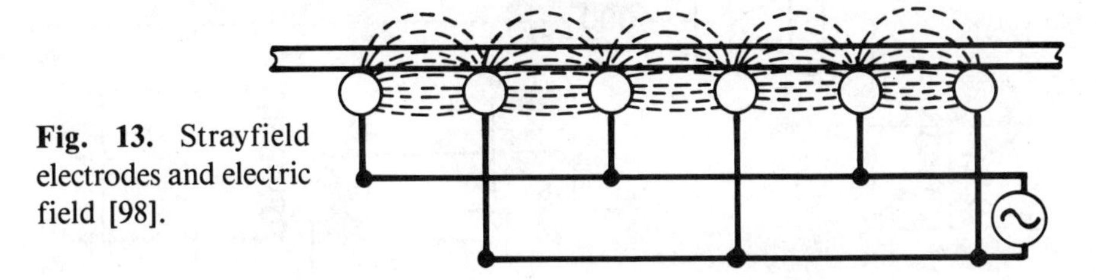

**Fig. 13.** Strayfield electrodes and electric field [98].

This type of applicator provides an electric field parallel to the layer or the belt of material to be treated, and therefore high power density. In particular, this system is used for treatment of fine products, up to thicknesses of approximately 10 mm; beyond this thickness, the garland configuration is generally preferred since it provides better homogeneity.

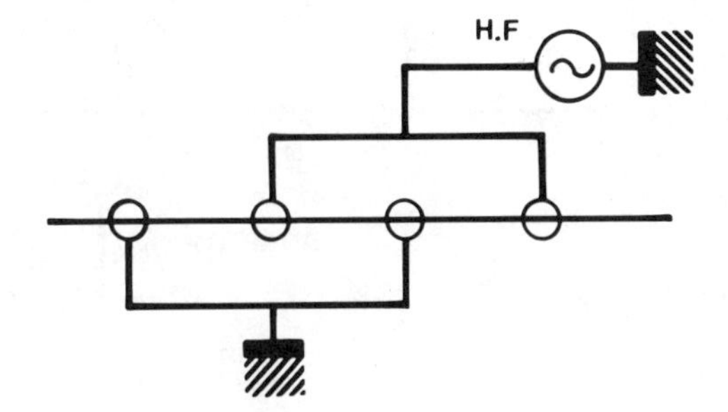


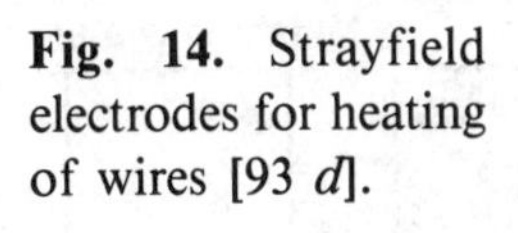

**Fig. 14.** Strayfield electrodes for heating of wires [93 *d*].

A variant of this system consists in transforming the "strayfield" rods into loops, through which the wire to be heated is passed.

### 3.2. The radio frequency generator

Heating can occur only if the frequency is sufficiently high. The generators used consist of self-excited oscillators, with one or more triodes with a filament heating system and a high voltage rectifier applying a very high anode voltage (5,000 to 15,000 V). These electronic generators can be compared to those used in electromagnetic induction heating, but the working frequency is much higher.

The maximum unit power with this type of generator is of the order of 600 kW HF, i.e. 850 kW from the line, due to a generation efficiency of the order of 70%. Generally, this radio frequency energy generation efficiency is between 55 and 70%, which is relatively low. However, the overall efficiency is very close to these values, since the energy is given off directly within the material to be heated, and since heating system inertia is very low (triode warmup time on starting);

generally, overall efficiency is of the order of 50%. Tube power is often given at the radio frequency side of the circuit, and the conversion efficiency must be taken into account in calculating the power absorbed from the line.

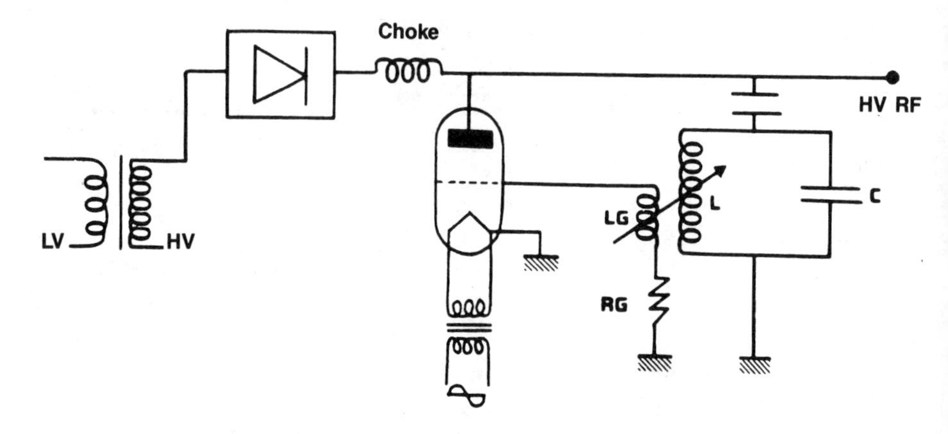


**Fig. 15.** Self-excited triode oscillator [89].

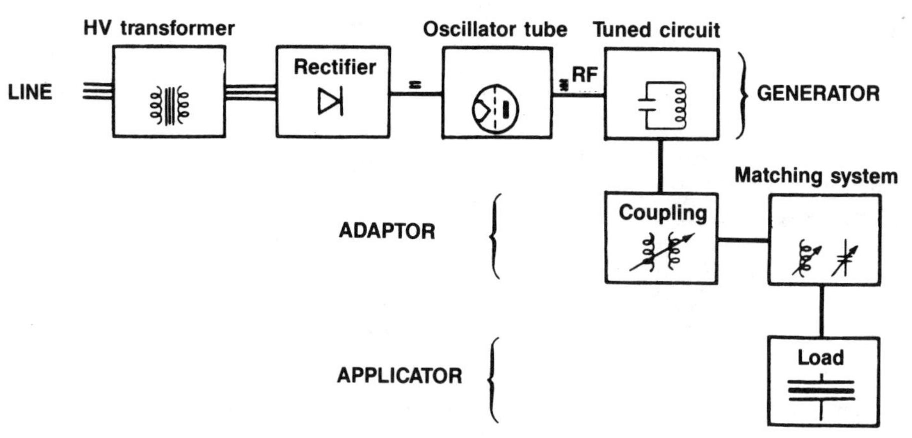


**Fig. 16.** Block diagram of a radio frequency dielectric heating installation [89].

The electron tubes are either water- or forced-air cooled. Natural cooling is often used for low powers. The energy lost in cooling the tubes can be recovered for heating of premises, thus noticeably improving process efficiency, or better still, recovered directly to provide added furnace heating.

The service life of the electron tubes is generally between 5,000 and 10,000 hours, and, on average, is between 6,000-7,000 hours.

To remain within the frequency tolerances imposed by telecommunications regulations, it is necessary to allow for modifications in capacitance which occur during treatment of a given body or different bodies. Equipment manufacturers must also guarantee a limit value for radio disturbance due to harmonics. Frequencies other than those imposed by regulations are sometimes used for low power equipment. In such cases, it is necessary to prevent propagation into space of the frequencies produced (Faraday cage protection, etc.); these devices can also be useful with the frequencies reserved for industrial applications [89].

### 3.3. Matching the generator to the load

As in induction heating, it is indispensable that the generator be electrically matched to the load. Generally, in induction heating this function is provided by matched transformers and capacitors. In dielectric heating, the load impedance must be matched to the oscillator tube anode, so that the tube emits maximum power with adequate efficiency. Matching is obtained by compensating the reactance of the load capacitor by a series or parallel inductance. Also, an impedance transformer can be used, if a radio frequency coaxial cable is used as a power conductor between the generator and the electrodes. Power adjustment can be provided by two systems. In the first, the electrode separation is adjusted by means of a moving electrode, and therefore the electric field can be adapted to specific requirements; in the second, electrical field variation is obtained by altering the electrode working voltage and also the coupling between the anode circuit and the oscillatory circuit.

In some cases, as for drying numerous materials in which the absorbed power decreases automatically with water content, heating is practically self-adjustable.

The connections between the power source and the electrodes must be as short as possible, since they produce electromagnetic radiation and Joule effect losses. The high voltages used lead to low currents, but, due to the radio frequency, the current concentrates in a very thin layer, which results in high Joule losses if the connection lengths are increased. To reduce electromagnetic radiation losses, conductors are, insofar as possible, concentric, and made from tubes and not wires, sudden bends being avoided. The generator should be as close as possible to the load electrodes [4], [89].

### 3.4. Handling system

In continuous heating systems, handling of parts plays an important role. Where the material itself forms the conveyor — belt or sheet products — efforts must be made to keep the product from separating from the electrodes when a strayfield system is used. Conversely, when the product is placed on a conveyor, the product material must not disturb the electric field (no special metal elements). As in the air gap case, the conveyor is then equivalent to a dielectric in series with the load, and loss angle must obviously be very low if it is not to overheat. It must also be capable of withstanding mechanical, thermal and, where necessary, the chemical stresses to which it is subjected.

### 3.5. Safety devices

Safety measures required by telecommunications personnel must be applied to radio frequency dielectric heating installations. Therefore, in both fields and for most countries, standards have been drawn up and are respected by equipment constructors, although the diversity of such regulations often causes problems when processing imported equipment. Therefore, this point must be carefully examined by potential purchasers (French standards are at present identical to British, German and American standards as far as protection of workers is concerned). Generally, these standards are less stringent for radio frequency than for UHF (microwaves). For example, for a power density in the vicinity of the equipment of 10 mW/cm$^2$, the total exposure time must be less than 8 hours per day, while no exposure to a power of 25 mW/cm$^2$ is permitted.

The thousands of installations in operation comply with these standards, and all necessary arrangements are taken by manufacturers (shielding, Faraday cage, grounding, etc.).

## 4. ADVANTAGES AND DISADVANTAGES OF RADIO FREQUENCY DIELECTRIC HEATING IN INDUSTRIAL PROCESSES

### 4.1. Advantages of radio frequency dielectric heating

The analysis of radio frequency dielectric heating characteristics has demonstrated that this method offers original properties which can result in significant advantages for an industrial process.

— *Direct transfer of energy to the product*

The energy is transferred directly to the product to be heated, which absorbs the energy of an electromagnetic wave without intermediate transfer mechanism. This type of heating therefore eliminates the constraints of conventional thermal laws, in which heat transfers take place through convection, radiation and conduction, with slow penetration of heat into the material; therefore, the energy transfer can be much faster, and superheating used with other processes to accelerate energy transfer, avoided.

— *Homogeneous heating within the product*

If the product is homogeneous and of regular shape, and the material thickness low with respect to the penetration depth which is generally the case at radio frequency, heating is homogeneous throughout the material.

— *Selective heating*

• Only the product treated is heated; its support and the applicator chamber remain cold, with the exception of thermal losses from the product to be heated, which remain limited. Unless the final temperature is high, overall thermal losses are very low.

• The various components of a complex material can heat differently. For example, when bonding materials, only the cement or glue is heated, or in drying, the water, enabling the wet parts to be heated very rapidly and to prevent burning or total dehydration of dry parts.

— *High power density*

The power density in the product to be heated is high; therefore, the potential increase in productivity is also high.

— *No thermal inertia*

The thermal inertia of a radio frequency dielectric heating system is very low, some tens of seconds or a few minutes, depending on power required for the electron tube to warm up on starting the installation.

— *Heating rate*

This property is a direct result of the previous properties and forms a potential source for increase in productivity.

— *Rather high efficiency*

The generator efficiency is rather low, 55 to 70%, and overall efficiency is very similar, 50 to 60% approximately. Compared with the overall efficiency of other heating systems (for example, steam heating), radio frequency dielectric heating often provides energy savings, or is at least comparable. When used for selective heating, the energy saving becomes very important.

— *Reduction in floor space*

This is a direct consequence of high power density.

— *Wide range of potential applications*

The coupling between applicator and product to be heated can vary widely, with a variety of series and parallel setups obtainable. The potential applications of this method of heating are therefore numerous and diverse.

— *Treatment quality*

The treatment quality obtained with this type of heating is generally excellent, due to the properties of homogeneity, selectivity and absence of overheating previously described.

— *Resolving difficult heating problems*

Some heating operations are difficult to perform through conventional thermal methods, and radio frequency dielectric heating may often be a simple remedy for such difficulties (heating thermal insulators, powder, thermofragile, thick products, selective heating of composite materials, requirement for homogeneous mass heating or to prevent superficial overheating, requirement for high quality, etc.).

The above advantages make radio frequency dielectric hysteresis heating an extremely attractive process; unfortunately, it has certain limits, both economic and technological, which have slowed its expansion.

### 4.2. Disadvantages of radio frequency dielectric heating

The disadvantages of radio frequency dielectric heating are basically economic, and sometimes, but more rarely, technological.

#### *4.2.1. Economic limits*

If the price of installed power is retained as an investment cost measurement criteria, which, of course is simplistic, but usable to give an initial estimate, a radio frequency dielectric heating installation costs four to six times more than a resistance furnace of the same power, the energy efficiencies being comparable in most cases, with a slight advantage in efficiency for dielectric heating in one instance, and for resistance heating in another.

For dielectric heating to impose itself on other techniques, its advantages must more than compensate for this financial handicap.

#### *4.2.2. Technological limits*

The principle of the procedure itself shows that it is suitable for heating formed parts, panels, belts and layers, or products providing regular and repetitive shapes, but that it is difficult to ensure uniform heating in parts of some shapes. In mixed, dielectric process heating systems, the requirements of both techniques are sometimes difficult to combine (for example the necessity of placing the electrodes appropriately, while intense agitation of the atmosphere is necessary).

For treatment of layers or wide belts, standing wave problems arise, and end effects may occur.

*Therefore*, for dielectric heating to compete with other techniques, it is necessary that, after examining all the possible alternatives, this appears as the only satisfactory technical solution to the problem or that it offers significant economic advantages: gain in productivity, improvement in quality increasing the market value of the product, low energy efficiency of competing techniques for all or part of the treatment proposed, reduction of intermediate stocks, etc. [689], [93 *d*].

## 5. INDUSTRIAL APPLICATIONS OF RADIO FREQUENCY DIELECTRIC HEATING

An analysis of the characteristics of radio frequency dielectric heating has demonstrated the extent of its potential applications in the industrial heating field, together with its drawbacks and, in particular, those of an economic nature. The decision to invest in radio frequency dielectric heating equipment must therefore be carefully considered, and this even more so than in other fields.

### 5.1. Selection of a radio frequency dielectric heating equipment

Radio frequency dielectric heating is suitable only for heating bodies which do not conduct electricity, on condition that their "loss factor" $\epsilon_r \tan \delta$ is sufficient, i.e., greater than 0.01 — 0.02. If these values are not available or cannot be obtained from suppliers, preliminary tests must be conducted with an equipment manufacturer or with suitably equipped research centers (textile industry research center, Électricité de France laboratories, etc.) to conduct a feasibility study. The materials to be heated must in particular be free of electrical conductors (carbon, metal inclusions, etc.) and comply with the conditions defined above for dielectric heating to be efficient.

Dielectric heating will most probably be suitable if the applications proposed belong to three major families.

### *5.1.1. Specific applications difficult to obtain with another technique*

Applications exist for which it is necessary to uniformly through heat a thick, low thermal conductivity material, or again, a thermally fragile material for which superficial overheating is unacceptable. A selective heating requirement for a composite material can also make dielectric heating a possible choice; similarly, heating of a powder product is easier than with other processes.

The total absence of product pollution can also be a determining factor [89].

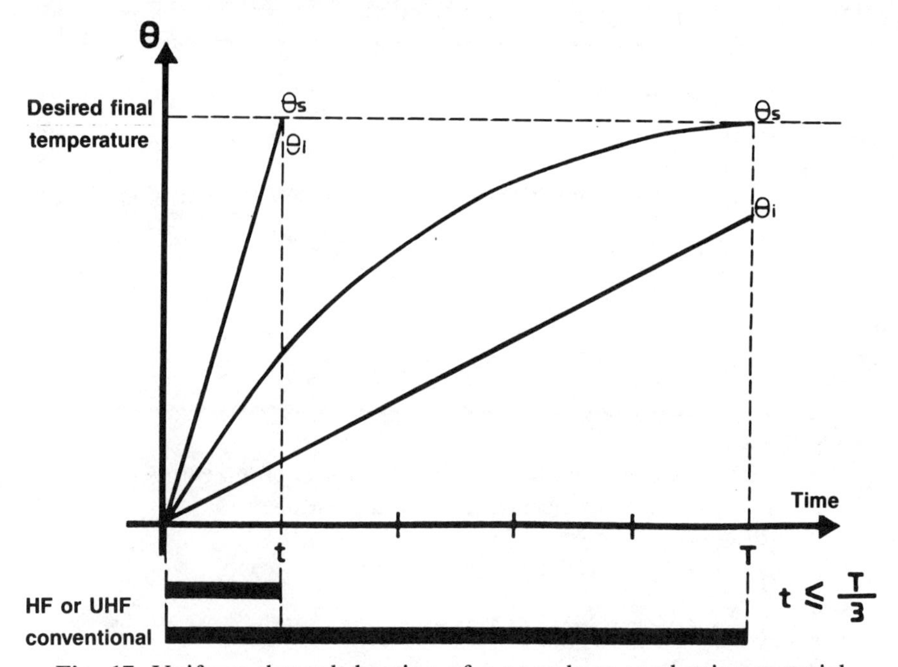


**Fig. 17.** Uniform through heating of a poor heat conducting material

### *5.1.2. Mixed applications, dielectric heating plus other processes*

The specific advantages of dielectric heating are used to improve another, more conventional process, in order to increase overall efficiency. In this manner, investment cost is much lower than with full dielectric heating, but productivity or quality is considerably improved. Generally, radio frequency power is 5 to 20% lower than the total power.

Radio frequency heating is, above all, used for rapid preheating of products, the heat then being maintained by a conventional method, sometimes hot air, or for added drying, or again to enhance migration of water towards the outside of the product at the start of or during drying or, on the contrary, for final drying which is difficult to obtain with the standard techniques (89).

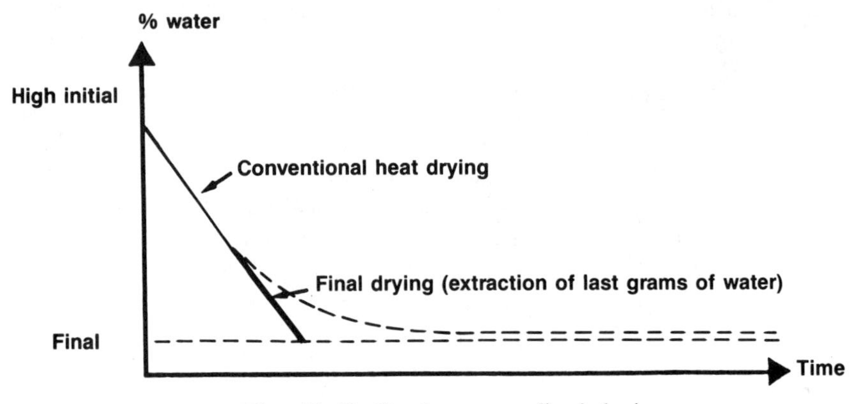


**Fig. 18.** Radio frequency final drying

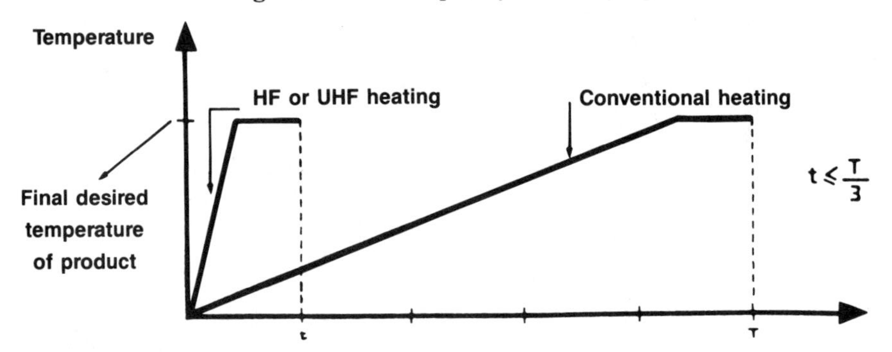


**Fig. 19.** Fast temperature buildup using dielectric heating. Holding using conventional heating.

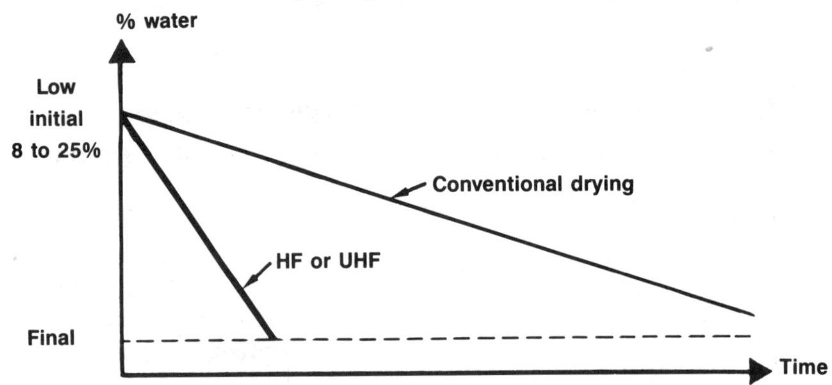


**Fig. 20.** Total drying using dielectric heating of a product with a low initial water content

### *5.1.3. Total heating where limited powers are involved*

A mixed heating system may be too costly when the added energy required for treatment is limited; for example total drying of products with a low initial water content, or heating of small quantities of high added value products. In such cases, dielectric heating can provide all the heating required.

### *5.1.4. Power to be installed and treatment time*

While the previous analyses show the applications of dielectric heating, an initial estimate of the power to be installed can be performed. The thermal variables, specific heat and latent heat, the weight of the product to be treated, the desired temperature rise and the temperature rise rate are known values. The electrical variables, loss angles and dielectric constants are available or have been estimated during preliminary tests.

Using the dissipated power expressions drawn up previously and an estimated efficiency (in general, a conservative value of 50% is used), it is possible to calculate radio frequency power and power demanded from the line, which is more or less double the former. However, in some cases, thermal losses may be relatively high (operation at high temperature of more than 150°C approximately), and a finer analysis is necessary to obtain better accuracies. However, it must not be overlooked that the material to be treated can impose power density limits. Especially with drying, it is necessary to ensure that an excessively fast temperature rise does not cause excessively fast vaporization of water, which could damage the product.

### *5.1.5. Final choice*

The calculated power provides the order of magnitude of the investments required and an appreciation of profitability.

However, the best type of applicator must be chosen (flat capacitor, garland or strayfield electrodes, etc.) and, in general where new applications are involved, it is necessary to conduct complete tests with the manufacturer or specialized research laboratories.

It is often necessary to provide a cooling system to extract water vapor or other vapors produced during drying or heating.

More detailed studies may be necessary to optimize the use of radio frequency heating and obtain the best efficiency, whatever the operating conditions (impedance matching and determination of power transmission components), while ensuring that the system dielectric strength is maintained.

## 5.2. Field of application of radio frequency dielectric heating

The analyses described above illustrate the wide range of applications of dielectric heating, either alone or combined with other techniques. Some of these applications have existed for a long time, others are being developed and others are yet to be imagined [1], [7], [8], [19], [68], [96].

The list below simply reviews the variety of existing or developing applications, while some are described in detail.

### *5.2.1. Existing drying applications*

— drying mass textile products (spools, rovings, skeins);
— final drying of paper;

— drying glass fibres and spools;
— drying water-based glues in the paper-cardboard industry;
— drying cigarettes;
— final dehydration of biscuits at outlet from baking ovens;
— wood and sawdust drying;
— drying silica;
— drying pharmaceutical products;
— honeycomb ceramic structure drying;
— rapid heating of paper before drying.

### *5.2.2. Existing heating applications*

— welding plastic materials and preheating prior to forming;
— firing foundry cores in casting;
— polymerization of fibre panels;
— glueing woods;
— printing and marking in the textile, leatherware and shoe industries;
— polymerization of latex coatings on textiles for manufacture of floor coverings;
— melting honey;
— heating rubber prior to vulcanization;
— welding glass formed sections, bonding multi-layer glass products.

### *5.2.3. Applications in the development stage*

— drying leathers and hides;
— drying powder products;
— drying various papers;
— drying water inks
— gelification of plastisols;
— drying pharmaceutical products;
— drying ceramics and refractories;
— sterilization and cooking foodstuffs.

Numerous other applications may be envisaged as a function of the specific properties of dielectric heating [20], [93 *e*].

## 5.3. Textile industry applications

### *5.3.1. Preheating and drying of mass materials*

Due to the poor thermal conductivity of fibres and the porous structure of materials, textiles are very good thermal insulators. Now, in convection heating, heating speed or drying speed of a fibrous mass depends, above all, on the rate of propagation of heat from the surface to the inside of the material.

Thermal exchanges through voluminous textile materials are therefore slow, and large quantities of hot air must be circulated over very long periods to obtain through heating. Therefore, heating and ventilation energy consumptions are high. For example, in drying, energy consumption is a minimum of 2.5 to 3 th (2.9 to 3.5 kWh thermal) per kilogram of evaporated steam in a boiler, to which

cooling energy must be added. Therefore, this situation favors the use of radio frequency dielectric heating, on condition that the geometry of the products is sufficiently regular for the electric field inside the mass to be uniformly distributed and the temperature homogeneous, since:
— the energy radiated is absorbed by the material and the penetration depth is generally sufficient at radio frequency for heating to be homogeneous;
— thermal losses are very low;
— heating and treatment times are generally reduced, a few tens of minutes against some hours to some tens of hours in a forced convection dryer;
— the heat given off in the product to be heated increases as a function of the humidity contained in the product. In drying, the material is therefore the source of a self-regulating process. The wet parts heat more quickly than the dry parts, and humidity tends towards a constant value which can be controlled by measuring the absorbed power (or if the voltage applied is constant, by measuring the anode current). This property is of interest, since over-drying forms an energy loss and causes unwanted modifications to the characteristics of fibres.

Under such conditions, dielectric heating installations are at present used for:
— heating bales of wool or cotton from producing countries. The raw wool contains impurities and grease, ensuring its cohesion at temperatures of than 20°C. Preheating between 30 and 35°C is necessary to enable the wool to be sorted and to accelerate elimination of grease and insolubles during washing. Regular heating of a bale of 1.2 to 2 $m^3$ and a weight of 150 to 400 kg calls for 2 to 3 days in a convection oven. A 30 kW radio frequency installation can be used to heat a bale of a weight of 150 kg every 4 minutes in a continuous tunnel oven, with a temperature rise of approximately 30°C. The energy consumed from the line is between 20 and 25 kWh/t of material. The operating frequency is generally 30.56 or 27.12 MHz; it is also possible to preheat bales of cotton or other raw textile materials;
— drying textile products in a mass, spools, skeins, roving, ribbons, etc.

The products treated vary widely: combed wool ribbons, polyester fibre cables, spools of viscous and glass fibres, ribbons of acrylic fibres, stopples of linen and hemp, spools of wool, cotton, polyester and mixed fibres, rayon blocks, stockings in packets, etc. Drying is often performed after dying [9], [54], [55], [61], [62], [93 *f*], [E], [V].

In most installations, the products are continuously treated in flat electrode equipped ovens. They are connected to the oscillator circuit of the radio frequency power generator, which is generally located over the treatment chamber, so as to keep the connections as short as possible, reducing electromagnetic radiation losses to a minimum.

The oven is equipped with a water vapor extractor. Fresh air introduced into the chamber can be preheated, either by recovery from the electronic generator cooling system, or any other thermal method. The distance between the electrodes and the load is adjustable.

In general, the frequencies used are 27.12 or 13.56 MHz. The radio frequency voltage applied to the electrode is between 5,000 V for low powers (5 to 10 kW) and 10,000 to 20,000 V for higher powers (at present, approximately 150 kW in the textile industry).

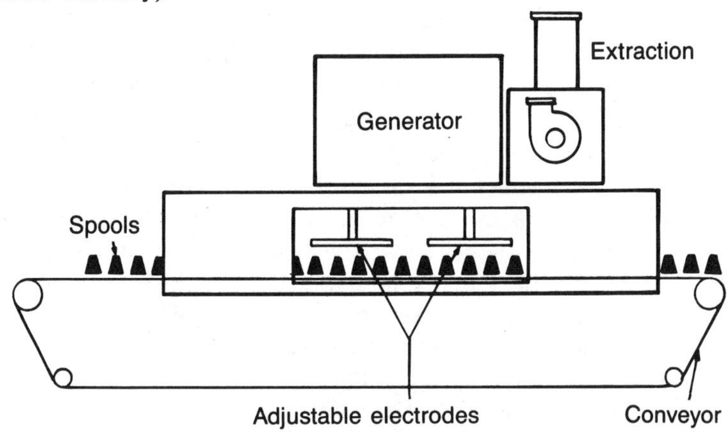


**Fig. 21.** Drying of textile spools [93 *f*].

For example, a textile spool drying, radio frequency installation of 30 kW at 13.56 MHz enables evacuation of 30 to 40 kg of water per hour, i.e. treatment of 125 kg per hour of wool (300 g per spool), dried from 45 to 18% or 300 kg per hour of acrylic textile (500 g per spool), dried from 12 to 2% of water. The specific consumption is 1.1 to 1.3 kWh/kg of water eliminated, and drying time is 20 to 30 minutes. Another 50 kW radio frequency installation (27.12 MHz, power at line 80 kW), associated with hot air of 80°C, can evacuate 65 kg water and dry spools of wool from 45 to 18% and spools of polyester from 46 to 8%. Drying time is between 15 and 45 minutes, versus 4 to 12 hours with the former conventional installation. Specific consumption is of the order of 1 to 1.3 kWh/kg of evacuated water. The main advantages noted by users of this type of spool dryer are:

— a reduction in overall drying costs of 40 to 60%;

— a considerable decrease in treatment time (some tens of minutes, against some hours);

— reduction of intermediate stocks;

— a product humidity factor dispersion of less than 0.5%;

— possibility of treating products of different shapes;

— high overall instantaneous energy efficiency;

— very low thermal inertia, 3 minutes approximately for startup;

— increase in product quality;

— better operational flexibility, which is highly appreciated in particular in contracting drying enterprises;

— low maintenance costs.

The economic analyses which have been performed on this type of equipment show that the investment return time is generally between 2 and 3 years.
At present, dielectric heating is used in industry for mass products other than spools, for example products in skeins or packets.

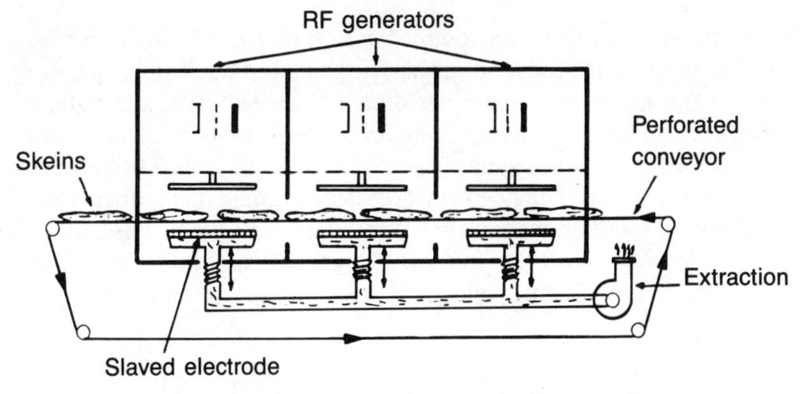


**Fig. 22.** Radio frequency dielectric drying of skeins for products in packets [93 *f*].

Equipment of this type is similar in general design to spool dryers. For example, a radio frequency (13.56 MHz, power absorbed from line 120 kW), radio frequency 85 kW nylon stocking dryer can treat 1,800 dozen pairs of stockings per hour in packets. These stockings, of which there are 150 different models, are first of all mechanically dried after dying to reduce their water content to 15-18%; then humidity is lowered by 3 to 4% in the radio frequency dryer. Selective heating, provided by the radio frequency dryer, provides the regular humidity profile which is indispensable for product quality, and prevents local burns, since overheating is impossible. Stockings are not twisted as in other conventional dryers and the operation required to eliminate twist is therefore dispensed with. The machine is very compact: 15 meters overall in length, and easy to handle. Investment return time was two years.

While mass product drying is well developed in the textile industry, much research has been conducted to improve the techniques mentioned above and to extend the range of applications and increase profitability.

As a result, new drying techniques for spool products are being developed. In the installations described above, the water vapor escapes freely from the wet materials placed on a conveyor, and is then evacuated by a low velocity heated-air draft at relatively low temperature, by recuperation from the electron tube cooling system or another thermal method. This system is not appropriate for relatively low density spools, through which steam can be diffused freely. In fact, in very compact spools, water vapor given off inside the spool cannot be eliminated through the thickness of the material as rapidly as it is produced; internal stresses which then occur can cause the spools to explode (drying of such spools using conventional convection methods is also extremely long).

A solution to this problem was proposed by the Textile Industry Research Center at Mulhouse. The water vapor is sucked through the spools as and when it forms, and can be vented freely at the surface. This concept is now being used industrially. The radio frequency drying phase is preceded by a preliminary mechanical elimination of water by centrifugal methods or suction. Under these conditions, the total drying time has been considerably reduced, 7 to 8 minutes to reduce the water content of a spool of cotton of 700 g from 80 to 8%, 2 to 4 minutes to lower that of a spool of acrylic thread of equivalent weight from 50 to 10%. The material temperature does not exceed 70°C throughout the operation, which improves quality (better whiteness, lower breakage factor on looms). However, to obtain a regular final humidity profile, it is possible to reverse the suction direction alternately. Further research is going on in other countries, in order to reduce spool drying time to around 10 minutes. In all cases, the overall efficiency will be around 60%, and much higher if the heat given off by the electron tubes and the latent heat of the water vapor are recovered.

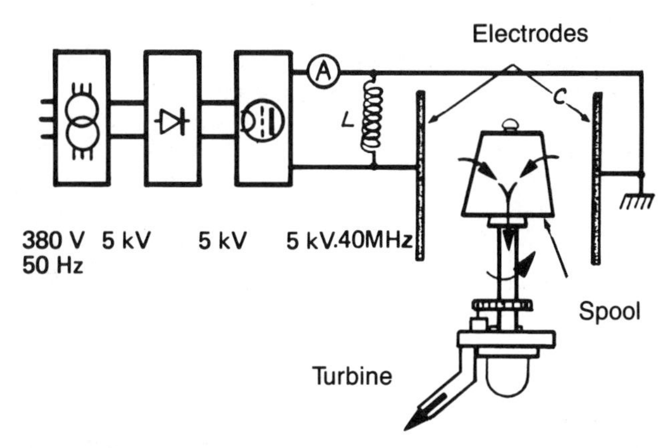


**Fig. 23.** Drying textile spools with water vapor. Draw-off through the spool [93 *f*].

Other research concerns drying products in the form of skeins.

While radio frequency heating is more appreciated for mass treatment of textiles, applications also apply to sheet products.

### *5.3.2. Drying and treatment of sheet textiles*

By using applicators consisting of divided field strayfield or garland electrodes, which are better adapted to the geometry of flat and thin products than plate electrodes, radio frequency dielectric drying can be used for drying and enabling operations requiring fast homogeneous action. The electric field created by this type of electrode is in more or less the same direction as the fabric or sheet of fibres to be treated [9], [54], [55], [61], [62], [84], [86].

Present applications are highly diversified: thermosetting rayon, heating latex used with glass fibre to produce tire carcasses, heating latex on the backs of carpets, production complex materials by back coating, thermal setting dyes and pigments, polymerization of coatings, drying products intended for the production of babies' bedding, binding non-woven polyester sheets with elastomer, thermoprinting by transfer of manufactured articles, thermosetting dyes on bulk fibres, dying by impression and setting of the dye on clothing and carpeting [97], [98], [E], [S], [V].

For example, thermosetting of bulk fibres combines radio frequency dielectric heating and moderate pressure. This process applies to most fibres, acrylics, polyamids, wool, nylon, etc., but involves problems for dying wool using chrome dyes, or dying polyesters, with plasto-soluble dyes, since, in the latter case, it is necessary to reach a high temperature, about 230°C. After dying, the fibre layers are slightly compressed by means of two conveyor belts, entering the setting chamber. The setting chamber consists of a glass enclosure between two pairs of radio frequency electrodes (these electrodes are flat plates, since the thickness of the fibre matting is such that the type of heating is intermediate between mass heating and thin sheet heating). The radio frequency electric fields has a double effect :

— a thermal effect, providing steam inside the fibre itself, resulting in swelling;
— activation of the dye molecules, by increasing the setting speed.

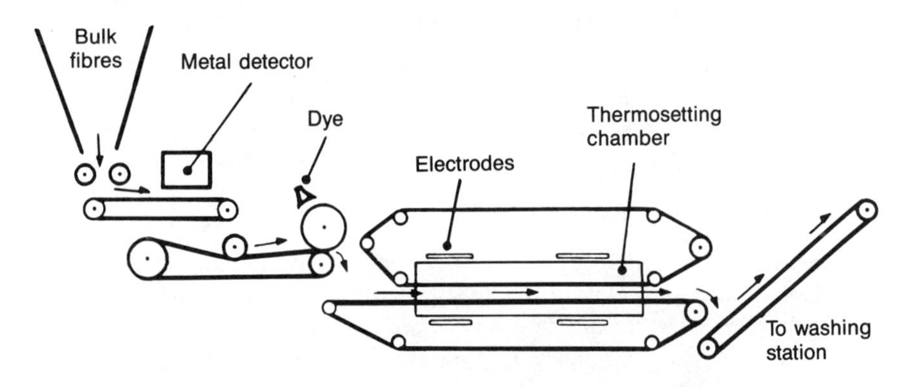


**Fig. 24.** Radio frequency thermosetting of dyes on bulk fibres [98], [X].

An installation for 200 kg per hour of dyed fibre consists of two radio frequency 16 kW generators (27.12 MHz, total power absorbed from line 50 kW). Temperature rise (to 100°C approximately) and the treatment process are extremely rapid, since the time spent by the product in the setting chamber is about 10 seconds. The installation is very compact: 4 meters for the setting chamber. After continuous mode washing, softening, rinsing and drying (the dryer used is one half as small as with the conventional process, and the air heating power is only 14 kW), the fibres pass through an opening machine and a double card, the productivity of which is increased by 5%. This increase is attributed to the fact that the radio frequency provides a more open material than discontinuous, autoclave dying processes. The energy saving is 30 to 45% [98], [X] with respect to a discontinuous conventional dying installation, even if equipped with a heat transfer.

In most other applications concerning sheet products, strayfield or garland electrodes are used. The type of each electrode configuration depends on the number of parameters, material thickness, type of fibre and humidity factor. Tests must often be performed to find the best adapted solution.

In spite of high investment cost, radio frequency dielectric heating techniques are developing in the textile industry for applications in which their special characterictics offer major advantages. They are often used in conjunction with other heating methods. Research at present being conducted in this field indicates that this development will continue to accelerate [E], [V].

## 5.4. Cardboard-paper industry applications

In paper production, drying is a fundamental operation. On a paper machine, the water is first of all eliminated by pressure, then by drying on steam-heated dryer rollers. The thermal energy required for drying is high and cannot, except in exceptional cases, be completely provided by radio frequency dielectric heating. Conversely, this technique may provide additional heating at carefully chosen locations on a paper machine [10], [11], [18], [30], [42 *b*], [54], [61], [73], [86].

### *5.4.1. Initial fast heating of paper machine*

Radio frequency dielectric heating is used for fast heating at beginning of drying. This installation then replaces the first dryer cylinders of the paper machine. The advantage of an installation of this type is the increase in productivity provided, but to date, this has not been widely used [T].

### *5.4.2. Humidity profile correction*

Dielectric heating humidity profile correction has been more widely developed. At end of drying, paper often has very important transverse humidity factor variations. Now, irregular distribution of humidity in the paper reduces quality:
— for glazed papers, variations in humidity result in surface irregularities;
— for coated papers, variations in the acceptable humidity cause irregular deposit of the glaze, unwanted migration of binders in the glaze, and therefore a reduction in surface quality, which causes problems during printing;
— for base paper, humidity irregularities cause variable shrinkage, entailing difficulties during rolling or storing;
— for printing papers, variations in humidity affect color absorption and the speed of printing presses.

In most installations, paper is over dryed to ensure that humidity peaks do not exceed a commercial tolerance. Sometimes the paper is rehumidified after super-drying. This system considerably increases production costs, and in particular energy costs and those related to overdimensioning of equipment (extra floor area, maintenance, etc.) and ever more so since the efficiency of the dryer cylinders at the end of the drying machine is low.

Conversely, and due to its selective effect, radio frequency dielectric heating can correct the humidity profile of the paper at end of drying, and can then be substituted for one or more steam heated or electrical (resistance heated) finishing rollers. Water, which has a "loss factor" approximately twenty times higher than that of dry paper, in fact absorbs much more energy than the dry parts, overheats rapidly and evaporates at the surface of the paper. Strayfield or garland electrodes are used in such cases.

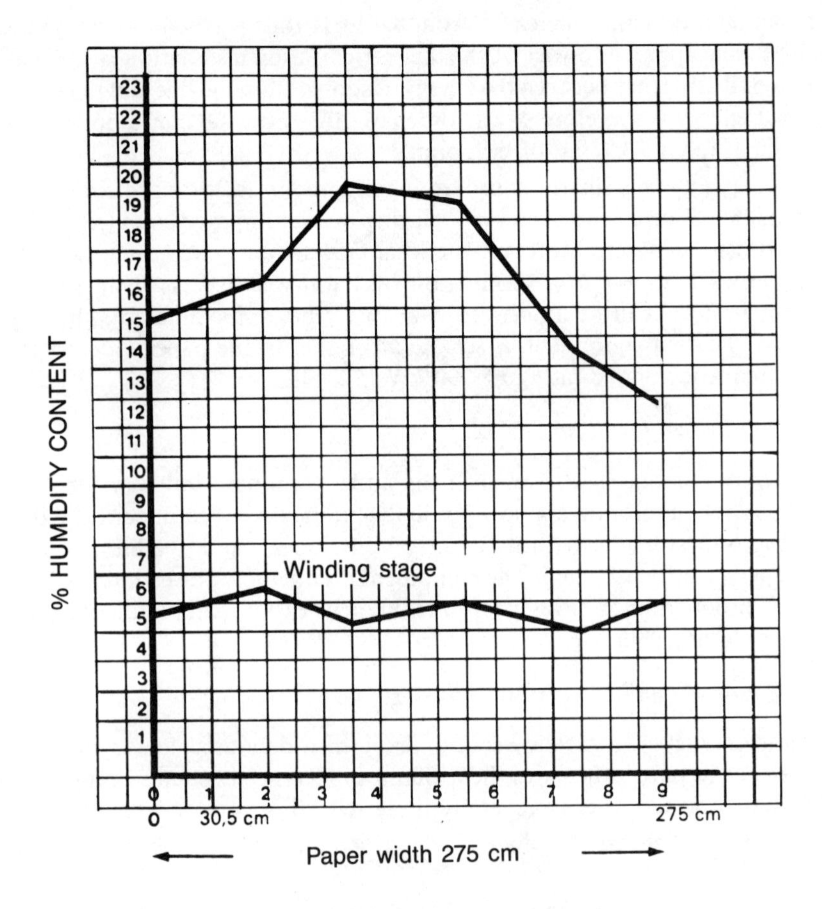


**Fig. 25.** Paper humidity profile correction [T]:
— top curve: humidity profile before dielectric drying;
— bottom curve: humidity profile after dielectric drying.

Air in contact with the paper surface must be heated to prevent recondensation of the water vapor on the surface. Preheating of this air can be provided by recovery from the electron tube cooling system.

Dielectric drying suppresses almost all humidity irregularities, since the humidity factor dispersion is of the order of ±0.5%, therefore considerably improving product quality. Humidity is not only regular across the paper, but also through it. Energy saving is noticeable, since over-drying is avoided, and the water to be eliminated is preferentially heated. Also, the productivity of an existing paper machine can be increased by adding this type of dryer, and an increase in production of 20 to 25% can be obtained, especially for heavy papers.

For example, a 600 kW radio frequency dielectric dryer, installed on a machine producing printing-writing quality paper with weights of 60 to 120 g/m$^2$ enables paper belts of 3.85 m wide to be treated at a rate of 500 m/min. approximately.

Machine production increased from 8.4 to 11 tons per hour, i.e. an increase of 30%. The apparent power of the dielectric heating installation is 1,000 kVA. The overall efficiency between the energy absorbed from the line and the effective thermal energy is therefore of the order of 70%. Specific consumption varies between 1 and 1.3 kW/kg of evaporated water [S].

Paper quality was clearly improved, and humidity factor dispersion was less than ±0.5%. Rejects are decreased, glazing quality improved and treatment by paper users facilitated. Moreover the total primary energy consumption of the machine with respect to a production of 1 ton was decreased by 10%. The investment costs can be amortized over two years approximately. The largest dielectric heating equipment in service at present in the paper industry has a radio frequency power rating of 900 kW [S].

### *5.4.3. Other applications*

Dielectric heating is also used in the paper-cardboard industry for various drying or polymerization applications in the production of cardboard and drying aqueous adhesive glues (production of envelopes, books, adhesive papers, wallpapers, cardboard, etc.). The equipment is either low power, a few kilowatts (for the production of envelopes), or some tens of kilowatts (adhesive papers, cardboards) [5], [30].

## 5.5. Applications in the wood industry.

Radio frequency dielectric heating has been widely used for drying woods, but, except in particular cases, heat pump or electric resistance drying, which offers much lower costs is generally preferred [5].

Conversely, dielectric heating is widely used in wood glueing, for two major applications :

— high production per run of wood products (furniture, windows, frameworks in wood cored plywood, etc.) by assembling the basic components;

— production of particle or fibre panel.

In both cases, the objective is to obtain accelerated hardening, and in some cases almost instantaneous hardening of the glue by polymerization. The operation is performed with moderate heating, and generally accompanied by elimination of water. The glues must be adapted to this type of heating, i.e. contain water which heats rapidly and selectively with respect to the wood. In the furniture industry, urea-formol or vinyl-type glues are used.

Relative dielectric constant and loss factor of woods and water glues

| | $\epsilon_r$ | $\tan \delta$ | $\epsilon_r \tan \delta$ |
|---|---|---|---|
| Glue prior to polymerization | 20 to 40 | 0.5 to 0.8 | 10 to 32 |
| Glue after polymerization | 3 to 6 | 0.04 to 0.1 | 0.12 to 0.6 |
| Dry wood (8 to 10% humidity) | 3 to 4 | 0.04 to 0.01 | 0.12 to 0.4 |

**Fig. 26.**

To facilitate heating of the glue, the electrodes must be correctly located. In fact, depending on the electrode position, the assembly consisting of the wood elements to be assembled-glued joint can be considered, from the dielectric heating viewpoint, as a series or parallel mounting, and power dissipation in both materials takes place differently (*see* paragraph 2.2).

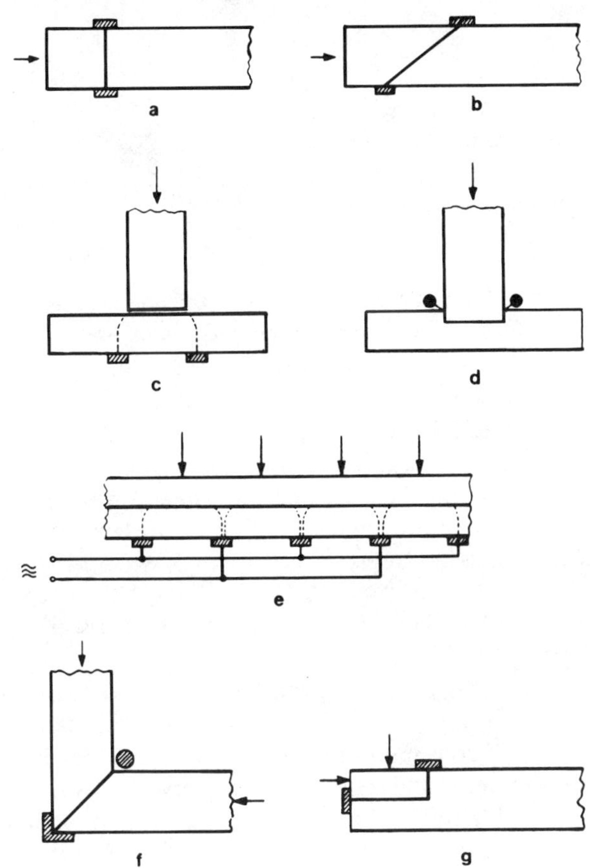


**Fig. 27.** Examples of correct electrode configurations for gluing [B]:
*a)* perpendicular joint;
*b)* beveled joint;
*c)* fillet glueing of wide joint;
*d)* glueing of embedded wide joint;
*e)* glueing of joint in stray field when the electrode cannot be applied to the joint itself;
*f)* mitred assembly of a butt glued frame;
*g)* corner assembly of a morticed frame.

Therefore, and whenever possible, the setup must be such that the power is dissipated essentially in the glue (parallel connection). However, forming of plywood demands a series configuration (dielectric heating is pertinent for this configuration as soon as the thickness reaches 8 mm [31].

When the wooden sections and glued joints are in series within the load, the electrodes are often series-parallel connected to reduce heating power and to offer better electrical insulation.

In most cases, heating is made under pressure, and the electrodes are often the press plates. The pressure applied is of the order of 3 kg/cm$^2$ for resinous woods and 10 to 15 kg/cm$^2$ for hard woods.

For given radio frequency generator power, drying time depends on several factors. The glue used, the area of the joint to be glued and the humidity factor of the wood are all major factors. The type of wood and the pressure applied are also important. For example, glueing time is much higher with high residual humidity woods than with a correctly dried wood, since the water contained in the wood absorbs part of the radio frequency energy. Water diffusion phenomena during drying and polymerization also require minimum treatment times.

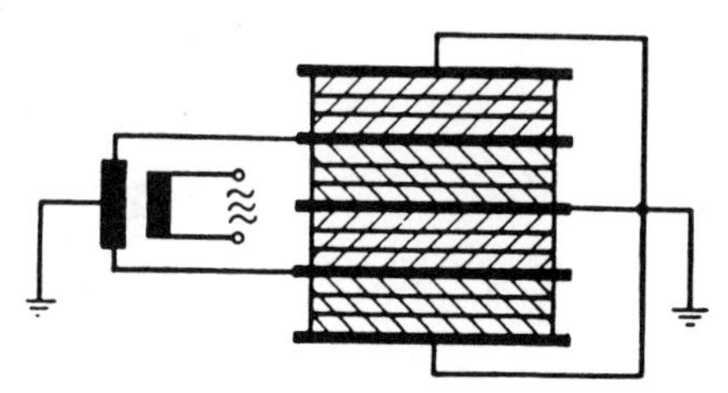

**Fig. 28.** Series-parallel connection of electrodes for heating wooden assemblies.

For a 4 kW radio frequency power generator, and a joint surface of the order of 1,000 cm², approximate glueing times are :
— resinous-to-resinous woods : 1 minute;
— hard-to-resinous woods: 1.5 minutes;
— hard-to-hard woods: 2 minutes.

Glueing time can however, be reduced by using a higher power generator, up to 10 kW radio frequency approximately, in the case described above. Another method of estimating glueing time is to relate power density to temperature rise times. Thus, for a power density of 5 W/cm³, the temperature rise time in the glue is of the order of 225°C per minute, and only 45°C per minute for a power density of 1 W/cm³. Tests are sometimes necessary to define production drying times.

These gluing times are much shorter than those obtained with other heating methodes, illustrating the advantage of dielectric heating.

Energy consumptions vary widely and depend on the quantity of water to be evaporated and also on the humidity of the wood. For instance, in producing plywood, almost 60 kWh/m³ are required to eliminate 3% of the water, with a temperature of 100°C at the glue. To eliminate 10% water, consumption is only doubled. A 100 kW radio frequency power generator can be used to glue approximately 3 m³ per hour with elimination of 3% of water.

In production of particles of chipboard panels, the need for dielectric heating is proportional to panel thickness.

With a single-stage press, daily production rates of :
— 100 m³ of panel, with a radio frequency power of 100 kW;
— 200 m³ of panel with a radio frequency power of 300 kW are obtained.

With a two-stage press, this production increases to :
— 300 m³ of panel for a radio frequency power of 300 kW.

A 900 kW radio frequency generator (1,450 kW demanded from line) produces fibre panels of 2.22 cm thick, with a wood fibre weight at the press output of 15.7 kg/m² on a six-level press, with a cycle of 14 min. 15 sec., including 2 min. 30 sec. of radio frequency heating [3], [4], [5], [45].

The generators used in wood gluing cover a wide power range, from 1 to 1,400 kW radio frequency approximately. Applications are various: gluing of curved facing for seats and backrests, windowframes, beams and frames, production of ground covering panels from wooden planking, assembly of radio or TV set cases and small cabinets by glueing, furniture production, etc. [31], [57].

## 5.6. Food industry applications

### *5.6.1. Ranges of application*

Dielectric heating has found several applications in the food industries: heating coffee or cocoa beans to facilitate roasting, drying sugar compressed in moulds, heating blocks of chocolate, destruction of parasites in flours, semolina, etc. and also in an associated industry, drying tobacco leaves [14], [16], [21].

However, these applications are limited, and final cooking of biscuits is the most important development.

### *5.6.2. Final cooking of biscuits*

Industrially produced biscuits are generally cooked in continuous conveyor ovens, heated by resistances or fossil fuels, techniques which call for lower investment costs than a radio frequency dielectric oven.

The core temperature at beginning of cooking is reached rapidly, and therefore dielectric heating is of little interest for this phase. Conversely, during the last stage of cooking, the product is surrounded by a more or less browned crust forming a thermal insulator. To reduce the water content in the biscuit to below 5%, two contradictory requirements must be satisfied; non-excessive browning of the outside surface of the biscuit and the most thorough, fastest internal drying possible. Now, increasing the temperature of the conventional oven to accelerate elimination of water may cause excessive browning, or even carbonization of the biscuits. Therefore, long ovens must be used to prevent excessive temperatures reducing product quality, and an increase in cooking time must be accepted. Because of this, conventional ovens can be 60 to 90 m long.

Dielectric heating, which enable the water content inside the product to be reduced independently of its surface browning, is obviously of interest; cooking time and oven length are both decreased. Moreover, since the power dissipated inside the product is proportional to the water content, the humidity profile tends to equalize across the conveyor, and the imperfections encountered in a conventional oven, for which fine adjustment is always a delicate matter, are therefore corrected; this reduction in finished product humidity and low dispersion enable better conservation [2], [21], [41], [A], [N], [0].

For example, in a biscuit production line in which final drying calls for elimination of 80 kg/hr, of water (reduction of humidity factor by 3% approximately), a radio frequency dielectric heating system 5 m long replaces 30 m of conventional oven. The radio frequency power is 80 kW, and the frequency is 27.12 MHz. Specific consumption is 1.2 to 1.4 kWh/kg of evaporated water, and primary energy saving is 30% approximately with respect to a conventional fossil fuel oven. Although equipment costs are twice as high with dielectric heating as compared to fossil fuel heating, the overall investment, allowing for site costs and buildings constructed, is practically the same in both cases, while the operating costs and qualitative advantages clearly favor dielectric heating.

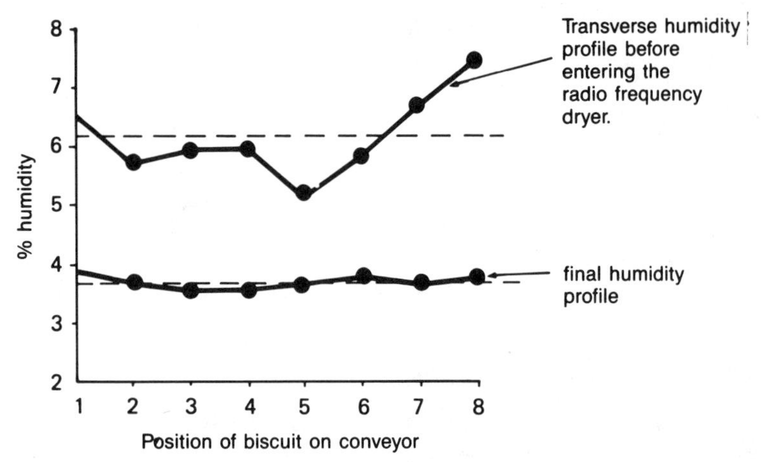


**Fig. 29.** Final cooking of biscuits and equalization of residual humidity [A].

## 5.7. Plastic forming industry applications

Doubtlessly, it is in the plastic forming industries that radio frequency dielectric heating is most widely used. In fact, it has not only enabled economic heating operations, but also enabled creation of new products in the leatherwork, toy, packaging, industrial parts, shoe-making, sports articles, household articles and furnishing industries, etc. [27], [50], [63], [K], [L], [M], [N], [P], [R].

**Fig. 30.** Diagram of a dielectric heating plastic welding machine: M, materials to be welded; E, top mobile electrode; T, fixed bottom electrode.

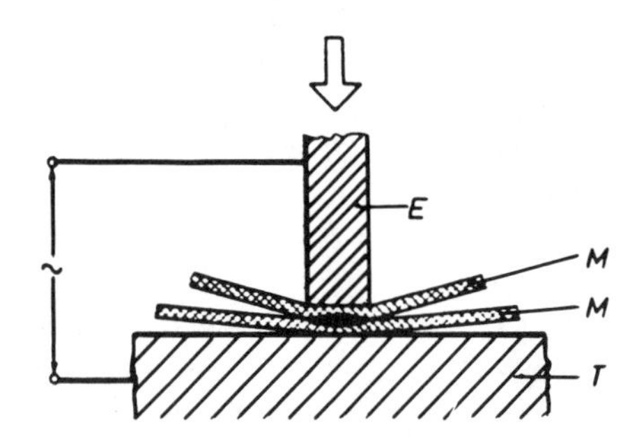


Two applications have particularly benefited from this growth: preheating heat-hardenable casting materials and welding. Some plastics cannot hovever, be heated by radio frequency dielectric hysteresis, since the "loss factor", $\epsilon_r \tan \delta$ is too low, less than 0.001 (polystyrene, polyethylene, polytetrafluorethylene, etc.).

### *5.7.1. Preheating thermosetting plastics*

Dielectric heating enables rapid through heating of thermosetting powders and pellets. Therefore, it is possible to place material in moulds, for which the uniform temperature throughout the mass is close to that of polymerization, and to press them to obtain the objects desired.

Water vapor and reaction products escape freely, and blisters are not formed. The mechanical power of the machines is reduced, since the pressure is transmitted uniformly. Moulding time is also strongly reduced, since the material is totally malleable and easy to handle, since the temperature remains relatively low. Also, mould wear is reduced and product quality improved.

These processes are widely used in the plastics industry for the production of various articles. Where production rates require, it is possible to construct continuous installations.

### *5.7.2. Welding of thermoplastics*

Some plastics such as PVC (polyvinyl chloride) have high loss factors in the radio frequency range. Therefore, they can be welded easily using radio frequency dielectric heating. Welding involves a heating phase, followed by a cooling phase both being performed under pressure [27].

A dielectric heated welding machine consists of three main units: the welding press, the electrodes and the radio frequency generator.

The welding press features all the items required for the operations: the device at the to required pressure, the frame which can withstand this pressure and the control devices. Generally, the machines are discontinuous (cooling time under pressure required), but special equipment enables continuous welding.

The materials to be assembled are placed on a slab or panel which behaves as a bottom electrode. The top electrode, of appropriate shape, not only enables pressure welding, but also embossing and cutting of the object to be produced. The electrodes, which are high precision machined, are generally of copper alloy or very hard aluminum. Unlike to other welding techniques in which the press plates are heated, the electrodes remain practically cold, since the energy is dissipated directly in the plastic material. Pressure cooling can therefore be very fast, since the heat is evacuated by conduction through the cold electrodes machined from high thermal conductivity metal alloys.

For low and medium powers, up to approximately 5 kW, the radio frequency generator is built into the welding machine. In this manner, connection lengths are reduced to a minimum. For higher powers, the generator is placed as near as possible to the welding machine. A generator can then supply one or more work stations.

The specific powers used vary from 50 to 200 W/cm$^2$ (i.e. 500 to 2,000 kW/m$^2$) of welded surface, or again, from 10 to 20 W/cm$^2$ of plastic material facing the electrodes. The temperature rise rate is very high, of the order of 3,000 to 5,000°C per second. For example, with a 1 kW radio frequency generator, it is possible to weld a joint 60 cm long and 3 mm wide, between two sheets of 0.3 mm thick PVC in one second.

With a new application, tests are generally conducted to determine the power density to be used and the heating and cooling times accurately. The power required and welding time depend on the surface to be welded, the thickness of the sheet and the width of the joint, and also on the nature of the material. For equal welding surface area and joint width, energy consumption per unit volume of material to be welded increases when the thickness decreases, due to the thermal conduction cooling effect of the electrodes.

Radio frequency dielectric heating welding applications for plastic materials are extremely varied: fabrication of clothing, inflatable objects, packages, boxes, leatherware objects, shoe-making, automobile upholstery, furniture, etc. [45], [K], [L], [M], [N], [P], [R].

## 6. CHARACTERISTICS OF MICROWAVE HEATING

In principle, microwave heating (or ultra high frequency heating) is identical to radio frequency dielectric hysteresis heating, and is characterized by the absorption of electromagnetic radiation by the body to be heated; therefore, this consists of submitting an electrically insulating body to an ultra high frequency, variable electrical field, so as to heat this body by dielectric hysteresis [12], [15], [37], [52], [53].

Since the frequency is higher — 300 to 30,000 MHz — than for radio frequency, heating is even faster. While with radio frequency, the heating effect can, in most cases, be attributed at least as much to ionic conduction as to bipolar rotation; in microwave heating, the heating effect is essentially due to dipolar rotation.

To make the use of microwaves possible in heating applications without disturbing telecommunications, the following frequency bands have been reserved for this purpose by the International Telecommunications Unions:

— 915±25 MHz (wavelength $\lambda \simeq 33$ cm);
— 2,450±50 MHz (wavelength $\lambda \simeq 12$ cm);
— 5,800±75 MHz (wavelength $\lambda \simeq 5$ cm);
— 22,125±125 MHz (wavelength $\lambda \simeq 1.35$ cm).

In France, as in most European countries, only the last three bands are authorized (the 915 MHz frequency requires special authorization).

Due to the high cost and limited power of equipment at frequencies 5,800 MHz and 22,125 MHz, only the 2,450 MHz frequency is used.

However, frequencies of 915 and 2,450 MHz are used in the United States. In Great Britain, frequencies of 896 MHz, and sometimes 915 MHz are used, in addition to the frequency of 2,450 MHz [12], [15], [48], [56].

### 6.1. Power dissipated in the load

The expression for the power dissipated in the load is similar to that drawn up for radio frequency heating. The specific power density is equal to:

$$\frac{dP_w}{dm} = 2\pi \varepsilon_0 f \varepsilon_r \operatorname{tg} \delta \, V^2,$$

$P_w$, dissipated power;
$m$, load volume;
$f$, frequency;

$\epsilon_0$, dielectric constant of vacuum;
$\epsilon_r$, relative permittivity;
tan $\delta$, loss angle tangent;
$V$, electric field.

or

$$\frac{dP_w}{dm} = 0.556 \times 10^{-12} f \, \varepsilon_r \, \text{tg} \, \delta \, V^2,$$

$dP_w/dm$ in Watts per cubic centimeter;
$V$ in Volts per centimeter;
$f$ in Hertz.

For equal voltage, the power density may be much higher than with radio frequency dielectric heating.

### 6.2. Penetration depth

As with radio frequency heating, the penetration depth is equal to:

$$d = \frac{\lambda_0}{4\pi}\left[\frac{\varepsilon_r(\sqrt{1+\text{tg}^2\,\delta}-1)}{2}\right]^{-1/2}.$$

When tan $\delta$ is low, which is generally the case, this expression becomes:

$$d = \frac{\lambda_0}{2\pi\sqrt{\varepsilon_r}\,\text{tg}\,\delta}, \qquad \text{or } d = \frac{4.77 \times 10^7}{f\sqrt{\varepsilon_r}\,\text{tg}\,\delta} \text{ in meters.}$$

The penetration depth is much lower than with radio frequency dielectric heating where it is not necessary to take this into account for most applications where heating can be considered as homogeneous.

The penetration depth is of the order of 5 to 10 cm for bodies having high water content, but may be much higher, several tens of cm, for other materials; generally, this is sufficient, even for through heating of relatively solid products, especially when it is thought that the electromagnetic waves completely surround the load penetrating it from all sides. Temperature homogenization is also obtained by conduction of heat within the load. In some cases, however, the power density must be limited, or special devices must be adapted, for example surface cooling to prevent overheating of the surface layer of the body to be heated [89].

### 6.3. Power density

With microwave heating, power density is often very high, 5 to 50 W/cm³. The values used, however, depend on numerous factors: dielectric characteristics of materials, the maximum temperature rise rate authorized for the product, temperature homogeneity between the core and the surface, etc.

Moreover, the heat is given off within the product itself, avoiding heat transfer times due to conduction, which, with poor conductors of heat, can be very long.

Therefore, treatment times can often be decreased in considerable proportions and, in numerous cases, divided by 10 to 100 or even more. For example, in mass thawing of foodstuffs, the time required to take the body from a temperature of -20 or -30°C to -4°C can be divided by 200. This is, of course, an exceptional case, but fully illustrates the possibilities of microwave heating.

### 6.4. Efficiency and specific consumptions

Efficiency on generation of microwave energy is between 55 and 65%, while overall efficiency varies from 50 to 60%. Generator powers are often given in microwave kilowatts, and therefore conversion efficiency must be taken into account when calculating the power absorbed from the line.

Microwave applications are extremely varied, and it is difficult to give general specific consumption figures.

## 7. MICROWAVE HEATING EQUIPMENT

Basically, a microwave heating installation consists of a microwave generator, a waveguide, the applicator itself and ancillary monitoring, handling and safety devices.

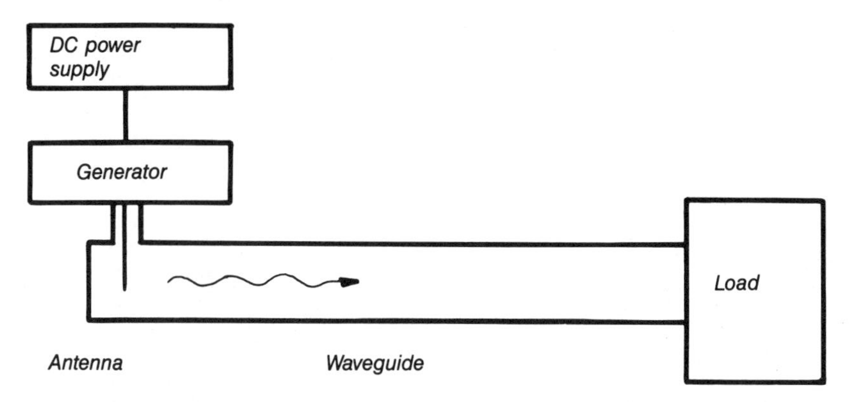


**Fig. 31.** Block diagram of a microwave heating installation [93 *a*].

Since the human body absorbs microwaves, it is critical that radiation leakage be kept at a very low level by the use of appropriate devices (metal shields, wave traps, door and other access switches, etc.). The permissible levels of radiation and the protection systems with which installations are to be equipped are defined in regulations [91], [93 *i*], [C], [D], [I], [J], [K], [U], [W].

### 7.1. Applicators

There are numerous types of microwave applicators:

— tunnel;

— slotted waveguide;

— radiating antenna waveguides;

— rectangular resonant and multimode cavities;
— circular resonant cavity;
— resonant ring;
— slow wave device;
— horn antenna.

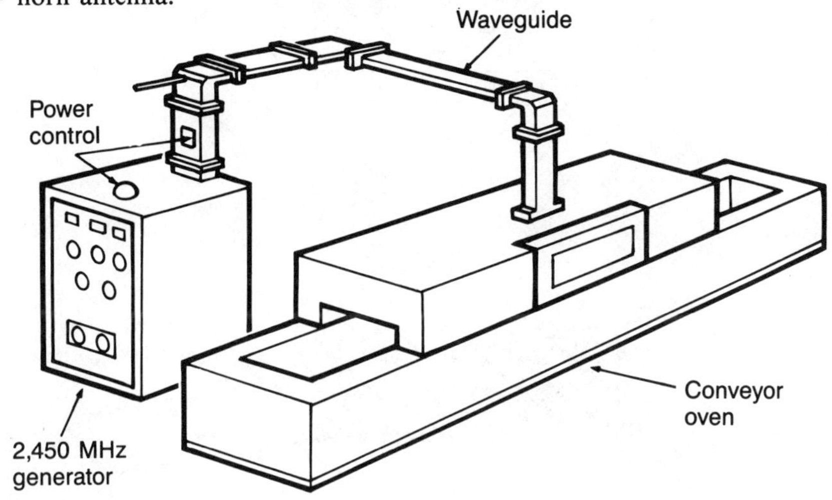


**Fig. 32.** Microwave heating installation [98].

These applicators are used as a function of the type of fabrication process, continuous or non-continuous and the nature and shape of the products [12], [13], [26], [40], [70].

### *7.1.1. Tunnel applicators*

This type of applicator consists of a tunnel which is supplied by the microwave generators, and in which the products move continually on an electrically insulated conveyor belt. This tunnel is fitted with an entry and exit hatch, which blocks electromagnetic wave leakage.

Tunnel applicators are, for example, used for thawing large masses or cooking foodstuffs. Similar devices are used to vulcanize rubber.

### *7.1.2. Folded, slotted waveguides*

Narrow slots are made in the center plane of a waveguide. The electric field is maximum in this plane, and a strip of dielectric material (fabric, paper, etc.) passing through the slots therefore absorbs a maximum of energy.

Generally, a single passage is not sufficient to apply sufficient energy to the material. Moreover, the electric field is partly attenuated over the width of the product to be treated, causing non-uniform heating. A folded, slotted waveguide is used to resolve these problems.

A water load must be placed at the end of the waveguide to absorb wave residual energy.

This applicator is suitable for continuous treatment of flat and thin products.

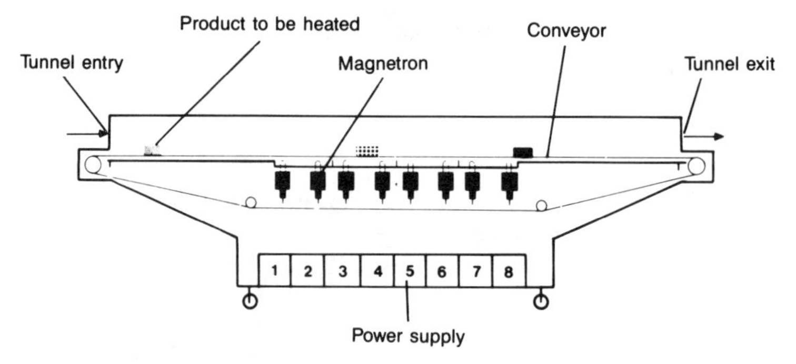


**Fig. 33.** Tunnel applicator [92].

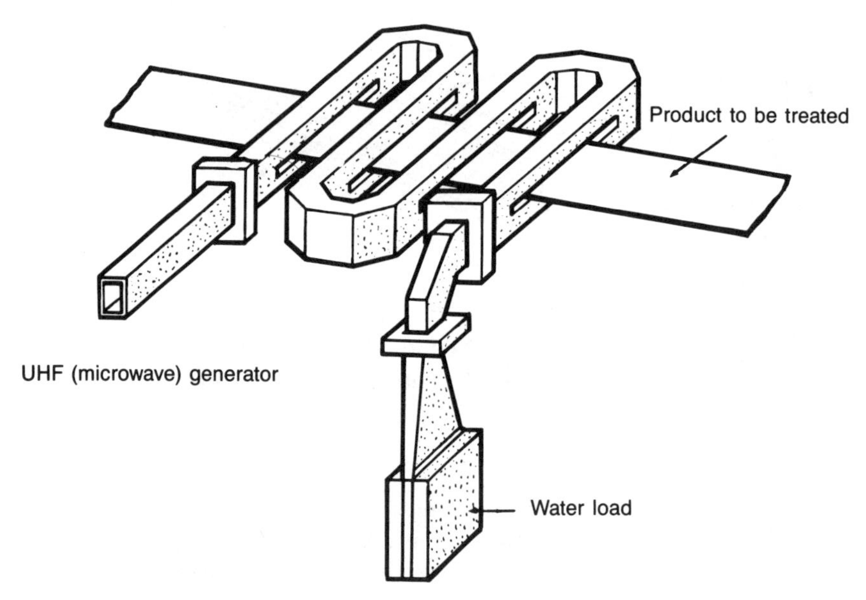


**Fig. 34.** Folded, slotted waveguide applicator [93 *d*].

### *7.1.3. Radiating slot waveguides*

A waveguide, through which an electromagnetic wave passes, has induced currents in its walls. If a slot is made in the wall of the waveguide which cuts the current lines, both edges offer a potential difference, and the slot behaves as an electrical dipole radiating into space. In order to radiate, the slots are placed on the longest dimension of the guide, and are either oblique or parallel to the waveguide, and also offset with respect to the center line (a slot in the center line does not radiate), or on the short dimension, either oblique or horizontal (a vertical slot does not radiate).

The product to be treated passes over the radiating slots. To improve efficiency, each slot can be equipped with a small correcting rod used to match slot impedance to that of space.

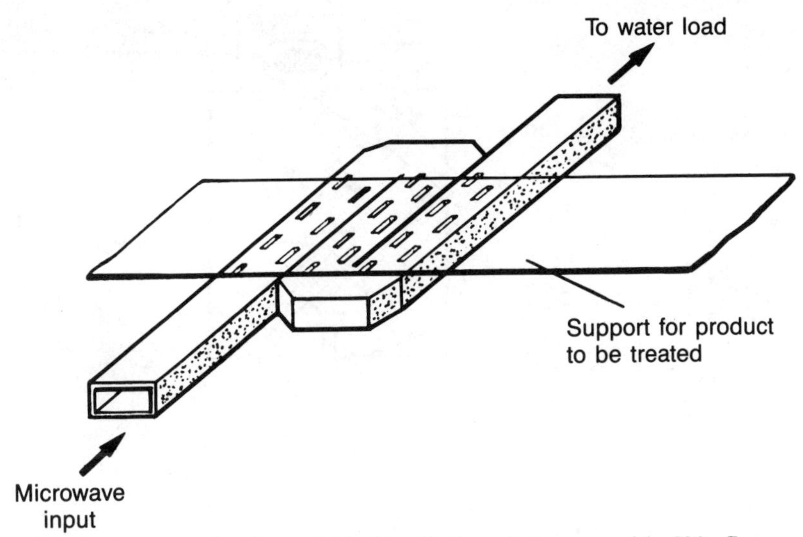


**Fig. 35.** Single or folded radiating slot waveguide [93 *d*].

As for the previous type of waveguide, the wave residual energy is absorbed by a water load.

This arrangement is used for continuous treatment of flat and thin products.

### *7.1.4. Rectangular and multi-mode resonant cavities.*

The waveguide opens into a resonant cavity in which the product to be placed is heated. This arrangement has the advantage that it does not depend on the geometric shapes of the products.

The waves reflect from the metal walls of the cavity and are absorbed by the product. Cavity dimensions are such that standing waves are generally formed. The condition for formation of standing waves in a rectangular cavity is that the equation:

$$\frac{1}{\lambda_r^2} = \left(\frac{m}{2a}\right)^2 + \left(\frac{n}{2b}\right)^2 + \left(\frac{p}{2c}\right)^2, \quad m, n, p \text{ integers}$$

must be satisfied by the three dimensions designated a, b and c and the wavelengths for which resonance occurs.

Each of the wavelengths satisfying this equation is known as the cavity inherent wavelength. A large number of discrete values exists for these inherent wavelengths, the three integers *m, n* and *p* define a mode, and the longest wavelength satisfying the above equation defines the fundamental mode.

To avoid insufficiently uniform distribution of the energy due to this phenomenon, the load can be placed on a rotating support, or a wave mixer consisting of a metal vane fan, can be used to alter wave reflection permanently.

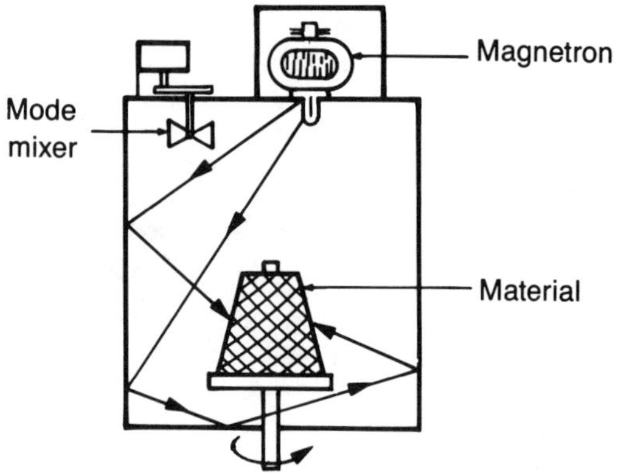

Fig. 36. Multimode resonant cavity [93 *d*].

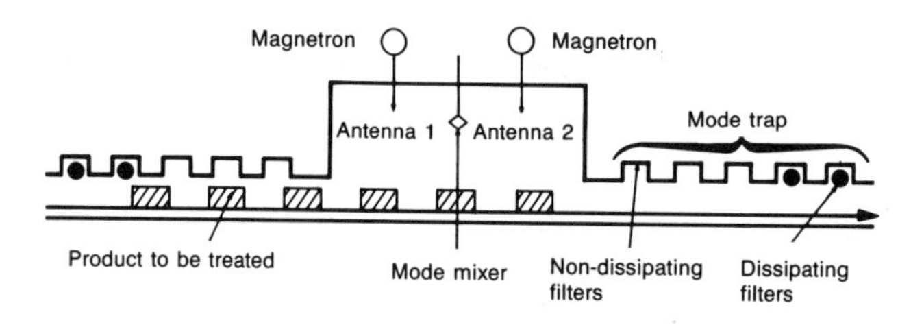


**Fig. 37.** Multi-mode cavity for continuous treatment [89].

These cavities are known as multi-mode cavities, and are widely used due to their relatively simple design. The dimensions of the rectangle are generally long with respect to the wavelength used.

The products treated are mostly mass products, foodstuffs, textile spools, etc.

### *7.1.5. Circular resonant cavities*

The applicators described above are unsuitable for treatment of fiber-shaped materials, since there is insufficient interaction with the electric field. In such cases, circular resonant cavities, the characteristics of which are that the electric field is parallel to the center line, are used.

Electrically, a resonant cavity can be analogous to a resonant circuit with series coil and resistance in parallel with a capacitance, the $Q$ of which is very high. Since the electrical field is high, it is possible to heat even slightly absorbant products in resonant cavities.

### *7.1.6. Other applicators*

Other applicators, such as horn antennae, resonant rings and slow waves exist, but have few industrial applications.

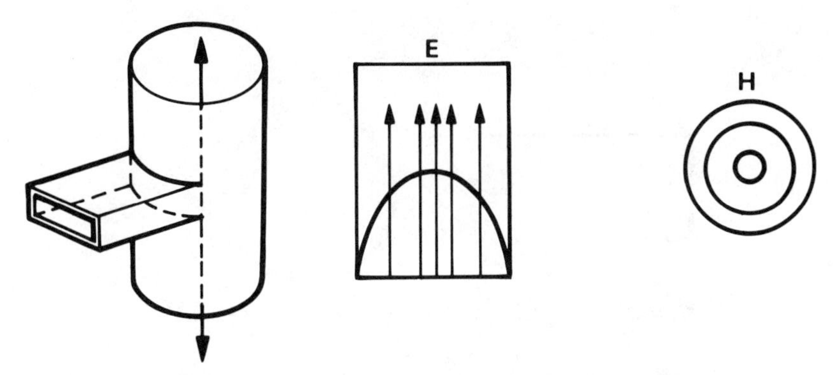


**Fig. 38.** Circular cavity applicator (single mode cavity) Coupling method and field distribution [89].

## 7.2. Waveguides

A waveguide is a metal product which channels electromagnetic waves emitted by the generator to prevent propagation in all directions. The waves are propagated through the inside dielectric by reflection from the metal walls. Usually, waveguides are of square, circular or coaxial cross-section. A mode is a special type of coupling between the generator and the load, and is characterized by its electromagnetic field configuration. The dimensions and types of waveguides define a cutoff wavelength $\lambda_c$ for one mode. Other modes for which the cutoff wavelength is greater than $\lambda_c$ are propagated within the waveguide, and the wavelength within the waveguide is given by:

$$\frac{1}{\lambda^2} = \frac{1}{\lambda_g^2} + \frac{1}{\lambda_c^2}.$$

In the opposite case, the electromagnetic fields decrease exponentially. Generally, waveguides are designed so that only one mode, known as the fundamental mode, propagates at operating frequency. For the 2,450 MHz frequency band, rectangular guides have, for example, a cross-section of 86.36 x 43.18 mm.

The fundamental mode electric field configuration in rectangular, circular and coaxial waveguides is given in figure 39 below.

In order to obtain optimum generator-load coupling, and therefore maximum efficiency, and prevent risks of damage to the generator, it is indispensable that the load be matched to transmission line impedance. Therefore, formation of incident and reflected standing waves must be prevented. Numerous techniques exist, but all consist of placing a non-dissipating object in front of the load, such that the load-matching device assembly does not produce a reflective wave [93 *a*].

Where this is difficult to do, for example in cases where the load changes noticeably with time, the generator is protected by a circulator, which branches the wave reflected by the load to an ancillary adapted load (*see* Figure 42).

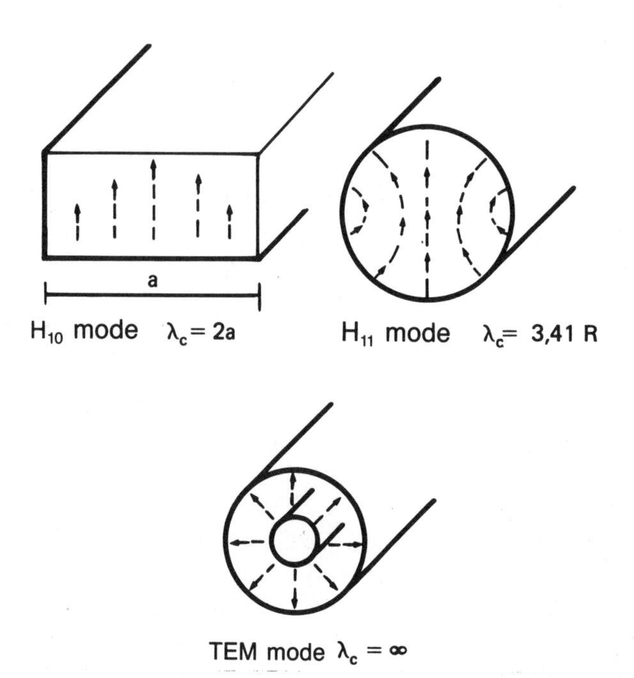


**Fig. 39.** Configuration of fundamental mode electric field in different waveguides [93 *a*].

## 7.3. Microwave generators

At the frequencies used in microwave heating, conventional triode oscillators are no longer suitable for the construction of power generators, and special generators such as magnetrons and klystrons are used.

### *7.3.1. Magnetrons*

The magnetron is a vacuum tube which functions like an oscillator, and consisting of two electrodes, a cylindrical central cathode, surrounded by a circular anode, in which oscillating cavities are formed. The interaction space is located between the anode and cathode. A high DC voltage is applied so as to take the cathode to a very high negative potential with respect to the anode.

When heated, the cathode emits electrons which are attracted by the positive anode. Moreover, a magnetic field is applied to the tube, generally through a permanent magnet, parallel to the cathode axis.

Due to the combined effect of the electric and magnetic fields, the electrons move away from the cathode along a curved path towards the anode. Each slot-cavity of the anode forms a resonant circuit at a precise frequency, the cavity walls representing the inductive part and the slot the capacitive part. The resonant frequency is determined by cavity and slot dimensions.

Emission of the anode current causes oscillations to appear in the slot and cavity combinations; in turn, these oscillations give rise to electric fields which are located in the interaction space where they alter the electron paths. Due to the effect of these microwave electric fields, the electrons, which follow a curved path from the cathode to the anode, are sometimes accelerated and sometimes decelerated, depending on the direction of the field in the region in which they are located at a given moment.

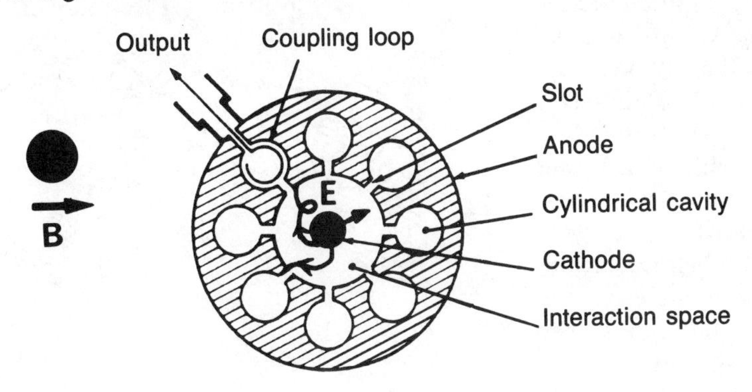


**Fig. 40.** Diagram of a cavity magnetron [89].

The accelerated electrons receive kinetic energy from the electric field, and follow paths returning them towards the cathode onto which they are projected, causing heat and emission of secondary electrons. Moreover, the slowed down electrons give off a part of their energy to the microwave field, are directed towards the anode and recover the energy which they lost to the microwave field from the DC field.

In this manner, most of the electrons describe a sequence of small cycloids and gradually lose their kinetic energy to the resonant cavities encountered along their path; finally, they strike the anode at low speed. There is an excessive energy in the tube, i.e. more than is necessary to maintain oscillation; this enables part of the energy to be recovered at one of the cavities by means of a loop or any other method.

The permanent mode which occurs in this manner in the magnetron is shown in figure 41; the electron clouds form spokes rotating around the cathode in phase with the electric field oscillations. Rotation of the electron beam takes places at constant angular velocity. A high quantity of energy is regularly converted from DC energy into microwave energy by the effect of the electric and electromagnetic fields, and can be applied to a load or load through a coupling [88].

This principle of operation applies not only to the cavity magnetron, but also to magnetrons of different structures.

Magnetrons operating at a frequency of 2,450 MHz generally have microwave powers of between 0.2 and 6 kW, and overall efficiency is of the order of 50 to 60%. At 915 MHz, it is possible to construct high power magnetrons of up to 25 kW microwave output approximately, with an overall efficiency of approximately 55 to 65%. Service life of magnetrons is between 3,000 and 6,000 hours. For low powers, air cooling is used (household ovens), while water cooling is used for higher powers [88], [89], [C].

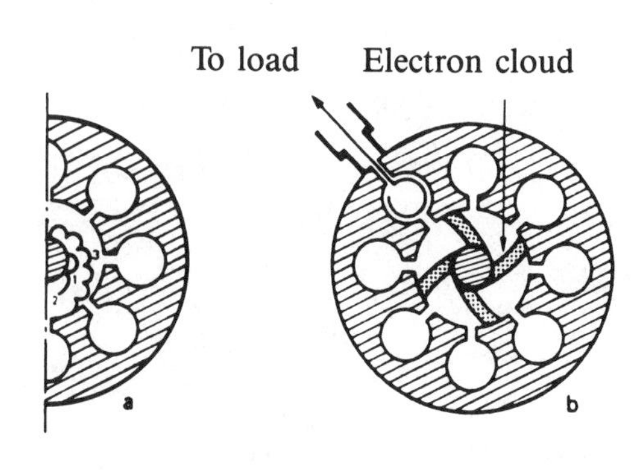


**Fig. 41.** *a)* Cross-section of a cavity magnetron showing:
1. path of an accelerated electron returned to the cathode
2. path of a secondary electron
3. path of a slowed down electron describing a cyclic path before reaching the anode.
*b)* Continuous operating cavity magnetron

Due to its relatively low cost, reliability and high stability, the magnetron is the most widely used microwave source for industrial applications; conversely, its power per unit is rather limited, which may be a drawback for some applications.

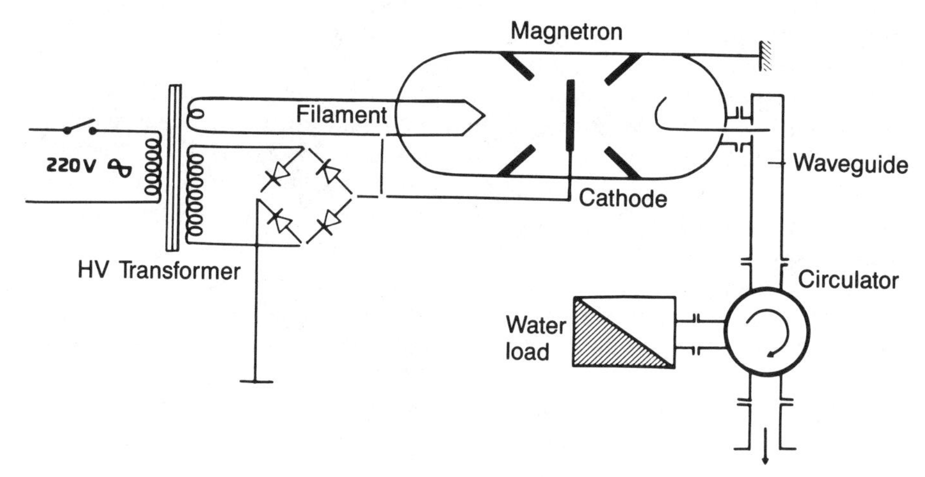


**Fig. 42.** Magnetron-power supply-waveguide assembly [89].

*7.3.2. Klystrons*

In contrast to the magnetron, the klystron has a linear structure. The electrons emitted by the electron gun cathode form a very fine beam, kept at constant diameter by a strong magnetic field parallel to the axis of the tube, and reach very high speeds due to the effect of the intense electric field.

The electron beam passes through a certain number of resonant cavities separated by cylinders known as "drift spaces". A microwave signal is applied to the first cavity, and the amplified output signal is extracted from the last cavity. The electron beam is picked up by a collector which dissipates the residual kinetic energy in the form of heat.

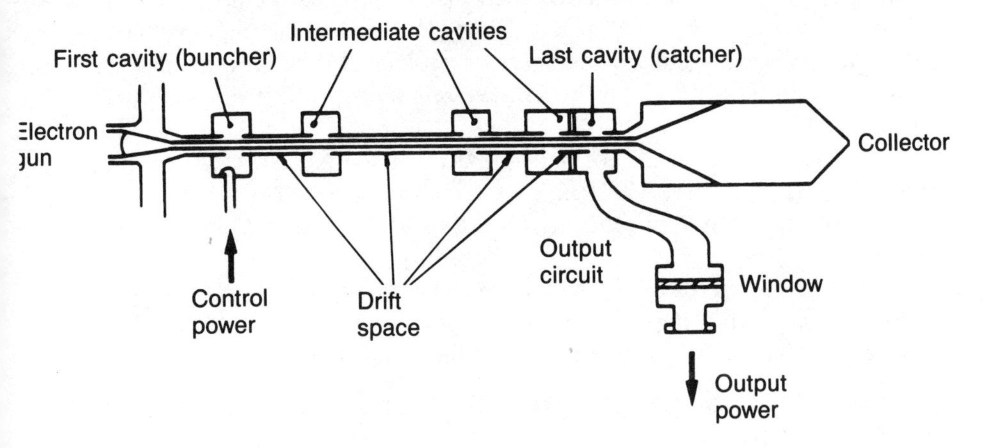


**Fig. 43.** Diagram of a klystron [88].

The amplification principle used in the klystron is based on the interaction of the microwave electric field in the resonant cavities, and the beam of electrons passing through the cavities. The microwave signal applied to the first cavity causes a microwave voltage to appear at the input of the first drift tube. This voltage produces an alternating electric field which accelerates passing electrons during the first half-cycle and slows them down during the second half-cycle, when the electric field reverses. Thus, the input signal velocity modulates beam. In the first drift tube, the electrons which were accelerated catch up with those which were slowed down, forming bunches of electrons. In this manner, the beam velocity modulation is converted into current modulation. The modulated beam induces alternating power on passage through the second cavity, and an electric field builds up at the input to the second drift tube. The electron beam is again velocity modulated then converted into current modulation during passage through the second drift tube, and so on. The high energy induced in the final cavity by the arrival of very dense bunches of electrons is extracted by coupling of the output to a waveguide, followed by a microwave window [88], [C].

The main interest of the klystron is its high power compared with that of the magnetron. For example, a five-cavity klystron, operating at 2.5 kV, produces a microwave power of 50 kW at 2,450 MHz, with an efficiency of 60%. The tube length is about one meter long. Klystron service life is high, of the order of 15,000 hours. Conversely, investment costs are high. Generally, klystrons are water cooled.

Klystrons are used mostly when the microwave power required exceeds 30 kW, or where it is difficult to parallel several magnetrons.

## 8. ADVANTAGES AND DISADVANTAGES OF MICROWAVE HEATING

The advantages of microwave heating are similar to those of radio frequency dielectric heating: direct transfer of energy to the product, homogeneous mass heating, selective heating, high power density, low thermal inertia, rapid heating, rather high efficiency and the possibility of resolving difficult heating problems. However, both types of heating differ in certain points, which will be described in paragraph 9.

Microwave heating leads to higher investment costs than radio frequency dielectric heating, since, when the cost per kilowatt installed is considered, a microwave heating system generally costs 5 to 10 times more than a resistance furnace.

Therefore, for microwave heating to be competitive with other techniques, it is indispensable that the technical or economic advantages provided largely compensate the heavy investment capital [66], [68], [69], [72], [93 *c*].

## 9. COMPARISON OF RADIO FREQUENCY AND MICROWAVE DIELECTRIC HEATING

Radio frequency and microwave dielectric heating first of all differ in the nature of the generators and applicators used, and the propagation modes for the electromagnetic waves. In particular, microwave heating calls for the installation of waveguides.

The penetration depth for a given material is generally less with microwave heating than with radio frequency heating, since the frequency is much higher; however, when relative permittivity and loss angle $\delta$ are taken into account, the above statement is modified. The behavior of the material must often be studied first so as to better define the values of $\epsilon_r$ and tan $\delta$, and therefore penetration depth, and also the transferrable power.

In general, the power density in a material may be higher with microwave heating, on condition that the change in the loss factor $\epsilon_r \tan \delta$ is not excessively unfavorable when frequency is increased. In particular, when $\epsilon_r \tan \delta$ is low, a higher frequency causes a higher power density, thus reducing the risk of arcing which would be caused by the use of high voltages.

The unit powers for radio frequency generators are much higher (up to 900 kW radio frequency) than those of microwave generators (maximum power of the order of 50 kW).

When product dimensions are very irregular, microwave heating is much easier to use than radio frequency heating, since the latter requires the use of electrodes fully adapted to the shape of the product, in order to obtain homogeneous heating, while microwave heating dispenses with the requirement for electrodes.

In fact, the microwaves surround the load and penetrate it from all sides not in contact with a wall or metal support.

Investment costs are high for both heating methods, but the cost is almost twice as high with microwave as with radio frequency heating, when considering the kilowatts delivered to the heated part and the auxilliaries.

Overall energy efficiencies are slightly higher with radio frequency heating (5 to 10%) than with microwave, but these are not absolute figures, and in some cases, may be inverted.

Therefore, radio frequency dielectric heating is more appropriate for materials of regular shape, of large dimensions and offering a relatively high "loss factor". Conversely, microwave heating is better adapted to the treatment of compact materials, complex shapes, and having a low dielectric "loss factor" [89].

The table in figure 44 below gives a summary of the comparative characteristics of both heating methods.

Comparative characteristics of radio frequency and microwave heating

| | Radio frequency | Microwave |
|---|---|---|
| Frequency | 13.56 MHz ($\lambda$=22.2 m)<br>27.12 MHz ($\lambda$=11.1 m)<br>40.68 MHz ($\lambda$= 7.4 m) | 915 MHz ($\lambda$=32.8 cm)<br>2,450 MHz ($\lambda$=12.2 cm)<br>5,800 MHz ($\lambda$= 5.2 cm)<br>22,125 MHz ($\lambda$= 1.4 cm) |
| $\epsilon_r$ tan $\delta$ | Must be relatively high | May be low |
| Principal origin of heating | Ionic conduction | Rotation of dipoles due to the effect of the electric field |
| Penetration depth | Deep | Medium or low |
| Product dimensions | May be large and variable | Small or medium |
| Product shapes | More or less regular | Any |
| High investment cost (investigate a mixture of dielectric + other heating methods) | Generator 60%<br>Applicator 40% | 1.5 to 2 times higher than with radio<br>generator 40%<br>applicator 60% |
| Tube service life | 5,000 to 10,000 hours | 2,000 to 5,000 hours (magnetrons)<br>15,000 hours (klystrons) |
| Generator maximum unit power | 900 kW radio frequency | 10 kW microwave at 2,450 MHz for magnetrons<br>50 kW microwave at 2,450 MHz for klystrons |

**Fig. 44.**

The above indications are simply intended as guidelines, and a thorough analysis must be performed in order to determine the type of dielectric heating, if applicable, best adapted to a given industrial application.

## 10. INDUSTRIAL APPLICATIONS OF MICROWAVE HEATING

Due to the high investment costs, microwave heating, as radio frequency dielectric heating, is above all intended for three applications:
— specific applications, difficult to perform with another technique;
— mixed microwave plus other process applications;
— applications involving low powers.

In spite of these limits, microwave heating is widely used. In the household equipment field, there are at present 20 million microwave ovens in use throughout the world, and in particular in the United States and Japan, while their use is at present developing in Europe. Power is low, less than 1 kW output, and generally of the order of 500 to 600 W output, but cooking times are very fast, in most cases a few minutes, compared with some tens of minutes with conventional ovens. In many cases, this heating method leads to high energy savings, in spite of the relatively low efficiency of microwave generation. In fact, only the foodstuffs are heated, the walls of the oven remain cold and heating is very rapid which more than compensates for the investment cost.

This gain in energy consumption is generally between 25 and 50%.

Generally, these ovens are used in the restaurant industry, for heating cooked dishes, defrosting and cooking. The power, which is generally higher than domestic ovens, is 1 to 3 kW output.

Moreover, microwaves, at much lower power levels, are used for medical applications which are not covered by this publication [93 *h*].

The industrial applications of microwave heating are rather numerous:
— cooking, defrosting, sterilization and drying of foodstuffs [13], [33];
— vulcanization of rubber [298], [29], [35];
— polymerization of synthetic resins [33];
— drying, cooking and sintering ceramic and refractory products [36], [44], [47], [65];
— melting and recovery waxes in steel casting [82], [98];
— drying glues on paper [33];
— manufacture of contact lenses [98]; — drying photographic films [24], [33], [36];
— drying and sterilization of pharmaceutical products [93 *c*];
— drying asbestos braids and polymerization of coatings on this material [94].

Numerous other applications are being developed [24], [25], [32], [39], [42 *a,b*], [67], [68], [74], [75], [76], [78], [81], [85], [90], [9 *c,e*], [E], [F], [G], [H].

### 10.1. Microwave heating applications in rubber and plastic working industries

Basically, these applications consist of heating prior to vulcanization of rubber and welding plastics.

#### *10.1.1. Rubber industries*

Microwave heating is mostly used for preheating operations prior to pressure moulding, and heating extruded sections up to vulcanizing temperature (hardening rubbers with sulphur based additives), this temperature being then maintained by passage through an electric resistance oven [28], [29], [35], [42 *c,d*], [59].

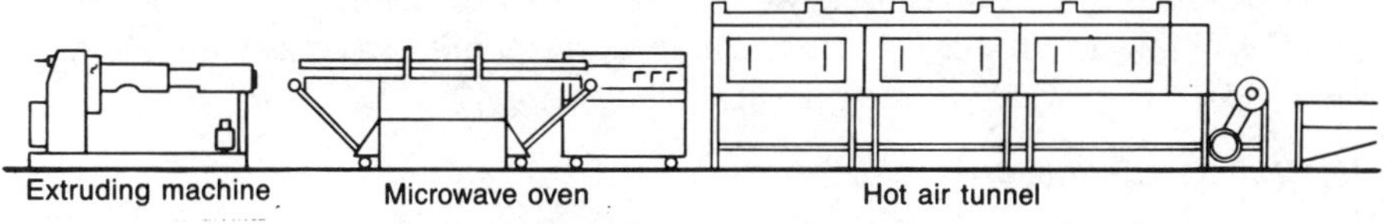


**Fig. 45.** Diagram of a rubber molded part extrusion line with microwave heating [98], [D].

A study of microwave heating of rubbers is relatively complex. Their electrical properties, and therefore energy absorption depend on many factors:
— behavior of basic polymers;
— addition of polar polymers;
— carbon-based additive content;
— nature of the different plasticizers;
— material temperature.

For example, pre-treatment relative permittivity of rubber can vary from 6 to 30 and tan $\delta$ can be between 0.01 and 0.15.

These values change during vulcanization and, at the end of this operation, the relative permittivity varies from 7 to 40 and tan $\delta$ from 0.03 to 0.25.

Therefore, each compound must be studied separately, and some mixtures are produced specially for microwave treatment.

Several successive operations are performed on a rubber molded part vulcanization system:
— plasticization, to soften the rubber by mechanical working; the temperature reached can vary between 75 and 110°C;
— extrusion enables formed sections to be obtained by passage of the rubber through a die;
— vulcanization involves two phases;
• heating the material up to vulcanizing temperature in a microwave oven;
• holding this temperature in a resistance oven, up to completion of vulcanization;
— cooling, which lowers the temperature to around 80°C, is followed by preparation of the products.

The advantage of microwave heating is the very rapid temperature rise of the rubber throughout its mass; with more conventional techniques, through heating is very slow, since rubber is a bad conductor of heat. Moreover, during temperature rise, the molded part passes through a plastic state; to obtain adequate dimensional stability, this passage must be as brief as possible, and this requirement is easy to obtain with microwave heating [77].

Generally, the microwave applicators used are of the 2,450 MHz resonant cavity or folded split waveguide type and, in some cases, slow wave applicators are used. A line of 5 kW microwave power and 20 kW for the resistance oven (8 kW demanded on average during the production phase, the total power being used only for the oven temperature rise) enables approximately 100 kg/hr. of rubber to be treated. 25 kW microwave power and 45 kW resistance oven lines are currently being constructed and handle approximately 500 kg/hr. of molded parts. Total consumption (microwave and resistance) of a line of this type is 0.15 to 0.17 kWh/kg of product, which generally corresponds to a considerable energy saving with respect to more conventional systems [98].

### *10.1.2. Plastic industries*

Some plastics have high dielectric losses over the microwave range, while at radio frequency, these losses are too low to heat the product. It is then necessary to use microwaves, and all the more so since the product $f \chi \epsilon_r \tan \delta$ is important in the dissipated power expression.

For polyethylene, the loss factor $\epsilon_r \tan \delta$, which is very low at radio frequency, approximately 0.0004, reaches 0.005 at 2,450 MHz. The product $f \chi \epsilon_r \tan \delta$ is then 1,000 times higher at 2,450 MHz than at 27.12 MHz, and heating polyethylene using microwave dielectric heating becomes possible. In general, it can be stated that microwave heating of plastics becomes possible as soon as the "loss factor" is greater than 0.001.

The most important applications in microwave heating in this field are welding plastic sheets and polymerization of resins on coated fabrics (araldite, epoxy, isobutyrate, etc.).

The welding machines are either fixed or portable. The latter are used to weld sheet to make tents, tarpaulins, greenhouses, footballs, swimming pools, packaging, etc. The power involved is low, a few hundred Watts at a frequency of 2,450 MHz. The welding rate is of the order of 1 m/min. and consumption approximately 1 kWh for 100 m of welding. These portable welding machines can also be used for field repairs of torn or punctured elements [44], [E], [F], [J].

## 10.2. Microwave applications in the food industries

Microwave applications are the most diversified and expected to develop widely in the food industries.

All food products in fact contain large quantities of water, and therefore offer high loss factors, $\epsilon_r \tan \delta$. Microwaves can be used for numerous thermal operations — cooking, sterilization, pasteurization, drying, thawing, etc. — when conventional methods give inadequate results, or in combination with other methods to improve their efficiency [1], [6], [13], [17], [33], [38], [42 *e*], [43], [51], [59], [93 *e*].

### *10.2.1. Thawing*

Deep freezing is an excellent method of conserving food products in their natural state. However, to work and convert such products, they must be thawed. Thawing has to take place at a temperature slightly lower than 0°C to prevent risk of contamination by pathogenic bacteria (regulations in each country define conditions) and alteration of product quality.

Thawing deep frozen foods requires several hours or several tens of hours when performed under these conditions, since the thermal conductivity of deep frozen products is low, and for the reasons stated above, it is impossible to work at positive ambient temperatures.

To increase the productivity of thawing operations and flexibility of the production system, it is therefore necessary to develop more rapid methods of thawing foodstuffs. Microwaves, which heat products directly in the mass, are an interesting alternative to conventional methods, since they dispense with the limits imposed by thermal conduction and ambient temperature [16], [49], [58], [80], [93 *g*].

Pure ice offers a very low "dielectric loss factor", of the order of 0.003, causing very limited heating; however, even at low temperature, foodstuffs have relatively high dielectric losses, since the crystalized water contains numerous substances, and it is therefore easily heated with microwaves. In particular, this applies to meat and fish, but also to numerous other foodstuffs such as butter or some vegetable products.

Relative permittivity and loss angle however vary with temperature, and both increase between the deep freezing temperature (-20°C) and the temperature at which pure ice melts (0°C) with a very high increase between — 5 and 0°C. This associated change of $\epsilon_r$ and tan $\delta$ has two consequences, the effects of which mutually reinforce each other. On the one hand, the penetration depth of microwaves in the product to be thawed decreases as the temperature rises; therefore, surface heating increases. On the other hand, surface heating tends to accentuate with increase in temperature, since the product $\epsilon_r$ tan $\delta$ increases. Excessive heating and even partial cooking or localized burns are to be feared on the surface or in the thin parts of the product.

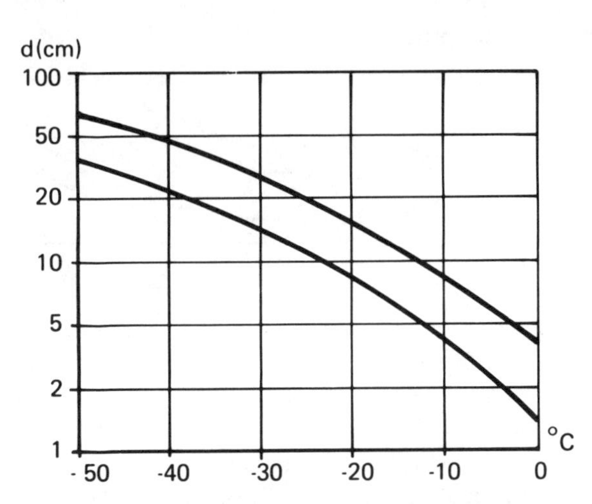


**Fig. 46.** Variation in penetration depth against temperature [44]:
— top curve: 915 MHz;
— bottom curve: 2,450 MHz.

Several methods of avoiding the risks of overheating exist. First of all, microwaves are used to heat the products from -20 to -4°C approximately, which reduces but does not eliminate the risks of surface overheating, but above all limits installed power, and therefore investment costs; the microwave energy is used only to apply the relatively limited sensitive heat required for heating the product, and not the latent heat of fusion of ice, which is much higher. To completely eliminate risk of overheating, the ambient temperature in the thawing tunnel is kept at very low temperature, between -20 and -10°C by an independent refrigerating unit; the supplementary energy consumption is very limited: approximately 5%, since cold air flows in a closed circuit, and the installation is very well insulated thermally. Another technique consists of spraying the product with liquid nitrogen. Final thawing is then obtained conventionally, for example in an ordinary cold room or in a tank heated by lukewarm water obtained from the electron tube cooling system. The product can also be worked directly without total thawing, which offers high energy savings and considerably reduces storage requirements. This is the most widely adopted solution in the food industry, and enables continuous production lines to be designed.

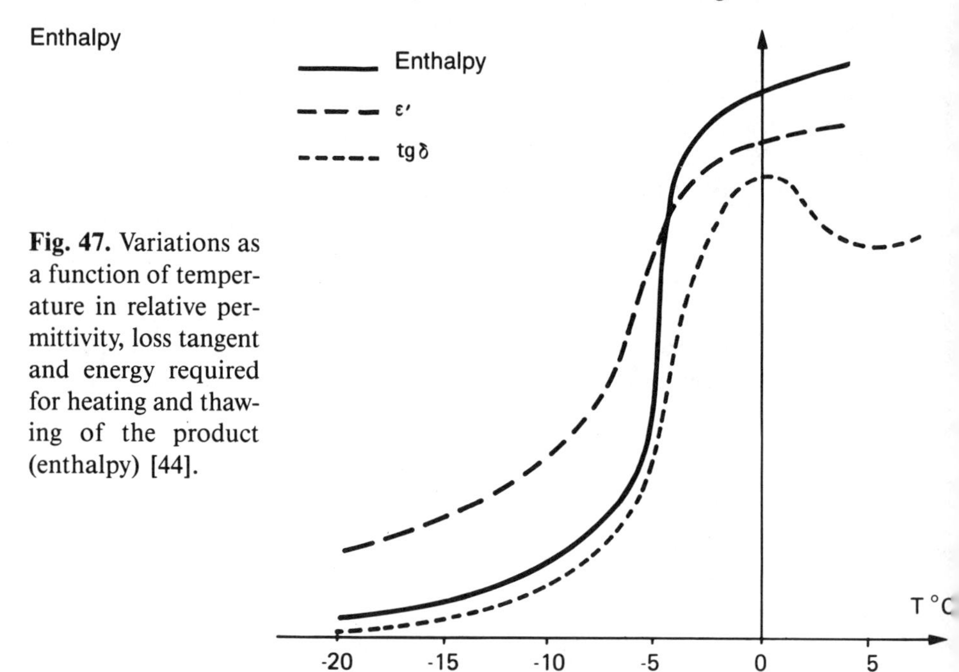


**Fig. 47.** Variations as a function of temperature in relative permittivity, loss tangent and energy required for heating and thawing of the product (enthalpy) [44].

This technique is used for thawing entire salmon, blocks of fish, pork and beef sides, blocks of chickens and other meat products, butter, etc., and can be extended to vegetable products; this is also being adapted for treatment of large masses such as complete quarters of beef [89], [W].

The equipment used are either continuous tunnels or static equipment for smaller production batches. In order to increase thawing capacity by simply adding modules, thawing tunnels are generally of modular design. The operating frequency is 2,450 MHz, and sometimes, in North and South America, 915 MHz; the applicator is a resonant cavity with several magnetrons located above and below the cavity, providing homogeneous distribution of heat in the products [U], [W].

To thaw beef from -20 to -5°C, the installed power required is of the order of 17 kW microwave output per ton/hour. For thawing to -2.5°C, the power required is 31 kW microwave per ton/hour, therefore practically double, which illustrates the importance of the final thawing temperature. As an initial approximation, these figures can also be applied to thawing fish. Specific consumption in this case is between 25 and 30 kWh per ton of thawed product, but recovery of energy from the electron tube cooling system can noticeably decrease this figure.

Another example is a 75 kW microwave output installation operating at 915 MHz (in the U.S.A.), which is used to thaw 2 tons per hour of butter from -12 to +2°C, with a consumption of 60 kWh approximately per ton treated [98].

The main advantages of microwave thawing are the thawing rate (10 to 20 minutes against some tens of hours in general), the flexibility provided for production organization, temperature homogeneity, suppression of exudates and superficial oxidization and improvement of biological and sensory qualities.

### *10.2.2. Drying of pasta*

With this application, microwaves are used in combination with conventional hot air drying. Located slightly after the start of the drying system, the microwave section considerably accelerates drying by enhancing migration of water towards the surface, from which it is drawn by the hot air flow. The length of the dryers and drying time are often divided by two to three, and overall energy consumption is greatly reduced. The product obtained is crustless, enabing very fast cooking of pasta [33].

For example, on a line producing 100 kg per hour of pasta, the installed power is 50 kW microwave output. On another line producing 1,500 kg per hour, the microwave power is only 60 kW. The section devoted to microwaves varies from one installation to the other. The frequency used is generally 915 MHz (in North America), since pasta is produced in thick layers. This technique is at present being developed in the United States on short pasta production lines.

### *10.2.3. Vacuum dehydration*

Microwave heating is used to dehydrate liquid foodstuffs and produce powders in a partial vacuum. This technique, which competes with lyophilization, dispenses with the necessity of deep freezing products and enables continuous operating installations to be designed [66], [87], [93 *g*].

Heat transfer partly controls the duration of the dehydration cycle. In lyophilization, the surface layer, which is already dry, forms an insulating thermal shield around the central and still frozen zone. In vacuum dehydration by expansion, an abundant foam forms on the surface of the product, which slows down thermal exchanges. This problem is resolved by microwave heating, which directly through heats the mass of the product.

Present installations are intended for the production of powder of small red fruit (rasberries, black currents, exotic fruit, etc.), a concentrated extract (up to

90% approximately of dried material) is placed in a chamber at a vacuum of 5 to 10 torr. This extract, submitted to the combined action of the vacuum and the microwaves, forms an abundant foam as it progresses to the back of the oven. It then forms a meringue of 10 to 20 cm thick, which is picked up by a rotary scraper and crushed to form instantaneously soluble pellets or powder. The applicator itself is relatively complex and consists of several resonant cavities. A 7.5 kW microwave power supply is used to evaporate 8 kg per hour of water, with a specific consumption from the line of 1.5 kWh/kg of water, to which the energy required to create the vacuum must be added. The working temperature varies between 35 and 60°C. Equipment of this type can evaporate 32 kg/hr. of water for a microwave power of 30 kW [W].

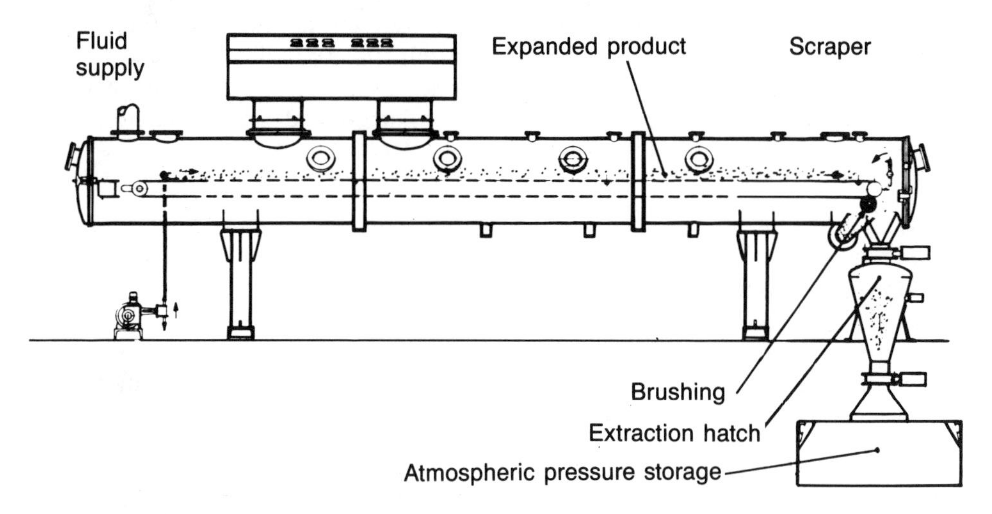


**Fig. 48.** Microwave equipment for vacuum dehydration of fruit juices [66].

This type of equipment can be used for drying aromatic products (fruit and vegetable juices, licorice, vanilla, etc.), natural colorants, aromatic plant extraction, grain products, proteins, eggs and solid products (mushrooms, strawberries, asparagus, etc.) [W].

### *10.2.4. Other applications*

Microwaves are used for numerous other applications in the food industries, precooking sardines, making fritters, drying and cooking potato chips, pasteurization and sterilization of liquid foodstuffs [33], [42 *e*], [46], [51], [53], [58], [60 *b*], [64], [79], [83], [94].

Numerous tests demonstrating the feasibility of microwave techniques have been conducted throughout the world. High investment costs and difficulties sometimes encountered in developing final industrial equipment have, however, slowed down the development of this heating method. Therefore, microwave heating often finds its applications during the development of new products, or when conventional techniques have given only partially satisfactory results.

### 10.3. Microwave applications in the textile industries

While radio frequency dielectric heating has developed in the textile industeries, microwaves have not as yet found many applications. Nevertheless, they are used to treat some high added-value products [23], [33], [93 *f*], [95]:
— grafting copolymers from different monomers, to provide basic textiles with new properties (high resistance to fire, dirt-proofing, etc.);
— grafting copolymers to modify the characteristics of the fibres for new applications;
— miscellaneous drying operations.

**Fig. 49.** Treatment of textile spools in a multimode cavity.

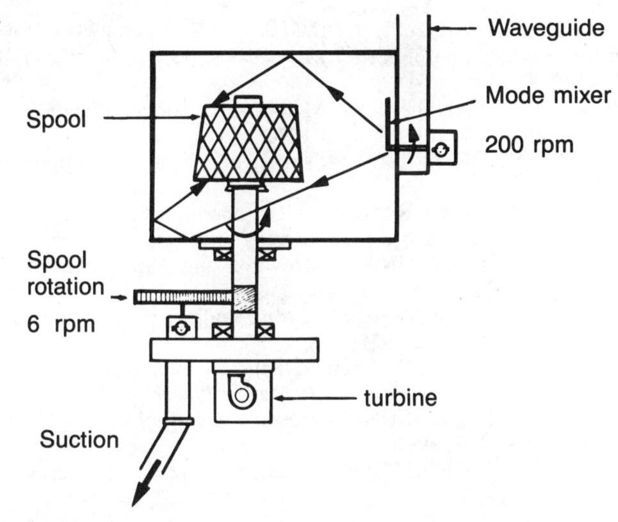


## 11. BIBLIOGRAPHY

[1] A. VON HIPPEL, *Dielectric materials and applications*, John Wiley and Sons, New York, 1954.
[2] J. M. HOLLAND, *La cuisson haute fréquence en biscuiterie*, Réunion annuelle du Biscuit and Cracker Institute, 9 avril 1963.
[3] T. H. HAFNER, Le chauffage diélectrique haute fréquence appliqué à la fabrication des panneaux agglomérés, *Revue BBC*, n° 10-11, 1964.
[4] H. LIND et F. POPERT, Le problème de l'adaptation dans les presses à panneaux agglomérés alimentées en haute fréquence, *Revue BBC*, n° 10-11, 1964.
[5] P. ADAMI, Quelques installations nouvelles de chauffage diélectrique haute fréquence, *Revue BBC*, n° 10-11, 1964.
[6] G. P. DE LOOR et F. W. MEIJBOOM, The dielectric constant of foods and other materials with high water contents at microwaves frequencies, *Journal of Food Technology*, n° 1, 1966.
[7] N. H. LANGTON, Le préchauffage diélectrique d'ébauches aux radio-fréquences, *Revue générale des caoutchoucs et plastiques*, n° 1, vol. 44, 1967.
[8] J. C. BLONDEL, Derniers développements du chauffage diélectrique appliqué à l'industrie du caoutchouc, *Revue générale des caoutchoucs et plastiques*, n° 6, vol. 44, 1967.
[9] H. G. GRASSMANN, Séchage haute fréquence dans l'industrie textile, *Chemiefasern*, n° 9, septembre 1967.

[10] M. D. PRESTON, Dielectric dryers can improve paper machine performance, *Paper Trade Journal*, n° 4, vol. 152, janvier 1968.
[11] F. CHURCH, How dielectric heating helps to control moisture content, *Pulp and Paper International*, février 1968.
[12] E. C. OKRESS, *Microwave power engineering*, Academic Press, 1968.
[13] C. A. LOFDAHL, Microwave food applications, status and potential, *Microwave Energy Applications News Letter* n° 4, vol. 1, 1968.
[14] T. HAFNER, Traitement des denrées alimentaires par chauffage haute fréquence, *Revue BBC*, n° 3, mars 1968.
[15] S. ADAM, *Microwave theory and applications*, Prentice Hall, 1968.
[16] J. B. VERLOT, Décongélation du poisson congelé ou surgelé, *la revue générale du froid*, n° 5, 1969.
[17] M. A. K. HAMID et R. J. BOULANGER, A new method for the control of moisture and insect infestation of grain by microwave power, *Journal of Microwave Power*, n° 1, vol. 4, 1969.
[18] S. F. GALEANO et R. A. MACK, Electromagnetic drying of lineboard in the RF Band, *Tappi*, n° 4, vol. 63, avril 1970.
[19] P. E. SHARP, Industrial advances in dielectric heating, *Electrical Review*, octobre 1970.
[20] P. Y. MAC CORMICK, *Unit operations-drying*, *Industrial and Engineering Chemistry*, n° 12, vol. 62, décembre 1970.
[21] CONGRÈS BFMIRA, Microwave and radiofrequency in the food industry, Londres, 22 septembre 1971.
[22] E. SIMMEN, Bases théoriques du chauffage haute fréquence des matériaux isolants, *Revue BBC*, novembre 1971.
[23] J. E. PENDERGRASS *et al.*, Deposition of finishes and dyes in material dried using microwave heating, *Journal of Microwave Power*, n° 3, vol. 7, 1972.
[24] E. W. STEPHANSEN, Microwave drying of coated films, *Journal of Microwave Power*, n° 3, vol. 7, 1972.
[25] VAN KOUGHNETT et W. WYSLOUZIL, A microwave dryer for ink lines, *Journal of Microwave Power*, n° 4, vol. 7, 1972.
[26] D. A. JOHNSTON et W. A. VOSS, Use of the TEM mode in microwave heating applicators, *IEEE Transcriptions* (USA), n° 8, vol. 20, 1972.
[27] J. P. BOULANGER *et al.*, Étude des caractéristiques diélectriques des matériaux en relation avec l'utilisation de la haute fréquence dans l'industrie de la chaussure, *Technicuir*, mars 1972.
[28] J. C. BRULE, Application du chauffage par hyperfréquence à la vulcanisation en continu des profilés, *Revue générale des caoutchoucs et plastiques*, n° 4, vol. 49, avril 1972.
[29] P. CHANET et M. MONCEL, Rôle de l'humidité des mélanges dans le chauffage diélectrique par ultra haute fréquence, *Revue générale des caoutchoucs et plastiques*, n° 4, vol. 49, avril 1972.
[30] R. M. PATE, Dielectric drying of paper and converted products, *American Paper Industry*, mai 1972.
[31] T. HAFNER, Collage et durcissement d'éléments de construction au moyen de chauffage HF, *Revue BBC*, n° 6, vol. 59, juin 1972.
[32] N. ANDERSON *et al.*, Microwave drying of paper, an experimental study, *Svenk Paperstidning Arg*, n° 16, vol. 75, septembre 1972.
[33] J. A. JOLLY, *Technical advances in operational microwave industrial process systems in North America*, Congrès de l'Union internationale d'Électrothermie, Varsovie, 1972.
[34] B. MICHEL et J. P. BOULANGER, L'utilisation de la haute fréquence dans la fabrication de la chaussure, *Technicuir*, n° 9, novembre 1972.
[35] H. E. SCHWARTZ *et al.*, Microwave curing of synthetic rubbers, *Journal of Microwave Power*, n° 3/4, vol. 8, 1973.

[36] M. GUERGA et B. HALLIER, *Le chauffage par micro-ondes, application au séchage des moules de fonderie*, Comité français d'Électrothermie, 5 avril 1973.
[37] E. CATIER, Le chauffage par micro-ondes, *L'Électricien*, octobre 1973.
[38] G. BARTHOLIN *et al.*, *Problèmes posés par l'industrie de transformation des légumes*, Centre technique des conserves et produits agricoles, janvier 1974.
[39] A. G. NORTH, Progress in radiation cured coatings, *Pigment and Resin Technology*, février 1974.
[40] A. C. METAXAS, Design of TM 010 resonant cavity as a heating device at 2,45 GHz, *Journal of Microwave Power*, n° 2, vol. 9, 1974.
[41] J. M. HOLLAND, Dielectric post baking in biscuit manufacture, *Baking Industries Journal*, février 1974.
[42] IMPI, Symposium international sur les applications des micro-ondes, Loughborough, G.-B., 11-12 septembre 1973 :
*a*) J. GERLING, *High power applications;*
*b*) P. L. JONES et J. LAWTON, *Comparison of microwave and radio frequency drying of paper;*
*c*) R. A. PETERSON, *A microwave preheater for giant tires;*
*d*) H. SCHWARTZ *et al.*, *Microwave curing of synthetic rubbers;*
*e*) D. W. ALEXANDER, *Process parameters for continuous microwave sterilization.*
[43] N. MEISEL, Apport d'énergie instantané au cœur des produits alimentaires, les micro-ondes, *Industries alimentaires et agricoles*, octobre 1974.
[44] L. THOUREL, *Les applications industrielles du chauffage par micro-ondes*, Journées électro-industrielles de Paris, EDF, novembre 1974.
[45] T. HAFNER, Fabrication de panneaux de construction au moyen du chauffage HF, *Revue BBC*, n° 1, vol. 62, janvier 1975.
[46] T. HAFNER, Traitement sans dommage par chauffage HF des denrées alimentaires à stocker, *Revue BBC*, n° 1, vol. 62, janvier 1975.
[47] B. HALLIER, *Le raffermissement des pâtes céramiques et autres applications des hyperfréquences*, Comité français d'Électrothermie, 29 avril 1975.
[48] D. A. COPSON, *Microwave Heating* (2nd edition), The Avi Publishing Cy, Westport, Connecticut, 1975.
[49] A. BOSSAVIT, Avant-projet d'un modèle mathématique pour l'étude de la décongélation rapide de produits alimentaires, *Documentation EDF*, septembre 1975.
[50] E. J. JOHNSTON, Basic principles in dielectric heating of synthetics, *Industrial Heating*, octobre 1975.
[51] N. MEISEL, Applications industrielles des micro-ondes dans les industries alimentaires, *Revue générale du froid*, janvier 1976.
[52] A. J. BERTEAUD, Les hyperfréquences et leurs applications, Presses Universitaires de France, *Que sais-je*, 1976.
[53] IMPI SYMPOSIUM, Louvain, 27-30 juillet 1976, *Ensemble de communications sur le chauffage par micro-ondes.*
[54] P. L. JONES *et al.*, *Radiofrequency drying of papers and textiles*, Congrès Union internationale d'Électrothermie, Liège, 11-15 octobre 1976.
[55] J. CHABERT et A. VIALLIER, *Applications des techniques de chauffage infrarouge, haute fréquence et micro-ondes au séchage et aux traitements thermiques d'ennoblissement des textiles*, Congrès Union internationale d'Électrothermie, Liège, 11-15 octobre 1976.
[56] W. R. TINGA, *Microwave process systems, advances and problems*, Congrès Union internationale d'Électrothermie, Liège, 11-15 octobre 1976.
[57] LE MARCHÉ DE L'INNOVATION, *Collage de bois lamellé par chauffage haute fréquence*, n° 23, novembre 1976.
[58] K. OGURA *et al.*, Current industrial microwave heating equipment; latest microwave heating equipment for food industries and microwave moisture measurement devices, *Toshiba Review*, n° 106, novembre 1976.
[59] P. M. KOHN, Microwave technology, penetrating CFI markets, *Chemical*

*Engineering* (New York), n° 1, vol. 84, janvier 1977.
[60] UNIVERSITÉ CLAUDE-BERNARD, Villeurbanne, Congrès sur les applications industrielles des micro-ondes et des hautes fréquences, 26 janvier 1977, en particulier :
*a*) N. MEISEL, *Séchage-décongélation par micro-ondes*,
*b*) P. VAROQUAUX, *Blanchiment, pasteurisation, stérilisation*.
[61] D. L. HODGETT *et al.*, *Drying by means of radiofrequency power*, Congrès IEE, Londres, 8-9 mars 1977.
[62] J. CHABERT, Nouveaux procédés de séchage thermique, *La technique moderne*, n° 3, vol. 69, mars 1977.
[63] M. LEFRANC et M. JOLION, *Application du chauffage haute fréquence à la gélification des plastisols*, Congrès EDF, Toulouse, 3 mars 1977.
[64] G. FAILLON *et al.*, New uses of microwave power in the food industry, *Journal of Microwave Power*, n° 1, vol. 12, mars 1977.
[65] I. TRIPSA *et al.*, Étuvage des noyaux par micro-ondes, *Fonderie*, n° 365, mars 1977.
[66] N. MEISEL, *La déshydratation par micro-ondes dans les industries alimentaires*, Comité français d'Électrothermie, 21-22 avril 1977.
[67] L. THOUREL, *Quelques possibilités d'économies offertes par l'emploi des micro-ondes dans l'industrie*, Comité français d'Électrothermie, 21-22 avril 1977.
[68] M. JOLION, *Vers des voies nouvelles d'applications industrielles des hautes fréquences et des hyperfréquences*, Comité français d'Électrothermie, 19-20 octobre 1977.
[69] L. THOUREL, Quelques possibilités d'économies offertes par l'emploi des micro-ondes dans l'industrie, le froid, *Le Conditionnement d'air et la climatisation*, n° 291, novembre 1977.
[70] R. G. BOSISIO *et al.*, A non contact temperature monitor for the automatic control of microwave ovens, *Journal of Microwave Power*, n° 4, vol. 12, décembre 1977.
[71] M. D. PRESTON, *Dielectric Heating*, The Electrification Council, 13e Congrès, Cincinnati, 6-8 février 1978.
[72] H. MELGAARD, *Saving dollars and energy with microwave heating*, The Electrification Council, 13e Congrès, Cincinnati, 6-8 février 1978.
[73] M. MAC CORMICK, The application of R.F. to paper drying, *Electrical Review*, n° 11, vol. 202, 1978.
[74] P. HARTELMEYER, les micro-ondes dans le séchage du plâtre, *Industries et Techniques*, n° 336, mars 1978.
[75] S. LEFEUVRE *et al.*, *Industrial materials drying by microwave and hot air*, Congrès IMPI, juin 1978.
[76] IMPI SYMPOSIUM, Ottawa 1978, Communications du Congrès, 28-30 juin 1978.
[77] M. JOLION, *L'utilisation des UHF dans la vulcanisation du caoutchouc*, Conférence à l'Institut français du Caoutchouc, 1978.
[78] T. HIROSE *et al.*, Microwave heating of multi-layered cigarettes, *Journal of Microwave Power*, n° 2, vol. 13, juin 1978.
[79] Y. TAKAHASHI *et al.*, Measurement of total milk solids by microwave heating, *Journal of Microwave Power*, n° 2, vol. 13, juin 1978.
[80] D. BIALOD *et al.*, Microwave thawing of food products using associated surface cooling, *Journal of Microwave Power*, n° 3, vol. 13, septembre 1978.
[81] Y. KASE et K. OGURA, Microwave power applications in Japan, *Journal of Microwave Power*, n° 2, vol. 13, juin 1978.
[82] J. HITCHON, *Utilisation d'un four à micro-ondes industriel pour l'opération de décirage*, 8e Congrès Fonderie de précision, Stockholm, 18-21 juin 1978 (*voir* également *Fonderie*, n° 383, vol. 33, novembre 1978).
[83] N. MEISEL, Les micro-ondes au service des techniques alimentaires, *Industries alimentaires et agricoles*, n° 9, septembre 1978.
[84] J. CHABERT *et al.*, Le séchage « tout électrique » des nappes textiles, *l'Industrie textile*, n° 1081 (septembre 1978).
[85] E. GORDON, Hautes fréquences et micro-ondes : de nouvelles applications dans l'industrie, *L'Usine nouvelle*, septembre 1978.

[86] J. CHABERT, *Séchage du papier et des fibres textiles, Comité français d'Électrothermie*, 12-13 *octobre* 1978.
[87] N. MEISEL, *Procédé Gigavac pour la fabrication en continu de poudres de fruits instantanément solubles*, Comité français d'Électrothermie, 12-13 octobre 1978.
[88] THOMSON-CSF, *Applications industrielles des micro-ondes*, décembre 1978.
[89] R. LE GOFF et M. JOLION, Travaux de recherche dans le domaine des rayonnements HF et UHF, *Documentation EDF*, 1978-1979.
[90] H. LE DOUSSAL, Application des ondes hyperfréquences à l'accélération de la prise des bétons réfractaires, *L'industrie céramique*, n° 724, janvier 1979.
[91] EDF, *Sécurité des fours à micro-ondes*, 74 références parues depuis 1969, avril 1979.
[92] RWE, *Die industriellen Elektrowärme-Verfahren*, RWE, Essen, 1979.
[93] IMPI, 14e Symposium International sur les applications des micro-ondes, Monaco, 12-15 juin 1979 (Documents Comité français d'Electrothermie) :
*a*) S. LEFEUVRE, *Les ondes électromagnétiques, définitions, propriétés, utilisations, génération;*
*b*) G. ROUSSY, *Les matériaux diélectriques;*
*c*) M. JOLION, *Intérêt de l'utilisation des ondes électromagnétiques très haute fréquence;*
*d*) J. P. PELISSIER, *Les applicateurs micro-ondes et hautes fréquences;*
*e*) R. LE GOFF, *Panorama des applications existantes;*
*f*) J. CHABERT, *les micro-ondes dans l'industrie textile;*
*g*) N. MEISEL, *Les micro-ondes dans les industries alimentaires;*
*h*) B. SERVANTIE, *Effets biologiques et applications médicales des micro-ondes;*
*i*) M. STUCHLY, *Normes de protection vis-à-vis des micro-ondes.*
[94] P. WAGNER, Les micro-ondes renouvellent la conception du chauffage, *L'Usine nouvelle*, n° 29, juillet 1979.
[95] S. AUSSUDRE, *Réticulation par micro-ondes des matrices organiques de tissus préimprégnés époxy-fibres de verre et époxy-fibres de graphite*, Comité français d'Électrothermie, 6-7 mars 1980.
[96] M. MANOURY, *Le liage par haute fréquence des matériaux isolants de récupération*, Comité français d'Électrothermie, 6-7 mars 1980.
[97] DOCUMENTATION ELECTRICITY COUNCIL, Londres, G.-B.
[98] DOCUMENTATION EDF, 92080 Paris La Défense.
[99] DOCUMENTATION RWE, Essen, RFA.

**List of research centers and equipment suppliers mentioned in this chapter:**

[A] STRAYFIELD, Redding, G.-B.
[B] CEM-BBC, 75008 Paris.
[C] THOMSON-CSF, 92100 Boulogne-Billancourt.
[D] LAMBDA-INTERNATIONAL, 95880 Enghien-les-Bains.
[E] INSTITUT TEXTILE DE FRANCE, 69130 Écully.
[F] UNIVERSITÉ CLAUDE-BERNARD, 69100 Villeurbanne.
[G] ÉCOLE NATIONALE SUPÉRIEURE D'ÉLECTRICITÉ, D'ÉLECTRONIQUE, D'INFORMATIQUE ET D'HYDRAULIQUE, 31000 Toulouse.
[H] ONERA, 31000 Toulouse.
[I] MASSER, 91170 Viry-Châtillon et Bruxelles, Belgique.
[J] SAREM, 69000 Lyon.
[K] SFAMO, 88370 Plombières-les-Bains.
[L] THIMONNIER, 69338 Lyon.
[M] SINTRA-SEF, 92700 Colombes.
[N] TOCCO-STEL, 91300 Massy.
[O] RADYNE, 75009 Paris.
[P] MHM ELECTRONIC, 78630 Orgeval.
[Q] MDP, 78470 Magny-les-Hameaux.

[R] HFI, 94700 Maisons-Alfort.
[S] SIEMENS, 93000 Saint-Denis.
[T] INTERTHEM, Londres, G.-B.
[U] ABC FOOD MACHINERY, Milton Keynes, G.-B.
[V] CENTRE DE RECHERCHES TEXTILES, 68100 Mulhouse.
[W] IMI, INDUSTRIES MICROONDES INTERNATIONALES, 78680 Epône.
[X] PLATT LONGCLOSE, Leeds, G.-B.

# Chapter 8

# Electric Arc Heating

## 1. PRINCIPLE OF ELECTRIC ARC HEATING

An electric arc is the result of current flow between two electrodes in an ionized gas environment (air or special gas).

This heating method permits very high powers and temperatures which, for normal furnaces, are about 3,000°C.

The first work performed on electric arc heating dates from 1880. Davy highlighted the phenomenon by creating an arc between two carbon electrodes using Volta batteries. Around 1887, Moissan invented the use of heat industrial applications, and in 1899, Heroult installed the first industrial furnace for the production of steel.

Heat transfer varies greatly with furnace design. In practice, the various types of arc furnace, unlike other electrothermal techniques, are very independent of the application.

However, two major arc furnace families exist:

— *Electric arc-heated melting furnaces*

The material to be heated is first loaded into the furnace, then melted by heat transferred from one or more electric arcs. As soon as melting is completed, the product is refined and poured. When these operations are complete another load is placed in the furnace. Basically, these furnaces are used for the production of steel and cast iron.

— *Arc resistance-heated reduction furnaces*

The product is reduced either continuously or discontinuously. These furnaces, which are sometimes known as submerged arc or arc resistance furnaces are more like conduction furnaces (direct current flow through load) than arc furnaces, since the load is heated by the Joule effect created by the current flowing through the load. Conversely, the construction and design of these furnaces are very similar to arc furnaces. These furnaces are thus dealt with in this chapter rather than the one devoted to "conduction heating", respecting the usual classification encountered in electrothermics.

Reduction furnaces are used in two fields:

- electrochemistry, for preparation of calcium carbide, phosphorous, etc.;
- electrometallurgy, for the production of ferro-alloys such as ferrosilicon, ferromanganese or cast iron in electric furnaces.

Also, the electric arc phenomenon is used for industrial operations other than melting or reduction. The most widespread applications of this heating method are the various arc welding processes. Electric arc heating is also used for heating reagent gases in the chemical industry, treatment often taking place in the arc itself.

Arc heating involves extremely high powers and electrical energy consumptions. As a function of the countries involved, the consumption of melting and reduction arc furnaces is generally between 30 and 60% of the total electrical energy consumption intended for industrial thermal usage (35% in France). In spite of this importance, the study of arc furnaces will be kept short, since the number of installations is relatively low (for example, there is often only one reduction furnace in a country or even a continent, due to the cost of such equipment) and furnaces are often specific to one application.

## 2. ARC HEATING CHARACTERISTICS

### 2.1. Formation of the arc and voltage drop across electrodes

Establishment of an electric arc requires striking; for example, this can be obtained by a current pulse, following contact between two electrodes taken to different potentials. The incandescent cathode then emits electrons which move towards the anode due to the effect of the electric field existing between the electrodes. These electrons encounter gas molecules, which are rendered conductive subsequent to ionization due to shock, allowing current to flow within the gas column. The ions, which are accelerated by the electric field, strike the cathode and heat it, thus maintaining electron emission [4], [55], [66].

The arc actually originates on a small part of the cathode known as the "cathode spot" and terminates on a small part of the anode known as the "anode spot" (with AC, these spots move at high speeds, more than 500 kilometers per hour, according to studies performed by Union Carbide engineers); this arc causes a voltage drop which can be divided into three zones:

— The cathode drop, of approximately 10 V, extends over a distance of about $10^{-6}$ m, independent of the length of the arc, and creates an electric field of $10^7$ to $10^8$ V/m. The arc is created by short-circuiting the electrodes. The resultant heating of the cathode lowers the electron extraction potential and facilitates build-up of the arc. Once the arc has struck, the ions bombard the cathode at high speed and take it to a sufficient temperature, between 3,600 and 4,000°C, to maintain electron emission. The material forming the cathode determines the value of the electron extraction potential, and therefore the voltage drop; this is slightly smaller for metal electrodes than for graphite or amorphous carbon electrodes. The use of a limited thermally conductive material reduces thermal losses by conduction due to the electrodes. The current density generally varies from 15 A/cm$^2$ for high diameter electrodes to 40 A/cm$^2$ for low diameters, but can reach 10,000 A/cm$^2$ at the cathode spot.

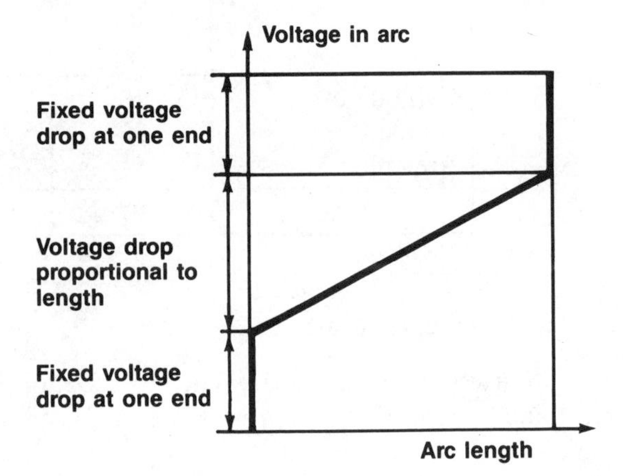


**Fig. 1.** Voltage drop in an electric arc.

— The anode drop is generally two to three times higher than the cathode drop. The anode is bombarded by electrons from the arc column; these are much more numerous than the ions (several hundred times more) and are much faster. The anode can reach temperatures of between 4,000 and 4,500°C, causing secondary emission from the anode. These electrons are responsible for the anode voltage drop.

— The voltage drop in the gas column depends on the length of the arc and the working environment (temperature and pressure inside furnace, nature of load, etc.). In the arc column, gas is ionized, but the positive charges practically balance out the negative charges, rendering the assembly overall neutral. A gas state of this type is known as a "plasma", and in this case, "arc plasma" (arc heating is in fact a special form of plasma heating, which is described in the next chapter). The electrons, due to their high mobility, are mainly responsible for electrical conductivity, and the positive ions compensate the electron charges. Due to internal collisions in the gas column, the arc temperatures can reach 6,000°C. This value is, however, simply a mean, and the temperature can rise to 15,000°C approximately in the center of the arc column.

The table in Figure 2 below gives an estimate of voltage and power distribution between the anode and cathode drops and the arc columns, for a given supply voltage. However, these proportions can vary rather widely.

Compared to resistance heating, arc heating is therefore related to the use of low voltages. The sum of the anode and cathode voltage drops gives the minimum voltage required to maintain the arc. This minimum voltage is of the order of 40 Volts, and in large melting arc furnaces, the voltage is generally between 40 and 700 V. For some chemical syntheses, long arcs require voltages of several kilovolts.

Power and voltage distribution in an electric arc [6 *a*].

| | Voltage (V) | Power (%) |
|---|---|---|
| Cathode drop | 10 | 10.8 |
| Anode drop | 30 | 32.4 |
| Arc column | 53 | 56.8 |
| Total | 93 | 100 |

**Fig. 2.**

## 2.2. The direct current arc

The arc itself is barrel-shaped. The cross-section of the arc increases with current and its resistance decreases. For example an order of magnitude of the voltage drop in an arc is:

— for low current (a few Amperes): 5,000 V/m;

— for high current (1,000 Amperes): 500 to 1,000 V/m.

The decrease in resistance is greater than the increase in current, so that the voltage drop decreases for constant arc length. At usual furnace currents, arcs therefore have a decreasing current-voltage static characteristic. The assembly behaves as a negative resistance, and the voltage-current curve is almost hyperbolic (area I of curves in Figure 3, or silent as opposed to hissing arcs).

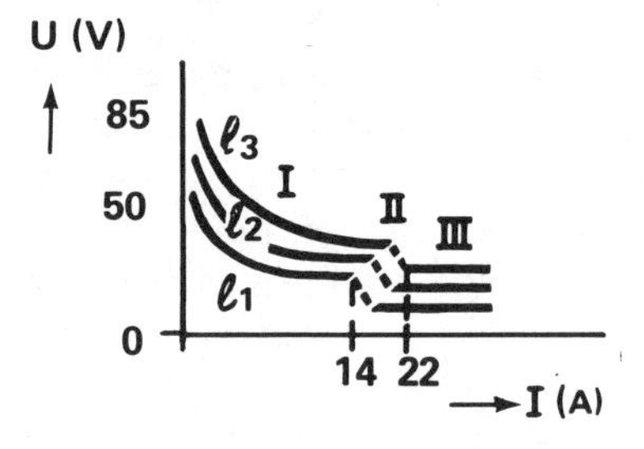


**Fig. 3.** Current-voltage curves of DC arc.

Beyond a certain current, the arc becomes unstable (area II of Figure 3) and suddenly reaches a very low positive resistance. In this area, the arc produces a hiss (area III of Figure 3 or hissing arc area). Each curve of Figure 3 corresponds to a given arc length l [33], [36], [55].

### *2.2.1. Silent arc zone*

In the silent arc zone, the voltage drop across the arc is expressed by Ayrton's formula:

$$U_0 = a + bl + \frac{c + dl}{I},$$

$U_0$, voltage in volts;
$I$, current in Amperes;
$l$, arc length in millimeters;
$a$, $b$, $c$, $d$, constants.
Constants $a$, $b$, $c$ and $d$ depend on the type of electrodes used, their dimensions and the environment in which the arc takes place (for a DC arc occurring between two graphite electrodes, $a$, $b$, $c$ and $d$ for example equal 39, 2, 12 and 20). Current $I$ is then expressed as:

$$I = \frac{c + dl}{U_0 - a - bl}.$$

Current I decreases when voltage $U_0$ increases, which provides a certain degree of self-regulation of power $U_0I$ dissipated in the arc. This self-regulation is, however, insufficient to ensure correct furnace operation. To ensure arc stability and adjustment, a resistance must be connected in series with the furnace circuit. Knowing curve $U_0 = f(I)$, it is possible to provide values for the adjusting resistor, to determine the highest value of this resistor for which the arc can subsist, and the greatest possible arc length, with the adjusting resistor constant.

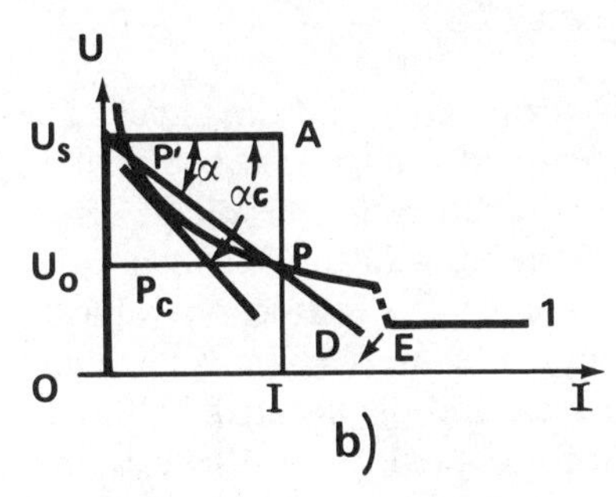


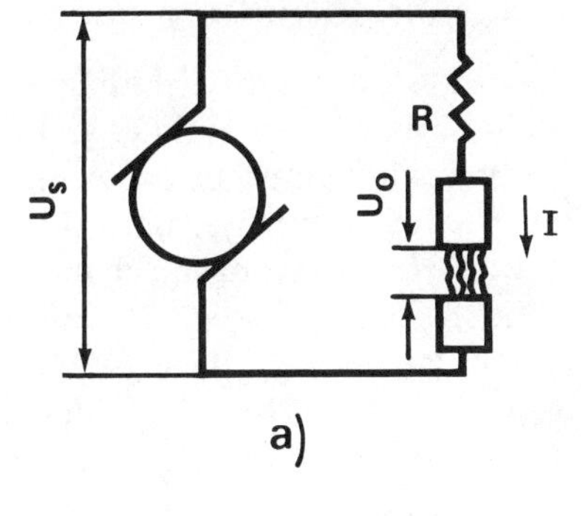


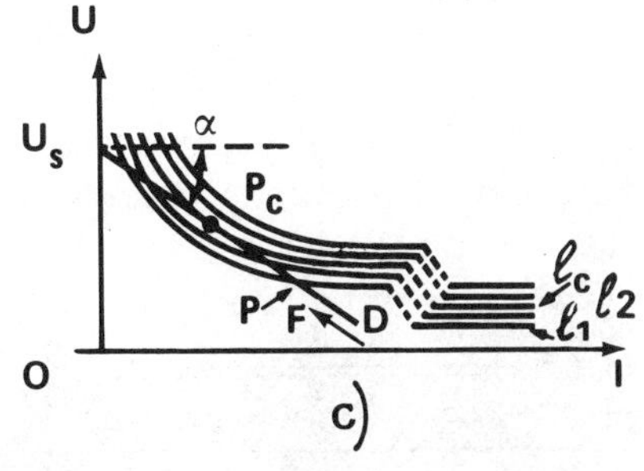


**Fig. 4.** DC arc stabilization:
a) stabilization and adjusting resistor;
b) stable arc and maximum adjusting resistor value;
c) maximum arc length for a given adjusting resistor value.

If figurative point $P$, the intersection of the straight line $D$ of equation $U_s - RI$ and the curve providing the voltage drop in the arc $U_0$ (according to

Ohm's law $U_0 = U_s - RI$) representing a stable operating point of the furnace ($P'$ is an unstable operating point), $\tan \alpha = AP/U_s A = RI/I = R$; graphically, resistance $R$ is therefore represented by $\tan \alpha$ (*fig.* 4*b*). The maximum value of $R$ for which the arc subsists is equal to $\tan \alpha_c$, $\alpha_c$ being obtained by taking the tangent of curve $U_0 = f(I)$ of the voltage drop in the arc (contact point $P_c$) through point $U_s$.

If adjusting resistor $R$ is constant, $\tan \alpha$ is constant and linear section $D$ is fixed. When the arc length increases, point $P$ moves along $D$ in the direction of arrow $F$ to arrive at contact $P_c$ (*fig.* 4*c*) with curve $l_c$ corresponding to the maximum arc length.

### 2.2.2. *Hissing arc zone*

Hissing arcs correspond to high currents. The curves of Figure 3 show that, for a given voltage $U_0$, current is undetermined (in this zone, the voltage is of the form $U_0 = a' + b'l$, $a'$ and $b'$ constant) for a DC arc. It is therefore necessary to regulate this current.

## 2.3. Alternating current arcs

Arcs can be DC or AC-supplied. In a DC arc, wear of both electrodes is unequal, since the anode temperature is much higher than that of the cathode. This dissymmetry, and above all the existence of AC line and energy losses, make AC preferable to DC.

### 2.3.1. *Characteristics of alternating current arc*

In an AC arc, each electrode is alternately the cathode and the anode. The supply voltage is sinusoidal, and the arc current and voltage vary as shown in the curves in figure 5 below. At the beginning of a half-cycle, the voltage across the electrodes increases until the arc strikes (striking voltage $U_{am}$, then drops to value $U_a$ (arc voltage); it then increases to value $U_{ext}$ (extinction voltage), while the supply voltage decreases, extinguishing the arc. The arc voltage and current are identical during the positive half-cycle and the negative half-cycle, insofar as the electrodes are identical, which, for example, is not the case in a three-phase arc furnace intended for melting scrap metal.

In practice, striking inertia and advance on extinction strongly distort the current curve, which deviates notably from the sinusoidal shape.

A series inductance is inserted in the circuit to stabilize the arc. If the arc is extinguished, the induced voltage $- L\mathrm{d}I/\mathrm{d}t$ facilitates reignition, and all the more so if the inductance causes a phase-shift between the current and voltage. The current modulus is given, considering it is sinusoidal and single-phase, by:

$$I = \frac{U}{\sqrt{(R_v + R_a)^2 + (X_v + X_a)^2}},$$

$U$, supply voltage (at transformer output);
$R_a$, arc resistance;

$R_v$, transformer resistance, supply lines;
$X_a$, arc reactance;
$X_v$, sum of installation reactances, other than arc (reactance of adjusting coil, supply conductors, transformer, etc.);
$R$, total resistance equal to $R_v + R_a$;
$X$, total resistance equal to $X_v + X_a$.

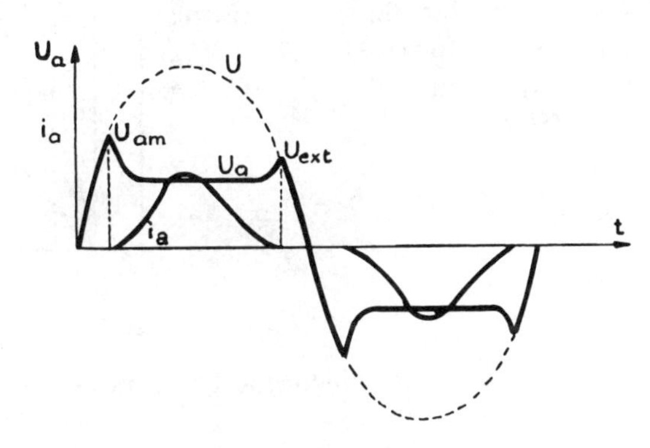


**Fig. 5.** Voltage and current evolution in an AC arc [55].

The values $R_v$ and $X_v$ depend on the furnace construction method. $R_a$ can vary between 0 (short-circuited electrode, for example on contact with the metal bath) and infinity (when the arc is extinguished). A semi-circle of diameter U/X can be considered as the circular diagram of the vector representing current (*Fig.* 6, curve 1); this is given by the expression:

$$I'_k = \frac{U}{X_v} = \frac{U}{\omega L}$$

$I'_k$, short-circuit current, loss resistance $R_v$ being neglected;
$L$, total installation inductance;
$\omega$, frequency in radians/s.

A scale indicating active power can be imposed on the vertical axis providing the current active component, and a scale representing reactive power imposed on the horizontal axis. The arc of the circle centered at the origin of radius $I'_k$ (*Fig.* 6, curve 2) also provides the power factor (cosine $\varphi$).

This type of diagram is used to study the electrical behavior of arc furnaces and to determine control parameters; in practice, arc furnace transformers have several voltage tappings (generally 6 to 14), and by plotting the curves corresponding to each voltage level, it is possible to define furnace operating conditions.

The following points of this diagram should be noted:

— point $K$, corresponding to the real short-circuit current $I_k$ defined by the equation $I_k = U/\sqrt{R^2 + X_v^2}$, being slightly less than $I'_k$ defined above;

— point $M$, obtained by plotting the tangent to the current semi-circle parallel to the vector, representing short-circuit current $I_k$ and corresponding to arc maximum power;

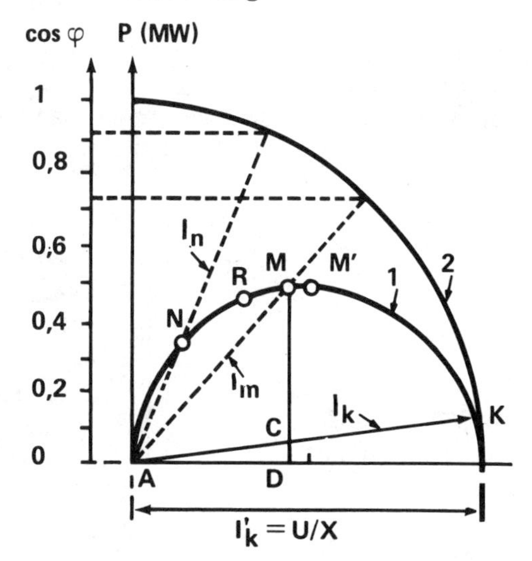


**Fig. 6.** Circular diagram of arc furnace operating current [55], [61 *e*].

— point $M'$, for which the total power absorbed by the installation is maximum;
— point $N$ corresponding to the rated power of the supply transformer;
— point $R$, for which installation efficiency is maximum.

On the power scale, segment $MC$ (together with any similar segment for another furnace operating point $M$) represents the furnace active power, segment $CD$ the resistive losses of the installation as a whole and segment $AD$ reactive power. The values can be determined for various values of operating voltage, and combined to form furnace characteristics. Figure 17 gives an example of adjustment of a Heroult UHP arc furnace (*see* paragraph 3.2.2.1.) intended for production of steels, and the corresponding curves.

Point $N$, which is characteristic of the supply transformer rated power, to the left of point $M$, represents the furnace maximum power; however, one should not conclude that the maximum arc power can never be attained. This high power has to be provided only over rather short periods, allowing an overload of 10 to 20%, depending on the transformer used, during these periods. However, these overload capabilities depend on the type of furnace.

### *2.3.2. Arc power*

The active power dissipated in the arc is equal to $R_a I^2$, i.e.:

$$P_a = \frac{R_a U^2}{(R_v + R_a)^2 + X^2}.$$

### *2.3.3. Efficiency*

The total power absorbed by the installation is equal to:

$$P = (R_v + R_a) I^2 = \frac{(R_v + R_a) U^2}{(R_v + R_a)^2 + X^2}.$$

Effective power $P_u$ represents the difference between arc power $P_a$ and thermal losses $P_c$ with $P_u$ equal to $P_a - P_c$. Efficiency $\eta$ is equal to $P_u/P$, i.e.:

$$\eta = \frac{P - R_v I^2 - P_c}{P}.$$

Generally, efficiency is high, and often between 70 and 80%. Thermal losses $P_c$ are relatively low, and efficiency $\eta$ is generally close to efficiency $\eta'$ defined by the ratio $P_a/P$, which is equal to $R_a/(R_v + R_a)$. The overall efficiency however depends on furnace operating conditions. Measured specific consumption values are given in the paragraphs describing the various applications.

### *2.3.4. Power factor*

The power factor, cosine $\varphi$ is equal to $P/UI$, i.e.:

$$\cos \varphi = \frac{R_v + R_a}{\sqrt{(R_v + R_a)^2 + X^2}}.$$

Power factor $\cos \varphi$ varies with the furnace operating point, and in the event of a short-circuit, drops to a very low value, of between 0.1 and 0.2.

Since operating point varies during furnace operation, it is therefore difficult to define an installation power factor other than the mean power factor during the operating cycle.

Appropriate connection of the power lines ensures low reactance. The power supply line length must be as short as possible, and some manufacturers have designed power transformers which tilt together with the furnace. However, circuit reactance can be too low, since the short-circuit current could become too high; the coil self-induction keeps the ratio between the short-circuit current $I_k$ and the rated current $I_n$ at a satisfactory value. The power factor and ratio $I_k/I_n$ are related by the equation:

$$\cos \varphi = \sqrt{1 - \frac{1}{(I_k/I_n)^2}} \quad [66].$$

Therefore, a high power factor, of the order of 0.95, would lead to a short-circuit current triple the rated current; in general, the short-circuit current must remain between 1.8 and 2.5 times the rated current, which, for nominal load, leads to operation at a power factor between 0.83 and 0.92. In the most modern furnaces, automatic regulation devices prevent the power factor from dropping below a predetermined threshold, providing a mean power factor of around 0.8. However, to ensure better productivity, in some furnaces other values are used (*see* UHP furnaces, paragraph 3.2.2.1.). Numerous devices are now available to limit line disturbances (*see* paragraph 3.2.6.).

### 2.4. Heat transfer in arc furnaces

The method by which the heat is transferred to the load depends basically on the type of arc furnace. The information given below essentially applies to Heroult arc furnaces intended for the production of steel. In these furnaces, the method by which the heat is transferred to the bath is closely related to arc formation, as described above. In particular, the arc moves during each half-cycle of the current on the electrode surfaces at a very high speed, of the order of 500 km/h on average, but may exceed 1,000 km/h. Under these conditions, heat transfer to the bath takes place as follows:

— the hot point of the electrode radiates onto the bath and the load;

— the plasma column transfers energy by radiation and convection to the load, to the bath and sometimes to the lining; when the current is maximum, the column is inclined and radiates towards the center of the furnace, improving heat transfer;

— the low point of the plasma column transfers heat directly to the bath on contact.

In spite of the very high arc temperature, the main heat transfer mechanism is not radiation but convection, which is responsible for 50 to 60% of the total energy transferred, due to high temperature gas jets projected at very high speeds against the load [61 *a*], [63], [E].

Arc movements also prevent local overheating and distribute the heat. The advantage of direct arc furnaces is closely related to the heat transfer method characterizing it; the slag over the bath is heated before the metal, and taken to higher temperature, rendering it highly reactive, and enabling metallurgical operations (refining, dephosphoration and desulphuration) to be performed.

In other types of arc furnace, heat transfer may take place differently (by radiation in a radiating arc furnace, by the Joule effect and thermal conduction in submerged arc furnaces, etc.).

## 3. VARIOUS TYPES OF ARC FURNACES AND THEIR APPLICATIONS

Due to the highly specific applications of these furnaces, only a brief description of each type is given.

### 3.1. Radiating arc melting furnaces

Furnaces of this type, which are known as Mazières or Detroit furnaces, are of similar design to radiating resistor melting furnaces described in the "resistance furnace" chapter.

Heating of the load is provided by radiation from an arc occurring between two graphite electrodes horizontally mounted along the furnace axis. The melting chamber is more or less spherical, since the arc length is low, and the heat source is practically pinpoint. There is no contact between the arc and the metal. A high amplitude vibration is applied to the furnace, so that the molten metal absorbs most of the heat stored in the refractory during its exposure to the arc radiation. The installed powers are generally a function of capacity and the type of metal.

Range of radiating arc steel melting furnaces [C].

| Capacity (kg) | Power (kW) | Melting time hot start (min) |
|---|---|---|
| 125 | 240 | 60 |
| 250 | 400 | 70 |
| 500 | 630 | 80 |
| 1,000 | 900 | 100 |

**Fig. 7.**

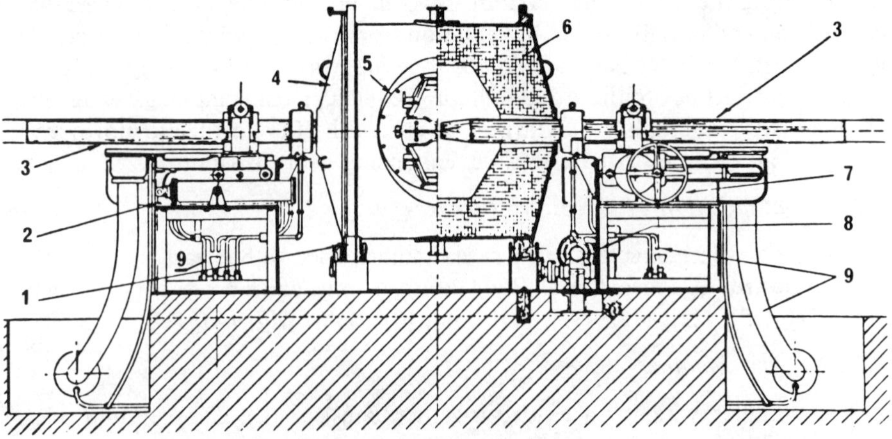

**Fig. 8.** Mazière type radiating arc furnace [63], [C].
1. rollers; 2. hydraulic servo motor; 3. electrodes; 4. housing; 5. loading door; 6. lining; 7. arc manual control; 8. pouring motor reduction gear unit; 9. piping.

The refractories used in these furnaces are of the same type as those in Heroult arc furnaces. The silica ramming mixture withstands 400 to 600 melts for cast iron. The operational characteristics of these furnaces are given in figure 9.

| Metal | Specific consumption | | |
|---|---|---|---|
| | Energy (kWh/t) | Electrodes (kg/t) | Refractories (kg/t) |
| Steel | 750 | 6 | 50 |
| Cast iron | 650 | 5 | 30 |
| Bronze | 350 | 3 | 35 |
| Copper-nickel | 500 | 4 | 40 |
| Monel | 650 | 5 | 40 |
| Nickel | 725 | 6 | 50 |

**Fig. 9.**

These simple, inexpensive furnaces are gradually being replaced by more flexible induction furnaces.

## 3.2. Direct arc melting furnaces

These furnaces are the direct result of the technique developed by Heroult around 1900, and which, since then, have been improved in capacity, power, specific power and therefore productivity. Basically, these furnaces are used for production of ordinary steels from scrap in small steelworks, special steels, steels and synthetic cast iron in casting foundries, etc.

Three-phase AC line are generally used to supply direct arc furnaces, and an arc is maintained between each electrode and the load.

Electrically, three-phase furnaces, which represent almost all direct arc furnaces in service throughout the world, are similar to star-connected systems, in which the metal is the neutral point. The metal is heated jointly by heat transfer to the hot point of the arc column in contact with the bath (convection, conduction and radiation) and by radiation from the hot point of the electrode and the arc onto the bath and load, together with the extremely powerful convection effect due to the arc column projecting high temperature gases against the load (*see* paragraph 2.4.). During one current half-cycle, the arc moves over the terminal surface of the electrode, enhancing heat distribution.

### *3.2.1. Composition of direct arc furnaces*

A direct arc furnace consists of a steel chamber, lined with refractory materials, vertical electrodes, a mechanical tilting device, a dome and a power supply system [14], [21], [25], [28], [32], [39], [45], [58], [67], [68], [70], [A], [C].

#### *3.2.1.1. The chamber*

The chamber, which is a welded together, consists of a 20 to 35 mm thick steel cylindrical or truncated cone structure, and a chamber bottom consisting of heavy stamped sheet metal, which is either elliptical or spherical. Welded horizontal and vertical reinforcements are provided to prevent deformation, and the top of the chamber generally terminates in a sand seal and a cooling water system. The chamber, together with its refractory lining, contains the load and the molten metal bath. In some cases, a magnetic coil, through which a very low frequency current flows (of the order of 1 Hz approximately) is installed in the bottom of the chamber, causing electromagnetic stirring of the molten metal, which accelerates melting, homogenizes the bath composition and causes a certain degree of outgassing [F].

The chamber is equipped with a 45° forward tilt mechanism to enable pouring, and often has a 10 to 15° backward tilt for bath cleaning. The tilting system generally consists of a cradle with two rail sections coupled to the chamber, and their braces mounted on horizontal or sometimes curved rails. Therefore, the cradle supports the chamber, and tilting is provided by hydraulic rams (or sometimes a purely mechanical system) placed in the pit, or inverted to prevent damage if the chamber is pierced. Some small furnaces tilt around the pouring spout on trunnions located on either side of the chamber.

**Fig. 10.** Tilting of a direct arc furnace [66].

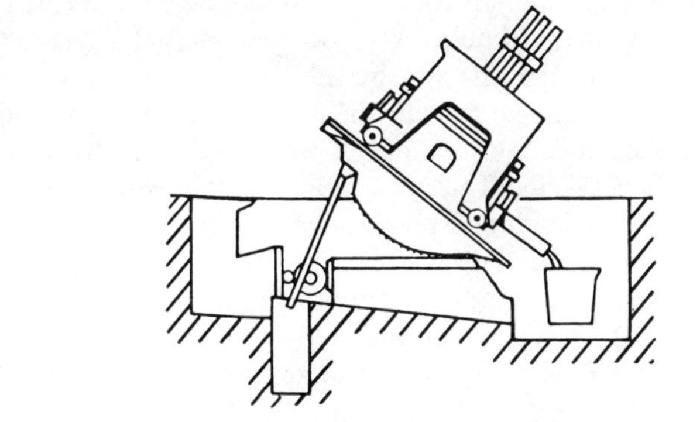

The pouring spout is located along the tilting direction at the front of the chamber, and a working door or gate used for cleaning and insertion of additives is located opposite the spout. In very low capacity furnaces (test furnaces), this door is sometimes used for loading (since the vault cannot be moved). To facilitate operations, some furnaces are fitted with side doors.

The working door, which is mechanically controlled by hydraulic or pneumatic rams, or manually controlled by levers, is generally water cooled.

**Fig. 11.** Diagram of a direct arc furnace [66].

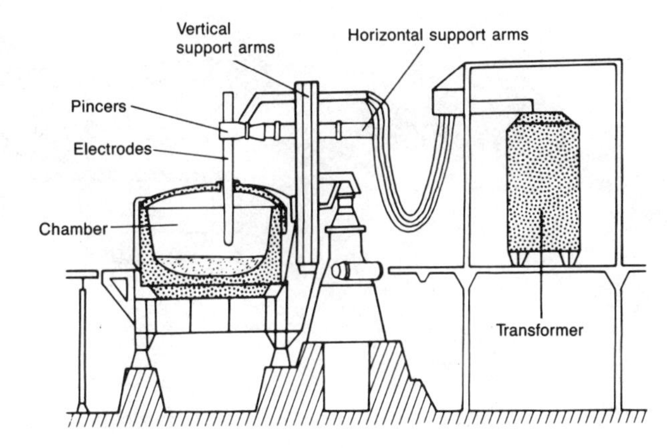


### *3.2.1.2. Vault and loading*

The furnace is capped by a vault resting on the chamber through a sand seal, or resting directly on the refractory material. The vault consists of a steel structure, the vault ring which is generally water cooled, holding the refractory bricks; frequently, this is spherical, offering better mechanical strength and longer service life. The vault has three electrode holes and, in some cases, a hole for venting fumes (this can also be achieved through the electrode holes). The hole peripheries are water cooled.

To facilitate loading, industrial furnaces are fitted with movable vaults. With arc furnaces, the loading system has a definite effect on productivity, since, if this operation is conducted rapidly, down-times are reduced and production rates increased. Mainly, two processes are used: overhead cranes and pivoting vaults.

With the first system, the crane straddles the chamber and carries the vault, which can be raised. The crane is mounted on rails, and when moved, carries the vault, thus completely uncovering the chamber. In the second process, the ram, vault and electrode arm assembly raises and pivots, to uncover the chamber during loading. Movable chamber furnaces, in which the chamber moves when the vault is raised to receive the load, can also be constructed. In small test furnaces, loading is often made through a side door.

Different systems are used to introduce solid loads, opening loading buckets, with flexible or formed quadrants, etc. (*fig.* 12).

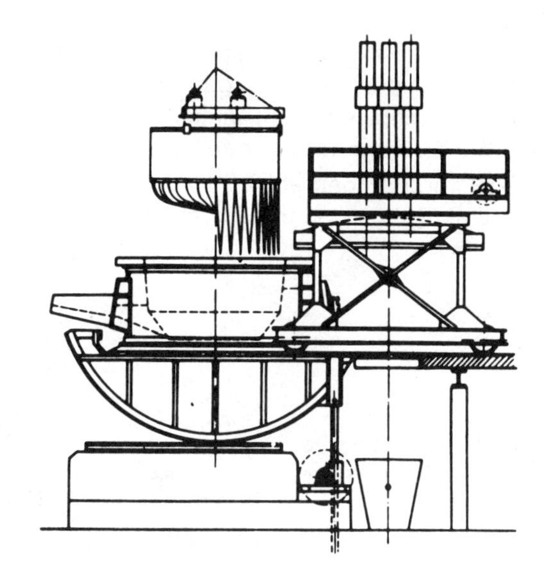

**Fig. 12.** Mobile gantry vault furnace.

### *3.2.1.3. Electrodes*

Electrodes must have low resistivity and high elasticity and mechanical strength. Three types of electrodes can be used in arc furnaces: graphite, amorphous carbon and self-baking electrodes.

#### *3.2.1.3.1. Graphite electrodes*

The manufacture of graphite electrodes was described in "Conduction heating" (graphitizing furnaces, paragraph 4.1). Because of their multiple advantages such as ease of machining, regularity, high current density capability, low risk of accidental carburation in melting baths, direct arc furnaces use only this type of electrode. The current density is a function of the diameter and varies from 30 to 50 A/cm$^2$ for small diameters and 15 to 25 A/cm$^2$ for large diameters. In the most powerful furnaces, the current carried can be up to 100,000 Amperes.

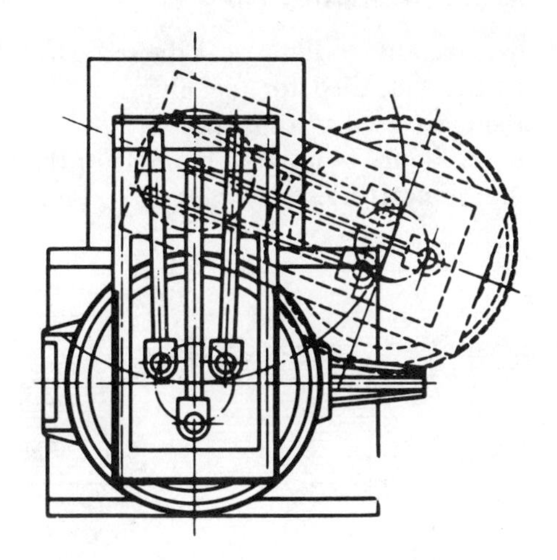

**Fig. 13.** Pivoting arm vault furnace [63].

Electrodes wear permanently, and must be renewed frequently. To achieve this, graphite electrodes are butt jointed, either by means of threaded cylindrical or twin-tapered graphite couplings, which screw onto the nipples of the electrodes, or screw directly into each other, the first solution being the most common. This operation can be performed with the furnace in operation, or outside the furnace.

*3.2.1.3.2. Amorphous carbon electrodes*

Amorphous carbon electrodes have a lower load capability than graphite electrodes, and can only handle current densities not exceeding 8 A/cm$^2$. For a given current, these electrodes are therefore heavier and of larger diameter than graphite electrodes. These electrodes have been more or less abandoned for graphite electrodes which are lighter, wear less rapidly and offer better mechanical qualities.

*3.2.1.3.3. Self-baking electrodes*

Self-baking electrodes, which are also known as Söderberg electrodes after their inventor, consist of a mixture of coke, anthracite, pitch and tar. The paste is inserted in a cylindrical sheet metal jacket. As the electrode descends into the furnace, the paste hardens and bakes due to the combined effect of furnace thermal losses and Joule losses. To compensate electrode wear, all that is required is to add paste and provide a sliding system for the electrodes.

Electrode diameters can reach 140 cm. Specific load varies between 6 A/cm$^2$ for a diameter of 140 cm and a current of 100,000 A to 10 A/cm$^2$ for a diameter of 25 cm and a current of 5,000 A.

Söderberg electrodes are only rarely used in direct arc furnaces, since they require frequent removal of the electrodes; conversely, they are used in most reduction furnaces (*see* paragraph 3.3.).

*3.2.1.4. Refractory lining*

Generally, to limit wear the refractory lining of a furnace must be the same as the slag used for metallurgical operations. Acid or basic refractories are therefore used as a function of furnace operation. Neutral refractories are also used. Figure 14 gives an example of the lining of a 4.8 m furnace, with acid or basic refractories (46), (63), (67), (68).

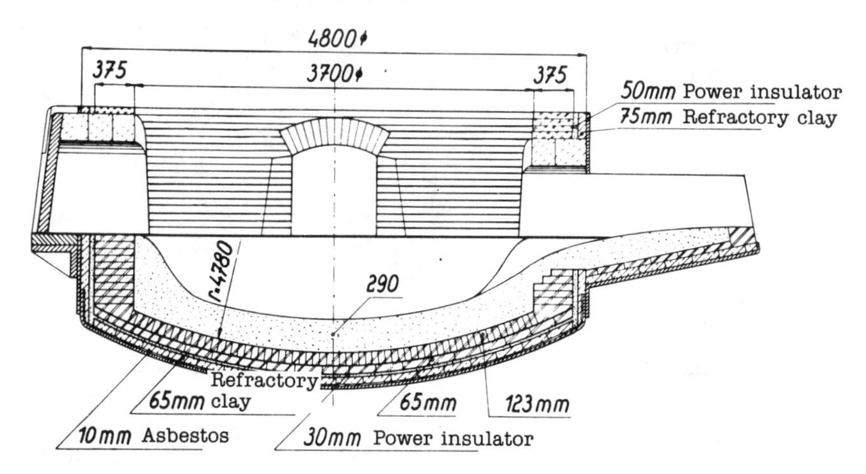


**Fig. 14.** Cross-section of furnace through pouring spout [63].

— *Acid lining*

This type of lining is used for production of pig iron and steels from scrap containing little phosphorous or sulphur, and is used mostly in casting foundries. Refractory and insulating bricks, together with the ramming mixture, are silica or alumino-silicate products. The service life of the ramming mixture lining for a furnace producing pig iron is between 250 and 600 melts, while that of the vault is between 100 and 300 melts (13 kg/t of metal for the chamber, 4.5 kg/t for the vault).

— *Basic lining*

Basic lining is mostly used for production of steels and, in particular, fine steels. Refractory bricks and insulators and the ramming mixture are magnesia or dolomite based (double carbonate of lime and magnesia). Sometimes the vault is made from alumino-silicate bricks, and more often from high alumina content bricks. With UHP furnaces (*see* paragraph 3.2.2.1), the chamber is often made from chromite-magnesia bricks; for these furnaces, the service life of the chamber bricks and that of the vault is between 150 and 160 pours (5 kg/t for a 64 t, 46 MVA furnace).

— *Lining technique evolution*

To improve the productivity of arc furnaces, lining techniques have evolved in two ways:

• rapid replacement of the worn refractory using an interchangeable chamber or cylinder;

• replacement of the refractory by cooled panels, which thus form a quasi-permanent lining, while water circulation structures replace the conventional refractory part; the service life of such structures is 1,000 pours approximately. Refractory consumption is therefore lower, but energy consumption increases by approximately 30 kWh/t.

**Fig. 15.** Lining with water circulation structure

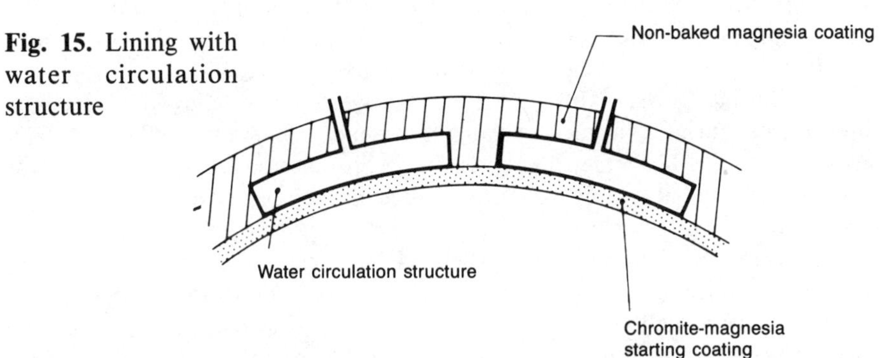


### *3.2.1.5. Electrical equipment*

The electrical power supply equipment for arc furnaces is, in principle, very simple. Due to the very high currents, equipment must have very high quality characteristics.

**Fig. 16.** Diagram of a direct arc furnace [63], [C]:
1. mechanical or electrically controlled isolating switch;
2. fuse;
3. potential transformers;
4. furnace remote control circuit-breaker;
5. high voltage current transformers;
6. choke;
7. remote controlled choke switch;
8. furnace transformer;
9. low voltage current transformers;
10. arc furnace.

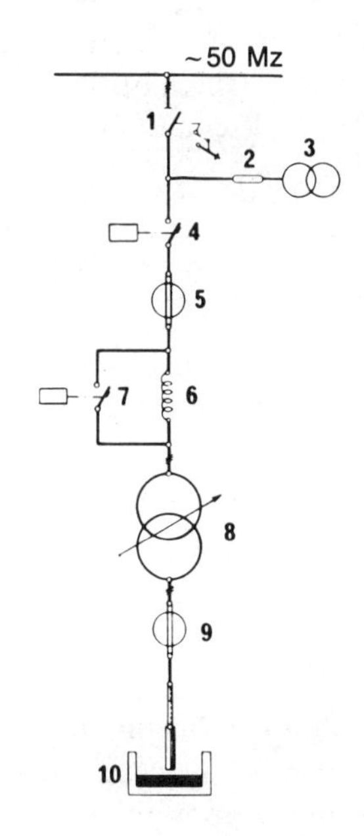

Transformers must accept overloads of approximately 20%. Depending on power, these can be supplied directly up to 220,000 Volts. The voltages at the secondary vary with furnace capacity: below 10 t, they are up to 200-250 V, while for larger furnaces, 100 to 450 V. Generally, transformers are equipped with stepped adjustable selector switches on the high tension line; in practice, the number of steps varies from 6 to 14. Oils, more often pyrolene, are used as cooling fluids [42], [43], [44], [53], [63].

The furnace circuit-breaker must provide a large number of tripouts (60 to 200 per day) and therefore must be overhauled frequently. Most often, a pneumatic compressed air circuit-breaker is used.

The flexible power cables must be enough long to allow for electrode movements, furnace tilting and vault movement. Generally, cables are water cooled and must be arranged so as to reduce installation reactance as much as possible (the spatial arrangement of the three phases must be as near as possible to an equilateral triangle).

The electrode clips must have surfaces large enough to prevent excessive overheating, and are water cooled; these must be made from a non-magnetic metal. For very high currents, more than 30 kA, these contacts are generally made from copper, protected from oxidation by coatings.

### *3.2.1.6. Regulation*

An electrode automatic regulation system is required to balance the phases, control electrode descent as and when the load melts, maintain arc length constant with respect to electrode wear and to stabilize arcs [26], [27], [35], [63].

Most modern regulation systems operate at constant impedance and affect the complete length of the arc, so as to keep the ratio between electrode voltage U and current I constant for each electrode. The arc mode is represented by the equation:

$$\boxed{U - AI = O,} \qquad \text{in which } A \text{ is a constant.}$$

If $U - AI$ is positive, the current increases and the electrode descends until this equation is satisfied. If $U - AI$ is negative, the current decreases, and the electrode rises until $U - AI = O$.

It is also possible to introduce a supplementary term into the above equation, which then becomes:

$$\boxed{U - AI - B = O,} \qquad \text{in which } A \text{ and } B \text{ are constants.}$$

Constant $B$ offers the advantage of enabling the electrodes to rise automatically in the event of a power cutoff ($U = O, I = O$).

Depending upon the manufacturers, electrode movement is provided either by hydraulic rams or electric motors. With the development of electronics (semi conductor signal amplifier) and computer control, all-electric systems are preferred. Modern regulation systems have very short response time, and also prevent oscillations [41], [59], [61 *c*], [72].

#### *3.2.1.7. Fume and dust trapping*

Arc furnaces are relatively slightly polluting. The quantity of dust produced varies from 2.5 to 15 kg/t, the average being between 10 and 13 kg/t. Treatment of gas given off from arc furnaces consists essentially in cleaning. The dust, which contains a high proportion of iron oxide, up to 80 to 90%, can be recovered to produce high iron content brickettes.

It is also possible to recover energy from fumes to preheat loads. The use of oxygen increases the volume of the gases produced, and therefore requires larger fume treatment systems.

### ***3.2.2. Control of direct arc furnaces and applications***

#### *3.2.2.1. Electrical control — UHP furnaces*

##### *3.2.2.1.1. Principle*

In particular, electrical control concerns the power and length of the arcs. Basically, power is increased by using higher voltage. However, arc power is not the only criteria to be taken into account; stability, power factor and length are also interdependent values. Electrical controls for an arc furnace are designed using a circular diagram, as shown in paragraph 2.3. Figure 17 provides an example of electrical control of a UHP arc furnace using this diagram [63].

Computerized control of arc furnaces is being more and more widely used (power demand, electrode position adjustment, etc.) [61 *d* and *e*], [45].

##### *3.2.2.1.2. Development of electrical control of arc furnaces*

Since the end of the 1960's, electrical control of arc furnaces has changed radically. In conventional arc furnaces, thin, low current, relatively long arcs supplied at rather high voltages were preferred. This type of operation provided excellent electrical efficiency due to the reduction of joule effect losses, and to rather small cross-section conductors, together with a high power factor, thus optimizing furnace electrical design [42], [47], [63].

Conversely, short, high current arcs are better for heat transfer and reduce refractory wear. However, the power factor is lower than in the previous case, about 0.7, while in conventional ovens, this is between 0.8 and 0.85 (it is, however, possible to install capacitors to rectify the power factor). Short arc furnaces therefore enhance thermal efficiency and installation productivity. Improvement in thermal efficiency also more than compensates for the reduction in electrical efficiency, thus increasing installation overall efficiency (decrease in specific consumption). Moreover, high current arcs demand less electrode movements for regulation, and the flicker phenomenon is reduced. The furnace operating sequence can then be described schematically as follows:

— *Well drilling*

The electrodes drill their wells in the load. The time taken is a function of the load density and arc power; generally, this is of the order of 5 minutes. The arc column is relatively long during this period, so as to enlarge well diameter and prevent damage to the furnace hearth.

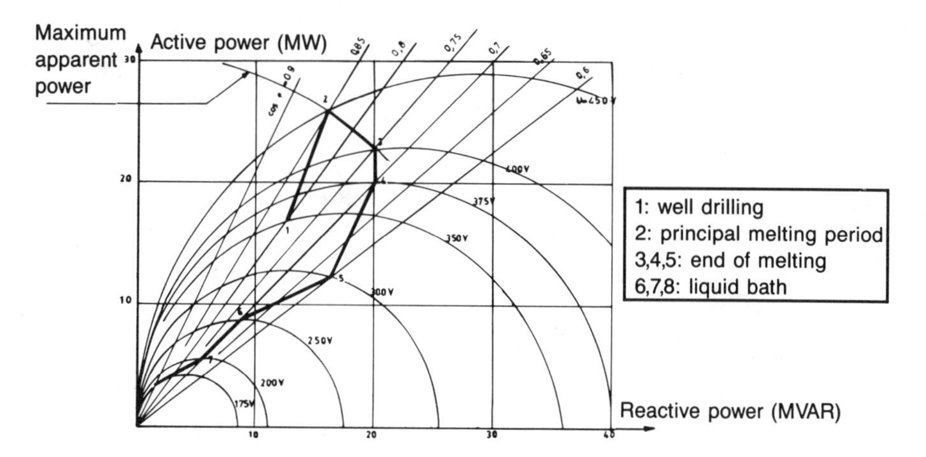


**Fig. 17.** Examples of electrical settings for a UHP furnace in the circular diagram [63].

— *Protected arc period*

During this period, arcs are completely surrounded by the load. They are relatively unstable, since scrap metal movements quench the arc and cause short-circuits; the current and voltage therefore fluctuate strongly. The power may be very high during this phase, since there is a risk of damage to the refractory materials. Modern furnaces therefore operate with short arcs and use high currents and voltages.

— *Calm bath period*

This period characterizes end of melting. During this period, the refractory may be damaged, since the arc radiates directly onto the walls; the arc must be set very short, and this period must be as short as possible.

— *Metallurgical operation period*

The power may often be strongly reduced during this period, which consists of providing the liquid metal with the desired chemical composition, and subjecting it to certain characteristic improving treatments.

*3.2.2.1.3. The UHP concept*

The evolution of the construction and operating characteristics of arc furnaces has led to the definition of UHP (*Ultra High Power,* later *Ultra High Productivity*) furnaces. Originally, the UHP concept concerned the use of much higher melting power levels than usual, and in particular, short, high current arcs during melting of the load when the refractories are most exposed to radiation [11], [12], [13 *a*], [16], [40], [63].

Other criteria were later added to this idea to define an operating mode for direct arc furnaces. When these conditions are respected, furnace productivity is maximum.

— *Specific power level*

The diagram in figure 18 gives three areas, corresponding to different specific power levels (kW per ton of capacity) for arc furnaces.

The specific power of UHP furnaces is 1.5 to 2 times higher than that of ordinary furnaces (RP on figure 18).

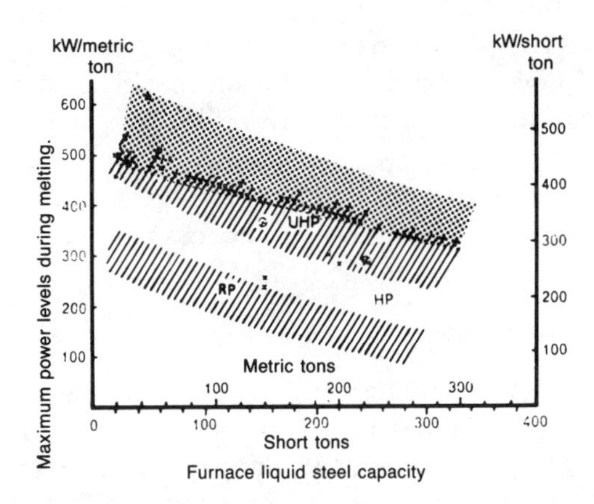


**Fig. 18.** Arc furnace specific power levels [40], [63], [E].

— *Mean to maximum power ratio*

In UHP operation, the mean power (energy consumed during the cycle, divided by the total furnace power on time) to maximum power ratio must be high, greater than 0.7.

— *Time utilization*

The time utilization is the ratio of the total powered time to the times separating two pours. In UHP, this should be greater than 0.7.

— *Adaptation of power and arc length to heat transfer conditions*

The arc length must be adjusted to ensure the highest possible productivity and the best heat transfer conditions. The power factor provides an indication of arc length.

— *Electrical balancing of three arcs*

Electrical balancing of three arcs (power balancing) balances the furnace and provides optimum operating conditions. Balancing is provided by the furnace control system, and in particular by the electrode position adjustment system.

*3.2.2.2. Metallurgical control of direct arc furnaces*

Due to the high temperatures reached by the bath surface slag, direct arc furnaces facilitate metallurgical operations such as refining, dephosphoration, desulphuration, adjusting and calming. This paragraph provides a brief review of these metallurgical operations which take place once the metal is molten.

*3.2.2.2.1. Metallurgical operations*

— *Dephosphorization*

Phosphorous is converted into phosphoric anhydride ($P_2O_5$) by oxidation; this combines with the lime contained in the slag (basic slag) to provide phosphate of lime. The phosphate thus formed is eliminated by cleaning to prevent later reduction of the phosphoric acid, which would increase the phosphorous content.

Oxidation is obtained by adding ore (iron oxide FeO), or in modern processes, by blowing gaseous oxygen into the bath. In more recent techniques, the lime powder is introduced in an oxygen nozzle, enabling rapid and thorough dephosphorization [63], [64].

— *Refining*

In fact, refining consists in decarburizing the metal. To adjust the carbon content of the steel, it is in fact preferable to decarburize the bath first of all, and then recarburize to the desired content. In modern installations, oxygen decarburizing is used. Elimination of carbon results in the giving off of carbon monoxide (CO) in the form of gas bubbles, which draw off the gases such as hydrogen dissolved in the liquid steel.

Other oxidizable elements such as silicon or aluminum are also eliminated during this phase.

— *Deoxidation*

After the oxidizing phase, the liquid bath contains a certain proportion of iron oxides which are distributed according to the laws of equilibrium between the slag and the steel. Deoxidation is initially performed on the slag, to which lime is added to reduce the iron oxide content and the content of reducing agent such as silicon and carbon.

At the end of the operation, deoxidation is completed by adding energetic deoxidants such as aluminum, silicon or titanium.

— *Desulphurization*

Sulphur, which is present in the steel in the form of iron sulphides and manganese sulphides, are eliminated by deoxidization through the slag. These sulphides, in fact, combine with the slag lime (and, where applicable, with calcium carbide) to produce calcium sulphide, if the residual iron oxide and manganese contents are sufficiently low. New nozzle-type desulphurization techniques, where oxygen and lime powder are blown in, have also been used in recent proceses.

— *Adjustment, calming and pouring*

Adjustment consists in setting the metal bath to the final composition or grade and adjusting its temperature. Calming prevents outgassing during pouring, in particuar due to reaction with oxygen and carbon, to produce carbon monoxide. Both these operations are generally performed simultaneously. The composition is adjusted by adding graphite and ferro-alloys, so as to maintain each element at the desired content. Calming is obtained by introducing aluminum in conjunction with other deoxidants (Si-Mn, Si-Ca, Fe-Si-Zr, etc.).

*3.2.2.2.2. Basic and acid operation*

The above sequence of operations and the conditions under which they take place depend basically on the product produced (pig iron, steel, etc.), the raw materials used and the desired quality. These processes form the metallurgist's art, and direct arc furnaces are one of his best tools.

Two major families of processes are used: basic operation and acid operation, named after the slags employed. Figure 19 gives the sequence in which the various operations are conducted for each process.

Arc furnace metallugical control [63].

| Melting<br>Adjustment | Melting<br>Refining<br>Adjustment<br>Calming | Melting<br>Dephosphorization<br>Refining<br>Adjustment<br>Calming | Melting<br>Dephosphorization<br>Refining<br>Desulphurization<br>Adjustment<br>Calming |
|---|---|---|---|
| Pig iron | Steel | Steel | Steel |
| Lining (acid) | | | |
| | Lining (basic) | | |
| | | 1 slag | 2 slags |

**Fig. 19.**

*3.2.2.3.Direct arc furnace applications*

*3.2.2.3.1. Production of steel*

For steel, basic operation, which was implictly described in the previous paragraph concerning metallurgical operations, is used to produce all steels from highly diverse loads, since the use of two slags makes all metallurgical operations possible. If desulphurization is not required, only one slag may be used. These processes are primarily intended for the production of fine and special steels.

Acid operation, which comprises neither dephosphorization nor desulphurization, is used essentially in casting foundries. Melting, refining and adjusting are very rapid, and the metal obtained is of excellent quality. In steel casting foundries, this process is substituted for the cupola — converter system for economic and technical reasons (better metal quality, less oxidation losses, etc.).

In steel-making applications, in addition to the production of high quality steels, ordinary steels ("concrete rod" type) are produced in "mini steel-making plants", installed near the consumption site, and using cheap scrap as raw material.

The direct arc furnace is also used for the production of steels of all types from prereduced ore in the form of pellets. This process, which is being developed in some countries, has not been applied in France [31], [44], [67], [68], [70].

*3.2.2.3.2. Production of pig iron*

Some large casting foundries use direct arc furnaces to produce pig iron. These are acid operation furnaces, and the process consists of single melting and adjustment. However, it is possible, if necessary, to desulphurize by injecting a mixture of nitrogen and calcium carbide, although, in this case, pocket desulphurization is the best technique.

For melting casting iron, arc furnaces compete with crucible induction furnaces. Basically, their advantage is the possibility of using low price scrap and power density; conversely, they call for rather high investment costs, discontinuous operation and complete furnace loads (holding furnaces are often required) and do not provide bath stirring. Therefore, an investment choice calls for a thorough study of the complete melting system, so as to evalute the feasibility of each solution.

*3.2.2.3.3. Other applications*

Production of steel and pig iron is the principal application of direct arc furnaces, and steel melting remains an essential part of the production system.

These furnaces can be used for other applications such as remelting ferromanganese for addition to a Thomas or oxygen-type converter, and also copper alloys; the last application, however, is being replaced by induction furnaces.

### *3.2.3. Range of direct arc furnaces*

Direct arc furnaces offer an extremely wide range of capacities and powers, since capacities range from a few hundred kilograms to more than 300 tons, and powers from some hundreds of kilowatts to 100,000 kW approximately. The largest furnace in the world, in service in the United States has a capacity of 400 "short tons", i.e. 360 metric tons approximately, with an apparent transformer power of 162 MVA.

The table in figure 20 provides a range of UHP arc furnaces (in the lower portion of the UHP range, presented in Figure 18). Some UHP furnaces are

1.1 to 1.3 times more powerful (for example 80 MVA for a 135 t furnace with a chamber diameter of 6.7 m, 18 MVA for a 20 t furnace with a chamber diameter of 4 m). For comparison, and for conventional furnaces of 20 tons, the specific power was only 450-500 kVA/t, while with UHP, this is 750-800 kVA/t, i.e. 1.6 times more.

The melting rate depends on the material melted, the temperature level, specific power and furnace control. During the melting phase, electricity consumption is between 420 and 450 kWh/t for steel (hot start), i.e., with a power of 100 kW, it is possible to obtain a melting rate of 2.2 to 2.4 t/h. For pig iron, the melting rate is more or less the same, since the load generally consists of scrap steel; the final temperature is lower than with steel, but the load must be recarburized, which gives comparable melting times.

Range of direct arc furnaces [C].

| Capacity (t) | Power (kVA) | Chamber diameter (m) | Electrode diameter (mm) | Specific power (kVA/t) |
|---|---|---|---|---|
| 3 | 2,600 | 2.2 | 200 | 866 |
| 6 | 4,800 | 2.8 | 250 | 800 |
| 12 | 9,600 | 3.2 | 350 | 800 |
| 20 | 15,000 | 4 | 400 | 750 |
| 40 | 24,000 | 4.6 | 450 | 600 |
| 70 | 60,000 | 5.2 | 500 | 560 |
| 90 | 48,000 | 5.8 | 550 | 540 |
| 130 | 68,000 | 6.4 | 600 | 520 |
| 180 | 90,000 | 7 | 600 | 500 |

**Fig. 20.**

This melting range must not, of course, be confused with furnace production, which depends essentially on the metallurgical operations subsequently performed. Thus, in modern steelworks and in basic operation, metallurgical operations represent 30 to 70% of the time required for melting itself (about 15 years ago, this time was often between 1 or 2 times melting time). More and more metallurgical operations are being performed in ladle, therefore increasing the furnace use factor.

### *3.2.4. Specific consumptions*

In addition to refractories, for which some figures were given in paragraph 3.2.1.4., arc furnaces consume energy, electrodes, water and often oxygen.

#### *3.2.4.1. Energy consumption*

Specific consumptions vary widely and depend on the type of raw materials, the products produced, temperature level, furnace specific power and control. Theoretically, the energy required to take one ton of steel to its melting point is 380 kWh; in practice, overall specific consumption varies between 500 and 1,000 kWh/t [63], [66].

The most frequently encountered overall specific consumptions occur for steel production:

— in acid operation, between 500 and 650 kWh/t, with oxygen blow through, towards 650 kWh/t with ore being added during metallurgical operations;

— in basic operation, between 600 and 700 kWh/t with oxygen blown through, and around 700-750 kWh/t with ore.

Specific consumption increases with the size and power of the furnace, and is between 500-550 kWh/t for large UHP steelwork furnaces.

Generally, the oxygen consumed in steel melting represents 5 to 10 cubic meters per ton, and 15 to 38 kWh/t when used for refining.

For producing synthetic pig iron from scrap and carbon, specific consumption is generally between 500 and 600 kWh/t (for example, this is 480 kWh/t for a UHP furnace of 20 t, and a power of 18 MVA, producing basic pig iron at around 1,450°C).

#### *3.2.4.2. Electrode consumption*

The figures published in this domain differ widely. The most currently used values are:

- iron melting: 3.8 to 8 kg/t;
- steel melting: 4 to 8.5 kg/t.

#### *3.2.4.3. Water consumption*

Water consumption increases with furnace size. Of the order of 26 $m^3$/hr. for a 10 ton furnace, it reaches 38 $m^3$/hr. for a 17 ton furnace, and 70 $m^3$/hr. for a 40 ton furnace.

### *3.2.5. **Energy recovery***

Direct arc furnaces are highly efficient, since efficiency during melting is between 85 and 90%, and overall efficiency is sometimes more than 50-60% (including the energy required for metallurgical operations). However, it is possible to recover fumes to preheat the load prior to insertion in the furnace, together with the cooling water for various parts of the furnace, for heating premises and production of hot sanitary water or water necessary for other industrial processes. Few results are at present available, but the efficiency gain should remain low, at about 5-10% [65].

### *3.2.6. **Power line disturbances created by direct arc furnaces***

Due to the high powers involved, but above all to sudden power variations, power feeding of arc furnaces calls for special precautions. Unwanted alterations in arc length subsequent to changes in the load position and the physical phenomena taking place in arcs, in fact cause continuous current and power fluctuations. Basically, these disturbances occur during the melting phase, and become very limited during refining. Therefore, unlike reduction furnaces, melting furnaces require special devices intended to limit disturbances.

In particular, these phenomena involve line voltage variations which may disturb other consumer equipment such as lighting installations, television sets and computers. In particular, incandescent lamps are highly sensitive to such voltage variations, since the light flux varies very quickly with supply voltage (flicker).

To remedy these drawbacks, auxiliary devices must be used, and the furnace must be connected to power lines that tolerate high short circuit power and therefore to high voltage lines.

The variation of voltage amplitude results basically from the reactive power variations of arc furnaces, and, as an initial approximation, can be expressed by the equation:

$$\boxed{\frac{\mathrm{d}U}{U} = \frac{\mathrm{d}Q}{P_k}} \qquad [51], [61\ b],$$

$\mathrm{d}U/U$ representing the relative voltage variation;
$\mathrm{d}Q$, variation in demanded reactive power;
$P_k$, line short-circuit power.

Flicker measurement methods and imposed limitations have not, up to the present, given rise to the adoption of a general set of regulations. Various methods have been proposed, and the one most widely used in France combines flicker level (flicker begins to become noticeable for a relative voltage variation of 0.3-0.35%), frequency (maximum sensitivity of the average individual, representing the users as a whole, is around 10 Hz) and the received flicker level.

Generally, in order not to disturb the line, the short-circuit power must be approximately 100 times higher than the nominal power of a conventional arc furnace, and 65 times higher than that of a UHP furnace.

Several types of auxiliary devices are used to limit voltage variations and therefore the flicker phenomenon:
— connection to a high voltage level;
— synchronous compensators (synchronous rotating machines) in parallel with the supply busbars;
— batteries of capacitors (these, however, do not offer rapid compensation, even when used with thyristor static contactors);
— stato-compensators, consisting essentially of reactances controlled by thyristors, and sometimes capacitors.

It is probable that, in the future, thyristor electronic control static systems will be developed and will be substituted for other systems. The requirement of connection to a sufficiently high voltage will, however, remain, since the use of a stato-compensator, for example, can reduce the line short-circuit power to furnace rated power ratio from 100 to 50 and 65 to 33. [13 *b,c,d*], [19], [34], [37], [51], [54], [60], [61 *b*], [72].

### 3.3. Submerged-arc furnaces

As described at the beginning of this chapter, heating in these furnaces is based more on direct current flow through the load than on arc heating, since the electrodes are submerged directly in the load. They are used mostly for the production of ferro-alloys (ferrosilicon, ferrochromium, ferromanganese, ferromolybdenum, ferrotungsten, ferrovanadium, etc.) and assimilated products (silicocalcium, silico-aluminum, silicomanganese, etc.) by reduction of some ores occurring in the form of oxides. The most widely used reducer is carbon (and sometimes silicon) and the reaction taking place in a reduction furnace is often (schematically) of the type:

$$MO + C + \text{electric energy} \rightarrow CO + M.$$

(where M is a metal).

However, products are not limited to ferro-alloys. Some of the products enabling this technique to be used are:

— calcium carbide ($CaC_2$) from lime ($CaO$) and coke, with carbon monoxide given off [10], [D];

— phosphorous, from a mixture of tricalcidiphosphate (or soda phosphate), carbon and quartz containing sand [15], [33];

— melted cement, from a mixture of bauxite, lime and carbon [15];

— pig iron in electric low shaft furnaces, from iron ore and coke [15], [17];

— refractory materials or insulators, such as silicon carbide, corundum, magnesia, mineral wool [17], [33];

— metals such as tin, copper, zinc, lead and manganese [1], [2], [30].

### *3.3.1. Submerged arc heating process*

With furnaces of this type, the complex behavior of the arc is of only secondary importance. Conversely, since the load is essentially heated by the Joule effect due to the current flowing through it, current flow in the molten bed is of fundamental importance, as well as the furnace-load electrical parameters (melting bed resistivity, electrode separation, power supply voltage).

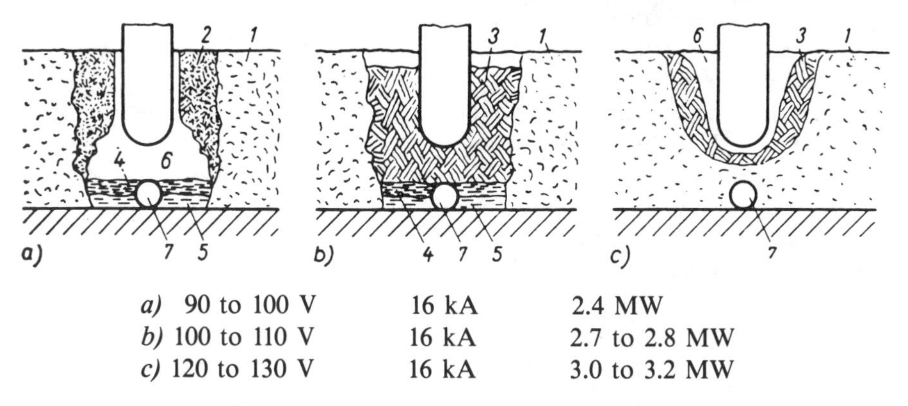


| | | |
|---|---|---|
| *a)* 90 to 100 V | 16 kA | 2.4 MW |
| *b)* 100 to 110 V | 16 kA | 2.7 to 2.8 MW |
| *c)* 120 to 130 V | 16 kA | 3.0 to 3.2 MW |

**1. Reserve of material forming the melting bed; 2. Melting bed with crust; 3. Paste type melting bed; 4. Liquid melting bed; 5. Liquid product; 6. Hollow space; 7. Pour hole**

**Fig. 21.** Optimum voltage at the electrodes of a reduction furnace [3].

Figure 21 represents various operating conditions for resistance arc furnaces. In case a), only the hottest parts of the melting bed provide conduction, since the voltage is too low, and the rest of the bed stays cold. Hard crusts form, which render the melting bed progress along the electrode towards the reaction zone more difficult, and prevent venting of gases formed during chemical reactions. In situation b), the voltages are close to their optimum values; the current density attained enhances softening of the melting bed and its renewal along the electrode. Finally, if the voltage becomes higher than an optimum value, case c), the part of the melting bed in the immediate vicinity of the electrode is overheated, and the latter is not sufficiently immersed in the load. Bath losses due to radiation and vaporization increase, decreasing efficiency (except when vaporization is necessary, such as in production of magnesium in a partial vacuum).

The melting bed resistivity can be varied, even when the voltage applied to the furnace or the electrode load possibility is limited (for example, by varying the carbon content of the load or the varieties of carbon used — metallurgical coke, lignite, mixture granularity, etc.).

In reduction furnaces, there is no anode or cathode voltage drop as in true arc furnaces. The specific powers are therefore much higher.

### *3.3.2. Submerged arc furnace composition*

#### *3.3.2.1. General design*

Constructionally, these furnaces belong to the direct arc furnace family; basically, a submerged arc furnace consists of:

— a steel chamber with a complex thermally insulated refractory lining;

— the electrodes, with their suspension, regulation and slippage system; electrode movement can be provided by a hydraulic or electromechanical system. Reduction furnace regulation can be of the constant current or constant impedance type;

— a furnace loading system; in continuous reduction furnaces, the load is applied permanently through fixed or moving filling ducts, and efforts are made to introduce the load near the electrodes or over the source points in which the melting bed has collapsed;

— a step-down voltage transformer, power cables and, in particular, the conductors of the transformer secondary, arranged so as to offer minimum reactance, and the parts in contact with the electrodes (for example, equilateral triangle arrangement of the electrodes and three single-phase transformers supplying them, known as the "knapsack" configuration);

— a gas and fume collection system; gas energy is often recovered to improve the furnace thermal balance.

Reduction furnaces are not tilting; in some cases, the chamber rests on a frame, which may be slowly rotated. Pouring of the product is made through one or more pour holes, located at the same level as the furnace hearth [8],[9], [18].

#### *3.3.2.2. Different types of submerged arcs*

Numerous types of submerged arcs exist, and different classification criteria can be used. The major classifications are:

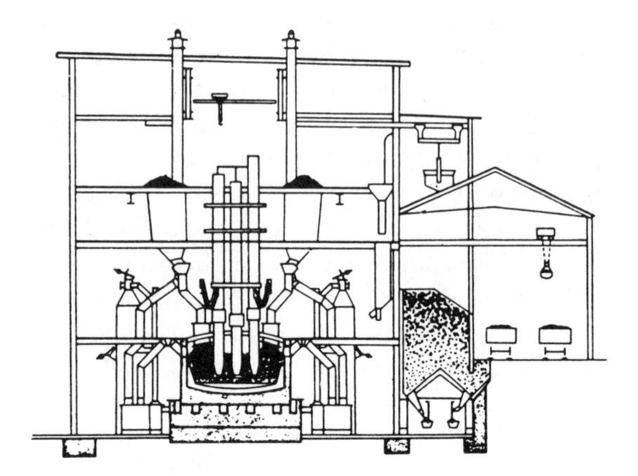

Fig. 22. Diagram of a three-phase submerged arc reduction installation [66].

— *Single or polyphase furnaces*

Numerous single-phase furnaces exist, with one electrode submerged in the load, while the other electrode consists of the carbon or graphite crucible. Due to the phase balancing requirement for high powers, most furnaces are three-phase supplied, the tops of the three electrodes forming an equilateral triangle as in conventional direct arc furnaces. Some furnaces also use 3, 4 or 6 electrodes in line.

— *Open or closed furnaces*

Closed reduction furnaces are used systematically for large dimension high power units. Closing of the furnace, which is obtained by a special water-cooled steel bell placed over the furnace, enables recovery and use of the gas produced during reduction. The energy recovered can also be used to improve the economic balance of the operation. An analysis of the composition of the gas produced and continuous measurement of temperatures and pressures beneath the furnace vault generally enables better control of furnace operating conditions. Treatment of the gases given off can strongly limit pollution.

Also, open furnaces exist in which the top surface of the load is in contact with the atmosphere. To reduce pollution, these furnaces can be equipped with smoke purifying installations. They are widely used in the simplest ferro-alloy production installations.

— *Continuous or discontinuous furnaces*

In discontinuous furnaces, the material to be melted is loaded into the furnace, then melted; as soon as melting is completed, the material is refined, then poured. Once these operations have been completed, a new load is introduced. Conversely, in continuous furnaces, the product to be melted is permanently loaded into the furnace and is subjected to the reduction operation.

**Fig. 23.** Single-phase submerged arc furnace for production of magnesium using the "Magnetherm" process [30]:
1. Central electrode;
2. Secondary circuit;
3. Hearth electrode;
4. Refractory lining;
5. Carbon lining;
6. Material feed;
7. Pouring hole;
8. Vacuum tube;
9. Sprinkler;
10. Condensation chamber;
11. Crucible.
The sprinkler [9] cools crucible [11].

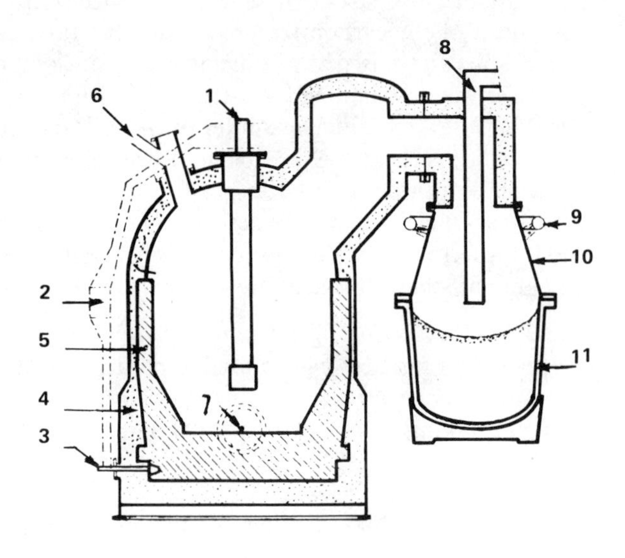


— *Fixed or rotating chamber furnaces*

Some furnaces are equipped with a chamber which rotates slowly during treatment, which may improve melting conditions. This arrangement is often used with open furnaces, but may also be used with closed furnaces.

— *Amorphous carbon or graphite self-baking electrode furnaces*

Self-baking (or Söderberg) electrodes are the most widely used in reduction furnaces. In some furnaces, they are replaced by amorphous carbon electrodes (e.g., production of silicon metal) and, more rarely, by graphite electrodes (*see* paragraph 3.2.1.3.).

### *3.3.3. Submerged-arc furnace applications*

#### *3.3.3.1. Production of ferro-alloys*

Submerged arc furnaces are used to produce most ferro-alloys and similar products. Generally, the most modern equipment consists of continuous closed reduction furnaces, which enable purification and recovery of gases produced. Furnace powers range from a few megawatts to several tens of megawatts.

#### *3.3.3.2. Production of pig iron*

In electric low shaft furnaces used to produce pig iron from ore and carbon, the production temperature is obtained by the Joule effect (in conventional blast furnaces, this is obtained by combustion of coke), and the gases which are given off do not contain nitrogen, but only the gases released by reduction ($CO$, $CO_2$, $H_2$, etc.). Therefore, this gas has a energy content of 1.5 times higher than that of blast furnace gases. These low shaft electric furnaces only require half or one-third of the carbon required with blast furnaces. Almost all the qualities of carbon can be used as a reducing agent, since submerged-arc furnaces, which are relatively low in height (of the order of 6 m, with 2 to 3 m for the melting bed) have few requirements in terms of resistance to crushing and abrasion of charcoal: it is therefore possible to use smaller grained charcoal and ore, while in a blast furnace, it is necessary to use high quality coke.

Low shaft electric furnace design is similar to that of closed reduction furnaces used for ferro-alloys. Furnace power is generally between 10 and 100 megawatts. At present, these furnaces are rarely used in Europe.

#### *3.3.3.3. Production of calcium carbide*

Reduction furnaces can be used to produce calcium carbide ($CaC_2$) from lime ($CaO$) and coke, with carbon monoxide being given off, according to the basic reaction taking place between 1,700 and 2,000°C:

$$CaO + 3\,C \rightarrow CaC_2 + CO.$$

Basically, calcium carbide production is related to acetylene chemistry. Therefore, due to the reduction in use of this chemical, production of calcium carbide has been strongly decreased. The largest furnaces have powers of the order of 50 MW, and are of the continuous closed type with gas recovery. Thermal efficiency may be very high, up to 85%.

#### *3.3.3.4. Production of phosphorous and alumina cement*

Phosphorous is produced in gas form in closed furnaces, then condensed and recovered, using the chemical reaction:

$$2Ca_3\,(PO_4)_2 + 6\,SiO_2 + 10C \rightarrow P_4 + 10CO + 6(CaO.SiO_2).$$

The alumina cement is obtained from a mixture of bauxite, lime, a little charcoal, and can be used up to temperatures exceeding 1,000°C.

#### *3.3.3.5. Production of magnesium*

Magnesium can be obtained by reduction of dolomite by silicon, using the simplified reaction:

$$2(MgO, CaO) + Si \rightarrow SiO_2, 2CaO + 2\,Mg.$$

This reaction takes place around 1,600°C, and magnesium is given off in vapor form and condensed to the liquid state in a condenser, then poured into a crucible in which it is solidified or collected in the liquid state. Bauxite is added to the mixture introduced into the furnace to promote heating by the Joule effect. A 4,500 kW furnace produces approximately 7 tons of magnesium per 24-hour cycle (*see* figure 23) [30].

### *3.3.4. Specific consumptions and advantages of submerged arc furnaces*

Published specific consumptions vary widely; in particular, these depend on the nature of the loads and the manner in which the gases given off are evaluated.

Figure 24 gives specific electricity consumptions for the production of various products.

Figure 25 provides examples of the characteristics of two closed-chamber reduction furnaces, the first producing pig iron and the second calcium carbide.

The values provided in Figures 24 and 25 are approximations only. However, they demonstrate that specific electricity consumptions are relatively high (however, they do not take into account the possibility of reusing gases given off during chemical reactions). The cost of electrical energy is therefore a determining factor in cost, since electrical energy sometimes represents 30 to 50% of product cost. Therefore, the choice of this type of process is highly sensitive to the relative cost of energies.

Conversely, the investment costs for submerged-arc furnaces are relatively limited with respect to those for competing processes, and production unit size is often more limited enabling an increase in production tool flexibility (for example, low shaft electric furnaces compared to blast furnaces).

Specific electricity consumption for submerged arc furnaces [5]

| | Specific consumptions (kWh/t) |
|---|---|
| Continuously produced products: | |
| Calcium carbide | 3,000 |
| Phosphorous | 10,000 to 15,000 |
| Pig iron | 2,500 to 3,000 |
| Ferromanganese 75 to 80% Mn | 3,500 to 4,000 |
| Silicomanganese 70% Mn, 20% Si | 6,000 |
| Ferrosilicon 45% Si | 5,500 to 6,000 |
| Ferrosilicon 75% Si | 10,500 to 11,000 |
| Ferrosilicon 90% Si | 14,500 to 15,000 |
| Discontinuously produced products: | |
| Silicon carbide | 10,000 to 12,000 |
| Corundum ($Al_2O_3$) | 3,000 to 4,000 |
| Alumina cement | 1,500 |
| Ferrochromium 70% Cr, 4 to 8% C | 6,500 |
| Ferrochromium 68% Cr, 1 to 4% C | 8,000 to 9,000 |
| Ferrochromium 0.02 to 0.10% C | 14,000 to 15,000 |
| Ferromanganese 80% Mn, 1% C | 8,000 to 9,000 |
| Ferromolybdenum 70% Mo, 1% C | 7,000 |
| Ferrotungsten 80% W, 1% C | 8,000 |
| Magnesium | 12,000 to 13,000 |

**Fig. 24.**

Under these conditions, the choice of this type of production process, which can serve very vast markets with a single production unit, should be the subject of exhaustive economic and technical analyses.

### 3.4. Consumable electrode vacuum arc melting furnace

These furnaces are used for vacuum melting of special metals and steels of very high quality by remelting of one electrode by heating of the arc. [45], [50], [56], [62], [A], [B].

#### *3.4.1. Functions and construction of a consumable electrode vacuum arc furnace.*

Vacuum consumable electrode arc furnaces are remelting furnaces. These furnaces have only one electrode, prepared in advance and consisting of a metal similar in composition to that obtained by remelting; the essential objective of the operation is to improve metal quality and prepare it for transformation.

Characteristics of reduction furnaces producing pig iron and calcium carbide [5]

| Product produced | Pig iron | Calcium carbide |
|---|---|---|
| Transformer power | 17,000 kVA | 43,500 kVA |
| Maximum absorbed power | 14,500 kW | 35,000 kW |
| Minimum absorbed power | 5,000 kW | 8,000 kW |
| Power factor | 0.80 to 0.85 | 0.80 to 0.85 |
| Primary voltage | 20 kV | 108 kV |
| Secondary voltage | 240 to 140 V | 240 to 80 V |
| Secondary maximum current | 62,000 A | 105,000 A |
| Self-baking electrode diameter | 1,350 mm | 1,400 mm |
| Daily production rate | 120 to 130 t/d | 285 t/d |
| Electricity specific consumption | 2,600 to 2,800 kWh/t | 2,940 kWh/t |
| Electrode specific consumption | 20 to 30 kg/t | 15 to 20 kg/t |
| Slag production | 400 kg/t | |
| Gas production | 880 $m^3$/t | 350 to 380 $m^3$/t |

**Fig. 25.**

This electrode is remelted due to the heat resulting from an arc between its end and the crucible (striking) or the molten metal. The arc is DC supplied, the electrode forming the cathode and the crucible and molten metal the anode.

The molten metal is collected in the copper crucible which is either water cooled or a mixture of molten sodium and potassium mixture cooled (in general, a mixture of sodium and liquid potassium of — 11 to 784°C, circulation being provided by an electromagnetic pump). Due to this cooling, there is no reaction between the molten metal and the walls. The DC power supply is obtained from silicon rectifiers. The feed voltage is between 25 and 50 V for most furnaces, while current can reach 40,000 A. Furnace capacities vary between a few hundred kilograms and about ten tons, but very high capacity furnaces exist, of the order of 50 tons and more, producing large ingots with diameters of more than 1.5 m and also furnaces capable of treating a few kilograms of metal.

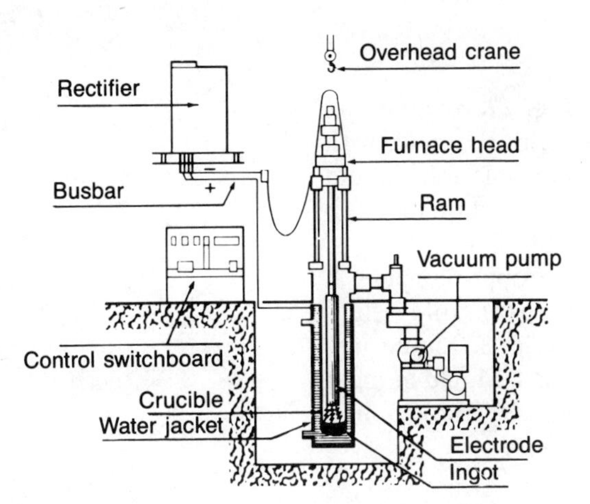


**Fig. 26.** Diagram of a consumable electrode vacuum arc furnace

The vacuum created in the furnace is generally between $5 \times 10^{-4}$ and 1 torr. Pumping facilities must often be very powerful, and therefore represent an important part of investment costs. An inert, low pressure atmosphere can also be created in the furnace.

The electrode is obtained by compacting powder (for very reactive metals such as titanium or zirconium), or by melting in induction or arc furnaces (often in a vacuum induction furnace), then poured to the appropriate dimensions.

Due to crucible cooling, the metal solidifies gradually in its own solid phase. Towards the end of melting, the current is gradually reduced to prevent shrinkage cracks. The formed ingot is easily removed from the crucible, since the metal contracts during cooling. In some furnaces, the liquid metal is poured directly through an orifice underneath the crucible into ingot moulds; this operation also takes place in a vacuum.

Electromagnetic systems (induction coils) are sometimes installed to mix the bath and rotate the arc so that the energy is better distributed.

### *3.4.2. Advantages and application of consumable electrode vacuum arc furnaces*

The basic interest of the consumable electrode arc furnace is that it enables thorough outgassing and, in particular, elimination of hydrogen due to the action of the vacuum and absence of contamination due to the crucible. The high arc temperature also causes evaporation of volatile impurities and elimination of oxides. Non-metallic residues generally float on the bath making them easily removable. The ingot solidifies gradually from the bottom of the crucible and the metal is fully homogeneous.

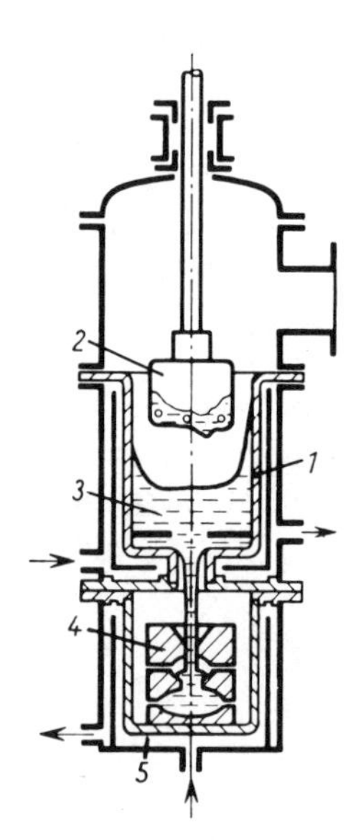


**Fig. 27.** Vacuum arc melting furnace with pour-off into mould or ingot maker [7], [15]:
1. Copper crucible; 2. Consumable electrode; 3. Liquid metal; 4. Mould or ingot maker; 5. Cooling water.

Consumable electrode vacuum arc furnaces therefore produce a very high quality pure metal; electron beam furnaces are the only better method. Outgassing is, in particular, better than with electroslag remelting, but, due to vacuum generating and power rectifying equipment, investment costs are higher than with this type of furnace, which, due to the use of a variable composition electroslag, enables better control of the metal refining conditions (see "conduction heating" section). Both processes are also complementary and are used as a function of the desired final characteristics of the metal produced.

Consumable electrode vacuum furnaces are used for two main application families:

— remelting of high reactive metals with high melting points, such as titanium, zirconium, hafnium, molybdenum, palladium, tantalum, tungsten and their alloys; this technique provides very high quality ingots, often from compacted powders or castings, directly vacuum poured into the moulds (titanium and uranium compound casting, etc.);

— improvement in the quality of special steels by vacuum remelting; the ingots obtained are often intended for forging of very high mechanical quality parts, since remelting improves strength and prevents segregations, thus increasing malleability and ductility. These parts are used in aviation, gas turbines, nuclear power stations, the petrochemical industry and, in general, in mechanical construction, etc.; for steel, specific consumption of electricity is of the order of 800 to 1,000 kWh/t.

### 3.5. Chemical reactors

The electric arc enables very high temperatures and power densities, to be obtained which provides temperature rises of the order of $10^{6}$ °C/s, considerably accelerating chemical reactions.

For example, with this molecule disassociation process, it is possible to obtain acetylene from hydrocarbons. Energy consumption is about 11 to 12 kWh/kg of acetylene, and power density 400 W/cm$^3$ in the furnace. Acetylene production has been practically abandoned, since the petrochemical industry has turned towards production of other major intermediate chemical products (in particular, ethylene). Unlike arc furnaces, these reactors call for high voltages across the electrodes.

Furnaces based on this principle and plasma furnaces, which provide even higher power densities in the electric arc (see "plasma heating") can, however, be used to obtain other chemical reactions.

## 3.6. Arc welding

Electric arcs are used as a heat source to simultaneously melt the edges of metal parts to be assembled, and sometimes to melt an added metal. This thermal effect is obtained by causing an arc between the electrode of the welding equipment and the part to be welded. This technique enables assembly operations (self welding) or recharging (addition of metal to worn parts).

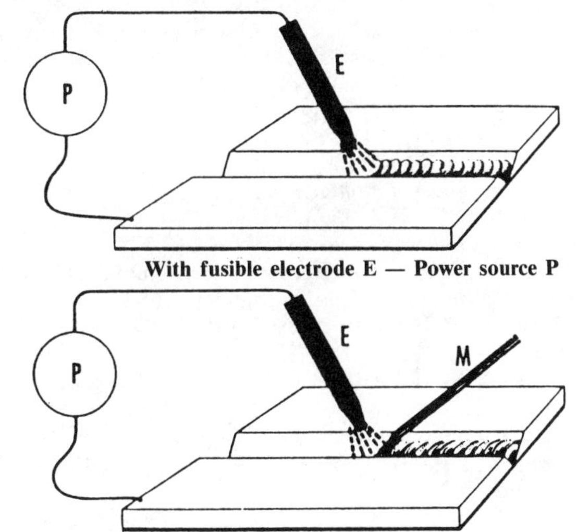


**Fig. 28.** Arc welding principle

Welding can be performed using AC or DC. The type of current depends on the type of electrodes used and the part to be welded. For DC, the welding electrode generally consists of the cathode; this begins to melt before the part which forms the anode. The inverse polarity is used only in special cases such as aluminum welding. The technique has given rise to numerous options, which differ in the nature of the electrodes, current, flux, atmosphere, degree of automation, current stabilization method and the slag elimination technique.

Arc welding is a widely used technique, and numerous publications have been devoted to this [20], [22], [23], [24], [29], [38], [48], [49], [52], [57], [69].

This application, in spite of its industrial importance, is mentioned for reference only. However, one of the essential evolutions in this technique, consisting of the development of protective atmosphere welding and, in particular, the T.I.G. and M.I.G. processes, should be noted. The T.I.G. (*Tungsten Inert Gas*) process uses tungsten electrodes and a protective gas composed of argon, helium or a mixture of both usually with no added metal. The M.I.G. (*Metal Inert Gas*) process uses similar conditions but with added metal. The advantage of these processes is the absence of fluxes, and therefore the elimination slag, and the high quality of the welds obtained. These processes are of major importance in the welding of stainless and special steels, and for welding non-ferrous metals and, in particular, light alloys.

## 4. BIBLIOGRAPHY

[1] BUNKER HILL AND SULLIVAN, Electrolytic Antimony, *Mining World* 4, n° 6, p. 3-9, 1942.

[2] LES TECHNIQUES DE L'INGÉNIEUR, *Fours de réduction de minerais d'étain*, vol. M 2300-9.

[3] G. VOLKERT *et al.*, *Stahl und Eisen*, 70, p. 369-371, 1950.

[4] M. H. DODERO, Essai de réhabilitation du courant continu en métallurgie électrothermique, *Bulletin de la Société française des électriciens*, n° 77, Vol. 7, p. 276-284, 1957.

[5] G. POHLE, *Jahrbuch der Elektrowärme*, 6 et 192, 1957.

[6] 4ᵉ Congrès International d'Electrothermie, Stresa, 1959 :
(*a*) W. E. SCHWABE, rapport 220;
(*b*) A. J. FORST, rapport 203.

[7] H. GRUBER, VDE Fachberichte (rapports techniques VDE), 21, 1960.

[8] M. MORKRAMER, Principes de dimensionnement des fours électriques de réduction, *Journal du four électrique*, n° 2 et 3, 1960.

[9] M. MARTINEZ, Proposition de calcul systématique des fours à arcs submergés, *Journal du four électrique*, n° 9, 1961.

[10] R. SEVIN, Le four à carbure de calcium de 30 000 kVA de Sadaci à Langerbrugge, *Journal du four électrique*, juin-juillet 1965.

[11] W. E. SCHWABE et C. ROBINSON, *Journal of Metals*, vol. 17, n° 1, p. 75-80, 1965.

[12] W. E. SCHWABE et C. ROBINSON, *Journal of Metals*, vol. 19, n° 4, p. 67-75, 1967.

[13] 6ᵉ Congrès International d'Electrothermie, Brighton (G.B.), 1968, en particulier :
(*a*) W. E. SCHWABE et C. G. ROBINSON, Experience with UHP electric furnace operation;
(*b*) P. G. KENDALL, The measurement of voltage fluctuations caused by arc furnaces;
(*c*) U. BUTTIKOFER, Rapport du Groupe de Travail Perturbations de l'UIE;
(*d*) L. DI STASI, Étude du fonctionnement des fours à arcs et à résistances en régime perturbé.

[14] *JOURNAL DU FOUR ÉLECTRIQUE*, Structure et fonctionnement des fours à arc, 5ᵉ leçon, août-septembre 1968.

[15] F. LAUSTER, *Manuel d'Électrothermie Industrielle*, Dunod éditeur, 1968.

[16] N. HERMONT, Les fours à arcs de grande puissance, *La technique moderne*, mars 1969.

[17] R. SEVIN, L'usine électrométallurgique de la Sandur Manganese and Iron Ores Ltd, Vyasankere (Inde), *Journal du four électrique*, janvier 1970.

[18] M. VOGEL, Lichtbogenerwarmung zur Stahlerzeugung und für Reduktionsprozesse, Elektrowärme, 1971, Heft 8, p. 453-456.

[19] P. MEYNAUD, Flicker et conditions de raccordement au réseau d'appareils produisant des variations rapides de tension, *Revue générale d'électricité*, novembre 1971.

[20] P. T. HOLDCROFT, Les procédés de soudage, Dunod éditeur, 1971.

[21] 7e Congrès International d'Électrothermie, Varsovie, 1972, section Fours à arcs.
[22] M. LE GUELLEC, Le rechargement dur par soudage et particulièrement avec les alliages spéciaux dans les industries mécaniques, *Soudage et techniques connexes*, vol. 26, n° 10, p. 373-384, octobre 1972.
[23] H. GERBEAUX, Le soudage des aciers inoxydables, *Soudage et techniques connexes*, vol. 26, n° 12, p. 455-462, décembre 1972.
[24] ESSERS *et al.*, Les caractéristiques de l'arc et du transfert de métal dans le soudage MIG au plasma, *Métal construction*, vol. 4, n° 12, décembre 1972.
[25] IRON AND STEEL CONGRESS, Amsterdam, février 1973, Metals Society, p. 57-93, communications sur les fours à arcs.
[26] N. EDEMSKY *et al.*, Entwicklung und Verbesserung automatischer Regler für 200 t Lichtbogenstahl-Schmelzöfen, *Elektrowärme*, p. 224-227, 1973.
[27] A. BUXBAUM, Elektronische Regeleinrichtung für die Elektrodenregelung von Lichtbogenöfen, Technische Mitteilungen, *AEG Telefunken* 63, p. 335-339, 1973.
[28] G. CHAUDRON et J. TROMBE, *Les hautes températures et leurs utilisations en physique et en chimie*, Masson éditeur, 1973.
[29] S. BELAKHOWSKY, *Théorie et pratique du soudage*, PYC éditeur, Paris, 1973.
[30] R. SEVIN, L'usine de magnésium silico-thermique de la Société française d'électrométallurgie, *Journal du four électrique*, n° 5, mai 1973.
[31] E. ELSNER et H. KNAPP, Electric Furnace Congress, AIM-CBEE, Liège, Belgique, novembre 1973, La fusion de l'éponge de fer dans les fours à arcs.
[32] W. E. SCHWABE et C. ROBINSON, Some considerations on large arc furnace design and operation, Electric Furnace Congress, Liège, novembre 1973.
[33] K. KEGEL *et al.*, Elektrowärme, *Theorie und Praxis*, Verlag Girardet éditeur, 1974.
[34] T. E. HARRAS, a static var compensation for flicker control on the Saskatchewan Power Corporation System, CEA Montreal, mars 1974.
[35] K. EICHACKER *et al.*, Exakte Lichtbogenregelung an einem 100 T, *Lichtbogenöfen Elektrowärme B*, p. 335-339, 1974.
[36] R. MALGAT, Cours d'Électrothermie, École de Thermique, Institut Français de l'Énergie, 1974.
[37] A. R. OLTROGGE, Arc furnace voltage can be critical, Electrical World, décembre 1974.
[38] N. RYKALINE, Les sources d'énergie utilisées en soudage, *Soudage et techniques connexes*, vol. 28, n° 11-12, décembre 1974.
[39] BONIS, Les fours à arcs en fonderie d'acier, *Métaux corrosion industrie*, n° 597, mai 1975.
[40] W. E. SCHWABE, The status of Ultra High Power Electric Furnace, I and SM, juillet 1975.
[41] G. STEINSMETZ, Produktionerhöhung and Kostenersparnis durch elektronische Regeleinrichtungen zur Elektrodenregelung an Lichtbogenöfen, *Elektrowärme B*, p. 96-99, 1975.
[42] H. G. LEU, Fours à arcs de haute capacité de fusion, *Revue Brown-Boveri*, n° 63, 1976.
[43] J. KREUZER, Transformateurs à forte intensité et inductances pour installations de redresseurs et de fours, *Revue BBC*, n° 63, 1976.
[44] W. E. SCHWABE, Electric furnace problems, design and operating requirements for UHP arc furnaces melting prereduced charge materials, 8e Congrès Union International d'Électrothermie, Liège, 1976.
[45] 8e Congrès Union Internationale d'Électrothermie, Liège, 1976, Section I, fours à arcs, et IV, Automatisation.
[46] F. D. JACKSON et E. L. BEDEL, Fused cast and rebonded fused grain refractories in electric arc and auxiliary steelmaking vessels, Industrial Heating, avril 1976.
[47] J. ANTOINE et J. DUMONT, Problèmes de l'arc dans l'élaboration d'acier au four électrique, *Journal du four électrique*, vol. 81, n° 2, p. 37-39, février 1976.
[48] M. ROBERT, les développements technologiques du procédé TIG en soudage des aciers inoxydables, *Revue de soudure*, n° 2, p. 55-64, 1976.

[49] C. WEISMAN et KEARNS, WELDING HANDBOOK, 7e édition, *American Welding Society*, vol. 1, 1976 et vol. 2, 1977.
[50] CAMERON IRON WORKS, From vacuum melting to finished forgings, Metallurgia and Metal Forming, juillet 1976.
[51] J. TOULEMONDE et D. BIRFET, Le flicker et les moyens de le compenser, *Hommes et fonderies*, décembre 1976.
[52] I. DREWS et J. KING, Schweissen und Schneiden, *VDI-Z*, n° 1-2, vol. 119, p. 65-71, janvier 1977.
[53] COMMISSION ELECTRO TECHNIQUE INTERNATIONALE (CEI), Sécurité dans les installations électrothermiques, 4e partie, règles particulières pour les fours à arc, 1977.
[54] E. WANNER et W. HERBST, Installations statiques de compensation de la puissance réactive pour fours à arc, *Revue BBC*, n° 64, 1977.
[55] LABORELEC, *Manuel d'électrothermie*, juin 1977.
[56] E. JARVIS, Vacuum arc melting furnaces, *Electrical Review*, vol. 201, n° 21, novembre 1977.
[57] L. MENDEL, *Manuel pratique de soudage à l'arc*, Bordas-Dunod éditeurs, 1976.
[58] U. BECKER-BARBROCK *et al.*, Installation de fours à arcs de grande puissance de fusion, Revue Brown-Boveri, n° 65, février 1978.
[59] F. FURRER, Régulation électronique d'électrodes pour fours à arc, *Revue BBC*, n° 65, février 1978.
[60] V. SEKULOVSKI, Four à arc et installation statique de compensation de puissance réactive pour une aciérie spéciale, *Revue BBC*, n° 65, février 1978.
[61] COMITÉ FRANÇAIS D'ÉLECTROTHERMIE, Journées d'Études des 5 et 6 avril 1978 :
(*a*) M. B. BOWMAN, Caractéristiques des fours à arcs comme instrument de fusion;
(*b*) X. HENRY et J. BERGEAL, Le stato-compensateur, un moyen pour limiter les perturbations sur les réseaux électriques;
(*c*) M. A. MAECHLER, Conception nouvelle de la régulation des électrodes d'un four à arcs;
(*d*) M. MAITRE, Ordonnancement et conduite optimisée des outils de production et de transformation d'une aciérie électrique;
(*e*) M. ROJON, Mesures électriques et thermiques sur les fours à arcs; conduite et contrôle par calculateur.
[62] R. PITT, Investment in nickel alloy production, Metallurgia and Metal Forming, novembre 1978.
[63] L. HALLOT, Les fours électriques de fusion et de maintien des métaux, Cours à l'École Supérieure de Fonderie, 1979.
[64] P. FREGEAC et J. ROUDIER, L'oxygène au four à arcs, *Journal du four électrique*, février et mars 1979.
[65] J. PAUTZ, Possibilités de récupération d'énergie sur les fours à arcs, *Journal du four électrique*, n° 10, décembre 1979.
[66] DOCUMENTATION EDF, 1980, (92) PARIS LA DÉFENSE.
[67] DOCUMENTATION Centre Technique des Industries de la Fonderie (CTIF) 75783 Paris, 1980.
[68] DOCUMENTATION Institut de Recherche de la Sidérurgie, 78100 Saint-Germain-en-Laye, 1980.
[69] DOCUMENTATION Institut de Soudure, 75880 Paris.
[70] DOCUMENTATION CYCLATEF (Association Technique de Fonderie), 75783 Paris.
[71] DOCUMENTATION LABORELEC, Bruxelles, Belgique.
[72] G. SEGUIER, *L'électronique de puissance*, Dunod Technique, 75000 Paris.

**List of equipment manufacturers and suppliers mentioned in this chapter:**

[**A**] CREUSOT-LOIRE, LECTROMELT, 75008 Paris et Pittsburg (USA).
[**B**] LEYBOOLD-HERAEUS-SOGEV, 91400 Orsay et Hanau (RFA).
[**C**] CEM-BBC, 75008 Paris et Baden (Suisse).
[**D**] Ing. Leone Tagliafferi, Milan (Italie).
[**E**] Union Carbide, 94150 Rungis et New York (USA).
[**F**] ASEA, 95340 Persan et Vasteras (Suède).

# Chapter 9

# Plasma Heating

## 1. PRINCIPLE OF PLASMA HEATING

While initial plasma research was conducted around 1920-1925, it was not until 1960, with the appearance of plasma cutting, cladding and welding, that industrial applications of this technique were significantly developed. Other applications, such as melting and chemical reactions also began to use this technique.

Plasma is a physical state of matter obtained by ionizing a gas; a plasma consists of positively charged ions and negatively charged free electrons, but remains electrically neutral. However, the fact that a plasma is ionized makes it an electrical conductor. Plasma is often considered as a fourth state of matter. There are two types of plasma:

— Plasmas in which the degree of ionization is close to unity, for example in thermonuclear fusion. The temperatures are very high, several million degrees. At present, there are no industrial applications for such plasmas.

— Partially ionized plasmas; the standard ionized gases, where ionization varies between a few percent and fifty percent maximum, are, by extension, still known as plasmas. The temperatures reached vary between 2,000 and 50,000°K. It is these plasmas, whose industrial applications are developing due to the technical possibilities offered by such temperature levels, that are discussed.

## 2. CHARACTERISTICS OF PLASMA HEATING

### 2.1. Conditions for the creation of a plasma

Conversion of a gas into a plasma requires that energy and the degree of ionization increase with temperature. The degree of ionization at a given temperature depends on the ionization potential of the gas and varies greatly from one gas to another. For polyatomic gases, the molecules can either first break down into atoms or be ionized directly.

Ionization potential and disassociation energy
for some gas molecules [52].

| | $H_2$ | $N_2$ | $O_2$ | CO | NO |
|---|---|---|---|---|---|
| Ionization energy (eV) | 15.6 | 15.5 | 12.5 | 14.1 | 9.5 |
| Disassociation energy (eV) | 4.4 | 9.7 | 5.1 | 9.6 | 5.3 |

**Fig. 1.**

The energy required to create a plasma is provided by various electrical sources, the principal being:

— the arc plasma generator, in which the ionization potential is provided by an electric arc (paragraph 3.1.1.);

— the high or ultra high frequency plasma generator, in which the ionization potential is provided by an electromagnetic field (paragraph 3.1.2.).

## 2.2. Thermal properties of plasmas and ranges of application

An industrial plasma consists of electrons, ions, energized moleclues, disassociated molecules and neutral molecules and atoms. Plasma state bodies behave as conductors of electricity. Due to the effect of the currents flowing through the gas stream and the resulting magnetic forces, the gas stream contracts; this results in a high increase in temperature, especially in the center of the gas stream. The plasma temperatures used in industry vary between 6,000 and 20,000°K; the brightness or luminosity of the plasma jet is therefore highly intense, and adequate protection must be provided. Pinching of the plasma can be used to shape and adapt it for the application.

In addition, plasma is sensitive to wall effects: a cold wall tends to contract a plasma. The walls of plasma generators of all types must be cooled to provide adequate service life, therefore reinforcing the contracting effect of the electromagnetic forces. This contraction accentuates the radial temperature gradient and increases the axial thermal effect.

The thermal properties of the ionized gas, (high thermal conductivity, a temperature favoring intense radiation), promotes thermal exchange coefficients better than those obtained with more conventional heating methods. Recombination of ions, electrons and atoms also causes significant quantities of heat to be given off.

The advantages of plasma heating are due to the following characteristics:

— power densities are high; therefore, plasmas can be used to reach specific powers of about $10^5$ W/cm$^2$, or expressed in power densities, powers of the orders of $10^5$ kW/m$^3$;

— these powers can be concentrated on very small areas or volumes: the minimum impact area of a plasma jet is about $10^{-3}$cm$^2$;

— thermal exchanges are highly accelerated and reaction speeds very high;

— under these conditions, the size of plasma furnaces can be much smaller than equipment using more conventional techniques;

— thermal inertia is very low, which can offer greater flexibility.

Plasmas can therefore improve the performance of some thermal production techniques, but, above all, and due to their specific properties, make possible applications which are delicate or impossible using other methods;
— applications requiring very high temperatures, such as the melting of highly refractory materials;
— highly endothermic reactions;
— reactions for which the kinetic energy is too slow at the temperatures used;
— reactions resulting from excitation of molecules and atoms, which can lead to the production of new products;
— reactions requiring high specific energy within a restricted volume (pyrolysis, volatilization);
— phase changes, modifying the physical properties of materials (atomization of metals and high specific surface refractory powders);
— reduction of transformation processes requiring several phases from the initial to the final product, in conventional processes, within a single stage [42].

Thermally, two types of plasma should be distinguished:
— thermal plasmas, in which the electron and heavy particle (ions, excited atoms) temperatures are more or less the same: 10,000 to 20,000°K. Since the pressure is close to atmospheric pressure or higher, collisions between the various particles submitted to intense agitation in the plasma are numerous, so that transformation of their kinetic energy into heat causes a considerable rise in temperature. Gas enthalpy is the main factor in such applications. This type of plasma is excellent for treating non-organic materials, both for heating and for reactions during heterogeneous phases at high temperature.
— cold plasmas (or luminescent discharges) in which the temperature of the heavy particles forming the gas is much less than that of the electrons; the ratio between these temperatures is generally between 0.01 and 0.1, with electron temperatures about 10,000°K. These plasmas operate at low pressure, between 0.1 and 100 torr approximately and the probability of particle collisions is lessened, thus the heat given off is much lower. The thermal aspect, while significant, is therefore secondary with these plasmas. Conversely, it is the particle excitation aspect which triggers and accelerates chemical reactions. The luminescent nature of these plasmas also permits some photochemical reactions. Cold plasmas are therefore fully adapted to the chemical reactions involving thermosensitive materials found in many organic compounds. Applications are also developing in surface treatment of metals, together with ion nitriding (ionic bombardment of the surface of a metal part due to the effect of an electric field); soon, carbonitriding and ionic hardening will also use luminescent discharge plasmas [28], [51 *b*], [C], [E].

This chapter discusses primarily plasma heating (i.e. thermal plasmas).

## 3. COMPOSITION OF PLASMA HEATING EQUIPMENT

Plasma heating equipment must provide two basic functions:
— production of the plasma;
— effective heating of the product for a given application.

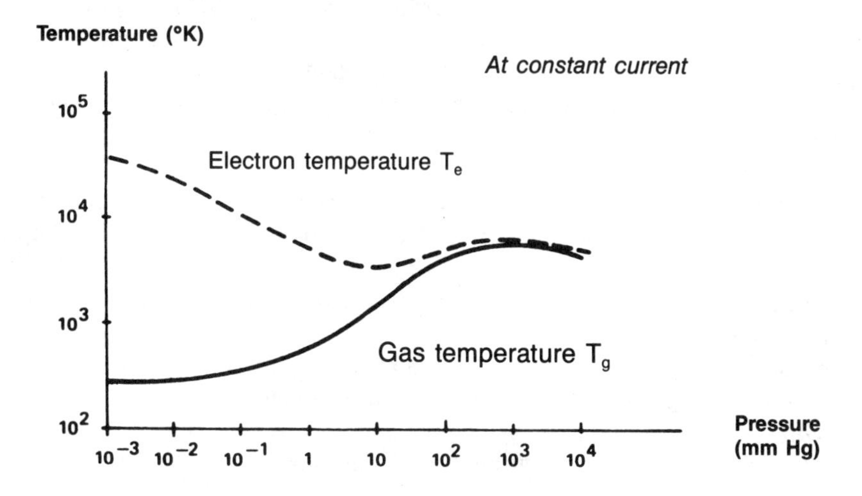


**Fig. 2.** Gas and electron temperatures in a plasma [52].

Basically, the equipment consists of:

— a plasma generator, comprising a plasma gas supply and an electrical power supply;

— an applicator, enabling the plasma created to be used (furnace, reactor, etc.). This device may be very small or even non-existent, when the plasma is applied directly to solid parts (welding, spray cladding, etc.);

— handling systems (load transport, plasma torch orientation, etc.);

— monitoring and regulation systems and the various support devices.

## 3.1. Plasma generators

### *3.1.1. Arc-plasma generators*

#### *3.1.1.1. Principle*

An arc-plasma generator can be supplied with AC or DC; creation of the plasma is basically the same.

Briefly, a DC arc plasma is obtained by passing a pressurized gas (argon, nitrogen, helium, etc.) through an electric arc operated and maintained between two electrodes: the cathode, generally consisting of a thoriated tungsten bar (thorium, which has a low electron extraction potential, enhances thermionic emission), and the anode, normally copper, whose end forms a nozzle. When the plasma generating gas is an oxidizing gas, air or oxygen, it is impossible to use tungsten cathodes which are rapidly destroyed by oxidation; most of the other materials usually used as cathodes run the same risk, and it is therefore difficult to operate arc plasma generators using oxidizing plasma generating gases. Research on zirconium cathodes may help resolve this problem.

Generally, the arc current is a function of the quantity of electrons released by thermionic emission from the cathode surface. These electrons are accelerated in the cathode voltage drop region by the intense electric field around the cathode; they therefore acquire high kinetic energy, enabling them to ionize neutral atoms by collision, and beforehand, if the plasma generating gas is polyatomic, disassociation of the molecules. The positive ions which form are accelerated in the opposite direction and strike the cathode; they give up their energy, enabling thermionic emission to be maintained.

Due to the high temperature levels in the plasma and the intensity of the ion and electron beams to which they are submitted, the cathode and the anode, which behaves as a nozzle or jet, must be cooled.

The pressurized plasma gas blows the arc through the small diameter jet. Supplementary pinching of the arc occurs in the jet due to contact with the cold walls of the anode, and the Lorenz force due to the magnetic field of the current in the plasma.

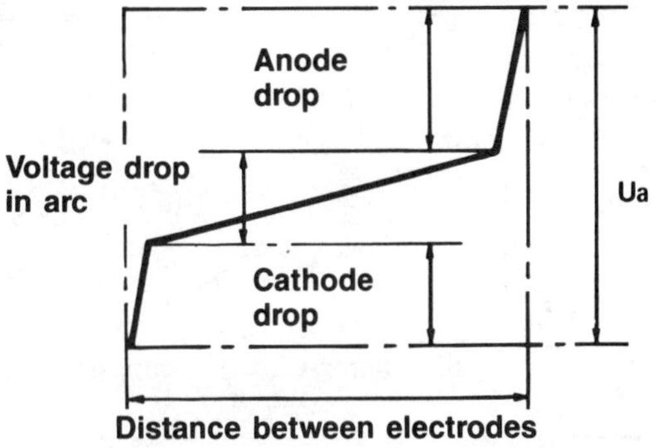

**Fig. 3.** Voltage drop in an electric arc [52].

The energy density increases, causing an increase in temperature and ionization level. The temperature of the central stream generally reaches 15,000 to 20,000°K. This cylindrical plasma jet is surrounded by a gas stream of lower temperature due to contact with the nozzle walls. The ionized gases are discharged from the jet at very high speed (several hundred meters per second) due to the combined effect of the high axial expansion due to the high temperatures and pinching of the jet resulting from electromagnetic forces and contact with the cold walls.

This type of plasma generator is often known as a plasma torch or blow torch by analogy with conventional equipment (oxyacetylene torch or TIG welding conventional arc torch).

The anode and cathode are separated by a few millimeters only. The generator is DC-supplied with high currents at a voltage of between 15 and 150 V. Only non-reactive gases are used to prevent electrode wear. Reactive products can only be introduced to the hot gas outside the electrode area. The plasma jet power can be adjusted by regulating the gas flow, voltage and arc current. Figure 5 provides the characteristics of some arc plasma generators.

For example, the characteristics of a 12 kW arc plasma generator are as follows:

— current: 600 A;

— nozzle diameter: 5 mm;

— plasma gas: argon;
— inter-electrode distance: 1.5 mm;
— cooling water flow rate: 25 l/min;
— gas flow rate: 30 l/min.

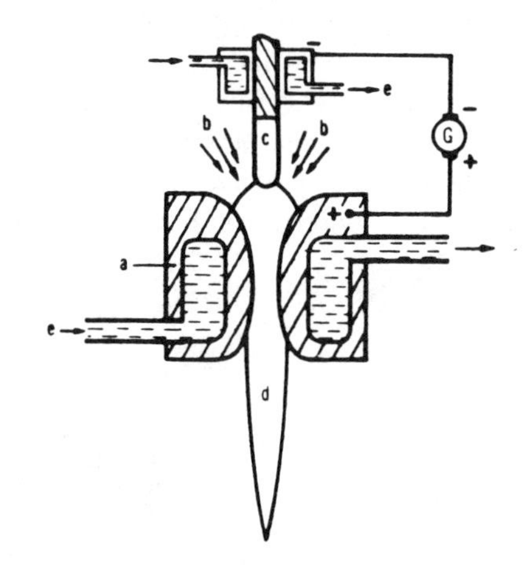


**Fig. 4.** Diagram of a blown arc plasma generator [52]:
c, cathode;
a, anode;
e, water;
b, plasma gas;
d, plasma jet;
g, DC generator.

Characteristics of some arc plasma generators [52].

| | Argon | Nitrogen | Hydrogen |
|---|---|---|---|
| Arc voltage (V) | 25 | 40 | 70 |
| Current (A) | 1,500 | 1,000 | 680 |
| Temperature (°K) | 10,000-20,000 | 15,000-25,000 | 15,000-30,000 |

**Fig. 5.**

These characteristics will, of course, vary from one application or manufacturer to another.

In a single-phase AC arc plasma generator, the electrodes act as the cathode and anode alternately, and must therefore be made from the same material.

### *3.1.1.2. Arc stabilization*

The arcs obtained in this type of plasma generator must be stabilized to regulate operation and to prevent excessive wear of the electrodes.

Numerous arc stabilization methods exist:

— Adjusted fluid flow stabilization (tangential injection of gas to form a vortex enhancing arc stability, introduction of gas along the cathode parallel to the latter, etc.);

— magnetic field stabilization (self-stabilization obtained by rotating the arc on annular electrodes, external magnetic field stabilization, in which case the arc is pinched and rotates around the anode);

— wall stabilization (the arc is stabilized due to the position of the anode with respect to the cathode);
— liquid wall stabilization (water vapor plasma using a graphite cathode and a copper anode with water cladding which simultaneously provides cooling and the required water vapor, and ethanol plasma which is based on a similar principle).

The most widely used stabilizing method, especially in small and medium power installations, combines adjusted fluid flow (vortex sheathed arc ([1]) between two electrodes) and careful positioning of the electrodes. This is by far the cheapest solution for this type of installation, and plasma generation efficiency is high.

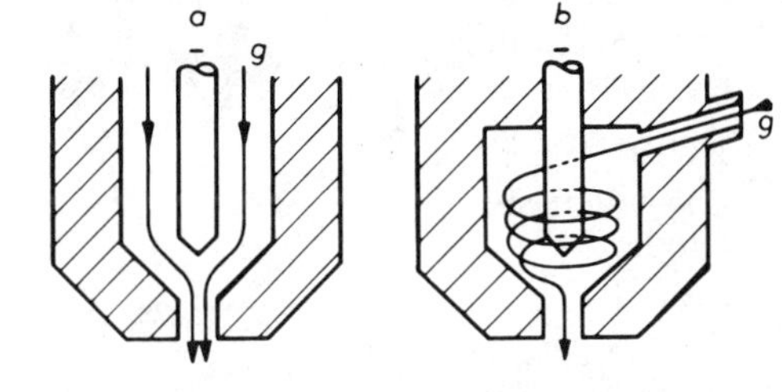


**Fig. 6.** Arc stabilization [2]: *a*, injection of plasma gas parallel to axis; *b*, tangential injection of plasma generating gas (vortex).

([1]) Vortex: centrifugal (spiral) flowing of gas by radial injection into an axial flow.

### *3.1.1.3. Various types of arc plasma generators*

— *Blown arc-plasma generators*

In the arc generators described so far, which are the simplest, the electric circuit is completed through the generator itself, containing the cathode and the anode. The plasma is blown through the anode nozzle. The plasma jet therefore does not conduct the current outside the nozzle, except in some cases and then only over short lengths.

To increase the power dissipated in the plasma, other plasma generators have been developed [2], [21], [43], [44], [52].

— *Transferred arc-plasma generators*

With this type of generator, the electrical circuit does not close between the generator electrodes, but between the generator cathode and the part or material to be treated which then behaves as an anode.

Electric current therefore flows through the plasma jet and even more energy is applied to the target material. The kinetic energy of the electrons bombarding the target over a small cross-section further accentuates this thermal effect. The maximum temperature reached is about 30,000°K, and power density is $2.5 \times 10^4$ W/cm$^2$.

To start the torch, a low power arc must be created between the cathode and anode. The nozzle is used to constrict the arc and stabilize it, and also create a guide channel to concentrate the energy. Electrically, the nozzle is used simply to create the arc. In the normal mode, the nozzle current density is zero, and therefore nozzle heating is more limited than with the blown arc method (however, radiation from part of the plasma jet is intense, and heat transfer due to convection high).

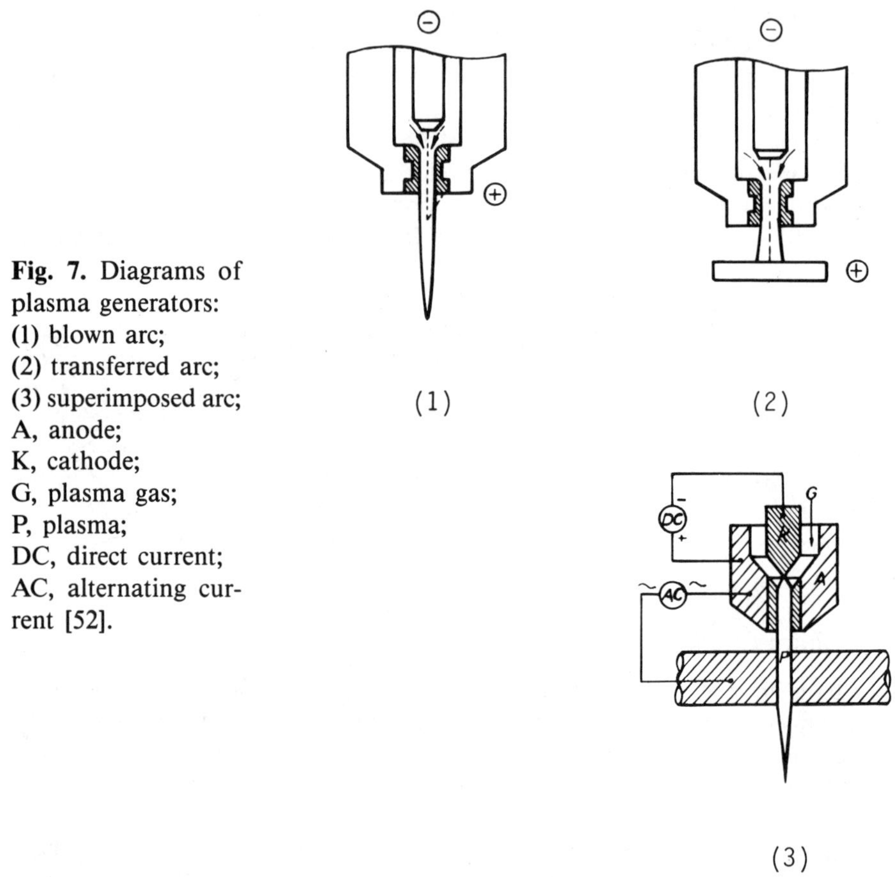


**Fig. 7.** Diagrams of plasma generators: (1) blown arc; (2) transferred arc; (3) superimposed arc; A, anode; K, cathode; G, plasma gas; P, plasma; DC, direct current; AC, alternating current [52].

— *Superimposed arc-plasma generator*

Even higher powers can be developed in the plasma jet using a superimposed arc. In this type of generator, the plasma jet is used as a conductor of electricity (plasma state resistor) through which an AC or DC current, much higher than that used to create the plasma, flows. Therefore, numerous variants of this basic principle exist [33 *a, e*], [43], [44].

A single-phase AC power arc-plasma generator can, for example, operate on startup as a blown arc generator. A DC power source is connected between the anode and cathode nozzle. When the arc strikes, the gas blows the plasma through the nozzle. As soon as the plasma reaches the target material to be treated, the superimposed AC power source is stabilized. The plasma jet can also be transferred to another jet; in this case, after striking, a plasma column builds up between both torches. This column forms a gas-state resistor of regular cross-section, through which the superimposed AC current flows and radiates uniformly. This concept is used in certain melting furnaces (*see* Fig. 41).

Other types of superimposed arc plasma generators have been designed:
— Superimposition of three-phase AC between three DC generators (*Fig.* 8).

Three blown arc plasma generators, operating in the DC mode, are arranged so that the jets produced intercept each other at the same point. The three plasma jets form three gas-state resistors which can be supplied with three-phase AC power connected to the anode of each of the generators. After striking the arc, the relative distance and position of the three generators can be adjusted to suit the application.

This type of device can be used to heat reagent fluids, melt bars of electrically conducting material, treat liquid or solid refractory charges, etc.
— Superimposition of an AC current between a pilot plasma generator and several clad electrodes (*Fig.* 9).

This device consists of a DC-supplied blown arc pilot plasma generator and three electrodes, the axes of which cross at a given point along the path of the plasma jet from the pilot generator. These three electrodes are supplied with three-phase AC. The plasma gas is taken via silica cladding to the electrodes around which it is injected.

Once the arc has struck, the electrodes can be withdrawn to their operating position, and their inclination with respect to the axis of symmetry of the system can be modified. The pilot torch can then be extinguished.
— Superimposition of an AC current between one or two plasma generators and a bath or mass behaving as a third electrode (*Fig.* 10).

Two jets, produced by two DC-plasma generators, form the two phases of a three-phase AC current superimposed on both jets. A bath or mass behaves as the third phase of this superimposed current. Convergence of both jets on a given point of the mass helps prime the system. The heat flux attained is $3 \times 10^4$ W/cm$^2$.

It is also possible to superimpose DC on the plasma jets, but these devices are of less practical interest.

Superimposed arc plasma generators enable very high powers since the power ratio of the superimposed current to that of the current creating the plasma can be up to 10.

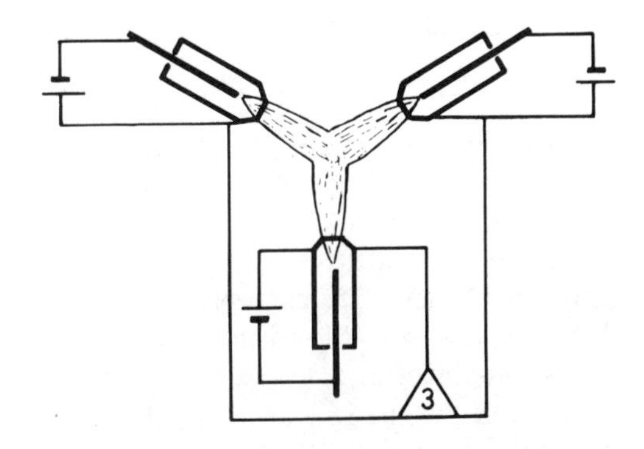


**Fig. 8.** Three-phase superimposition [42].

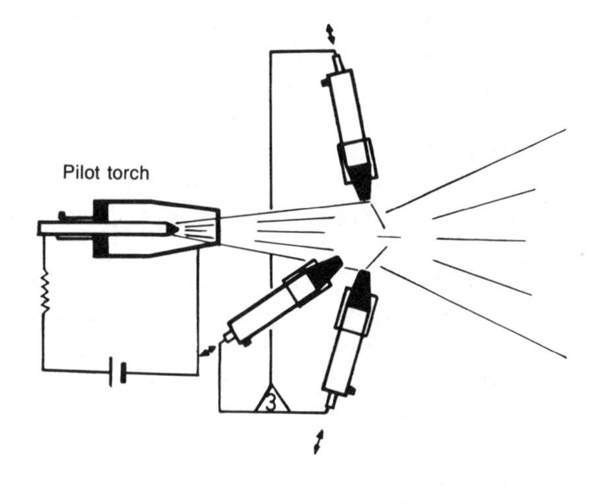


**Fig. 9.** Three-phase arc between clad electrodes [43].

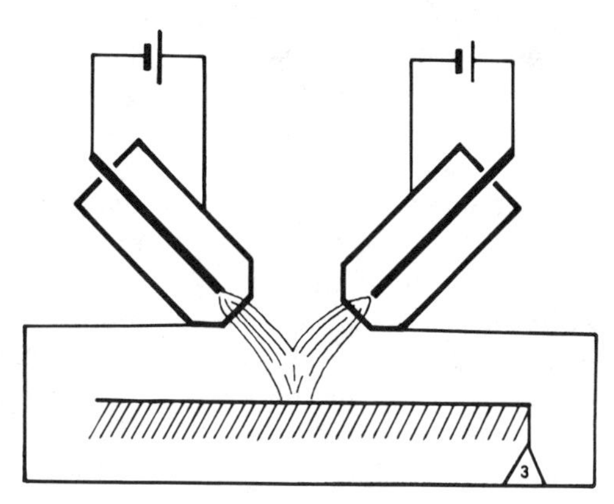


**Fig. 10.** Three-phase superimposition between two torches and a bath [43].

In addition, the efficiency of a plasma of this type is much better and approaches unity; the overall thermal efficiency depends, of course, on torch utilization conditions.

In general, the power of plasma generators is adjusted by varying the superimposed current.

### *3.1.1.4. Efficiency of arc plasma generators*

In its simpler versions, the efficiency of an arc plasma generator is rather low. The very high losses due to electrode cooling must be added to losses due to rectification of the AC (5 to 10% depending on the type of rectifier). In fact, electrode cooling losses vary between 30 and 60% depending on the type of generator. These figures apply to blow- or transferred-arc plasma generators; however, it is possible to improve efficiency considerably, especially for high energy applications.

To limit the drawback due to the use of rectifiers, it is possible to use AC so that both electrodes behave symmetrically. It is also possible to use a consumable electrode generator which operates at very high current; the anode resistivity is then high, its surface area low and therefore current density is high. Since the anode is only slightly cooled if at all, the material forming it vaporizes. A cloud develops around the anode which can participate in the reaction to be produced but also causes pollution. The efficiency of this generator is higher than those described above but the energy gain must be compared to electrode costs.

The best method of obtaining very high efficiency is to use a superimposed arc-plasma generator. Most of the power is, in fact, dissipated due to the Joule effect in the plasma gas column as a result of the superimposed AC, and the overall plasma generation efficiency is then around 90%.

#### *3.1.1.5. Arc-plasma applications*

The arc-plasma generator is by far the most widely used in industry. The technology is now well mastered and a wide range of generators is available. Powers used vary from a few watts to about 10 megawatts, and the temperatures reached enable treatment of most refractory materials.

Powders can be introduced in the generator at the nozzle, in the area in which the temperature is highest, which enables coatings by spraying of very refractory materials. Protective gas can also be injected around the plasma through a nozzle concentric with the main nozzle, which protects the plasma impact area on the target material.

Plasma generator efficiency, which is relatively low for the simplest versions (about 40 to 50%) can be increased considerably by using techniques such as superimposition of AC power, where it can reach 70 to 90%, excellent for applications involving high powers.

However, this generator has some disadvantages. The operating mode requires the use of neutral or reducing gases. However, oxidizing gases may be introduced by special devices such as auxiliary nozzles. The electrodes pollute the plasma gas, which may exclude certain applications where very pure plasma is required.

Arc-plasma generator applications are therefore numerous and varied — cutting, welding and melting metals, spraying metallic or non-metallic high melting point coatings, activation of chemical reactions, etc.

### *3.1.2. Electromagnetic field plasma generators*

#### *3.1.2.1. Principle*

The energy of disassociation and the ionization potential are provided by an electromagnetic field. There are three types of electromagnetic field generators: the induction plasma generator, the capacitive generator and the microwave generator. Only the induction generator is used in industrial applications, the others at the moment being confined to research laboratories. The induction plasma generator is also known (as is the capacitive generator) as a high frequency or radiofrequency generator [7 *e*], [22].

Induction plasma generator operation is very simple. The ionized gas, which is electrically conductive, is induction heated. This gas, which is injected into the top end of a 15 to 100 mm diameter vertical quartz tube, is pre-ionized on starting of the torch by an auxiliary metal or graphite electrode inserted into the tube; this auxiliary electrode is induction-heated and causes the gas temperature to rise, its ionization potential to drop, and the torch to ignite. Pre-ionization can also be obtained in a partial vacuum, of the order of $10^{-2}$ mbars. The pre-ionized gas passes through the magnetic field of a solenoid through which the RF current flows. Due to the effect of the induced currents, the gas heats and its ionization factor increases. The center of the plasma reaches a high temperature, a function of the radiofrequency power applied.

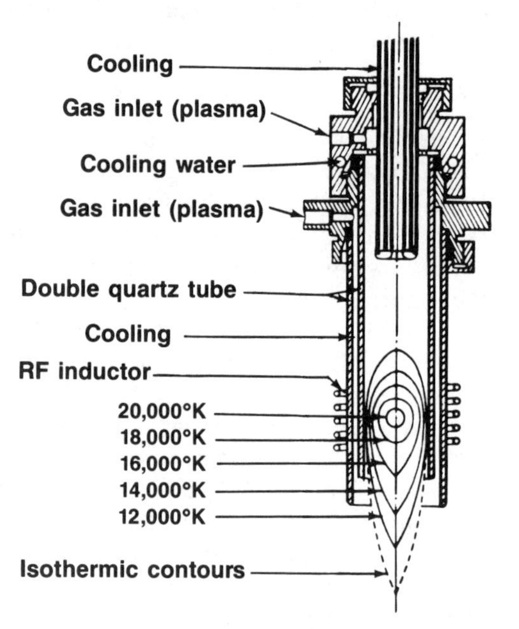


**Fig. 11.** Radio frequency plasma generator [52]:
— diameter of silica tube: 20 mm;
— power: 5 kW.

### *3.1.2.2. Induction plasma generator characteristics*

For a power of 8 kW and an argon rate of 20 l/min, the maximum temperature of the plasma center is about 20,000°K: if the radiofrequency power is dropped to 1.5 kW, the temperature only reaches 800°K (52°C). Generally, the induction coil consists of 2 to 8 turns of water-cooled copper. The quartz tube must be gas cooled or, for high powers, water cooled. In most cases, plasma stabilization is provided by tangential injection at the apex of the gas torch which flows spirally into the quartz tube.

The power applied to the plasma is a function of the diameter of the gas column, the type of gas and frequency; therefore, as in all induction heating methods, power depends on the electrical and magnetic characteristics of the body to be heated and on frequency. Since a plasma can be analagous to a non-magnetic conductor of electricity, resistivity and frequency are the determining factors in transferred power, current penetration depth and power transmission efficiency (*see* "Electromagnetic induction heating"). The resistivity of the gas column is generally high, but drops as the temperature increases, in proportion to the rate of ionization increase, as shown in Figure 12 (this is 10 to 20 times higher than for graphite, and 500 to 1,000 times higher than copper). Therefore, it is necessary to operate at radio frequencies between 1 and 30 MHz with quartz tube diameter between 15 and 85 mm.

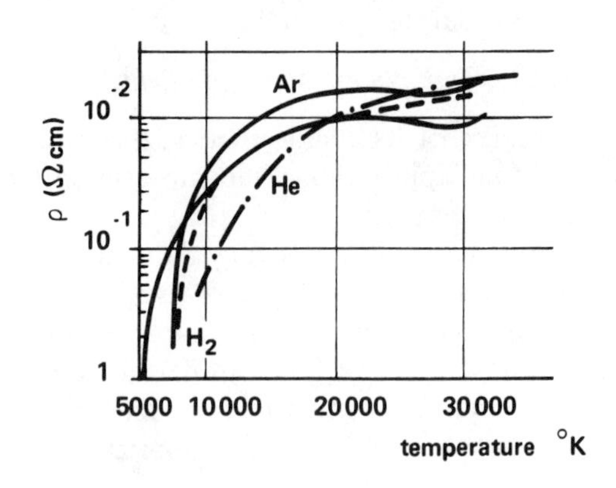


**Fig. 12.** Variation in resistivity of ionization gas versus temperature [52].

Various gases such as argon, nitrogen, helium, hydrogen, air or mixtures thereof in variable proportions are used as plasma gases. Figure 13 below gives the characteristics of some argon generators.

Characteristics of argon induction plasma generators.

| Diameter (mm) | RF power (kW) | Argon rate (l/min) |
|---|---|---|
| 15 to 40 | 4 to 5 | 15 to 20 |
| 15 to 60 | 8 to 10 | 25 to 35 |
| 15 to 85 | 12 to 20 | 50 to 60 |

**Fig. 13.**

Plasma generator power is generally low, a few kilowatts to a few tens of kilowatts, and the maximum powers currently reached are about 1 megawatt.

The basic quality of this type of generator is the pureness of the plasma obtained. The absence of electrodes (except on startup) provides purity equal to that of the plasma gas introduced. In addition, the absence of electrodes enables any gas to be used.

However, by comparison induction plasma generators have some disadvantages:

— the efficiency of the plasma generator compared to the power provided by the inductor is about 60 to 70%; the efficiency of the RF generator is between 50 and 60%, the power ratio available in the plasma to that provided by the

line is only 0.30 to 0.40. The plasma generation efficiency is therefore low, excluding many thermal applications from the inductive plasma field of applications;
— investment costs are much higher than with arc plasma generators;
— maximum power is relatively low, limiting use to low quantity products (very high purity products) or to laboratory work;
— torch utilization lacks flexibility since it is almost impossible to modify its position during operation.

#### *3.1.2.3. Inductive plasma applications*

Because of its characteristics, the inductive plasma torch has only limited industrial applications; the most important are activation of endothermic chemical reactions:
— zirconia, niobium or ruby monocrystal growth (the Verneuil method which does not seem to have had major industrial developments);
— the production of cyanhydric acid from methane and nitrogen, of cyanogen by injection of carbon powder in a nitrogen plasma, of metallo-organic compounds (tetra-ethyl of lead, zinc dimethyl, etc.);
— depositing thin films by vaporization.

### 3.2. Plasma gases

The first element in choosing a plasma gas is the type of atmosphere desired for the application: inert or neutral, reducing or oxidizing. If the generator used does not permit the use of oxidizing gases (arc plasma generators), they are introduced in the plasma jet through an auxiliary device, and the plasma gas is then a neutral gas.

However, the use of a plasma gas has an important effect on generator efficiency and on the thermal characteristics of the plasma; if the reactive gas is fixed, it is not always desireable to use it as a plasma gas [2], [21], [23], [52].

A detailed study of their thermodynamic properties will determine the relative advantages of different gases.

Figure 14 gives the energy efficiency of some gases versus temperature; diatomic gases (nitrogen, hydrogen and oxygen) and monoatomic gases (argon, helium) differ widely. For monoatomic gases, the energy density first increases slowly up to the ionization threshold, then very rapidly through the ionization zone. With diatomic gases, energy increases from a much lower temperature and rapidly reaches high values which are much higher than for monoatomic gases. This higher enthalpy in diatomic gases is basically due to the energy of disassociation of the molecules which does not exist with monoatomic molecules. This aspect is critical for heat transfer to the load. In contact with the load, on the one hand, the electrons and ions recombine, and on the other, the molecules reform. Diatomic gas plasma can therefore not only provide ionization energy but also disassociation energy. Power density is therefore higher with diatomic gases.

However, compared to that required by a monoatomic gas, the energy required for a diatomic gas to create a plasma is much higher. Figure 15 gives an example of the variation, as a function of temperature, of the power ratio for the gas carried in the plasma for argon and nitrogen. Therefore, it is much more difficult to generate a plasma with a diatomic gas such as nitrogen. Some plasma gases also use argon as a priming gas, and another gas, hydrogen or nitrogen, during the production phase.

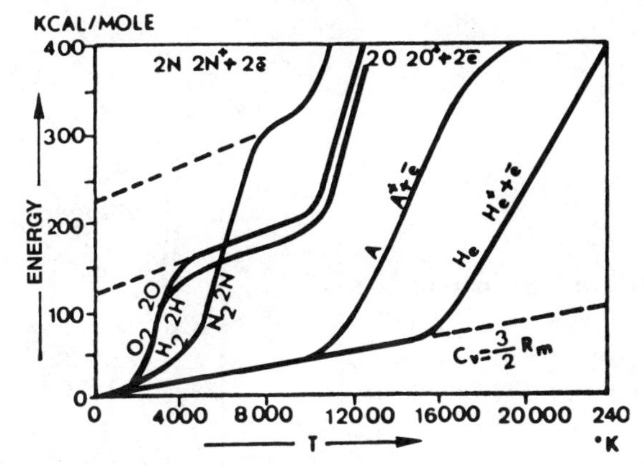


**Fig. 14.** Energy versus temperature for different gases [52].

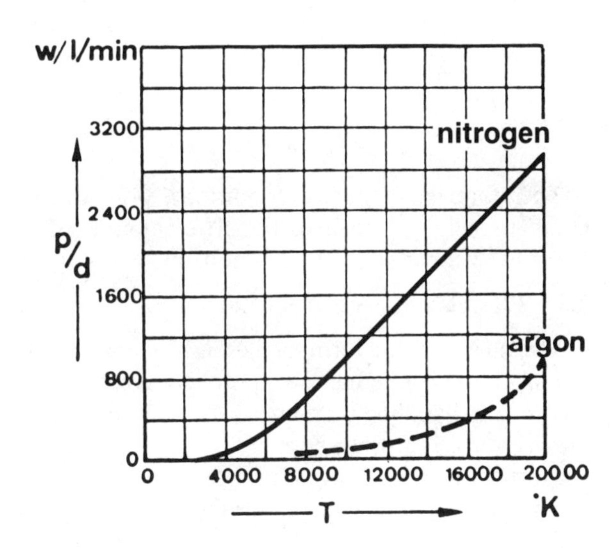


**Fig. 15.** Temperature reached versus electric power ratio (W) to gas flow rate (l/min) [52].

In transferring heat to the load, the thermal conductivity of the plasma is also critical; Figure 16 gives a variation of this value as a function of temperature for different gases. The thermal conductivity of diatomic gases shows a pronounced maximum at the disassociation temperature of the molecule, and subsequently maintains a high value, generally higher than that of monoatomic gases (with the exception of helium, whose thermal conductivity is very high, but as with monoatomic gases, has the drawback of having low energy density; it is also expensive). Diatomic gases therefore offer better heat transfer to the load.

In practice, gas mixtures are widely used, since they offer properties which are noticeably different from their original components, and therefore lower costs.

For example, the thermal conductivity of argon mixed with a low proportion (approximately 10 to 20%) of nitrogen or hydrogen, gives an increase of 10 to 20 times over the diatomic gas disassociation zone, and may be 2 to 3 times higher than the argon ionization zone.

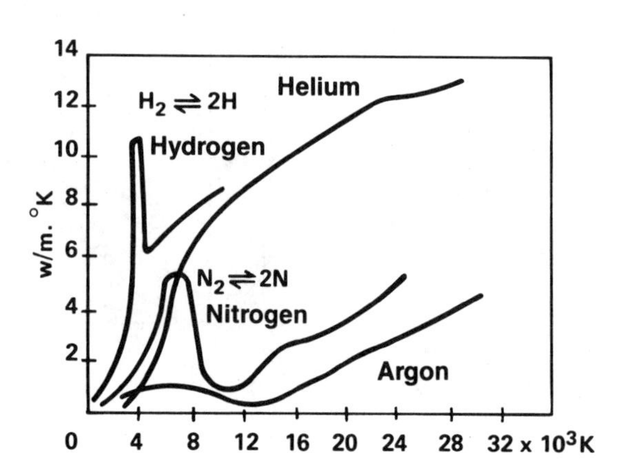


**Fig. 16.** Thermal conductivity of plasma versus temperatures in degrees Kelvin [52].

Therefore, atmospheric conditions, power densities and thermal transfer, combined with the costs of the different gases, determine the best gas mixture to be used for a given application.

### 3.3. Plasma furnaces or reactors

Plasma generators form torches or blow torches which are then used for various applications. For some, the equipment consists essentially of the torch and its control and handling systems especially for cutting, welding and surface spraying applications. For other applications, the generator is a heat source for a more complex system such as a furnace or reactor.

The furnace or reactor consists of a chamber in which plasma generators are used in different ways:

— *blown-arc generator*, without arc transfer to the charge; this technique is used for melting and treating electrically non-conductive bodies;

— *superimposed arc generator*; the powers dissipated can be very high, and this technique, which offers high efficiency, is of interest in applications where high energy levels are required.

Due to continuous introduction of plasma gas into the furnace, it is generally necessary to provide the furnace with a pressure regulation system.

Many types of plasma furnaces and reactors have been conceived in research laboratories; some have reached the prototype installation stage, and some are currently in use in industry.

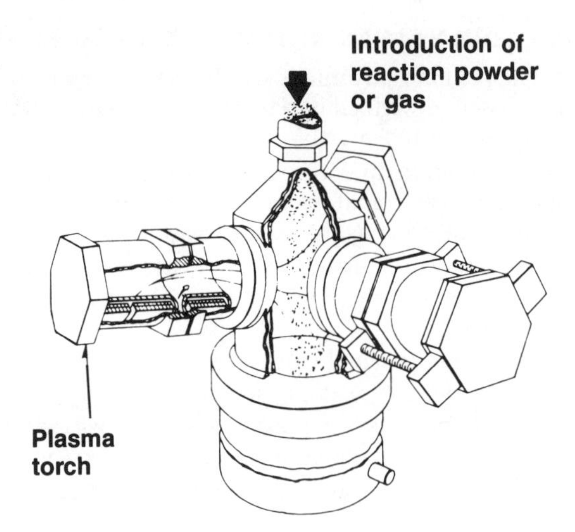


**Fig. 17.** Three-phase plasma furnace for chemical reactions [32].

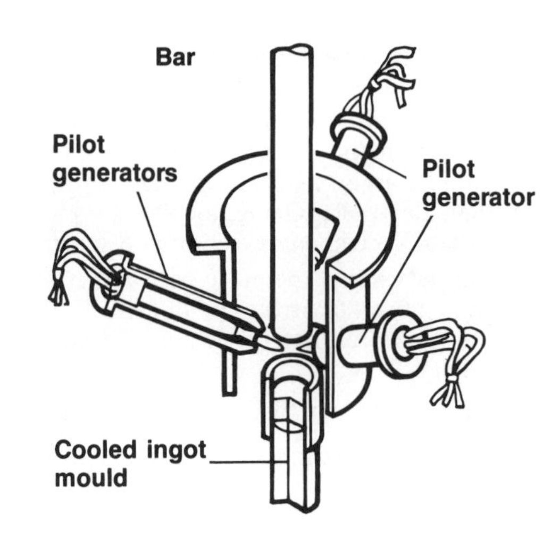


**Fig. 18.** Remelting of steel using three-phase plasma [G].

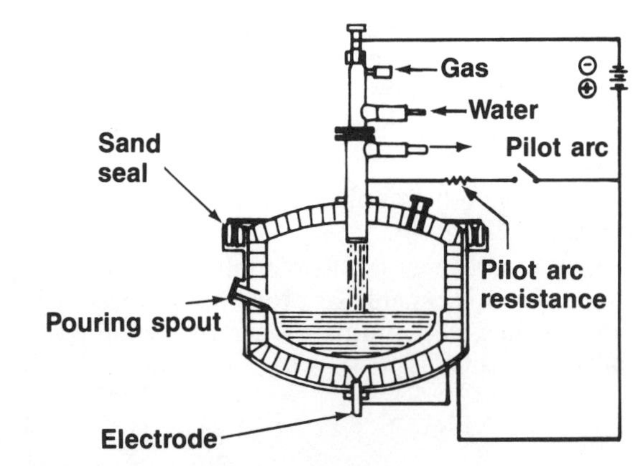


**Fig. 19.** Arc plasma furnace for melting metals (hearth furnace), [52].

## 4. ADVANTAGES AND DISADVANTAGES OF PLASMA HEATING

The primary advantage of plasma heating is in its special thermal properties. These were described in paragraph 2.2 "Thermal properties of plasmas and field of application"; briefly, they are:

— very high temperature in plasma jets, enabling treatment of highly refractory materials, the transformation of which is difficult using other techniques, or creation of chemical reactions requiring very high temperatures;

— high power densities which, if necessary, can be concentrated on very small surfaces or volumes, favoring highly endothermic reactions and accelerating processes;

— very high thermal conductivity, providing very high reaction speeds thereby reducing the size of equipment;

— adaptability of the plasma gas composition: the atmosphere can be neutral, reducing or oxidizing, and better adapted to the requirements of the application;

— high gas ejection rates whereby the mechanical action of the plasma can reinforce its thermal effect.

In some cases, combinations of these characteristics can reduce the number of material production stages. New materials can also be produced using this technique.

The specific advantages of the various applications can be seen from their descriptions.

Conversely, in terms of investment, plasma heating is more expensive than conventional methods. Gas consumption can weigh heavily in operating costs. For example, argon as a plasma gas is responsible for more of the operating costs than electricity. Generally, gas costs in plasma generators are equal to or higher than electric energy costs and must be carefully supervised.

Plasma heating is therefore used either for applications in which there are practically no alternatives or, due to its thermal characteristics, to improve fabrication of products for which conventional processes offer relatively low productivity.

## 5. INDUSTRIAL APPLICATIONS OF PLASMA HEATING

Some current plasma heating applications have numerous installations in industry:

— cutting metals [10]

— welding [2], [6], [13];

— spray coating [1].

Other applications are less widespread but nevertheless significant in industry: melting special steels and refractory materials, and in certain chemical reactions, spheroidization of powders, production of monocrystals and, very recently (1979), production of sponge iron [7], [15], [16 *c*], [33 *f*], [34], [36], [38], [39], [42], [47], [48], [55].

Numerous other applications have been the subject of laboratory studies but are as yet unproven.

Finally, aerospace industry applications should be mentioned. The arc plasma torch, which is capable of providing gas jets at hypersonic speeds and at high temperatures, can simulate operating conditions for supersonic aircraft and ballistic missiles [20], [51 *c*].

## 5.1. Metal cutting

Arc plasma metal cutting was introduced in France around 1965. Since then, it has been adapted by many metal industries [10].

### *5.1.1. Mechanism and characteristics of metal cutting using arc plasma*

— *Principle*

Plasma metal cutting involves the use of a generator in which the arc is transferred onto the workpiece. The circuit anode is the actual workpiece, and the generator nozzle anode is only used to start the arc, which is produced at radio frequency between the anode and the cathode.

The plasma jet cuts by melting the metal at impact point then ejecting molten metal. The gas flow rate must be high, 20 to 100 l/min (5 to 20 times higher than that used in welding) so that expulsion of the melted metal is complete, providing a clean cut. The plasma jet is surrounded by an annular gas layer which is relatively cold and non-ionized. This protective jacket of inert gas prevents oxidation around the plasma jet impact area [10], [17].

— *Plasma gases*

The gases generally used are argon-hydrogen, nitrogen-hydrogen or argon-nitrogen-hydrogen mixtures.

The arc is often started using pure argon, being easier with a monoatomic gas (*see* paragraph 3.2.). Attempts have been made to use air or water vapor as plasma gases but few industrial applications use this principle. Other generators enable water to be injected directly into the torch nozzle.

The plasma gases used for metal cutting must have the following characteristics:
— very high enthalpy and thermal conductivity; the quantities of heat transferred to the parts and the transfer rates are high, and cutting therefore rapid. Diatomic gases such as hydrogen and nitrogen have both these properties (*see* paragraph 3.2), and mixtures are widely used in cutting;
— high molecular mass; the plasma jet can then easily expel the molten metal over a wide thickness, with limited gas flow rate; nitrogen and argon meet this requirement;
— very low chemical reactivity; reaction of nitrogen with the nozzles, though slow, causes more frequent replacement than with argon. Moreover, the reaction of nitrogen with the oxygen contained in air causes formation of nitrous gases. The reducing nature of hydrogen reduces or prevents oxide formation on cut surfaces; the cut surfaces are much cleaner than those obtained with inert gases such as argon.

The complementary qualities of the various gases explain the use of mixtures. Adding hydrogen to argon or nitrogen increases heat transfer to the workpiece, while permitting relatively low arc voltages (high arc voltages are required for pure hydrogen). Ternary mixtures are becoming more widely used; argon facilitates striking and arc stability, hydrogen provides high enthalpy and nitrogen high kinetic energy.

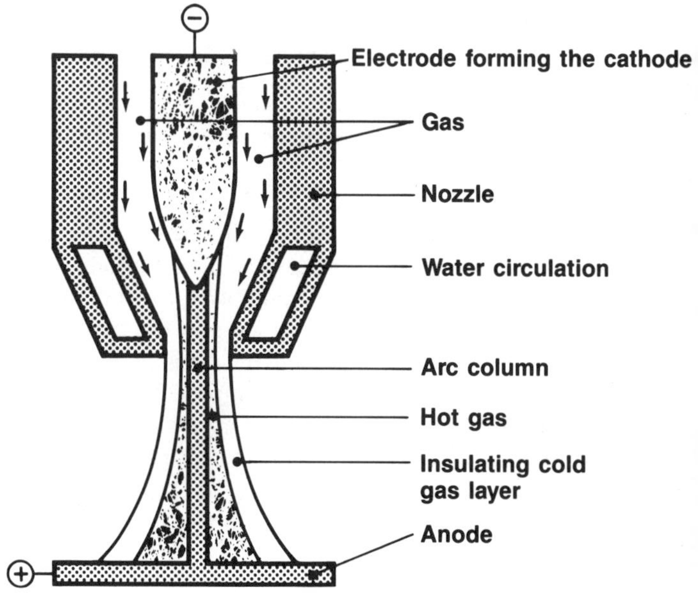


**Fig. 20.** Diagram of a transferred arc plasma generator for metal cutting [10].

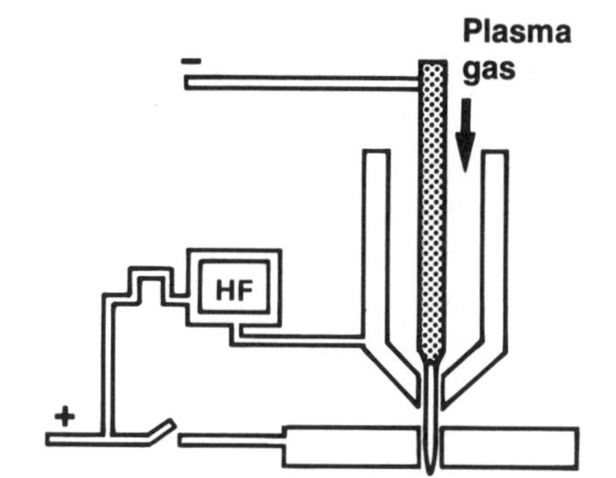


**Fig. 21.** Composition of metal cutting equipment [10].

The most widely used mixtures are argon and hydrogen (80% Ar, 20% $H_2$), argon and nitrogen (60% Ar, 40% $N_2$) and a ternary mixture of argon, nitrogen and hydrogen (35% Ar, 35% $N_2$ and 30% $H_2$).

— *Aspect of cutting*

The energy available for cutting is obtained from two sources: the plasma jet and the anode portion of the cutting arc.

The anode point is located in the center of the cut so that, above this region, melting is provided by the plasma jet due to the Joule effect and by radiation from the arc column. Below the anode point, cutting takes place by means of the molten metal at a temperature 100 to 200°C higher than its melting point, and by the plasma whose temperature gradually decreases.

**Fig. 22.** Aspect of a plasma jet cut.

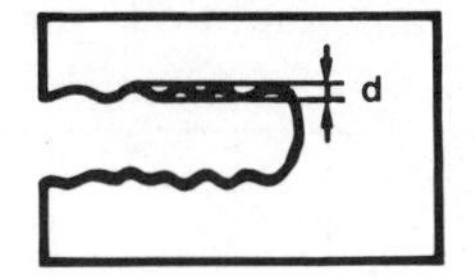


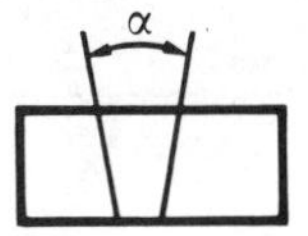


This cutting mechanism results in a special cutting shape, which is wider at its apex than at its base [10], [21].

The area thermally affected by plasma cutting is small, and always less than that produced by oxyacetylene cutting. Figure 23 gives characteristics when cutting stainless steel.

The thermally affected area is much smaller than with welding. Therefore, no machining is required to weld parts previously cut by plasma jet.

Zone thermally affected by plasma cutting titanium-stabilized 18/8 stainless steel [10].

| Thickness cut (mm) | Power used (kW) | Cutting rate (m/h) | Thermally affected zone (mm) |
|---|---|---|---|
| 10 | 35 | 120 | 0.3 |
| 20 | 35 | 36 | 1.5 |

**Fig. 23.**

— *Cutting rate*

Cutting rate depends on the electrical power, the type of metal, the composition and flow rate of the gas mixture, and also on the diameter of the nozzle and the torch-workpiece distance.

Increasing the electric power increases the cutting rate, or enables larger thicknesses to be cut without decreasing cutting rate. Installed power ratings are commonly between 20 kW (no-load voltage: 130 V) and 100 kW (no-load voltage: 400 V).

The characteristics of some cutting equipment are given in Figure 24.

Enrichment of the plasma gas mixture with hydrogen increases the plasma temperature, allowing greater thicknesses to be cut. Also, gas flow rate must increase as cut thickness increases.

As the nozzle-workpiece distance increases, the plasma jet tends to widen and become irregular in shape, resulting in a more jagged cut.

Finally, the nozzle diameter varies according to the thicknesses cut and the desired cut width. The cut width is generally about twice the nozzle diameter. For example, a nozzle diameter of 3 mm can cut thicknesses between 5 and 30 mm.

Plasma cutting conditions for different materials [10].

| Material | Thickness (mm) | Electric power (kW) | Current (A) | Gas flow rate (l/m) argon + 35% hydrogen mixture | Cutting rate (m/h) (adequate cut) |
|---|---|---|---|---|---|
| Stainless steel | 40 | 30 | 350 | 50 | 12 |
| | 100 | 100 | 700 | 100 | 10 |
| Cast iron | 5 | 26 | 240 | 50 | 30 |
| | 20 | 28 | 260 | 50 | 12 |
| Aluminum | 45 | 30 | 350 | 50 | 18 |
| | 120 | 100 | 700 | 100 | 10 |

**Fig. 24.**

### *5.1.2. Plasma cutting applications and advantages*

Arc plasmas can be used to cut most metals. However, their technical and economic interest varies as a function of the nature of the metals and the characteristics of the parts to be cut.

Plasma arc cutting is basically used for high alloy steels such as stainless or refractory steels, light alloys and copper alloys.

For high alloy steels, plasma cutting competes with iron powder oxygen cutting where conventional oxygen cutting is not possible. Due to the cutting rate, the use of the plasma torch has increased gradually, despite high investment costs; it provides high productivity and cleanliness, considerably reducing pollution and improving working conditions (in all cases, however, personnel must be protected against intense light radiation).

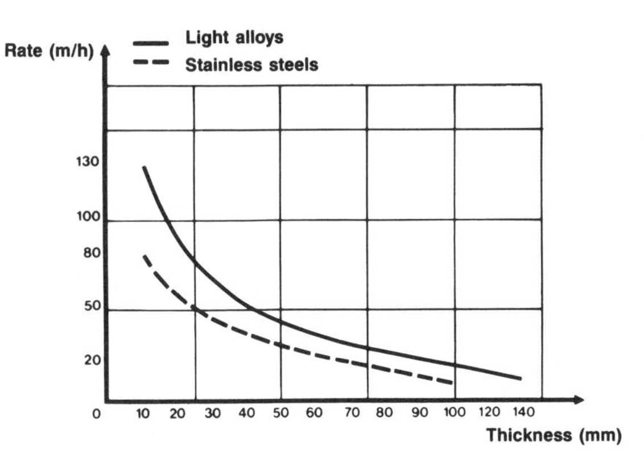


**Fig. 25.** Cutting stainless steels and light alloys using a 100 kW plasma torch [10].

Plasma cutting is, of course, applicable to other steels — mild steels, carbon steels, low alloy steels, etc. — but is more expensive than oxygen cutting. However, its use may be profitable when cutting speed becomes a determining factor.

The curves of Figure 26 show that plasma cutting ordinary steels less than 25 mm thick, can be much faster than oxygen cutting.

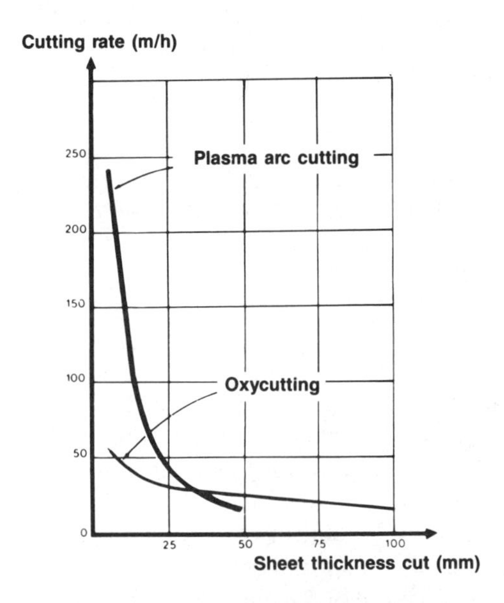


**Fig. 26.** Comparison of rates obtained for straight cuts in plasma and oxycutting [10].

Greater thicknesses may be cut due to increased plasma torch power densities.

For light alloys and other non-ferrous metals (copper alloys, magnesium, nickel, etc.), plasma cutting offers high productivity, better quality and greater flexibility than mechanical processes.

In addition to cutting for incorporation into metal assemblies by mechanical welding methods and similar techniques, arc plasma cutting is also used for other applications, such as cutting jets in steel or pig iron casting, cutting discards in foundries for recycling in melting furnaces, repair of chemical reactors and other assemblies by cutting weak or damaged portions of walls with plasma jets and replacing them with new metal, etc.

Mobile plasma cutting equipment is available with manual torches, as are fixed machines of various types including plasma torches which can be mounted on oxycutting machines. Arc plasma cutting machines enable single or multiple cuts. Automatic controls, either magnetic or electronic, are also used.

### 5.2. Plasma welding

Plasma welding started to appear in industry around the mid 1960's [2], [6], [11], [12], [13], [30], [31].

#### *5.2.1. Principles and characteristics of plasma welding*

Arc plasma welding is similar to the cutting process described above; in fact, it is a fusion welding process using a transferred arc with the fusion bath acting as the anode. The gas flows are much lower in welding, since high flow rates would form a continuous groove in the weld. Generally, flow rates vary between one and five l/min; they cannot provide adequate protection of the weld area (melted metal, solid weld bead, metal edges taken to high temperature) so protection is provided by a non-ionized cold neutral annular gas jet surrounding the plasma jet. The flow rate for this protective gas is about 15 to 20 l/min. in conventional applications, and may be higher in special cases such as titanium welding [2], [21], [24], [26].

— *Plasma gases*

In welding, argon is the most widely used plasma gas. It is also possible to use an argon-hydrogen mixture, with a small amount of hydrogen, at least for stainless steels, nickel and copper-nickel alloys. The addition of hydrogen improves heat transfer to the part, but its percentage in the mixture must be low so that there is no outgassing of the weld bead (affinity of some metals for hydrogen).

The same atmosphere can be used for the plasma jet and the protective gas jet, but it is also possible to use different combinations, argon in the center and an argon-hydrogen mixture for annular protection and vice versa.

If the argon-hydrogen mixture is not appropriate, it is necessary, for example, to use an argon-helium mixture. For welding titanium and zirconium, the gases used must be totally free of hydrogen. For non- or low-alloyed steels, it is possible to use $CO_2$ as a protective gas.

The choice of the plasma gas and the protection gas obviously affects welding rate and the shape of the melted area.

— *Weld shape and welding rate*

In general, the weld bead is similar to that shown in Figure 27. The top flared part can be compared to a conventional arc crater, which is shallow, and whose formation is essentially due to the effect of hot gases surrounding the arc column.

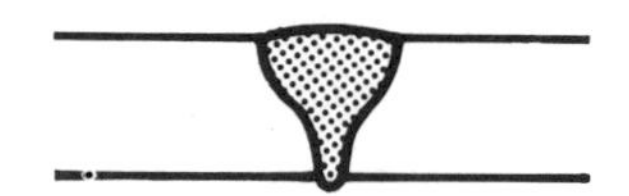

**Fig. 27.** Shape of an arc plasma weld.

The lower cylindrical part is melted only by the plasma jet itself. The width of the crater at the top of the joint therefore depends not only on the plasma jet and its characteristics (heat flux, thermal and electrical conductivity of the ionized gases, electron beam intensity in the plasma, etc.), but also on the properties of the metal (conductivity, specific heat, thermal diffusivity, etc.) and the nozzle characteristics.

It is possible to alter the width of the weld seam by using a nozzle having two orifices situated on either side of the main orifice. These gas injections, which are much colder than the main jet, lower the temperature of the gas layer on each side of the joint, and flatten the plasma jet by ovalizing it. The melted area is therefore narrower and more elongated, and the welding rate increased. The separation between the two orifices must not, however, be too small, so as to prevent the formation of channels.

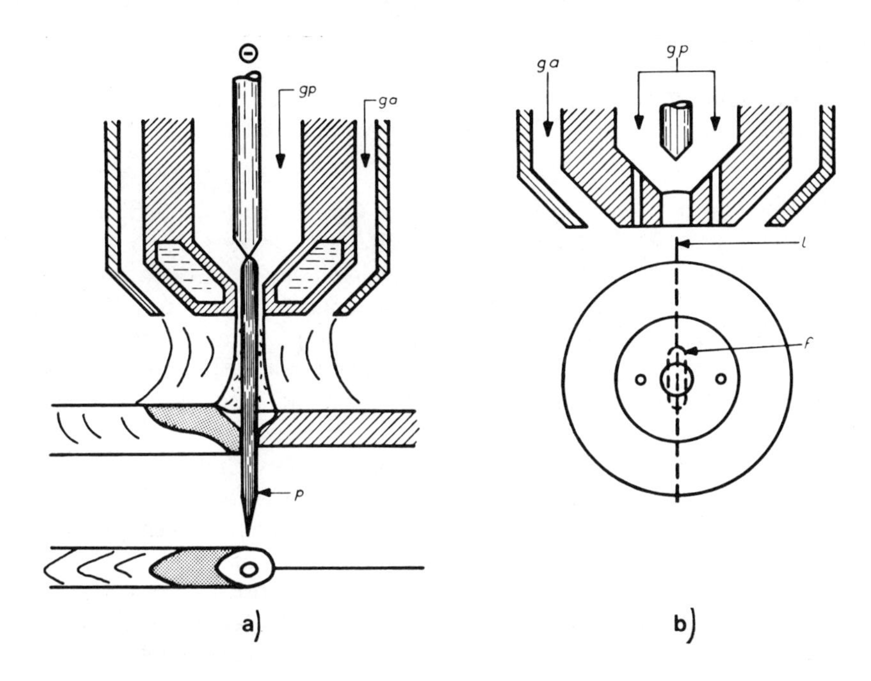


**Fig. 28.** Operation of plasma torch in continuous welding [2]:
*gp*, plasma gas;
*ga*, annular protection gas;
*p*, plasma jet passing through the hole formed in the joint;
*l*, weld line;
*f*, oval plasma jet shaped by orifice;
a, single orifice nozzle;
b, two-orifice nozzle.

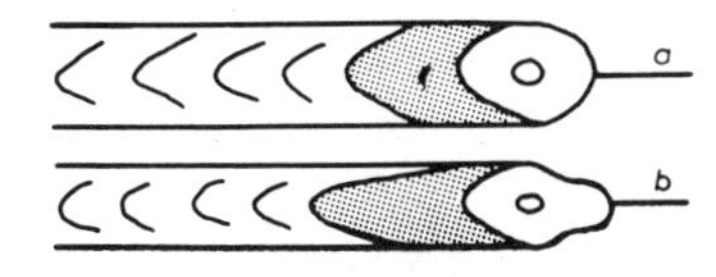


**Fig. 29.** Weld made with nozzle [2]:
- top: single orifice;
- bottom: multi-orifice.

The plasma gas also has a marked effect on the characteristics of the weld seam. The narrowest welds are obtained by using argon for the plasma gas and for the protective jet. The high temperature gradient in the argon plasma in fact gives a small cross-section arc column with a gas layer at the periphery where the temperature is relatively low. Conversely, the welding rate is low with this monoatomic gas due to limited thermal performance.

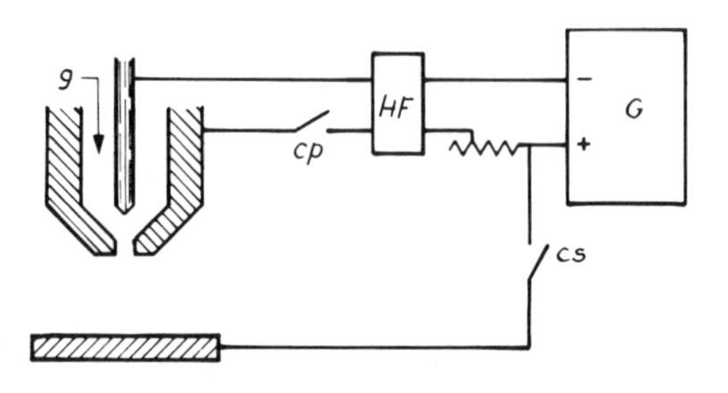


**Fig. 30.** Simplified diagram of a plasma welding station [52]: *g*, gas; *cp*, pilot arc contactor; *cs*: welding contactor; HF, high frequency oscillator.

If a small proportion of hydrogen is added to the annular jet and the plasma gas is argon, some of the hydrogen disassociates and ionizes in the hottest layers touching the plasma; the weld bead widens and the welding rate increases. Introduction of hydrogen into the plasma accenutates both effects. The use of an argon-hydrogen mixture as a plasma gas and argon as a protective gas is of little interest: since argon ionizes easily, the annular jet is ionized over a wide cross-section and the plasma expands widening the weld bead and causing relatively low welding rates despite the added hydrogen.

The welding rate also depends on other parameters such as electrical power, type of metal and thickness. For example, with a 6 kW torch, it is possible to weld 3 and 5 mm thick stainless steel parts at respective rates of 100 and 35 cm/min. without added metal. With added metal (melting of wire by the plasma), welding speeds change to 61 and 20 cm/min for thicknesses of 3 and 6 mm. For spheroidal graphite cast iron, preheated to 300°C before welding, the rate is 22 cm/min for a thickness of 6 cm and a power of 4.5 kW. With titanium and a 5 kW argon torch, welding rate is 30 cm/min for a thickness of 5 mm.

### *5.2.2. Advantages and applications of plasma welding*

Welding stainless steels and some special metals such as titanium often involve problems difficult to resolve with standard techniques such as the TIG (Tungsten Inert Gas) process, which uses a conventional arc and protects the welding area by an inert gas such as argon. Arc plasma welding has enabled welds impossible or difficult to obtain with other techniques [4], [5], [13], [14], [25], [A], [B], [D].

In terms of power density performance, the width of the weld bead and the welded thickness-weld bead width ratio, arc-plasma welding is situated between the TIG process on the one hand, and the electron beam or laser process on the other.

In practice, arc-plasma welding is better suited than the TIG process for stainless steel and special steels thicker than 3-4 mm. The maximum weldable thickness is about 8 mm, but recent developments have raised this limit to 12 mm.

Another application is microplasma welding. This process is used to weld stainless and special steels, but also almost all grades of steel which can be welded using the TIG process (when between 0.01 and 0.5 mm) at lower costs and with more flexibility than the TIG process, which is difficult to implement at low powers.

Compared performances of plasma, electron beam and TIG welding [21].

| | TIG process | Plasma | Electron beam | |
|---|---|---|---|---|
| | | | 30 kV | 150 kV |
| Maximum power density ($W/cm^2$) | $5x10^4$ | $5x10^5$ | $10^6$ | $10^9$ |
| Minimum welding surface ($mm^2$) | $10^{-1}$ | $10^{-2}$ | $10^{-4}$ | $10^{-5}$ |
| Weld seam maximum depth-to-width ratio | 1/2 | 2/1 | 10/1 | 20/1 to 40/1 |
| Linear consumption of energy for a given thickness (J/cm) | 21,000 | 7,500 | 1,450 | — |

**Fig. 31**

| PLASMA | | | TIG | | |
|---|---|---|---|---|---|
| Preparation | Number of runs | Th. mm | Preparation | Number of runs | Th. mm |
|  | 1 | 2,5 |  | 1 | 2,5 |
|  | 1 | to 7 |  | 1 or 2 | 3 |
|  60° 5 | 2 | > 7 < 12 |  75° 1,5 | 4 to 5 | 7 |

**Fig. 32.** Comparison of TIG and plasma processes [2].

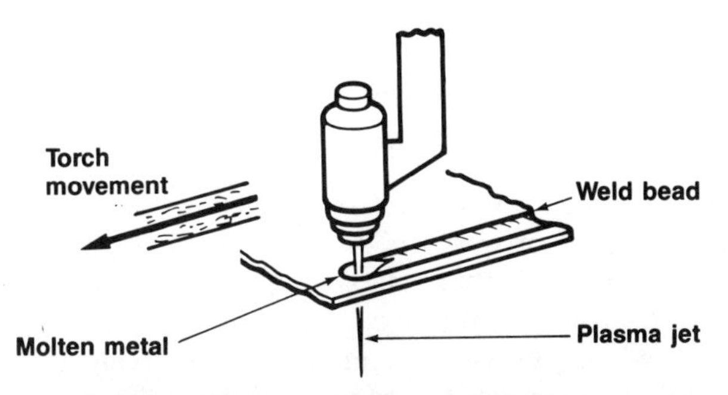


**Fig. 33.** Welding plasma torch [27].

Microplasma is particularly attractive because of its stability, a characteristic which is difficult to obtain with the TIG process. The powers of microplasma welding torches vary from a few tens to hundreds of Watts (arc current: 0.2 to 15 A at 35 to 50 V). The corresponding torches are light and mobile. Generators are usually of the transferred arc type, but blown arc generators are used for welding electrically non-conductive parts and special metal assemblies (metal gauzes and delicate brazing work).

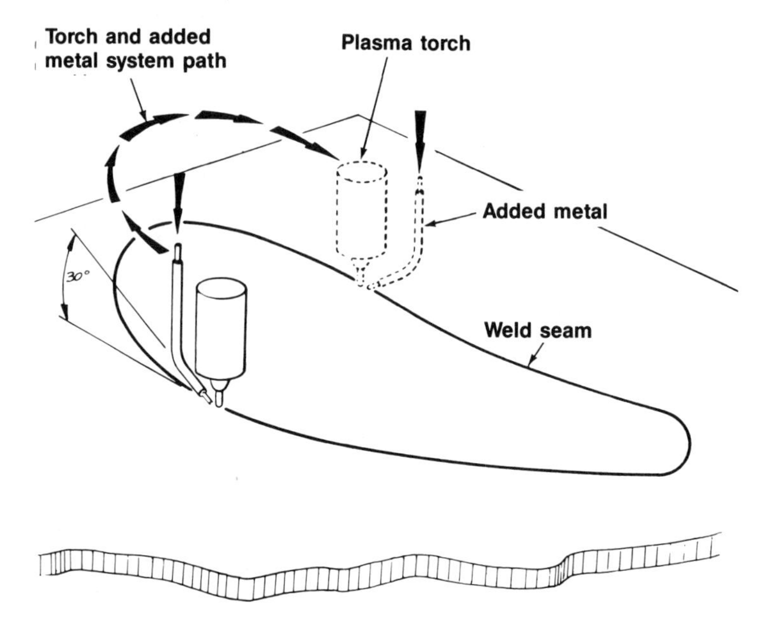


**Fig. 34.** Plasma welding installation [27].

Spheroidal graphite cast-iron plasma welding has given adequate results but as yet is not widely used in industry [16].

Also, plasma jets are used for hard resurfacing of mechanical parts [8], [29].

Plasma welding installations (with the exception of microplasma torches) generally consist of automatic machines having a rigorous welding process control system.

## 5.3. Plasma coating and forming

### *5.3.1. Principles and characteristics of plasma spray*

Since the beginning of the 1960's, arc plasmas have been used to spray substrates with refractory powders. Arc-plasma torches, which provide very hot, high ejection speed gases, are in fact capable of melting and spraying powders of highly refractory materials.

The fluidized powder, which is mixed with a carrying gas, is accelerated so that it can penetrate the plasma and be driven by it. The powder grains are taken to a temperature close to their melting point and sprayed at high speed onto the substrate. If blown-arc plasmas are the most widely used in this application, induction plasmas in some cases offer major advantages [1], [9].

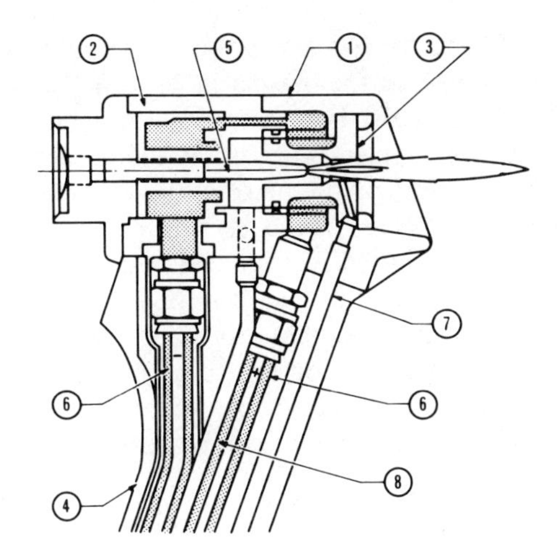


**Fig. 35.** Cross-section of a metallization gun (blown arc) [52], [53]:
1, flange;
2, body;
3, anode;
4, handle;
5, tungsten cathode;
6, DC supply through water-cooled flexible coaxial cables;
7, plastic powder feed hose;
8, plastic gas feed hose.

Similar processes are used to produce refractory objects by spraying powders onto forms. When the forms are removed, the parts require very little finishing.

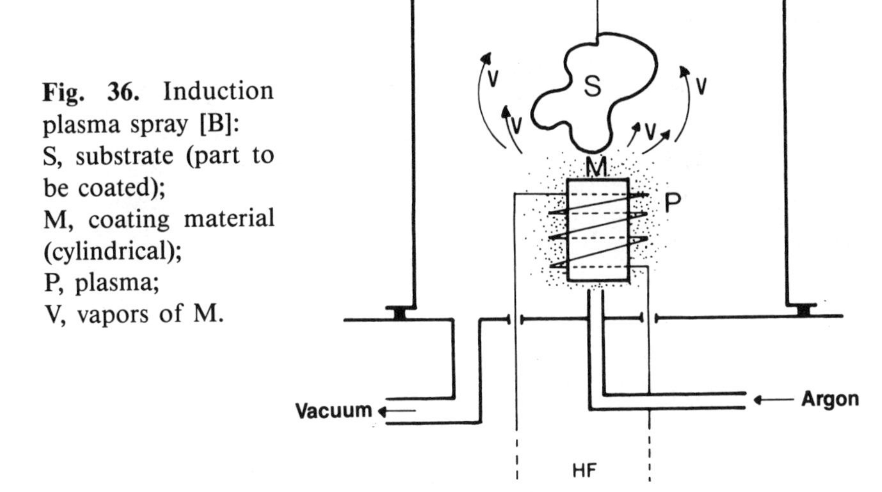


**Fig. 36.** Induction plasma spray [B]:
S, substrate (part to be coated);
M, coating material (cylindrical);
P, plasma;
V, vapors of M.

— *Injection of particles in plasma gas*

Powders can be injected into the plasma using different methods; with blown-arc generators:

— directly into the nozzle channel for bodies or powdered mixtures such as carbides and oxides; this method, which offers excellent efficiency by weight, is the most widely used;

— via injectors in the plasma flame, just at the outlet from the nozzle, which enables injection distances to be chosen as a function of the melting point of

the materials and the gas and powder flow rates. Inclination of the injection nozzle with respect to the plasma jet axis can be adjusted so that the time the powder remains in the hot gases varies with its thermal characteristics; this technique is easily implemented, but it is difficult to make the particles penetrate into the hottest part of the plasma due to its high viscosity;
— around the periphery of the jet, which allows only limited temperatures (this technique has been used to spray nylon particles);
— by benefiting from the Maecker effect; a low pressure zone exists at the tip of the cathode. The gases and solids injected into this area are drawn into the core of the jet. This method provides prolonged staying times in the plasma, and is excellent for materials having very high temperatures and melting points, such as oxides and zirconium compounds.

Characteristics of some materials used in plasma coating [21].

| | Melting point °C | Vaporization temperature °C | Heat of fusion J/g | Thermal conductivity W/cm X °C |
|---|---|---|---|---|
| $SiO_2$ | 1,713 | 2,590 | 235 | 0.0125 |
| MgO | 2,800 | 3,600 | 1,930 | 0.335 |
| $ZrO_2$ | 2,715 | 3,730 | 2,020 | 0.0209 |
| $Al_2O_3$ | 2,050 | 2,980 | 1,448 | 0.293 |
| $Cr_2O_3$ | 2,275 | 3,000 | — | — |
| W | 3,370 | 5,527 | 184 | 1.63 |
| Cr | 1,903 | 2,680 | — | 0.67 |
| Fe | 1,539 | 3,070 | 50.3 | 0.71 |
| Al | 670 | 2,327 | 323 | 2.18 |
| Bi | 271 | 1,560 | — | 0.084 |

**Fig. 37.**

The most widely used plasma gases are argon, nitrogen and nitrogen-hydrogen mixtures but final choice depends on the desired atmosphere. The gas ejection speed is generally very high, 100 to 1,500 m/sec permitting very dense coatings [23].

To reduce the costs inherent with argon or hydrogen, water-stabilized generators using steam as the plasma gas have been developed. The powder is injected into the nozzle just before its tip. Figure 38 shows the characteristics of such a generator which is capable of providing coatings 0.06 to 30 mm thick.

With induction plasma generators, the powder injection mechanism can be similar to that used with the arc plasma method. Generators operating on other principles also exist. The material to be sprayed is shaped by a solenoid connected to a RF generator or a hollow cylinder placed inside a water-cooled copper solenoid, which is itself connected to the RF generator. After having created a vacuum of about $10^{-6}$ torr in the chamber, a low argon current is applied to the coil and the coil circuit is closed. Plasma appears, and the material vaporizes gradually due to the intense bombardment of the plasma particles. The material is then deposited on the substrate. The material to be sprayed is only slightly heated, and along with the fact that the electrode does not contaminate, constitute the two major advantages of this type of plasma [B].

Characteristics of a water-stabilized plasma generator (Corundum spray) [54], [F].

| | |
|---|---|
| Power | 160 kW |
| Mean spraying capacity | 30 kg/h |
| Maximum spraying capacity | 56 kg/h |
| Water consumption | 10 l/h |
| Water pressure | 8 to 10 bars |
| Graphite loss at cathode (diameter 13 mm) | 2 mm/h |
| Anode rotation (diameter 150 mm) | 2,800 tr/mn |
| Maximum plasma temperature | 30,000°C |

**Fig. 38.**

— *Depositing rate*

The depositing rate depends on many things: powder flow rate, particle size, type of plasma gas, electrical power, etc.

Particle size is basically a function of the thermal characteristics of the powder and the porosity desired; generally, this varies from 1 to 100 $\mu$m, larger sizes being used to obtain porous coatings. Particle dispersion must be low in order to obtain adequate deposit homogeneity.

Figure 39 gives examples of the depositing rate as a function of materials, power, nozzle-substrate distance and the plasma gas.

Characteristics of some plasma powder spray equipment [21].

| Powdered material | Gas | Distance to target (cm) | Powder flow rate (g/min) | Power (kW) |
|---|---|---|---|---|
| TiC | Argon | 7.5 | 14.5 | 19 |
| ZrC | Argon | 7.5 | 11.2 | 17.5 |
| TaC | Argon | 7.5 | 14.4 | 19.5 |
| $UO_2$ | Argon | 10 | 29.5 | 20 |
| $ZrO_2$ | Argon | 10 | 25 | 25 |
| Steel | Argon | 10 | 12.7 | 10 |
| W | $N_2+H_2$ | 7.5 | 50 | 30 |

**Fig. 39.**

### *5.3.2. Advantages and applications of plasma coating (surfacing)*

Flame torches (oxyacetylene, oxyhydric, oxypropane) are used to deposit oxide coatings and numerous other chemical compounds, but their use is limited to depositing materials which do not require temperatures higher than 2,500°C. At present, there is a requirement for coatings based on high melting point materials, not only to benefit from their refractory properties, but also the special characteristics of certain oxides, carbides, nitrides and borides. The economic motive is clear: deposit a coating on a relatively low value part or substrate, to provide it with very high surface properties. Plasma generators are excellent for this.

In fact, these generators can spray a wide variety of materials onto highly differing substrates. There are few limitations concerning the choice of materials, which, to be sprayable, need only meet the following requirements:
— possess a stable liquid phase;
— possess a liquid phase over a temperature range wide enough to avoid premature vaporization.

Most industrial materials meet these requirements (for example, *see* Figure 37). The materials most widely used in plasma spraying are relatively high melting point metals (nickel, nickel-chromium, titanium, copper), and especially the refractory metals (tungsten, molybdenum and tantalum), oxides (alumina, titanium oxide, zirconia and zirconates, chrome oxide, magnesia), carbides (tungsten, chromium and titanium carbide), nitrides, borides and cermets, etc. Mixtures of various materials are often used provided their melting points do not differ widely [1], [9], [19], [37], [45], [46], [51].

The substrates can also be very different: metals and ceramics, but also plastics, textiles, glass, semiconductors, etc. They must withstand the added heat from particles, but it is always possible either to cool them (causing sudden hardening of the sprayed particles) or to use induction plasmas, and in particular cold plasmas, when the substrate is highly sensitive to heat. The body to be coated or surfaced must also withstand the treatment environment and the spraying gases. The surface to be coated must be carefully prepared, since its cleanliness and surface condition are controlling factors in coating adherence.

If these conditions are respected, plasma coating offers high quality and properties interesting to many fields, since it provides low porosity (but it is also possible to produce porous coatings), excellent adhesion and high resistance to oxidation. The specific qualities are a function of the coating material.

Some of the applications for such coatings are described below:
— Anti-wear coatings, using tungsten and chromium carbides or special nickel, chromium, boron, and silicon-based alloys; these provide a high degree of hardness and wear resistance for parts such as forging dies, valve seats, valves, furnace rollers, conveyor screws, drive rollers, and turbine and fan vanes.
— Anti-corrosion coatings, using zirconium silicate, alumina, zirconia, zirconates, etc.; these are used to protect moulds, shells, ladles, pyrometric rod sheaths or floats against corrosion due to molten light alloys, and also ingot-making machines in metalwork.

— Production of thermal barriers, used to provide a temperature drop through their thickness sufficient to protect the substrate from thermal destruction and corrosion. These coatings (several millimeters) are used for parts constantly subjected to intense heat differentials (combustion chambers and the bottoms of petrol engine pistons, diesel engines and compressors); the temperature drop can exceed 400°C. For short durations, a thin coating of 0.5 to 1 mm of zirconia or alumina forms an efficient barrier (aerospace industries, casting moulds, etc.).
— The production of electric barriers and electronic components using alumina, copper, barium titanate, ferrite (semiconductors, submerged motors), ceramics, etc.
— Variable friction coefficient coatings consisting of complex mixtures of different materials.
— Production of parts such as crucibles, nacelles, tubes and furnace components by spraying onto forms.

## 5.4. Production of special steels and refractory materials

Throughout the world, arc plasmas are used to produce special steels and refractory metals. Their general design is similar to that of direct arc or radiating arc furnaces.

Figure 40 shows an industrial melting furnace for the production of special steels. The three torches are located at the apex of an equilateral triangle (in the same manner as the electrodes in an arc furnace). They are superimposed arc generators, with superimposed AC. The plasma jet first bores a well in the charge with the molten metal being used to support the arc. Heat is transferred to the rest of the charge by conduction from the plasma jet impact area, and by radiation from the gas column onto the well walls. The furnace refractory lining is therefore less important than with arc furnaces.

Unlike graphite electrode arc furnaces, current stability is adequate (variation of 2%). There is no short-circuit during melting, and it is possible to apply maximum power to the furnace without risking an instantaneous overload of the power feed circuit.

Therefore, gas-type electrodes have the advantage of not requiring an electrode regulator [21], [52].

Moreover, there is no risk of the bath being contaminated by the electrode as in conventional arc furnaces. The gas current and the magnetic field also provide intense currents in the bath. Therefore, this type of furnace can compete with the vacuum furnace for producing special, very high quality steels.

However, plasma furnace development in the steel-making industry has been very limited and only a few industrial furnaces are in service in Eastern European countries. A typical example is a 5-ton furnace in operation since 1970, is equipped with a DC argon-plasma torch of 3,500 kW producing 4 to 6 t/hr of liquid metal. Higher capacity furnaces were later put into service; for example, a 10-ton capacity furnace equipped with three plasma torches, bias-mounted in the furnace walls. The power demand varies from 7,000 to 8,500 kW, and the voltage between 200 and 600 V. Production ranges from 7 to 10 t/hr and specific consumption is about 650 kWh/t. Furnace capacities continue to increase, and 50-ton, 24 megawatt furnaces will soon enter service [7 *a, b, c, d*], [18], [33 *b, d*], [35], [38], [40], [42].

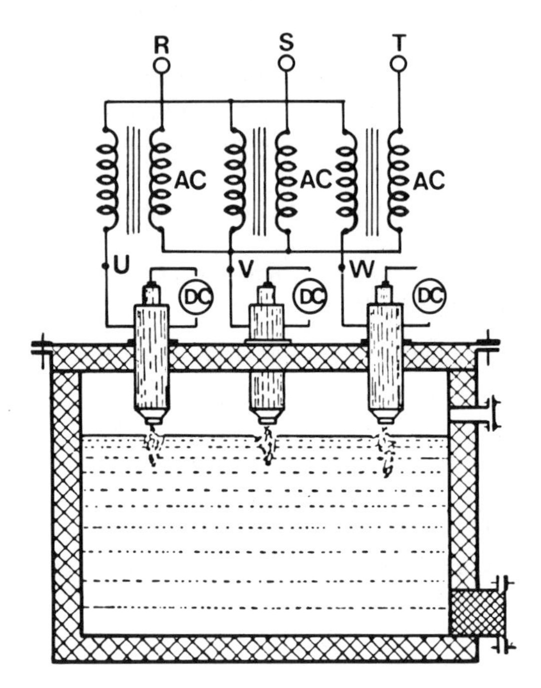


**Fig. 40.** Three-phase arc plasma furnace for making of special steels.

Other types of plasma furnaces can be used for the production of special steels or making refractory metals (tantalum, niobium, tungsten, zirconium, etc.). These furnaces are similar to an arc or radiant resistor furnace; heating is provided by radiation from a plasma column onto the metal to be melted which is in a rotating furnace (Fig. 41). This concept, laboratory tested, has yet to be applied in industry [16 *a*], [22].

It is currently difficult to judge the success of the arc plasma furnace in this field since it competes with furnace types which have worked well and are familiar to users (arc, induction, vacuum arc, electron beam furnaces, etc.).

### 5.5. Melting non-metallic refractory materials

A radiating resistor furnace with a plasma gas column can also be used for melting non-metallic refractory metals, since, unlike superimposed arc furnaces, it is not necessary that the material be a conductor of electricity. An arc plasma generator, superimposed between two torches, is placed at each end of the cylindrical furnace. The furnace, with water-cooled the outside walls, rotates at 50 to 500 rpm around the plasma column. The radiated heat melts the load which is then both mixed and centrifuged. The part of the load in contact with the furnace's cooled wall remains solid and forms a refractory lining which insulates the furnace. Since the material forming the lining is identical to the load, all risks of contamination are avoided and the product is very pure. This furnace is used to produce refractory oxides (zirconium oxide, thorium oxide, etc.) and other refractory materials [15 *a*], [33 *d*], [35].

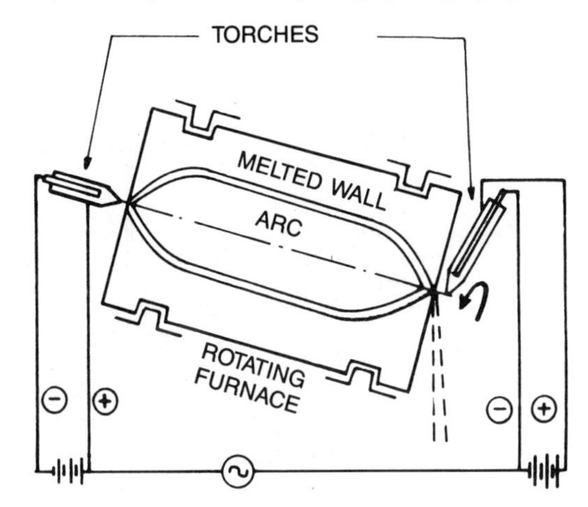


**Fig. 41.** Radiating plasma resistor furnace for melting refractory materials (arc superimposed between two pilot torches) [49].

### 5.6. Plasma production of sponge iron

The diagram in Figure 42 below shows the principle of a sponge iron manufacturing process using arc plasma heating to reform the gas (conversion of $CO_2$ into CO) produced by ore reduction (the reformed gas is a mixture of carbon monoxide, hydrogen and nitrogen). A small part of this gas is used, after $CO_2$ is removed, as a plasma gas. The arc plasma is used to heat the rest of the recycled gas and a fuel (natural gas, liquified petroleum gas, fuel, fluidized coal, etc.) used as a reducing agent.

This reaction takes place in a reduction chamber which is used only to mix the gases and to allow them sufficient time for the reaction to be as complete as possible [50].

The temperature in the plasma reaches 4,000 to 5,000°K. The generator power is 7,500 kW for production of 70,000 t/year of sponge iron.

This plasma furnace has replaced reforming furnaces in which the fuel used as a gas reducing agent obtained from the reduction furnace was coke, heated by direct current flow through its mass. Electricity consumption was reduced by 15 to 30%, depending on the fuel used as a reducing agent, but coke has been replaced by a gas or liquid hydrocarbon. The future of this process is unknown.

Research is being conducted in the use of plasma furnaces enabling direct passage from concentrated ores to liquid metal without agglomeration and without using coke [50], [5].

### 5.7. Particle spheroidization

Particle spheroidization is obtained by partial melting, and formation into spheres on cooling due to the effect of internal material tensions. Existing industrial applications use arc plasma generators, but inductive plasmas can also be used.

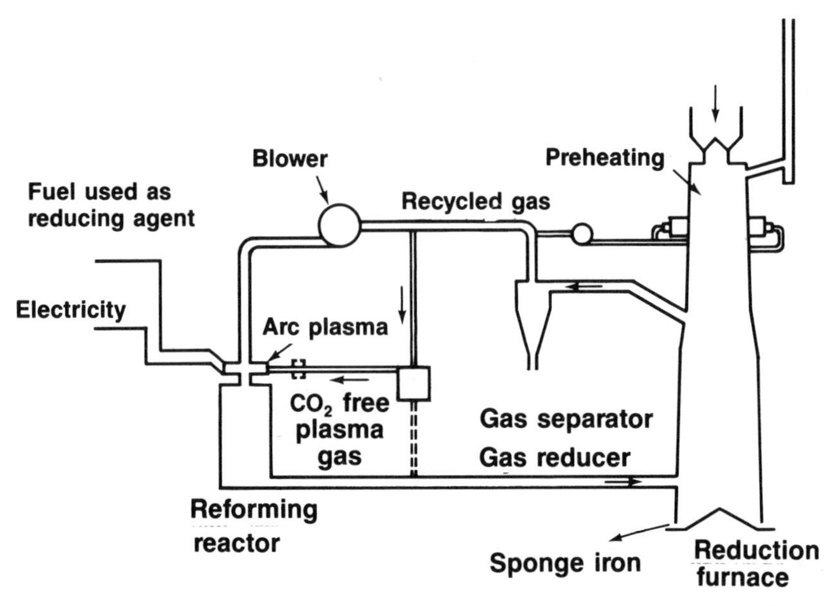


**Fig. 42.** Diagram of a sponge iron production installation using arc plasma to reheat gases prior to reforming [50].

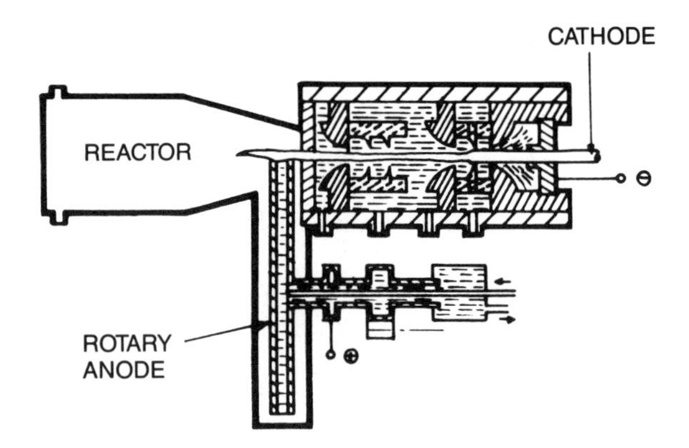


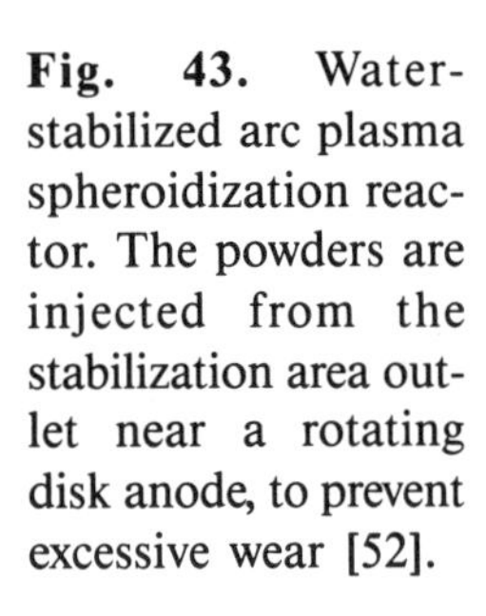

**Fig. 43.** Water-stabilized arc plasma spheroidization reactor. The powders are injected from the stabilization area outlet near a rotating disk anode, to prevent excessive wear [52].

The main applications of spheroidized products are electrostatic copying, abrasives (shot blasting), catalysers, controlled porosity materials and nuclear reactors.

Numerous materials can be spheroidized such as refractory oxides ($Al_2O_3$, $SiO_2$, $ZrO_2$, MgO, etc.), carbides or some metals.

Industrial or semi-industrial installations exist for the production of spheroidized magnetite and alumina. Magnetite and alumina are produced by water-stabilized arc plasma equipment with rates of 100 and 40 kg/h respectively and specific energy consumptions ranging from 2.3 to 3 and 5 to 5.8 kWh/kg [32], [33 *c, d*], [35], [42].

### 5.8. Production of powder refractories

Plasma furnaces can be used to obtain refractories (oxides, carbides, nitrides, etc.) in the form of ultra-fine powders. The powders are used for a wide range of applications — high density ceramic coatings, pigments, catalysers, thickeners, etc. The powder qualities obtained by the plasma method generally differ from those obtained by conventional methods [3], [7 *b*], [15], [32], [33 *c, d*].

By causing oxygen/chloride reactions, it is possible to obtain pigments ($TiO_2$ from $TiCl_4$), ultra-pure silica ($SiO_2$ from $SiCl_4$)or oxide mixtures (chromium-titanium or chromium-aluminum).

For example, it is possible to produce "titanium white" which is used as a pigment, in a three-phase plasma furnace (*Figures* 17 and 45) using the following chemical reaction:

$$TiCl_4 + O_2 \rightarrow TiO_2 + 2Cl_2.$$

This reaction calls for a temperature of more than 1,500°C, easily obtained with an arc-plasma generator. A two kW installation can produce variable quantities of titanium oxide as a function of reaction conditions.

Production of titanium oxide in a 400 kW arc plasma reactor (G).

| | Production conditions | |
|---|---|---|
| | 1 | 2 |
| Plasma gas | Argon<br>30 l/min | Nitrogen 67% — Argon 33%<br>18 l/min |
| Reactive (reagent) gases<br>oxygen<br>titanium chloride | <br>65 m³/h<br>350 kg/h | <br>100 m³/h<br>550 kg/h |
| Axis temperature<br>at 30 mm from axis<br>at 60 mm from axis<br>on walls (80 mm from axis) | 1,720°C<br>1,750°C<br>1,680°C<br>900°C | 2,000°C<br>1,900°C<br>1,650°C<br>1,100°C |
| Titanium oxide production rate | 164 kg/h | 260 kg/h |

**Fig. 44.**

A 1,000 kW industrial installation in Great Britain seams to be the only one in existence.

The three torches are located along the edges of a triangular-based pyramid. The three gas jets converge at the apex of the pyramid which is the neutral point of the three-phase power supply. The AC current superimposed in the ionized turbulent gas jet varies as a function of the distance between the electrodes, environmental gases and the degree of ionization of the plasma, etc.

### 5.9 Chemical reactions

The production of titanium oxide is in fact a chemical reaction. The plasma reactor described here is fully suited for other chemical applications (*see* Figure 17). Due to its high enthalpy, the plasma enables very high reaction temperatures. Therefore, plasma furnaces are most suitable for endothermic reactions whose direction is enhanced at high temperatures [7 *b, e*], [15], [32], [33 *c, d*], [35], [41].

**Fig. 45.** Diagram of a reactor for the production of powder refractories [G].

The preparation of reformed gas, previously described for the production of sponge iron, is also an example of plasma used for chemical reactions.

A lot of research has been devoted to this field, but for the moment and with the exception of the applications mentioned above, it has not led to many industrial installations. However, this situation may well change over the years to come.

## 6. BIBLIOGRAPHY

[1] H. S. INGHAM et A. P. SHEPARD, *Flame spray handbook*, vol. 3, Metco, 1965.
[2] P. DEMARS, Le soudage Plasma, *Communication à la société des ingénieurs soudeurs*, octobre 1967.
[3] J. SUNNEN *et al.*, Das Plasma Verfahren und seine Anwendung, *ARCOS Sonderbroschure*, Bruxelles 1968.
[4] D. WUNDERLICH, Advanced welding techniques and the high nickel alloys, *Revue de soudure*, n° 3, 1970.
[5] K. E. DORSCHU, Weldability of a new ferritic stainless steel, *Welding Journal*, n° 9, septembre 1971.
[6] P. T. HOULDCROFT, *Les procédés de soudage*, Dunod, 1971.
[7] CONGRÈS INTERNATIONAL D'ÉLECTROTHERMIE, Varsovie, 1972 :
*a*) C. ASADA *et al.*, *The industrial plasma-induction furnace;*
*b*) H. R. P. SCHOUMAKER, *Fours à plasma; données d'exploitation;*
*c*) N. J. BORTNITCHUK et E. N. EKZHANOWA, *Installation de fusion à plasma sous vide;*
*d*) V. DEMBOVSKY, *Plasmaöfen für die Herstellung hochschmelzender Metalle und Legierungen;*
*e*) J. GALLIKER, *Plasma-Erzeugung mit Hochfrequenz.*

[8] M. LE GUELLEC, Le rechargement dur par soudage et particulièrement avec alliages spéciaux dans les industries mécaniques, *Soudage et techniques connexes*, octobre 1972.
[9] G. AUBIN *et al.*, Les revêtements de surface par projection au plasma, *Traitements de surface*, n° 116, octobre 1977.
[10] M. VAGNARD, *Les chalumeaux à plasma d'arc pour le coupage des métaux*, Colloque du Comité français d'électrothermie, 5-6 avril 1973.
[11] P. DEMARS *et al.*, Les nouvelles possibilités du procédé plasma en soudage, *Soudage et techniques connexes*, vol. 27, n° 5-6, 1973.
[12] M. SCHULTZ, Utilisation des plasmas d'arc en soudage, *Entropie* n° 53, octobre 1973.
[13] S. BELAKHOWSKY, *Théorie et pratique du soudage*, Pyc Édition, 1973.
[14] B. BLANCHET, Le soudage du titane, *Revue de métallurgie*, n° 1, janvier 1974.
[15] A. BELL, *Techniques and applications of plasma chemistry*, John Wiley and Sons, New York, 1974.
[16] COLLOQUE DU COMITÉ FRANÇAIS D'ÉLECTROTHERMIE, 2 mai 1974 :
*a*) J. AUBRETON *et al.*, *Le four à plasma d'arc adapté à la fusion des métaux;*
*b*) P. BOUVARD, *Soudage à arc plasma des fontes à graphite sphéroïdal;*
*c*) M. BILLARD *et al.*, *Nouveau procédé de recuit destiné aux tréfilés minces.*
[17] J. LAWTON et P. J. MAYO, Fluid flow aspects in the formation and removal of dross during plasma-torch cutting, *Journal of Engineering, Materials and Technology*, vol. 96, n° 3, juillet 1974.
[18] U. CHANDRA, Entwicklung eines leistungsstarken Wechselstrom-Plasmabrenners, *Elektrowärme International*, vol. 32, n° 4, août 1974.
[19] M. VILLAT, Application de couches de protection contre l'usure sur les composants de réacteurs, *Revue technique Sulzer*, 1974.
[20] H. HUGEL, *Arc heater for high enthalpy plasma flows*, IEE, Londres, 9-12 septembre 1974.
[21] R. KEGEL *et al.*, *Elektrowärme, Theorie und Praxis*, Verlag Girardet, 1974.
[22] J. MONTUELLE, Application du four à plasma à la purification des métaux et alliages, *Journal du four électrique*, n° 1, janvier 1975.
[23] H. S. INGHAM et A. J. FABEL, Comparison of plasma flame spray gases, *Welding Journal*, vol. 54, n° 2, février 1975.
[24] J. C. METCALFE et B. C. QUIGLEY, Heat transfer in plasma-arc welding, *Welding Journal*, vol. 54, n° 3, mars 1975; et Keyhole stability in plasma-arc welding, *Welding Journal*, vol. 54, n° 11, novembre 1975.
[25] D. R. WOODFORD et J. NORRISH, Development of plasma welding, *Metal Construction*, vol. 7, n° 6, juin 1975.
[26] G. JELMONIRI *et al.*, Welding characteristics of the plasma, MIG process, *Metal Construction and British Welding Journal*, vol. 7, n° 11, novembre 1975.
[27] D. J. BEAUCHAMP, Plasma welding of aero-engine materials, IEE, *Conference Publication* n° 133, Londres, novembre 1975.
[28] H. DE CATHEU, Nitruration ionique, *Comité français d'électrothermie*, avril 1976.
[29] R. VENNEKENS, Plasma, MIG surfacing, *Revue de soudure*, n° 5, 1976.
[30] C. WEISMANN, *Welding Handbook*, U.S.A., 1976 (American Welding Society).
[31] L. MENDEL, *Manuel pratique de soudage à l'arc*, Bordas-Dunod, 1976.
[32] M. G. FEY, Electric arc (plasma) heaters for the process industries, *Industrial Heating*, juin 1976.
[33] CONGRÈS DE L'UNION INTERNATIONALE D'ÉLECTROTHERMIE, Liège, octobre 1976 :
*a*) I. DE VYNCK, *Four à plasma-arc utilisant la superposition de courants intenses sur un jet de plasma soumis à l'action d'un champ magnétique axial;*
*b*) C. I. MEYERSON, *Plasma steelmaking in ceramic crucible furnaces of up to* 30 *t capacity;*
*c*) P. D. JOHNSTON, *Plasma reactors for material processing;*
*d*) H. R. P. SCHOUMAKER, *Plasma heating ovens;*

*e)* F. PIETERMAAT, *Étude des paramètres influençant la vitesse de rotation de l'arc à plasma dans un champ magnétique axial;*
*f)* G. HETHERINGTON et J. A. WINTERBURN, *Electrical methods for production of fused quartz.*

[34] T. J. FOX et J. E. HARRY, *Surface heat treatment using a plasma torch with a rectangular jet*, Congrès IEE, Londres, 8 mars 1977.

[35] H. R. P. SCHOUMAKER, Fours à chauffage par plasma, *Journal du four électrique*, vol. 82, n° 4, avril 1977.

[36] P. FAUCHAIS, Développement des fours à plasma en vue d'applications industrielles, Colloque du Comité français d'électrothermie, octobre 1977.

[37] A. BORIE et P. FAUCHAIS, *Utilisation d'un générateur à plasma d'arc pour la protection de revêtements de surface*, Colloque du Comité français d'électrothermie, avril 1978.

[38] J. AUBRETON *et al.*, *Les fours à plasma en métallurgie extractive*, Colloque du Comité français d'électrothermie, Versailles, avril 1978.

[39] C. BONET, *Le four à plasma en minéralurgie et métallurgie extractive*, Congrès SEE, Issy-les-Moulineaux, avril 1978.

[40] J. AMOUROUX *et al.*, Étude technico-économique de la production d'acier au four à plasma, *Revue générale d'électricité*, vol. 87, n° 6, juin 1978.

[41] P. FAUCHAIS et *al.*, *La chimie des plasmas et ses applications à la synthèse de l'acétylène à partir des hydrocarbures et du charbon*, Congrès SEE, Gif-sur-Yvette, juin 1978.

[42] J. AUBRETON et P. FAUCHAIS, Les fours à plasma, *Revue générale de thermique*, vol. 17, n° 200, août 1978.

[43] J. AUBRETON *et al.*, *Les fours à plasma*, Laboratoire de Thermodynamique, U.E.R. des Sciences, Université de Limoges, 1978.

[44] P. FAUCHAIS *et al.*, *Rappel général sur les plasmas*, Laboratoire de Thermodynamique, U.E.R. des Sciences, Université de Limoges, 1978.

[45] V. VESELY, La technique et les procédés technologiques de projection à chaud en Tchécoslovaquie, *Matériaux et technique*, novembre-décembre 1978.

[46] R. BENSIMON et E. HEDDE, Techniques récentes de projection plasma en anti-corrosion, *Matériaux et technique*, novembre-décembre 1978.

[47] P. R. SAVAGE, La technique du plasma est apte à la récupération des métaux, *Chemical Engineering*, n° 5, 1979.

[48] A. I. W. MOORE, La Pera utilise le réchauffage au plasma pour l'usinage ultra-rapide, *Metalworking Production*, n° 1, 1979.

[49] D. YEROUCHALMI, Four rotatif à haute température chauffé axialement par un plasma d'arc pour la fusion et l'élaboration de produits ultra-réfractaires et de métaux, *Journal du four électrique*, n° 3 et 4, mars-avril 1979.

[50] S. SANTEN, Plasmared, a new technology for the manufacture of sponge iron, *Metallurgia and Metal Forming*, vol. 46, n° 12, décembre 1979.

[51] COMITÉ FRANÇAIS D'ÉLECTROTHERMIE, Versailles, 6-7 mars 1980 :
*a)* P. FAUCHAIS *et al.*, *Les projections de revêtements protecteurs par plasma et les dépôts par ion plating pour la maîtrise des états de surface;*
*b)* P. COLLIGNON, *Perspectives et développement de la nitruration ionique;*
*c)* M. LABROT, *Exploitation et perspectives d'utilisation des générateurs de plasma de grande puissance de l'Aérospatiale.*

[52] Documentation EDF, 92080 Paris-La Défense.

[53] DOCUMENTATION SOCIÉTÉ NOUVELLE DE MÉTALLISATION, 75013 Paris.

[54] DOCUMENTATION ENODIM, 63910 Chignat-Vertaizon.

[55] INDUSTRIES ET TECHNIQUES, Acier suédois, la plasma dès 1981, n° 43, septembre 1980.

**List of equipment manufacturers and suppliers mentioned in this chapter:**

[**A**] SAF, 95310 Saint-Ouen-l'Aumône.
[**B**] CEA, 38000 Grenoble.
[**C**] CFI, 93100 Montreuil.
[**D**] SARRAZIN, 78800 Houilles.
[**E**] KLÖCKNER, 75008 Paris et Cologne (R.F.A.).
[**F**] SKODA-EXPORT, 75000 Paris et Prague (Tchécoslovaquie).
[**G**] ARCOS, Bruxelles, Belgique.

# Chapter 10

# Electron Beam Heating

## 1. PRINCIPLE OF ELECTRON BEAM HEATING

The idea of using electron beams for heating dates from the 1900's and the work of von Pirani. However, some forty years elapsed before progress in vaccum techniques and optoelectronics could provide the high vacuums (about $10^{-4}$ Torr) required for the propagation of the electron beam and control of its path and characteristics, and thereby make it practical for industrial applications.

In principle, electron beam heating is very simple. The kinetic energy of the electron beam is converted into heat when the body to be heated, acting as target, is struck.

In industry, electron beam heating is used for welding, melting, special metal alloys, vaporizing metals and surface deposit coatings, heat treating of metals and micromachining. Another electron beam application is polymerization, crosslinking and grafting organic materials, in particular coatings onto all substrates (wood, metal, etc.); in this case, it is the radiochemical effect of electron irradiation and not the thermal effect of the electron beams which is used.

## 2. CHARACTERISTICS OF ELECTRON BEAM HEATING

The physics of electron beams in a vacuum were thoroughly studied during the development of electron microscopes and electron tubes. These studies were completed during the development of industrial applications involving the use of beams at much higher powers.

Complete trajectory calculations, which are analytically intractable, require highly sophisticated digital techniques [1], [2], [4], [8], [14].

### 2.1. Electron beam energy

In electron beam heating, it is necessary to "extract" electrons from an appropriate surface, the cathode, provide them with kinetic energy by accelerating them in an electric field (speed is about 85,000 km/sec for a voltage of 20 kV) and focus them on a target.

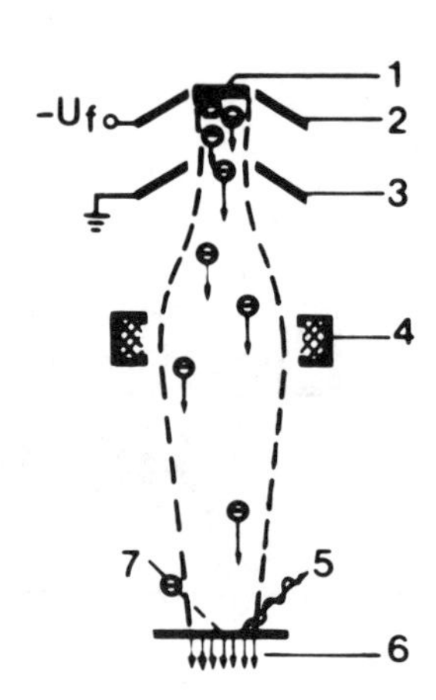


**Fig. 1.** Electron beam heating equipment [H]:
1, cathode; 2, Wehnelt electrode (control electrode);
3, anode; 4, electromagnetic lens;
5, secondary electromagnetic radiation, X rays;
6, electron beam penetration;
7, reflected and secondary electrons.

When the electrons hit the target, their kinetic energy is converted into usable heat. A small part of the energy is, however, lost in the form of secondary electron emission and X-rays from which protection must be provided by a well-designed working chamber (emission of X-rays is very low when the voltage is limited, approximately 0.1 percent of the energy being re-emitted for an acceleration voltage of 20 kV; in such cases, the walls provide sufficient protection).

The kinetic energy of the electron beam is approximately equal to:

$$\boxed{W = 0.5\, nm\, V^2 = ne\, U,}$$

$n$, number of electrons emitted;
$m$, electron mass;
$e$, electron electric charge;
$V$, speed of the electrons arriving at the positive electrode;
$U$, acceleration voltage.

If $n_s$ is the number of electrons emitted per unit of time, the maximum thermal power which can be collected is expressed as follows:

$$\boxed{P = n_s e\, U = UI.}$$

The electron mass is very low (approximately 0.9 X $10^{-30}$ kg). Its electrical charge is 1.6 X $10^{-19}$ coulombs, and, in an electron beam, energy is often measured in electron-volts; this quantity represents the energy acquired by an electron accelerated by 1 V (1 eV = 1.6 X $10^{-19}$ J).

### 2.2. Power density

Generally, the electron beam is electrostatically and electromagnetically focused. The energy may be highly localized as in welding or machining, distributed uniformly as in melting or diffused as in surface hardening. The impact surface is therefore very small, and it is possible to obtain very high target power densities with an electron beam, generally of the order of $10^2$ to $10^9$ W/cm$^2$, depending on the application. The following power densities are currently used in industrial applications:
— melting: $10^3$-$10^4$ W/cm$^2$;
— welding: $10^6$-$10^8$ W/cm$^2$;
— evaporation: $10^4$-$10^5$ W/cm$^2$;
— surface hardening: $10^6$-$10^8$ W/cm$^2$;
— hardening/tempering: $10^2$-$10^3$ W/cm$^2$;
— machining: $10^7$-$10^9$ W/cm$^2$.

### 2.3. Effect of electron beam on the target-penetration depth

Accelerated electrons acquire high kinetic energy. Electrons of 100 keV reach speeds of 160 000 km/sec. When the target is hit, energy is transferred to the material to be heated at the impact point; its temperature increases very rapidly. The penetration depth in the material depends on the energy of the electron and the characteristics of the material to be heated. The energy lost by an electron beam passing through a metal is given by the Thomson-Whiddington law:

$$W_0^2 - W_x^2 = k\rho \frac{N}{A} x,$$

$x$, distance covered in centimeters;
$W_x$, electron energy in electron-volts after a path of length $x$;
$\rho$, metal density;
$N$, metal atomic number;
$A$, metal atomic weight;
$k$, constant equal to 7.75 X $10^{11}$ for all metals.

Electron beam penetration is therefore proportional to the increase in its original energy [43].

The energy applied by the electron beam has several consequences:
— heating and sometimes melting the material;
— heat transfers through conduction, radiation and sometimes convection, as a function of the residual atmosphere; these heat transfers have practically no effect during temperature rise due to the very high heating rate;
— possibility of ejection of droplets and local vaporization of the metal;
— thermo-electronic emission from the hot zone;
— collision between the primary electrons and the target atoms, with emission of X and light photons;
— sometimes deep penetration of the beam and the thermal effect: a beam of specific power 3 X $10^7$ W/cm$^2$ can pass through steel plate 10 cm thick.

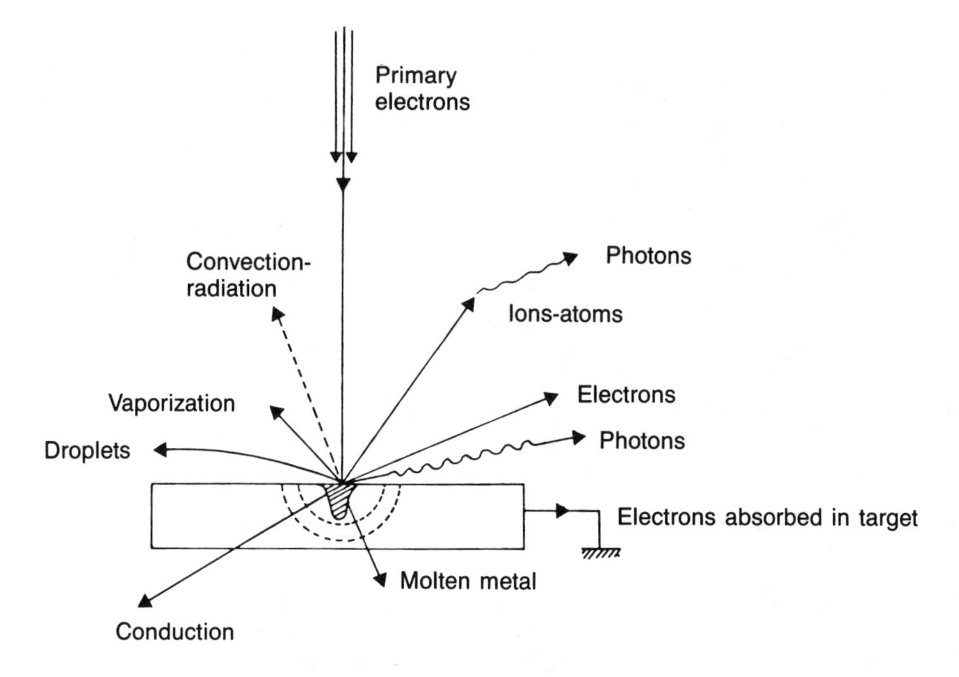


**Fig. 2.** Effect of an electron beam on the target [43].

The thermal effect of electron beam is used in various ways to perform numerous industrial operations [1], [2], [3], [43].

## 3. COMPOSITION OF AN ELECTRON BEAM INSTALLATION

Basically, an electron beam installation consists of:

— an electron gun and its various peripherals (high voltage supply, heater, etc.) consisting of:

- an electron emitting cathode,
- an acceleration electrode (anode),
- a beam-forming electrode (Wehnelt), which can be biased to modulate and adjust the electron flow,
- an electromagnetic focusing coil, used to concentrate the beam on different points of the surface to be heated,
- an electromagnetic deflection coil, capable of making the electron beam follow any path;

— a control system, generally computerized and capable of controlling and monitoring the various operating parameters of the equipment;

— a work chamber in which the products to be treated are located;

— the gun and work chamber vacuum equipment.

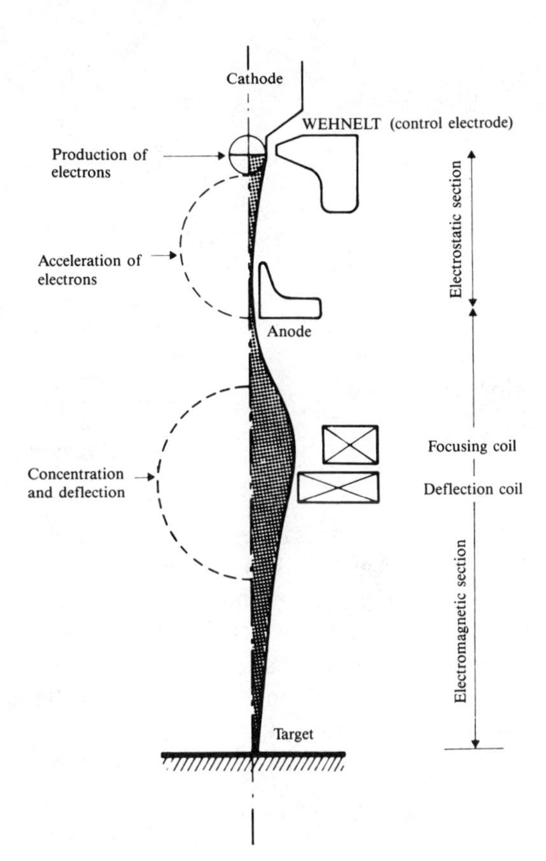


**Fig. 3.** Composition of an electron beam installation [A].

## 3.1. Cathode

### *3.1.1. Electron emission*

The cathode maintains electron emission. In the balanced state the electrons in the cathode are near the reference energy or Fermi level, which, in electrically conducting materials, is located in the center of the conduction band.

For the cathode to emit electrons, an energy level higher than the potential barrier must be applied to them because of the crystalline network attraction of the electrons; the released electrons in fact form a space charge which remains around the cathode and is maintained by the cathode's positive charge.

The energy required to extract electrons from the metal or the output work is equal to the difference between the vacuum level $W_0$ and the Fermi energy $W_F$; this is a characteristic of the cathode material. Materials with low work functions are used for cathodes.

The energy required to extract electrons can be provided by various methods. With electron beam heating, thermionic emission is used.

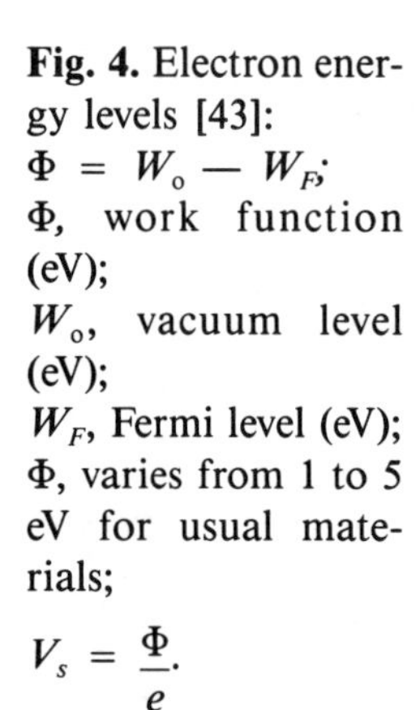

**Fig. 4.** Electron energy levels [43]:
$\Phi = W_o - W_F$;
$\Phi$, work function (eV);
$W_o$, vacuum level (eV);
$W_F$, Fermi level (eV);
$\Phi$, varies from 1 to 5 eV for usual materials;

$V_s = \frac{\Phi}{e}$.

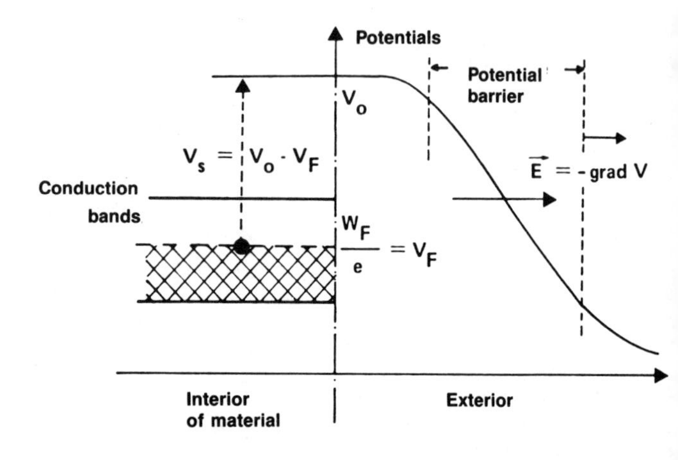


When cathode temperature increases, the conducting electron energy increases; when it becomes greater than the potential barrier, the electrons escape and form a negative space charge around the cathode. If a positive electric field is applied, the electrons are then extracted from the space charge.

There are three types of electron emission: emission limited by the space charge, emission limited by thermal saturation and Schottky-effect emission; the last is not used in electron guns.

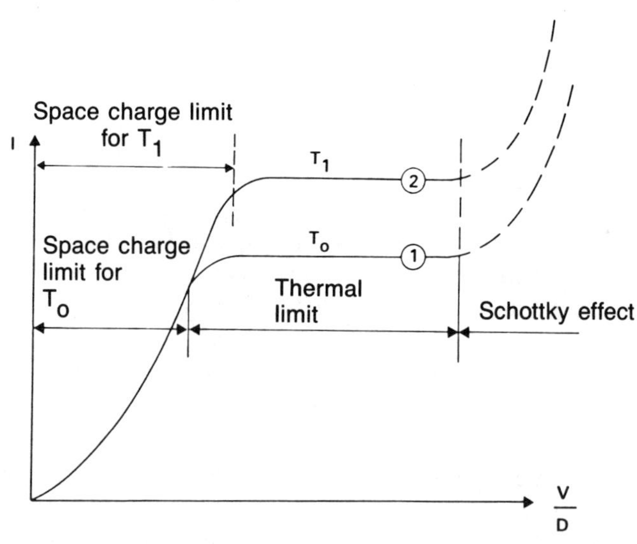


**Fig. 5.** Electron emission from cathode: emitted current $I$ as a function of electric field $V/D$, $V$ being the applied voltage and $D$ the cathode-anode distance [43].

When the electric field is relatively low, the current density essentially depends on the voltage applied and the anode-to-cathode distance and little on temperature. When the electric field increases, all electrons in the cloud created at temperature T are extracted; the current density depends on cathode temperature and type but not electric field. Emission due to thermal saturation provides a uniform cathode current density.

### *3.1.2. Composition of the cathode*

The cathode is made from refractory material (tantalum, tungsten), refractory oxide (thoriated tungsten) or rare earths (lanthane hexaboride); it is directly heated by the joule effect, or indirectly by auxiliary electron beam for high powers. In the first case, the cathode takes the form of a filament, and that of a disk in the second. If the emission current is more than 500 mA, a disk cathode heated by electron beam is used systematically; service life is between 50 and 300 hours but can be much shorter; for example, in welding, it is only a few hours. Thermionic emission becomes significant at about 1,600°C. The pressure at the cathode is about $10^{-5}$ torr [(1)].

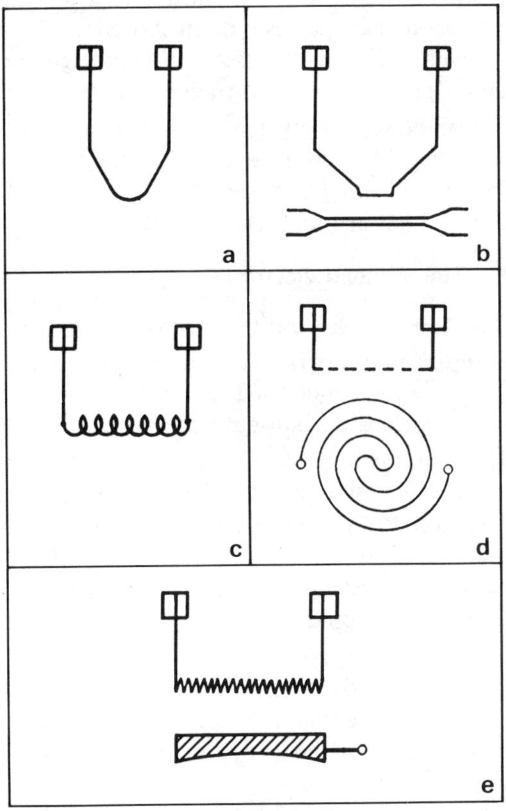

**Fig. 6.** Various types of cathode [59]:
a) hairpin;
b) ribbon;
c) coil;
d) filament;
e) electron beam indirectly heated disk.

[1]) Note: The standard unit of pressure is the pascal. However, in publications devoted to electron beam heating, the unit used is the torr.
These units are related by the following equations:
1 torr = 1 mmHg = $1.33 \times 10^{-3}$ bar = $1.33 \times 10^{-2}$ Pa.

In electron guns, thermionic emission is limited to two modes: space charge limited and thermal saturation limited. In the first mode, the current is independent of temperature of the cathode and depends only on the acceleration voltage (proportional to $V^{3/2}$), but electron emission is less uniform at the cathode surface than in the second mode.

### 3.2. The anode

The anode is taken to a high voltage, 20 to 200 kV depending on applications and gun design. The electric field created around the cathode enables extraction of a certain number of electrons from the surrounding electron cloud. These electrons form a beam, and a flow is built up between the cathode and anode which contains a central bore allowing the electron beam to pass. After acceleration due to the effect of the electric field created by the potential difference between the cathode and anode, the electrons are directed towards the target at constant speed. The diameter of the anode diaphragm through which the electron beam passes is often chosen to offer very low conductance for gases, and to allow different pressures inside the gun and work chamber, thus reducing vacuum pump power and therefore cost.

A high power density demands a high acceleration voltage. However, the higher the voltage, the more penetrating the X-rays generated at the load surface. Adequate shielding is then required to stop the X-rays. This secondary emission becomes high at an acceleration voltage of 50 to 60 kV.

### 3.3. The Wehnelt electrode

The Wehnelt electrode has a double role. Its configuration, combined with that of the anode, defines the shape of the beam. Moreover, its bias with respect to the cathode can be used to modify the electric field in front of the cathode, and therefore the beam current and power.

If the Wehnelt is taken to the same potential as the cathode (diode electron gun) and emission current $I$ is limited by the space charge, the current is proportional to the 3/2 power of the acceleration voltage. The electron beam power, equal to $UI$, is proportional to the 5/2 power of the voltage, and the variation in acceleration voltage is then used to adjust the thermal power. This arrangement is used for welding and melting applications.

The Wehnelt can also be taken to a lower potential, which is adjustable with respect to the cathode (triode operation). The power is then adjusted by varying the Wehnelt potential; the electron and current energy parameters of the beam then become independent. This design is used primarily for welding and heat treatment applications.

The anode and Wehnelt electrode form the electrostatic part of the geometry of the Wehnelt electrode, the anode and the electric field determine the electron paths and speed. In the electrostatic part of the gun, the electric field and electron beam assembly provide axial symmetry.

### 3.4 Beam focusing and deflection

In the electrostatic part, the electrons are focused at the "cross-over point"; however, the beam has a tendency to split due to the mutual repulsion of the electrons. One or more electromagnetic lenses are then required to focus the beam, using electromagnetic forces, to obtain the desired specific power at the load surface. The induction created provides an axis of symmetry; the electromagnetic lenses consist of coils surrounding the beam. The focal length of the lens is set by the coil current and, to prevent alterations to the electromagnetic image on the load, must not fluctuate.

The electron gun can also be fitted with deflection coils creating induction perpendicular to the axis of symmetry. In practice, two magnetic fields perpendicular to each other are often created. Due to the action of the two frequency- and amplitude-variable fields, it is possible to make the beam follow any curve and send it in the desired direction.

### 3.5. Various types of electron guns

Various types of electron guns exist and are characterized by the shape of their electrodes, the acceleration voltage, the type of deflection, and the vacuum maintained at the cathode and throughout the rest of the installation [4], [8], [58], [59], [A], [G], [H].

— *Axial symmetry guns*

The axial (or Pierce) gun is derived directly from the electron microscope, but has had many modifications to become an industrial tool; different variants of this gun exist:

— Pierce diode gun; the acceleration voltage which may reach 150 kV is used to adjust gun power. Gun design is therefore relatively simple; it is used mostly in welding and machining;

— Pierce triode gun; the power is adjusted by biasing the Wehnelt electrode, which varies the number of electrons allowed to escape from the cathode. For welding the beam is finely focused, and mediumly focused for melting;

— multiple chamber gun; this gun is similar to the above but has several chambers which isolate the gun from the work chamber. These chambers are separated by other chambers offering high impedance to gas flow allowing the beam to pass. Each chamber is separately vacuum pressurized, and the pumping power and equipment costs are lower. The pressure can reach $10^{-2}$ torr in the work chamber without the pressure in the cathode chamber reaching a critical value (the pressure at the cathode is between $10^{-5}$ and $10^{-4}$ torr).

— *Transverse beam guns*

This type of gun is used for the high powers required in melting. The beam is deflected by an angle greater than 90°, and the cathode is fully protected against metal vapor and spray. The cathode is located in the work chamber; therefore, powerful vacuum pumps are required.

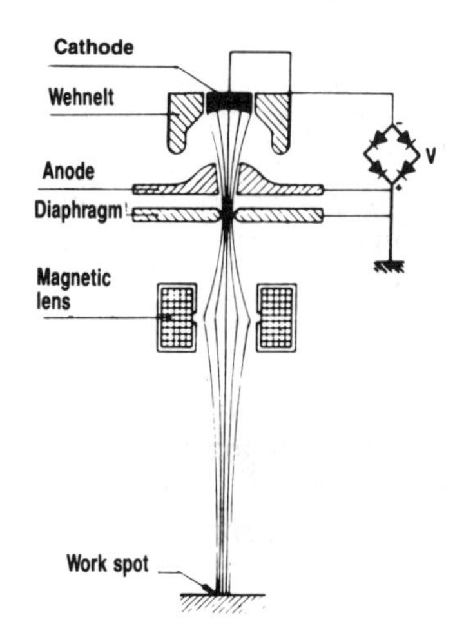


Pierce diode gun [G]

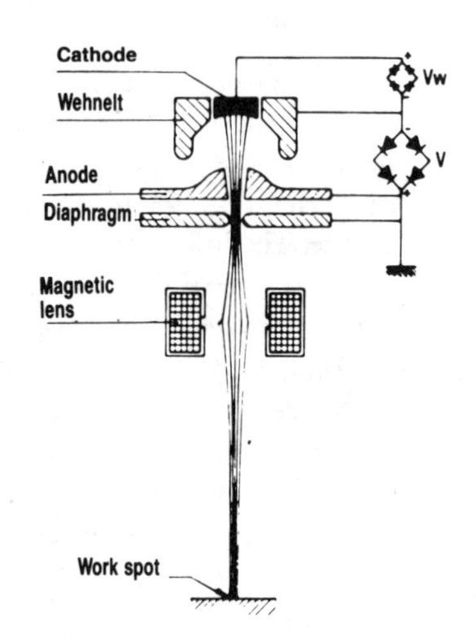


Pierce triode gun [G]

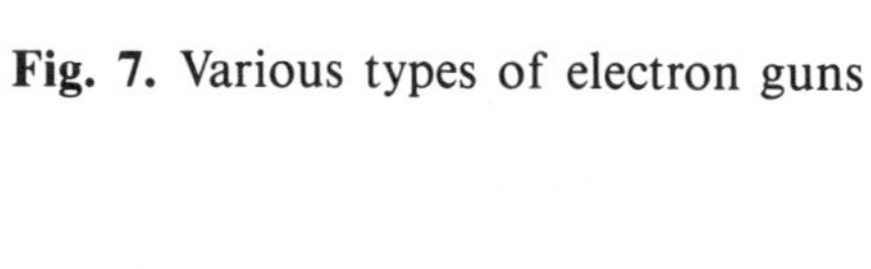

**Fig. 7.** Various types of electron guns

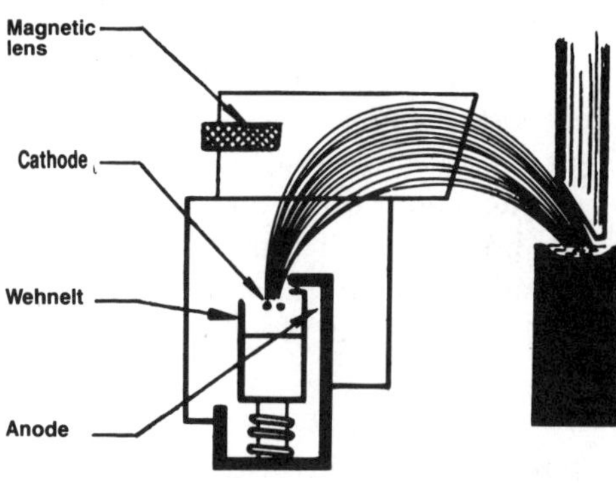


Transverse beam gun [H]

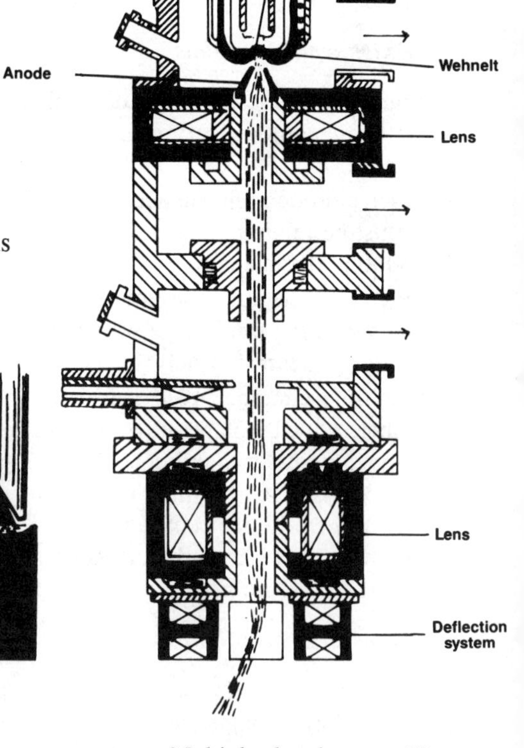


Multiple chamber gun [4].

## 4. ADVANTAGES AND DISADVANTAGES OF ELECTRON BEAM HEATING

Electron beam heating offers several advantages justifying its industrial use:
— very high power density, variable according to the application;
— vacuum operation giving very high product quality;
— very high temperatures can be reached;
— very accurate location of the thermal effect;
— treatment speed;
— can be used in radiochemistry;
— sometimes high energy savings;
— pollution-free.

Conversely, investment costs are very high, and maintenance requires a favorable technological environment.

These advantages and disadvantages can be seen in the present and future development of this technique. The various points and their relative importance are analysed in detail in the description of each application.

## 5. ELECTRON BEAM HEATING INDUSTRIAL APPLICATIONS

Electron beam heating requires high investment costs, often 5 to 15 times more per installed kilowatt than with more conventional processes. This roughly shows that electron beam heating applications would not have been developed without specific advantages:
— performance of industrial operations which are impossible or difficult with other techniques;
— very important increase in productivity, compensating the heavy investment costs;
— increase in quality for products in which safety is of prime importance;
— creation of new products with better characteristics than those produced using conventional methods.

The following paragraphs briefly describe the main industrial applications of electron beam heating [4], [9], [16], [31], [43], [45], [49].

### 5.1. Melting metals and semiconductors

Electron beam melting is the oldest application of this technique; its properties are as follows:
— the operation is performed in a high vacuum, between $10^{-4}$ and $10^{-2}$ torr, which offers very high metallurgical quality (outgassing of the metal is excellent) and is used to produce metals which react strongly with oxygen or other atmospheric gases;
— the load need not necessarily consist of bars, but may consist of waste such as scrap, enabling recycling of discarded, high value materials;
— power density is high, providing high melting speeds and enabling high melting point alloys to be produced;
— deflection of the electron beam prevents local concentrations of energy, and therefore local overheating, thus increasing product quality;

— beam stability and adjustment of beam parameters makes accurate process control possible — melting speeds, temperature, overheating, etc. — and automation of the process;

— precise localization of the thermal effect enables water-cooled copper ingot moulds to be used, which eliminates the risks of reaction between the molten metal and the crucible refractories;

— the capability, within certain limits, of keeping metal fluid without melting.

### *5.1.1. Electron beam melting installations*

The guns used in melting are either axial or transverse:

— *Pierce axial gun*

The disk-shaped cathode, generally made from tungsten or tantalum, is heated by secondary electrode as soon as the power exceeds 30 kW. Service life is about 100 hours. In most cases, the gun behaves as a diode in the space charge mode, and the power is adjusted by the electron acceleration voltage. A magnetic lens concentrates the beam on the opening of a diaphragm separating the gun and the melting chamber. In this chamber, the concentration and deflection coils can control the beam, adjustments being manual or automatic. The beam is mediumly focused.

Several separate pumping chambers are located between the cathode and the work chamber. This arrangement protects the gun from metal spray and vapor from the melting chamber, and reduces the overall pumping work. Pressure within the working chamber is about $10^{-2}$ torr which is generally sufficient, but if necessary can be lowered; at the cathode, it is $10^{-4}$ to $10^{-5}$ torr.

— *Transverse gun*

With the transverse gun (Stauffer-Temescal type), the beam is deflected by an angle greater than 90° and sometimes up to 180°C. The cathode consists of two tungsten filaments, the length of which is adapted to furnace power, and located at a very short distance from the anode. The cathode is located in the melting chamber; very powerful pumps must be used to provide sufficient vacuum pressure. Cathode service life is about 50 hours.

Industrial furnaces have one or more guns of the axial or transverse type.

Gun power and furnace power are both highly variable. The maximum power of the guns used for melting is about 400 to 500 kW, for both axial and transverse types. In most cases, guns of lower unit powers are used, and the total furnace power can then reach 1,000 kW, and even several thousand kW [13], [34], [44], [H], [J].

### *5.1.2. Electron beam melting applications*

Electron beam furnaces have been used for numerous years for melting refractory metals such as tantalum and niobium, but also for melting other metals such as tungsten, molybdenum, zirconium and hafnium [44], [H], [J].

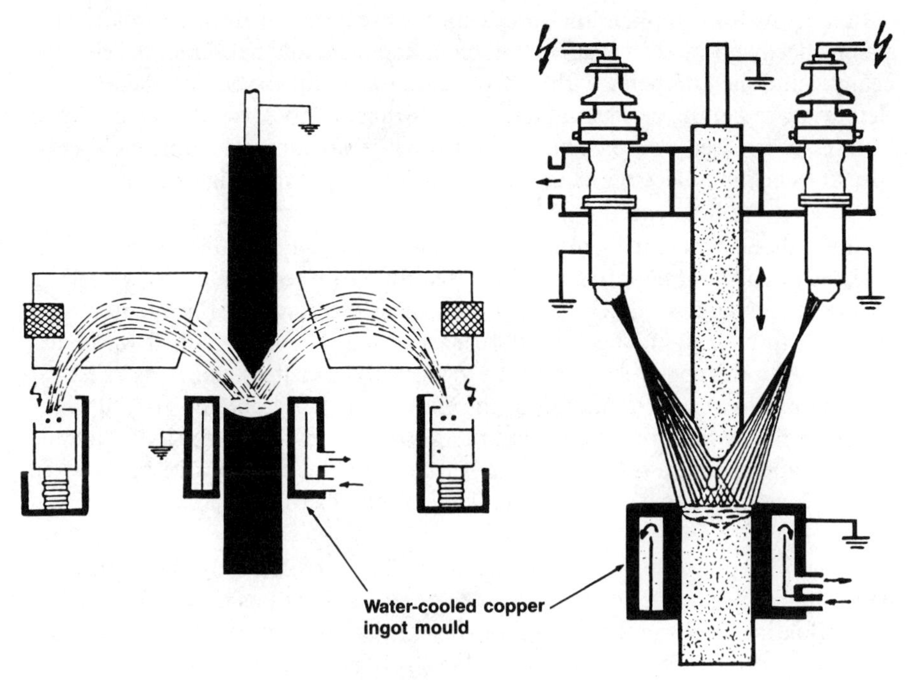


**Fig. 8.** Diagram of electron beam melting furnaces [H].

This type of furnace is also used in the nuclear materials industry for melting plutonium and other metals.

Electron beam also seems to be used for remelting and pouring titanium in the following applications: production of ingots from scrap, granulates and other waste with extremely low metal loss, production of fluid metal for casting, etc.

The special steel industries are using electron beam techniques for the production of very high quality alloys. However, difficulties can occur for alloys containing high partial vapor pressure components such as chromium, since they then have a tendency to evaporate.

In all these applications, electron beam is used either for melting or remelting alloys produced by another technique; in the second case, this technique provides high purity and quality.

Several electron beam melting installations can be described:

— a 10 kW furnace, used for melting and centrifugal pouring of plutonium alloys for the production of cardiac stimulators [44];

— a 1 200 kW furnace, equipped with six guns of 200 kW each, for production of tantalum and niobium ingots, with a melting rate of 500 kg/hr for tantalum and 330 kg/hr for niobium; the four diffusion pumps have a capacity of 300,000 l/sec.;

— a 4,000 kW furnace, equipped with 22 electron guns, capable of producing 120 t/day of high quality stainless steel; doubtlessly, this is the most powerful electron beam melting furnace ever built [12].

In most melting applications, and with the exception of those in which there are practically no valid alternatives (niobium, tantalum and hafnium), the electron beam technique competes with other melting techniques such as vacuum arc, electroslag remelting or vacuum induction furnaces; moreover, these techniques can, in some cases, complement each other like vacuum induction melting of special steels and electron beam remelting, to obtain very high quality ingots for forging applications.

The choice of these techniques depends on numerous variables — compared investment costs, which often preclude electron beam techniques (for remelting titanium, it is approximately 20% more expensive than with the vacuum arc), materials to be melted (vacuum arc furnaces cannot be used to melt waste and require electrodes constructed from the material to be melted), metal losses (either by evaporation or by elimination of the ends of the ingot obtain), partial vapor pressure of alloy components, extent of local overheating, the necessity of holding metal temperature (in the vacuum arc furnace, the metal electrode melts when current flows through it), crucible service life and risk of destruction by the heat source, etc.

In melting applications, electron beam can be used differently to obtain small bars of refractory or special metals. The method consists of using a narrow melted zone (floating zone) in a bar, and moving this slowly at constant speed from one side of the bar to the other. This floating zone purification process is used in the semiconductor industry, but is less widely used than radiofrequency induction heating [H].

Finally, the special advantages of electron beam melting, despite high investment costs, is excellent not only for the production of special metals but also for very high quality steels and semiconductors.

### 5.2. Metallization and vacuum deposition

Metallization is a surface coating operation where vaporized metal is deposited on a substrate. An electron beam is used to evaporate the metal to be deposited in a vacuum and the atoms emitted by the metal source condense on the substrate. In general, there are two variants of this process: evaporation and ionic deposition [6], [8], [28], [43], [51], [H].

With the ionic deposition method, the substrate is negatively charged with respect to the metal to be evaporated (some kilovolts), and the chamber contains an inert gas, generally argon at low pressure (about $10^{-2}$ torr), while with evaporation there is no polarization and the air pressure is very low, $10^{-3}$ to $10^{-4}$ torr. In both cases, the power density is generally higher than in melting, $10^4$ to $10^6$ W/cm$^2$, and in particular for high melting point metals.

In evaporation, the atoms emitted by a metallic source propagate linearly in all directions, and only the surfaces facing the source are coated. A variant in the process consists of using low pressure argon, $10^{-3}$ to $10^{-1}$ torr; collision of the atoms emitted by the source with the argon molecules redistributes the propagation directions and increases the homogeneity and thickness of the deposits, and also enables coating holes and other surfaces which do not face the source directly. Evaporation provides the highest deposit speeds, up to 5 $\mu$m per minute, but is unsuitable for complex shapes.

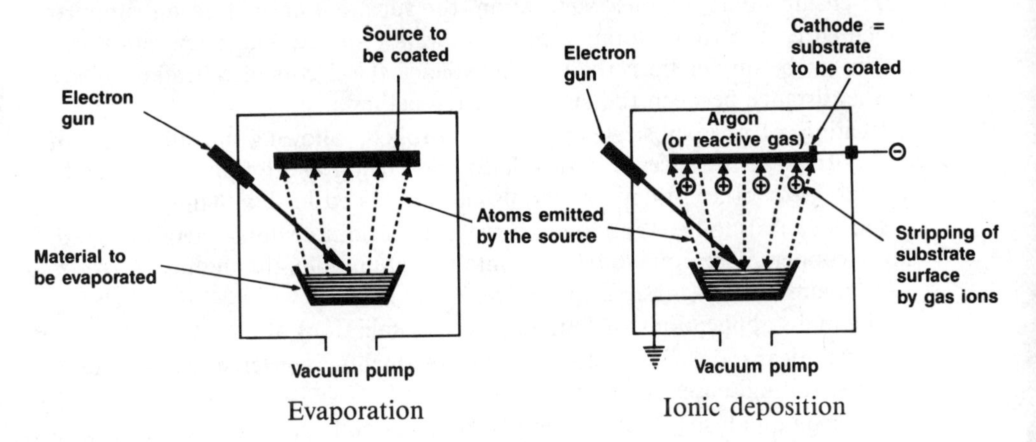


**Fig. 9.** Principle of electron beam metallization [51].

With ionic deposition, a luminescent discharge occurs in the gas, and a small fraction of the vapor atoms from the source and the gas is ionized. Treatment is done in two phases. The gas ions first bombard the substrate and strip it; then, without interrupting ionic beam, the deposit itself is made by evaporating the source. Ion stripping offers excellent deposit adherence. Since only a small number of the atoms emitted are ionized, approximately 2%, it would seem that these propagate as in evaporation, and not due to the effect of electrical forces. The deposit rate can be up to 20 $\mu$m/min. The ionic deposition is used for electrically conductive substrates, but may be extended to dielectric materials by operating at radio frequency.

In the strict sense of the term, metallization designates vapor coating by a metal, but this depositing technique also applies to metal alloys, metal oxides and other compounds (nitrides, carbides) when reactive gases are used.

Generally, there are two types of electron beam metallization applications, depending on the power of the equipment and the industry:

— the low power field, in general less than 20 kW, characteristic of the optical and electronic industries;

— the high power field, sometimes up to several hundred kilowatts, which mainly applies to the metallurgical industries.

This division, however, is gradually disappearing as electron beam metallization develops, for applications such as coating large plates of glass with special products, or mechanical components with compounds intended to improve mechanical or chemical characteristics (increase in resistance to corrosion or wear, reduction of friction coefficients, etc.).

### *5.2.1. Low power domain*

The guns used are either axial or transverse. In some installations, the material to be evaporated forms the anode and is heated by the electrostatically deflected electron beam. The substrata can be secured to a rotating platform, which enables metallization of various parts without disturbing the vacuum.

To ensure homogeneous evaporation, the substrata are placed on supports which either revolve or rotate around the vapor source. The evaporation rate is adjusted by varying the cathode heater voltage, the electron acceleration voltage or the distance between the anode and the cathode.

Heating of the substrate can be controlled, allowing metallization of thermofragile substances. Plastic films, glass, nylon and teflon mouldings or mouldings of other organic materials can be treated in this manner.

The materials to be evaporated are usually precious or noble metals — gold, silver, copper, nickel, molybdenum, tantalum, aluminum, titanium — or oxides — titanium and magnesium oxide, etc.

The main applications of limited power installations are:

— the optical industries: anti-reflection treatments, interference filters, laser mirrors, tinting lenses for glasses;

— electronic industries: production of resistors and capacitors for integrated circuits, interconnections on hybrid circuits and electrical contacts;

— mechanical industries: improvement in hardness by depositing titanium nitride or carbide, self-lubrication (molybdenum sulphide-based compound), aesthetic effects (the color of titanium nitride varies from metallic grey-white to deep golden yellow), increase in corrosion resistance (deposits of cadmium, aluminum, etc.), reduction of the friction coefficient by depositing an iron-molybdenum-cobalt complex coating, multiple use deposits such as coatings on jet engine turbine vanes using a cobalt-chromium-aluminum-yttrium compound.

Surface treating mechanical parts often require powers between those used in optics and electronics, and high power equipment required for continuous coating.

The main recognized advantages of this technique are high evaporation speeds, very high purity in the coatings obtained and the possibility of evaporating most metals.

### *5.2.2. High power domain*

High power industrial installations are used to deposit thin metal films on large surfaces — steel, glass, plastic film belts, etc.

In most cases, transverse beam or multiple chamber axial symmetry guns are used. Transverse beam guns are similar to those used for low powers, but beam deflection is electromagnetic to avoid the risks of discharge onto the metal plates. In multiple chamber guns, the beam is focused first by an electromagnetic lens, then deflected 90° towards the product to be evaporated by a second lens located in a water-cooled copper crucible.

With this type of gun, the cathode cannot be contaminated by metal vapors.

These high power installations are used for coating sheet steel with thin protective metal films, generally aluminum films. Within this field, the process competes with electrolytic tin plating (conventional white iron). A 300 kW installation can evaporate approximately 30 kg of aluminum per hour.

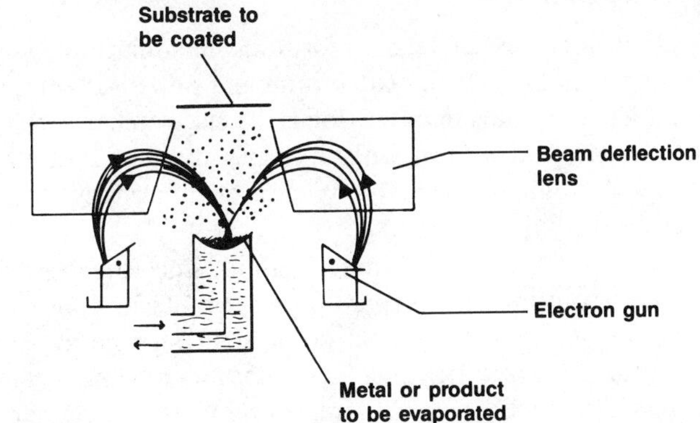


**Fig. 10.** Electron beam metallization [H].

A 1,000 kW installation in the United States treats sheets up to 1 150 mm wide (thickness limited to 0.90 mm) at a rate of 5 m/sec. Due to a vacuum hatch at the entry and exit from the spraying equipment, these installations can operate continuously. Their advantage is that an expensive metal such as tin is replaced by a cheaper metal such as aluminum. Other processes intended to replace tin, however, compete with the electron beam aluminum coating process, and it is difficult to forecast how well this technique will do.

Depositing metals other than aluminum has also been studied. In Japan, a coating unit for ionic depositing of zinc onto steel belts with a capacity of 450,000 t/year is being studied.

With respect to electrolytic coatings, the energy saving provided by vapor phase deposits may be high. Thus, for chrome deposits, electrolysis requires 24 kWh/kg, while 5 kWh/kg is sufficient for a vapor phase deposit using electron beam heating.

Other industrial installations enable continuous deposits of metal films or dielectrics on large surfaces such as glazing, facings, plastic film or paper, etc. The powers involved can exceed 100 kW, with a product feed rate of 5 m/sec. The interest in vapor phase coating techniques using electron beam heating is due basically to specific characteristics of this process:

— ability to evaporate all metals and substitute relatively cheap for costlier metals;

— ability to deposite various metals in layers of different thicknesses on two surfaces of the same substrate;

— ability to deposite several layers of different materials in a single operation and create complex structure coatings;

— high evaporation rate;

— high purity.

Despite of high investment costs and competition from other processes (cathode spraying, electrolytic coatings, etc.) vapor phase deposits using electron beam heating as an energy source should continue to develop.

### 5.3. Machining

Electron beams can be used for some micromachining operations; for example, drilling, milling, cutting, etc. Equipment powers are low, generally between 0.01 and 1 kW. The guns intensely focus the electron beam which can have a diameter between 10 and 30 um with a clearly defined contour. The power densities obtained are high, $10^5$ to $10^9$ W/cm$^2$, enabling very fine machining operations such as drilling and cutting.

This is a thermal machining method, since the desired effect is obtained by local evaporation at the beam impact point. To limit the thermal effect in the zone surrounding the impact point, it is possible to use pulse heating ($10^{-2}$ to $10^{-5}$ s) for localized treatment, or a fast scan of the beam at a speed of several meters per second over the area to be treated, if larger.

Technical accuracy is such that very fine holes can be drilled. The depth to diameter ratio is 10 in most cases but can be up to 20.

Some guns operate according to a principle different from that described above. The pressure around the part to be machined is approximately $10^{-1}$ Torr, much higher than in normal installations. The residual atmosphere is ionized due to incident electrons or reflected secondary electrons. Therefore, a luminescent discharge mode is established. These guns are used for machining non-metallic materials.

Electron beam machining is used mostly in the electronic industry. High precision machining operations are sometimes required to produce different circuits and components and especially resistors and capacitors. Electron beam has replaced mechanical machining, which is much less accurate. There are industrial production lines for the manufacture of high precision resistors (thin film resistors).

Very fine holes can also be drilled in materials such as quartz, sapphire and special glasses.

These applications not only developed in the electronic industries, but also in the optical, watch-making and micromechanical industries.

Within these fields, accuracy and results are such that there is practically no technological competition, with the exception of the laser [2], [8], [9], [39], [43], [55].

### 5.4. Welding

Originally, electron beam techniques were reserved for welding high melting point metals very sensitive to oxygen or other gases such as zirconium, tantalum or beryllium. These metals were intended primarily for the nuclear and aircraft industries. Knowledge of the advantages of this welding technique and improvements made to it have led to its use in other industries, including production of cheap, large production parts in the automobile industry.

#### *5.4.1. Electron beam welding mechanism*

When the beam hits the metal, most of the kinetic energy of the electrons is converted into heat to make the weld.

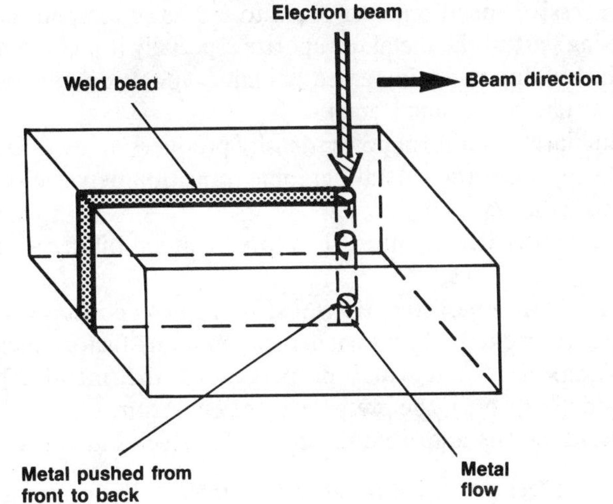


**Fig. 11.** Electron beam welding mechanism [24].

A small quantity of electrons escapes from the workpiece (back scattered electrons) or produces X-rays, from which protection is required.

A cavity filled with molten metal forms in the electron beam impact zone. When the beam is moved, it heats the front of the cylinder more, and the molten metal is driven behind the beam, creating a solid surface which melts in turn; behind the beam, the metal solidifies and forms a narrow weld over the complete height of the part [1], [11], [14], [24], [26], [27].

If the electron beam is concentrated at the junction of both parts, first placed edge-to-edge, melting provides material continuity over the complete thickness, and after cooling, a weld seam is formed.

This welding technique implies special electron beam features. The high specific power beams used in machining, or the relatively low power beams with uniform distribution within a straight cross-section as used in melting, cannot be used in edge-to-edge or butt welding [24].

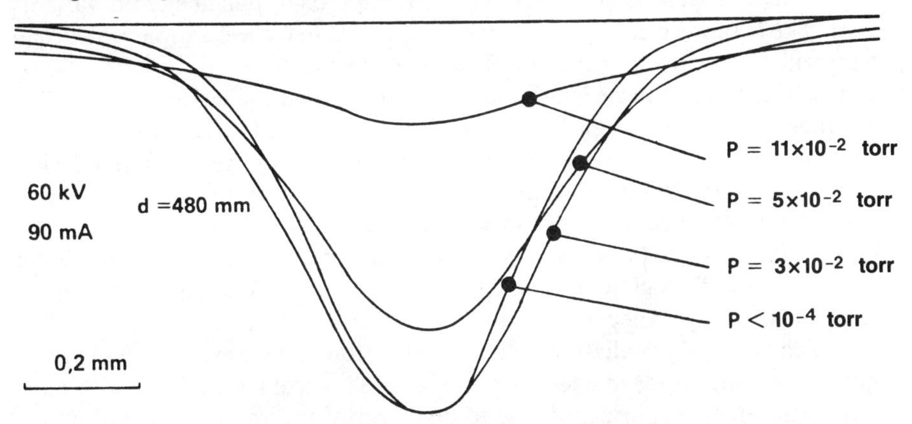


**Fig. 12.** Power density distribution in a beam section at different pressures [24].

Excessive specific power leads to excessive temperatures in the melted area causing part of the metal to vaporize especially if it contains volatile components. If the welding rate is increased to limit evaporation, the weld bead can have weak areas that cause small cracks.

Similarly, a uniform power density produces an excessive temperature gradient and suppresses the crystallographic transition between the melted zone and the base metal.

The power density must therefore be much higher at the center of the beam than at the edges.

The beam characteristics must be kept as constant as possible in order to guarantee weld quality; they depend on many factors such as welding rate, weld penetration, the thermal properties of the metal (specific heat, thermal conductivity, etc.), the pressure in the enclosure, etc. Figure 13 gives an example of weld depths and welding speeds for steel and for a given beam power.

### *5.4.2. Electron beam welding equipment*

Electron beam welding equipment consists of an electron gun, a welding chamber, pump units, a siting system, a part movement mechanism and various monitoring and regulation devices [24], [25], [41], [A], [E], [H], [I].

Electron beam power and welding characteristics [24], [42].

| Beam power (kW) | 3 | 30 | 100 | 100 |
|---|---|---|---|---|
| Welding rate (m/min) | 8 | 0.5 | 1.5 | 0.35 |
| Weld thickness (mm) | 2 | 50 | 100 | 200 |

**Fig. 13.**

#### *5.4.2.1. Electron guns*

The electron guns used in welding are always axial, but numerous variants exist. The cathode can be directly or indirectly heated, acceleration voltage can vary widely and the gun may be fixed or movable, etc.

The triode gun is more widely used in welding than the diode type. Welding installations can be grouped as a function of acceleration voltage:

— "low voltage" installations where the acceleration voltage is less than 30 kV; they do not require X-ray precautions but their power is limited and the weld seam depth-width ratio cannot exceed 10 to 20;

— "medium voltage" installations where acceleration voltage is between 30 and 80 kV; they enable high weld bead depth-to-width ratios of about 40 but require X-ray precautions. They are the most widely used guns;

— "high voltage" installations where acceleration voltage exceeds 80 kV; these guns have some disadvantages, such as high molten metal turbulence within the weld bead, metal vaporization due to high energy electron impact, and X-rays which require costly protection. The design of the gun-chamber assembly also varies widely to meet the application.

#### 5.4.2.2. *Welding chambers*

If the electron beam encounters gas molecules, electrons will collide causing path modification; this widens the beam and decreases the specific power, limiting its use in air. Moreover, to protect the cathode from oxidation and to ensure adequate voltage, the pressure in the gun must be very low. Electron beam heating is thus often performed in a vacuum chamber which contains the electron gun and the tools used to move the parts to be welded. Three major types of installation are used:

— *Overall vacuum chamber systems*

The parts to be welded are placed in a thick, 15 to 30 mm, mild steel or stainless steel vacuum chamber whose volume varies according to the parts, and ranges from a few litres to over 10 $m^3$. Pumping equipment is used to create the vacuum which may be primary, (between $10^{-1}$ and $10^{-3}$ torr) or secondary (about $10^{-4}$ torr).

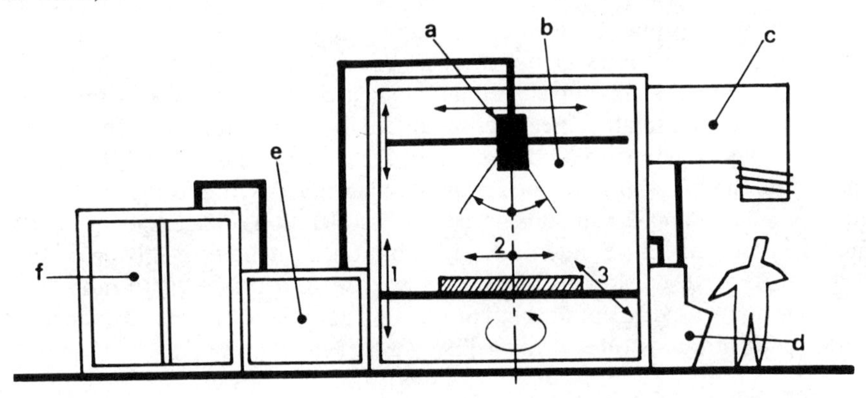


a: movable gun
b: chamber
c: pump unit
d: control console
e: power feed
f: electronic control and monitoring cabinet
1-2-3 Movement of parts along the three axes.

**Fig. 14.** A secondary vacuum welding station [24].

High vacuum equipment is used to obtain very precise weld seams with a very high depth-to-width ratio, or for welding metals which are very sensitive to air. To increase productivity in these installations, very powerful vacuum pumps create the vacuum required in a chamber of several cubic meters within a few minutes.

If the weld specifications are less stringent or the metals to be welded less sensitive to air, such as nickel, steel, copper, aluminum and their alloys, a primary vacuum installation will suffice in most cases; especially if the acceleration voltage is high since the gas molecules present in the chamber will have less effect on the beam. The pumping time required to create a partial vacuum is then about half that required to obtain a "secondary" vacuum.

The electron gun may be movable; the machine then uses a "secondary" vacuum. Simultaneous movement of the gun and the use of beam deflection enables highly complex welds. Part dimensions are limited by chamber size.

The electron gun can also be placed outside the chamber. An independent pumping unit provides a secondary vacuum of between $10^{-4}$ and $10^{-5}$ torr in the gun. The parts to be welded are placed in a chamber in which a primary or secondary vacuum is created. The chamber and gun are coupled through a seal. Poor conductance of gas molecules in the hole separating the chamber and gun permit welding, even when a pressure difference exists between them. This type of machine where parts may be placed on a rotating platform and the chamber may be very small, permits very high welding rates suitable for mass production.

Vacuum-chamber electron-beam welding systems have many applications and constitute the majority of installations.

— *Normal pressure welding systems*

Using a gun appropriately constructed, it is possible to weld in a chamber with normal atmospheric pressure. A pump maintains pressure between $10^{-4}$ and $10^{-5}$ torr in the cathode chamber. The beam then passes through several chambers where the pressure is gradually increased. Each chamber is separated from the next by a small orifice, a poor conductor of gas molecules. Interaction of gas molecule electrons, however, diffuses and rapidly widens the beam. Therefore, the parts to be welded must be placed very close to the end of the gun, 15 to 20 mm maximum, and a very high acceleration voltage (150 to 250 kV) must be used; this causes intense X-ray radiation, demanding costly protection systems. Beam dispersion also limits the weld seam depth-to-width ratio. This system is rarely used except for high production automatic treatment of identical parts for continuous tube welding [29], [33], [59], [H].

— *Local vacuum chamber systems*

When the parts to be assembled are large, or it is impossible to move them, local vacuum chambers permit electron beam welding without having to create a large vacuum area. Generally, the enclosure in which the operation takes place is small; it is equipped with suction-type sealing systems, sliding seals, tapes alone or in combination [10], [24], [25], [38], [50], [A], [E].

The local vacuum chamber may be fixed and the sealing principle used comparable to that of a suction cup. The welding chamber, fitted with seals which even seal the non-machined surfaces, is placed around the parts to be welded. The primary vacuum is then created in a few minutes, or even seconds, and the weld made.

Movable vacuum chamber machines consist either of a fixed base forming the suction unit and a chamber, which can move with respect to the base, or a chamber which moves directly on the part to be welded. Generally, these machines are used for linear welds on flat or curved surfaces.

There are many variations: sliding or fixed ribbon, sliding and mobile chamber for orbital welding, etc. Generally, they are easily transported from one work station to another, and the electron acceleration voltage can be relatively low so that no X-rays are produced. This highly flexible machine is being used to weld heat transfer tubes, thick steel plates (reactor tanks, turbine rotors, etc.), and to make circular or very long welds.

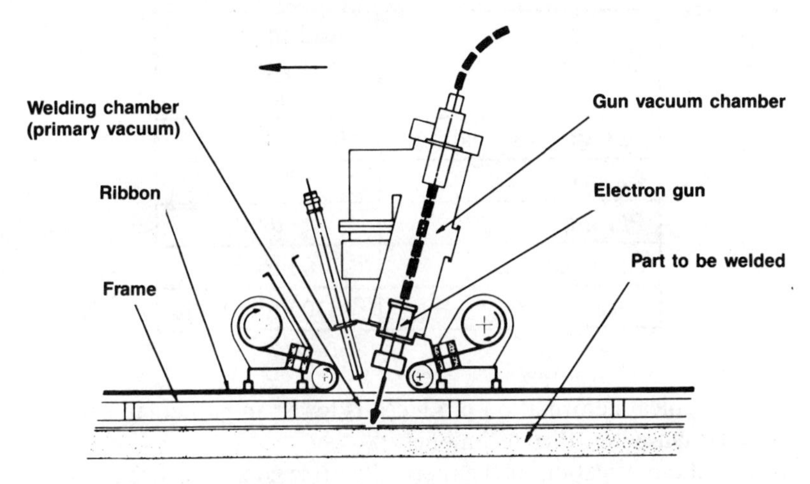


**Fig. 15.** Fixed ribbon vacuum chamber [25].

*5.4.2.3. Other components of an electron beam welding installation*

The precision of the welding operation calls for the use of sighting systems to center the beam on the joint. This operation is performed during welding by optical sighting devices, TV cameras or electronic systems using back-scattered electrons known as reflectrons.

Also, machines are equipped with various devices to move the gun or the part either manually or automatically, control and regulate the vacuum, welding rate, high voltage stability and beam power.

***5.4.3. Electron beam welding characteristics***

Welding techniques are many, and some are relatively cheap, especially in terms of investment costs. Therefore, the advantages and drawbacks of electron beam welding must be examined to better define potential applications [24], [47], [57].

Basically, the advantages of beam welding are:

— deep penetration in a single run; value is related to beam characteristics, welding rate and type of metal (see *Fig.* 13). Welds of more than 200 mm can be made in a single run;

— a very high weld bead depth-to-width ratio, up to 40 approximately; this property is related to the very high power density ($10^6$ to $10^8$ W/cm$^2$) of the electron beam and the ability to produce very fine beams.

— the ability to weld high melting point metals, different metals in direct contact, or parts of different mechanical characteristics;

— possibility of forming welds impossible with other techniques (easily oxidized metals, light alloys, etc.);

— straight melted edges;
— enhanced metallurgical qualities:

Minimum cross-section of weld bead in different welding processes [59].

| Welding method | Bead minimum cross-section ($mm^2$) |
|---|---|
| Autogenous welding | 1 |
| Electric arc | $10^{-1}$ |
| Plasma | $10^{-2}$ |
| Electron beam | $10^{-4}$ to $10^{-5}$ |
| Laser | $10^{-5}$ to $10^{-6}$ |

**Fig. 16.**

• the mechanical deformation produced is lower; in particular, shrinkage is much lower than with other techniques,

• structural modification of the thermally affected zone is less pronounced; this zone is also kept small with minimum internal tensions,

• the zone welded is outgassed on melting, risks of contamination are practically nil and the mechanical strength of the seam is increased;

— this process is suitable for automation and high production due to the high welding rate;
— some post-welding operations, such as machining, can be dispensed with;
— material savings (due to decrease in welding and machining losses) and energy savings may be high; for small thicknesses, of the order of 15 mm, the energy saving is approximately 30% with arc welding, while for large thicknesses (more than 90 mm), it is 50% with "narrow gap" welding, the most competitive technique.

All these advantages explain the rapid development of electron beam welding, which is without doubt the most widespread application of this method of heating. This development has, however, been slowed by the large investment costs involved. Some special techniques also cause problems:
— in some cases, channel formation on the front and back of the weld bead;
— irregular penetration in fillet welds (fluctuation of seam bottom);
— spraying of melted metal.

These problems, however, rarely prevent the use of electron beam welding.

### *5.4.4. Electron beam welding applications*

Electron beam welding applications vary widely. They were originally reserved for welding special metals which, due to their high oxidability, could not be welded using conventional processes. This welding technique can now be applied to highly diversified metals and sectors of economic activity, and is developing in two directions:

— mass production of relatively simple components; in fact, due to its characteristics (fast execution, no consumables, automatability, etc.) it can lead to cost reductions; especially high rate welding of automobile gearbox and transmission components, etc.;
— assembly of thick parts whose production by more conventional processes involves technological, metallurgical, and economic problems; for example, welding thick turbine rotors or nuclear reactor vessels where thicknesses can reach 200 to 300 mm.

Electron beam welding is keeping its place in leading industries where quality is critical (nuclear, aeronautic, aerospace, micromechanical industries, etc.) [15], [17], [18], [19], [21], [23], [36], [37], [56], [57[.

The following products are obtained using this technique:
— nuclear fuel rods and their cladding;
— satellite shafts for automatic gearboxes, ring-gear twin-bevel gear-sliding dog-splined support assemblies, gearbox pinions, axle spindles and other light components in the automobile industry;
— built-on vane flux rectifiers, turbine wheels, compressor rotors in titanium alloy, main spar of combat aircraft in light alloy, etc., in the aviation industry;
— tubes on plates (heat transfers), cylinder assemblies, mechanically welded assemblies, etc., in boilerwork;
— steam turbine drive gears, rifle barrels, bi-metal circular saws, assembly of sheet metal, etc.

## 5.5. Surface heat treatments

The use of the electron beam for surface heat treating of metals is recent. Induction heating or other, more conventional techniques satisfy such requirements more economically. However, for some complex parts or with irregular treatment surfaces, it is difficult to obtain regular reproducible results. Electron beam treatment can then be justified, both from the technical and economic viewpoints [20], [46], [48], [53], [57], [A].

### *5.5.1. Electron beam heat treatment*

This technique is used for surface hardening steels. Due to the very high power density of the electron beam ($10^6$ to $10^8$ W/cm$^2$), the metal surface is taken to austenitization temperature very quickly without heating the remaining metal. Surface hardening is not, as in conventional processes, obtained by contact with a coolant, but by heat flow from the superficially heated area to the colder area. The shorter the heating period, the smaller the heated area, the faster the cooling, and the more efficient the self-hardening mechanism.

During the first attempts to use the electron beam for heat treatment, the electron beam "scanned" the surface to be treated at high speed. Applications of this method have, however, remained limited since the curves followed by the beam often intersect, and therefore cause hot spots which can create micromelting on the metal surface.

Currently the surface to be treated is divided into a finite number of elementary zones, and scanning is accomplished by deflecting the electron beam from one point to another. A strictly controlled quantity of energy is applied to each point over a given time and precisely defines the thermal cycle of the affected part of the workpiece. In machines currently being used, each surface can be divided into 500 elementary zones, and the beam time at each point can be varied from 50 to 100 $\mu$s. These characteristics will of course get better as electron beam heat treatment equipment evolves.

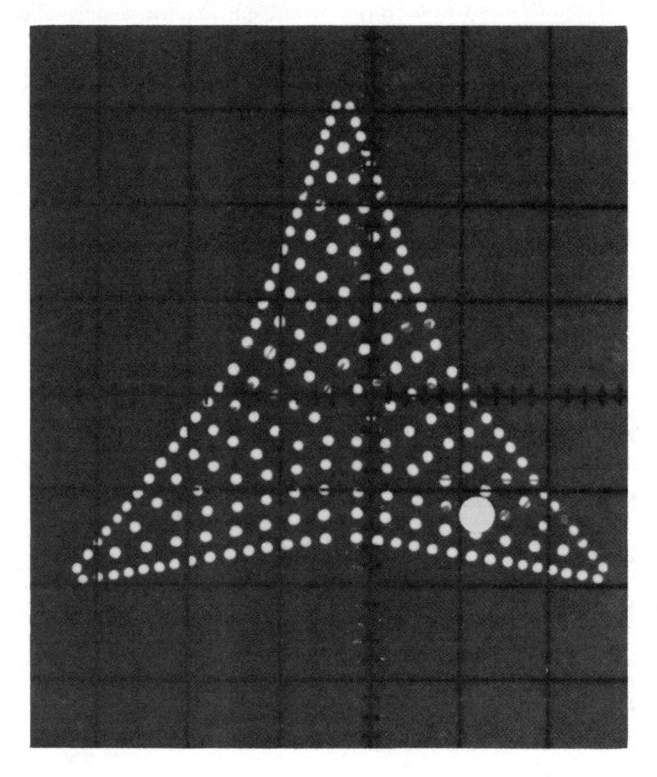

**Fig. 17.** Discrete focal points of a surface for electron beam treatment [48].

Distribution of the elementary areas over the surface treated is not necessarily uniform, so that the treatment effect can be varied according to part shape and provide very fine adjustment of the mechanical characteristics.

### *5.5.2. Electron beam heat treatment equipment and their applications*

A heat treatment machine consists of:

— an electron gun with an electromagnetic focusing coil which concentrates the beam as required, and an electromagnetic deflection coil capable of producing two perpendicular magnetic fields with a very wide bandpass; the characteristics of this coil are fundamental in obtaining adequate heat treatment;

— a computer, capable of controlling and monitoring the various operating parameters of the equipment;

The computer controls beam deflection, the pitch separating two consecutive points and the dwell time on each point;

— a vacuum chamber, together with its pumping system required for automatic system operation.

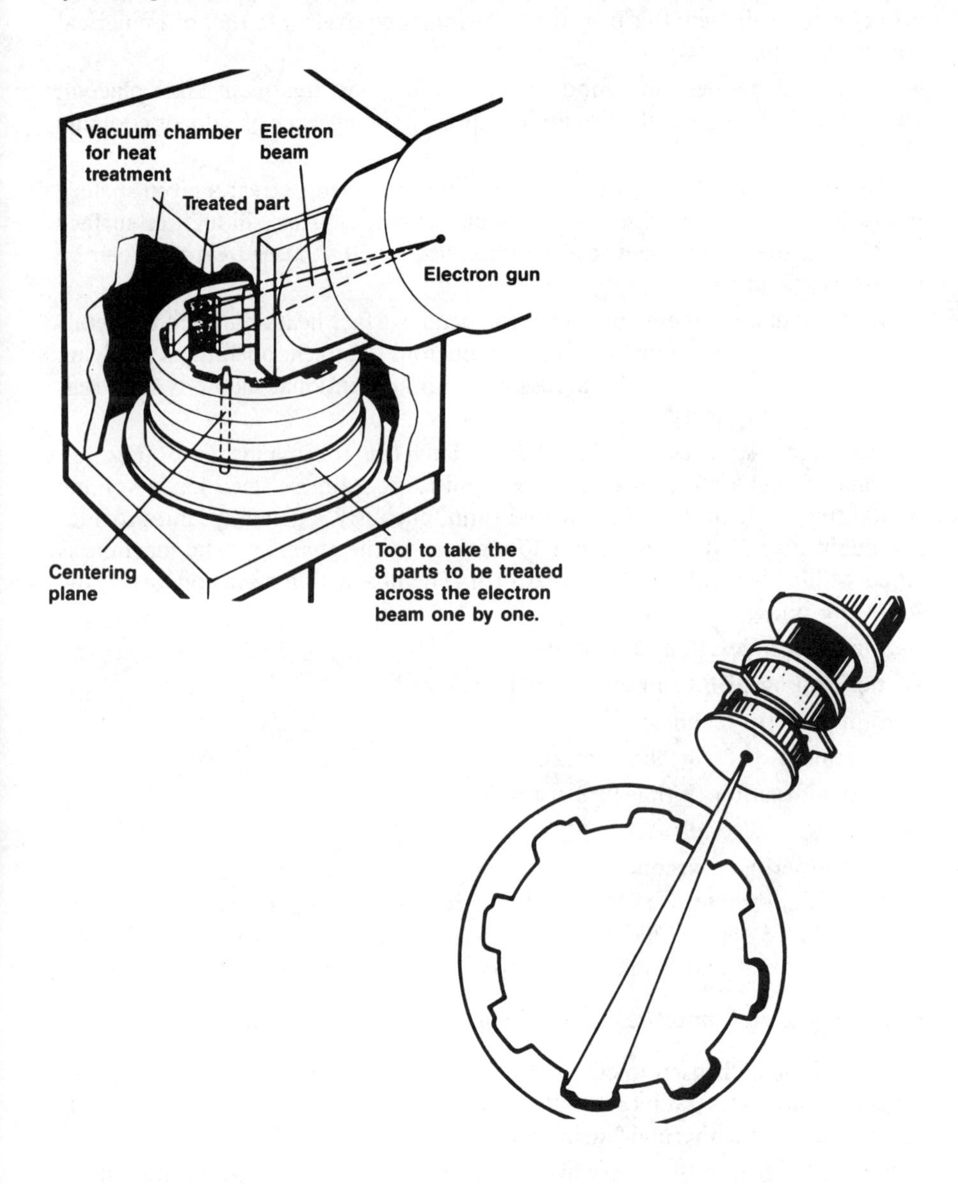


**Fig. 18.** Surface heat treatment of a clutch cam [48].

The advantages of electron beam surface heat treatment are:

— low parts distortion after treatment, eliminating re-work sometimes necessary with more conventional techniques;

— easy treatment of complex parts, treatment accuracy and reproducibility;
— process flexibility and adaptability; just alter the computer program to change energy profile applied; this provides a flexibility equivalent to that of numerical control machine tools;
— quality, cleanliness and good working conditions; treatment takes place in a vacuum providing excellent treatment quality and absence of salts or coolants eliminates pollution;
— very high process energy efficiency; energy required to treat a given surface is much smaller than with conventional techniques. Also, induction surface hardening, already efficient, can be improved by 50%; laser treatment can be improved by up to 90%.

Due to high investment costs, electron beam surface heat treatment of metals cannot replace conventional processes. Numerous treatment operations involving complex parts, or which are impossible with conventional methods, however, are economically feasible.

For example, surface hardening of drive clutch cam roller guide ramps (*Fig.* 18). The machine chamber, enclosed in 25 mm thick stainless steel, has a volume of 30 litres; three parts can be treated simultaneously. The pump unit provides a vacuum of $5 \times 10^{-2}$ torr within 10 seconds in the chamber. The machine is fitted with a six-position rotary platform. The production cycle consists of the following phases:
— platform indexation: 2 seconds;
— positioning of tool in chamber: 1 second;
— pumping: 10 seconds;
— treatment of 8 surfaces: 16 seconds;
— part indexation (8 times): 8 seconds;
— venting: 4 seconds;
— tool lowering: 1 second.

Under these conditions, the equipment treats 255 parts per hour. Electron gun power is 42 kW (60 kV, 700 mA) and is designed for heating the combined part surfaces prior to hardening in the static mode.

## 5.6. Curing and modification of organic material properties

Electron beam is used to cure paints, varnishes or other organic coatings on different substrates and to modify organic compound characteristics. This application is not thermal but radiochemical [7], [23], [45], [52].

Electron energy must be very high, and therefore the acceleration voltage must be between 150 and 300 kV and sometimes even higher.

### *5.6.1. Radiochemical effect of an electron beam*

High energy electrons, strongly accelerated by voltages greater than 150 kV, cause ionization and excitation of the target atoms and molecules. Three basic phenomena may appear in the exposed material:

— *Radical polymerization*

Vinylic monomers, which are irradiated at ordinary temperatures, can be polymerized by the electron beam.

— *Polymer cross linking*

In linear macromolecules, exposure to radiation breaks some bonds, and reactions then take place between the chains creating molecule "bridges". Polyethylene, polypropylene, polystyrene and polyacrylates can be cross linked in this manner.

— *Grafting*

A radiated linear polymer has active areas on the chain which, in the presence of a monomer, are grafted onto the initial polymer. Grafted co-polymers are obtained in this manner.

Catalysts are often added to the products to be treated. Oxygen inhibits such reactions, and therefore must occur in an inert atmosphere. A detailed study of these reactions is extremely complex and has been the subject of many works. The reactions are being used more widely inndustry and are giving rise to new materials and manufacturing processes [5], [7], [22], [40].

### *5.6.2. Composition of an irradiator*

An irradiator consists of four parts:
— electron gun;
— high voltage generator;
— treatment chamber;
— control and monitoring devices and support devices.

— *The electron gun*

The electron gun consists of the standard components but the acceleration voltage is much higher. The end of the gun is covered with a very fine aluminum or titanium sheet 15 to 25 $\mu$m thick held by a grid of thin copper wires. Since the acceleration voltage is very high, the electrons pass through this sheet or "window" with very low absorption, 10% approximately at 150 kV, 7% at 300 kV. The accelerator and the electron source are maintained in a vacuum; thus the window must withstand atmospheric pressure and be coated because of overheating caused by the electrons.

— *High voltage generator*

The high voltage (300 kV, 20 to 50 mA) is provided by a rectifier supplied from a frequency converter (400 to 2,000 Hz).

The accelerator and the high voltage power supply are contained in metal chambers filled with a protective gas (sulphur hexafluoride $SF_6$) at a pressure of 1.5 to 3 bars, providing electrical insulation and enabling a compact product to be obtained.

— *Treatment chamber*

The strip or sheet of threads to be treated move continuously very close to the radiator window to prevent electron dispersion due to molecule collisions in the treatment chamber atmosphere. Therefore, irradiation takes place in an inert gas, generally nitrogen, or a nitrogen-carbon dioxide mixture. The treatment chamber is made of pressurized concrete which contains the X-rays resulting from the molecule collisions.

However, new irradiators are being developed which use lower acceleration voltages of between 100 and 150 kV and require less protection (in particular, no concrete chamber).

— *Auxiliaries*

These auxiliaries comprise control and monitoring devices, safety devices, pumping units and inert gas generators.

### *5.6.3. Industrial applications of electron beam irradiation*

Electron beam irradiation is used for curing coatings such as paints, varnishes, lacquers, plasters, inks, etc., on different substrates and modification of organic materials [23], [31], [52], [60], [C], [D], [F].

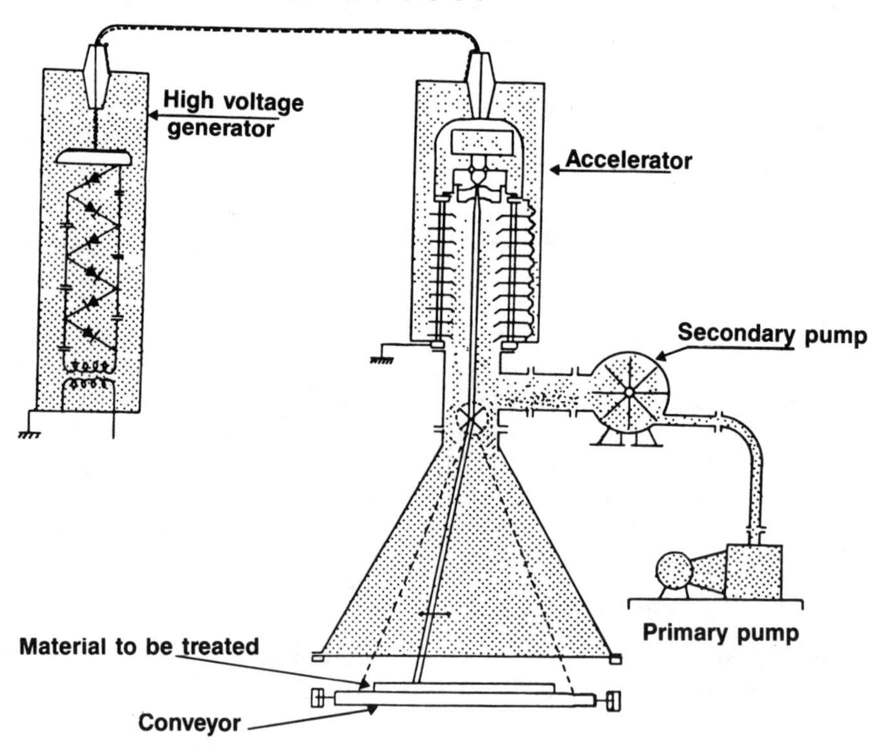


**Fig. 19.** Diagram of an irradiator [B], [D].

— *Curing of paints and varnishes*

Unlike ultraviolet radiation, where the reaction is also radiochemical, electron beams can be used to treat pigmented products, due to relatively deep penetration through even the most opaque coatings. This property is used for:

— Curing coatings on wood

Treatment times are extremely short and the coating quality excellent. Since thicknesses are relatively small, curing by ultraviolet radiation is sometimes preferred because of lower investment costs (5 to 20 times).

— Hardening coatings on metals

The technique has been used some in this area but development has been slowed by a lack of data concerning coating durability. Power is in the tens of kilowatts.

Electron beam heating can be used to cure coatings on other media (construction materials, coatings on plastics, etc.).

The advantages are short treatment time (the product feed speed can be more than 50 m / min. for wood and 250 m / min. for metal), compact installations (at 120 m / min., the length of a conventional oven for baking paints on sheet metal is about 90 m versus 3 m for an electron beam irradiator), energy savings (the process consumes 20 to 50 times less energy than a conventional forced convection process) and low pollution since the coatings involved do not contain solvents.

Moreover, treatment is performed at ambient temperature which is excellent for fragile materials such as wood, paper, plastics, etc.

The coating quality is also improved due to better mechanical and chemical strength and greater brilliance.

Electronic irradiation, however, requires radiosensitive products which are still expensive and whose coating durability has yet to be proven. Moreover, it can only be used on flat products. Finally, investment costs are high which can only be recovered by mass production sometimes exceeding market requirements.

— *Cross linking of plastic sheaths on cables*

Electron beam is used to cross link plastic sheets (PVC, polyethylene, etc.) on electrical and telephone cables. The treatment improves temperature behavior, resistance to aging and cracking under tension. Electron-beam-cross linked cables are, for example, used as elements in floor heating systems. Generally, this type of cable is used whenever operating conditions are less than ideal [45], [54], [F], [N].

— *Fabrication of heat-shrinkable polyethylene films and sheets*

Irradiation causes cross linking of the polymers which, combined with biorientation of chains, improves the film properties — better impact resistance, better shape retention improved brilliance and transparency due to a higher drawing factor.

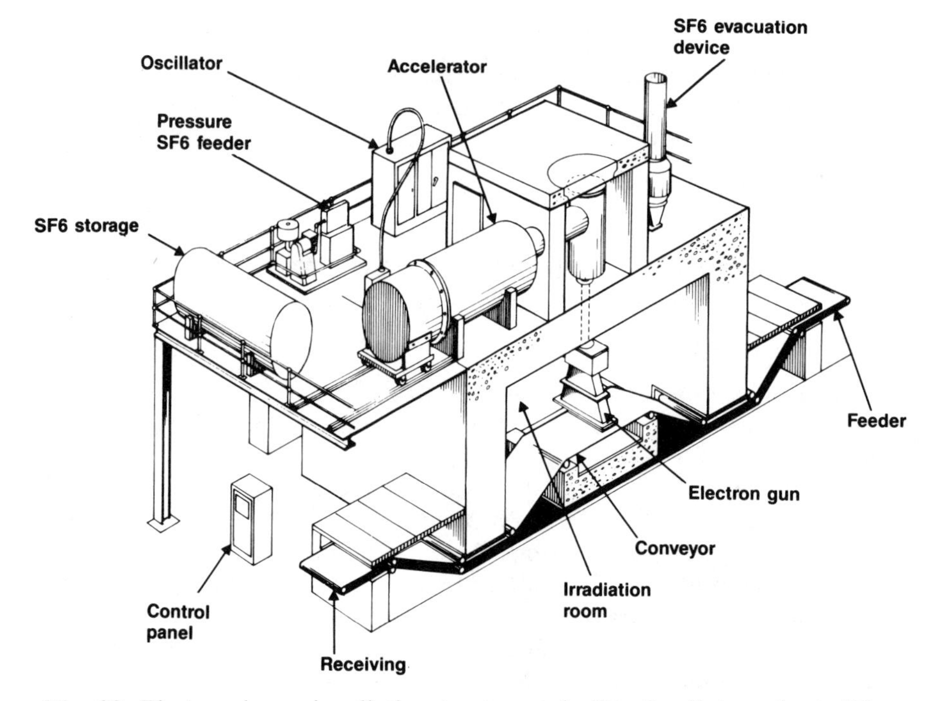


**Fig. 20.** Electron beam irradiation treatment facility for flat products [F].

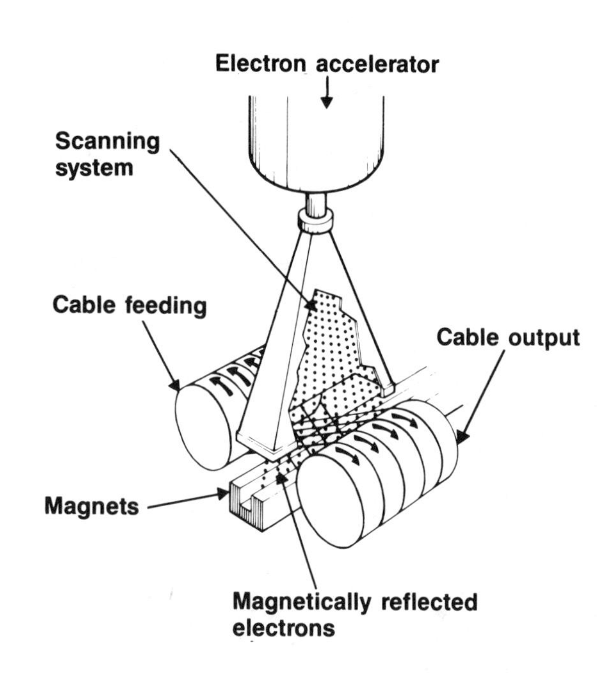


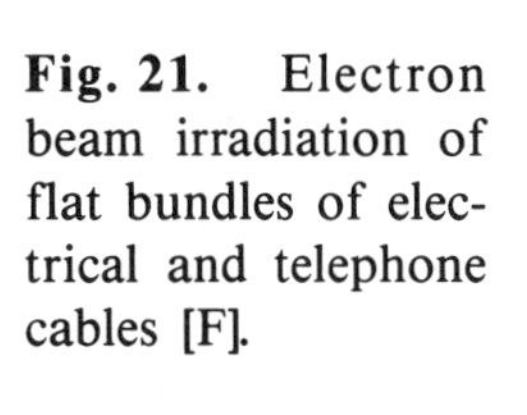

**Fig. 21.** Electron beam irradiation of flat bundles of electrical and telephone cables [F].

— *Miscellaneous applications*

Many other applications have been or are being developed: production of electrically conductive plastics, sterilization of surgical equipment, manufacture of polyethylene — and polypropylene-based adhesives by acrylic acid grafting, cross linking of elastomers, neutralization of steel-work effluents ($SO_2$ and $NO_2$), grafting of acrylic acid onto textile fibres (polyester, polypropylene, etc.) to improve dye affinity, fire and dirt resistance, washability, moisture take-up and appearance (permanent crease) [5], [7], [30], [32], [45], [F], [K], [L], [M].

Generally, these applications are the result of long research.

## 6. BIBLIOGRAPHY

[1] M. STOHR, Brevet français n° 1141535 relatif au soudage par faisceau d'électrons, 1956.

[2] C. K. CRAWFORD. Electron Beam Machining, *Introduction to Electron Beam Technology*, John Wiley, New York, 1962.

[3] WEINRYB, Étude de la rétrodiffusion des électrons, *Métaux, Corrosion et Industrie*, n° 476, 1965.

[4] M. VON ARDENNE *et al.*, Stand und Entwicklungstendenzen der Elektronenstrahltechnologie, *Die Technik*, mai 1966.

[5] HALL *et al.*, Activated gas Plasma, surface treatment of polymers for adhesive bonding, *Journal of applied Polymer Science*, vol. 13, p. 2085-2096, 1969.

[6] M. OLETTE *et al.*, Perspectives de la protection de l'acier par métallisation sous vide, *Corrosion-Traitements-Protection-Finition*, n° 3, avril-mai 1969.

[7] K. H. MORGENSTERN, *Radiation vulcanization*, 98th meeting of the division of Rubber Chemistry, Chicago, 1970.

[8] PIETERMAAT, Bombardement électronique et applications, *Usinages par procédés non conventionnels*, Masson, 1971.

[9] CAST, INSA, *Les applications industrielles du bombardement électronique*, INSA Lyon, mars 1972.

[10] R. ROUDIER *et al.*, Perspectives actuelles du soudage sur site de tôles épaisses, *Soudage et techniques connexes*, n° 5-6, vol. 26, juin 1972.

[11] G. SAYEGH et P. DUMONTE, Analyse des phénomènes qui se produisent dans un canon de soudage par bombardement électronique, *Revue de la soudure*, Bruxelles, n° 4, vol. 28, 1972.

[12] INDUSTRIAL HEATING, *Electron beam melting of new type stainless steel at revolutionary plant*, août 1972.

[13] UNION INTERNATIONALE D'ÉLECTROTHERMIE, *Reduction, refining and remelting processes*, Congrès de Varsovie, septembre 1972.

[14] SCIAKY, Le soudage par faisceau d'électrons, *Documentation Sciaky*, 1973.

[15] Y. DE BONY, Le soudage d'aluminium en fortes épaisseurs, *Revue de l'aluminium et de ses applications*, n° 415, février 1973.

[16] G. CHAUDRON et F. TROMBE, *Les hautes températures et leurs utilisations en physique et en chimie*, Masson, 1973.

[17] H. MONTIGNI, Le soudage par bombardement électronique, *l'Usine nouvelle*, numéro mensuel, mai 1973.

[18] S. BELAKHOWSKY, *Théorie et pratique du soudage*, Pyc-édition, 1973.

[19] J. GAUTHIER, Applications du soudage par faisceau d'électrons aux pièces mécaniques, *Soudage et techniques connexes*, n° 9-10, vol. 27, juin 1973.

[20] W. HILLER *et al.*, Nouvelles possibilités du traitement thermique par bombardement d'électrons, *Traitement thermique*, n° 80, 1973.

[21] J. COLLOMB *et al.*, Le soudage par faisceau d'électrons appliqué aux fabrications industrielles de combustibles nucléaires, *Soudage et techniques connexes*, n° 11-12, vol. 27, novembre 1973.

[22] G. GAUSSENS et J. DUCHEMIN, Optimisation des compositions thermodurcissables pour différents subjectiles, *Revue générale des caoutchoucs et plastiques*, n° 11, 1973.
[23] A. G. NORTH, Progress in radiation cured coatings, *Pigment and Resin Technology*, *février* 1974.
[24] P. DUMONTE et G. SAYEGH, *Soudage par faisceau d'électrons, applications industrielles et perspectives d'avenir*, Congrès du Comité français d'électrothermie, Versailles, 2 mai 1974.
[25] R. ROUDIER, *Le soudage par faisceau d'électrons avec chambres à vide locales et portables*, Congrès du Comité français d'électrothermie, Versailles, 2 mai 1974.
[26] D. DUPOUX, *Electron beam welding, relationship between the parameters of the beam and the mechanical features of the gun*, Advances in Welding Processes, 3e Conférence, Harrogate, 5 mai 1974.
[27] P. DUMONTE et G. SAYEGH, *Effect of accelerating voltage on the penetration in electron beam welding*, Advances in Welding Processes, 3e Conférence, Harrogate, 5 mai 1974.
[28] C. OTIS PORT, Métallisation sous vide des pièces en plastique, *Modern Plastics International*, n° 12, 1974.
[29] R. J. LANYI *et al.*, Non-vacuum electron-beam welding, an advanced metal joining technique, *Westing house Engineer*, n° 4, vol. 34, octobre 1974.
[30] M. R. CLELAND *et al.*, Use of high-power electron beam radiation for the treatment of municipal and industrial wastes, *Documentation Radiation Dynamics*, Westbury (New York).
[31] K. H. MORGENSTERN, *Processing with radiation*, Chemtech, octobre 1974.
[32] K. H. MORGENSTERN, Appraisal of the advantages and disadvantages of gamma, electron and X-ray sterilization, *Radiation Dynamics*, Westbury (New York), 1974.
[33] J. H. FINK, Analysis of atmospheric electron beam welding, *Welding Journal*, n° 5, vol. 54, mai 1975.
[34] P. MERRIEN et A. BARBIER, Moulage du titane, *Matériaux et Techniques*, n° 5, 1975.
[35] J. PITROU, Le Soudage par faisceau d'électrons, *Matériaux et Techniques*, novembre 1975.
[36] B. SAVORNIN, Applying electron beam welding in nuclear engineering, *Nuclear Engineering International*, n° 240, vol. 21, mars 1973.
[37] T. E. BURNS, Applications of electron beam welding, *Metal Construction*, n° 6, vol. 7, juin 1976.
[38] J. C. GOUSSAIN et Y. LE PENVEN, Étude du soudage par faisceau d'électrons avec pistolet portable de tubes sur plaques d'échangeurs, *Soudage et techniques connexes*, n° 7-8, août 1976.
[39] D. VON DONEBECK et K. H. STEIGERWALD, Electron beam machining; the process and its industrial applications, *IEEE Conference*, Publication n° 133, Londres, novembre 1975.
[40] MILEO, Composition de revêtements réticulables par bombardement électronique, *Revue de l'Institut français du pétrole*, nos 4, 5, 6, 1976.
[41] C. WEISMANN, *Welding Handbook*, 7e édition (American Welding Society), 1976.
[42] G. SAYEGH, *Le soudage par canon à électrons de* 100 kW *d'acier de fortes épaisseurs*, 8e Congrès International d'Électrothermie, Liège, 1976.
[43] D. BIAU, *Applications industrielles du bombardement électronique*, Étude EDF non publiée, 1977.
[44] H. STEPHAN, Present position of electron beam melting and casting, *Elektrowärme International*, n° 3, juin 1977.
[45] BUSINESS WEEK, *Radiations bright new future*, étude parue dans le numéro de juillet 1977.
[46] C. L. GILBERT, Computerized control of electron beam for precision surface hardening, *Industrial Heating*, janvier 1978.

[47] R. G. WILCOX et W. R. BERGER, Factors affecting the quality of electron beam welding, *Metal Progress*, n° 4, vol. 113, avril 1978.
[48] J. BURKETT et G. SAYEGH, *Nouvelles possibilités des faisceaux d'électrons dans le traitement thermique des métaux*, Congrès Comité français d'électrothermie, 5-6 avril 1978.
[49] S. SCHILLER *et al.*, Leistungsfachige Elektronenstrahltechnik in Industrieöfenbau (faisceaux électroniques de grande puissance pour fours industriels), *Elektrowärme International* n° 4, août 1978.
[50] P. MARTIN *et al.*, Matériel de soudage par faisceau d'électrons de forte puissance pour utilisation en vide local, *Soudage et techniques connexes*, n° 9, vol. 32, septembre 1978.
[51] P. LAPERROUSSAZ, Traitements de surface, des possibilités nouvelles avec les dépôts sous vide, *l'Usine nouvelle*, n° 3, 18 janvier 1971.
[52] P. LAPERROUSAZ, Traitement des polymères par irradiation, une nouvelle chance aux USA, *l'Usine nouvelle*, n° 13, 29 mars 1979.
[53] G. SAYEGH et J. BURKET, Principe et application de l'emploi des faisceaux d'électrons commandés par minicalculateurs dans le traitement thermique de l'acier, *Traitement thermique* n° 136, juin 1979.
[54] J. H. BLY et R. S. LUNIEWSKI, Dynamitron high power industrial accelerators in the wire and cable industry, *Documentation Radiation Dynamics*, Westbury (New York) 1979.
[55] D. COUE, Perçage par faisceau d'électrons : 5 000 trous par seconde, *l'Usine nouvelle*, n° 37, septembre 1979.
[56] K. MARVIN, Thermal processing of steel strip grades at new computerized cold rolling mill related to specific applications, *Industrial Heating*, septembre 1979.
[57] G. SAYEGH, *Comparaison du faisceau d'électrons et du faisceau laser dans le travail, le soudage et le traitement thermique des métaux*, Comité français d'électrothermie, avril 1980.
[58] DOCUMENTATION EDF, 75080 Paris-La Défense.
[59] DOCUMENTATION LABORELEC, Bruxelles, Belgique.
[60] M. MOREL, L'irradiation par faisceau d'électrons, *Revue du bois*, n° 11, 1975.

**List of equipment manufacturers and suppliers mentioned in this chapter:**

[A] SCIAKY, 94400 Vitry-sur-Seine.
[B] SAMES, 38000 Grenoble.
[C] STEINEMANN, 94150 Rungis et Saint-Gallen, Suisse.
[D] CAPRI (CEA), 91400 Saclay.
[E] CLOVER, 93210 La Plaine-Saint-Denis.
[F] RADIATION DYNAMICS, Westbury, New York, 11590, U.S.A.
[G] TECMETA, 74000 Annecy.
[H] LEYBOLD-HERAEUS-SOGEV, 91400 Orsay.
[I] SAF, 95310 Saint-Ouen-l'Aumône.
[J] ALCATEL-HEURTEY, 75823 Paris.
[K] ENERGY SCIENCE, Burlington, Massachusetts 01803, U.S.A.
[L] HIGH VOLTAGE ENGINEERING, Burlington, Massachusetts, 01803, U.S.A.
[M] VEECO, 91400 Orsay.
[N] ACOME, 75008 Paris.

# Chapter 11

# Laser Heating

## 1. PRINCIPLE OF LASER HEATING

Laser heating is very new with laboratory developments dating from the early 1960's. However, lasers were soon used in industries such as electronics, micromechanical engineering, watchmaking, etc.

The word laser stands for "light amplification by stimulated emission of radiation".

This concept is most easily explained by comparing it with standard light sources. Light emission is due to a change in the path of electrons which, due to the effect of applied energy (thermal, in the case of an incandescent source), leave their normal orbit for a higher level, then return to their original orbit, emitting energy equal to the difference between levels (emission of photons).

Light emitted by the standard sources occupies an important part of the electromagnetic spectrum (wavelengths of visible light are between 0.4 and 0.8 $\mu$m approximately) and consists of radiation of different wavelengths (corresponding to photons of different energies).

With most sources, the energy emitted can be represented by a continuous curve as a function of the wavelength, although important concentrations often occur around certain lines. Moreover, the waves emitted are phase-shifted with respect to each other and are propagated in all directions.

This type of emission has several consequences:

— it is impossible to cause an energy receptor to resonate, enabling a very high energy transmission efficiency to be reached (as in the case of radiofrequencies).

— although the light beam can be concentrated by means of optical systems, emission remains diffuse and the impact area is relatively large thus limiting power density and accuracy.

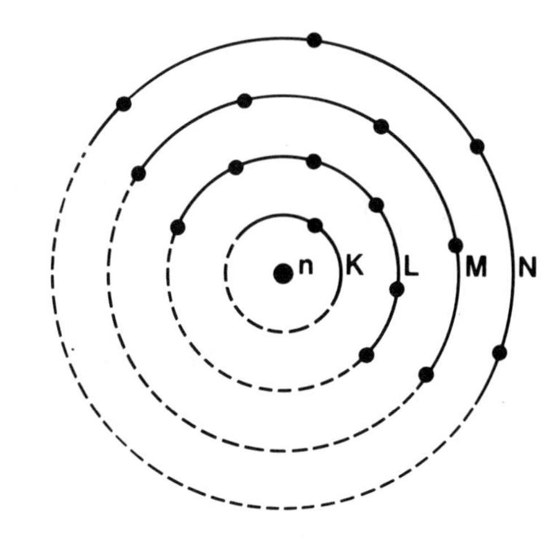


**Fig. 1.** Schematic representation of an atom and its central core n and the various electron paths K, L, M and N. A clearly defined energy level corresponds to each path and increases from the center towards the periphery. Return of electrons taken to a higher energy level $E_z$ by an excitation source to a lower level $E_x$ results in emission of a light ray, the frequency of which is given by the equation:

$$E_z - E_x = hf,$$

$h$ is Planck's constant ($6.624 \times 10^{-34}$ J/s); $f$, frequency.

Laser emission overcomes these limits. In fact, a laser emits over a very short period or continuously, a monochromatic, linear, extremely intense light beam. The maximum power density of a laser beam is of the order of $10^{15}$ W/cm$^2$. In addition, all oscillations are in phase (coherent light source) [1], [3], [4], [5], [6], [25], [47].

The following are necessary for lasers:

— corpuscular population, such as electrons; each element of this population is at its own energy level in the material. Distribution of the population is normal if the population density decreases with distance from the basic level (fundamental level). This occurs with bodies in a state of equilibrium at a given temperature.

A population is inverted if a higher energy level is more populated than the level immediately below. This population is then in a state of unstable equilibrium and tends to return spontaneously in a disordered manner to the normal state, which is the general case with light emission. This return can also be stimulated and is then accompanied by coherent emission; this is the laser effect.

Certain materials are used to activate corpuscular populations favorable to the laser effect, such as chromium-doped alumina monocrystals (rubies), neodymium-doped glasses, some semiconducting materials, gas mixtures, etc..

— a method of inverting this population; to take this population to a higher energy level, an energy source must be used which acts as a pump taking a certain number of particles from a lower energy level to a higher energy level. For solid bodies, pumping is obtained by optical methods (for example, xenon flash), or electrical discharge for gases (other pumping techniques exist such as chemical reactions but are not used in industrial applications);

— a method of stimulating population return to the normal state, thus enabling emission of coherent energy; inversion and stimulation of the population to return it to the fundamental state are separated by a fraction of a second, and the constantly present spontaneous emission causes forced emission of coherent radiation. For operation to be continuous, the amplifier thus created is converted into an oscillator by adding a feedback loop, which reapplies part of the output energy to the laser system. The reapplied photons cause in-phase emission of more photons at the same frequency, the same energy and along the same direction.

To ensure amplification and resonance, the laser system is placed between two parallel or slightly curved mirrors. One of these mirrors is partially transparent to radiation (0.3 to 50% depending on the type of laser) and the other totally reflecting. If the distance between the two reflecting walls is equal to an integer of the half wavelength of the laser emission, a standing wave occurs in the space between the mirrors which behaves as a resonant cavity and is amplified by successive reflection from the mirrors (this device is sometimes known as the Fabry-Perot interferometer). If amplification reaches a threshold sufficient to compensate for inevitable losses due to diverse absorptions, a beam of monochromatic coherent light is rapidly formed, traverses the partially radiation-transparent mirror, and is directed towards the target.

The properties of laser beams result directly from their emission mode and type:

— very low light beam divergence; for a plane source with all points emitting in phase, most of the energy is radiated within a cone, the apex angle of which, known as emitter divergence, has a value of $1.22\lambda/d$, where $\lambda$ is the emission wavelength and d the source diameter. For a ruby laser, the emitting wavelength is 0.7 um approximately with an emissive surface of 1 cm; application of the above formula gives a divergence of the order of $10^{-4}$ radians. A laser of this type therefore emits a beam which, without an interposed optical system, will illuminate a circle of 1 m diameter only at a distance of 10 km;

— the beam can be focussed by optical devices; the diameter of the light spot obtained at the source is up to several wavelengths, depending on the optical system used;

— the stimulated emission phenomenon can be triggered very quickly; the excitation phase corresponds to storage of energy which, restored in a few nanoseconds, leads to very high instantaneous powers. For example, a pulse laser with a pulse width of 30 ns ($30\times10^{-9}$s) provides an instantaneous power of

$10^9$ W; i.e., 1 000 MW (present power of a nuclear power station) if the energy emitted during the pulse is 30 joules ($8.33 \times 10^{-6}$ kWh);

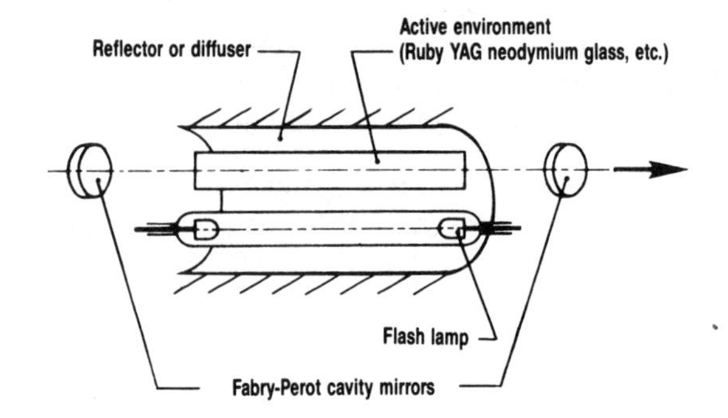


**Fig. 2.** Diagram of a solid laser [E].

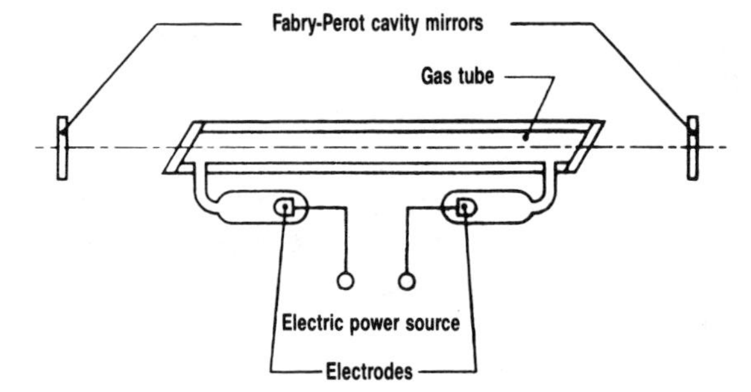


**Fig. 3.** Gas laser diagram [E].

— the power density is very high, the laser beam can be concentrated on a very small surface (a few wavelengths), and pulse instantaneous power is very high. It is therefore possible to obtain extremely high power densities, of the order of $10^{14}$ to $10^{15}$ Watts/cm$^2$, by concentrating the pulse defined in the preceding example on a circular surface of 10 $\mu$m diameter (approximately 15 times the wavelength of a laser ruby), a power density of $10^{15}$ W/cm$^2$ is obtained. In practice, the specific powers used are often lower, of the order of $10^{10}$ W/cm$^2$, which is still very high.

Essentially, it is the very high power density and the ability to concentrate the beam on a very small area which justifies the use of laser heating despite high investment costs.

However, investment costs reduce the scope of laser beam applications as a heating method to those for which no other technique is available. For example, they are used in micromachining, drilling, milling or cutting in microengineering, electronics, etc. However, they are more widely used in other industries, including some which are not usually classified as leading industries, for high-speed cutting of a wide range of materials (plastic transformation industries, metalwork, textile and glass industries, etc.). Some applications have been used in industries using mass production techniques such as the automobile industry (surface heat treatments) [16], [18], [25], [26], [37], [38], [40], [49].

Also, laser applications exist for which purposes are non-thermal (medicine, guidance, telemetry, decoration, etc.).

## 2. COMPOSITION OF LASER HEATING INSTALLATIONS

Basically, a laser heating installation consists of a laser generator, the beam focusing optical system and handling and control support systems.

Lasers can be classified as a function of various criteria such as the active material used, the type of pumping, continuous or pulse operation, etc. The classification used below is based on the type of active material and analysis of the different variants.

### 2.1. Types of lasers

#### *2.1.1. Optically pumped solid lasers*

The oldest of these, the ruby laser, is a pulse laser. The ruby consists of a chromium-doped alumina monocrystal. Pumping energy is very high and must be provided in the form of pulses. Excitation is obtained by optical pumping from the discharging of a bank of capacitors into a xenon arc lamp surrounding the ruby crystal.

Radiation is emitted in pulses. Each pulse in fact consists of a pulsetrain consisting of several hundred very short pulses with a wavelength of 0.6943 $\mu$m (red part of visible spectrum). Pulsetrain width is a few tens of nanoseconds, and instantaneous power is very high, of the order of $10^8$ to $10^9$ W [1], [4], [5], [6], [47], [B], [C], [E].

To obtain these very high instantaneous powers, special devices such as Kerr cells are often used (giant pulse lasers). This cell behaves as a special switch, which, by reversing its polarity, releases the energy accumulated in the active element during pumping, precisely when a predetermined value is reached.

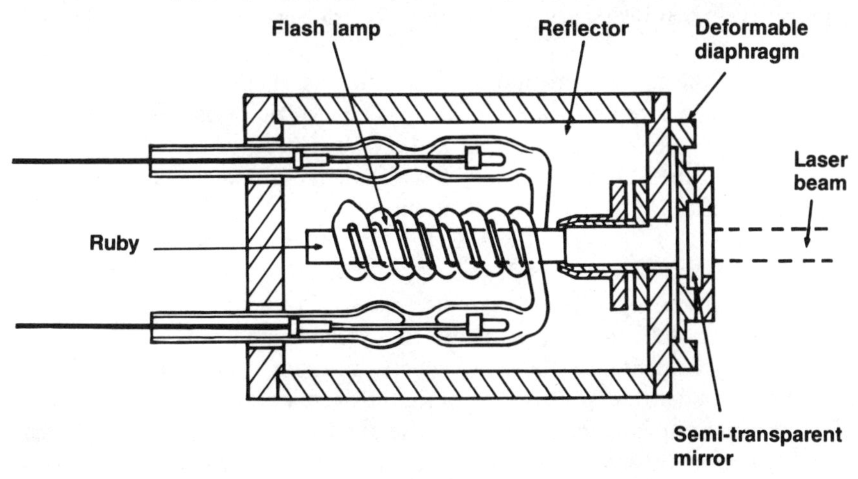


**Fig. 4.** Diagram of a ruby laser [47].

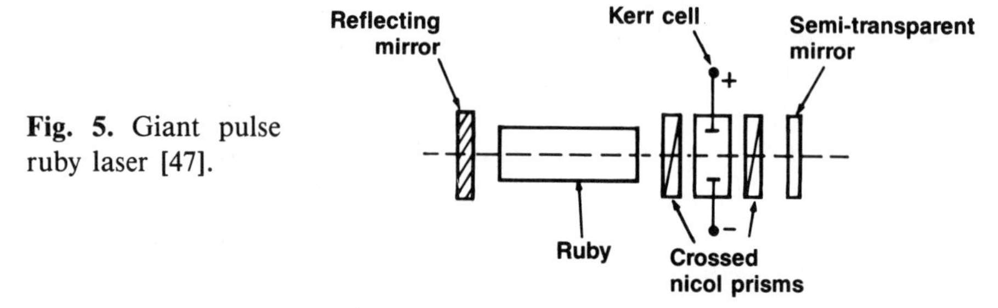


**Fig. 5.** Giant pulse ruby laser [47].

The energy conversion efficiency in a ruby laser is very low, of the order of 0.1%. Mean unit power is a few watts (as compared with instantaneous power, which is very high).

Other types of solid lasers, such as a neodymium-doped glass laser or YAG lasers (yttrium, aluminum garnet) exist.

The YAG laser may be continuous or pulsed, and due to the favorable characteristics of the active material pumping can be provided by a continuous light source such as a krypton arc, thus forming a continuously emitting solid laser. Wavelength is equal to 1.06 $\mu$m, which is in the infrared. Efficiency, although still very low, 1% approximately, is much higher than with the ruby laser. The mean maximum power of a continuously operating YAG laser is about 1,000 watts.

### *2.1.2. Radiofrequency discharge pumped gas lasers*

With a gas laser, the active element is a pure gas or a mixture of gases under low pressure (some tens of torrs); the gases most widely used are a helium-neon mixture, argon, krypton and, most commonly, carbon dioxide.

Pumping energy is provided by a RF generator which maintains an electrical discharge in the tube (see *Fig.* 3 and 14). Generally, gas lasers operate continuously, but, if necessary, can be designed for pulse operation [1], [3], [4], [5], [8], [B].

For example, the active element of a helium neon laser consists of 90% helium and 10% neon, in a quartz tube 1 m long. The electric discharge is provided by generators of some tens of watts, at 27.12 MHz. Energy emission takes place on the neon red line. Due to the very low efficiency (between 0.1 and 0.5‰), the mean power available is very low, about 0.1 to 0.3 W. For a $CO_2$ laser, the gas can consist of a mixture of 8% $CO_2$, 75% He and 17% $N_2$.

At present, only carbon dioxide lasers can enable relatively high powers to be obtained with acceptable energy conversion efficiencies. A wide range of powers, between 100 and 20,000 watts, is available and sufficient to meet numerous requirements. Efficiency, which is about 10%, becomes more acceptable for industrial applications, since the overall efficiency, depending on the use to which the laser beam is put in the production operation considered, begins to approach the standard industrial efficiencies, and can even, in some cases, exceed them due to the specific operating characteristics of the lasers. The drawback inherent in carbon dioxide lasers is the long wavelength, 10.6 $\mu$m in the long infrared, which is strongly reflected by metals, thus limiting applications [8], [9], [16], [46], [A], [B], [F], [G].

Carbon monoxide lasers (CO) have high efficiency, of the order of 60%, but, for safety reasons, have been of limited industrial use.

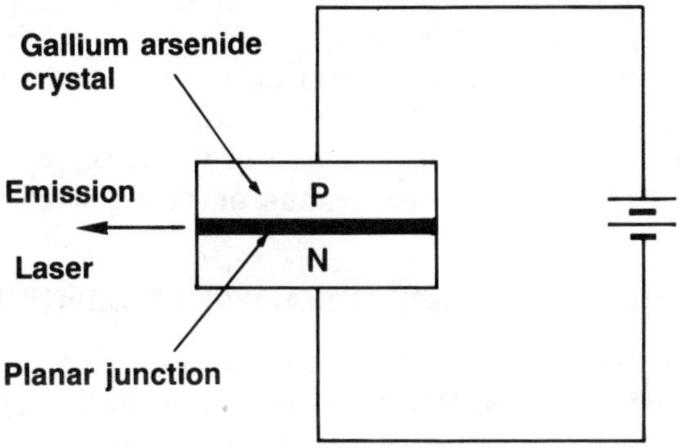

**Fig. 6.** Semiconductor (solid state) laser [47].

### *2.1.3. Semiconductor lasers (solid state laser)*

A semiconductor or solid state laser consists of a P-N junction, composed of diffused zinc in a gallium arsenide crystal. Both surfaces of the junction are perfectly polished and parallel. A current of several hundred amps per square meter of junction can produce a laser emission which is generally located in the infrared. The current required is much higher than that supportable by a junction in the continuous mode. Therefore, this is a pulse laser, and must be cooled by a cryogenic liquid; there are practically no industrial applications.

### *2.1.4. Liquid laser*

The laser effect can be obtained from organic liquids. These lasers, which are not at present used in industry, can vary wavelength which could be interesting in some applications.

## 2.2. Optical components

In principle, they are similar to those encountered in conventional optics (lenses, mirrors, etc.), but demand very high machining accuracy.

The components must be mounted to protect them from vapor spray or pieces of the part treated.

For lasers such as $CO_2$ lasers, with a long wavelength (10.06 $\mu$m), the standard glass optical components are totally opaque, and it is necessary to use potassium chloride, germanium, tellurium and cadmium, zinc selenide or diamond-based materials.

Energy absorption by such components is not negligible, and all precautions must be taken to allow them to cool normally.

## 2.3. Handling and positioning systems

Laser equipment consists mainly of machine tools whose components have been machined to extremely tight tolerances.

To prevent unwanted movements which adversely affect accuracy and cost, and to use the optical properties of lasers, laser beams are deflected via mirrors and prisms.

Also, it is necessary to control and regulate all the equipment operating parameters:

— mirror alignment in cavity;
— gas mixture and pressure in cavity (gas laser);
— discharge and beam power output;
— beam movements with respect to the workpiece;
— where necessary, the temperature of the various components (mirrors, window, etc.).

## 3. ADVANTAGES AND DISADVANTAGES OF LASER HEATING

Essentially, laser heating permits very high power densities which can be localized on very small surfaces, as shown in Figure 7, where the laser is compared to other heating processes. Very high temperatures of several thousand degrees (much higher in laboratories) are easily attained. Finally, the process is totally non-polluting and thermal or mechanical constraints are extremely low.

Comparison of power densities and minimum impact areas of different heating methods [48].

| Energy source | Maximum power density ($W/cm^2$) | Minimum area heated ($cm^2$) |
|---|---|---|
| Welding flame | $5 \times 10^4$ | $10^{-2}$ |
| Arc-plasma | $10^5$ | $10^{-3}$ |
| Electron beam | $5 \times 10^9$ | $10^{-7}$ |
| Laser | $10^{15}$ | $5 \times 10^{-8}$ |

**Fig. 7.**

The technology most like the laser is electron beam heating. The technologies sometimes compete. The advantage of the laser is that there is no need for a vacuum chamber or protection against X-rays. Conversely, its energy efficiency, when compared with that of electronic beam, is very low, and the powers developed much more limited. This competition only occurs in some special cases demanding thorough analysis [46 *b*].

In terms of investments, the laser is costly especially when the low energy conversion efficiency is taken into account. Its use is thus limited to specific applications practically impossible with other techniques. Some interesting attempts have been made however to adopt laser technology to mass production of relatively low added value parts, by optimum exploitation of its properties in order to compensate for the heavy initial investment [16], [22], [24], [25], [26], [30], [32], [33], [36], [37], [38], [40], [44].

## 4. INDUSTRIAL APPLICATIONS OF LASER HEATING

Laser beams are used in medicine (treatment of retinal separations, etc.) and parts measurements (clock-making), in amusements (laser beams used in shows), guidance and telemetry, but industrial applications are developing:
— cutting various materials;
— machining, and in particular micromachining;
— special welding operations;
— surface heat treatment.

The development of the laser accompanies the introduction of new, highly refractory materials which are difficult to weld, machine and cut, and ever-increasing miniaturization in electronics. The performances sought are very high energy concentrations on the smallest possible surfaces, absence of mechanical forces, absence of pollution due to foreign matter and high operating flexibility enabling processes to be automated easily. The laser meets all these criteria [20], [24], [25], [26], [27], [34], [36], [37], [40], [45], [49].

The power required to melt or vaporize the most current materials, including those encountered in the leading industries, is given in Figure 8 below.

Energy required to melt and vaporize different materials [36].

| Materials | Melting and vaporization energy |
|---|---|
| Metal melting | 1 to 15 kJ/cm$^3$<br>Aluminum: 2.5 kJ/cm$^3$<br>Tungsten: 12.5 kJ/cm$^3$ |
| Vaporization, decomposition and depolymerization of materials such as quartz, plastics, ceramics and woods | 2 to 4 kJ/cm$^3$ |
| Vaporization of metals | 30 to 80 kJ/cm$^3$ |

**Fig. 8.**

Wavelength 10.6 $\mu$m is highly absorbed by most non-metallic materials. $CO_2$ lasers, which offer the advantage of relatively high efficiency, are therefore used for these materials. Conversely, for metals which strongly reflect this wavelength, either this technique is used but with much higher powers and adapted energy transfer methods, or solid active element lasers.

Figure 9 shows the penetration effect of a pulsed laser beam. The minimum hole diameter is between 1 and 2 times the wavelength. The penetration depth corresponding to one pulse is very low, of the order of 0.01 to 0.05 $\mu$m depending on the material, for radiation at a wavelength of 0.7 $\mu$m and high pulse repetition rate enable deep penetration. The first pulse must be powerful enough to raise the temperature of the material above the threshold at which absorption increases. The time between two elementary pulses must be long enough to enable vapor formed by the material in the impact area to escape, but short enough to prevent the temperature dropping below that at which the material absorbs best [17].

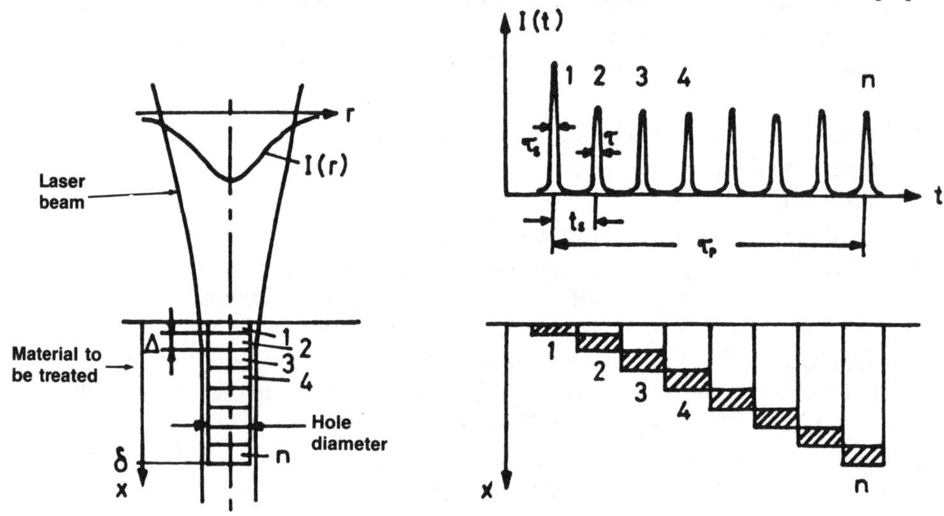


**Fig. 9.** Laser beam penetration mechanism [17]: *I*(*r*), power density distribution within the beam; *I*(*t*), beam intensity against time.

Laser characteristics [36]

| Laser | Wavelength (μ m) | Mean power (W) | Pulse energy (J) and frequency (Hz) | Electric efficiency (%) | Usual applications |
|---|---|---|---|---|---|
| Argon | 0.351-0.529 | 2-20 | Continuous | <0.1 | Film drilling |
| YAG | 1.06<br>1.06 | 10-1,000<br>20-50 | Continuous<br>0.5 J at 50 Hz | 1<br>1 | Cutting<br>Drilling diamonds and rubies |
| $CO_2$ | 10.6 | 500 | Continuous | 10 | Cutting non-metallic materials |
| $CO_2$ | 10.6 | 20,000 | Continuous | 10 | Welding metals |
| $CO_2$ | 10.6 | 2,000 | 2 J at 1 kHz | 10 | Machining ceramics |

**Fig. 10.**

This penetration mechanism is used for drilling and cutting operations. Conversely, for surface heating applications, the evaporation phenomenon must be avoided by limiting power density.

Figure 10 gives the characteristics of some of the lasers used in industry.

### 4.1. Cutting

$CO_2$ lasers are used for cutting a wide range of materials: stainless steel, plastics, fabrics, ceramics, wood, semiconductors and quartz. Cutting rate and width depend on the material treated and its thickness. Material is removed by evaporation at the impact point of the focus beam [9], [26], [31], [34], [35], [38], [46 *a*], [49].

For metals, which strongly reflect the $CO_2$ laser wavelength as previously described, it would seem that metal reflectivity drops when a critical threshold has been reached, enabling the high power laser to be used in industry. System efficiency can also be improved by combining the effect of a laser and an oxygen jet (inert gas jets are also used in other cases for purely protective purposes, either to protect the material or the optical system). Figure 12 gives current cutting rate values obtained with $CO_2$ lasers [36].

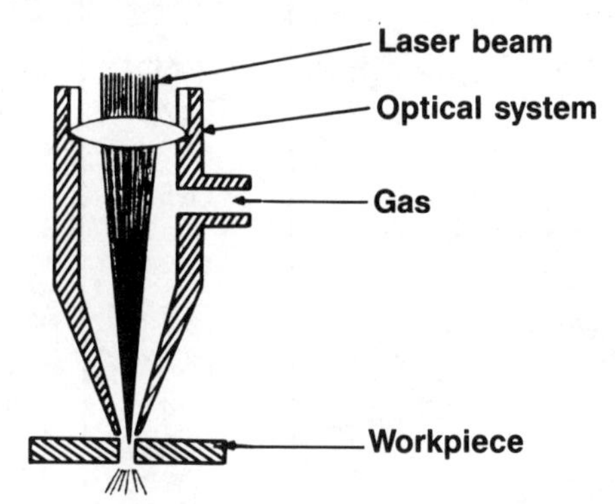


**Fig. 11.** Diagram of laser and gas jet cutting system [46 *a*].

For non-metallic materials, cutting is generally made in air. For example, a 300 W $CO_2$ laser can cut 1.5 mm thick formica at a rate of 9 m/min, with a cut width of 0.8 mm. The cut thickness for these materials may be high, 20 mm for perspex and 200 mm for polyurethane with the laser described above. Laser cutting is of particular interest for very hard materials such as metal carbides (tungsten carbide) or very fragile materials such as glass, and when extremely complex cutting profiles are required [A], [B], [C], [D].

### 4.2. Welding

Ruby, YAG or carbon dioxide lasers are used primarily for welding. The weld seam depth-to-width ratio is relatively limited, of the order of 10 maximum, since the beam energy is absorbed by a very thin layer. The power density must not be too high in order to prevent evaporation of the metal which also reduces melting front penetration. Inert gas spray is often used to prevent oxidation [9], [10], [11], [12], [13], [21], [23], [27].

Laser welding is used for microwelding in microelectronics and micromechanical engineering (welding fine wires to solid conductors, welding sintered material to stainless steel to produce automatic watch components) and to perform welding operations which, due to geometry, assembly or the type of parts to be welded, is difficult with other methods (metal ceramic welding, welding metals on glass, aluminum, epoxy resins, glass fibre-based materials) [19], [28], [29], [32], [38], [41], [46 *b*], [49].

Cutting and welding characteristics of $CO_2$ lasers [36].

| Material | Thickness (mm) | Speed (m/min) | Area affected by thermal effect (mm) | Power (kW) |
|---|---|---|---|---|
| Cutting: | | | | |
| Paper | 70 g/m$^2$ | 600 | 0.13 | 0.4 |
| Textile | 450 g/m$^2$ | 50 | 0.25 | 0.3 |
| Quartz | 2 | 1 | 0.25 | 0.4 |
| | 1.5 | 3 | 0.25 | 0.4 |
| Composite materials, fibreglass, epoxy resins, | 12.7 | 4.6 | 0.63 | 20 |
| plywood | 17 | 0.5 | — | 0.5 |
| | 25.4 | 1.5 | 1.50 | 8.0 |
| Tool steel | 3.0 | 1.7 | 0.20 | 0.4 |
| Mild steel | 1.2 | 4.6 | 0.20 | 0.4 |
| Stainless steel | 2.5 | 1.27 | 0.25 | 0.4 |
| | 4.7 | 1.27 | 2.00 | 20.0 |
| Titanium | 1.0 | 7.50 | 0.50 | 0.6 |
| Welding: | | | | |
| Mild steel | 2.0 | 0.20 | 1.50 | 0.7 |
| Stainless steel | 1.0 | 0.38 | 1.50 | 0.6 |
| | 6.0 | 0.60 | 0.50 | 6.0 |
| | 20.0 | 1.27 | 3.30 | 20.0 |

**Fig. 12.**

Figure 12 gives the characteristics of welds obtained with $CO_2$ lasers [36].

The heat applied around the weld is low and causes minimum shrinkage and tension.

### 4.3. Micromachining

Laser beams can be used for micromachining operations by removal of material, for drilling, milling, straightening, grooving or adjustment, and to make engravings on various supports.

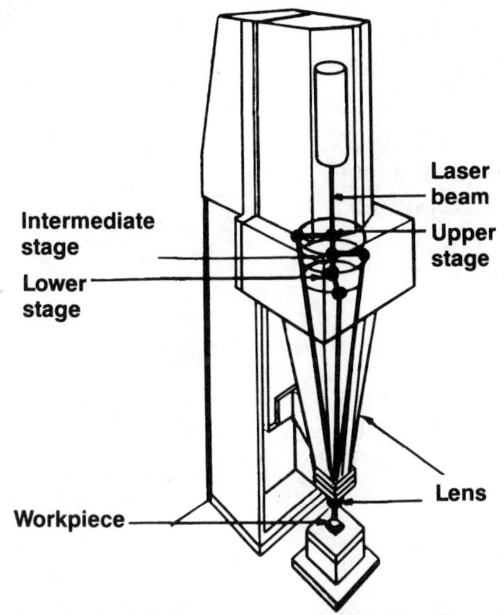


**Fig. 13.** Diagram of a laser machining unit [D].

All materials, metals, ceramics, industrial stones, crystals, plastics, etc., can be treated using this technique. Material is generally removed by evaporation; pulse lasers, and in particular giant pulse lasers, are often used to prevent interaction between the laser beam and the vapors [2], [7], [16], [17], [20], [34], [35], [40].

Through or blind holes having a depth-to-diameter ratio of more than 10 can be obtained, even for very small diameter holes of the order of 0.005 mm. By deflecting the laser beam, it is possible to cut openings of any shape, for example tapered blind holes. The minimum practical hole diameter is of the order of 1 or 2 wavelengths; i.e. 1 to 2 $\mu$m for a YAG laser and 10 to 20 $\mu$m for a $CO_2$ laser. The rate is, for example, 40 mm/s for drilling 200 $\mu$m diameter holes in ceramic or glass materials with $CO_2$ lasers of a power of 30 W.

Laser drilling is used mainly for treating refractory metals such as titanium, zirconium and tungsten, hard insulators such as diamonds, quartz and ceramics and semiconductors. Applications involve drilling rubies and diamonds, drilling holes in fine diaphragms, balancing wheels and tuning forks in clock-making and production of masks for integrated circuits [14], [37], [41], [44], [49].

Due to removal of a defined and controlled quantity of material, it is also possible to machine three-dimensional shapes, to adjust, align or balance mechanical, electromechanical, electrical and electronic systems or to engrave characters on surfaces of any type (production of high precision resistors, ball-bearings for pivots, instrument mountings, capacitors, engraving of printing cylinders, etc.) [15], [46 *a*], [49].

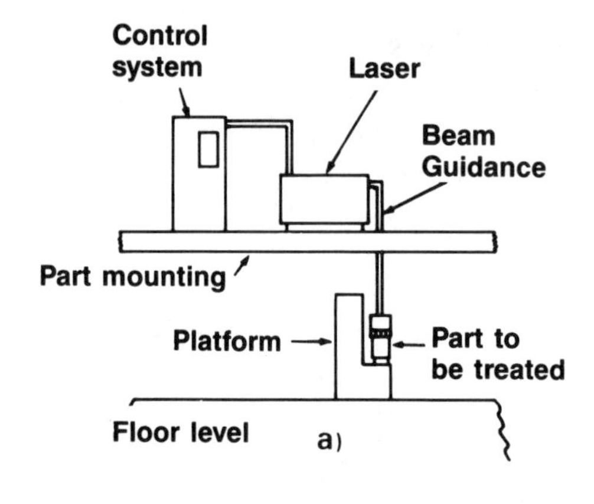


Fig. 14. Installation for surface hardening of diesel engines using a $CO_2$ laser with a power of 5 kW and a beam diameter of 1.9 cm, and a zinc selenide semi-transparent mirror [39], [F].

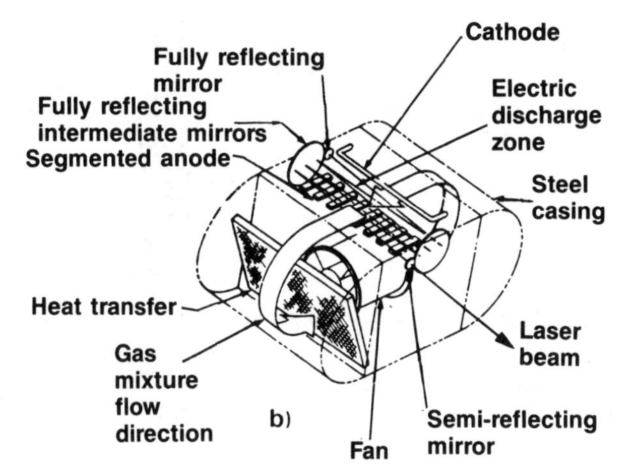


## 4.4. Surface heat treatment of metals

Lasers are now being used to surface harden steels. They develop high power densities and surface temperature over an elementary area of a part which can be heated very rapidly. For example, for 0.5% carbon steel, a $10^4$ W/cm$^2$ laser can raise the surface layer temperature to 940°C over a thickness of 0.5 mm in 50 ms. The cooling rate is the order of $10^{3}$°C/s, which is much lower than the critical cooling speed needed to harden the metal [16], [26], [36], [46 *a* and *b*], [49].

To harden a given surface, the laser scans it at a rate of between 0.5 and 2 m/min. The diameter of the impact area, which is higher than with other laser applications, is a few millimeters in most cases. Scanning is provided by mirror systems [36], [39], [42], [43].

The advantage of this surface hardening process is its accuracy, quality (absence of deformation) and flexibility, enabling complex-shaped parts to be treated easily. Due to its characteristics, this technique competes with electron beam heat treatment. It offers the advantage of not requiring vacuum equipment; conversely, energy efficiency is low and available powers are limited. Electron beam on the other hand requires high investment costs [46 *b*].

Research is also being conducted to obtain modifications of metallic surface properties by surface melting and adding alloying elements with laser beams [36], [46 *a*].

## BIBLIOGRAPHY


[1] T. P. MELA, *Introduction to masers and lasers*, Chapman and Hall Ltd, 1967.
[2] M. COHEN et B. A. UNGER, Laser machining of thin films and integrated circuits, *The Bell System Journal*, vol. 47, n° 3, mars 1968.
[3] A. L. BLOOM, *Gas lasers*, John Wiley and Sons, 1968.
[4] B. A. LENGEYL, *Introduction à la physique du laser*, Eyrolles, 1968.
[5] SMITH et SOROKLIN, *The laser*, MacGraw Hill, 1969.
[6] W. KLEEN et R. MULLER, *Laser*, Springer Verlag, Berlin, New York, 1969.
[7] J. L. HAKAUSON, Laser machining thin film electrode arrays on quartz crystal substrates, *Journal of Applied Physics*, vol. 40, n° 8, 1969.
[8] A. M. ROBINSON et D. C. JOHNSON, A carbon dioxide laser bibliography, *IEEE Journal of Quantitative Electronics*, vol. 6, n° 10, octobre 1970.
[9] J. ADAMS, *Large scale cutting and welding with $CO_2$ lasers*, Welding Institute, 1971.
[10] J. HOUBART, L'univers futuriste du soudage, *Industries et techniques*, n° 224, mars 1973.
[11] E. L. BAARDSEN, High speed welding of sheet steel with $CO_2$ laser, *Welding Journal* n° 4, vol. 52, avril 1973.
[12] S. BELAKHOWSKY, *Théorie et pratique du soudage*, Pyc-Édition, septembre 1973.
[13] D. T. SWIFT-HOOK, Penetration welding with lasers, *Welding Journal*, n° 11, vol. 52, novembre 1973.
[14] B. A. BAKEWELL, Assesment of holes drilled by laser beam, Conference on Electrical Methods of Machining, Forming and Coating, *IEE Conference Publication* n° 133, Londres, novembre 1975.
[15] R. CHILD et B. F. SCOTT, Rectification of figured scales on thin metallic films using a shaped laser beam, *IEE Conference Publication* n° 133, Londres, novembre 1975.
[16] A. S. KAYE et J. H. MEGAW, Multikilowatt $CO_2$ lasers and their applications, *IEE Conference Publication* n° 133, Londres, novembre 1975.
[17] J. STEFEN, High precision laser machining, *IEE Conference Publication* n° 133, Londres, nowembre 1975.
[18] G. CHAUDRON et F. TROMBE, *Les hautes températures et leur utilisation en physique et en chimie*, Masson, 1973.
[19] P. SALMON, Recherches pour l'actualisation du soudage, *Industries et Techniques*, n° 265, septembre 1974.
[20] E. D. SEAMAN et E. V. LOCKE, Metal working capability of a high power laser, Congrès S.A.E., San Diego, California (USA), octobre 1974.
[21] N. RYKALINE, Les sources d'énergie utilisées en soudage, *Soudage et techniques connexes*, n° 11/12, vol. 28, décembre 1974.
[22] A. CLEUET et A. MAYER, Risques liés à l'utilisation industrielle des lasers, *Techniques industrielles*, n° 66, mars 1975.
[23] P. T. HOULCROFT, laser and electron beam for welding engineering components, *Metal Construction and British Welding Journal*, n° 10, vol. 7, octobre 1975.
[24] J. HOUBART, Les lasers décollent, *Industries et techniques*, n° 317, mai 1976.
[25] J. P. MEGAN et I. J. SPALDING, High power continuous lasers and their applications, *Physics in Technology*, n° 5, vol. 7, septembre 1976.
[26] E. M. BREINAN *et al.*, Processing materials with lasers, *Physics to-day*, n° 11, vol. 29, novembre 1976.
[27] WELDING INSTITUTE, *Laser welding applications* 1970-1975, 1976.
[28] C. WEISMAN, *Welding Handbook*, 7e édition, vol. 1, 1976.
[29] R. W. TURNER et A. B. TOWSEND, Welding uranium with a multikilowatt, continuous wave, carbon dioxide laser welder, *Energy Research and Development Administration (U.S.A.)*, juin 1977.

[30] C. DESFORGES, Le laser, un outil comme les autres, *l'Usine nouvelle*, n° 37, septembre 1977.

[31] LE GÉNIE CIVIL, *Les méthodes modernes de découpage du béton*, janvier 1977.

[32] G. SAYEGH, Utilisation et perspectives des faisceaux lasers de haute puissance dans le travail des métaux, *Soudage et techniques connexes*, vol. 31, mai 1977.

[33] E. CATIER, Les lasers industriels en quête de kilowatts, *Automatique et applications industrielles*, n° 59, août 1977.

[34] J. F. DELORME, Le laser, un outil de production dans l'industrie des caoutchoucs et plastiques, *Caoutchoucs et plastiques*, n° 574, vol. 54, octobre 1977.

[35] COHERENT RADIATION, Laser machining of plastics and rubber, *Documentation de constructeur*, 1977.

[36] T. K. ALLEN *et al.*, *The current status of lasers as industrial tools*, UKAEA, Lulhal Laboratory, 1977.

[37] LASER INSTITUTE OF AMERICA, *Guide for materials processing by lasers*, 1977.

[38] D. G. ANDERSON, What designers should know about laser welding and cutting, *Mechanical Engineering*, n° 6, vol. 100, juin 1978.

[39] INDUSTRIAL HEATING, *Laser treating of cylinder liners for diesel engines to increase wear resistance*, septembre 1978.

[40] J. F. READY, *Industrial applications of lasers*, Academic Press, 1978.

[41] N. RYKALIN, *Laser machining and welding*, Pergamon Press, 1978.

[42] A. MULOT et J. P. BADEAU, Influence de la structure initiale et de la composition chimique sur les caractéristiques des couches durcies obtenues par trempe superficielle, bombardement électronique et laser, *Traitement thermique*, n° 136, juin 1979.

[43] G. STAHLI, Possibilités et limites du durcissement superficiel rapide de l'acier, *Traitement thermique*, n° 136, juin 1979.

[44] E. A. METZBOWER, *Applications of lasers in material processing*, Congrès ASM, Washington, 1979.

[45] N. G. NILLSON et K. G. SVANTESSON, The role of free carrier absorption in laser annealing of silicon, *Journal of Physics*, n° 13, 1980.

[46] COMITÉ FRANÇAIS D'ÉLECTROTHERMIE, Journées d'étude des 6-7 mars 1980 :

*a*) J. BAUJOIN, *Usinage par laser à $CO_2$;*

*b*) G. SAYEGH, *Comparaison du faisceau d'électrons et du faisceau laser dans le travail, le soudage et le traitement thermique des métaux.*

[47] DOCUMENTATION EDF, 75080 Paris La Défense.

[48] DOCUMENTATION LABORELEC, Bruxelles, Belgique.

[49] M. KIMURA *et al.*, *Flexible manufacturing system complex provided with laser*, août 1980, Tokyo et Cité des Sciences de Tsukuba.

**List of equipment manufacturers and suppliers mentioned in this chapter:**

[A] CILAS, 91460 Marcoussis.
[B] LASAG, 75008 Paris et Thur, Suisse.
[C] SAF, 95310 Saint-Ouen-L'Aûmone.
[D] LLOYD INDUSTRIEL, 75003 Paris.
[E] CGE, 91460 Marcoussis.
[F] GTE SYLVANIA, 95380 Louvres et U.S.A.
[G] COHERENT RADIATION, 94250 Gentilly et Palo Alto, U.S.A.

# Alphabetical Index

P



Q



R



S